Construction Site Work, Site Utilities, and Substructures Databook

Other *Databooks* by Sidney M. Levy

Construction Building Envelope and Interior Finishes Databook
Construction Databook
MEP Databook

Construction Site Work, Site Utilities, and Substructures Databook

Sidney M. Levy

McGraw-Hill

New York Chicago San Francisco Lisbon London Madrid
Mexico City Milan New Delhi San Juan Seoul
Singapore Sydney Toronto

Library of Congress Cataloging-in-Publication Data

Levy, Sidney M.
    Construction site work, site utilities, and substructures databook / Sidney M. Levy.
        p.   cm.
    Includes index.
    ISBN 0-07-136021-2
    1. Building sites. 2. Foundations. I. Title.

TH375.L48   2001
624—dc21   2001030146

McGraw-Hill

A Division of The **McGraw·Hill** Companies

1 2 3 4 5 6 7 8 9 0   KGP/KGP   0 7 6 5 4 3 2 1

ISBN 0-07-136021-2

The sponsoring editor for this book was Larry S. Hager and the production supervisor was Sherri Souffrance. It was set in ITC Century Light by TechBooks.

Printed and bound by Quebecor/Kingsport.

McGraw-Hill books are available at special quantity discounts to use as premiums and sales promotions, or for use in corporate training programs. For more information, please write to the Director of Special Sales, Professional Publishing, McGraw-Hill, Two Penn Plaza, New York, NY 10121-2298. Or contact your local bookstore.

This book is printed on acid-free paper.

Contents

Introduction

The *Construction Site Work, Site Utilities, and Substructures Databook* provides the builder, the project manager, the construction superintendent, design consultants, and facility managers with a one-source reference guide to soils, excavation and site work, substructures, underground piping systems, landscaping and site improvement components, installation techniques, and product specifications. No need to consult numerous volumes when all of the most frequently required data is readily at hand in one book.

Valuable information such as typical working ranges for excavating and hoisting equipment can be extracted in order to select the right equipment for the job. A section on metrification will prove useful as the U.S. construction market slowly adjusts to this soon-to-be-universal system.

Complete specifications and recommended installation instructions for concrete and corrugated metal pipe, ductile iron, cast pipe and fittings along with non-metallic piping will ensure that a proper installation can be effected. Need to convert the Latin name for those trees or shrubs on the landscaping plan to common names? Just look in Section 7.

Much of the material in the *Databook* has been gleaned from manufacturers' sources and trade association–furnished information; some of this information is proprietary in nature, but much is generic.

How many times during project meetings, field visits, or conversations with specialty contractors is it convenient to have a concise source of site-work-related information handy? The *Construction Site Work, Site Utilities, and Substructures Databook* will go a long way in satisfying those needs.

I selected those products and equipment that, in my 40 years in the industry, appear to be those for which reference material is so often required, always needed "yesterday," and so elusive to find in a hurry.

For those experienced construction professionals, the *Construction Site Work, Site Utilities, and Substructures Databook* may serve as a "refresher" course, and for those new to the industry, it offers a simpler way to become familiar with the complex and often bewildering array of products on the market.

I hope you find the *Databook* a worthwhile addition to your construction library.

Sidney M. Levy

Sitework

Contents

1.0.0 Investigation

Site work involves working with various types of soils and dealing with the unexpected—even in the presence of extensive soil test borings.

Even prior to commencing construction, a thorough investigation of the site, both visually and after a review of available geotechnical reports, the contractor will be more prepared for what lies ahead.

1. Does a visual inspection of the site reveal any clues to the composition and consistency of the soil?
2. Are there rock outcroppings? If so, what is the nature of the rock?
3. Is there any indication of the presence of ground water close to the surface of the site?
4. Do any remains of abandoned subsurface structures appear in areas where excavation will be required?
5. Do any structures require demolition in areas where new structures are to be built or where underground utilities are to be installed?
6. Are any utilities absent that might be required during construction (i.e., water, electric power, telephone lines, sanitary and storm sewers, or gas mains)? Are any of these utilities in areas where new construction will be required and are to be relocated?
7. What do the soil test borings reveal?

Analyzing a typical soil test boring should start with a look at the consistency of the soil (as reported on the report), the presence or absence of rock or any other underground obstructions, the level at which water was observed, and the blow count (an indication of soil-bearing capacity). The blows per foot also reveal the plasticity of the soil.

1.1.0 Glossary of Terms

AASHTO   American Association of State Highway and Transportation Officials.

AASHTO T-180   American Association of State Highway and Transportation Officials standard for the modified Proctor test.

AASHTO T-99   American Association of State Highway and Transportation Officials standard for the standard Proctor test.

Aeolian deposits   Wind-deposited materials such as sand dunes or other silty-type materials.

Aggregate (coarse or fine)   Crushed rock, sand, or gravel that has been graded and may be used as backfill material.

Air gap reading   The nuclear density meter test procedure that allows for cancellation of error in reading due to the chemical composition of the soil tested.

Alluvium   Material that has been deposited by streams that may no longer exist or that form existing floodplains.

Amplitude   The distance an oscillating body moves in one direction from its neutral axis to the outer limit of travel.

Aquifer   A geologic formation that provides water in sufficient quantities to create a spring or well.

ASTM   American Society for Testing and Materials.

ASTM D 1557   American Society for Testing and Materials standard for the modified Proctor test.

ASTM D 698   American Society for Testing and Materials standard for the standard Proctor test.

Backfill   Materials used to refill a cut or other excavation, or the act of such refilling.

Backscatter   A method of nuclear density meter soil testing in which the radiation source is placed in contact with the soil surface and density readings taken from the reflected radiation, the principle being that dense materials absorb more radiation than materials that are not as dense.

Bank   A mass of soil that rises above the normal earth level. Generally any soil that is to be dug from its natural position.

Bank-run gravel (run of bank gravel)   Gravel as it is excavated from a bank in its natural state.

Bank-yards   The measurement of soil or rock taken before digging or disturbing from its original position.

Base   The course or layer of materials in a road section on which the actual pavement is placed. This layer may be composed of many different types of materials, ranging from selected soils to crushed stone or gravel.

Base course   A layer of material selected to provide a subgrade for some load-bearing structure (such as paving) or to provide some for drainage under a structure above.

Berm   An artificial ridge of earth. This term is generally applied to the slide-slopes of a road bed.

Binder   A material that passes through a No. 40 U.S. standard sieve that is used to fill voids or hold gravel together.

Borrow pit   An excavation from which fill material is taken.

Boulder   A rock fragment with a diameter larger than 12 inches (304.8 mm).

Capillary action   The cohesive, adhesive, or tensive force that causes water that is contained within soil channels to rise or depress on the normal horizontal plane or level.

Cemented soil   Soil in which particles are held together by a chemical agent, such as calcium carbonate.

Centrifugal force   The pulling force of an eccentric weight when put in rotary motion that may be changed by varying the rotational speed and/or mass of the eccentric and/or center of gravity (shape) of the eccentric weight.

Clay   A cohesive mineral soil consisting of particles less than 0.002 mm in equivalent diameter, a soil textural class, or a fine-grained soil with more than 50 percent passing through a No. 200 sieve that has a high plasticity index in relation to its liquid limit.

Clean   Free of foreign material. When used in reference to sand or gravel, it means the lack of a binder.

Cobble   A rock fragment, generally oblong or rounded, with an average dimension ranging from 3 inches (75 mm) to 12 inches (305 mm).

Cohesion   Shear resistance of soil at zero normal stress; also, the quality of some soil particles to attract and stick to like particles; sticking together.

Cohesionless soil   A soil that when air-dried in an unconfined space has little cohesion when submerged.

Cohesive material   A soil having properties of cohesion.

Cohesive soil   A soil that when in an unconfined state has considerable strength when air-dried and submerged.

Compacted yards   The cubic measurement of backfill after it has been placed and compacted in fill.

Compaction   A process to decrease voids between soil particles when subjected to the forces applied by special equipment.

Compressibility   The property of a soil to remain in a compressed state after compaction.

Contact reading   A reading by a nuclear density meter when the bottom of the meter is in full contact with the compacted material to be tested.

Core   A cylindrical sample of an underground formation, cut and raised by a rotary hollow bit drill.

Crown   The center elevation of a road surface used to encourage drainage.

Datum   Any level surface used as a plane of reference to measure elevations.

Density   The mass of solid particles in a sample of soil or rock.

Double amplitude   The distance an oscillating body moves from its neutral axis to the outer limit of its travel in opposite directions.

Dry soil   Soil that does not exhibit visible signs of moisture content.

Dynamic linear force   The force pounds per inch (lb/in.) seen by the soil as produced by a vibratory roller. Calculated by dividing the centrifugal force by the width of the compacting surface(s).

Eccentric   A mass of weight off-balanced to produce centrifugal force (lb) and being part of the exciter unit that produces vibration.

Elasticity   Properties that cause soil to rebound after compaction.

Embankment   A fill whose top is higher than the adjoining natural compaction.

End result specifications   Compaction specifications that allow results instead of method specifications to be the determining factor in the selection of equipment.

Exciter   The component of a vibratory compactor that creates centrifugal force by means of a power-driven eccentric weight.

Fines   The smallest soil particles (less than 0.002 mm) in a graded soil mixture.

Fissured soil   Soil material that has a tendency to break along definite planes of fracture with little resistance.

Foot or shoe   The bottom part of a vibratory impact rammer contacting the soil.

Frequency   The rate at which a vibrating compactor operates, usually expressed in vibrations per minute (VPM).

Glacial till   Unstratified glacial materials deposited by the movement of ice and composed of sand, clay, gravel, and boulders in any proportion.

Grade   Usually defined as the surface elevation of the ground at points where it meets a structure; also, surface slope.

Grain distribution curve   A soil analysis graph showing the percentage of particle size variations by weight.

Granular material   A type of soil whose particles are coarser than cohesive material and do not stick to each other.

Granular soil   Gravel, sand, or silt with little or no clay content. It has no cohesive strength, cannot be molded when moist, and crumbles easily when dry.

Gravel   Round or semi-round particle of rock that pass through a 3-inch (76.2 mm) sieve and be retained by a No. 4 U.S. standard sieve [approximately $\frac{1}{4}$ inch (6.35 mm)]. It is also defined as an aggregate, consisting of particles that range in size from $\frac{1}{4}$ inch (6.35 mm) to 3 inches (76.2 mm).

Gumbo   Clays that are distinguished in the plastic state by a soapy or waxy appearance and great toughness.

Hardpan   Soil that has become rocklike because of the accumulation of cementing minerals, such as calcium carbonate, in the soil.

Impervious   Resistance to movement of water.

In-situ   The natural, undisturbed soil in place.

Internal friction   The soil particle's resistance to movement within the soil mass. For sand, the internal friction is dependent on the gradation, density, and shape of the grain and is relatively independent of the moisture content. For a clay, internal friction varies with the moisture content.

Layered system   Two or more distinctly different soil or rock types arranged in layers.

Lift   A layer of fill as spread or compacted. A measurement of material depth. The amplitude of a rammer's shoe. The rated effective soil depth a compactor can achieve.

Liquid limit   The water content at which the soil changes from a plastic to a liquid state.

Loam   A soft, easily worked soil that contains sand, silt, clay, and decayed vegetation.

Loess   A uniform aeolian deposit of silty material having an open structure and relatively high cohesion because of the cementation of clay or marl.

Marl   Calcareous clay that contains from 35 to 65 percent calcium carbonate.

Muck   Mud rich in humus or decayed vegetation.

Mud   Generally, any soil containing enough water to make it soft and plastic.

Optimum moisture content   Water content at which a soil can be compacted to a maximum-unit dry-unit weight.

Organic clay/soil/silt   Clay/soil/silt with high organic content.

Pass   A working trip or passage of an excavating, grading, or compaction machine.

Peat   A soft, light swamp soil consisting mostly of decayed vegetation.

Perched water table   A water table of generally limited area that appears above the normal free-water elevation.

Plastic   A property of soil that allows the soil to be deformed or molded without cracking or causing an appreciable volume change.

Plasticity index   The numeric difference between a soil's liquid limit and its plastic limit.

Plastic limit   The lowest water content of a soil, at which the soil just begins to crumble when rolled into a cylinder approximately $\frac{1}{8}$ inch (3.17 mm) in diameter.

Proctor modified   A moisture–density test of more rigid specifications than the standard Proctor test. The basic difference is use of a heavier weight dropped from a greater distance in laboratory determinations.

Proctor standard   A test method developed by R. R. Proctor for determining the density–moisture relationship in soils. It is almost universally used to determine the maximum density of any soil so that specifications may be properly prepared for field construction requirements.

Quicksand   Fine sand or silt that is prevented from stabilizing by a continuous upward movement of underground water.

Relative compaction   The dry unit of weight of soil compared to the maximum unit weight obtained in a laboratory compaction test and expressed as a ratio.

Silt   A soil composed of particles between 0.00024 inches (0.006 mm) and 0.003 inches (0.076 mm) in diameter.

Soil   The loose surface material of the earth's crust.

Specific gravity   The ratio of weight in air of a given volume of solids at a stated temperature to the weight in air of an equal volume of distilled water at the stated temperature.

Stabilize   To make soil firm and prevent it from moving.

Static linear force   The force in pounds per inch (lb/in.) seen by the soil as produced by a nonvibratory roller. Calculated by dividing the dead weight of the compactor by the width of the compacting surface(s).

Subbase   The layer of selected material placed to furnish strength to the base of a road. In areas where construction goes through marshy, swampy, unstable land, it is often necessary to excavate the natural material in the roadway and replace it with more stable materials. The material used to replace the unstable natural soils is generally called subbase material, and when compacted is known as the subbase.

Subgrade   The surface produced grading native earth, or inexpensive materials that serve as a base for a more expensive paving.

VPM   Vibrations per minute, derived by the rate of revolutions the exciter makes each minute.

1.1.1 Soil Classification Systems

Soils can be classified in several different methods and categories. The Tyler System uses opening per lineal inch of wire screen to determine particle size. For example, according to this system, a No. 20 mesh has 20 openings per lineal inch of screen, which equates to a sieve size of 0.0328 inches (0.833 mm).

The Unified Soil Classification System, the most widely used classification system, uses letters to designate soil types within three major groups: coarse-grained, fine-grained, and highly organic soils.

- *Coarse-grained soil*   Includes gravel, sands, and mixtures of the two. The letter *G* denotes gravel and the letter *S* denotes sand. In mixtures, the first letter indicates the primary constituent, e.g., GS. Both gravel and sand are further divided into four groups:

  - *Well graded*   Designated by the letter *W.*
  - *Poorly graded*   Designated by the letter *P.*
  - *Dirty with plastic fines*   Designated *P.*
  - *Dirty with nonplastic silty fines*   If it will pass through a No. 200 sieve, it is designated by the letter *M.*

  The coefficient of uniformity (Cu) is computed from data taken from a grain size distribution curve.

- *Fine-grained soils*   These soils are further divided into inorganic silts (M), inorganic clays (c), and organic silts or clays (O). Each group is further divided into soils having liquid limits lower than 50 percent (L) and those with liquid limits higher than 50 percent (H). For example, an inorganic silt with liquid limit lower than 50 percent would be designated *ML.*

- *Highly organic soils*   This group is identified by the letters *Pt,* for peat, which is characteristic of materials in this grouping.

1.1.2 Unified Soil Classification

TABLE 2.		UNIFIED SOIL CLASSIFICATION	
MAJOR DIVISIONS		**SYMBOL**	**TYPICAL NAMES**
GRAVELS (More than half of coarse fraction larger than #4 sieve size)	Clean gravel	GW	Well-graded gravel, gravel-sand mixtures
		GP	Poorly graded gravel, gravel-sand mixtures
	Gravel with fines	GM	Silty gravels, gravel-sand-silt mixtures
		GC	Clayey gravels, gravel-sand clay mixtures
SANDS (More than half of coarse fraction smaller than #4 sieve size)	Clean Sands	SW	Well-graded sands, gravelly sands
		SP	Poorly graded sands, gravelly sands
	Sands with fines	SM	Silty sands, sand-silt mixture
		SC	Clayey sands, sand-clay mixtures
FINE-GRAINED SOILS (More than half of material is smaller than #200 sieve)	Silts & Clays (LL < 50)	ML	Inorganic silt & very fine sands, clayey fine sands/silts with slight plasticity
		CL	Inorganic clays of low to medium plasticity, sandy/gravelly clays
		OL	Organic silts and organic silty clays of low plasticity
	Silts & Clays (LL > 50)	MH	Inorganic silts, fine sandy/silty soils, elastic silts
		CH	Inorganic clays, medium to high plasticity, fat clays
	Highly organic soils	PT	Peat and highly organic soils

1.1.3 OSHA Soil Classification

OSHA uses a soil-classification system as a means of categorizing soil and rock deposits in a hierarchy of stable rock, Type A soil, Type B soil, and Type c soil, in decreasing order of stability. Maximum allowable slopes are set forth, according to the soil or rock type.

Soil or rock type	Maximum allowable slope for excavation less than 20 feet
Stable rock	Vertical (90 degrees)
Type A soil	3/4:1 (53 degrees)
Type B soil	1:1 (45 degrees)
Type C soil	1½:1 (34 degrees)

A short-term maximum allowable slope of 1½ H:1V (63°) is allowed in excavations in Type A soil that are 12 ft (3.67 m) or less in depth. Short-term maximum allowable slopes for excavations greater than 12 ft (3.67 m) in depth shall be ¾ H:1V (53°)

Note: Consult OSHA for definition of *short-term.*

Type A: A cohesive soil with an unconfined compressive strength of 1.5 tons per square foot (144 kPa) or greater. Cohesive soils can be categorized as silty, clay, sandy clay, clay loam, and cemented soils. No soil is classified Type A if:

1. The soil is fissured.
2. The soil is subject to vibration from heavy traffic or pile driving.
3. The soil has previously been disturbed.
4. The soil is part of a sloped, layered system, where the layers dip into the excavation on a slope of 4 horizontal to 1 vertical.
5. The material is subject to other factors that tend to make it less stable.

Type B: A cohesive soil with an unconfined compressive strength of greater than 0.5 tons per square foot (48 kPa), but not less than 1.5 tons per square foot (144 kPa). This classification applies to cohesionless soils, including angular gravel (similar to crushed rock), silt, silt loam, sandy loam, and in some cases, silty clay loam and sandy clay loam. This classification also applies to previously disturbed soils, except those that would be classified as Type C or soil that meets the unconfined compressive strength or cementation requirements for type A, but is fissured or subject to vibration; dry rock that is not stable; or material that is part of a sloped, layered system, where the layers dip into the excavation on a slope less steep than 4 horizontal to 1 vertical, but only if the material would otherwise be classified Type B.

Type C: A cohesive soil with an unconfined compressive strength of 0.5 tons per square foot (48 kPa) or less, generally consisting of granular soils (including gravel, sand and loamy sand, submerged soil, soil from which water freely seeps, submerged rock that is not stable, or material in a sloped, layered system, where the layers dip into the excavation on a slope of 4 horizontal to 1 vertical (or steeper).

OSHA, in 1926.652 Appendix B, lists standards, interpretations, and illustrations of simple, single, multiple benches, and the use of trench support and shield systems for 20-foot (maximum) excavation depths. OSHA pages 186.8 and 186.9 of Appendix B contain diagrams that depict benched excavations for various types of excavations. For a complete explanation of excavations and trench-protection requirements, refer to the entire text of OSHA CFR 1926.652 in Appendix B.

1.1.4 OSHA Simple Slope and Single- and Multiple-Bench Diagrams

1.

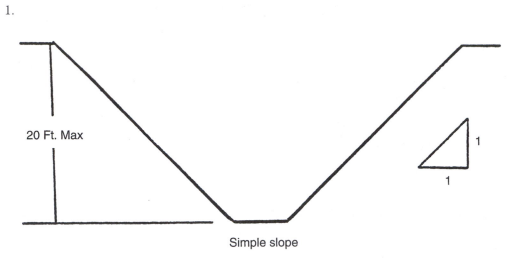

Simple slope

2. All benched excavations 20 feet or less in depth shall have a maximum allowable slope of 1 : 1 and maximum bench dimensions as follows:

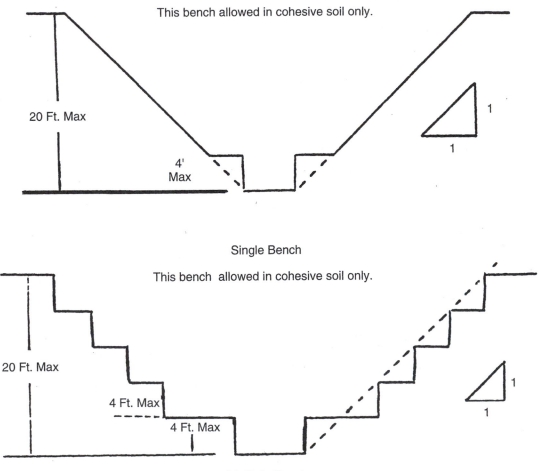

Single Bench

Multiple Bench

3. All excavations 20 feet or less in depth that have vertically sided lower portions shall be shielded or supported to a height at least 18 inches above the top of the vertical slide. All such excavations shall have a maximum allowable slope of 1 : 1.

1.1.5 OSHA Simple Slope and Vertical-Sided Trench-Excavation Diagrams

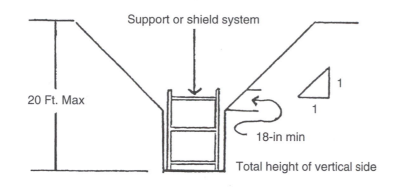

Vertical-Sided Lower Portion

4. All other sloped excavations shall be in accordance with the other options permitted in §1926.652(b).

1.1.6 Excavations Made in Type C Soil

1. All simple slope excavations 20 feet or less in depth shall have a maximum allowable slope of $1\frac{1}{2}:1$.

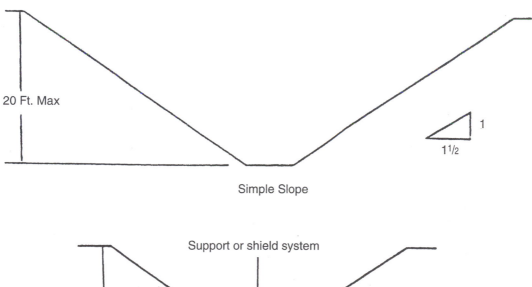

Simple Slope

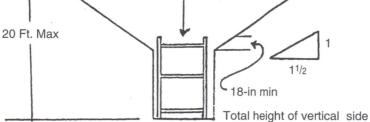

Vertical-Sided Lower Portion

1.2.0 Definition of Soil by Grain Size

Sieve size	Corresponding soil classification
12" (304.8 mm) or more	Boulders
3" (76.2 mm) to 12" (304.8 mm)	Cobbles
¾" (19.05 mm) to 3" (76.2 mm)	Coarse gravel
No. 4 to ¾" (19.05 mm)	Fine gravel
No. 4 to No. 10	Coarse sand
No. 10 to No. 40	Medium sand
No. 40 to No. 200	Fine sand
Passing through No. 200	Silt and clay fines

1.2.1 Uniform Building Code Standard 18-1: Soils Classification

Based on Standard Method D 2487-69 of the American Society for Testing and Materials.
Extracted, with permission, from the *Annual Book of ASTM Standards,* copyright American Society for
Testing and Materials, 100 Barr Harbor Drive, West Conshohocken, PA 19428

See Sections 1801.2 and 1803.1, *Uniform Building Code*

SECTION 18.101 — SCOPE

This standard describes a system for classifying mineral and organomineral soils for engineering purposes based on laboratory determination of particle-size characteristics, liquid limit and plasticity index.

SECTION 18.102 — APPARATUS

Apparatus of an approved type shall be used to perform the following tests and procedures: Preparation of soil samples, liquid limit test, plastic limit test and particle-size analysis.

SECTION 18.103 — SAMPLING

Sampling shall be conducted in accordance with approved methods for soil investigation and sampling by auger borings, for Penetration Test and Split-barrel Sampling of Soils, and for Thin-walled Tube Sampling of Soils.

The sample shall be carefully identified as to origin by a boring number and sample number in conjunction with a job number, a geologic stratum, a pedologic horizon or a location description with respect to a permanent monument, a grid system or a station number and offset with respect to a stated center line.

The sample should also be described in accordance with an approved visual-manual procedure. (A soil which is composed primarily of undecayed or partially decayed organic matter and has a fibrous texture, dark brown to black color, and organic odor should be designated as a highly organic soil, PT, and not subjected to the classification procedures described hereafter.)

SECTION 18.104 — TEST SAMPLE

Test samples shall represent that portion of the field sample finer than the 3-inch (76 mm) sieve and shall be obtained as follows:

Air dry the field sample; weigh the field sample; and separate the field sample into two fractions on a 3-inch (76 mm) sieve. Weigh the fraction retained on the 3-inch (76 mm) sieve. Compute the percentage of plus 3-inch (76 mm) material in the field sample and note this percentage as auxiliary information. Thoroughly mix the fraction passing the 3-inch (76 mm) sieve and select test samples.

SECTION 18.105 — PRELIMINARY CLASSIFICATION PROCEDURE

Procedure for the determination of percentage finer than the No. 200 (75 μm) sieve is as follows:

1. From the material passing the 3-inch (76 mm) sieve, select a test sample and determine the percentage of the test sample finer than the No. 200 (75 μm) sieve. (This step may be omitted if the soil can obviously be classified as fine-grained by visual inspection.)

2. Classify the soil as coarse-grained if more than 50 percent of the test sample is retained on the No. 200 (75 μm) sieve.

3. Classify the soil as fine-grained if 50 percent or more of the test sample passes the No. 200 (75 μm) sieve.

SECTION 18.106 — PROCEDURE FOR CLASSIFICATION OF COARSE-GRAINED SOILS (MORE THAN 50 PERCENT RETAINED)

Select test samples from the material passing the 3-inch (76 mm) sieve for the determination of particle-size characteristics, liquid limit and plasticity index. Determine the cumulative particle-size distribution of the fraction coarser than the No. 200 (75 μm) sieve.

Classify the sample as *gravel,* G, if 50 percent or more of the coarse fraction [plus No. 200 (75 μm) sieve] is retained on the No. 4 (4.75 mm) sieve. Classify the sample as *sand,* S, if more than 50 percent of the coarse fraction [plus No. 200 (75 μm) sieve] passes the No. 4 (75 mm) sieve.

If less than 5 percent of the test sample passed the No. 200 (75 μm) sieve, compute the coefficient of uniformity, C_u, and coefficient of curvature, C_z, as given in Formulas 18-1-1 and 18-1-2:

$$C_u = \frac{D_{60}}{D_{10}} \qquad (18\text{-}1\text{-}1)$$

$$C_z = \frac{(D_{30})^2}{D_{10} \times D_{60}} \qquad (18\text{-}1\text{-}2)$$

in which D_{10}, D_{30} and D_{60} are the particle size diameters corresponding respectively to 10, 30 and 60 percent passing on the cumulative particle size distribution curve.

Classify the sample as well-graded gravel, GW, or well-graded sand, SW, if C_u is greater than 4 for gravel and 6 for sand, and C_z is between 1 and 3. Classify the sample as poorly graded gravel, GP, or poorly graded sand, SP, if either the C_u or the C_z criteria for well-graded soils are not satisfied.

If more than 12 percent of the test sample passed the No. 200 (75 μm) sieve, determine the liquid limit and the plasticity index of a portion of the test sample passing the No. 40 (425 μm) sieve in accordance with approved methods.

Classify the sample as silty gravel, GM, or silty sand, SM, if the results of the limits tests show that the fines are silty, that is, the plot of the liquid limit versus plasticity index falls below the "A" line (see Plasticity Table 18-1-A) or the plasticity index is less than 4.

Classify the sample as clayey gravel, GC, or clayey sand, SC, if the fines are clayey, that is, the plot of liquid limit versus plasticity index falls above the "A" line and the plasticity index is greater than 7.

If the fines are intermediate between silt and clay, that is, the plot of liquid limit versus plasticity index falls on or practically on the "A" line or falls above the "A" line but the plasticity index is in the range of 4 to 7, the soil should be given a borderline classification, such as GM-GC or SM-SC.

If 5 to 12 percent of the test sample passed the No. 200 (75 μm) sieve, the soil should be given a borderline classification based on both its gradation and limit test characteristics, such as GW-GC or SP-SM. (In doubtful cases the rule is to favor the less plastic classification. Example: A gravel with 10 percent fines, a C_u of 20, a C_z of 2.0, and a plasticity index of 6 would be classified as GW-GM rather than GW-GC.)

1.3.0 Section 18.107—Procedure for Classification of Fine-Grained Soils (50 Percent or More Passing)

From the material passing the 3-inch (76 mm) sieve, select a test sample for the determination of the liquid limit and plasticity index. The method for wet preparation shall be used for soils containing organic matter or irreversible mineral colloids.

Determine the liquid limit and the plasticity index of a portion of the test sample passing the No. 40 (425 μm) sieve.

Classify the soil as inorganic clay, C, if the plot of liquid limit versus plasticity index falls above the "A" line and the plasticity index is greater than 7.

Classify the soil as inorganic clay of low to medium plasticity, CL, if the liquid limit is less than 50 and the plot of liquid limit versus plasticity index falls above the "A" line and the plasticity index is greater than 7. See area identified as CL on the Plasticity Chart of Table 18-1-A.

Classify the soil as inorganic clay of high plasticity, CH, if the liquid limit is greater than 50 and the plot of liquid limit versus plasticity index falls above the "A" line. In cases where the liquid limit exceeds 100 or the plasticity index exceeds 60, the plasticity chart may be expanded by maintaining the same scales on both axes and extending the "A" line at the indicated slope. See areas identified as CH on the Plasticity Chart, Table 18-1-A.

Classify the soil as inorganic silt, M, if the plot of liquid limit versus plasticity index falls below the "A" line or if the plasticity index is less than 4, unless it is suspected that organic matter is present in sufficient amounts to influence the soil properties, then tentatively classify the soil as organic silt or clay, O.

If the soil has a dark color and an organic odor when moist and warm, a second liquid limit test should be performed on a test sample which has been oven dried at 110°C ± 5°C for 24 hours.

Classify the soil as organic silt or clay, O, if the liquid limit after oven drying is less than three fourths of the liquid limit of the original sample determined before drying.

Classify the soil as inorganic silt of low plasticity, ML, or as organic silt of low plasticity, ML, or as organic silt or silt-clay of low plasticity, OL, if the liquid limit is less than 50 and the plot of liquid limit versus plasticity index falls below the "A" line or the plasticity index is less than 4. See area identified as ML and OL on the Plasticity Chart, Table 18-1-A.

Classify the soil as inorganic silt of medium to high plasticity, MH, or as organic clay or silt-clay of medium to high plasticity, OH, if the liquid limit is more than 50 and the plot of liquid limit versus plasticity index falls below the "A" line. See area identified as MH and OH on the Plasticity Chart of Table 18-1-A.

In order to indicate their borderline characteristics, some fine-grained soils should be classified by dual symbols.

If the plot of liquid limit versus plasticity index falls on or practically on the "A" line or above the "A" line where the plasticity index is in the range of 4 to 7, the soil should be given an appropriate borderline classification such as CL-ML or CH-OH.

If the plot of liquid limit versus plasticity index falls on or practically on the line liquid limit = 50, the soil should be given an appropriate borderline classification such as CL-CH or ML-MH. (In doubtful cases the rule for classification is to favor the more plastic classification. Example: a fine-grained soil with a liquid limit of 50 and a plasticity index of 22 would be classified as CH-MH rather than CL-ML.)

TABLE 18-1-A—SOIL CLASSIFICATION CHART

	MAJOR DIVISIONS		GROUP SYMBOLS	TYPICAL NAMES
COARSE-GRAINED SOILS More than 50% retained on No. 200 (75 μm) sieve*	GRAVELS 50% or more of coarse fraction retained on No. 4 (4.75 mm) sieve	CLEAN GRAVELS	GW	Well-graded gravels and gravel-sand mixtures, little or no fines
			GP	Poorly graded gravels and gravel-sand mixtures, little or no fines
		GRAVELS WITH FINES	GM	Silty gravels, gravel-sand-silt mixtures
			GC	Clayey gravels, gravel-sand-clay mixtures
	SANDS More than 50% of coarse fraction passes No. 4 (4.75 mm) sieve	CLEAN SANDS	SW	Well-graded sands and gravelly sands, little or no fines
			SP	Poorly graded sands and gravelly and sands, little or no fines
		SANDS WITH FINES	SM	Silty sands, sand-silt mixtures
			SC	Clayey sands, sand-clay mixtures
FINE-GRAINED SOILS 50% or more passes No. 200 (75 μm) sieve[1]	SILTS AND CLAYS Liquid limit 50% or less		ML	Inorganic silts, very fine sands, rock flour, silty or clayey fine sands
			CL	Inorganic clays of low to medium plasticity, gravelly clays, sandy clays, silty clays, lean clays
			OL	Organic silts and organic silty clays of low plasticity
	SILTS AND CLAYS Liquid limit greater than 50%		MH	Inorganic silts, micaceous or diatomaceous fine sands or silts, elastic silts
			CH	Inorganic clays of high plasticity, fat clays
			OH	Organic clays of medium to high plasticity
	Highly Organic Soils		PT	Peat, muck and other highly organic soils

[1]Based on the material passing the 3-inch (76 mm) sieve.

1.4.0 U.S.A./Metric Sieve Sizes

This chart shows the various sieve-size openings and their metric conversions.

U.S.A. Sieve Series and Equivalents—A.S.T.M. E-11-87

Sieve Designation		Sieve Opening		Nominal Wire Diameter	
Standard (a)	Alternative	mm	in (approx. equivnts.)	mm	in (approx. equivnts.)
125 mm	5"	125	5.00"	8.00	.3150"
106 mm	4.24"	106	4.24"	6.40	.2520"
100 mm	4"(b)	100	4.00"	6.30	.2480"
90 mm	3.5"	90	3.50"	6.08	.2394"
75 mm	3"	75	3.00"	5.80	.2283"
63 mm	2.5"	63	2.50"	5.50	.2165"
53 mm	2.12"	53	2.12"	5.15	.2028"
50 mm	2"(b)	50	2.00"	5.05	.1988"
45 mm	1.75"	45	1.75"	4.85	.1909"
37.5 mm	1.5"	37.5	1.50"	4.59	.1807"
31.5 mm	1.25"	31.5	1.25"	4.23	.1665"
26.5 mm	1.06"	26.5	1.06"	3.90	.1535"
25.0 mm	1"(b)	25.0	1.00"	3.80	.1496"
22.4 mm	7/8"	22.4	0.875"	3.50	.1378"
19.0 mm	3/4"	19.0	0.750"	3.30	.1299"
16.0 mm	5/8"	16.0	0.625"	3.00	.1181"
13.2 mm	.530"	13.2	0.530"	2.75	.1083"
12.5 mm	1/2"(b)	12.5	0.500"	2.67	.1051"
11.2 mm	7/16"	11.2	0.438"	2.45	.0965"
9.5 mm	3/8"	9.5	0.375"	2.27	.0894"
8.0 mm	5/16"	8.0	0.312"	2.07	.0815"
6.7 mm	.265"	6.7	0.265"	1.87	.0736"
6.3 mm	1/4"(b)	6.3	0.250"	1.82	.0717"
5.6 mm	No. 3-1/2(c)	5.6	0.223"	1.68	.0661"
4.75 mm	No. 4	4.75	0.187"	1.54	.0606"
4.00 mm	No. 5	4.00	0.157"	1.37	.0539"
3.35 mm	No. 6	3.35	0.132"	1.23	.0484"
2.80 mm	No. 7	2.80	0.11"	1.10	.0430"
2.36 mm	No. 8	2.36	0.0937"	1.00	.0394"
2.00 mm	No. 10	2.00	0.0787"	.900	.0345"
1.70 mm	No. 12	1.70	0.0661"	.810	.0319"
1.40 mm	No. 14	1.40	0.0555"	.725	.0285"
1.18 mm	No. 16	1.18	0.0469"	.650	.0256"
1.00 mm	No. 18	1.00	0.0394"	.580	.0228"
850 μm	No. 20	0.850	0.0331"	.510	.0201"
710 μm	No. 25	0.710	0.0278"	.450	.0177"
660 μm	No. 30	0.600	0.0234"	.390	.0154"
500 μm	No. 35	0.500	0.0197"	.340	.0134"
425 μm	No. 40	0.425	0.0165"	.290	.0114"
355 μm	No. 45	0.355	0.0139"	.247	.0097"
300 μm	No. 50	0.300	0.0117"	.215	.0085"
250 μm	No. 60	0.250	0.0098"	.180	.0071"
212 μm	No. 70	0.212	0.0083"	.152	.0060"
180 μm	No. 80	0.180	0.0070"	.131	.0052"
150 μm	No. 100	0.150	0.0059"	.110	.0043"
125 μm	No. 120	0.125	0.0049"	.091	.0036"
106 μm	No. 140	0.106	0.0041"	.076	.0030"
90 μm	No. 170	0.090	0.0035"	.064	.0025"
75 μm	No. 200	0.075	0.0029"	.053	.0021"
63 μm	No. 230	0.063	0.0025"	.044	.0017"
53 μm	No. 270	0.053	0.0021"	.037	.0015"
45 μm	No. 325	0.045	0.0017"	.030	.0012"
38 μm	No. 400	0.038	0.0015"	.025	.0010"
32 μm	No. 450		0.00126"	.0011	
25 μm	No. 500		0.00098"	.001	
20 μm	No. 635		0.00079"	.0008	

(a) These standard designations correspond to the values for test sieve apertures recommended by the International Standards Organization Geneva, Switzerland.
(b) These sieves are not in the fourth root of 2 Series, but they have been included because they are in common usage.
(c) These numbers (3-1/2 to 400) are the approximate number of openings per linear inch but it is preferred that the sieve be identified by the standard designation in millimeters or microns (1000 microns = 1 mm.)

1.5.0 Geotechnical Investigations

GEOTECHNICAL INVESTIGATION REQUIREMENTS

I.   PURPOSE

The primary purpose of the geotechnical investigation is to assist in identifying the key soil strength parameters for design of the steel foundation elements. In addition to the above, such studies are useful for the following reasons:

Resistance Piers:

- To locate the depth of firm bearing stratum for end bearing support of the underpinning pier.

- To establish the location of any weak soil zones in which column stability of the pier shaft must be considered.

- To determine if there are any barriers to installing the pier such as rubble fill, boulders, zones of chert or other similar rock, voids or cavities within the soil mass, any of which might require pre-drilling.

- To do a preliminary evaluation of the corrosion potential of the foundation soils as related to the performance life of the steel pier.

Helical Foundation Piers/Tiebacks/Anchors:

- To locate the depth and thickness of the soil stratum suitable for seating the helical plates of the pier and to determine the necessary soil strength parameters of that stratum.

- To establish the location of weak zones, such as peat type soils, in which column stability of the pier for compression loading situations may need to be considered.

- To locate the depth of the groundwater table (GWT).

- To determine if there are any barriers to installing the piers such as rubble fill, boulders or zones of cemented soils, chert or similar conditions, which might require pre-drilling.

- To do a preliminary evaluation of the corrosion potential of the foundation soils as related to the performance life of the steel pier.

Rock Anchors:

- To establish the depth and location of the rock.

- To determine the type of rock and competence of the rock to support the rock anchor.

- To determine the presence of groundwater.

- To do a preliminary analysis of corrosion potential of the rock and overburden soil for designing a corrosion protection system.

The geotechnical investigation generally consists of four phases: (1) Reconnaissance and Planning, (2) Test Boring and Sampling Program, (3) Laboratory Testing, and (4) Geotechnical Reporting. A brief description of the requirements and procedures, along with the required soil

Figure 1.5.0 *(By permission: Atlas Systems, Inc., Independence, Missouri)*

parameters used in designing manufactured steel foundation products is given in the following sections.

II.   RECONNAISSANCE AND PLANNING

Reconnaissance and planning includes: (1) reviewing the structural load requirements and size of the structure and whether the project is new construction or structure repair, (2) a review of the general soil and geologic conditions in the proximity of the site, (3) a site visit to observe topography and drainage conditions, rock outcrops if present, placement of borings, evidence of soil fill, including rubble and debris and evidence of landslide conditions.   The planning portion includes making a preliminary determination of the number and depth of each boring as well as determining the frequency of soil sampling for laboratory testing and the requesting of marking of all utilities in the zone in which borings will be conducted. Indicated below are guidelines for determining the number of borings and the depth to which the boring should be taken based on the project type.

A. Minimum Number Of Borings

Whether the project involves underpinning/repair of an existing structure or new construction, boring(s) should be made at each location where a pier or series of piers are to be placed. The recommended minimum number of borings necessary to establish a foundation soil profile is given below:

- Residential Home - One (1) boring.

- Commercial Building - One (1) boring for every 50 to 100 lineal feet for multistory-story structures, and every 100 to 150 lineal feet for other commercial buildings, warehouses and manufacturing buildings.

- Communication Towers - One (1) boring for each location of a pier anchor cap and One (1) boring at the tower center foundation footing.

- Sheet Pile/Earth Stabilization for Earth Cuts - One (1) boring for every 200 to 400 feet of project length.

B. Depth Of Boring(s)

The depth of each boring will vary depending on the project type, magnitude of foundation loads and area extent of the project structure. Some general guidelines for use in estimating required boring depth is given below:

- Residential Home - At least 15 feet into good bearing stratum, generally N > 8 to 10   (See Section III-A, Page A4 for a description of Standard Penetration Test and "N" values.)

- Commercial Building - At least 20 feet into good bearing stratum (generally N> 15)

- Communication Towers - Minimum of 35 feet for towers over 100 feet tall and at least 20 feet into a suitable bearing stratum (typically medium dense to dense for sands and stiff to very stiff for clays) for anchor piers.   The suitable bearing stratum should have a minimum value of "N" of 18 for sands and a minimum of 15 for cohesive soils.

- Sheet Piling/Earth Stabilization - boring should be taken to a depth that is at least as deep as the structure (sheet pile, etc.) to be anchored or until a suitable stratum is reached for seating the helical plates of the tiebacks (generally medium or denser sand or stiff clays).

Figure 1.5.0   *(cont.)*

1.5.1 Test Boring and Sampling

III. PROGRAM

A. Method Of Boring And Frequency Of Sampling

The most common method used in soil exploration is the use of truck mounted continuous flight auger systems together with the use of Standard Penetration Test (SPT) equipment. The continuous flight augers can be either hollow stem or solid stem. The equipment and procedure used are in conformance to ASTM D 1586. *Figure 1* shows a typical drill truck with a hollow stem auger being used to continuously sample the soil. The advantage of use of the hollow stem auger is to permit the sampler rod for the SPT to be inserted through the auger rather than have to remove the

Figure 1. Rotary Auger Boring Operation

auger stems each time an SPT is conducted. As the auger rotates, the soil moves to the surface and based on visual observation is classified as to type and condition and recorded on a log sheet at the depth of the auger stem. The result of this auger advance process is to establish the vertical sequence of the soil substrata. Generally a sample is recovered and visually classified at each 5-ft. interval of depth or at a change in soil type. This classification continues to the bottom of the boring. During the advance of the auger in the borehole, the driller also records the depth to any observed ground water table (GWT). The presence of a GWT will influence the unit weight and strength characteristics of the soil.

B. Standard Penetration Test And Sampling

The Standard Penetration Test (SPT) consists of driving the sampler with a 140-pound hammer and counting the number of blows (30 inch drops of the 140-pound hammer) applied in each of three 6" increments of penetration. The SPT, "N" value is then determined by adding together the total number of blows for the last 12" of penetration. The split spoon sampler consists of a 2" OD hollow (split spoon) sampler tube mounted on the bottom of a steel rod. *Figure 2* shows the SPT being conducted and *Figure 3* shows the split spoon sampler opened showing the soil sample contained in the sampler. These

Figure 2. Standard Penetration Test (SPT)

Figure 1.5.1 *(By permission: Atlas Systems, Inc., Independence, Missouri)*

blow counts are shown on the field-boring log at the depth the SPT was conducted. If the sampler is driven less than 18", such as would occur in highly resistant material, the blows for each complete 6" and for each partial increment is shown on the boring log. Portions of the recovered samples from each SPT are placed in sealed bags for the purpose of conducting laboratory tests.

Figure 3. Recovered Sample from Split Spoon Sampler (SPT)

The STP results are the most widely used soil parameter to guide the selection of soil strength for the design of helical foundation piers. The N values also can assist in determining the depth of installation requirements for resistance piers. Values of soil friction angle (ϕ) and cohesion (c) can be selected through correlation with the SPT, "N" values. References are provided on page A12 at the end of this section to guide the user in selecting correlated values.

C. Rock Coring And Quality Of Rock Measurement

When bedrock is encountered, and rock anchors are a design consideration, a continuous rock core must be recovered to the depth or length specified. Typical rock anchors may be seated 20 ft. or 30 ft. into the rock formation. The same type of truck drill rig is used for the rock coring operation, however, hardened steel or a diamond coring bit is used at the end of the core barrel to assist in cutting through the rock to recover the sample. Water under pressure is forced down the barrel and into the bit to carry the rock dust out of the hole as the water is circulated.

In addition to conducting compressive tests on the recovered rock core samples (See *Table 1*, below.); the rock core is examined and measured to determine the rock competency (soundness or quality). The rock quality designation (RQD) is the most commonly used measure of rock quality and is defined as:

$$\text{RQD} = \frac{\Sigma \text{ \underline{Length of intact pieces of core (>100 mm.)}}}{\text{Length of core advance}}$$

The values of RQD range between 0 and 1.0 where an RQD of 0.90 or higher is considered excellent quality rock.

Figure 1.5.1 *(cont.)*

1.5.2 Laboratory Testing

IV. LABORATORY TESTING OF RECOVERED SOIL SAMPLES

Every recovered sample from the field boring and sampling program is inspected visually and given a visual description as to its color, condition and type. (See *Table 2*.) In addition to this visual classification, a representative number of samples are selected to conduct the following tests:

- Water Content – measures the amount of moisture in the soil.

- Particle Size Analysis -- measures the distribution of particle sizes within the soil sample.

- Liquid Limit (LL), Plastic Limit (PL) and Plastic Index (PI) – applies to cohesive types of soil and is a measure of the relative stiffness of the soil and potential for expansion.

Figure 1.5.2 *(By permission: Atlas Systems, Inc., Independence, Missouri)*

- Strength Characteristics – in some instances undisturbed soil samples are recovered in the field using a thin wall (Shelby) tube. These recovered samples are tested either in triaxial or direct shear tests to determine directly the friction angle (ϕ) and the cohesion (c) of the soil. For cohesive (clay) soil samples, an unconfined compression (q_u) is often conducted. The cohesion of the clay sample is then taken to be one-half of q_u.

The results of the above tests are used to classify the soil as to type and condition. The most widely used classification system is the Unified Soil Classification System (USCS). This classification system divides soils between cohesionless (sands, gravels, silty sands, etc.) and cohesive (clays, silty or sandy clays, etc.) types. This classification is an important consideration in choosing the number and sizes of the helical plates since the strength mechanism for cohesionless and cohesive soils are quite different. The classification system also recognizes that soils occur in a mixed condition where the soil may exhibit both cohesionless and cohesive characteristics. *Table 2*, page A10 gives an introduction to USCS classification and the references at the end of this section provides a full description of the USCS.

Figure 1.5.2 *(cont.)*

1.5.3 Mechanical Properties of Various Rocks

TABLE 1.   MECHANICAL PROPERTIES OF VARIOUS ROCKS

Rock	Young's Modulus at Zero Load (10^5 kg/cm^2)	Bulk Density (g/cm^3)	Porosity (percent)	Compressive Strength (kg/cm^2)	Tensile Strength (kg/cm^2)
Granite	2 - 6	2.6-2.7	0.5-1.5	1,000-2,500	70-250
Microgranite	3 - 8				
Syenite	6 - 8				
Diorite	7-10			1,800-3,000	150-300
Dolerite	8-11	3.0-3.05	0.1-0.5	2,000-3,500	150-350
Gabbro	7-11	3.0-3.1	0.1-0.2	1,000-3,000	150-300
Basalt	6-10	2.8-2.9	0.1-1.0	1,500-3,00	100-300
Sandstone	0.5-8	2.0-2.6	5 - 25	200-1,700	40-250
Shale	1-3.5	2.0-2.4	10 - 30	100-1,000	20-100
Mudstone	2 - 5				
Limestone	1 - 8	2.2-2.6	5 - 20	300-3,500	50-250
Dolomite	4-8.4	2.5-2.6	1 - 5	800-2,500	150-250
Coal	1 - 2			50-500	20-50
Quartzite		2.65	0.1-0.5	1,500-3,000	100-300
Gneiss		2.9-3.0	0.5-1.5	500-2,000	50-200
Marble		2.6-2.7	0.5-2	1,000-2,500	70-200
Slate		2.6-2.7	0.1-0.5	1,000-2,000	70-200

Note: 1. For the igneous rocks listed above Poisson's ratio is approximately 0.25.
       2. For a certain rock type, the strength normally increases with increase in density and increase in Young's modulus. (After Farmer, 1968)
       3. Taken from *"Foundation Engineering Handbook"* by Winterkorn and Fong, Van Nostrand Reinhold, pg. 72.

Figure 1.5.3 *(By permission: Atlas Systems, Inc., Independence, Missouri)*

1.5.4 Geotechnical Report

The geotechnical report provides a summary of the findings of the three phases detailed above and also contains recommendations on options for a foundation together with the recommended soil related design values. Included in this report are the results of the laboratory testing of the soil samples and borings logs providing a visual summary of the vertical profile of foundation soils at the project site. *Figure 4* gives the boring log generated from the field exploration program as shown in *Figures 1* through *3*. A review of this boring log indicates the following:

- The total depth of the boring was 74.3 ft. Except for the upper one-half foot, the soil layers were all lean or lean to fat clay with some variations in color and stiffness down to the depth of 64 ft. At 64 ft., a shale stratum was encountered which was in a highly weathered condition. Shale is a rock but in a weathered condition such as noted on the boring log, it is probable that the helical plates of a pier could be set 1 ft. to 3 ft. into the shale stratum.

- Standard Penetration Tests (SPT) were conducted at each 5-ft. interval of depth down to the bottom of the boring. From the SPT, N column on the boring log, it is noted that the stiffness (or strength) of the lean clay is fairly consistent from depth 30 ft. to 64 ft. (N ranged from 9 to 14). The upper part of this stratum (around 35 ft. to 40 ft. is where the helical plates would be seated).

- Moisture contents were taken on the recovered split spoon samples from the SPT. Again, below about 25 ft., the moisture content of the soil was fairly consistent (ranging between 23-1/2 to 26-1/2 percent). This low variation in moisture content is consistent with the consistent range of N values.

- Liquid Limit and Plastic Limit tests were also conducted on the recovered split spoon samples. The average LL = 45; the average PL = 20, resulting in a PI = 25. These results indicate that the in-situ moisture content of the lean clay ($\cong$ 25%) from 30 ft. to 60 ft. is just above the Plastic Limit (20%). As the in-situ moisture content approaches the Plastic Limit, the clay soil will become stiffer (higher cohesion).

- The boring log also shows a column for unconfined compression strength for the clay type soil. The values indicated in this column were determined in the field using a hand held penetrometer device inserted into the split spoon sampler at the end of the exposed soil sample. This is not recognized as an accurate type of strength test for clay type soil but does provide a general order of magnitude of strength and also allows a comparison of strengths between various depths in the boring log. At a depth of 35 ft., the unconfined compression strength, q_u = 5000 psf. The cohesion of the soil, c, is taken to be 1/2 of qu or c = 2500 psf. This is generally a higher value than would be determined through correlation of the cohesion through the N value. From standard correlation charts, the cohesion, c, would likely be between 1500 and 1800 psf.

- In the bottom left hand corner of the boring log of *Figure 4*, it is indicated that no ground water table (GWT) was identified at the site.

- The fourth column in *Figure* 4 shows the soil classification symbols to help describe the soils. *Table 2*, Page A10 gives an abbreviated listing of the Unified Soil Classification system.

Figure 1.5.4 *(By permission: Atlas Systems, Inc., Independence, Missouri)*

1.5.5 Boring Logs

LOG OF BORING No. B-1

CLIENT: ATLAS SYSTEMS, INC.		DATE: 6-22-99	#02995604	RIG: CME 75
SITE: 1026B South Powell Road Independence, Missouri		PROJECT: ATLAS LOAD TEST		

GRAPHIC LOG	DESCRIPTION	DEPTH ft.	USCS SYMBOL	NUMBER	TYPE	RECOVERY in.	SPT – N BLOWS/ft	WATER CONTENT %	DRY UNIT WT. pcf	UNCONFINED STRENGTH qu psf	ATTERBERG LIMITS LL, PL, PI
	6" GRAVEL				PA						
	LEAN CLAY, silty trace organics, gray brown, trace dark brown and red brown, medium (Possible Fill)		CL	1	SS	14	7	34.1		2000*	45,21,34
		5			HS						
	LEAN CLAY, calcareous, trace sand and limestone gravel dark brown, brown, very stiff (Possible Fill)										
		10	CL	2	SS	6	5	18.6		7000*	45,23,22
					HS						
		15	CL	3	SS	24	9	24.1		5500*	
					HS						
	LEAN CLAY, trace silt, gray brown, trace dark gray, red brown and dark brown, stiff to very stiff										
		20	CL	4	SS	24	10	22.3		3500*	44,20,24
					HS						
		25	CL	5	SS	24	5	27.6		2500*	
					HS						
	LEAN CLAY, silty, gray brown, trace dark brown, stiff to very stiff										
		30	CL	6	SS	24	19	26.5		5000*	42,18,24
					HS						
	Trace limonites at 34.0'	35	CL-CH	7	SS	24	14	23.5		5000*	
					HS						
	LEAN TO FAT CLAY, gray brown, trace dark brown, very stiff										

Figure 1.5.5 (*By permission: Atlas Systems, Inc., Independence, Missouri*)

GRAPHIC LOG	DESCRIPTION	DEPTH ft.	USCS SYMBOL	NUMBER	TYPE	RECOVERY in.	SPT – N BLOWS/ft	WATER CONTENT %	DRY UNIT WT. pcf	UNCONFINED STRENGTH q_u psf	ATTERBERG LIMITS LL, PL, PI
	LEAN TO FAT CLAY, gray brown, trace dark brown, very stiff	40	CL-CH	8	SS	24	13	24.3		5000*	48,20,26
					HS						
	LEAN CLAY, silty, gray brown, brown, trace dark brown, red brown and gray, medium to stiff	45	CL	9	SS	24	11	24.5		3000*	
					HS						
	Trace gravel at 49.0'	50	CL	10	SS	24	10	26.3		3000*	46,21,25
					HS						
		55	CL	11	SS	24	13	24.7		3500*	
					WB						
	Trace gravel at 59.0'	60	CL	12	SS	24	9	25.7		1500*	
					WB						
	***SHALE, highly weathered, trace silty clay and gravel olive brown, gray trace brown	65		13	SS	16	55	21.1		9000*	
					WB						
	***SHALE, highly weathered, trace clay and black coal, very dark gray, gray	70		14	SS	6	44/3"	37.6			
					WB						
	***SHALE, highly weathered, calcareous, gray 74.3 BOTTOM OF BORING			15	SS		50/1"	13.1		3000*	

WATER LEVEL OBSERVATIONS, ft. NONE – WD   NONE - AB

* Calibrated Hand Penetrometer
** CME 140H SPT automatic hammer
*** Classification estimated from disturbed samples. Core samples and petrographic analysis may reveal other rock types.
The stratification lines represent the approximate boundary lines between soil and rock types: in-situ, the transition may be gradual.

Figure 1.5.5 *(cont.)*

1.5.6 Interpreting Soil-Test Boring Logs

Typical auger boring, spoon sampling report.

The blow count reveals that it took:

1 blow to drive a 140-pound hammer 6 inches.

3 blows were required to drive a 140-pound hammer 12 inches.

5 blows were required to drive a 140-pound hammer 18 inches.

8 blows were required to drive a 140-pound hammer to a depth of 24 inches.

16 blows were required to drive the hammer to a depth of 5 feet.

100 blows were required to drive the hammer to 10 feet 5 inches.

As far as water level is concerned, the use of water during the coring operation did not allow the Geotech to ascertain groundwater levels with certainty. When nonwater operations are used, groundwater levels are so indicated.

1.5.7 Classification Terminology Used in Conjunction with Test Borings

COMPONENT GRADATION TERMS

MATERIAL	FRACTION	SIEVE SIZE
GRAVEL	COARSE	3/4" TO 3"
	FINE	NO. 4 TO 3/4"
SAND	COARSE	N0. 10 TO NO. 4
	MEDIUM	NO. 40 TO NO. 10
	FINE	NO. 200 TO NO. 40
FINES		PASSING NO. 200

FINES FRACTION

PLASTICITY	PI	NAME	SMALLEST THREAD DIA ROLLED
NON-PLASTIC	0	SILT	NONE
SLIGHT	1-5	Clayey SILT	1/4"
LOW	5-10	SILT & CLAY	1/8"
MEDIUM	10-20	CLAY & SILT	1/16"
HIGH	20-40	Silty CLAY	1/32"
VERY HIGH	>40	CLAY	1/64"

RELATIVE DENSITY OR CONSISTENCY TERMS

NON-PLASTIC SOILS		PLASTIC SOILS	
BLOWS/FT	DENSITY	BLOWS/FT	PLASTIC SOILS
0-4	V. LOOSE	<2	V. SOFT
4-10	LOOSE	2-4	SOFT
10-30	M. DENSE	4-8	M. STIFF
30-50	DENSE	8-15	STIFF
>50	V. DENSE	15-30	V. STIFF
		>30	HARD

PROPORTIONAL TERMS

PROPORTIONAL TERM	PERCENT BY WEIGHT
AND	35-50
SOME	20-35
LITTLE	10-20
TRACE	1-10

BEDROCK WEATHERING CLASSIFICATION

GRADE	SYMBOL	DIAGNOSTIC FEATURES
Fresh	F	No visible signs of decomposition or discoloration. Rings under hammer impact.
Slightly Weathered	WS	Slight discoloration inwards from open fractures, otherwise similar to F.
Moderately Weathered	WM	Discoloration throughout. Weaker minerals such as feldspar decomposed. Strength somewhat less than fresh rock but cores cannot be broken by hand or scraped by knife. Texture preserved.
Highly Weathered	WH	Most minerals somewhat decomposed. Specimens can be broken by hand with effort or shaved with knife. Core stones present in rock mass. Texture becoming indistinct but fabric preserved.
Completely Weathered	WC	Minerals decomposed to soil but fabric and structure preserved (Saprolite). Specimens easily crumbled or penetrated.
Residual Soil	RS	Advanced state of decomposition resulting in plastic soils. Rock fabric and structure completely destroyed. Large volume change.

1.6.0 Soil Compaction

Soil compaction has been practiced by man for thousands of years.

The first attempts at earthen dams and irrigation ditches demonstrated the value of compaction by adding strength and some measure of protection against moisture damage. Early tamped earth buildings depended on thorough compaction for stability.

However, it was not until road building became a highly developed art, during the period of the Roman Empire, that the importance of soil compaction was fully appreciated. Roman roads, which are still in use, were built with careful attention to subsoil conditions and with thorough compaction of gravel and clay base. (Figure 1)

The Roman road builders knew that their cut stone road surfaces were only as good as the foundation on which they rested.

Some years ago, most compaction was done only on large construction sites, such as roads and airports. Very large, heavy machinery was used.

Only in the last few decades has the importance of **confined area** compaction been recognized.

With the introduction of self-contained portable rammers and vibratory plates, confined area compaction became practical.

Now soil compaction is commonly specified for building foundations, trench backfills, curbs and gutters, bridge supports, slab work, driveways, sidewalks, cemeteries and other confined area work.

Properly done, soil compaction adds many years to the useful life of any structure by increasing foundation strength and greatly improving overall stability.

Figure 1 - Roman Road

1.6.1 What Is Soil?

Soil is any natural material found on the surface of the earth except for embedded rock, organic plant and animal material.

Soil may be divided into four major groups according to particle size. As shown in Figure 2, the major groups are:

Clay — with particle sizes 0.00024" (.006 mm) and smaller;

Silt — with particle sizes ranging from 0.00024" (.006 mm) — 0.003" (.076 mm);

Sand — with particle sizes ranging from 0.003" (.076 mm) — 0.08" (2.03 mm);

Gravel — with particle sizes ranging from 0.08" (2.03 mm) — 3" (76.2 mm).

Figure 1.6.1 *(By permission: Wacker Corporation, Menomonee Falls, Wisconsin)*

1.6.2 Soil Classification According to Size

Figure 2 - Soil Classification According to Size

ASTM	Clay	Silt	Fine Sand	Coarse Sand	Gravel			
AASHTO	Clay	Silt	Fine Sand	Coarse Sand	Fine Gravel	Medium Gravel	Coarse Gravel	Boulders

Sieve Size: 270, 200, 140, 60, 40, 20, 10, 4, 1/2", 3/4", 3"

Particle Sizes mm: .001 .002 .003 .004 .006 .008 .01 .02 .03 .04 .06 .08 .1 .2 .3 .4 .6 .8 1.0 2.0 3.0 4.0 6.0 8.0 10 20 30 40 60 80

Particle Sizes in: .00004 .0002 .003 .01 .015 .04 .08 .2 .35 .50 .75 1.0 3.0

1.6.3 AASHTO Group Classification

The American Society for Testing and Materials (ASTM) and the American Association of State Highway and Transportation Officials (AASHTO) classify soil as **granular** or **cohesive** on the basis of a sieve analysis of the soil. See Figure 3.

Granular soil consists mainly of sands and gravels. **Cohesive soil** consists mainly of silts and clays.

In granular soil, the particles are held in position due to the frictional force that exists at the contact surfaces. In the dry state, granular soil particles can be easily separated and identified. In a moist state, a granular material, such as sand, may be formed to desired shapes but will crumble easily as soon as it is disturbed.

In cohesive soil, the molecular attraction between soil particles is the force which holds the soil in place. As these particles are very small in size, high in number, and densely arranged, the cohesive force within the soil is very high. Cohesive soils are very hard in the dry state. When moist, they are plastic and can be molded or rolled to almost any shape.

General Classification	Granular Material							Cohesive Material More than 35% of Total Sample Passing No. 200			
Group Classification	A-1				A-2						A-7
	A-1-a	A-1-b	A-3	A-2-4	A-2-5	A-2-6	A-2-7	A-4	A-5	A-6	A-7-5 A-7-6
Sieve Analysis % Passing											
No: 40 _____	30 max	50 max	51 min								
No. 200 _____	15 max	25 max	10 max	35 max	35 max	35 max	35 max	36 min	36 min	36 min	36 min

Figure 3 - AASHTO Group Classification

1.6.4 How to Test for the Correct Amount of Moisture

The moisture content of a soil, either granular or cohesive, is a critical factor in the workability of that soil.

Moisture acts as a lubricant between soil particles. Too little moisture will not allow soil particles to move into a dense arrangement. Too much moisture will saturate a soil taking up space which would normally be filled by soil particles.

Figure 1.6.4 *(By permission: Wacker Corporation, Menomonee Falls, Wisconsin)*

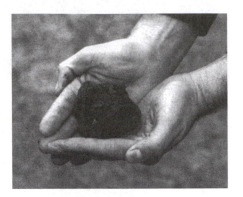

Figure 4 - Simple Soil Moisture Test

A simple method of testing whether the soil contains the right amount of moisture for compaction, is to take a handful of the soil to be compacted and squeeze it into the size and shape of a tennis ball and drop it on the ground from about 1 foot. See Figure 4. Uniform gravel or mostly sandy soil do not react well to this test.

At optimum soil moisture — The ball breaks apart into a small number of fairly uniform fragments.

If too dry — The soil does not form into a ball at all, and moisture must be added to the soil.

If too moist — The soil does not break apart (unless the soil is very sandy), and soil should be allowed to dry if possible.

1.6.5 Grain Size Distribution

Since a soil may contain different particle sizes, it is useful to know the amount of each size present in the soil. To do this, a sample of soil is dried, crumbled to separate the particles and then run through a series of standard sieves of different sizes. The amount of soil retained on each sieve is noted and calculated as a percentage of the total sample weight. The percentages obtained are plotted against sieve sizes to give a **Grain Distribution Curve** for the soil under investigation as shown in Figure 5.

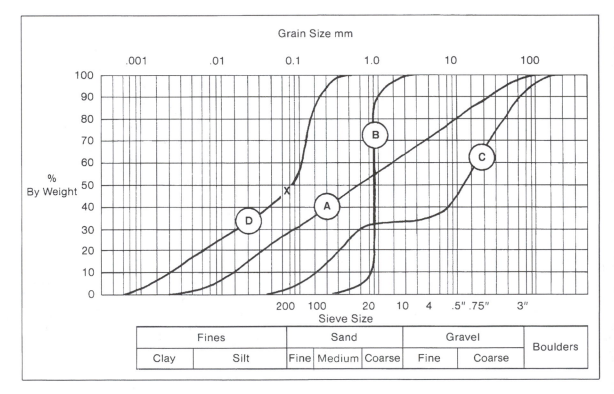

Figure 5 - Grain Distribution Curves

Figure 1.6.5 *(By permission: Wacker Corporation, Menomonee Falls, Wisconsin)*

The shape of curve so obtained gives an indication of the **gradation** of the soil. A "well graded" soil is defined as a soil which contains a broad range of grain sizes. A well graded soil is distinguished by a curve with a fairly uniform incline similar to Curve A in Figure 5.

A "poorly graded" or "uniform" soil is a soil that contains limited range of grain sizes. A steep curve is characteristic of this soil as shown in Curve B in Figure 5.

A soil that is missing certain particle sizes will have a curve with a horizontal portion as indicated in Curve C in Figure 5. Such soil is termed as "gap graded".

Point X on Curve D in Figure 5 shows that 48% by weight of that soil is finer than #200 sieve, meaning a very cohesive soil.

A well graded soil compacts to higher density than a poorly graded soil, and, therefore, has a higher load bearing capacity. This is because the finer grains can be vibrated or compacted into the cavities between the larger particles. If the fine grains were not present, those cavities would stay unfilled, resulting in air voids and lower load bearing capacity of the soil.

Response to Moisture

The response of soil to moisture is important, since the soil has to carry the load year round, rain or shine. Rain, for example, may transform soil into a plastic state or even into a liquid form. In these forms, the soil has little or no load bearing capacity.

Soil Classification Systems

Various soil classification systems exist to indicate the quality of soil as construction material. These classification systems take into consideration particle sizes, grain size distribution, and the effect of moisture on the soil. One of the soil classification systems is the Unified Soil Classification System (USC). A summary of USC is indicated in Figure 6. The meaning of the System Code Letters is also indicated in Figure 6 to make the Chart simple and easy to understand.

Group Symbol	Brief Description	Suitable as Construction Material
GW	Well graded gravels	Excellent
GP	Poorly graded gravels	Excellent to Good
GM	Silty gravels	Good
GC	Clayey gravels	Good
SW	Well graded sands	Excelllent
SP	Poorly graded sands	Good
SM	Silty sands	Fair
SC	Clayey sands	Good
ML	Inorganic silts of low plasticity	Fair
CL	Inorganic clays of low plasticity	Good to Fair
OL	Organic silts of low plasticity	Fair
MH	Inorganic silts of high plasticity	Poor
CH	Inorganic clays of high plasticity	Poor
OH	Organic clays of high plasticity	Poor
PT	Peat, mulch and high organic soils	Not Suitable
Code:	G = Gravel	W = Well Graded
	S = Sand	P = Poorly Graded
	M = Silt	L = Low Liquid Limit
	C = Clay	H = High Liquid Limit
	O = Organic	PT = Peat

Figure 6 - Unified Soil Classification

Figure 1.6.5 *(cont.)*

1.6.6 What Is Soil Compaction?

Soil compaction is the process of applying energy to loose soil to consolidate it and remove any voids, thereby increasing the density and consequently its load-bearing capacity.

Why is soil compaction necessary?

Nearly all man-made structures are ultimately supported by soil of one type or another. During the construction of a structure, the soil is often disturbed from its natural position by excavating, grading or trenching. Whenever this occurs, air infiltrates the soil mass and the soil increases in volume. Before this soil can support a structure over it or alongside it, these voids must be removed in order to be a solid mass of high strength soil.

Residential construction as well as commercial construction can benefit from Soil Compaction.

Soil Compaction provides the following benefits:

Increases Load Bearing Capacity - Air voids in the soil cause weakness and inability to carry heavy loads. With all soil particles squeezed together, larger loads can be carried by the soil because the soil particles support each other better.

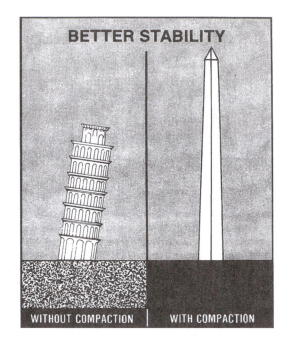

Reduces Water Seepage - Compacted soil reduces water penetration. Water flow and drainage can then be brought under control.

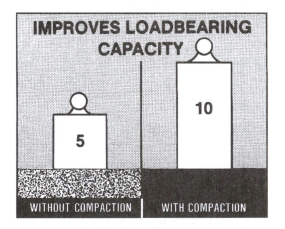

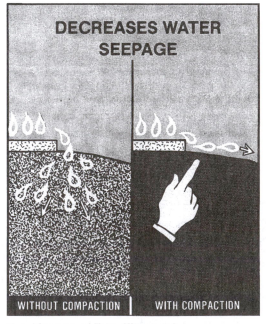

Prevents Soil Settlement - If a structure is built on uncompacted or on unevenly compacted soil, settlement of soil occurs causing the structure to deform. If the settlement is more pronounced at one side or corner, cracks or complete failure can result.

Figure 1.6.6 *(By permission: Wacker Corporation, Menomonee Falls, Wisconsin)*

Reduces swelling and contraction of soil - If air voids are present, water may penetrate the soil to fill the air voids. The result will be a swelling action of the soil during the wet period and a contraction action during the dry season.

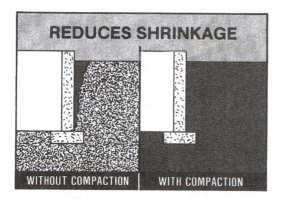

Prevents frost damage - Water expands and increases its volume upon freezing. This action often causes pavement heaving and cracking of walls and floor slabs. Compaction reduces these water pockets in the soil.

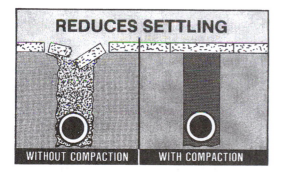

In summary, compaction should be used every time the soil is disturbed. Effective compaction means densely packed soil without any voids.

1.6.7 Methods to Compact Soil

Three major methods are used to compact soil.

1.) **Static force** - Compaction is achieved using a heavy machine whose weight squeezes soil particles together without the presence of vibratory motion. **Example**: A static roller. See Figure 7.

Figure 7 - Static Roller

2.) **Impact force** - Compaction comes from a ramming shoe alternately striking and leaving the ground at a high speed, literally "kneading" the ground to increase its density. **Example**: A rammer. See Figure 8.

Figure 8 - Wacker Rammer

3.) **Vibration** - Compaction is achieved by applying a high frequency vibration to the soil. **Example**: A vibratory plate. See Figure 9.

Figure 9 - Wacker Vibratory Plate

Figure 1.6.7 *(By permission: Wacker Corporation, Menomonee Falls, Wisconsin)*

1.6.8 Choosing the Correct Method

Granular soils are best compacted by **vibration**. This is because the vibration action reduces the frictional forces at the contact surfaces, thus allowing the particles to fall freely under their own weight. At the same time, as soil particles are set in vibration, they become momentarily separated from each other, allowing them to turn and twist until they can assume a position that limits their movements. This settling action and repositioning of particles is compaction. All the air voids that were previously present in the soil mass are now replaced by solidly packed soil.

Cohesive soils are best compacted by **impact force**. Cohesive soils do not settle under vibration, due to natural binding forces between the tiny soil particles.

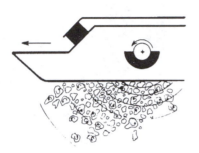

These soils tend to lump, forming continuous laminations with air pockets in between.

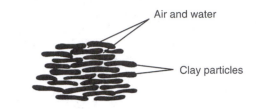

Clay particles especially present a problem because of their extremely light weight which causes the clay to become very fluid when excess moisture is present. Also, clay particles have a flat, pancake shape appearance which prevents them from dropping into voids under vibration. Therefore, cohesive soils, such as silt and clay, are effectively compacted using **impact force** which produces a shearing effect that squeezes the air pockets and excess water to the surface and moves the particles closer together.

Combinations of impact force and vibration are also used. For example, large vibratory plates and vibratory rollers combine static weight with vibration to achieve compaction.

1.6.9 Soil Testing

Compacted Soil is measured in terms of **Density** in pounds per cubic foot (lb./cu. ft.).

For Example:
Loose soil may weigh 100 pounds per cubic foot. After compaction, the same soil may have a density of 120 pounds per cubic foot. This means that by compaction, the density of the soil is increased by 20 pounds per cubic foot (PCF).

Laboratory Testing

To determine the density value of a soil from a given job site, a sample of the soil is taken to a soil test lab and a **Proctor Test** is performed. See Figure 10.

The purpose of a Proctor Test is twofold. The Proctor Test (1) measures and expresses the density attainable for any given soil as a standard; and (2) determines the effect of moisture on soil density.

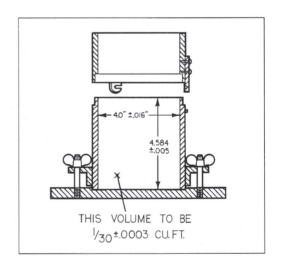

Figure 10 - Cylindrical Mold
for Performing Proctor Soil Tests

1.6.10 Standard and Modified Proctor Tests

In this test, a sample of soil is compacted in a standard container 4" dia x 4.59" high which has 1/30 cubic foot capacity. The container is filled in 3 layers. Each soil layer is compacted using a 5.5 lb. weight which is lifted through a distance of 12" and dropped 25 times evenly over each soil layer, yielding a soil sample which has received a total of 12,375 ft. lb. of energy per cubic foot, determined as follows:

1 ft. x 5.5 lb. x 25 drops x 3 layers x 30 =
12,375 ft. lb./cu. ft.

After striking off the surface of the container, the soil sample is weighed immediately after the test (wet weight) and then weighed again after drying the soil in an oven (dry weight). The difference between the wet and dry weights represents the weight of water that was contained in the soil. The density of the dry soil can now be expressed in terms of lb. per cubic foot. The amount of water or moisture may also be expressed **as a percentage** of the dry weight.

The procedure is repeated, adding different amounts of water to the soil with each repetition and the soil weights as well as the percentages of moisture are recorded as described previously.

Example: For a given 1/30 cu. ft. soil sample:
Wet weight = 4.6 lb.
Dry weight = 4.0 lb.
Weight of water lost = 0.6 lb.

We then calculate:
Soil Dry Density = 4.0 lb. / 1/30 cu. ft. =
120 lb./cu. ft.
% Moisture = 0.6 lb. / 4.0 lb. x 100 = 15%

Plotting the data on a graph, a curve similar to the one shown in Figure 11 is obtained.

The curve is referred to as the **Moisture-Density Curve** or **Control Curve**.

Conclusions

1. At a certain moisture, the soil reaches a maximum density when a specific amount of compaction energy is applied.
2. The maximum density reached under these conditions is called **100% Proctor density**.
3. The moisture value at which maximum density is reached is called **optimum moisture**.
4. When compacting soil at above or below optimum moisture, and using the same compacting effort, the density of the soil is less than when compacted at optimum moisture.
5. The 100% Proctor value thus obtained in a laboratory test is used as a basis **for comparing the degree of compaction of the same type of soil on the job site.**

Example:

For the soil under investigation, 100% Proctor represents a density of 120 lb. per cubic foot. (Refer to Figure 11). Assuming the same soil is compacted to a dry density of 115 lb. per cubic foot. The degree of compaction for the soil is then expressed as:

$$\frac{115}{120} \times 100 = 96\%$$

In other words, the soil is being compacted to 96% Proctor.

The above lab test for soil density was developed by R.R. Proctor, a Field Engineer for the City of Los Angeles, California, back in the early 1930's. It is now universally accepted throughout the construction industry and is known as the **Standard Proctor Test**.

The trend to construct heavy structures, such as Nuclear Power Plants and jet runways, has increased the demand for tougher compaction specifications. For those structures, a **Modified Proctor Test** was developed. The principles and procedures for both tests are very similar.

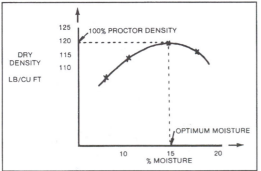

Figure 11 - Moisture Density Curve

Figure 1.6.10 *(By permission: Wacker Corporation, Menomonee Falls, Wisconsin)*

1.6.11 Requirements for Standard and Modified Proctor Density Tests

Specifications	Standard Proctor	Modified Proctor
Weight of the Hammer	5.5 lb.	10 lb.
Distance of Drop	12 inches	18 inches
Number of Soil Layers	3	5
Number of Drops on Each Layer	25	25
Volume of Test Container	1/30 cu ft.	1/30 cu ft.
Energy Imparted to Soil	12,375 ft. lb. per cu. ft.	56,250 ft. lb. per cu. ft.

Each soil behaves differently with respect to maximum density and optimum moisture. Therefore, each type of soil will have its own unique control curve. See Figure 13.

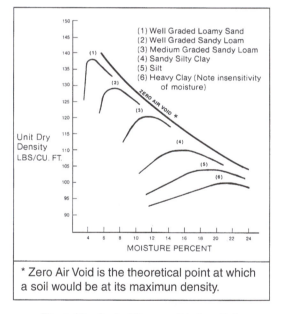

(1) Well Graded Loamy Sand
(2) Well Graded Sandy Loam
(3) Medium Graded Sandy Loam
(4) Sandy Silty Clay
(5) Silt
(6) Heavy Clay (Note insensitivity of moisture)

ZERO AIR VOID *

Unit Dry Density LBS/CU. FT.

MOISTURE PERCENT

* Zero Air Void is the theoretical point at which a soil would be at its maximun density.

Figure 13 - Control Curves of Various Soils

The Proctor Test is usually conducted in the laboratory, not on job sites.

It is quite possible that a soil may be compacted to more than 100% Proctor, for example, to 104%. This is because the 100% Proctor value is obtained by using a specific amount of energy during compaction. If more energy is put into the soil, higher densities are to be expected.

The Standard Proctor Test has been adopted as AASHTO, Standard T-99 and ASTM, Standard D 698. Modified Proctor has been adopted as AASHTO, Standard T-180 and ASTM, Standard D 1557, respectively.

On establishing the Proctor curve for the soil and determining its 100% density, the architect/engineer is now in a position to specify the percent Proctor to which the soil must be compacted. Actual compaction then takes place in the field.

Field Testing Methods

There are two major methods used for field compaction testing today.

1.6.12 The Sand Cone Method

A hole about 6" wide and 6" deep is dug into the compacted soil. The soil removed from the hole is weighed, then completely dried and weighed again. The amount of water lost, divided by the dry weight gives the % moisture in the soil. A cone and jar apparatus containing special fine-grained uniform sand is placed over the hole, and the hole is filled with sand. See Figure 14. The jar is weighed before and after filling the hole, and in this way, the volume of sand required to fill the hole is determined.

Dividing the dry weight of soil removed by the volume of sand required to fill the hole gives the density of the compacted soil in lb./cu. ft.. The density so obtained is compared to the maximum density from a Proctor Test and the relative Proctor Density is obtained. The sand cone method has been in use for a long time and is well known and accepted.

Figure 1.6.12 *(By permission: Wacker Corporation, Menomonee Falls, Wisconsin)*

However, several common mistakes are sometimes made by field testers, that can render the sand cone test inaccurate.

The use of uniform sand assumes that the sand is not compactible. However, since sand particles are not completely round, job site vibration can compact the sand and the test will show a lower density than actual.

Several types of sand are used in sand cone testing, but each sand cone device is calibrated to use only one type of sand. Errors can result if the wrong type of sand is used in a particular test device.

Since a sand cone test takes several hours to perform, testing after each compaction pass is unfeasible, and too much or too little compaction is a possibility.

1.6.13 The Nuclear Method

The Nuclear Density/Moisture Meter operates on the principle that dense soil absorbs more radiation than loose soil. The Nuclear Meter is placed directly on the soil to be tested and is turned on. See Figure 15. Gamma rays from a radioactive source penetrate the soil, and, depending on the number of air voids present, a number of the rays reflect back to the surface. These reflected rays are registered on a counter; and the counter reading visually registers the soil density in lb. per cu. ft.

This density is compared to the maximum density from a Proctor Test and the relative Proctor Density is obtained as before.

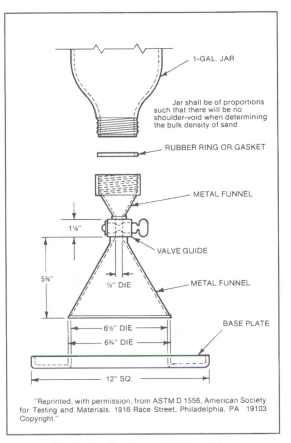

"Reprinted, with permission, from ASTM D 1556, American Society for Testing and Materials. 1916 Race Street, Philadelphia, PA 19103 Copyright."

Figure 14 - Sand Cone Testing Device

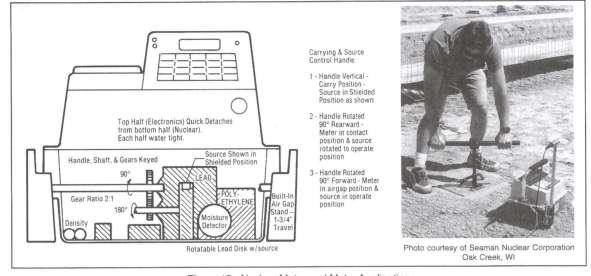

Figure 15 - Nuclear Meter and Meter Application

Figure 1.6.13 *(By permission: Wacker Corporation, Menomonee Falls, Wisconsin)*

The Nuclear Method has gained in popularity because it is accurate and fast — test results are obtained in 3 minutes — and the soil is not disturbed. Newer Nuclear Meters incorporate "quick indicator modes" for instantly checking the density after each pass the equipment makes. In addition, the Nuclear Meter method helps to quickly establish optimum compactor usage by eliminating over compaction, equipment wear and abuse, and wasted operator time. The initial cost of a Nuclear Meter can be upward of several thousand dollars; however, the time saving per test is considerable when compared to other methods.

The latest development in soil and asphalt density testing is called Density On The Run. With Density On The Run, the nuclear density meter is actually attached to the roller and a display screen is positioned near the operator for a continuous, visual read-out of the density measure. So the operator can precisely check the density of soils or hot asphalt without stopping to take a reading or rolling for miles with poor compaction results.

This compactor/density meter combination reduces roller hours, prevents over compaction, and encourages more efficient and productive man hours.

Figure 16 - Nuclear Density Meter on Soil and Asphalt
Photos courtesy of Seaman Nuclear Corporation Oak Creek, WI
Figure 1.6.13 *(cont.)*

1.6.14 Equipment Types and Selection

How does one choose the right compactor for the job? The answer is not always straightforward or simple, because a number of factors must be considered, mainly, soil type, physical conditions of the job site, compaction and specifications to be met.

The above factors must be evaluated with two purposes in mind. First, to determine which machines are able to do the job, and secondly, to recommend the one which will do the job most economically.

Let's discuss each factor separately.

Soil Type

As stated earlier, soil may be granular or cohesive in nature. See Figure 17.

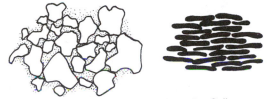

Figure 17 - Granular and Cohesive Soils

Granular Soils

For **granular** soils, compaction by vibration is most effective and economical. Vibration decreases friction between soil particles allowing them to rearrange themselves downward into a tightly packed configuration, eliminating all air voids. The effect of vibration penetrates deep into the soil, meaning that large layers of soil may be compacted, which contributes to the economy of the compaction process.

Vibratory plates are the machines commonly specified for use on granular soils because they are dependable, relatively inexpensive and very productive.

Vibratory rollers are used where even higher production rates are necessary.

The various granular soils have different **Natural Resonant Frequencies**, defined as that frequency which causes the greatest soil particle motion.

The **smaller** the soil particle, the **higher** the natural frequency; the **larger** the particle, the **lower** the natural frequency. See Figure 18.

That is why a lightweight vibratory plate, of 183 lb., with a high frequency of 6250 vibrations per minute and low amplitude, is the best compactor for fine and medium sands. Other vibratory compactors with lower frequencies and higher amplitude are necessary for coarse sands, gravels, and mixes containing more cohesive particles.

For optimum compaction, a plate with a frequency approximately equal to the natural frequency of the soil particle mix being compacted should be used.

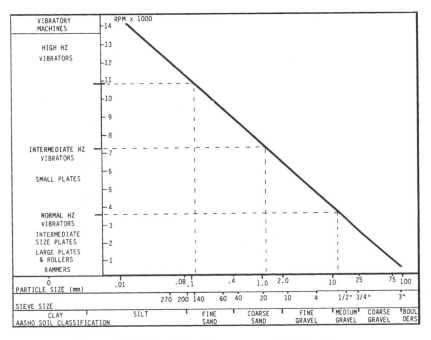

Figure 18 - Vibration Frequency - Particle Size

Figure 1.6.14 Choosing compaction equipment based on soil conditions. (*By permission: Wacker Corporation, Menomonee Falls, Wisconsin*)

1.6.15 Pea Gravel

A Word About Pea Gravel

One common misconception about Pea Gravel, a granular soil, is that it is not compactable and, therefore, does not require compaction.

Pea Gravel **is** compactable because the stones are not all perfectly round, making them subject to settling if compaction is not performed.

See Figure 19: Whether the soil settles 1/2 inch or 4 inches does not matter, as there is no support for the structure above it either way. So, be sure that Pea Gravel is always compacted.

Slurry Mixes also tend to settle in the same manner, and these mixes should NOT be specified **without** compaction.

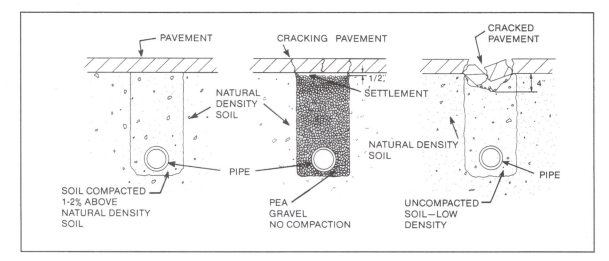

Figure 19 - Uncompacted Pea Gravel as well as Uncompacted Soil will Settle and Cause Damage to Support Structures

1.6.16 Cohesive Soils

For **cohesive** soils, impact type machines must be used. The impact force produces a shearing effect in the soil, which binds the pancake shaped particles together, squeezing air pockets out to the surface.

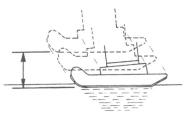

A high shoe lift of the ramming machine is very desirable in order to provide a high impact energy and to make the forward advance possible. A high rammer speed, in the range of 500–700 impacts per minute, also creates a vibratory action that is desirable with granular as well as with cohesive soils. Vibratory trench rollers with special cleated drums also perform well on cohesive soils because of their shearing action.

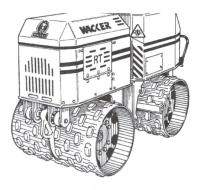

Thus, as a general rule:

For **granular soils**, the first choice should be a vibratory plate or smooth drum vibratory roller. See Figure 20. Your second choice could be a rammer for narrow trench applications. See Figure 21.

For **cohesive soils**, a rammer or vibratory trench roller should be used.

Figure 1.6.16   (*By permission: Wacker Corporation, Menomonee Falls, Wisconsin*)

1.6.17 Matching the Machine to the Job

WP 1550A Vibratory Plate

RSS 800A Vibratory Roller

Figure 20

BS 600 Vibratory Rammer

DS 720 Diesel Rammer

Figure 21

Physical Conditions of the Job Site

In a trench or next to a foundation wall, the space available often determines the machine model.

A 6" wide utility trench necessitates using a rammer with a shoe size not exceeding 6" wide.

A 24" wide trench with granular fill may be compacted using a rammer or vibratory plate, with the vibratory plate being the faster machine. If there is no room at the end of the trench to turn a unidirectional plate around, then a reversible vibratory plate should be used.

On the other hand, when compacting the granular base for a large warehouse or driveway, only a large gasoline or diesel vibratory plate or vibratory roller will provide the area capacity needed to do the job in a reasonable time.

Figure 1.6.17   *(By permission: Wacker Corporation, Menomonee Falls, Wisconsin)*

1.6.18 Specifications to Be Met

Many compaction specifications are still written as the "method" type which specify the type of machine to use, the soil depth or lift, and the number of passes. Machine selection in this case, is dictated by the specifications.

Many method type specifications indicate ignorance as to the "state of the art" of modern vibratory compaction equipment. For instance, it is completely unrealistic to specify a vibratory plate with 8000 lb. of centrifugal force to run over lifts of 4", and specify a minimum of 20 passes.

These type of specs not only waste many man-hours, but drastically increase the maintenance cost of compactors and, most of all, do not achieve compaction.

Most specifications, however, are the "end result" type, allowing the use of any equipment which will achieve the specified Proctor Density.

Soil is compacted in layers which are called "lifts", and most manufacturers rate their equipment as to the maximum lift each machine can compact under ideal conditions.

In the case of "end result" specs, recommend a machine according to the soil type and job physical dimensions, and be sure it has a lift rating greater than the depth of soil layer to be compacted.

However, for modern compaction machines, the lift to be compacted should not be less than 1/3 the maximum rated lift. If thin layers must be specified, a lighter machine must be used to achieve proper compaction.

During any compaction process, it is very important that the soil be at or as close to **optimum moisture** as possible as this will insure achieving the density required and expending the least amount of energy and making the minimum number of passes of the equipment.

Performing compaction on soil that is so dry that the compaction effort raises a cloud of dust is a waste of time and money. Such dry soil (See Figure 22) will not accept compaction energy, and even many passes will not compact the soil to an acceptable density.

All Wacker compaction equipment is designed and rated to provide 95% or better Standard Proctor Density with three to four passes, when the soil moisture is near optimum.

Figure 22 - Soil Is Too Dry For Good Compaction Results

Soil Should not be Overcompacted

As soon as the specified density is reached, compaction should be stopped. If the machine is continued to be run over a compacted area, soil particles will start to move sideways, under the effect of continued compactor pressure; thus breaking up a stable soil which results in a decrease in density.

If possible, soil testing should be started immediately after the first and each pass thereafter, continuous monitoring will eliminate the possibility of damaging the machine.

Summary

When recommending equipment for a compaction job, the soil types have to be taken into consideration.
Soil Type:

Cohesive Soils	— Use a rammer or cleated vibratory trench roller.
Granular Soils	— Use a vibratory plate or roller. — Rammers can also be used.
Mixed Soils	— Use any rammer or trench roller. Some faster vibratory plates and rollers can handle mixed materials.

Physical Dimensions and Restrictions of the Site

Match the size of the machine to the job.

Wacker offers optional narrow shoes for rammers and special narrow plates when compacting in extremely narrow areas.

Reversible vibratory plate are available for use in trenches and open areas where turning is impossible or inconvenient.
Specifications: Density requirements, site size, optimum moisture, and number of passes — Match the equipment to these special requirements.

Figure 1.6.18   *(By permission: Wacker Corporation, Menomonee Falls, Wisconsin)*

1.6.19 Vibratory Rammers

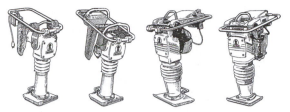

Figure 23 - Vibratory Rammers

Rammers produce an **impact force** which is necessary for the compaction of **cohesive soils**. See Figure 23.

Wacker rammers are classified as **vibratory impact rammers**. That is due to their high number of blows per minute, ranging from 450 to 800. At this high impact rate, vibration is induced in the soil. Because of their vibratory action, coupled with impact, vibratory impact rammers can also be used on **granular and mixed soils**.

An efficient rammer should provide:
1. **High impact power** (ramming shoe must come off the ground 2"–3")
2. **Good balance** (easy to guide and good shock isolation to reduce operator fatigue)
3. **Durability** (to withstand the high stresses created in the rammer)
4. **Easy maintenance**

Certain points should be considered when purchasing or comparing rammers. See Figure 24.

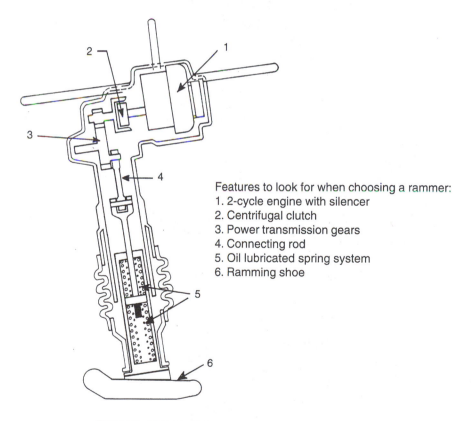

Features to look for when choosing a rammer:
1. 2-cycle engine with silencer
2. Centrifugal clutch
3. Power transmission gears
4. Connecting rod
5. Oil lubricated spring system
6. Ramming shoe

Figure 24 - Rammer Cutaway

Figure 1.6.19 *(By permission: Wacker Corporation, Menomonee Falls, Wisconsin)*

Two-Cycle Engine

Lubrication of the moving parts is provided by oil mixed with the fuel. A 4-cycle engine with the lubricating oil placed in the sump does not allow for hard impact jumps and side tilting of the machine because when tilted, the oil runs towards the cylinder and splashes against the piston causing heating, foaming and lubrication problems. Therefore, a 2-cycle engine is necessary to allow for the up and down jumping action, tilting when operating on slopes and transportation in the horizontal position. The 2-cycle engine has no valves and is light in weight.

Centrifugal Clutch

A centrifugal clutch allows easy engine starting with the ramming system disengaged, and also allows the engine to idle during short work stoppages.

Power Transmission Gears

The power which is developed by the engine must be transferred to the ramming system to produce 450 to 800 impact blows per minute. This is the function of the power transmission gears.

In all Wacker rammers, hardened gears machined from forgings do the job. Belt power transmissions are used in some rammers to reduce the cost of manufacturing; however, transmission via belts can prove to be quite costly to the end user in terms of increased maintenance and down time.

Ramming System

Also called spring system. It has two functions. One is to store the energy developed by the engine and then release it to the shoe during the downward stroke. The second is to work as an elastic buffer between the oscillating ramming shoe motion and the circular rotation of the upper mechanical components, that is the gears, clutch and engine.

Wacker spring systems are comprised of two sets of springs, with a piston in between. Each set of springs consists of two or three separate coil springs placed inside each other. Therefore, the total number of springs in any Wacker rammer is four or six springs depending on machine size. Lesser number of springs, or no springs at all, reduce the efficiency and impact power.

Oil Bath Lubrication

High quality rammers incorporate a sealed oil bath lubrication system. The oil is splashed throughout the machine, providing reliable and continuous lubrication for all internal parts. A periodic oil change is all the attention that an oil lubricated machine needs. Daily or weekly greasing is eliminated. Grease lubrication is only used in rammers which were designed decades ago.

Silencer

Equipment owners and operators as well as government agencies do not tolerate noisy equipment; therefore, most rammers are equipped with engine silencers as standard equipment.

High Impact Force

A **high impact** machine is desirable so that deeper soil lifts may be compacted. High impact force is possible only with a well designed spring system and a long shoe stroke.

A short shoe stroke reduces the impact force and prevents machine advance on slopes and unleveled soil.

Figure 1.6.19 *(cont.)*

1.6.20 Vibratory Plates

Vibratory plates apply high frequency, low amplitude vibrations to the ground, and are used mainly for compacting granular soils such as sand and gravel; mixes of granular and cohesive soil; and asphalt mixes, both hot and cold. See Figure 25.

Figure 25 - Wacker Vibratory Plate

Usually powered by small gasoline or diesel engines up to about 10 hp., some vibratory plates are available with electric motors for use where noise or fumes must be minimized.

A vibratory plate consists essentially of two masses, upper and lower. The upper mass includes the engine, centrifugal clutch and engine console. The lower mass,

which includes the base plate with the vibration producing exciter unit rigidly bolted onto it.

Vibratory plates are generally high production machines, in terms of volume of soil compacted (cubic yards per hour), because of their fast forward speed, deep effective lifts and large plate contact areas. See Figure 26.

Figure 26 - Two Wacker Gasoline Vibratory Plates

Rubber shockmounts isolate the upper mass from the vibrating lower mass. The power transmission from the engine to the exciter is achieved using V-belts. The guiding handle can be mounted on the upper or lower mass and is usually rubber shockmounted to reduce operator fatigue. Figure 27 shows the main components of a vibratory plate.

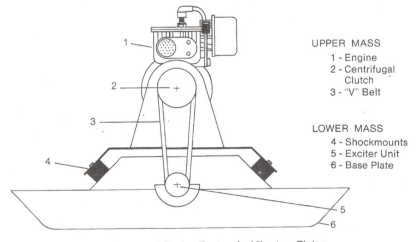

UPPER MASS
1 - Engine
2 - Centrifugal Clutch
3 - "V" Belt

LOWER MASS
4 - Shockmounts
5 - Exciter Unit
6 - Base Plate

Figure 27 - Important Design Factors for Vibratory Plates

Figure 1.6.20   *(By permission: Wacker Corporation, Menomonee Falls, Wisconsin)*

Important Design Factors for Vibratory Plates

The total static weight, the exciter design, the exciter frequency, and the positioning of engine/exciter all play an important part in the efficiency and performance of the vibratory plate.

Static Weight

The static weight of a small vibratory plate (150-300 lb. weight class) is usually negligible compared to the centrifugal force that is generated in the exciter. Here, the vibratory force is the dominant force which acts on soil particles during the compaction process. For larger vibratory plates (above 300 lb.), the vibration action, as well as the static weight have a combined effect on soil particles. The total effect is to vibrate and squeeze soil particles together to achieve compaction.

Exciter Design

The exciter unit of any vibratory plate can be thought of as the "heart" of the machine. See Figure 28.

Exciter units operate on the principle of turning an unbalanced "eccentric" weight at high speed to produce centrifugal force. This centrifugal force causes the machine to vibrate, move forward and compact the soil.

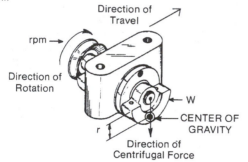

Figure 28 - Exciter Unit of a Wacker Vibratory Plate

The formula showing the centrifugal force produced by an exciter is:

Centrifugal Force = w x r x rpm² x K
Where
w = weight of the eccentric
r = radius from center of shaft to center of gravity of the eccentric weight
rpm² = speed of the exciter unit square
K = A constant factor = $\frac{2\pi}{60}$

The fact that centrifugal force produced varies with the square power of the exciter speed is of practical importance in the operation of a vibratory plate for two reasons.

If the engine is overspeeded just a little bit, centrifugal force increases a lot; which will overload the exciter bearings.

If the engine runs underspeed just a little bit, centrifugal force will be much too low, causing poor performance, slow forward speed and low compaction effort.

Therefore, it is extremely important that any vibratory plate engine be set to the manufacturer's recommended speed with a tachometer.

Exciter units can be either oil or grease lubricated. On an oil lubricated exciter, it is important that the exact amount of oil specified be contained in the unit. Too little oil will cause the bearings to burn up; too much oil will allow the exciter weights to churn the oil, causing foaming, overheating and poor performance.

Exciter Frequency

Each soil particle size responds differently to the various exciter frequencies. Laws of physics state that a small mass responds favorably to rapid vibration and that a larger mass responds favorably to slower vibration. Therefore, an attempt should be made to match the frequency of the exciter to the dominant particle size of a soil mix.

As the exciter frequency approaches the resonant frequency of particles, sympathetic vibration occurs and soil particles vibrate with maximum amplitudes.

Engine/Exciter Layout

The relative position of the exciter mounting to the engine is also an important design factor.

A centrally mounted exciter (Figure 29) is one placed in the direct center of the base plate, directly under the engine. This provides uniform amplitudes at the front and rear of the plate.

Figure 1.6.20   *(cont.)*

A front mounted exciter (Figure 29) is placed at the front of the base plate, and the engine is mounted in the rear. The amplitude at the front of the base plate is larger than that at the back. The result is faster forward speed and the ability to compact soil with a certain amount of cohesive material content. This design also allows for lower overall center of gravity which contributes to the stability of the machine.

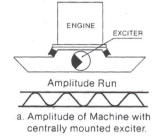

a. Amplitude of Machine with centrally mounted exciter.

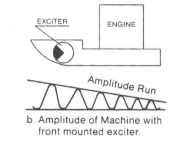

b. Amplitude of Machine with front mounted exciter.

Figure 29 - Types of Vibratory Plates

Reversibility

Some larger vibratory plates are unique due to the fact that they are **reversible**. The exciter system of a reversible vibratory plate has two eccentric weights that revolve in opposite directions. (See Figure 30).

These eccentric weights are arranged in a way that will move the plate in the opposite direction every time the relative position of one eccentric is changed 180° with respect to the other. This is done by a special spring and cam changing device that insures 180°° change in relative position with each shift without changing the direction of rotation of the two eccentric weights.

The change of direction of travel of a reversing plate occurs instantaneously at full shaft speed, without having to bring the plate to a neutral position or stop.

Modern designers use hydraulic power to change the eccentrics in infinite increments from full forward to full reverse.

This design allows a reversible plate to do spot compaction, i.e., no forward/reverse motion but the full centrifugal force of all eccentrics is used for compaction only. No energy is wasted to propel the machine.

The latest vibratory plate design features 2 pairs of reversible eccentrics in the exciter housing. These plates are fully steerable at the touch of a finger. And for additional operator safety -- some of these steerable plates are remote controlled. The operator stands on top of a trench and the unit works down in the trench. This safety feature may save costly trench shoring in many places.

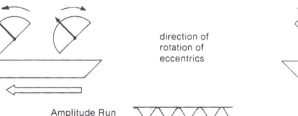

direction of rotation of eccentrics

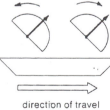

direction of travel

Figure 30 - Schematic of a Reversible Vibratory Plate

Diesel Power

Diesel powered vibratory plates offer several advantages over similar gasoline models. One is economy of operation in terms of both lower overall fuel consumption and lower cost per gallon of diesel fuel versus gasoline. Secondly, maintenance costs are also lower because diesels have no spark plugs, carburetor, or electrical ignition components to service.

Third, one can expect longer engine life due to the sturdier construction of diesels.

A diesel powered plate has higher initial cost, although its lifetime cost per hour of operation is lower than that of a gasoline engine.

Figure 1.6.20 *(cont.)*

1.6.21 Rollers

In general, rollers may be classified as static, vibratory, sheepsfoot, or pneumatic tire.

Static rollers (Figure 31) rely on their intrinsic weight to achieve compaction. Their use for soil compaction is steadily diminishing with the introduction of vibratory rollers because the static roller must have high intrinsic weight in order to handle even moderate soil lifts. The heavy static weight means higher component costs and increased size which make handling and transportation difficult.

Static rollers, however, are still used for asphalt rolling as they provide the desirable smooth surface.

Figure 31 - Static Roller

Vibratory rollers (Figure 32) have exciter weights in one or more drums and provide vibration action (dynamic force) in addition to the static weight, vibratory rollers produce superior compaction, particularly on granular soil because the vibratory impulses break up the frictional force between soil particles, thus allowing deeper layers of soil to vibrate and settle. The vibratory action permits the use of larger lifts and provides quick and effective particle rearrangement i.e., compaction.

Figure 32 - Wacker Vibratory Roller
For Use on Soil or Asphalt

Sheepsfoot rollers, static or vibratory, (Figure 33) have drums with many protruding studs, each similar in shape to a sheepsfoot, that provide a kneading action on the soil. The total force becomes concentrated on the small protruding sheepsfoot.

These machines can effectively compact cohesive soils as they break hard soil lumps and homogenize the soil into a dense layer. Sheepsfoot rollers are sometimes used for drying areas saturated with water because they create multiple indentations in the soil, increasing the exposed surface area, thus speed up drying.

Figure 33 - Wacker Sheepsfoot Roller

On **Pneumatic tired rollers,** (Figure 34) the combination of gross static weight, number of wheels, tire size, inflation pressure and travel speed, all have a bearing on the compaction performance. The pneumatic wheels are usually arranged so that the rear ones will run in the spaces between the front ones; theoretically leaving no ruts. Pneumatic tired rollers are mainly used for compacting granular soil bases as well as for compacting asphalt surfaces.

Because of their static roller design, Pneumatic tired rollers are primarily surface compactors with effective compaction depths up to 6"

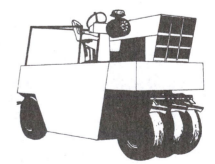

Figure 34 - Pneumatic Tired Roller

1.6.22 Other Compacting Rollers

There are other specialized rollers, such as the segmented pad and sanitary landfill type compactors; however, these are outside the scope of this booklet.

For confined area soil compaction, the double-drum vibratory rollers dominate the roller field. See Figure 35.

Figure 35 - Wacker Double Drum Walk-behind Roller

In a roller design such as the Wacker Model W74, both drums vibrate. While one drum is in its upper motion, the other drum is moving downwards and hitting the ground; as shown in Phases A & C in Figure 36. This means that there is always one drum in contact with the ground, and that the impact force is always directed downwards. The horizontal forces in each drum (Phases B & D) are equal and opposite in direction, and, therefore, they cancel each other out. See Figure 36.

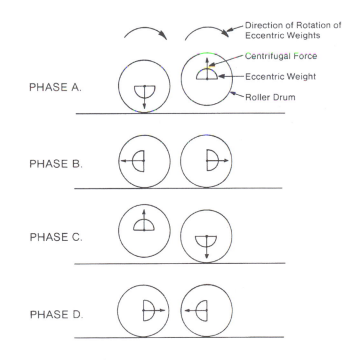

Figure 36 - Direction of Centifugal Forces from a Wacker Model W74 Vibratory Roller

Figure 1.6.22   *(By permission: Wacker Corporation, Menomonee Falls, Wisconsin)*

Because of the absence of a horizontal force, the forward motion of the machine has to be maintained through a direct gear drive, which propels both drums in both a forward and a reverse mode.

When selecting a walk-behind double drum vibratory roller for confined area soil compaction, the following specifications and parameters should be considered.

- Drum width for suitability to the application.
- Static weight for ease of handling and transport.
- Centrifugal force output for deep soil compaction.
- Dynamic linear force which is the centrifugal force per inch of the drum width.

$$\text{Dynamic Linear Force} = \frac{\text{Centrifugal force}}{\text{Number of drums x Drum width}}$$

For the Wacker roller, Model W74, the Dynamic Linear Force is:

$$\frac{9000 \text{ lb.}}{2 \times 29.5 \text{ in.}} = 152.5 \text{ lb./in.}$$

- Static linear force which is the static force per inch of the drum width.

$$\text{Static Linear Force} = \frac{\text{Static Weight}}{\text{Number of drums x Drum width}}$$

For the Wacker Roller, Model W74, static linear force is:

$$\frac{1863 \text{ lb.}}{2 \times 29.5 \text{ in.}} = 31.6 \text{ lb./in.}$$

Using Wacker roller, Model W74, as an example, the dynamic linear force is shown to be 5 times the static linear force. For this reason, a small vibratory roller can achieve higher compaction results than a static roller 5 times its size.

Other considerations in the selection of a roller compactor are adequate engine horsepower for reserve power in handling tough applications and grades; the sealing of all moving components from dust and moisture and the availability of a water sprinkling system for blacktop application (water is used to lubricate and clean the drum surfaces). Also keep in mind; the frequency of maintenance, accessibility, of components for service and maintenance, and the ready availability of parts and service.

The advantages of using vibratory rollers are: they are used for larger area compaction where width and speed are beneficial, they are used on inclines where vibratory plates do not have enough gradability, in trenches and around footings where the roller can propel itself without requiring lifting equipment to place it as well as on rough and uneven ground where the positive drive makes for easier travel. See Figures 37 and 38.

The designs of roller compactors are many and varied. Often the application dictates the design used. It is well to remember that it is not always the largest design that produces the most work.

Figure 37 - Double Drum Walk-behind Compacting Soil

Figure 38 - Double Drum Walk-behind Compacting Asphalt

Figure 1.6.22   *(cont.)*

1.6.23 Equipment Maintenance

Poor maintenance is the main cause of premature construction equipment failure. Because soil compaction equipment operates exclusively under conditions of heavy dust and vibration, proper machine maintenance is vital for long life.

The following are maintenance and repair tips which apply to different types of compaction equipment.

Rammers

Dust is rated as enemy number one for every engine. Under normal operating conditions, the elements will not require cleaning and should not be removed from the machine. If the elements do become plugged with dirt, the engine will begin to lose power. In this case, the air cleaner elements can be removed and cleaned as described below. Replace an element if it becomes so plugged with dirt it can no longer be cleaned.

Erratic operation or drop in rammer performance is usually caused by clogged outlet ports at engine exhaust. If this happens, clean ports and inspect the muffler. If the clogging problem occurs frequently, check fuel/oil mix. The correct ratio is 50:1. Too much oil accelerates carbon buildup and promotes spark plug fouling. To little oil causes excessive wear because of lack of lubrication.

In-tank fuel filters should be periodically cleaned with a solvent and the tank flushed to remove any sediment.

Lubrication of the rammer's spring system (lower unit) should be checked. This is done from the outside through an oil sight glass. Change system oil, first after 50 operating hours, then after each 300 operating hours.

Damaged ramming shoes and broken mufflers are often caused by loose bolts. Always tighten plow bolts on the shoe and inspect all nuts holding the muffler on engine.

The operator must know how to operate the rammers so that the shoe hits parallel to the soil surface. Not on its toe, nor on its heel. This prolongs the service life of the shoe.

Avoid overthrottling the rammer. This damages the shoe and contributes to erratic operation and difficult handling. Best compaction is achieved when the rammer produces smooth, rhythmic tamping action.

Care should be taken when loading and unloading rammers; springs may break, if machine is dropped from a high distance on hard ground.

The 2-cycle, air-cooled rammer engine always takes a minute or two to warm up on initial start. It is during this period that carburetors are often mis-adjusted because the machines do not appear to be running properly. This practice should be avoided. However, the Wacker rammer engine has a tinker-proof carburetor and adjustments are eliminated.

Vibratory Plates

Vibratory plates are relatively simple machines and usually require little maintenance.

As with rammers, the air filter must be kept functional at all times. Inspect and clean the air filter every 8 operating hours or more often as necessary. Replace the air filter cartridge on gasoline engine driven units, as recommended in the Service Manual. Change oil in the oil bath type air filter, for diesel engines, every 8 operating hours or more frequently when operating under dusty conditions. **Never run an engine without an air filter!**

Check and change engine oil as recommended for each model. Use good quality detergent oil of the recommended viscosity to suit different seasons.

Check and change exciter oil. Do not overfill exciter with oil. Overfilled exciter causes excessive heating and foaming of oil which may result in power loss (as most of engine energy would be expended in moving and splashing the oil unnecessarily); belt overheating and failure; loss of lubricating effect of exciter oil as a result of excessive heating or bearing failure.

Figure 1.6.23 (*By permission: Wacker Corporation, Menomonee Falls, Wisconsin*)

Follow instructions in the service manual on recommended filter changes and hydraulic fluid types.

If a gasoline engine loses power, and if all other normal components such as spark plugs, points, high-tension cable, etc, check all right, then the engine combustion chamber must be cleaned of combustion deposits. Note here that unleaded gasoline may be used in vibratory plate engines with some added advantages such as a cleaner spark plug and valves, and less contamination of lubricating oil.

Inspect rubber shockmounts under engine console or at handle and replace those that are defective, i.e., cracked or hardened.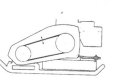

Check belt tension to prevent slipping (slack belts) and to prevent overloading of bearings and shockmounts (tight belts).

All large plates are equipped with lifting eyes for crane handling. Do not lift machines at other points to avoid breakage of machine components.

Keep base plate clean and free of soil accumulation. Dirt accumulation only increases the load unnecessarily, resulting in exciter oil overheating and a drop in performance.

Rollers

Wacker rollers are hydraulically driven.

Inspect and change engine oil regularly. Replace fuel filter at recommended intervals. This is a standard procedure for any internal combustion engine which should not be neglected.

The other type of drive systems used on Wacker rollers are hydraulic or hydrostatic transmission. A hydraulic system makes use of a fixed displacement pump supplying hydraulic fluid flow to steering cylinders, drum drive motors, and exciter drive motors. Since the output of the pump is continuous, control valves are required allowing operator control of the various machine functions.

By contrast, the hydrostatic system has an infinitely variable drive. It depends on a variable displacement hydraulic pump to provide totally smooth speed control from full forward, through neutral, to full reverse.

A hydraulic system is fairly easy to maintain. Cleanliness is No. 1 when it comes to servicing hydraulic systems. **Keep dirt and other contaminants out of the system.** Key maintenance problems are usually: not enough oil in system, incorrect oil, clogged or dirty filters, and loose lines.

Figure 1.6.23 *(cont.)*

1.6.24 Compaction Costs

As cost is a major factor in any construction project, the cost of compaction must be taken into consideration.

The examples below are an indicator of how cost of compaction can be calculated and how machine performance greatly affects these costs. Because machine prices are subject to change, the example is a comparison between two vibratory Plates.

The price information is hypothetical but the performance figures and formulae are **actual** and may be used by job estimators to forecast costs of compaction.

Descrpition	Unit of Measure	Machine "A"	Machine "B"
Weight	lb.	183	267
Plate width	in.	21	21
Centrifugal force	lb.	2850	4700
Lift	in.	10	13
Speed	ft./min	66	87
Fuel consumtion	qt/h	1.6	1.7
Production rate	cu yd/h	200	366
Hypothetical list price	$	1300	1800

Let us assume that we want to compact the granular backfill around the outside foundation walls of a high rise building. The building is rectangular, 400 x 300 ft. The width of trench around the building is 10 ft. The depth of trench is 10 ft. (See Figure 39).

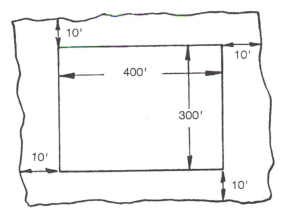

Figure 39 - Jobsite Layout

Volume of soil to be compacted:
= 2 [(300 x 10 x 10) + (420 x 10 x 10)]
= 2 x (30,000 + 42,000)
= 2 x 72,000
= 144,000 cu. ft.
or
= 144,000 cu. ft.
    27
= 5333 cu. yd.

By using the maximum effective lift for each machine, 95-100% Proctor Density is expected to be reached after 3 passes provided that moisture is close to optimum.

To determine the total cost of owning and operating the two vibratory plates for this job, assume that:

1. Machines have a life expectancy of 3000 hours. (4 hours per day, 250 working days per year, 3 years).
2. Depreciation is determined by pro-rating the price of the machine over their 3000 hour life.
3. Maintenance cost is 80% of the depreciation cost.
4. Fuel cost is $1.20 per gallon.*
5. Labor cost is $17.00 per hour.*

*These costs will vary.

Figure 1.6.24 *(By permission: Wacker Corporation, Menomonee Falls, Wisconsin)*

Using the above assumptions, the depreciation cost is:

Machine A: $\dfrac{\$1300}{3000 \text{ hr.}} = \$0.43$ per hour

Machine B: $\dfrac{\$1800}{3000 \text{ hr.}} = \$0.60$ per hour

The maintenance cost is:

Machine A: $0.43 x 80% = $0.34 per hour

Machine B: $0.60 x 80% = $0.48 per hour

Fuel costs are:

Machine A: $\dfrac{1.6 \text{ quarts per hour}}{4 \text{ quarts per gallon}}$ x $1.20 per gallon
= $.48 per hour

Machine B: $\dfrac{1.7 \text{ quarts per hour}}{4 \text{ quarts per gallon}}$ x $1.20 per gallon
= $.51 per hour.

Thus the total for machine depreciation, maintenance and fuel is:

Machine A: $0.43 per hour
0.34 per hour
0.48 per hour
Total: $1.25 per hour

Machine B: $0.60 per hour
0.48 per hour
0.51 per hour
Total : $1.59 per hour

Using Vibratory Plate Machine "A"

Production rate per hour   200 yd³/h

Production rate per 3 passes $\dfrac{200}{3} = 66.7$ yd³/h

Time needed for compacting 5333 yd³ $\dfrac{5333}{66.7} = 80$ hours

Based on the equipment costs of — $1.25 per hour; and labor costs of — $17.00 per hour;
Total cost of using Machine "A" for this job: ($17.00 + $1.25) x 80 hours = $1460.

Using Vibratory Plate Machine "B"

Following the same steps as before:
Production rate per hour   366 yd³/h

Production rate per 3 passes $\dfrac{366}{3} = 122$ yd³/h

Time needed for compacting 5333 yd³ $\dfrac{5333}{122} = 43.7$ hours

Based on equipment cost of $1.59 per hour and labor cost of $17.00 per hour:
The total cost of using Machine "B" for this job: ($17.00 + $1.59) x 43.7 hours = $812.38
Summarizing preceding calculations:

	Machine "A"	Machine "B"
Equipment costs (Depreciation, maintenance, and fuel) per hour:	$1.25	$1.59
Labor costs per hour:	+ $17.00	+ $17.00
Total costs per hour:	$18.25	$18.59
Time required to finish job (hours)	x 80 h	x 43.70 h
Total cost for this job:	$1460.00	$812.38
Savings by using Machine "B":		**$647.62**

Figure 1.6.24   *(cont.)*

Note that:

1. The hourly cost of depreciation, maintenance and fuel for Machine "B" is 27.2% greater than that of Machine "A" ($1.59 vs. $1.25).
2. Machine "B" required a little more than half the time needed by Machine "A" to finish the job (43.7 hours vs. 80 hours).

The above should not lead us to believe that Machine "B" is costly to own and operate. Its high rate of production offsets the higher cost of initial investment.

Since labor constitutes the major expense of any Confined Area Compaction operation, it is, therefore, always advisable to recommend the most productive machine, rather than the least expensive one.

The smaller machine costs less in terms of ownership but more in labor. As one combines both equipment and labor costs, the total cost to finish the job is $1460 for the small machine as compared to $812.38 for the larger machine. That is to say, the smaller machine, "A", costs almost twice as much to finish this job as the larger machine, "B". This final cost is what determines the profit or loss of the compaction process.

In general, for each compaction situation, one should look at the cost of equipment, as well as the cost of labor involved, and then try to strike a balance between equipment and labor costs.

Conclusion

This booklet has reviewed soil types, soil compaction tests and the factors that influence the selection of compaction equipment. Also, the general principles and applications of rammers, vibratory plates and rollers were covered.

With this information, one should now have a better understanding of soil compaction and the equipment connected with it and should also be in a position to be able to recommend the right equipment for the job.

If additional confined area compaction application assistance is needed, please consult with the Wacker Sales Engineering Department for expert technical advice; it is free and your inquiries are always welcome.

Figure 1.6.24 *(cont.)*

Excavating and Hoisting Equipment

Contents

2.0.0 Changing Roles of Construction Equipment

Changes in construction excavating equipment over the past several decades have provided the contractor with more manufacturers from whom to chose, and the introduction of a line of compact excavators has added a whole new dimension to working in confined spaces. The steer-skidder is now available with such a wide range of attachments that it has assumed the role of workhorse in and around construction sites. The fitting of lasers to excavating equipment has opened the door to more exacting tolerances in both rough and finish grading. One manufacturer of such equipment indicates that the payback is not so much in more controlled operations, but in the savings of imported fill or porous drainage material since the grade achieved via laser guided equipment is more consistent. As satellite-operated ground positioning equipment (GPS) becomes adapted to excavating equipment, the limits of more exacting line and grade achievement can only be imagined. Although the diagrams and specifications presented in this section of the book represent only a few of the many equipment manufacturers on the market today, their operational performance and configurations will be somewhat similar to competing brands.

2.1.0 Primary Use for an Excavator

The primary use for an excavator is digging. There are all types of buckets available, many designed for very specific applications. Specialized attachments make it possible to do a number of tasks faster, more efficiently and more economically.

Buckets are designed for different uses. They also have two different capacity measurements, struck capacity and heaped capacity.

Struck capacity is the volume inside the bucket enclosure itself without consideration for any material retained by the spillplate or bucket teeth.

Heaped capacity is the volume inside the bucket enclosure plus the volume of material

SAE BUCKET RATING

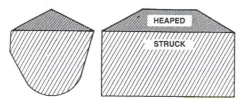

heaped above the strike-off plane, having an angle of repose of 1:1. Material retained by the spillplate or bucket teeth is not considered.

Typical excavator buckets include:
General purpose
Digging
Ditch cleaning/digging
Excavation
Utility
Trapezoidal
Weed cutting
Rock
Extreme service
Mass excavation
Rock ripping
Trenching
V-type excavation
Clamshell
Tilt

Booms and dipper sticks can also be considered as part of the attachment family. On many machines you can select different length booms and sticks, depending on the application.

Not all excavators can accommodate all types of buckets. It is important to size the bucket to

the machine. Bucket teeth also play an important role in the machine's application and will directly impact machine productivity. There are specific teeth for the various buckets and bucket applications.

Other attachments that make excavators such good work tool platforms include:
Quick couplers
Thumbs
Grapples:
Construction
Trash
Scrap
Logging
Hydraulic hammers
Vibratory plate compactors
Compaction wheels
Shears
Crushers
Pulverizers
Rock drills
Sorters
Mowers
Tree harvesters
Stump cutters
Augers
Brooms
Multi-processor

Make certain you size the attachment to the machine's capacities. If an attachment is too

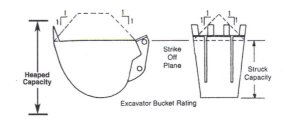

Excavator Bucket Rating

big for the machine you stand a good chance of damaging the machine as well as failing to achieve production goals. If you have doubts, consult your equipment dealer or manufacturer.

BACKHOE LOADERS:

One of the most popular tool carriers is the backhoe loader unless you want to call it the loader backhoe, or the tractor/loader/backhoe (often referred to as a TLB), or a backhoe. Actually the backhoe is an attachment used on skid-steer loaders...

The popularity of backhoe loaders on job sites continues to increase with the introduction of the many attachments that make them a virtual "Swiss Army Knife." With the recent introduction of the "toolcarrier" concept on the loader end of the machine and the addition of quick

Figure 2.1.0 *(By permission:* Construction, *ACP/Construction, Norcross, Georgia; Feb. 2000)*

couplers to the backhoe end, the backhoe loader has become a must for most contractors. It has become a "can-do-anything" machine.

Typical loader end attachments include:
- Quick coupler
- Buckets:
  - General purpose
  - Light material
  - Multipurpose
  - Heavy-duty
  - Side dump
- Pallet forks
- Hydraulic angle blades
- Hydraulic brooms
- Asphalt cutter
- Loader rake

Many of the attachments designed for compact wheel loaders can be used on the backhoe loader. Before doing this, consult with the equipment and/or attachment dealer.

Typical backhoe end attachments include:
- Quick coupler
- Ditch digging bucket
- Heavy-duty bucket
- Trenching bucket
- Extreme service bucket
- Hydraulic hammer
- Vibratory compaction plates
- Extendable reach sticks
- Ripper
- Jaw bucket
- Ripper bucket
- Auger
- Cold planer
- Concrete removal bucket

Many of the attachments designed for mini or compact excavators can be used on the current backhoe loaders. Always consult with your dealer before you do.

WHEEL LOADERS:

Wheel loaders make great attachment carriers. In fact, it's the wheel loader that becomes the tool-carrier with some design changes. The two machines are not the same. Wheel loaders are designed for load and carry operations and are production oriented. Most wheel loaders are designed with Z-bar linkage for greater breakout and greater lifting capacity. The Z-bar linkage will affect overall dump height when compared to the common parallel linkage used on the tool-carriers.

Buckets are the most common attachment for wheel loaders. They come in a wide array of configurations for applications ranging from general to very specific. As with excavators, loader buckets have two volume measurements, struck and heaped. The only difference in the way the capacities are calculated is that on the loader bucket the heaped material is at a 2:1 angle of repose.

Bucket capacity must be matched to the machine. Adding a quick coupler could cut down on the bucket size a machine can handle depending on its hydraulic system capacity. Always check with your dealer to make certain the bucket you are selecting fits the machine on which you will use it.

Typical wheel loader attachments include:
- Quick coupler
- Buckets:
  - General purpose
  - Coke
  - Coal
  - Coal seam
  - Wood-chip
  - Refuse
  - Light material
  - Loose material
  - Snow
  - Fertilizer
  - Sand and gravel
  - Rock
  - Crushed rock
  - Skeleton rock
  - Slag
  - Bonus
  - Multipurpose
  - Side dump
  - Control discharge
  - Top clamp
  - Grab and grip
  - Block handling
  - Stone sieve
  - High dump
- Tire loader
- Material handling arm
- Pallet fork
- Top clamp pallet fork
- Mill yard fork
- Log/lumber fork
- Logging fork
- Core fork
- Straight blade
- Angle blade — manual or hydraulic
- V-plow
- Reverse plow — manual or hydraulic
- Clamp rake

Figure 2.1.0   (*cont.*)

Clearing rake
Loader rake
Block handling fork
Hydraulic broom
Breaker tine
Boom clearing rake
Pickup sweeper
... And I'm certain that there are even more than the ones listed.

TRACK LOADERS:

Track loaders don't have the mobility that wheel loaders do but they do have excellent stability and great flotation. Track loaders can go places wheel loaders can't and are still extremely versatile machines. With the attachments available, the track loader's value on a job site continues to increase. If you need a loader on a job site that destroys tires, such as a demolition site, you need to consider the advantages of a track loader.

Buckets and blades make up the vast majority of attachments used with track loaders. Again, bucket capacity is given as struck and heaped.

Typical track loader attachments include:
Quick coupler
Buckets:
General purpose
Landfill
Landfill multipurpose
Skeleton rock
Coal
Wood-chip
Fertilizer
Multipurpose
Side dump
Demolition
Blades:
Trim
Straight
Angle — manual or hydraulic
Forks
Material handling arm
Loader rake

TOOLCARRIERS:

Toolcarriers are designed specifically to accommodate the use of attachments. Typically they come with integral quick couplers for fast tool exchange; have increased lift height and reach; provide parallel lift from ground level to maximum height; are plumbed for multiple tool functions; and provide excellent operator visibility.

They usually also have positive carry position for load stability and retention and return-to-work kickouts. They

are designed for multiple tool or attachment use and have the balance, power and hydraulics needed to deliver jobsite performance.

At the present time tool carriers feature the widest selection of attachments. Typical attachments for the toolcarriers can be broken into buckets, forks, blades and miscellaneous.
Buckets:
General purpose
Loose material
Light material
Multipurpose
High dump
Side dump
Grading
Grab and grip
Fertilizer
Top clamp
Wood-chip
Refuse
Forks:
Log or lumber forks
Wide frame forks
Stinger fork
Utility pallet forks
Pulpwood
Sorting
Pallet fork
Core fork
Blades:
Angle blade — manual or hydraulic
Straight blade
One-way snow blade
V-plow
Reversible plow — manual or hydraulic
Miscellaneous:
Material handling arm
Asphalt cutter
Hydraulic broom
Hydraulic hammer
Tire loaders
Loader rake

There are probably dozens of other attachments that are available.

TELESCOPIC HANDLERS:

Telescopic handlers are probably the most recent entry into the attachment carrier market. Originally designed for use in the masonry market, to load and unload bricks and blocks from trucks, these machines have become another jobsite must.

The telescopic handler was designed for forks. There are a large number of forks to consider. In addition to forks, there are other attachments available that make the telescopic handler a very nice tool to have on just about any job site.

Figure 2.1.0 *(cont.)*

SCRAPERS:

What kind of attachment is there for a scraper? Caterpillar has an auger attachment available for six models of the scraper line. The auger attachment is a self-loading system that offers an alternative to conventional, push-pull or elevating scrapers. An independent hydrostatic system powers the auger, which is located near the center of the bowl. The auger is available in diameters of 4 feet 4 inches, 5 feet, and 5 feet 6 inches.

SKID-STEER LOADERS:

It gets harder and harder to decide which piece of equipment is the king of attachments. If you're looking for the most versatile piece of equipment money can buy it will probably be the skid-steer loader.

The skid-steer loader left the farm for the city while still a newcomer to the industry. It was probably one of the best-kept secrets in the industry. It wasn't until recently that it has been given the attention and use that it deserves. If any machine was ever born to be a work tool platform, it's the skid steer loader. It can sweep and load and drill and till and grapple and doze and trench and plane and landscape and grind and lift and compact and saw and cut and mow and Wow! It can do a lot...

Some, but not all, of the attachments available for skid-steer loaders include:
  General purpose bucket
  Four-in-one bucket
  Combination bucket
  Dirt bucket
  Angle broom rotary cutter

Mower
Auger
Backhoe
Wheel saw
Concrete mixer
Grader
Stump grinder
Tiller
Planer
Snow blade
Dozer blade
Scarifier
Utility grapple
Tilt-tatch
Pallet forks
Snow blower
Landplane
Landscape rake
Snow bucket
Industrial grapple
Power rake
Trencher
Vibratory roller
Sweeper
Angle broom
Pickup broom
Breaker
Tracks

Figure 2.1.0   (*cont.*)

2.2.0 Typical Specifications for Compact Excavators

KOBELCO

COMPACT EXCAVATOR LINE

GENERAL SPECIFICATIONS

Model		15SR	25SR	35SR	45SR	70SR	115SRDZ	135SRLC
Operating weight	Lbs	3680	5830	7700	10,275	16,400	31,750	30,840
Bucket capacity	Cu Ft	1.57	2.8	3.9	4.94	0.14-0.37yd^3	0.44-0.91yd^3	0.44-0.91yd^3
Engine make		Yanmar	Yanmar	Yanmar	Yanmar	Isuzu	Isuzu	Isuzu
Engine horsepower	HP	14.3	21	25.9	37	54	84	94
Engine type	Diesel	D	D	D	D	D	D	D
Maximum digging depth	Ft-In	7-1	8-6	10-2	11-9	14-9	18-1	19-9
Vertical wall digging depth	Ft-In	5-2	6-11	8-3	9-4	13-0	16-2	17-2
Maximum dumping clearance	Ft-In	8-6	10-8	11-4	13-1	17-9	20-4	21-1
Minimum dumping clearance	Ft-In	3-6	4-4	4-5	5-3	6-9	6-11	5-5
Maximum digging height	Ft-In	11-11	14-9	16-0	18-4	17-9	28-4	29-1
Maximum digging reach	Ft-In	12-11	15-3	17-0	19-3	22-0	27-4	28-9
Maximum digging reach at ground	Ft-In	12-6	14-10	16-8	18-10	21-7	26-11	28-5
Bucket width	In	18	20	24	25.6	26.8	24	24
Bucket digging force	Lbs	2,944	4,675	6,064	8,003	11,903	18,100	19,000
Arm crowding force	Lbs	2,205	3,506	4,410	5,512	7,718	13,892	12,569
Gradeability	Degree	30	30	30	30	35	35	35
Counterweight overside	Ft-In	0-0	0-0	0-0	0-0	0-0	5.1"	5.1"
Travel speed	MPH	1.3/2.6	1.7/3.0	1.6/2.9	1.8/2.7	2.1/3.4	2.2/3.7	2.2/3.7
Auxiliary flow	GPM	7.6	12.8	17.4	14.53	17.4/35	33.3/66	33.3/66

Figure 2.2.0 (*Courtesy of Kobelco America, Stafford, Texas*)

2.2.1 Compact Excavators with 1/4- and 1/2- Cubic Yard Buckets

PERFORMANCE

Bucket capacity range	0.23~0.53 yd³ (0.18~0.40 m³)
Travel speed	3.4/2.1 mph (5.3/3.4 km/h)
Drawbar pulling force	14,500 lbs. (6,591 kg)
Bucket digging force	11,900 lbs (5,400 kg)
Arm digging force	8,630 lbs (3,900 kg)
Swing speed	12.5 rpm
Gradeability	35 (70%)

HYDRAULIC SYSTEM
Unit: psi (kg/cm²)

Main hydraulic pumps	2 variable displacement
Max discharge flow	2 x 17.4 US gal/min (2 x 66 1/min)
Max discharge pressures:	
Boom, arm & bucket	4,270 (300)
Propel circuit	4,270 (300)
Blade circuit	2,845 (200)
Control circuit	500 (35)
Swing circuit	3,560 (250)
Control valves	6 spool

DIMENSIONS

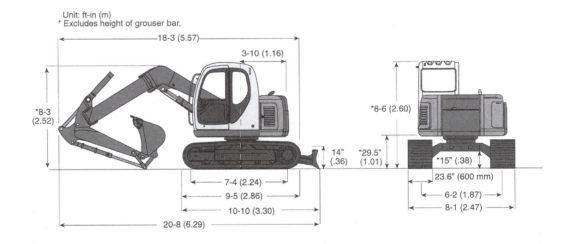

Unit: ft-in (m)
* Excludes height of grouser bar.

Figure 2.2.1 (*Courtesy: Kobelco America, Stafford, Texas*)

2.2.2 Compact Excavators with Small Buckets (2.8 Cubic Feet)

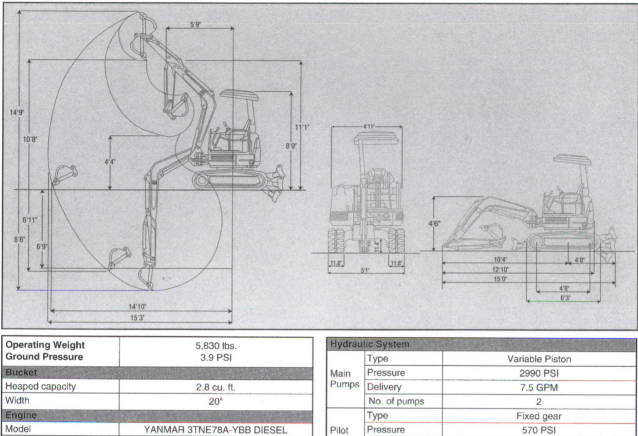

Operating Weight	5,830 lbs.
Ground Pressure	3.9 PSI
Bucket	
Heaped capacity	2.8 cu. ft.
Width	20"
Engine	
Model	YANMAR 3TNE78A-YBB DIESEL
Rated output	SAE NET 21 / 2,350 RPM
Total Displacement	73.47 cu. in.
No. Cylinders	3
Transportation Dimensions	
Overall length	15' 0"
Overall width	5' 1"
Overall height	8' 9"
Digging Force	
Bucket force	4,675 lbs.
Stick force	3,506 lbs.
Track	
Type	Rubber
Overall length	6' 3"
Overall width	5' 1"
Shoe width	11.8"
Capacities	
Fuel tank	9.25 U.S. Gallon
Hyd. System Capacity	13.5 U.S. Gallon
Hyd. Tank Capacity	9.5 U.S. Gallon
Swing Mechanism	
Max. swing speed	8.8 RPM

Hydraulic System		
Main Pumps	Type	Variable Piston
	Pressure	2990 PSI
	Delivery	7.5 GPM
	No. of pumps	2
Pilot Pump	Type	Fixed gear
	Pressure	570 PSI
	Delivery	2.8 GPM
	No. of pumps	1
Aux. Port	Pressure	2560 PSI
	Delivery	12.8 GPM
Travel Motors	Type	Axial Piston, 2-Speed
	No. of motors	2
Maneuverability		
Boom Swing Angle		85°L/55°R
Boom Offset		Left 2' 1" / Right 2' 3"
Front Swing Radius		5' 9"
Tail Swing Radius		**0' 0"**
Travel speed		1.7 / 3.0 MPH
Drawbar pull		4,630 lbs.
Gradeability		30°
Ground clearance		11.4"

Note: *Due to our policy of continual product improvement, all designs and specifications are subject to change without advance notice.*

Figure 2.2.2 (*Courtesy of Kobelco America, Stafford, Texas*)

2.2.3 Compact Excavators with Small Buckets (3.9 Cubic Feet)

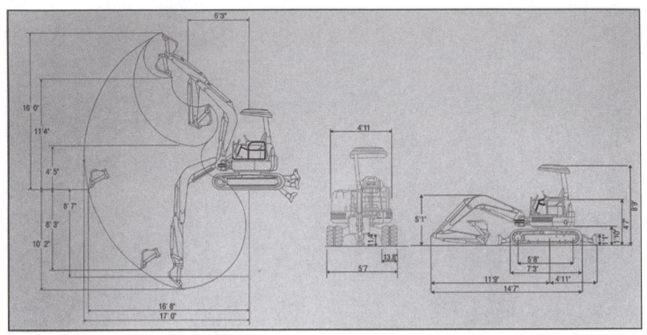

Operating Weight Ground Pressure	7,700 lbs. 4.4 PSI
Bucket	
Heaped capacity	3.9 cu. ft.
Width	24"
Engine	
Model	YANMAR 3TNE84-YBD DIESEL
Rated output	25.9 SAE NET HP / 2,350 RPM
Total Displacement	91.29 cu. in.
No. Cylinders	3
Transportation Dimensions	
Overall length	16' 8"
Overall width	5' 7"
Overall height	8' 9"
Digging Force	
Bucket force	6,064 lbs.
Stick force	4,410 lbs.
Track	
Type	Rubber
Overall length	7' 3"
Overall width	5' 7"
Shoe width	13.8"
Capacities	
Fuel tank	9.25 U.S. Gallon
Hyd. System Capacity	13.5 U.S. Gallon
Hyd. Tank Capacity	9.5 U.S. Gallon
Swing Mechanism	
Max. swing speed	8.5 RPM

Hydraulic System			
Main Pumps	Type	Variable Piston	
	Pressure	2990 PSI	
	Delivery	9.9 GPM	
	No. of pumps	2	
Pilot Pump	Type	Fixed gear	
	Pressure	570 PSI	
	Delivery	2.8 GPM	
	No. of pumps	1	
Aux. Port	Pressure	2560 PSI	
	Delivery	17.4 GPM	
Travel Motors	Type	Axial Piston, 2-Speed	
	No. of motors	2	
Maneuverability			
Boom Swing Angle	85˚L / 55˚R		
Boom Offset	Left 2' 5" / Right 2' 1"		
Front Swing Radius	6' 3"		
Tail Swing Radius	**0' 0"**		
Travel speed	1.6 / 2.9 MPH		
Drawbar pull	6,725 lbs.		
Gradeability	30˚		
Ground clearance	11.4"		

Note: *Due to our policy of continual product improvement, all designs and specifications are subject to change without advance notice.*

Figure 2.2.3 (*Courtesy of Kobelco America, Stafford, Texas*)

2.2.4 Reach of Compact Excavators

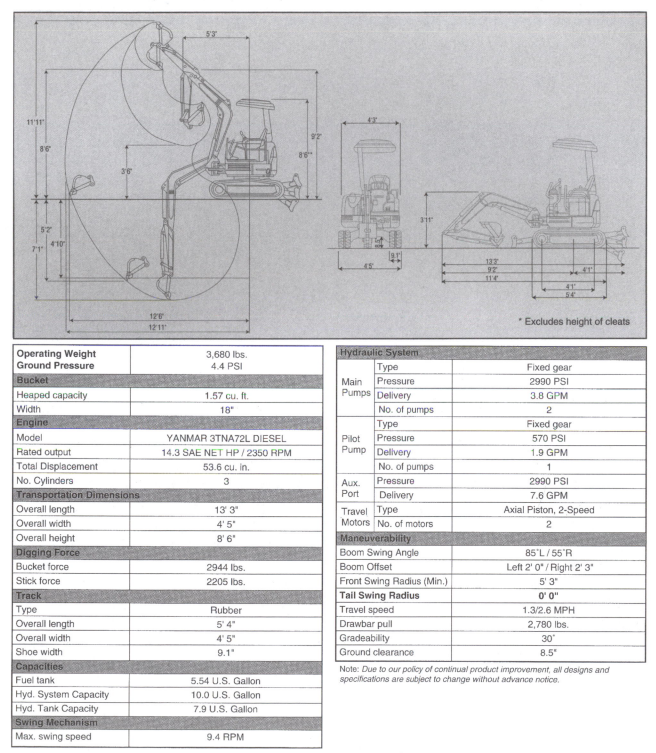

* Excludes height of cleats

Operating Weight	3,680 lbs.
Ground Pressure	4.4 PSI
Bucket	
Heaped capacity	1.57 cu. ft.
Width	18"
Engine	
Model	YANMAR 3TNA72L DIESEL
Rated output	14.3 SAE NET HP / 2350 RPM
Total Displacement	53.6 cu. in.
No. Cylinders	3
Transportation Dimensions	
Overall length	13' 3"
Overall width	4' 5"
Overall height	8' 6"
Digging Force	
Bucket force	2944 lbs.
Stick force	2205 lbs.
Track	
Type	Rubber
Overall length	5' 4"
Overall width	4' 5"
Shoe width	9.1"
Capacities	
Fuel tank	5.54 U.S. Gallon
Hyd. System Capacity	10.0 U.S. Gallon
Hyd. Tank Capacity	7.9 U.S. Gallon
Swing Mechanism	
Max. swing speed	9.4 RPM

Hydraulic System			
Main Pumps	Type		Fixed gear
	Pressure		2990 PSI
	Delivery		3.8 GPM
	No. of pumps		2
Pilot Pump	Type		Fixed gear
	Pressure		570 PSI
	Delivery		1.9 GPM
	No. of pumps		1
Aux. Port	Pressure		2990 PSI
	Delivery		7.6 GPM
Travel Motors	Type		Axial Piston, 2-Speed
	No. of motors		2
Maneuverability			
Boom Swing Angle			85°L / 55°R
Boom Offset			Left 2' 0" / Right 2' 3"
Front Swing Radius (Min.)			5' 3"
Tail Swing Radius			**0' 0"**
Travel speed			1.3/2.6 MPH
Drawbar pull			2,780 lbs.
Gradeability			30°
Ground clearance			8.5"

Note: *Due to our policy of continual product improvement, all designs and specifications are subject to change without advance notice.*

KOBELCO

Figure 2.2.4 (*Courtesy of Kobelco America, Stafford, Texas*)

2.2.5 Compact Excavators with Left and Right Offsets

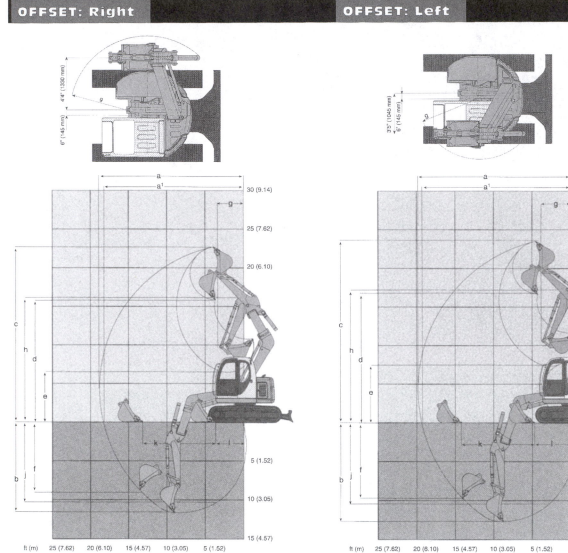

<table>
<tr><td colspan="4">WORKING RANGES</td><td>Unit: ft-in (m)</td></tr>
</table>

	OFFSET: RIGHT		
a	Max digging reach	18-10	(5.75)
a¹	Max digging reach at ground level	18-4	(5.60)
b	Max digging depth*	11-6	(3.51)
c	Max digging height*	22-6	(6.86)
d	Max dumping clearance*	15-8	(4.79)
e	Min dumping clearance*	6-5	(1.96)
f	Max depth of bucket hinge pin*	8-8	(2.64)
g	Min swing radius	6-9	(2.05)
h	Height at min swing radius*	16-6	(5.03)
j	Digging depth for 2.4m flat bottom	10-2	(3.11)
k	Horizontal digging distance - stroke	9-5	(2.88)
l	Horizontal digging distance - minimum	3-10	(1.16)

* Excludes height of grouser bar

	OFFSET: LEFT		
a	Max digging reach	19-10	(6.06)
a¹	Max digging reach at ground level	19-5	(5.92)
b	Max digging depth*	12-6	(3.82)
c	Max digging height*	23-5	(7.14)
d	Max dumping clearance*	16-8	(5.07)
e	Min dumping clearance*	7-4	(2.24)
f	Max depth of bucket hinge pin*	9-7	(2.93)
g	Min swing radius	5-1	(1.56)
h	Height at min swing radius*	17-5	(5.31)
j	Digging depth for 2.4m flat bottom	11-3	(3.42)
k	Horizontal digging distance - stroke	9-4	(2.86)
l	Horizontal digging distance - minimum	4-11	(1.49)

* Excludes height of grouser bar

Figure 2.2.5 (*Courtesy of Kobelco America, Stafford, Texas*)

2.2.6 Kobelco Full Range of Excavators

KOBELCO

HYDRAULIC EXCAVATORS

Model	70SR	115SRDZ	135SRLC	SK150LC MARK IV	ED180	SK200LC MARK IV	235SRLC	SK220LC MARK IV	SK270LC MARK IV	SK300LC MARK IV	SK400LC MARK IV
GENERAL											
Operating weight with bucket (lb)	16,400	31,750	30,840	35,800	41,800	45,900	53,580	55,800	61,997	73,300	100,900
Bucket capacity range SAE/PCSA (cu yd)	0.14-0.37	0.44-0.91	0.44-0.91	0.45-1.36	0.5-1.25	0.67-1.67	0.67-1.67	0.75-1.88	0.875-2.00	0.88-2.62	1.5-4.0
ENGINE											
Make and model	ISUZU A4JB1	ISUZU 4BGIT	ISUZU 4BGIT	CUMMINS 4BTA3.9	CUMMINS 4BTA3.9	CUMMINS 6BT5.9	MITSUBISHI 6D34-TE1	CUMMINS 6BTA5.9	CUMMINS 6BTA5.9	CUMMINS 6CTA8.3	CUMMINS MTA11
Displacement (cu in)	169	264	264	239.3	239.3	359	358	359	359	505	661
SAE NET Horsepower	54	84	94	103	103	141	142	175	175	238	306
Rated RPM	2,100	2,050	2,200	2,200	2,200	2,200	2,000	2,200	2,200	1,900	2,000
WORKING RANGES											
Arm length (ft-in)	6-9	8-0	9-6	10-0	10-0	9-8	9-8	9-9	11-2	10-4	11-4
Bucket digging force (lb)	11,903	18,100	19,000	22,000	22,000	*29,000	*30,000	*33,400	*33,400	*45,000	*55,000
Arm crowding force (lb)	7,718	13,892	12,569	15,400	15,400	*22,710	*23,152	*25,800	*24,030	*35,720	*45,860
Ground level reach (ft-in)	21-7	26-11	28-5	30-1	30-1	31-9	31-8	33-3	34-8	35-11	38-9
Digging depth (ft-in)	14-9	18-1	19-7	21-5	21-4	21-11	22-0	23-0	24-1	24-3	25-7
Dumping height (ft-in)	17-9	20-4	21-2	21-0	21-2	22-2	27-7	22-5	23-4	23-5	24-10
Vertical wall digging depth (ft-in)	13-0	16-2	17-3	18'-1"	17-11	19-8	20'	19-10	20-8	20-4	22-9
Max. digging depth at 8' level bottom (ft-in)	13-10	17-5	19-2	20-9	20-8	21-2	21'-4"	22-4	23-6	23-8	25-2
UNDERCARRIAGE											
Track overall length (ft-in)	9-5	11-9	12-3	13-1	12-7	14-8	15-4	15-4	16-3	16-3	17-11
Travel speed (mph)	3.4	3.7	3.7	3.7	3.7	4.3	3.1	4.3	3.2	3.4	3.4
Drawbar pull (lb)	14,500	23,108	23,108	35,230	42,556	44,660	52,920	50,420	59,840	59,500	87,300
Gradeability (%)	70	70	70	70	70	70	70	70	70	70	70
Ground clearance (in)	15.0	18.0	18.0	18.9	20	18.3	18.0	18.3	18.5	18.5	19.3
Ground pressure (psi)	3.3	5.43	5.12	5.3	5.9	4.69	5.3	5.55	5.7	6.82	7.68
Track shoes (in)	23.6	23.6	23.6	23.6	23.6	31.5	31.5	31.5	31.5	31.5	35.4
SWING STRUCTURE											
Swing speed (RPM)	12.5	11.7	11.7	10.8	10.8	13.0	11.0	12.0	12.0	10.5	9.0
Tail swing radius (ft-in)	3-10	4-8	4-8	8-0	8-0	8-10	5-3	9-4	9-4	10-6	11-4
Counterweight overhang overside (ft-in)	0-0	5.1"	5.1"	4-3	4-3	3-6	0-0	3-9	3-9	4-11	5-4

* Power Boost engaged

Note: Due to our policy of continual product improvement, all designs and specifications are subject to change without advance notice.

KOBELCO

Figure 2.2.6 *(Courtesy of Kobelco America, Stafford, Texas)*

2.2.7 Kobelco SK300LC Excavator Specifications

 DIMENSIONS

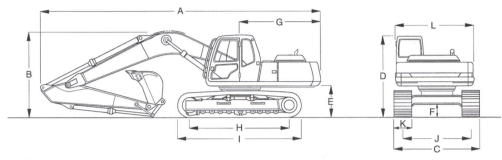

Dimensions

Unit:ft (m)

	Arm length	8-4 (2.55)	10 - 4 (3.14)	13 - 5 (4.1)
A	Overall length	35 - 11 (10.94)	35 - 7 (10.85)	35 - 8 (10.88)
B	Overall height (to top of boom)	11 - 7 (3.54)	10 - 8 (3.25)	11 - 1 (3.39)
C	Overall width	11 - 2 (3.40)	11 - 2 (3.40)	11 - 2 (3.40)
D	Overall height (to top of cab)	10 - 0 (2.89)	10 - 0 (3.06)	10 - 0 (3.06)
E	Ground clearance of rear end*	3 - 10 (1.17)	3 - 10 (1.17)	3 - 10 (1.17)
F	Ground clearance*	18.5" (0.47)	18.5" (0.47)	18.5" (0.47)
G	Tail swing radius	10 - 6 (3.20)	10 - 6 (3.20)	10 - 6 (3.20)
H	Tumbler distance	13 - 2 (4.01)	13 - 2 (4.01)	13 - 2 (4.01)
I	Overall length of crawler	16 - 3 (4.95)	16 - 3 (4.95)	16 - 3 (4.95)
J	Track gauge	8 - 6 (2.60)	8 - 6 (2.60)	8 - 6 (2.60)
K	Shoe width	31.5" (800mm)	31.5" (800mm)	31.5" (800mm)
L	Overall width of upper structure	10 - 0 (3.06)	10 - 0 (3.06)	10 - 0 (3.06)

*Excludes height of grouser bar.

Figure 2.2.7 (*Courtesy of Kobelco America, Stafford, Texas*)

2.2.8 SK300LC Reach and Working Ranges

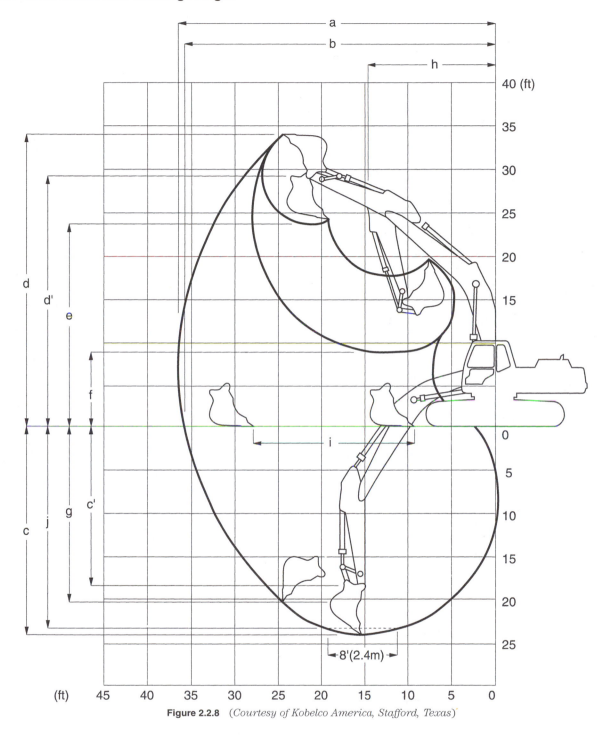

Figure 2.2.8 *(Courtesy of Kobelco America, Stafford, Texas)*

WORKING RANGES

Unit: ft-in (m)

Range	Arm	8 - 4 (2.55)	Standard 10 - 4 (3.14)	13 - 5 (4.1)
a — Max. digging reach		34 - 9 (10.60)	36 - 6 (11.13)	39 - 3 (11.98)
b — Max. digging reach at ground level		33 - 10 (10.40)	35 - 11 (10.94)	38 - 9 (11.80)
c — Max. digging depth		22 - 3 (6.79)	24 - 3 (7.38)	27 - 4 (8.34)
c' — Max. depth of bucket hinge pin		16 - 9 (5.10)	18 - 5 (5.64)	21 - 10 (6.65)
d — Max. digging height		33 - 1 (10.09)	33 - 8 (10.27)	34 - 6 (10.51)
d' — Max. height of bucket hinge pin		28 - 6 (8.68)	29 - 2 (8.88)	29 - 10 (9.09)
e — Max. dumping clearance		22 - 10 (6.95)	23 - 5 (7.15)	24 - 3 (7.40)
f — Min. dumping clearance		10 - 5 (3.17)	8 - 6 (2.59)	5 - 4 (1.62)
g — Max. vertical wall digging depth		17 - 8 (5.39)	20 - 4 (6.19)	22 - 11 (6.98)
h — Min. front swing radius		15 - 2 (4.61)	14 - 10 (4.53)	14 - 11 (4.54)
i — Horizontal digging stroke at ground level		14 - 0 (4.26)	18 - 4 (5.60)	23 - 11 (7.28)
j — Digging depth at 8'(2.4 m) level bottom		21 - 8 (6.61)	23 - 8 (7.22)	26 - 11 (8.21)
Bucket capacity SAE/PCSA heaped cu yd (m³)		2.09 (1.60)	1.83 (1.40)	1.57 (1.20)

*Excludes height of grouser bar.

Digging Force

Unit:lb (kg)

Arm length ft-in (m)	8 - 4 (2.55)	10 - 4 (3.14)	13 - 5 (4.1)
Bucket digging force	*45,000 (20,400)	*45,000 (20,400)	40,300 (18,300)
Arm crowding force	*43,340 (19,200)	*35,720 (16,200)	26,460 (12,000)

*Power boost engaged, available only for 8' 4" and 10' 4" arms.

Figure 2.2.8 *(cont.)*

2.2.9 Excavator with Bulldozer Blade Attachment

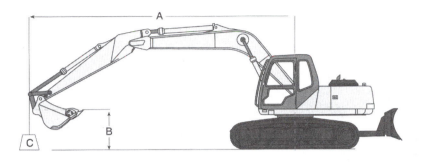

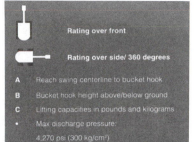

Rating over front

Rating over side/ 360 degrees

A   Reach swing centerline to bucket hook
B   Bucket hook height above/below ground
C   Lifting capacities in pounds and kilograms
•   Max discharge pressure:
     4,270 psi (300 kg/cm²)
•   Track shoe: 23.6" (600 mm) Triple grouser

LIFTING CAPACITIES

A / B / C		ED180 Arm: 10'0" Bucket: 0.68 cu yd (0.59 m3) SAE heaped – 933 lb (423 kg)									
		5' (1.5 m)		10' (3.0 m)		15' (4.5 m)		20' (6.1 m)		25' (7.6 m)	
B	**C**										
20'	lb							*6400	*7040		
(6.1 m)	kg							*2900	*3193		
15'	lb							*7000	6970		
(4.6 m)	kg							*3100	3162		
10'	lb			*14630	*14630	*9600	*9600	*8100	6650	*6300	4570
(3.0 m)	kg			*6636	*6636	*4300	*4300	*3700	3016	*2800	2073
5'	lb			*20600	17560	*12600	9760	*9960	6310	7650	4410
(1.5 m)	kg			*9344	7965	*5700	4427	*4518	2862	3470	2000
Ground	lb	*6900	*6900	*18500	15100	*15370	9250	10490	6040	7510	4280
Level	kg	*3100	*3100	*8392	6849	*6972	4196	4758	2740	3407	1941
–5'	lb	*12200	*12200	*22100	17040	16070	9030	10340	5900		
(–1.5 m)	kg	*5500	*5500	*10024	7729	7289	4096	4690	2676		
–10'	lb	*18100	*18100	*23400	17240	*15330	9060	10390	5940		
(-3.0 m)	kg	*8200	*8200	*10614	7820	*6954	4110	4713	2694		
–15'	lb	*25600	*25600	*19200	17780	*12900	9400				
(-4.6 m)	kg	*11600	*11600	*8709	8065	*5800	4264				

Notes:
1.   Do not attempt to lift or hold any load that exceeds these rated values at their specified load radii and heights.
2.   Lifting capacities assume a machine standing on a level, firm, and uniform supporting surface. Operator must make allowance for job conditions such as soft or uneven ground, out of level conditions, side loads, sudden stopping of loads, hazardous conditions, inexperienced personnel, weight of various other buckets, lifting slings, attachments, etc.
3.   The previous rated loads are in compliance with SAE Hydraulic Excavator Lift Capacity Standard J 1097. They do not exceed 87% of hydraulic lifting capacity or 75% of tipping load. Rated loads marked with an asterisk (*) are limited by hydraulic capacity rather than tipping load.
4.   Operator should be fully acquainted with the operator's manual before operating this machine. Rules for safe operation of equipment should be followed at all times.
5.   Capacities apply only to the machine as originally manufactured and normally equipped by KOBELCO America Inc.

DIGGING FORCE

Unit lb (kg)

ARM LENGTH ft-in (m)	10-0	(3.06)
Bucket digging force	22,000	(10,000)
Arm crowding force	15,440	(7,000)

Figure 2.2.9 (*Courtesy of Kobelco America, Stafford, Texas*)

2.2.10 Kobelco Mass Excavator Dimensions

 DIMENSIONS

Unit: ft-in (mm)

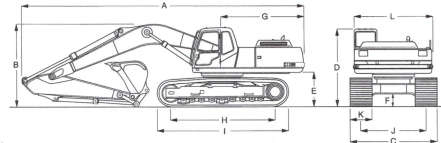

Dimensions

		Standard			Unit:ft (m)
	Arm length	**9 - 10** (3.0)	**11 - 4** (3.45)	**13 - 0** (3.95)	**16 - 1** (4.9)
A	Overall length	38 - 8 (11.78)	38 - 6 (11.73)	38 - 7 (11.77)	38 - 9 (11.82)
B	Overall height (to top of boom)	11 - 10 (3.61)	11 - 3 (3.43)	11 - 8 (3.56)	13 - 9 (4.18)
C	Overall width	12 - 0 (3.65)	12 - 0 (3.65)	12 - 0 (3.65)	12 - 0 (3.65)
D	Overall height (to top of cab)	10 - 6 (3.21)	10 - 6 (3.21)	10 - 6 (3.21)	10 - 6 (3.21)
E	Ground clearance of rear end*	4 - 4 (1.325)	4 - 4 (1.325)	4 - 4 (1.325)	4 - 4 (1.325)
F	Ground clearance*	19.3" (0.49)	19.3" (0.49)	19.3" (0.49)	19.3" (0.49)
G	Tail swing radius	11 - 4 (3.46)	11 - 4 (3.46)	11 - 4 (3.46)	11 - 4 (3.46)
H	Tumbler distance	14 - 4 (4.37)	14 - 4 (4.37)	14 - 4 (4.37)	14 - 4 (4.37)
I	Overall length of crawler	17 - 11 (5.46)	17 - 11 (5.46)	17 - 11 (5.46)	17 - 11 (5.46)
J	Track gauge	9 - 0 (2.75)	9 - 0 (2.75)	9 - 0 (2.75)	9 - 0 (2.75)
K	Shoe width	35.4" (900mm)	35.4" (900mm)	35.4" (900mm)	35.4" (900mm)
L	Overall width of upperstructure	10 - 8 (3.25)	10 - 8 (3.25)	10 - 8 (3.25)	10 - 8 (3.25)

*Excludes height of grouser bar

Figure 2.2.10 (*Courtesy of Kobelco America, Stafford, Texas*)

2.2.11 Operating Range of the SK400LC Mass Excavator

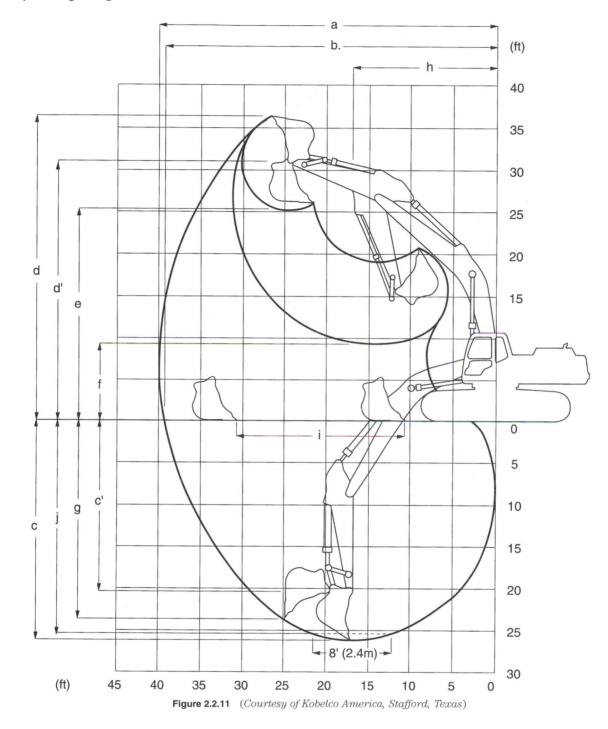

Figure 2.2.11   *(Courtesy of Kobelco America, Stafford, Texas)*

Operating Weight: ....107,500 lbs
Engine Output: ....306 hp (SAE net)
Bucket Range:..........4.5 - 5.5 cu. yd.
Arm Force: (7'10" Arm) ....56,825 lbs
Arm Force: (11'4" Arm) ....48,250 lbs
Bucket Force:
  5.0 - 6.0 cu. yd: ....58,770 lbs
  4.0 - 4.5 cu. yd: ....62,318 lbs

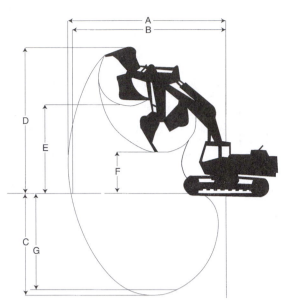

DIMENSIONS		
	7'10" Arm	Severe Duty 11'4' Arm
A	35'3"	38'0"
B	34'6"	37'0"
C	21'10"	24'0"
D	33'8"	34'0"
E	20'11"	23'1"
F	10'6"	7'10"
G	20'2"	24'2"

LIFTING CAPACITY CHART										
			10 Feet		15 Feet		20 Feet		25 Feet	
			Front	Side	Front	Side	Front	Side	Front	Side
7'0" Arm	20' (6.1m)	lbs							20,730	20,730
	15' (4.6m)	lbs					26,186	26,180	22,640	19,870
Bucket Weight:	10' (3.0m)	lbs			*42,420	41,660	30,750	26,720	25,090	18,630
5,200 lbs	5' (1.5m)	lbs			*49,450	37,700	34,830	24,610	27,440	17,460
Shoe: 36" (900mm)	Ground level	lbs	*32,020	*32,020	*52,390	35,880	37,420	23,240	28,680	16,610
Triple Grouser	-5' (-1.5m)	lbs	*48,890	*48,890	*52,060	35,440	38,070	22,650	28,220	16,200
	-10' (-3.0m)	lbs	*69,000	*69,000	*48,810	35,880	36,360	22,790		
	-15' (-4.6m)	lbs	*57,300	*57,300	*41,640	37,180	30,830	23,660		

Notes:

1. All figures conform to PCSA Standards No. 3 and SAE J1097. Hoist cylinder pressure (4980 psi).
2. Lifting capacities shown do not exceed 87% of machine hydraulic capacity, or 75% of machine tipping load.
3. Figures with asterisks (*) indicate that lifting capacity is limited by hydraulic capacity rather than tipping load.
4. Lifting capacities shown should not be exceeded.
5. Weight of all lifting accessories must be considered part of the load.
6. Lifting capacities assume the machine is standing level on a firm, uniform supporting surface. The user must make allowances for unfavorable job conditions such as soft or uneven ground, or sudden stopping of loads.
7. The operator should be fully acquainted with the Kobelco Operator and Maintenance Instruction Manual before operating the machine.
8. Capacities apply only to the machine with attachments manufactured by Jewell Manufacturing.

Figure 2.2.11 *(cont.)*

2.3.0 Komatsu Excavator Specifications

HYDRAULIC
EXCAVATORS

Model	Flywheel Horsepower	Operating Weight (lb)	Bucket Capacity	Digging Depth	Maximum Reach (cu yd)
PC95	72.5	20,084	.18–.46	13'1"	21'10"
PC120-6	87	26,530	.48–1.0	18'1"	26'10"
PC128UU	84	28,800	.24–.80	16'0"	23'10"
PW170-6	102	38,600	0.5–1.48	17'9"	28'5"
PC150LC-6	105	39,503	.50–1.12	19'7"	29'5"
PC200-6 Custom	128	43,880	.62–1.5	21'9"	32'5"
PC200LC-6	133	46,970	.62–1.5	21'9"	32'5"
PC210LC-6	145	51,180	.62–1.25	21'9"	32'5"
PC220LC-6	158	53,364	.75–1.75	22'3"	33'5"
PC250LC-6	158	60,795	1.0–2.0	22'0"	33'5"
PC300LC-6	232	72,312	1.0–2.5	24'3"	36'5"
PC300HD-6	232	79,785	1.0–2.5	23'5"	36'5"
PC400LC-6	306	95,147	1.25–3.0	25'6"	39'5"
PC400HD-6	306	97,665	1.25–3.0	25'0"	39'5"
PC750-6	444	162,480	3.5–5.25	27'8"	44'6"
PC750LC-6	444	171,070	3.5–5.25	27'8"	44'6"
PC750SE-6	444	170,850	4.0–6.0	23'0"	39'11"
PC750-6 SHOVEL	444	163,980	5.0–5.9	11'3"	32'8"
PC1000-1	542	209,400	4.1–7.1	30'6"	50'2"
PC1000LC-1	542	227,740	4.1–7.1	30'6"	50'2"
PC1000SE-1	542	213,850	7.45	26'2"	45'9"
PC1000-1 SHOVEL	542	216,100	7.2–9.2	12'10"	35'11"
PC1100-6	611	227,100	4.5–6.5	30'8"	50'4"
PC1100LC-6	611	244,710	4.5–6.5	30'8"	50'4"
PC1100SE-6	611	229,280	8.5	25'11"	46'2"

KOMATSU

Figure 2.3.0 *(Courtesy of Komatsu American International Co.)*

2.3.1 Komatsu Wheel Loader Specifications

WHEEL
LOADERS

Model	Flywheel Horsepower	Operating Weight (lb)	Bucket Capacity (cu yd)	Dumping Clearance	Breakout Force (lb)
WA120-3	100	18,110	1.7–2.25	8'11"	16,280
WA180-3	110	20,585	2.25–2.9	8'11"	20,550
WA180-3 PARALLEL LIFT	110	22,840	2.25–2.50	9'0"	20,254
WA250-3	127	24,790	2.5–3.5	9'1"	27,780
WA250-3 PARALLEL LIFT	127	27,540	2.50–3.0	9'4"	24,575
WA320-3	162	29,443	3.0–4.2	9'4"	28,770
WA380-3	189	38,765	3.4–5.25	9'7"	34,316
WA420-3	215	41,976	4.1–6.0	10'1"	39,904
WA450-3	264	49,350	4.7–6.8	10'1"	42,730
WA500-3	315	65,164	5.6–7.2	10'5"	52,140
WA600-1	430	90,456	7.5	10'11"	84,746
WA600-1 LOAD & CARRY	430	98,888	9.8	9'7"	73,193
WA700-1	641	151,082	11.1	13'3"	116,183
WA800-2	789	204,124	13.7	15'3"	152,120
WA900-1	828	208,340	17.0	15'3"	136,910
WA900-1 HIGH LIFT	828	212,520	15.0	17'1"	144,496

Figure 2.3.1   *(Courtesy of Komatsu American International Co.)*

2.3.2 Motor Graders—Various Moldboard Sizes

MOTOR GRADERS,
CRUSHERS, AND DUMP TRUCKS

MOTOR GRADERS

Model	Flywheel Horsepower	Operating Weight (lb)	Drive System	Moldboard	Overall Length
GD530A-2C	144	30,256	TORQUE PS 4WD	12' x 26" x 7/8" STD	27'7"
GD530AW-2C	144	30,556	TORQUE PS AWD	12' x 26" x 7/8" STD	27'7"
GD650A-2C	166	32,795	TORQUE PS 4WD	12' x 26" x 7/8" STD	27'9"
GD650AW-2C	166	33,095	TORQUE PS AWD	12' x 26" x 7/8" STD	27'9"
GD670A-2C	204	34,272	TORQUE PS 4WD	12' x 26" x 1" STD	27'9"
GD670AW-2C	204	34,572	TORQUE PS AWD	12' x 26" x 1" STD	27'9"
GD750A-1	225	43,000	TORQUE PS 4WD	14' x 27" x 1" STD	30'3"
GD825A-2	280	58,250	DIRECT PS 4WD	16'2" x 31" x 1"	32'10"

CRUSHERS

Model	Flywheel Horsepower	Rated Output (kW)	Crushing Capacity	Operating Weight (lb)
BR300J/310JG	153	114	48–144 U.S. TN/HR 43–130 U.S. TN/HR	61,730 66,140

DUMP TRUCKS

Model	Flywheel Horsepower	Empty Weight (lb)	Capacity (ton)	Capacity (cu yd)	Max Travel Speed (mph)
HD325-6	488	63,270	40	31.4	43.5
HD785-3	1050	149,716	100	74.0	39.8

KOMATSU

Figure 2.3.2 *(Courtesy of Komatsu American International Co.)*

2.3.3 Bulldozers by Komatsu—1.65- to 90-Yard Capacity

CRAWLER
DOZERS

Model	Flywheel Horsepower	Operating Weight (lb)	Blade Type	Blade Capacity (cu yd)
D31E-2	70	14,640	POWER-ANGLE-TILT	1.65
D31P-2	70	15,850	POWER-ANGLE-TILT	1.70
D32E-1	70	15,900	POWER-ANGLE-TILT	1.6
D32P-1	70	17,330	POWER-ANGLE-TILT	1.9
D37E-5	75	15,080	POWER-ANGLE-TILT	1.95
D37P-5	75	16,450	POWER-ANGLE-TILT	2.11
D38E-1	80	17,060	POWER-ANGLE-TILT	1.9
D38P-1	80	17,800	POWER-ANGLE-TILT	1.9
D39E-1	90	18,800	POWER-ANGLE-TILT	2.2
D39P-1	90	19,190	POWER-ANGLE-TILT	2.3
D41E-6	105	23,170	POWER-ANGLE-TILT	3.4/3.8
D41P-6	105	24,430	POWER-ANGLE-TILT	3.4/3.8
D58E-1B	130	32,830	POWER-ANGLE-TILT	3.7
D58P-1B	130	35,251	POWER-ANGLE-TILT	4.2
D65EX-12	190	40,960	STRAIGHT-TILT/SEMI-U	5.09/7.34
D65PX-12	190	42,500	STRAIGHT-TILT	4.83
D85E-21	225	55,340	STRAIGHT-TILT/U-DOZER	6.8/11.1
D85P-21	225	60,100	STRAIGHT-TILT	7.7
D155AX-3	302	85,160	SEMI-U/U-DOZER	11.4/15.3
D275A-2	405	111,800	SEMI-U/U-DOZER	16.7/20.0
D375A-3	525	149,180	SEMI-U/U-DOZER	24.2/28.8
D475A-2	770	214,740	U-DOZER/SEMI-U	33.5/45.0
D575A-2SD	1150	314,200	SUPER DOZER	90.0
D575A-2SR	1050	290,980	U-DOZER/SEMI-U	70.8/58.0

KOMATSU

Figure 2.3.3 (*Courtesy of Komatsu American International Co.*)

2.4.0 Hammer Selection for Demolition and Rock Breaking

Hammer Selection Guide

Proper hammer selection is key to successful hammer operation. Starting the selection process requires identifying how and where the hammer will be used and required production. If a hammer is being replaced, what brand and model is it? How did the hammer perform in the application? For large quarry hammers, will the hammer be used in primary benching or in secondary boulder breaking? What carrier is being considered and what are the machine's hydraulic flows and pressures? What boom and stick will be used? Will a quick coupler be used? Consult the Common Rock Strength table (at right) to identify the compressive strength of the material being broken.

Select two or three possible hammers using carrier weight class as reference. Validate the carrier hydraulic flows and pressures with those of the hammers selected. Eliminate hammers outside the carrier specs. Evaluate the use of smaller hammers on longer boom/stick combinations. How might a quick coupler affect the use? Did the previous hammer yield problems or marginal production? Compare its energy rating and weight to the candidate hammers. If a larger hammer needs to be considered, use only CIMA energy ratings. Avoid using a generic class. Select the larger of the candidates if the hammer will be used in primary benching. Examine the material's cohesive strength and fracturability, identify production requirements from the productivity tables in this section and then identify the hammer most compatible. Will the hammer require special modifications, i.e., steel mill, underwater, tunneling, etc.?

Once a hammer is chosen, make sure the proper bracket is selected. Be sure the proper oil is being used for the ambient temperature. Use grease especially formulated for hammer use, not generic. Once the hammer is installed on a carrier, measure and record actual back pressure, flows and pressures.

HAMMER APPLIATIONS

Sewer and Water — A hammer can be used on pockets of rock that slow down production. They are also good for breaking up old concrete pipes, manholes, etc.

Road Construction — A hammer is an essential

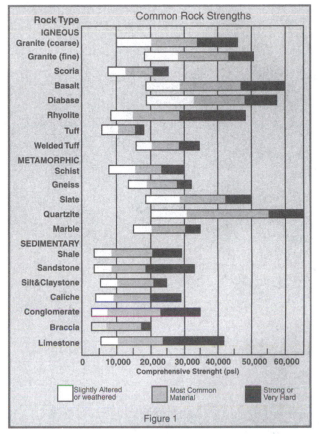

Figure 1

tool during improvements and upgrading. The hammer works well on removing existing curbs, traffic islands, ramps, or sections of concrete. With special tools, it can cut asphalt.

Bridge Renewal — Hammers are increasingly used to remove old bridge surfaces, railing supports, abutments, retaining walls, etc.

Demolition — The hammer-equipped excavator is often a key helper in industrial demolition. It can break up fallen wall sections and floor sections as well as handling loading ramps, foundations, or other brick and concrete structures near ground level.

Mining — Hammers can break oversized material to avoid secondary blasting, and size riprap.

Trenching/Primary Excavation — In soft or layered materials, the hydraulic hammer with a moil or chisel can be cost effective.

Tunneling — This work has traditionally been performed by tunnel boring machines or the drill and blast method. As hydraulic hammers have improved in power and reliability, they have been proven an economical altyernative.

Direct Quarrying — In many types of limestone, direct quarrying with hydraulic hammers can prove cost effective, especially where blasting is prohibited. See hydraulic hammer productivity graphs for production estimates.

The material for hammer selection and hammer applications is courtesy of Caterpillar Performance Handbook Edition 30.

Figure 2.4.0 (*By permission of* Construction, *ACP/Construction, Norcross, Georgia; Feb. 2000*)

2.5.0 Skid-Steer Loaders

Manufacturer	Model	Operating Capacity	HP	Dump Height (*lifting height)
ASV Inc	HD4520	2,800	115.0	114"*
	MD70	2,200	70.0	105"*
	MD2800	2,350	70.0	109"*
	MD70T	2,200	75.0	105"*
	MD2810	2,350	75.0	109"*
	HD4500	2,800	80.0	114"*
Bobcat	453	700	15.7	71.2"
	553	950	22.5	82"
	751	1,250	38.0	84.6"
	753	1,350	43.5	84.6"
	763	1,500	46.0	86.4"
	773	1,750	46.0	91"
	863	1,900	73.0	91.5"
	873	2,400	73.0	98.4"
	963	3,000	105.0	98.7"
Case	1825B	885	27.0	76"
	1838	1,300	51.0	84"
	1840	1,400	50.0	84"
	1845C	1,750	56.0	89.5"
	75XT	2,200	57.0	90"
	85XT	2,400	69.0	92.3"
	90XT	2,450	74.0	94.5"
	95XT	3,150	74.0	95"
Caterpillar	216	1,350	49.0	85"
	226	1,500	54.0	85"
	228	1,500	54.0	85"
	236	1,750	59.0	94"
	246	2,000	74.0	94"
	248	2,000	74.0	94"

First there was one now there are many. In fact the current count is that there are 18 manufacturers producing around 111 models of skid-steer loaders in North America. Today's skid steers are radically different than earlier models. Operating capacities range from 600 pounds to well over 3,000 pounds with engine horsepower ratings from 15.7 horsepower to 115 horsepower.

Manufacturers have developed a vast array of attachments for these machines making them among the most versatile pieces of equipment on a job site. In addition to every type of bucket imaginable, common attachments include augers, hammers, trenches, backhoes, vibratory compactors, tillers, rakes, brooms, cold planners and more.

It is important to match the attachment with the machine capacities for optimum performance.

The newest skid steers even address operator comfort with ergononically designed cabs, user-friendly controls and improved visibility.

Manufacturer	Model	Operating Capacity	HP	Dump Height (*lifting height)
Compact Technologies	1300D	1,350	49.5	89"
	1500D	1,500	60.0	89.8"
	1500DX	1,500	60.0	89.8"
	1750D	1,750	86.0	94"
	2000D	2,000	86.0	94.1"
	2000DX	2,000	86.0	94.1"
	2300D	2,300	86.0	94.1"
Daewoo	DSL601	1,300	42.0	86.7"
	1550XL	1,500	52.2	86.4"
	1760XL	1,700	63.0	91.5"
	2060XL	2,000	62.0	93.5"
Deere	3375	700	17.0	74.8"
	240	1,500	46.0	88.8"
	250	1,750	61.0	90.2"
	260	2,200	69.0	102"
	270	2,650	77.0	102"
Gehl	SL3725	950	39"	86.3"
	SL3825	1,000	28.0	86.3"
	SL4625	1,350	43.0	90.9"
	SL4625DX	1,350	43.0	90.9"
	SL4635	1,425	44.0	87.2"
	SL4635DX	1,425	44.0	87.2"
	SL4835DX	1,625	57.0	87.2"
	SL4835SX	1,625	57.0	87.2"
	SL5635DX	1,800	60.0	93.7"
	SL5635SX	1,800	60.0	93.7"
	SL6635DXT	2,300	80.0	93.7"
	SL6635SXT	2,300	80.0	93.7"

Figure 2.5.0 *(By permission of* Construction, *ACP/Construction, Norcross, Georgia; Feb. 2000)*

Manufacturer		Model	Operating Capacity (*estimated)	HP (*est.)	Dump Height (*lifting height)
Hecla Industries	Lahman	20G	600	20.0	69"
	Patriot	200G	600	20.0	69"
	Lahman	25G	800	24.0	70"
	Lahman	25D	800	24.0	70"
	Patriot	250D	800	24.0	70"
	Patriot	250G	800	24.0	70"
	Lahman	30D	1,200	33.3	86"
	Patriot	300D	1,200	33.3	86"
	Lahman	30G	1,200	38.6	86"
	Patriot	300G	1,200	38.6	86"
	Lahman	40D	1,400	41.0	86"
	Patriot	400D	1,400	41.0	86"
	Lahman	50D	1,800	56.0	96"
	Patriot	500D	1,800	56.0	96"
Hydra-Mac		1300	1,300	37.0	90"
		1700D	1,700	43.0	95"
		2050	2,050	70.0	92.8"
		2250	2,250	70.0	95"
		2550	2,550	70.0	N/A
		2650D	2,600	80.0	115"
		3250	3,250	100.0	115"
ICC Manufacturing Co		DL6000	1915*	60.0	91"
		DL7000	2,500*	72.0	97"
JCB		165 Robot - III	1,433	47.0	85"
		185 Robot - III	1,874	73.0	104"
		1105 Robot - III	2,300	78.0	123"

Manufacturer	Model	Operating Capacity (*estimated)	HP (*est.)	Dump Height (*lifting height)
LMC Corp Trackmaster	75	2,500*	75.0	91"
Mustang	2042	1,350	43.0	87.5"
	2050	1,550	49.0	90"
	2050 Hi-Flow	1,550	49.0	90"
	2060	1,750	62.0	92.7"
	2060 Hi-Flow	1,750	62.0	92.7"
	2070	2,100	75.0	92.7"
	2070 Hi-Flow	2,100	75.0	92.7"
New Holland	L250	600	18*	74.8"
	L255	700	15.5	74.8"
	LS140	1,380	30.0	87.9"
	LS150	1,400	36.0	87.9"
	LS160	1,750	40.0	90"
	LS170	1,765	50.0	90"
	LS180	2,555	60.0	100"
	LS190	2,800	75.0	100"
Ramrod	550	550	16*	N/A
	750	750	16*	N/A
	900T-G	900	20.0	48"
	900T-D	900	20.0	48"
Takeuchi	TL26	1,875*	61.0	93"
Thomas	T-95S	900	19.8	69"
	T-103S	1,000	33.5	76.8"
	T-135S	1,350	43.0	90"
	T-153S	1,500	52.0	89.8"
	T-173HLS-III	1,700	52.0	106.8"
	T-245HDS	2,400	83.0	96"
	T-245HDK	2,400	87.0	96.8"

Figure 2.5.0 (*cont.*)

2.6.0 Gradall 48-inch Wide Excavating Bucket

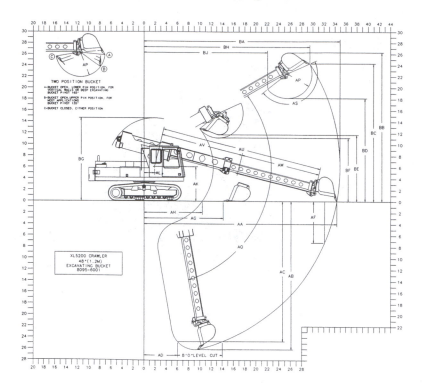
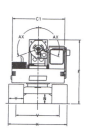
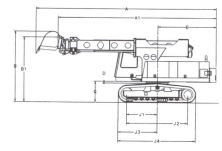

Shown with 8095-6001 48" (1.22m) excavating bucket
Metric units m unless otherwise noted.

A Overall length with bucket: 30'10" (9.4)
A1 Overall length without bucket: 27'1" (8.2)
B Overall height with bucket: 11'10" (3.6)
B1 Overall height without bucket: 10'10" (3.3)
C1 Width of upperstructure: 9'0" (2.7)
D Minimum clearance, upperstructure to undercarriage: 4" (102mm)
E Swing clearance, rear of upperstructure: 10'0" (3.0)
F Top of cradle to groundline: 10'10" (3.3)
G Clearance, upperstructure to groundline: 42" (1.1)
J1 Axis of rotation to centerline of drive sprockets: 5'3" (1.6)
J2 Nominal distance between centerlines of drive sprockets and idlers: 10'5" (3.2)
J3 Axis of rotation to end of track assembly: 6'8" (2.0)
J4 Nominal overall length of track assembly: 13'3" (4.0)
K Width of crawler with 23.6" (600mm) pads: 9'6" (2.9)
Width of crawler with 31.5" (800mm) pads: 10'2" (3.1)
N Ground clearance (per SAE J1234): 19" (483mm)
V Track gage, roller centerline to roller centerline: 7'6" (2.3)
Y Width of crawler track assembly: 31.5" (800mm)
AA Maximum radius at groundline (165° pivot): 34'4" (10.5)
AB Maximum digging depth (165° pivot) – end: 26'1" (7.9)
AC Maximum depth for 8' level cut – end: 24'8" (7.5)
AD Minimum radius of 8' level cut at depth "AC" – end: 6'0" (1.8)
AF Maximum depth of vertical wall which can be excavated: 7'3" (2.2)
AG Minimum level cut radius with bucket flat on groundline: 14'2" (4.3)
AH Minimum radius at groundline: 10'5" (3.2)
AK Boom pivot to groundline: 6'1" (1.8)
AL Boom pivot to axis of rotation: 22.5" (571mm)
AP Bucket tooth radius: 52" (1.3)
AQ Boom pivot angle: 30° Up & 80° Down

AS Bucket pivot angle: 135° & 165°
AU Maximum telescoping boom length (boom pivot to bucket pivot): 28'8" (8.7)
AV Minimum telescoping boom length (boom pivot to bucket pivot): 14'8" (4.5)
AW Telescoping boom travel: 14'0" (4.3)
AX Boom tilt angle (both sides of center): 110°
BA Maximum radius of working equipment (165° pivot): 34'11" (10.6)
BB Maximum height of working equipment: 26'4" (8.0)
BC Maximum bucket tooth height: 24'4" (7.4)
BD Minimum clearance of bucket teeth with bucket pivot at maximum height: 18'2" (5.5)
BE Minimum clearance of fully curled bucket at maximum boom height (165° pivot): 11'9" (3.6)
BF Minimum clearance of bucket teeth at maximum boom height: 11'2" (3.4)
BG Maximum height of working equipment with bucket below groundline: 14'8" (4.5)
BH Radius of bucket teeth at maximum height (165° pivot): 29'5" (9.0)
BJ Minimum radius of bucket teeth at maximum bucket pivot height (165° pivot): 22'1" (6.7)

Transport dimensions without attachment
Length: 27'1" (8.2)
Height: 10'10" (3.3)
Width: on 23.6" (600mm) pads - 9'6" (2.9)
on 31.5" (800mm) pads - 10'2" (3.1)

Units shown may have optional equipment

Figure 2.6.0 *(By permission of JLG Industries)*

2.6.1 Gradall 36-inch Wide Excavating Bucket

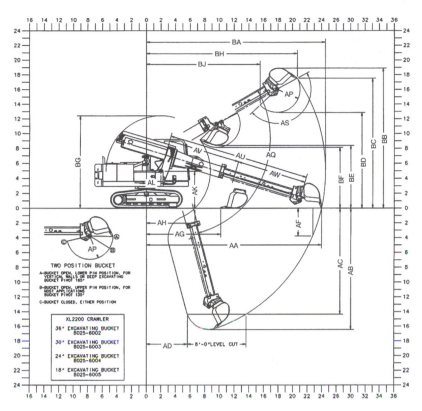

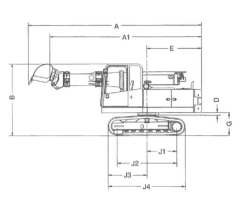

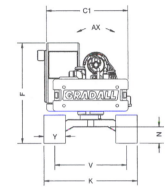

Shown with 8025-6002 36" (914mm) excavating bucket

A   Overall length with bucket: 22'1" (6.7)

A1   Overall length without bucket: 19'4" (5.9)

B   Overall height with bucket: 9'3" (2.8)

C1   Width of upperstructure: 7'5" (2.3)

D   Minimum clearance, upperstructure to undercarriage: 3" (76mm)

E   Swing clearance, rear of upperstructure: 7'0" (2.1)

F   Top of cab to groundline: 9'3" (2.8)

G   Clearance, upperstructure to groundline: 35" (889mm)

J1   Axis of rotation to centerline of drive sprockets: 3'10" (1.2)

J2   Nominal distance between centerlines of drive sprockets and idlers: 7'7" (2.3)

J3   Axis of rotation to end of track assembly: 5' (1.5)

J4   Nominal overall length of track assembly: 9'10" (3.0)

K   Width of crawler (standard): 8'6" (2.6)
Width of crawler (optional): 8'2" (2.5)

N   Ground clearance (per SAE J1234): 18" (457mm)

V   Track gauge, roller centerline to roller centerline: 6'6" (2.0)

Y   Width of crawler track assembly (standard): 23.6." (600mm)
Width of crawler track assembly (optional): 19.7" (500mm)

AA   Maximum radius at groundline (165° pivot): 24'0" (7.3)

AB   Maximum digging depth: 16'5" (5.0)

AC   Maximum depth for 8' level cut: 14'4" (4.4)

AD   Minimum radius for 8' level cut at depth "AC": 5'8" (1.7)

AF   Maximum depth of vertical wall which can be excavated: 3'9" (1.1)

AG   Minimum level cut radius with bucket flat on groundline: 10'4" (3.1)

AH   Minimum radius at groundline: 6'10" (2.1)

AK   Boom pivot to groundline: 4'11" (1.5)

AL   Boom pivot to axis of rotation: 24" (609mm)

AP   Bucket tooth radius: 3'2" (965mm)

AQ   Boom pivot angle: 30° Up & 75° Down

AS   Bucket pivot angle: 135° & 165°

AU   Maximum telescoping boom length (boom pivot to bucket pivot): 19'4" (5.9)

AV   Minimum telescoping boom length (boom pivot to bucket pivot): 9'11" (3.0)

AW   Telescoping boom travel: 9'5" (2.9)

AX   Bucket tilt angle: 360° (continuous)

BA   Maximum radius of working equipment (165° pivot): 24'6" (7.5)

BB   Maximum height of working equipment: 18'11" (5.8)

BC   Maximum bucket tooth height: 17'6" (5.3)

BD   Minimum clearance of bucket teeth with bucket pivot at maximum height: 12'10" (3.9)

BE   Minimum clearance of fully curled bucket at maximum boom height (165° pivot): 8'5" (2.6)

BF   Minimum clearance of bucket teeth at maximum boom height: 8'2" (2.5)

BG   Maximum height of working equipment with bucket below groundline: 12'8" (3.9)

BH   Radius of bucket teeth at maximum height (165° pivot): 20'8" (6.3)

BJ   Minimum radius of bucket teeth at maximum bucket pivot height (165° pivot): 15'7" (4.7)

Metric units are meters (m) unless noted

Machines shown may have optional equipment

Figure 2.6.1   *(By permission of JLG Industries)*

2.6.2 Gradall with Articulating 30-inch Excavating Bucket

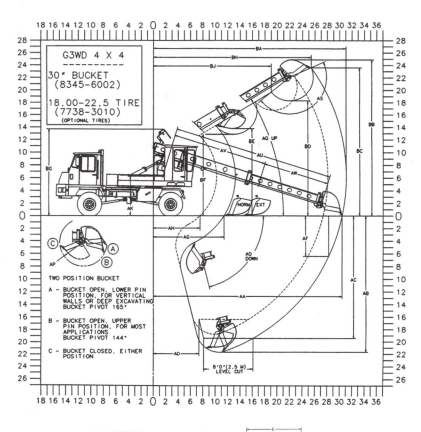

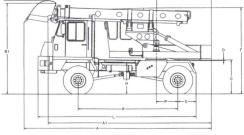

ATTACHMENTS

8345-6002 30" (760 mm) Excavating bucket

8345-6003 24" (610 mm) Excavating bucket

8345-6006 36" (910 mm) Excavating bucket

8345-6005 30" (760 mm) Pavement removal bucket

8345-6026 60" (1520 mm) Ditching bucket

8345-6019 66" (1680 mm) Ditch cleaning bucket

8345-6015 Single-tooth ripper

8345-6004 8' (2.4 m) Grading blade

8345-5008 Material handling extension

8345-5002 4' (1.2 m) Boom extension

8345-6032 Guardrail cleanout attachment

8345-5011 Fixed thumb grapple

Specifications subject to change without notice.

Some illustrations may show options.

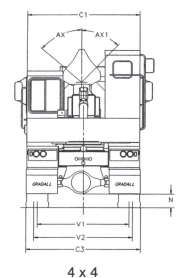

4 x 4

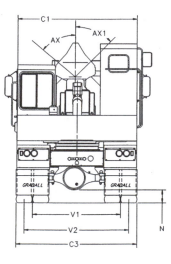

4 x 2

Figure 2.6.2 (*By permission of JLG Industries*)

2.6.3 Gradall Bucket and Blade Options

	Cu. yd.	m³
8025-6005 18" (457mm) Excavating bucket	3/16	.15
8025-6004 24" (610mm) Excavating bucket	1/4	.31
8025-6003 30" (762mm) Excavating bucket	3/8	.29
8025-6002 36" (914mm) Excavating bucket	1/2	.38

	Cu. yd.	m³
8025-6001 60" (1.5m) Ditching bucket	3/4	.61
8025-6012 60" (1.5m) Constant radius ditching bucket	5/8	.53

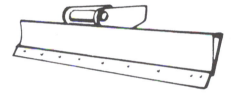

8025-6013 8' (2.4m) Grading blade with bolt on edge.

ATTACHMENTS
Buckets fabricated of steel plate, with high strength, low alloy cutting edges and wear strips. Standard attachments available for wide range of applications. Capacities shown are in heaped cu. yd.

8025-6006 30" (760mm) Pavement removal bucket

TWO-POSITION BUCKET

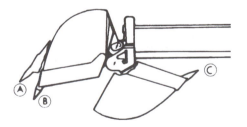

A: Bucket open, lower pin position, for vertical walls or deep excavating. Bucket pivot 165°.

B: Bucket open, upper pin position, for most applications. Bucket pivot 135°.

C: Bucket closed, either pin position.

Figure 2.6.3   (*By permission of JLG Industries*)

2.6.4 Specifications on Wheeled and Crawler Gradalls

Excavators	XL 2200 Crawler	XL 2300 Wheeled	G3WD Series E Wheeled Highway Speed	XL 3200 Crawler	XL 4100 Wheeled Highway Speed	XL 4200 Crawler	XL 5100 Wheeled Highway Speed	XL 5200 Crawler
Operating Range & Performance								
Engine Std	Cummins 4BT3.9	Cummins 4BT3.9		Deere 4045T	Cummins 6BT5.9	Cummins 6BT5.9	Cummins 6BTA5.9	Cummins 6BT5.9
HP @ RPM	93/2,200	110/2,200		114/2,200	148/2,200	148/2,200	174/2,200	174/2,200
Maximum Digging Depth	16'5" (5 m)	15'3" (4.6 m)	22' (6.7 m) †	20'7" (6.1 m)	22'6" (6.7 m)	23'3" (7.1 m)	25'4" (7.7 m)	26'1" (7.9 m)
Maximum Radius at Groundline	24'10" (7.3 m)	23'8" (7.2 m)	30'6" (9.3 m) †	27'7" (8.4 m)	30'1" (9.2 m)	30'3" (9.2 m)	34'3" (10.4 m)	34'4" (10.5 m)
Maximum Loading Height	12'10" (3.9 m)	14'0" (4.2 m)	18'1" (5.51 m) †	15'6" (5 m)	16'8" (5 m)	16'2" (4.9 m)	18'10" (5.7 m)	18'2" (5.5 m)
Max. Work. Ht. Boom Below Grade	13'5" (4.1 m)	13'9" (4.1 m)	13'2" (4.0 m) †	12'6" (3.8 m)	13'6" (4.1 m)	12'10" (3.9 m)	15'4" (4.7 m)	15'4" (4.5 m)
Boom Telescope	9'5" (2.9 m)	9'5" (2.9 m)	10'6" (3.2 m)	10'6" (3.2 m)	12'6" (3.8 m)	12'6" (3.8 m)	14'0" (4.3 m)	14'0" (4.3 m)
Boom Tilt	360° 3.9 rpm	360° 3.9 rpm	95°	220°	220°	220°	240°	240°
Bucket Pivot Angle	135° & 165°	135° & 165°	145°/165°	135° & 165°	135° & 165°	135° & 165°	135° & 165°	135° & 165°
Rated Boom Force	17,280 lbs (79.2 kN)	17,280 lbs (79.2 kN)	16,371 lbs (72.9 kN)	20,150 lbs (89.2 kN)	21,650 lbs (96.4 kN)	21,650 lbs (96.4 kN)	23,340 lbs (103.9 kN)	23,340 lbs (103.9 kN)
Rated Bucket Breakout Force	17,030 lbs (75.8 kN)	17,030 lbs (75.8 kN)	12,710 lbs (56.6 kN)	17,670 lbs (78.2 kN)	18,900 lbs (84 kN)	18,900 lbs (84 kN)	21,900 lbs (97.5 kN)	21,900 lbs (97.5 kN)
Maximum Lift Capacity	7,035 lbs (3,190 kg)	7,035 lbs (3,190 kg)	7,040 lbs (3,193 kg)	9,345 lbs (4,239 kg)	10,680 lbs (4,845 kg)	11,615 lbs (5,270 kg)	13,340 lbs (6,051 kg)	13,620 lbs (6,051 kg)
Weights & Dimensions								
Operating Weight	26,990 lbs (12,242 kg)	29,720 lbs (13,481 kg)	34,200 lbs (15,516 kg)	34,660 lbs (15,722 kg)	44,851 lbs (20,475 kg)	44,320 lbs (20,104 kg)	55,175 lbs (25,023 kg)	53,800 lbs (24,263 kg)
Swing Clearance	7'0" (2.1 m)	7'0" (2.1 m)	7'6" (2.29 m)	7'6" (2.3 m)	9'3" (2.8 m)	9'3" (2.8 m)	10'0" (3 m)	10'0" (3 m)
Transport Length w/Bucket	22'1" (6.7 m)	22'1" (6.7 m)	24'9" (7.54 m)	24'6" (7.5 m)	27'11" (8.5 m)	27'6" (8.4 m)	30'10" (9.4 m)	30'10" (9.4 m)
Transport Height w/Bucket	9'3" (2.8 m)	10'4" (3.1 m)	11'6" (3.51 m)	10'7" (3.2 m)	11'10" (3.6 m)	11'0" (3.3 m)	12'8" (3.9 m)	11'10" (3.6 m)
Transport Width w/Bucket	8'6" (2.6 m)	8'6" (2.6 m)	8' (2.44 m)	8'6" (2.6 m)	8' (2.4 m)	9'6" (2.9 m)	9'0" (2.7 m)	10'2" (3.1 m)
Drive	4x4	4x4	4x2 or 4x4		6x4 or 6x6		6x4 or 6x6	
Gross Vehicle Axle Rating		36,200 lbs (16,420 kg)			59,200 lbs (26,847 kg)		66,000 lbs (29,958 kg)	
Wheelbase		9'8.5" (2.5 m)	13' (3.96 m)		14'3" (4.3 m)		14'8" (4.5 m)	
Undercarriage								
Engine - Std			Cummins 6BTA5.9		Cummins 6BT5.9		Cummins 6CTA8.3	
HP @ RPM			190/2,500		200/2,500		240/2,100	
Travel Speed-MPH	2.34-3.44 mph (3.8-5.5 km/hr)	15 mph (24 km/hr)	55 mph (88.5 kph)	19.33 mph (31.53 km/hr)	52 mph (83 kph)	15.2-24 mph (2.5-3.7 kph)	51 mph (82 kph)	14-2.2 mph (2.2-3.5 kph)
Track Length	9'10" (3 m)			11'9" (3.6 m)		13'9" (3.9 m)		13'3" (4 m)
Pad Width	19.7" (500 mm) / 23.6" (600 mm)			19.7" (500 mm) 23.6" (600 mm)		23.6" (600 mm) / 31.5" (800 mm)		23.6" (600 mm) / 31.5" (800 mm)
Ground Bearing Pressure	6.86 psi (47.1 KPa)/5.8 psi (40 KPa)			7.1 psi (49 KPa) 6.0 psi (41 KPa)		6.9 psi (48 KPa)/5.4 psi (37 KPa)		8.1 psi (56 KPa)/6.2 psi (43 KPa)
Ground Clearance	18" (475 mm)	14" (355 mm)	11" (279 mm)	21" (533 mm)	10" (254 mm)	20" (508 mm)	10" (254 mm)	19" (482 mm)
Attachments								
Trenching Buckets	✓	✓	✓	✓	✓	✓	✓	✓
Excavating Buckets	✓	✓	✓	✓	✓	✓	✓	✓
Ditch Cleaning Buckets	✓	✓	✓	✓	✓	✓	✓	✓
Dredging Buckets	✓	✓	✓	✓	✓	✓	✓	✓
Pavement Removal Buckets	✓	✓	✓	✓	✓	✓	✓	✓
Grading Blades	✓	✓	✓	✓	✓	✓	✓	✓
Rippers	✓	✓	✓	✓	✓	✓	✓	✓
Telestick	✓	✓	✓	✓	✓	✓	✓	✓
Boom Extensions	✓	✓	✓	✓	✓	✓	✓	✓
Air Hammers	✓	✓	✓	✓	✓	✓	✓	✓
Hydraulic Hammers	✓	✓	✓	✓	✓	✓	✓	✓
Augers	✓	✓	✓	✓	✓	✓	✓	✓
Grapples	✓	✓	✓	✓	✓	✓	✓	✓

† Indicates Slide-o-Boom Extension Fully Extended

It is Gradall policy to continually improve its products. Therefore, designs, materials and specifications are subject to change without notice and without incurring any liability on units already sold. Units shown may have optional equipment.

Figure 2.6.4 *(By permission of JLG Industries)*

2.7.0 Articulating Boom Lift

REACH SPECIFICATIONS

Platform Height . . . . 150 ft. (45.72 m)
Maximum Horizontal Outreach . . . . . . 79 ft. 3 in. (24.15 m)
Up & Over Clearance @ Max. Horiz. Reach . 42 ft. 7 in. (12.98 m)
Below Ground Reach . . . . 30 ft. 4 in. (9.25 m)
Maximum Up and Over Clearance . . . . 80 ft. (24.38 m)
Horizontal Outreach @ Max. Up & Over . . . 72 ft. 4 in. (22.05 m)
Swing . . . . 360° Continuous
Platform Capacity (Unrestricted) . . . . 500 lbs. (227 kg)
Platform Capacity (Restricted) . . . . 1,000 lbs. (454 kg)
Platform Rotator . . . . 165°Hyrdaulic

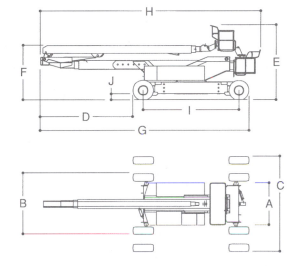

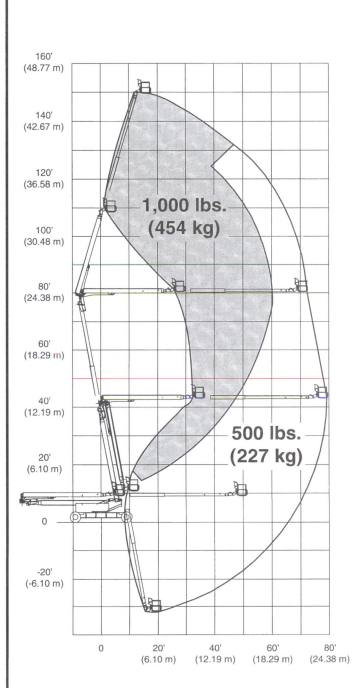

DIMENSIONAL DATA

A. Platform Size . . . . 36 x 96 in. (0.91 x 2.44 m)
B. Overall Width (Axles retracted) . . . . 11 ft. 6 in. (3.51 m)
C. Overall Width (Axles extended) . . . . . . 18 ft. 0 in. (5.49 m)
D. Tailswing (stowed) . . . . 17 ft. 7 in. (5.37 m)
   Tailswing (fully elevated) . . . . zero
E. Stowed Height . . . . 14 ft. 0 in. (4.27 m)
F. Stowed Height (Platform tilted) . . . . 10 ft. 6 in. (3.20 m)
G. Stowed Length . . . . 39 ft. 5 in. (12.01 m)
H. Stowed Length (Platform tilted) . . . . . . 41 ft. 7 in. (12.67 m)
I. Wheelbase . . . . 18 ft. 0 in. (5.50 m)
J. Ground Clearance . . . . 12 in. (0.30 m)
Gross Vehicle Weight . . . . 57,000 lbs. (25,855 kg)

CHASSIS

Drive Speed . . . . 2.7 mph (4.3 kph)
Gradeability 31%
Tires . . . . 445/65R22.5 Foam Filled
Turning Radius (outside)
   w/axles retracted . . . . 27 ft. 0 in. (8.23 m)
   w/axles extended . . . . 30 ft. 1 in. (9.17 m)
Turning Radius (inside)
   w/axles retracted . . . . 14 ft. 10 in. (4.52m)
   w/axles extended . . . . 14 ft. 7 in. (4.45 m)

POWER SYSTEM

Engine . . . . Cummins 4B3.9C 76 hp Diesel Engine
Fuel Tank Capacity . . . . 68 gal. (257.4 L)
Hydraulic Reservoir . . . . 124 gal. (469.3 L)
Main Pump Output (three section gear pump)
   Drive . . . . 25 gpm (95 liters/min) @ 2500 rpm
   Lift and Hi Drive . . . . 15 gpm (57 liters/min) @ 2500 rpm
   Steer . . . . 9 gpm (34 liters/min) @ 2500 rpm
Brakes . . . . . . Automatic spring applied, hydraulically released
Manual Descent . . . . Lever actuated hand pump

Figure 2.7.0 *(By permission of JLG Industries)*

2.7.1 JLG Semi-track Lift Specifications

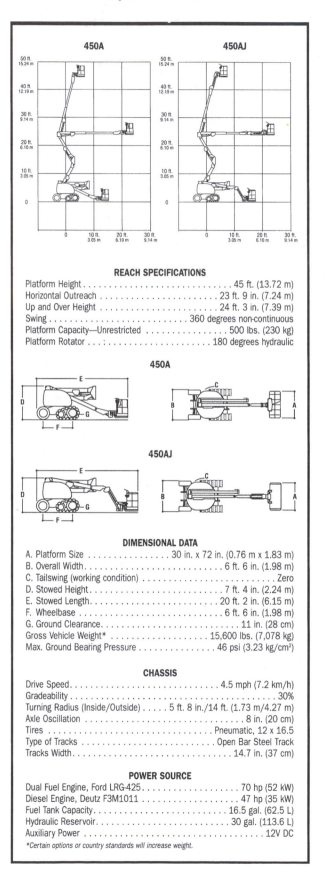

450A

450AJ

REACH SPECIFICATIONS

Platform Height . . . . 45 ft. (13.72 m)
Horizontal Outreach . . . . 23 ft. 9 in. (7.24 m)
Up and Over Height . . . . 24 ft. 3 in. (7.39 m)
Swing . . . . 360 degrees non-continuous
Platform Capacity—Unrestricted . . . . 500 lbs. (230 kg)
Platform Rotator . . . : . . . . 180 degrees hydraulic

450A

450AJ

DIMENSIONAL DATA

A. Platform Size . . . . 30 in. x 72 in. (0.76 m x 1.83 m)
B. Overall Width . . . . 6 ft. 6 in. (1.98 m)
C. Tailswing (working condition) . . . . Zero
D. Stowed Height . . . . 7 ft. 4 in. (2.24 m)
E. Stowed Length . . . . 20 ft. 2 in. (6.15 m)
F. Wheelbase . . . . 6 ft. 6 in. (1.98 m)
G. Ground Clearance . . . . 11 in. (28 cm)
Gross Vehicle Weight* . . . . 15,600 lbs. (7,078 kg)
Max. Ground Bearing Pressure . . . . 46 psi (3.23 kg/cm²)

CHASSIS

Drive Speed . . . . 4.5 mph (7.2 km/h)
Gradeability 30%
Turning Radius (Inside/Outside) . . . . . 5 ft. 8 in./14 ft. (1.73 m/4.27 m)
Axle Oscillation . . . . 8 in. (20 cm)
Tires . . . . Pneumatic, 12 x 16.5
Type of Tracks . . . . Open Bar Steel Track
Tracks Width . . . . 14.7 in. (37 cm)

POWER SOURCE

Dual Fuel Engine, Ford LRG-425 . . . . 70 hp (52 kW)
Diesel Engine, Deutz F3M1011 . . . . 47 hp (35 kW)
Fuel Tank Capacity . . . . 16.5 gal. (62.5 L)
Hydraulic Reservoir . . . . 30 gal. (113.6 L)
Auxiliary Power . . . . 12V DC

Certain options or country standards will increase weight.

Figure 2.7.1 *(By permission of JLG Industries)*

2.7.2 Material Handler by Gradall

MATERIAL HANDLER

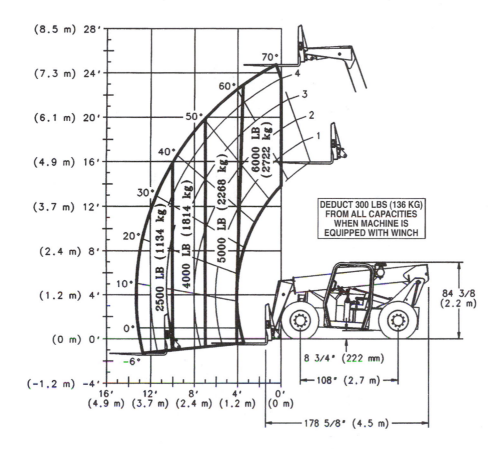

DEDUCT 300 LBS (136 KG) FROM ALL CAPACITIES WHEN MACHINE IS EQUIPPED WITH WINCH

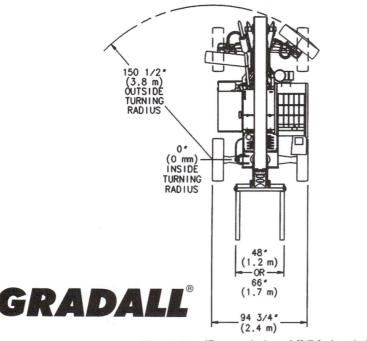

GRADALL®

Figure 2.7.2 *(By permission of JLG Industries)*

2.7.3 Platform Lifts—JLG Specifications

2033E3

DIMENSIONAL DATA

A. Platform Height—Elevated ....20 ft. (6.1 m)
B. Platform Height—Lowered...................................3 ft. 3.5 in (1.0 m)
C. Platform Railing Height* ....3 ft. 3.5 in. (1.0 m)
D. Overall Height ....6 ft. 7 in. (2.01 m)
E. Platform Size ....30 x 94 in. (0.76 x 2.39 m)
F. Platform Extension ....3 ft. (91 cm)
G. Overall Width ....2 ft. 9 in. (84 cm)
H. Overall Length ....8 ft. 4 in. (2.54 m)
I. Wheelbase ....6 ft. 1 in. (1.85 m)
J. Ground Clearance ....3.0 in. (8.0 cm)
Platform Capacity ....750 lbs. (340 kg)
Capacity on Deck Extension ....250 lbs. (113 kg)
Lift/Lower Speed..34/28 seconds
Maximum Drive Height..Full Height
Gross Vehicle Weight** ....3,700 lbs. (1,678 kg)

CHASSIS

Drive Speed—Lowered ....2.75 mph (4.42 kph)
Drive Speed—Elevated..............................0.5 mph (0.8 kph)
Gradeability...25%
Turning Radius (Inside) ....2 in. (5 cm)
Turning Radius (Outside)...............................80 in. (1.68 m)
Tire Size/Type ....16 x 5 Non-Marking Solid
Brakes.......................................Automatic Rear Wheel Parking

POWER SYSTEM

Batteries...4x6V, 220 amp-hr
Charger..Automatic 24V DC
Electric Motor ....4.5 hp (3.4 kW) Series Wound
Pump..Single Section Gear
Hydraulic Reservoir...4.0 gal. (15.2 L)

*special heights available, consult factory
**certain options or country standards will increase weight

2046E3

DIMENSIONAL DATA

A. Platform Height—Elevated ....20 ft. (6.1 m)
B. Platform Height—Lowered...................................3 ft. 3.5 in (1.0 m)
C. Platform Railing Height* ....3 ft. 3.5 in. (1.0 m)
D. Overall Height ....6 ft. 7 in. (2.01 m)
E. Platform Size ....42 x 94 in. (1.07 x 2.39 m)
F. Platform Extension ....3 ft. (91 cm)
G. Overall Width ....3 ft. 10 in. (1.17 m)
H. Overall Length ....8 ft. 4 in. (2.54 m)
I. Wheelbase ....6 ft. 1 in. (1.85 m)
J. Ground Clearance ....3.5 in. (9.0 cm)
Platform Capacity ....1,000 lbs. (454 kg)
Capacity on Deck Extension ....250 lbs. (113 kg)
Lift/Lower Speed..36/28 seconds
Maximum Drive Height..Full Height
Gross Vehicle Weight** ....3,940 lbs. (1,787 kg)

CHASSIS

Drive Speed—Lowered ....2.25 mph (3.62 kph)
Drive Speed—Elevated..............................0.5 mph (0.8 kph)
Gradeability...25%
Turning Radius (Inside) ....16 in. (41 cm)
Turning Radius (Outside)...............................95 in. (2.41 m)
Tire Size/Type ....16 x 5 Non-Marking Solid
Brakes.......................................Automatic Rear Wheel Parking

POWER SYSTEM

Batteries...4 x 6V, 220 amp-hr
Charger..Automatic 24V DC
Electric Motor ....4.5 hp (3.4 kW) Series Wound
Pump..Single Section Gear
Hydraulic Reservoir...4.0 gal. (15.2 L)

*special heights available, consult factory
**certain options or country standards will increase weight

Figure 2.7.3 *(By permission of JLG Industries)*

2.7.4 Bucket Lifts

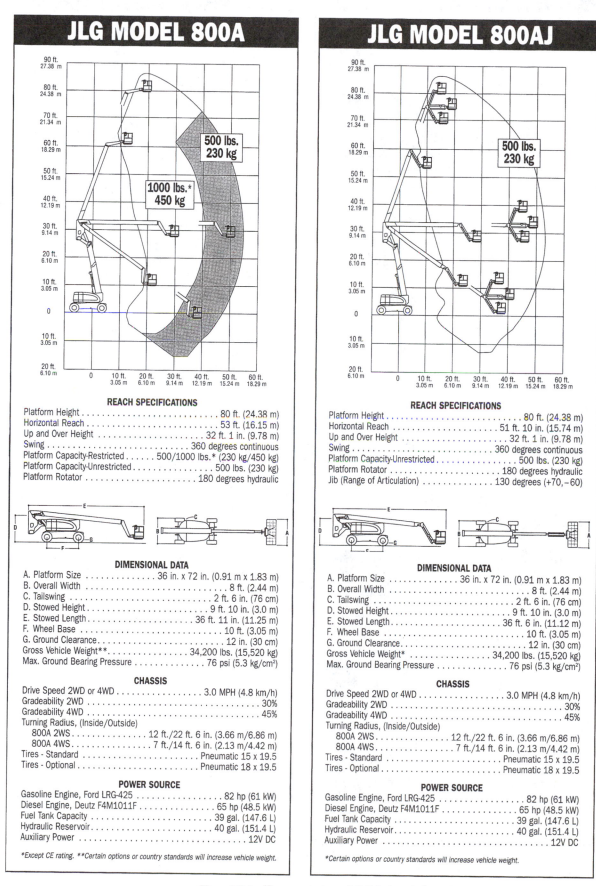

JLG MODEL 800A

500 lbs. 230 kg

1000 lbs.* 450 kg

REACH SPECIFICATIONS

Platform Height . . . . 80 ft. (24.38 m)
Horizontal Reach . . . . 53 ft. (16.15 m)
Up and Over Height . . . . 32 ft. 1 in. (9.78 m)
Swing . . . . 360 degrees continuous
Platform Capacity-Restricted . . . . . . 500/1000 lbs.* (230 kg/450 kg)
Platform Capacity-Unrestricted . . . . 500 lbs. (230 kg)
Platform Rotator . . . . 180 degrees hydraulic

DIMENSIONAL DATA

A. Platform Size . . . . 36 in. x 72 in. (0.91 m x 1.83 m)
B. Overall Width . . . . 8 ft. (2.44 m)
C. Tailswing . . . . 2 ft. 6 in. (76 cm)
D. Stowed Height . . . . 9 ft. 10 in. (3.0 m)
E. Stowed Length . . . . 36 ft. 11 in. (11.25 m)
F. Wheel Base . . . . 10 ft. (3.05 m)
G. Ground Clearance . . . . 12 in. (30 cm)
Gross Vehicle Weight** . . . . 34,200 lbs. (15,520 kg)
Max. Ground Bearing Pressure . . . . 76 psi (5.3 kg/cm²)

CHASSIS

Drive Speed 2WD or 4WD . . . . 3.0 MPH (4.8 km/h)
Gradeability 2WD 30%
Gradeability 4WD 45%
Turning Radius, (Inside/Outside)
    800A 2WS . . . . 12 ft./22 ft. 6 in. (3.66 m/6.86 m)
    800A 4WS . . . . 7 ft./14 ft. 6 in. (2.13 m/4.42 m)
Tires - Standard . . . . Pneumatic 15 x 19.5
Tires - Optional . . . . Pneumatic 18 x 19.5

POWER SOURCE

Gasoline Engine, Ford LRG-425 . . . . 82 hp (61 kW)
Diesel Engine, Deutz F4M1011F . . . . 65 hp (48.5 kW)
Fuel Tank Capacity . . . . 39 gal. (147.6 L)
Hydraulic Reservoir . . . . 40 gal. (151.4 L)
Auxiliary Power . . . . 12V DC

*Except CE rating. **Certain options or country standards will increase vehicle weight.*

JLG MODEL 800AJ

500 lbs. 230 kg

REACH SPECIFICATIONS

Platform Height . . . . 80 ft. (24.38 m)
Horizontal Reach . . . . 51 ft. 10 in. (15.74 m)
Up and Over Height . . . . 32 ft. 1 in. (9.78 m)
Swing . . . . 360 degrees continuous
Platform Capacity-Unrestricted . . . . 500 lbs. (230 kg)
Platform Rotator . . . . 180 degrees hydraulic
Jib (Range of Articulation) . . . . 130 degrees (+70, –60)

DIMENSIONAL DATA

A. Platform Size . . . . 36 in. x 72 in. (0.91 m x 1.83 m)
B. Overall Width . . . . 8 ft. (2.44 m)
C. Tailswing . . . . 2 ft. 6 in. (76 cm)
D. Stowed Height . . . . 9 ft. 10 in. (3.0 m)
E. Stowed Length . . . . 36 ft. 6 in. (11.12 m)
F. Wheel Base . . . . 10 ft. (3.05 m)
G. Ground Clearance . . . . 12 in. (30 cm)
Gross Vehicle Weight* . . . . 34,200 lbs. (15,520 kg)
Max. Ground Bearing Pressure . . . . 76 psi (5.3 kg/cm²)

CHASSIS

Drive Speed 2WD or 4WD . . . . 3.0 MPH (4.8 km/h)
Gradeability 2WD 30%
Gradeability 4WD 45%
Turning Radius, (Inside/Outside)
    800A 2WS . . . . 12 ft./22 ft. 6 in. (3.66 m/6.86 m)
    800A 4WS . . . . 7 ft./14 ft. 6 in. (2.13 m/4.42 m)
Tires - Standard . . . . Pneumatic 15 x 19.5
Tires - Optional . . . . Pneumatic 18 x 19.5

POWER SOURCE

Gasoline Engine, Ford LRG-425 . . . . 82 hp (61 kW)
Diesel Engine, Deutz F4M1011F . . . . 65 hp (48.5 kW)
Fuel Tank Capacity . . . . 39 gal. (147.6 L)
Hydraulic Reservoir . . . . 40 gal. (151.4 L)
Auxiliary Power . . . . 12V DC

Certain options or country standards will increase vehicle weight.

Figure 2.7.4 (*By permission of JLG Industries*)

JLG 600S

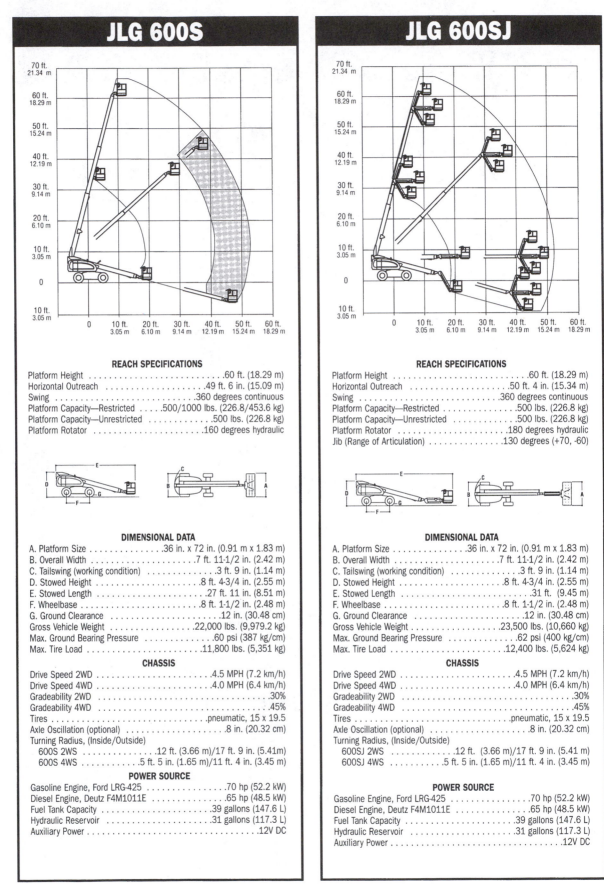

REACH SPECIFICATIONS

Platform Height . . . . .60 ft. (18.29 m)
Horizontal Outreach . . . . .49 ft. 6 in. (15.09 m)
Swing . . . . .360 degrees continuous
Platform Capacity—Restricted . . . . .500/1000 lbs. (226.8/453.6 kg)
Platform Capacity—Unrestricted . . . . .500 lbs. (226.8 kg)
Platform Rotator .160 degrees hydraulic

DIMENSIONAL DATA

A. Platform Size . . . . .36 in. x 72 in. (0.91 m x 1.83 m)
B. Overall Width . . . . .7 ft. 11-1/2 in. (2.42 m)
C. Tailswing (working condition) . . . . .3 ft. 9 in. (1.14 m)
D. Stowed Height . . . . .8 ft. 4-3/4 in. (2.55 m)
E. Stowed Length . . . . .27 ft. 11 in. (8.51 m)
F. Wheelbase . . . . .8 ft. 1-1/2 in. (2.48 m)
G. Ground Clearance . . . . .12 in. (30.48 cm)
Gross Vehicle Weight . . . . .22,000 lbs. (9,979.2 kg)
Max. Ground Bearing Pressure . . . . .60 psi (387 kg/cm)
Max. Tire Load . . . . .11,800 lbs. (5,351 kg)

CHASSIS

Drive Speed 2WD .4.5 MPH (7.2 km/h)
Drive Speed 4WD .4.0 MPH (6.4 km/h)
Gradeability 2WD 30%
Gradeability 4WD 45%
Tires . . . . .pneumatic, 15 x 19.5
Axle Oscillation (optional) . . . . .8 in. (20.32 cm)
Turning Radius, (Inside/Outside)
    600S 2WS . . . . .12 ft. (3.66 m)/17 ft. 9 in. (5.41m)
    600S 4WS . . . . .5 ft. 5 in. (1.65 m)/11 ft. 4 in. (3.45 m)

POWER SOURCE

Gasoline Engine, Ford LRG-425 . . . . .70 hp (52.2 kW)
Diesel Engine, Deutz F4M1011E . . . . .65 hp (48.5 kW)
Fuel Tank Capacity . . . . .39 gallons (147.6 L)
Hydraulic Reservoir . . . . .31 gallons (117.3 L)
Auxiliary Power . . . . .12V DC

JLG 600SJ

REACH SPECIFICATIONS

Platform Height . . . . .60 ft. (18.29 m)
Horizontal Outreach . . . . .50 ft. 4 in. (15.34 m)
Swing . . . . .360 degrees continuous
Platform Capacity—Restricted . . . . .500 lbs. (226.8 kg)
Platform Capacity—Unrestricted . . . . .500 lbs. (226.8 kg)
Platform Rotator .180 degrees hydraulic
Jib (Range of Articulation) . . . . .130 degrees (+70, -60)

DIMENSIONAL DATA

A. Platform Size . . . . .36 in. x 72 in. (0.91 m x 1.83 m)
B. Overall Width . . . . .7 ft. 11-1/2 in. (2.42 m)
C. Tailswing (working condition) . . . . .3 ft. 9 in. (1.14 m)
D. Stowed Height . . . . .8 ft. 4-3/4 in. (2.55 m)
E. Stowed Length .31 ft. (9.45 m)
F. Wheelbase . . . . .8 ft. 1-1/2 in. (2.48 m)
G. Ground Clearance . . . . .12 in. (30.48 cm)
Gross Vehicle Weight . . . . .23,500 lbs. (10,660 kg)
Max. Ground Bearing Pressure . . . . .62 psi (400 kg/cm)
Max. Tire Load . . . . .12,400 lbs. (5,624 kg)

CHASSIS

Drive Speed 2WD .4.5 MPH (7.2 km/h)
Drive Speed 4WD .4.0 MPH (6.4 km/h)
Gradeability 2WD 30%
Gradeability 4WD 45%
Tires . . . . .pneumatic, 15 x 19.5
Axle Oscillation (optional) . . . . .8 in. (20.32 cm)
Turning Radius, (Inside/Outside)
    600SJ 2WS . . . . .12 ft. (3.66 m)/17 ft. 9 in. (5.41 m)
    600SJ 4WS . . . . .5 ft. 5 in. (1.65 m)/11 ft. 4 in. (3.45 m)

POWER SOURCE

Gasoline Engine, Ford LRG-425 . . . . .70 hp (52.2 kW)
Diesel Engine, Deutz F4M1011E . . . . .65 hp (48.5 kW)
Fuel Tank Capacity . . . . .39 gallons (147.6 L)
Hydraulic Reservoir . . . . .31 gallons (117.3 L)
Auxiliary Power . . . . .12V DC

Figure 2.7.4 *(cont.)*

2.7.5 Range of Platform Lifts with Specifications

Model	Platform Height (Elevated)	Overall Height	Platform Size	Overall Width	Overall Length	Platform Capacity	Lift/Lower Time	Maximum Drive Height	Drive Speed Lowered	Gradeability	Turning Radius (Inside)	Turning Radius (Outside)	Tire Size/Type	Power Source	Gross Vehicle Weight*
E2 Series Scissor Lifts															
1932E2	19 ft. (5.79 m)	6 ft. 6 in. (1.98 m)	31 x 64 in. (0.79 x 1.63 m)	32 in. (81 cm)	5 ft. 8.5 in. (1.74 m)	500 lbs. (230 kg)	27/26 sec.	Full Height	2.5 mph (4 kph)	20%	6 in. (15 cm)	68 in. (1.73 m)	12.5 x 4 Non-Marking Solid	4 x 6V 220 amp-hr	2,620 lbs. (1,188 kg)
2032E2	20 ft. (6.1 m)	6 ft. 7.5 in. (2.02 m)	31 x 84 in. (0.79 x 2.13 m)	32 in. (81 cm)	7 ft. 5 in. (2.26 m)	750 lbs. (340 kg)	31/31 sec.	Full Height	2.7 mph (4.3 kph)	25%	39 in. (99 cm)	100 in. (2.54 m)	16 x 5 Non-Marking Solid	4 x 6V 220 amp-hr	3,930 lbs. (1,783 kg)
2646E2	26 ft. (7.92 m)	6 ft. (1.83 m)	45 x 84 in. (1.14 x 2.13 m)	46 in. (1.17 m)	7 ft. 5 in. (2.26 m)	750 lbs. (340 kg)	46/41 sec.	Full Height	2.25 mph (3.62 kph)	25%	44 in. (1.12 m)	111 in. (2.82 m)	16 x 5 Non-Marking Solid	4 x 6V 220 amp-hr	4,380 lbs. (1,987 kg)
3246E2	32 ft. (9.75 m)	6 ft. 5 in.** (1.96 m)	45 x 84 in. (1.14 x 2.13 m)	46 in. (1.17 m)	7 ft. 8 in. (2.34 m)	700 lbs. (318 kg)	50/57 sec.	Full Height	2.0 mph (3.2 kph)	25%	45 in. (1.14 m)	115 in. (2.92 m)	16 x 5 Non-Marking Solid	24V DC 4.5 hp (3.4 kw) electric	6,200 lbs. (2,812 kg)
E3 Series Scissor Lifts															
1532E3	15 ft. (4.57 m)	6 ft. 3.5 in. (1.92 m)	30 x 64 in. (0.76 x 1.63 m)	32.5 in. (83 cm)	5 ft. 11.5 in. (1.82 m)	600 lbs. (272 kg)	20/20 sec.	Full Height	2.5 mph (4.03 kph)	25%	2 in. (5 cm)	66 in. (1.68 m)	14 x 4.5 Non-Marking Solid	4 x 6V 220 amp-hr	2,630 lbs. (1,193 kg)
1932E3	19 ft. (5.79 m)	6 ft. 7.5 in. (2.02 m)	30 x 64 in. (0.76 x 1.63 m)	32.5 in. (83 cm)	5 ft. 11.5 in. (1.82 m)	500 lbs. (230 kg)	22/28 sec.	Full Height	2.5 mph (4.03 kph)	25%	2 in. (5 cm)	66 in. (1.68 m)	14 x 4.5 Non-Marking Solid	4 x 6V 220 amp-hr	2,900 lbs. (1,315 kg)
2033E3	20 ft. (6.1 m)	6 ft. 7 in. (2.01 m)	30 x 94 in. (0.76 x 2.39 m)	33 in. (84 cm)	8 ft. 4 in. (2.54 m)	750 lbs. (340 kg)	34/28 sec.	Full Height	2.75 mph (4.42 kph)	25%	2 in. (5 cm)	80 in. (2.0 m)	16 x 5 Non-Marking Solid	4 x 6V 220 amp-hr	3,700 lbs. (1,678 kg)
2046E3	20 ft. (6.1 m)	6 ft. 7 in. (2.01 m)	42 x 94 in. (1.07 x 2.39 m)	46 in. (1.17 m)	8 ft. 4 in. (2.54 m)	1,000 lbs. (450 kg)	38/28 sec.	Full Height	2.25 mph (3.62 kph)	25%	16 in. (41 cm)	95 in. (2.41 m)	16 x 5 Non-Marking Solid	4 x 6V 220 amp-hr	3,940 lbs. (1,787 kg)
2646E3	26 ft. (7.92 m)	7 ft. 4 in. (2.24 m)	42 x 94 in. (1.07 x 2.39 m)	46 in. (1.17 m)	8 ft. 4 in. (2.54 m)	750 lbs. (340 kg)	54/39 sec.	Full Height	2.25 mph (3.62 kph)	25%	16 in. (41 cm)	95 in. (2.41 m)	16 x 5 Non-Marking Solid	4 x 6V 220 amp-hr	4,370 lbs. (1,982 kg)
2658E3	26 ft. (7.92 m)	7 ft. 4 in. (2.24 m)	54 x 94 in. (1.37 x 2.39 m)	58 in. (1.47 m)	8 ft. 4 in. (2.54 m)	1,000 lbs. (450 kg)	56/39 sec.	Full Height	2.25 mph (3.62 kph)	25%	16 in. (41 cm)	100 in. (2.54 m)	16 x 5 Non-Marking Solid	4 x 6V 220 amp-hr	4,680 lbs. (2,123 kg)
Electric Drive Scissor Lifts															
3369 Electric	33 ft. (10.06 m)	6 ft. 1 in.** (1.85 m)	61 x 106 in. (1.55 x 2.69 m)	69 in. (1.75 m)	9 ft. 1 in. (2.77 m)	1,000 lbs. (450 kg)	75/60 sec.	Full Height	3.0 mph (4.8 kph)	25%	5 ft. (1.52 m)	12 ft. 5 in. (3.78 m)	205/75R15 Foam Filled	4 x 6V 220 amp-hr	8,000 lbs. (3,629 kg)
3969 Electric	39 ft. (11.89 m)	6 ft. 7 in.** (2.01 m)	61 x 106 in. (1.55 x 2.69 m)	69 in. (1.75 m)	9 ft. 1 in. (2.77 m)	750 lbs. (340 kg)	75/60 sec.	32 ft. (9.75 m)	3.0 mph (4.8 kph)	25%	5 ft. (1.52 m)	12 ft. 5 in. (3.78 m)	205/75R15 Foam Filled	8 x 6V 225 amp-hr	8,800 lbs. (3,992 kg)
Rough Terrain Scissor Lifts															
25RTS	25 ft. (7.62 m)	6 ft. 10 in.*** (2.08 m)	71 x 126 in. (1.8 x 3.2 m)	90 in. (2.29 m)	12 ft. 5 in. (3.78 m)	2,000 lbs. (907 kg)	55/35 sec.	Full Height	3.5 mph (5.6 kph)	35%	12 ft. 6 in. (3.81 m)	18 ft. 1 in. (5.51 m)	12 x 16.5 Loader Lug	35 hp DF	7,600 lbs. (3,447 kg)
26MRT	26 ft. (7.92 m)	7 ft. 5 in. (2.26 m)	54 x 94 in. (1.37 x 2.39 m)	69 in. (1.75 m)	8 ft. 9 in. (2.67 m)	1,000 lbs. (450 kg)	38/35 sec.	Full Height	4.5 mph (7.2 kph)	35%	8 ft. 1 in. (2.46 m)	15 ft. 4 in. (4.67 m)	26 X 12 X 12 Lug Tread	28 hp Diesel	5,500 lbs. (2,495 kg)
33RTS	33 ft. (10.06 m)	7 ft. 6 in. (2.29 m)	71 x 126 in. (1.8 x 3.2 m)	90 in. (2.29 m)	12 ft. 5 in. (3.78 m)	1,500 lbs. (680 kg)	60/35 sec.	Full Height	3.5 mph (5.6 kph)	35%	12 ft. 6 in. (3.81 m)	18 ft. 1 in. (5.51 m)	12 x 16.5 Loader Lug	22 hp Diesel	8,200 lbs. (3,720 kg)
40RTS	39 ft. 6 in. (12.04 m)	8 ft. 2 in.*** (2.49 m)	71 x 126 in. (1.8 x 3.2 m)	90 in. (2.29 m)	12 ft. 5 in. (3.78 m)	1,000 lbs. (450 kg)	65/35 sec.	Full Height	3.5 mph (5.6 kph)	35%	12 ft. 6 in. (3.81 m)	18 ft. 1 in. (5.51 m)	12 x 16.5 Loader Lug	35 hp Diesel	9,200 lbs. (4,173 kg)
400RTS	40 ft. (12.19 m)	7 ft. 2.5 in.** (2.2 m)	71 x 168 in. (1.8 x 4.27 m)	90 in. (2.29 m)	14 ft. 5 in. (4.39 m)	2,000 lbs. (907 kg)	60/60 sec.	Full Height	3.5 mph (5.6 kph)	35%	13 ft. 2 in. (4.01 m)	18 ft. 9 in. (5.72 m)	12 x 16.5 Loader Lug	70 hp Diesel	13,500 lbs. (6,124 kg)
500RTS	50 ft. (15.24 m)	8 ft. 0.5 in.*** (2.45 m)	71 x 168 in. (1.8 x 4.27 m)	90 in. (2.29 m)	15 ft. 5 in. (4.7 m)	2,500 lbs. (1,134 kg)	72/72 sec.	22 ft. (6.71 m)	3.5 mph (5.6 kph)	35%	13 ft. 2 in. (4.01 m)	18 ft. 9 in. (5.72 m)	12 x 16.5 Loader Lug	70 hp Diesel	15,300 lbs. (6,940 kg)

Figure 2.7.5 (*By permission of JLG Industries, Inc., McConnellsburg, Pa.*)

2.7.6 Range of Bucket Lifts with Specifications

Electric Articulating Boom Lifts

Model	Platform Height	Horizontal Outreach	Up-and-Over Height	Platform Capacity (Unrestricted)	Platform Size	Overall Width	Tailswing	Stowed Height	Stowed Length	Drive Speed (2WD)	Gradeability (2WD)	Turning Radius (Inside)	Turning Radius (Outside)	Power Source	Gross Vehicle Weight*
30electric	30 ft. (9.14 m)	13 ft. 5 in. (4.09 m)	16 ft. 7 in. (5.05 m)	500 lbs. (230 kg)	26 x 48 in. (0.66 x 1.22 m)	69 in. (1.75 m)	Zero	6 ft. 7 in. (2.01 m)	15 ft. 7 in. (4.75 m)	4.0 mph (6.4 kph)	25%	4 ft. 5 in. (1.35 m)	11 ft. 10 in. (3.61 m)	48V DC 8 x 6V 245 amp-hr	4,770 lbs. (2,164 kg)
35electric	35 ft. (10.67 m)	20 ft. 6 in. (6.25 m)	17 ft. 11 in. (5.46 m)	500 lbs. (230 kg)	26 x 48 in. (0.66 x 1.22 m)	69 in. (1.75 m)	Zero	6 ft. 7 in. (2.01 m)	17 ft. (5.18 m)	3.0 mph (4.8 kph)	25%	4 ft. 3 in. (1.3 m)	11 ft. 10 in. (3.61 m)	48V DC 8 x 6V 245 amp-hr	9,500 lbs. (4,309 kg)
n35electric	35 ft. (10.67 m)	20 ft. 6 in. (6.25 m)	17 ft. 11 in. (5.46 m)	500 lbs. (230 kg)	26 x 48 in. (0.66 x 1.22 m)	59 in. (1.5 m)	5 in. (13 cm)	6 ft. 7 in. (2.01 m)	17 ft. (5.18 m)	2.5 mph (4.0 kph)	25%	5 ft. 3 in. (1.6 m)	11 ft. 10 in. (3.61 m)	48V DC 8 x 6V 245 amp-hr	10,200 lbs. (4,627 kg)
40electric	40 ft. (12.19 m)	20 ft. 6 in. (6.25 m)	20 ft. (6.1 m)	500 lbs. (230 kg)	26 x 48 in. (0.66 x 1.22 m)	69 in. (1.75 m)	Zero	6 ft. 7 in. (2.01 m)	17 ft. 7 in. (5.36 m)	3.0 mph (4.8 kph)	25%	4 ft. 3 in. (1.3 m)	11 ft. 10 in. (3.61 m)	48V DC 8 x 6V 370 amp-hr	10,700 lbs. (4,854 kg)
n40electric	40 ft. (12.19 m)	20 ft. 6 in. (6.25 m)	20 ft. (6.1 m)	500 lbs. (230 kg)	26 x 48 in. (0.66 x 1.22 m)	59 in. (1.5 m)	5 in. (13 cm)	6 ft. 7 in. (2.01 m)	17 ft. 4 in. (5.28 m)	2.5 mph (4.0 kph)	25%	5 ft. 3 in. (1.6 m)	11 ft. 10 in. (3.61 m)	48V DC 8 x 6V 370 amp-hr	11,830 (5,366 kg)

E300 Series Articulating Boom Lifts

Model	Platform Height	Horizontal Outreach	Up-and-Over Height	Platform Capacity (Unrestricted)	Platform Size	Overall Width	Tailswing	Stowed Height	Stowed Length	Drive Speed (2WD)	Gradeability (2WD)	Turning Radius (Inside)	Turning Radius (Outside)	Power Source	Gross Vehicle Weight*
E300A	30 ft. (9.14 m)	20 ft. (6.1 m)	13 ft. 1 in. (3.99 m)	500 lbs. (230 kg)	30 x 48 in. (0.76 x 1.22 m)	48 in. (1.22 m)	Zero	6 ft. 7 in. (2.0 m)	17 ft. 4 in. (5.28 m)	3.0 mph (4.8 kph)	25%	5 ft. (1.52 m)	10 ft. 2 in. (3.1 m)	48V DC 8 x 6V 370 amp-hr	14,200 lbs. (6,441 kg)
E300AJ	30 ft. (9.14 m)	20 ft. (6.1 m)	13 ft. 1 in. (3.99 m)	500 lbs. (230 kg)	30 x 48 in. (0.76 x 1.22 m)	48 in. (1.22 m)	Zero	6 ft. 7 in. (2.0 m)	17 ft. 10 in. (5.44 m)	3.0 mph (4.8 kph)	25%	5 ft. (1.52 m)	10 ft. 2 in. (3.1 m)	48V DC 8 x 6V 370 amp-hr	14,300 lbs. (6,487 kg)
E300AJP	30 ft. (9.14 m)	20 ft. 6 in. (6.25 m)	13 ft. 1 in. (3.99 m)	500 lbs. (230 kg)	30 x 48 in. (0.76 x 1.22 m)	48 in. (1.22 m)	Zero	6 ft. 7 in. (2.0 m)	18 ft. 5 in. (5.6 m)	3.0 mph (4.8 kph)	25%	5 ft. (1.52 m)	10 ft. 2 in. (3.1 m)	48V DC 8 x 6V 370 amp-hr	14,800 lbs. (6,713 kg)

Multi-Power Articulating Boom Lifts

Model	Platform Height	Horizontal Outreach	Up-and-Over Height	Platform Capacity (Unrestricted)	Platform Size	Overall Width	Tailswing	Stowed Height	Stowed Length	Drive Speed (2WD)	Gradeability (2WD)	Turning Radius (Inside)	Turning Radius (Outside)	Power Source	Gross Vehicle Weight*
M45A	45 ft. (13.72 m)	22 ft. 8 in. (6.91 m)	24 ft. 7 in. (7.49 m)	500 lbs. (230 kg)	30 x 60 in. (0.76 x 1.52 m)	69 in. (1.75 m)	Zero	6 ft. 6 in. (1.98 m)	18 ft. 8 in. (5.69 m)	3.2 mph (5.2 kph)	30%	24 in. (61 cm)	10 ft. 4 in. (3.15 m)	48V DC 8 x 6V 370 amp-hr	12,800 lbs. (5,806 kg)
E45A	45 ft. (13.72 m)	23 ft. 1 in. (7.04 m)	24 ft. 7 in. (7.49 m)	500 lbs. (230 kg)	30 x 60 in. (0.76 x 1.52 m)	69 in. (1.75 m)	Zero	6 ft. 7 in. (2.0 m)	18 ft. 8 in. (5.69 m)	3.2 mph (5.2 kph)	30%	24 in. (61 cm)	10 ft. 4 in. (3.15 m)	48V DC 8 x 6V 370 amp-hr	12,750 lbs. (5,783 kg)

* Certain options or country standards will increase weight

Figure 2.7.6 (By permission of JLG Industries, Inc., McConnellsburg, Pa.)

VP Series Self Propelled Vertical Personnel Lifts

Model	Working Height	Platform Height	Maximum Drive Height	Platform Capacity	Platform Size (L x W)	Overall Length	Overall Width	Drive Speed Lowered	Drive Speed Elevated	Turning Radius	Gradeability	Break Over Angle (Grade)	Platform to Wall (Front)	Platform to Wall (Sides)	Gross Weight (Fixed Size Platform)
10VP	16 ft. 6 in. (5.03 m)	10 ft. 6 in. (3.2 m)	Full Height	350 lbs. (159 kg)	25.5 x 26 in. (64 x 66 cm)	4 ft. 4.5 in. (1.33 m)	32 in. (81 cm)	2 mph (3.2 kph)	0.5 mph (0.8 kph)	Zero	15%	15%	5 in. (13 cm)	3 in. (8 cm)	930 lbs. (425 kg)
15VP	21 ft. (6.40 m)	15 ft. (4.57 m)	Full Height	350 lbs. (159 kg)	25.5 x 26 in. (64 x 66 cm)	4 ft. 4.5 in. (1.33 m)	32 in. (81 cm)	2 mph (3.2 kph)	0.5 mph (0.8 kph)	Zero	15%	15%	2.5 in. (6 cm)	3 in. (8 cm)	1,355 lbs. (619 kg)*
20VP	25 ft. 9 in. (7.85 m)	19 ft. 9 in. (6.0 m)	Full Height	350 lbs. (159 kg)	25.5 x 26 in. (64 x 66 cm)	4 ft. 4.5 in. (1.33 m)	32 in. (81 cm)	2 mph (3.2 kph)	0.5 mph (0.8 kph)	Zero	15%	15%	Zero	3 in. (8 cm)	1,910 lbs. (873 kg)**

* For CSA add 190 lbs. (86 kg)
** For CSA add 280 lbs. (127 kg)

Optional Gull Wing Platform 26 x 28 in. (66 x 71 cm) – Optional Narrow Platform 24.5 x 22 in. (62 x 56 cm) – Optional Step-In Molded Platform 26 x 26 in. (66 x 66 cm) – Optional Extension Platform 49 x 26 in. (1.24m x 66 cm)

AM Series Vertical Personnel Lifts

Model	Working Height	Platform Height	Platform Capacity	Platform Size (L x W)	Overall Height Stowed	Overall Width Stowed	Overall Length Stowed	Overall Height Tilted Back	Overall Length Tilted Back	Ground Entry Height	Outrigger Footprint (Length)	Outrigger Footprint (Width)	Distance to Wall (Front/Side)	Gross Vehicle Weight (AC)	Gross Vehicle Weight (DC)
15AM	21 ft. (6.4 m)	15 ft. (4.57 m)	350 lbs. (159 kg)	25.5 x 26 in. (64 x 66 cm)†	6 ft. 5.5 in. (1.97 m)	29 in. (74 cm)	4 ft. 4.5 in. (1.33 m)	n/a	n/a	19 in. (48 cm)	n/a	n/a	0 in./1.5 in. (0 cm/4 cm)	820 lbs. (372 kg)	910 lbs. (413 kg)
20AM	26 ft. (7.92 m)	20 ft. 5 in. (6.22 m)	350 lbs. (159 kg)	25.5 x 26 in. (64 x 66 cm)†	6 ft. 5.5 in. (1.97 m)	29 in. (74 cm)	3 ft. 11 in. (1.19 m)	n/a	n/a	19 in. (48 cm)	5 ft. 2 in. (1.57 m)	4 ft. 5 in. (1.35 m)	8 in./13.5 in. (20 cm/34 cm)	715 lbs. (324 kg)	855 lbs. (388 kg)
25AM	31 ft. (9.45 m)	25 ft. 2 in. (7.67 m)	350 lbs. (159 kg)	25.5 x 26 in. (64 x 66 cm)†	6 ft. 5.5 in. (1.97 m)	29 in. (74 cm)	4 ft. 1 in. (1.24 m)	n/a	n/a	19 in. (48 cm)	5 ft. 2 in. (1.57 m)	4 ft. 5 in. (1.35 m)	5 in./13.5 in. (13 cm/34 cm)	800 lbs. (363 kg)	915 lbs. (415 kg)
30AM	36 ft. (10.97 m)	29 ft. 10 in. (9.09 m)	350 lbs. (159 kg)	25.5 x 26 in. (64 x 66 cm)†	6 ft. 5.5 in. (1.97 m)	29 in. (74 cm)	4 ft. 4 in. (1.32 m)	n/a	n/a	19 in. (48 cm)	5 ft. 8 in. (1.73 m)	5 ft. 1 in. (1.55 m)	5 in./17.5 in. (13 cm/44 cm)	870 lbs. (395 kg)	940 lbs. (426 kg)
36AM	42 ft. (12.8 m)	36 ft. 3 in. (11.05 m)	300 lbs. (136 kg)	25.5 x 26 in. (64 x 66 cm)*	8 ft. 8 in. (2.64 m)	29 in. (74 cm)	4 ft. 6.5 in. (1.38 m)	6 ft. 3.5 in. (1.92 m)	9 ft. (2.74 m)	19 in. (48 cm)	6 ft. 7 in. (2.01 m)	6 ft. 7 in. (2.01 m)	14 in./26.5 in. (36 cm/67 cm)	1,015 lbs. (460 kg)*	1,130 lbs. (513 kg)**
41AM	47 ft. (14.33 m)	40 ft. 9 in. (12.42 m)	300 lbs. (136 kg)	25.5 x 26 in. (64 x 66 cm)*	8 ft. 8 in. (2.64 m)	29 in. (74 cm)	4 ft. 9.5 in. (1.46 m)	6 ft. 5 in. (1.96 m)	9 ft. (2.74 m)	19 in. (48 cm)	6 ft. 7 in. (2.01 m)	6 ft. 7 in. (2.01 m)	11 in./26.5 in. (28 cm/67 cm)	1,070 lbs. (485 kg)†	1,210 lbs. (549 kg)††

* For CSA add 215 lbs. (98 kg)
** For CSA add 145 lbs. (66 kg)
† For CSA add 170 lbs. (77 kg)
†† For CSA add 125 lbs. (57 kg)

Optional Gull Wing Platform 26 x 28 in. (66 x 71 cm) – Optional Narrow Platform 24.5 x 22 in. (62 x 56 cm) – Optional Step-In Molded Platform 26 x 26 in. (66 x 66 cm)

Figure 2.7.6 (cont.)

Model	Platform Height	Horizontal Outreach	Platform Capacity (Restricted)	Platform Capacity (Unrestricted)	Platform Size	Overall Width	Tailswing	Stowed Height	Stowed Length	Drive Speed (2WD/4WD)	Gradeability (2WD/4WD)	Turning Radius (Inside)	Turning Radius (Outside)	Power Source	Gross Vehicle Weight*
40H Telescopic Boom Lifts															
40H	40 ft. (12.19 m)	33 ft. (10.06 m)	1000 lbs. (450 kg)	500 lbs. (230 kg)	36 x 60 in. (0.91 x 1.52 m)	95 in. (2.41 m)	2 ft. 10 in. (86 cm)	7 ft. 10 in. (2.39 m)	25 ft. 2 in. (7.67 m)	4.5/4.0 mph (7.2/6.4 kph)	25/40%	11 ft. 11 in. (3.6 m)	16 ft. 6 in. (5.03 m)	70 hp DF 42 hp Diesel	11,600 lbs. (5,262 kg)
600 Series Telescopic Boom Lift															
600S	60 ft. (18.29 m)	49 ft. 6 in. (15.09 m)	1000 lbs. (450 kg)	500 lbs. (230 kg)	36 x 72 in. (0.91 x 1.83 m)	96 in. (2.44 m)	3 ft. 9 in. (1.14 m)	8 ft. 5 in. (2.57 m)	27 ft. 11 in. (8.51 m)	4.5/4.0 mph (7.2/6.4 kph)	30/45%	12 ft. (3.66 m)	17 ft. 9 in. (5.41 m)	70 hp DF 65 hp Diesel	22,000 lbs. (9,979 kg)
600SJ	60 ft. (18.29 m)	50 ft. 4 in. (15.34 m)	500 lbs. (230 kg)	500 lbs. (230 kg)	36 x 72 in. (0.91 x 1.83 m)	96 in. (2.44 m)	3 ft. 9 in. (1.14 m)	8 ft. 5 in. (2.57 m)	31 ft. (9.45 m)	4.5/4.0 mph (7.2/6.4 kph)	30/45%	12 ft. (3.66 m)	17 ft. 9 in. (5.41 m)	70 hp DF 65 hp Diesel	23,500 lbs. (10,660 kg)
660SJ	66 ft. (20.12 m)	56 ft. 9 in. (17.3 m)	500 lbs. (230 kg)	500 lbs. (230 kg)	36 x 72 in. (0.91 x 1.83 m)	96 in. (2.44 m)	3 ft. 9 in. (1.14 m)	8 ft. 5 in. (2.57 m)	33 ft. 6 in. (10.21 m)	4.5/4.0 mph (7.2/6.4 kph)	30/45%	12 ft. (3.66 m)	17 ft. 9 in. (5.41 m)	70 hp DF 65 hp Diesel	25,500 lbs. (11,567 kg)
601S	60 ft. (18.29 m)	49 ft. (14.93 m)	1000 lbs. (450 kg)	500 lbs. (230 kg)	36 x 72 in. (0.91 x 1.83 m)	96 in. (2.44 m)	3 ft. 9 in. (1.14 m)	8 ft. 4 in. (2.54 m)	27 ft. 5 in. (8.36 m)	3.0 mph (4.81 kph)	45%	11 ft. 5 in. (3.48 m)	17 ft. 8 in. (5.38 m)	70 hp DF 65 hp Diesel	21,500 lbs. (9,752 kg)
70H & 80H Series Telescopic Boom Lifts															
70H	70 ft. (21.34 m)	59 ft. 11 in. (18.26 m)	1000 lbs. (450 kg)	500 lbs. (230 kg)	36 x 60 in. (0.91 x 1.52 m)	96 in. (2.44 m)	3 ft. 7 in. (1.09 m)	9 ft. 2 in. (2.79 m)	30 ft. 7 in. (9.32 m)	3.5/3.0 mph (5.6/4.8 kph)	25/35%	12 ft. 7 in. (3.8 m)	18 ft. 4 in. (5.59 m)	70 hp DF 47 hp Diesel	28,600 lbs. (12,973 kg)
80H	80 ft. (24.38 m)	70 ft. 7 in. (21.51 m)	1000 lbs. (450 kg)	500 lbs. (230 kg)	36 x 96 in. (0.91 x 2.44 m)	102 in. (2.59 m)	3 ft. 7 in. (1.09 m)	9 ft. 9 in. (2.97 m)	35 ft. 3 in. (10.74 m)	3.5/3.0 mph (5.6/4.8 kph)	25/35%	19 ft. 8 in. (6.0 m)	20 ft. 2 in. (6.15 m)	70 hp DF 70 hp Diesel	36,200 lbs. (16,420 kg)
Extendable Telescopic Boom Lifts															
80HX	80 ft. (24.38 m)	70 ft. 7 in. (21.51 m)	1000 lbs. (450 kg)	500 lbs. (230 kg)	36 x 96 in. (0.91 x 2.44 m)	96 in. (2.44 m)	2 ft. 11 in. (89 cm)	9 ft. 7 in. (2.92 m)	33 ft. 7 in. (10.24 m)	3.5/2.8 mph (5.6/4.5 kph)	25/35%	17 ft. (5.18 m)	23 ft. 2 in. (7.06 m)	70 hp DF 70 hp Diesel	31,600 lbs. (14,334 kg)
100HX	100 ft. (30.48 m)	60 ft. (18.29 m)	1000 lbs. (450 kg)	500 lbs. (230 kg)	36 x 96 in. (0.91 x 2.44 m)	102 in. (2.59 m)	2 ft. 10 in. (86 cm)	9 ft. 10 in. (3.0 m)	36 ft. 2 in. (11.02 m)	2.8 mph (4.5 kph)	20%	14 ft. 3 in. (4.34 m)	22 ft. 10 in. (6.96 m)	70 hp DF 70 hp Diesel	36,200 lbs. (16,420 kg)
110HX	110 ft. (33.53 m)	60 ft. (18.29 m)	1000 lbs. (450 kg)	500 lbs. (230 kg)	36 x 96 in. (0.91 x 2.44 m)	102 in. (2.59 m)	2 ft. 10 in. (86 cm)	9 ft. 10 in. (3.0 m)	38 ft. 7 in. (11.76 m)	2.8 mph (4.5 kph)	20%	14 ft. 3 in. (4.34 m)	22 ft. 10 in. (6.96 m)	70 hp DF 70 hp Diesel	41,100 lbs. (18,643 kg)
120HX	120 ft. (36.58 m)	60 ft. (18.29 m)	500 lbs. (230 kg)	500 lbs. (230 kg)	36 x 72 in. (0.91 x 1.83 m)	102 in. (2.59 m)	2 ft. 10 in. (86 cm)	10 ft. 4 in. (3.15 m)	35 ft. 11 in. (10.95 m)	2.75 mph (4.4 kph)	40%	12 ft. 10 in. (3.91 m)	18 ft. 10 in. (5.74 m)	70 hp Diesel	43,800 lbs. (19,868 kg)

* Certain options or country standards will increase weight

Figure 2.7.6 (cont.)

HA Articulating Boom Lifts

Model	Platform Height	Horizontal Outreach	Up-and-Over Height	Platform Capacity (Restricted/Unrestricted)	Platform Size	Overall Width	Tailswing	Stowed Height	Stowed Length	Drive Speed (2WD/4WD)	Gradeability (2WD/4WD)	Turning Radius (Inside)	Turning Radius (Outside)	Power Source	Gross Vehicle Weight*
34HA	34 ft. 2 in. (10.41 m)	20 ft. 1 in. (6.12 m)	13 ft. (3.96 m)	500 lbs. (230 kg)	36 x 60 in. (0.91 x 1.52 m)	78 in. (1.98 m)	Zero	6 ft. 5.5 in. (1.97 m)	16 ft. 8.5 in. (5.09 m)	4.5/3.5 mph (7.2/5.6 kph)	25/35%	8 ft. 8 in. (2.64 m)	13 ft. 8 in. (4.17 m)	32 hp DF / 28 hp Diesel	9,800 lbs. (4,445 kg)

450 Series Articulating Boom Lifts

Model	Platform Height	Horizontal Outreach	Up-and-Over Height	Platform Capacity (Restricted/Unrestricted)	Platform Size	Overall Width	Tailswing	Stowed Height	Stowed Length	Drive Speed (2WD/4WD)	Gradeability (2WD/4WD)	Turning Radius (Inside)	Turning Radius (Outside)	Power Source	Gross Vehicle Weight*
450A	45 ft. (13.72 m)	23 ft. 9 in. (7.24 m)	24 ft. 3 in. (7.39 m)	500 lbs. (230 kg)	30 x 72 in. (0.76 x 1.83 m)	78 in. (1.98 m)	Zero	7 ft. 4 in. (2.24 m)	20 ft. 2 in. (6.15 m)	4.5/2.25 mph (7.2/3.6 kph)	30/40%	5 ft. 8 in. (1.73 m)	14 ft. (4.27 m)	70 hp DF / 47 hp Diesel	14,600 lbs. (6,623 kg)
450AJ	45 ft. (13.72 m)	24 ft. 6 in. (7.47 m)	24 ft. 1 in. (7.34 m)	500 lbs. (230 kg)	30 x 72 in. (0.76 x 1.83 m)	78 in. (1.98 m)	Zero	7 ft. 4 in. (2.24 m)	22 ft. (6.71 m)	4.5/2.25 mph (7.2/3.6 kph)	30/40%	5 ft. 8 in. (1.73 m)	14 ft. (4.27 m)	70 hp DF / 47 hp Diesel	15,600 lbs. (7,076 kg)

600 Series Articulating Boom Lifts

Model	Platform Height	Horizontal Outreach	Up-and-Over Height	Platform Capacity (Restricted/Unrestricted)	Platform Size	Overall Width	Tailswing	Stowed Height	Stowed Length	Drive Speed (2WD/4WD)	Gradeability (2WD/4WD)	Turning Radius (Inside)	Turning Radius (Outside)	Power Source	Gross Vehicle Weight*
600A	60 ft. (18.29 m)	39 ft. 7 in. (12.07 m)	26 ft. 5 in. (8.05 m)	1000/500 lbs. (450/230 kg)	36 x 72 in. (0.91 x 1.83 m)	96 in. (2.44 m)	Zero	8 ft. 5 in. (2.57 m)	26 ft. 5 in. (8.05 m)	3.6/4.0 mph (5.8/6.4 kph)	30/45%	11 ft. 5 in. (3.48 m)**	17 ft. 8 in. (5.38 m)**	70 hp DF / 65 hp Diesel	20,700 lbs. (9,390 kg)
600A-N	60 ft. (18.29 m)	39 ft. 7 in. (12.07 m)	26 ft. 5 in. (8.05 m)	1000/500 lbs. (450/230 kg)	36 x 72 in. (0.91 x 1.83 m)	84 in. (2.13 m)	5 in. (13 cm)	8 ft. 5 in. (2.57 m)	26 ft. 5 in. (8.05 m)	3.6 mph (5.8 kph)	30%	12 ft. 2 in. (3.71 m)	16 ft. 6 in. (5.03 m)	70 hp DF / 65 hp Diesel	21,200 lbs. (9,616 kg)
600AJ	60 ft. (18.29 m)	39 ft. 9 in. (12.12 m)	26 ft. 5 in. (8.05 m)	500 lbs. (230 kg)	36 x 72 in. (0.91 x 1.83 m)	96 in. (2.44 m)	Zero	8 ft. 5 in. (2.57 m)	28 ft. 11.5 in. (8.82 m)	3.6/4.0 mph (5.8/6.4 kph)	30/45%	11 ft. 5 in. (3.48 m)**	17 ft. 8 in. (5.38 m)**	70 hp DF / 65 hp Diesel	22,100 lbs. (10,025 kg)
600AJ-N	60 ft. (18.29 m)	39 ft. 7 in. (12.07 m)	26 ft. 5 in. (8.05 m)	500 lbs. (230 kg)	36 x 72 in. (0.91 x 1.83 m)	84 in. (2.13 m)	5 in. (13 cm)	8 ft. 5 in. (2.57 m)	28 ft. 11.5 in. (8.82 m)	3.6 mph (5.8 kph)	30%	12 ft. 2 in. (3.71 m)	16 ft. 6 in. (5.03 m)	70 hp DF / 65 hp Diesel	22,450 lbs. (10,183 kg)

800 Series Articulating Boom Lifts

Model	Platform Height	Horizontal Outreach	Up-and-Over Height	Platform Capacity (Restricted/Unrestricted)	Platform Size	Overall Width	Tailswing	Stowed Height	Stowed Length	Drive Speed (2WD/4WD)	Gradeability (2WD/4WD)	Turning Radius (Inside)	Turning Radius (Outside)	Power Source	Gross Vehicle Weight*
800A	80 ft. (24.38 m)	53 ft. (16.15 m)	32 ft. 1 in. (9.78 m)	1000/500 lbs. (450/230 kg)	36 x 72 in. (0.91 x 1.83 m)	96 in. (2.44 m)	2 ft. 6 in. (76 cm)	9 ft. 10 in. (3.0 m)	36 ft. 11 in. (11.25 m)	3.0 mph (4.8 kph)	30/45%	12 ft. (3.66 m)**	22 ft. 6 in. (6.86 m)**	70 hp DF / 65 hp Diesel	34,200 lbs. (15,513 kg)
800AJ	80 ft. (24.38 m)	51 ft. 10 in. (15.8 m)	32 ft. 1 in. (9.78 m)	500 lbs. (230 kg)	36 x 72 in. (0.91 x 1.83 m)	96 in. (2.44 m)	2 ft. 6 in. (76 cm)	9 ft. 10 in. (3.0 m)	36 ft. 6 in. (11.13 m)	3.0 mph (4.8 kph)	30/45%	12 ft. (3.66 m)**	22 ft. 6 in. (6.86 m)**	70 hp DF / 65 hp Diesel	34,200 lbs. (15,513 kg)

150HAX Articulating Boom Lifts

Model	Platform Height	Horizontal Outreach	Up-and-Over Height	Platform Capacity (Restricted/Unrestricted)	Platform Size	Overall Width	Tailswing	Stowed Height	Stowed Length	Drive Speed (2WD/4WD)	Gradeability (2WD/4WD)	Turning Radius (Inside)	Turning Radius (Outside)	Power Source	Gross Vehicle Weight*
150HAX	150 ft. (45.72 m)	79 ft. 3 in. (24.16 m)	80 ft. (24.38 m)	1000/500 lbs. (450/230 kg)	36 x 96 in. (0.91 x 2.44 m)	138 in. (3.51 m)	17 ft. 7 in. (5.36 m)	14 ft. (4.27 m)	39 ft. 5 in. (12.01 m)	2.7 mph (4.3 kph)	31%	14 ft. 10 in. (4.52 m)†	27 ft. (8.23 m)	76 hp Diesel	57,000 lbs. (25,855 kg)

* Certain options or country standards will increase weight
** 2-wheel steer
† Axles retracted

Figure 2.7.6 (cont.)

2.8.0 Crane Nomenclature

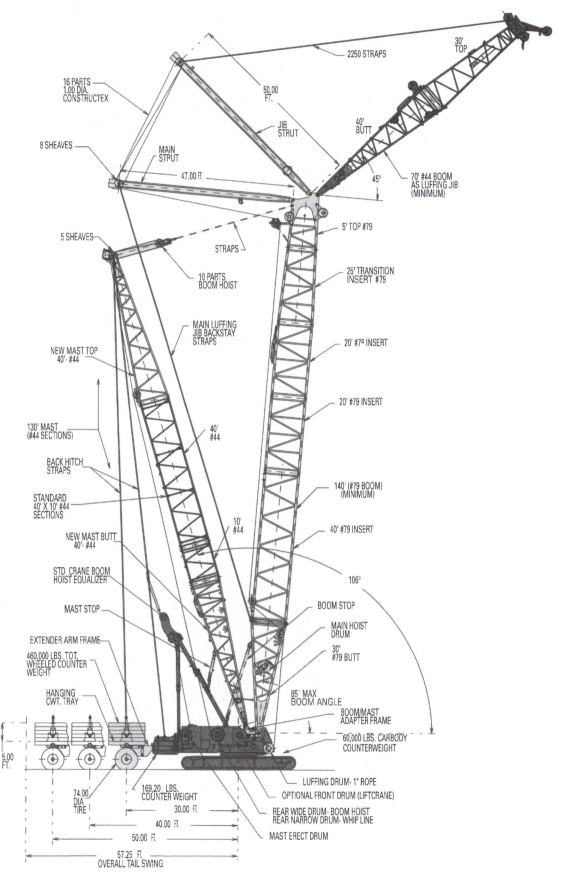

Figure 2.8.0 *(By permission: Manitowoc Cranes, Inc., Manitowoc, Wisconsin)*

2.8.1 Manitowoc's Model 111—80-ton Crane

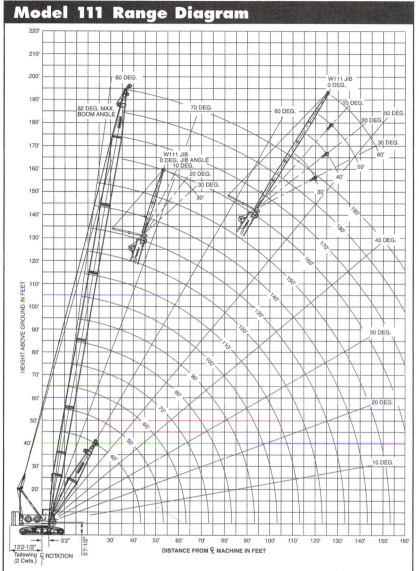

Model 111 Range Diagram

SWING SPEED: 3.0 RPM   **TRAVEL SPEED:** 1.5 MPH   **GRADEABILITY:** 30%

Reference Capacities
Model 111 • Boom No. W111

Liftcrane Capacities (thousands of pounds) - 360° Rating

Radius (feet)	Boom Length (feet)					
	40	70	100	130	160	190
10	160.0					
15	139.5	137.9				
20	86.0	85.8	85.4			
24	65.5	65.2	64.8	64.3		
30	48.0	47.6	47.2	46.6	46.1	
34	40.6	40.2	39.7		38.7	29.6
40	32.8	32.4	31.9	31.4	30.8	27.9
50		24.2	23.7	23.1	22.5	21.9
60		19.1	18.6	18.1	17.5	16.9
70		15.6	15.1	14.5	13.9	13.3
80			12.5	12.0	11.4	10.7
90			10.6	10.0	9.4	8.8
100			9.1	8.5	7.9	7.2
110				7.2	6.6	6.0
120				6.2	5.6	5.0
130				5.4	4.7	4.1
140					4.0	

Fixed Jib Capacities (thousands of pounds) - 360° Rating

Radius (feet)	30 Foot Jib, 3° Offset			60 Foot Jib, 3° Offset		
	Boom Length (feet)					
	120	150	180	120	150	170
30	20.0					
40	20.0	20.0	20.0	16.0	16.0	16.0
45	20.0	20.0	20.0	16.0	16.0	16.0
50	20.0	20.0	20.0	16.0	16.0	16.0
55	20.0	19.7	19.3	16.0	16.0	16.0
60	18.2	17.6	17.0	16.0	16.0	16.0
65	16.3	15.6	15.0	16.0	15.8	15.3
70	14.6	14.0	13.4	14.8	14.1	13.7
75	13.3	12.6	12.0	13.3	12.7	12.3
80	12.1	11.4	10.7	12.1	11.5	11.0
85	11.0	10.3	9.7	11.1	10.4	10.0
90	10.1	9.4	8.8	10.1	9.5	9.0
95	9.3	8.6	7.9	9.3	8.7	8.2
100	8.5	7.9	7.2	8.6	7.9	7.5
105	7.9	7.2	6.5	7.9	7.3	6.8
110	7.3	6.6	5.9	7.3	6.7	6.2
115	6.7	6.1	5.4	6.8	6.1	5.6
120	6.3	5.6	4.9	6.3	5.6	5.1
125	5.8	5.1	4.4	5.8	5.2	4.7
130	5.4	4.7	4.0	5.4	4.7	4.3
135		4.3	3.7	5.0	4.4	3.9
140		4.0	3.3	4.7	4.0	3.5
145		3.7	3.0	4.4	3.7	3.2
150		3.4		4.1	3.4	
155		3.1		3.8	3.1	

Shipping Weights and Dimensions

Component	Weight	Length	Width	Height
Upperworks, carbody, and boom butt	85,550 lbs.	41'2"	12'0"	11'9"
Crawlers (2), each	14,940 lbs.	19'4"	35.5"	43.3"
Boom top	2,290 lbs.	20'10"	4'4"	5'0"
Inner counterweight	28,500 lbs.	12'0"	2'8"	6'1"
Outer counterweight	15,500 lbs.	12'0"	2'3"	6'1"
10' jib insert with pendants	232 lbs.	10'3"	27.5"	21.5"
75-ton hook block with four sheaves	1,899 lbs.	–	–	–
12-ton weight ball	740 lbs.	–	–	–

Figure 2.8.1 (*By permission: Manitowoc Cranes, Inc., Manitowoc, Wisconsin*)

2.8.2 Manitowoc's Model 222—100-ton Crane

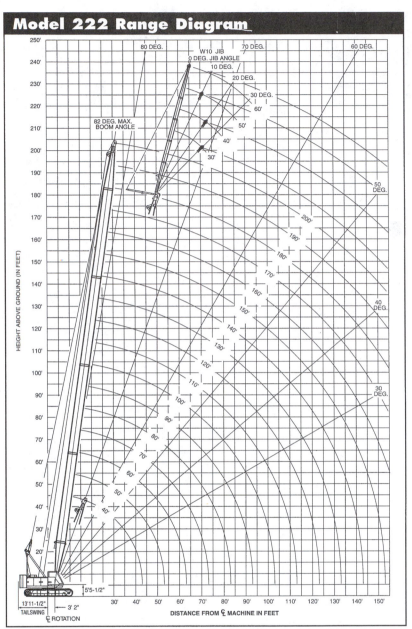

SWING SPEED: 3.0 RPM   **GRADEABILITY:** 30%   **TRAVEL SPEED:** 1.5 MPH

Shipping Weights and Dimensions

Component	Weight	Length	Width	Height
Upperworks, carbody, and boom butt	92,818 lbs.	49'5"	12'0"	12'0"
Crawlers (2), each	19,079 lbs.	22'4"	4'3"	3'7"
Boom top	2,267 lbs.	25'0"	5'0"	5'4"
Inner counterweight	28,500 lbs.	12'0"	1'9"	6'0"
Outer counterweight	24,800 lbs.	12'0"	1'7"	6'0"
10' Boom insert with pendants	824 lbs.	10'4"	50"	53.5"
20' Boom insert with pendants	1,349 lbs.	20'4"	50"	53.5"
40' Boom insert with pendants	2,444 lbs.	40'4"	50"	53.5"
30' Basic jib assembly with pendants	1,505 lbs.	30'0"	21.5"	3'6"
10' Jib insert with pendants	232 lbs.	10'0"	21.5"	21.5"
100-ton hook block with five sheaves	2,400 lbs.	–	–	–
12-ton weight ball	610 lbs.	–	–	–

Reference Capacities
Model 222 • Boom No. W222

Liftcrane Capacities (thousands of pounds) - 360° Rating

Radius (feet)	Boom Length (feet)					
	40	70	100	130	160	200
10	200.0					
15	148.2	138.5				
20	106.9	98.9	90.5			
24	84.3	80.5	74.1	66.4		
30	61.6	61.2	58.4	53.6	47.4	
34	50.6	51.7	50.7	46.3	42.8	25.5
40	35.8	41.7	41.1	39.2	36.4	23.8
45		35.8	35.3	34.5	31.5	22.8
50		31.3	30.8	30.2	27.6	22.0
55		27.7	27.2	26.6	24.6	21.3
60		24.8	24.3	23.6	22.0	19.3
65		22.0	21.8	21.2	19.9	18.2
70		18.2	19.8	19.3	18.8	16.3
75			18.2	17.6	17.0	14.7
80			16.7	16.1	15.5	13.3
85			15.4	14.8	14.2	12.1
90			14.2	13.6	13.0	11.1
95			12.6	12.6	12.0	10.1
100				11.7	11.1	9.2
110				10.2	9.6	7.7
120				8.2	8.1	6.4
130					6.8	5.3
140					5.5	4.4
150					4.2	3.5
155					3.5	3.1

Fixed Jib Capacities (thousands of pounds) - 360° Rating

Radius (feet)	30 Foot Jib, 0° Offset			60 Foot Jib, 0° Offset		
	Boom Length (feet)					
	120	150	190	120	150	190
24	20.0					
26	20.0					
28	20.0					
30	20.0	20.0		16.9		
32	20.0	20.0		16.7		
34	20.0	20.0	18.5	16.7	15.5	
36	20.0	20.0	17.7	16.6	15.2	
38	20.0	20.0	17.2	16.6	15.4	13.4
40	20.0	20.0	17.2	16.4	15.4	13.4
45	20.0	20.0	17.1	16.2	15.2	13.2
50	20.0	20.0	16.6	16.0	15.0	13.1
55	20.0	20.0	16.6	15.8	14.8	13.0
60	19.4	18.7	16.6	15.5	14.7	12.9
65	17.3	16.6	15.7	15.4	14.5	12.8
70	15.6	14.9	13.9	15.0	14.3	12.7
75	14.1	13.4	12.5	14.2	13.5	12.6
80	12.8	12.1	11.2	12.9	12.2	11.3
85	11.7	11.0	10.1	11.8	11.1	10.1
90	10.7	10.0	9.1	10.9	10.1	9.2
95	9.9	9.2	8.2	10.0	9.3	8.3
100	9.1	8.4	7.4	9.2	8.5	7.5
105	8.4	7.7	6.7	8.5	7.8	6.8
110	7.8	7.1	6.1	7.9	7.2	6.2
115	7.3	6.5	5.5	7.3	6.6	5.6
120	6.7	6.0	5.0	6.8	6.1	5.1
125	6.3	5.6	4.5	6.4	5.6	4.6
130	5.9	5.1	4.1	5.9	5.2	4.2
135		4.7	3.7	5.5	4.8	3.8
140		4.4	3.4	5.2	4.4	3.4
145		4.0	3.0	4.8	4.1	3.1
150		3.7	2.7	4.5	3.8	2.8
155		3.4	2.4	4.3	3.5	2.5
160		3.2	2.1	4.0	3.2	2.2
165			1.9		3.0	1.9
170			1.7		2.7	1.7
175			1.4		2.5	1.5
180			1.2		2.3	1.3
185					2.1	

Figure 2.8.2 *(By permission: Manitowoc Cranes, Inc., Manitowoc, Wisconsin)*

2.8.3 Manitowoc Model 777—169-ton Crane

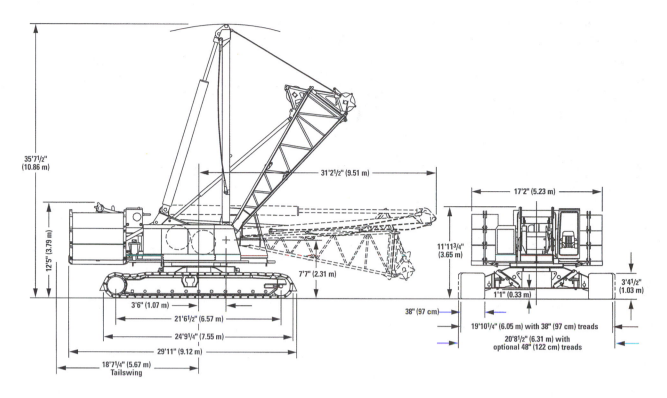

UPPERWORKS

ROTATING BED: High-strength fabricated steel rotating bed is mounted on 84½" (2.15m) diameter turntable bearing.

DRUMS: Two full-width drums are standard, both 30⅛" (77cm) wide and 19½" (49.5cm) in diameter. Each drum is antifriction bearing mounted and driven independently by a variable-displacement hydraulic motor and a planetary reduction. External contracting brakes mounted to the drum shafts are spring set and hydraulically released. A bi-directional redundant brake and a drum rotation indicator are included for each hoist.

BOOM HOIST: Independent boom hoist is provided by two double-acting hydraulic cylinders connected to the mast. Boom hoist provides full range of boom angles from horizontal to 88°, with or without load.

SWING SYSTEM: Independent swing is powered by a fixed-displacement hydraulic motor coupled to an internal brake and planetary reduction. Swing lock can be engaged with upperworks parked at any angle.

OPERATOR'S STATION: Insulated steel cab mounted to the left front corner of rotating bed is equipped with a sliding door, and large-safety glass windows on all sides and in roof. Signal horn, space heater, dome light, sun visor and shade, fire extinguisher, fan, and wipers for front and roof windows are standard.

ENGINE: Cummins 6CTA8.3-C260 diesel engine rated 260 HP (195kW) at 2200 RPM provides independent power for all operations through a multi-hydraulic-pump transmission. Power system includes ether starting aid, transmission-disconnect clutch, high-silencing muffler, hydraulic-oil cooler, radiator, and fan.

CONTROLS: Modulating electronic-over-hydraulic controls provide infinitely variable speed response that is directly proportional to control lever movement. Controls include Manitowoc's exclusive EPIC® system, which provides microprocessor-driven control logic, pump control, on-board diagnostics, and service information. Block-up limit control is standard for main and auxiliary hoists.

LOWERWORKS

CARBODY: High-strength steel fabrication connects rotating bed to crawlers. FACT™ connection system permits fast attachment and removal of crawlers. Optional self-assembly system includes four hydraulic jacks that raise entire upperworks and carbody off the transport trailer, then support the unit during crawler installation and removal.

CRAWLERS: Crawlers are 24'9¼" (7.55m) long steel fabrications with 38" (97cm) wide cast-steel treads and sealed low-maintenance intermediate rollers. Each crawler is powered independently by a variable-displacement hydraulic motor mounted on carbody. The motors are connected to planetary reductions on the crawlers by telescoping shafts. This permits the crawlers to be removed without opening their hydraulic circuits. Crawlers provide ample tractive effort for counter rotation with full rated liftcrane capacities.

Figure 2.8.3 (*By permission: Manitowoc Cranes, Inc., Manitowoc, Wisconsin*)

2.8.4 Manitowoc Model 777 Series-2—176-ton Crane

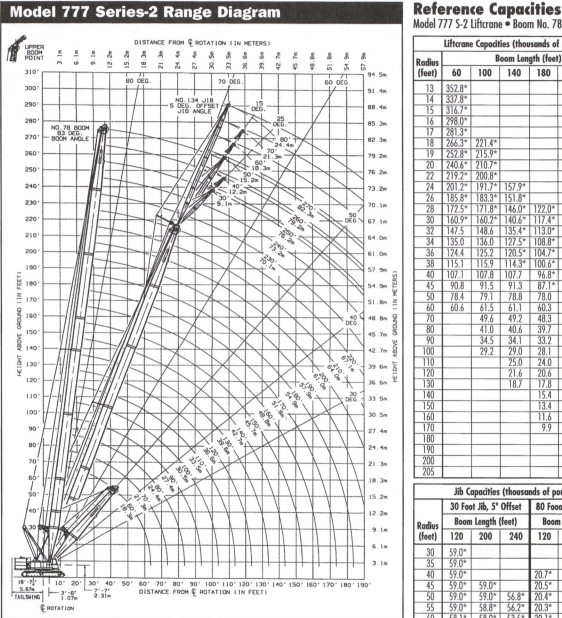

Model 777 Series-2 Range Diagram

SWING SPEED: 2.7 RPM   GRADEABILITY: 30%
TRAVEL SPEED: 1.33 MPH   BOOM HOIST SPEED: 270' boom, 0° to 82°, 1 minute, 40 seconds

Shipping Weights and Dimensions				
Component	Weight	Length	Width	Height
Upperworks, carbody, and boom butt	82,000 lbs.	49'9"	11'0½"	11'0¼"
Crawlers (2), each	28,700 lbs.	24'9¼"	4'3¾"	3'4½"
Boom top	8,000 lbs.	40'7"	7'4¼"	6'0¼"
Counterweight tray	31,000 lbs.	17'2"	8'1"	1'7"
Upper side counterweights (4 for S-1; 6 for S-2), each	18,500 lbs.	3'5"	7'4"	2'4"
Carbody counterweights (2, only on S-2), each	22,000 lbs.	6'4"	5'1"	2'5"
10' boom insert with pendants	1,660 lbs.	11'0"	7'4"	6'8"
20' boom insert with pendants	2,460 lbs.	21'0"	7'4"	6'8"
40' boom insert with pendants	4,360 lbs.	41'0"	7'4"	6'8"

© 1998 Manitowoc/98-002

Reference Capacities
Model 777 S-2 Liftcrane • Boom No. 78 • 360° Rating

Radius (feet)	Liftcrane Capacities (thousands of pounds) Boom Length (feet)					
	60	100	140	180	220	270
13	352.8*					
14	337.8*					
15	316.7*					
16	298.0*					
17	281.3*					
18	266.3*	221.4*				
19	252.8*	215.9*				
20	240.6*	210.7*				
22	219.2*	200.8*				
24	201.2*	191.7*	157.9*			
26	185.8*	183.3*	151.8*			
28	172.5*	171.8*	146.0*	122.0*		
30	160.9*	160.2*	140.6*	117.4*		
32	147.5	148.6	135.4*	113.0*	105.1*	
34	135.0	136.0	127.5*	108.8*	102.4*	
36	124.4	125.2	120.5*	104.7*	96.6*	
38	115.1	115.9	114.3*	100.6*	91.5*	
40	107.1	107.8	107.7	96.8*	86.9*	
45	90.8	91.5	91.3	87.1*	77.5*	66.7*
50	78.4	79.1	78.8	78.0	70.0*	59.6*
60	60.6	61.5	61.1	60.3	58.1*	49.3*
70		49.6	49.2	48.3	47.0	41.8*
80		41.0	40.6	39.7	38.4	35.5*
90		34.5	34.1	33.2	31.8	30.1
100		29.2	29.0	28.1	26.7	25.0
110			25.0	24.0	22.6	20.9
120			21.6	20.6	19.3	17.5
130			18.7	17.8	16.4	14.7
140				15.4	14.0	12.3
150				13.4	12.0	10.2
160				11.6	10.2	8.4
170				9.9	8.6	6.7*
180					7.2	5.2*
190					6.0	
200					4.8	
205					4.3	

	Jib Capacities (thousands of pounds)					
	30 Foot Jib, 5° Offset			80 Foot Jib, 5° Offset		
Radius (feet)	Boom Length (feet)			Boom Length (feet)		
	120	200	240	120	180	220
30	59.0*					
35	59.0*					
40	59.0*			20.7*		
45	59.0*	59.0*		20.5*		
50	59.0*	59.0*	56.8*	20.4*	20.7*	
55	59.0*	58.8*	56.2*	20.3*	20.6*	20.7*
60	58.1*	58.0*	53.5*	20.1*	20.5*	20.6*
65	55.7	53.2	49.0*	20.0*	20.4*	20.5*
70	50.2	47.7	45.1*	19.8*	20.2*	20.4*
75	45.6	43.0	41.5*	19.7*	20.1*	20.3*
80	41.6	38.9	37.6	19.6*	20.0*	20.2*
85	38.1	35.4	34.1	19.4*	19.9*	20.1*
90	35.1	32.4	31.0	19.0*	19.8*	20.0*
100	30.0	27.2	25.8	18.2*	19.6*	19.8
110	25.9	23.1	21.7	17.5*	19.3*	19.1
120	22.5	19.7	18.3	16.8*	18.7*	18.5*
130	19.7	16.9	15.5	16.1*	18.1*	17.2
140	17.2	14.5	13.1	15.5*	16.4	14.7
150		12.4	11.0	14.9*	14.3	12.7
160		10.6	9.2	14.3*	12.5	10.8
170		9.0	7.6	13.2	10.9	9.3
180		7.6	6.2	11.8	9.5	7.9
190		6.4	4.9	10.5	8.3	6.6
200		5.2			7.1	5.5
210		4.2*			6.1	4.5
220					5.2	
230					4.4	

*Capacity based on structural competence

Figure 2.8.4 (*By permission: Manitowoc Cranes, Inc., Manitowoc, Wisconsin*)

2.8.5 Manitowoc Model 777—220-ton Truck Crane

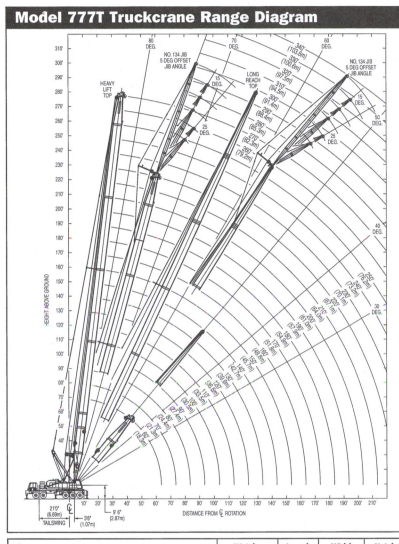

Model 777T Truckcrane Range Diagram

Preliminary Capacities
777T Truckcrane • Boom No. 78T

Liftcrane Capacities (thousands of pounds) - 360° Rating

Radius (feet)	Nominal Boom Length (feet)					
	60	100	150	200	250	300
14	353.7					
15	331.6					
18	278.5	241.2				
20	251.3	225.8				
22	228.8	211.9				
24	209.8	199.6				
26	193.5	188.7	161.1			
28	178.6	178.7	155.1			
30	165.2	166.0	148.0			
32	153.6	154.3	138.9	117.9		
36	134.3	135.1	123.9	110.7		
40	119.0	119.8	112.0	104.3	80.7	
45	103.7	104.7	100.2	95.5	76.8	58.8
50	91.6	92.6	91.8	86.3	73.0	57.2
60	69.9	72.5	75.8	72.5	62.4	54.0
70		57.7	61.1	59.7	53.6	46.9
80		47.2	50.7	49.2	46.8	40.4
90		39.2	42.9	41.4	39.8	35.4
100		32.9	36.9	35.3	33.7	31.3
110			32.0	30.5	28.8	26.9
120			28.1	26.5	24.8	22.9
130			24.8	23.2	21.5	19.5
140			21.9	20.4	18.7	16.7
150			19.4	18.0	16.3	14.3
160				15.9	14.2	12.2
170				14.1	12.4	10.4
180				12.5	10.8	8.8
190				11.0	9.4	7.3
200				9.6	8.1	6.1
210					6.9	4.9
220					5.9	3.8
230					4.9	2.9
240					4.0	2.0
250					3.1	1.2

Capacities above heavy line are for heavy-lift boom top. Capacities below line are for long-reach boom top.

Fixed Jib Capacities (thousands of pounds) - 360° Rating

Radius (feet)	30 Foot Jib, 5° Offset			80 Foot Jib, 5° Offset		
	Nominal Boom Length (feet)					
	150	200	280	150	200	260
40	59.0					
50	59.0	59.0		20.6		
60	59.0	58.7	41.5	20.3	21.8	
70	56.9	50.4	38.7	20.1	21.6	18.5
80	47.2	42.4	35.0	19.8	21.3	17.8
90	39.2	36.3	32.0	19.6	21.0	17.1
100	33.0	31.2	29.3	19.2	20.7	16.4
110	31.3	26.8	26.3	18.5	20.4	15.8
120	27.3	24.8	22.3	17.8	19.8	15.2
130	24.0	22.1	18.9	17.2	17.6	14.4
140	21.1	19.2	16.0	16.6	15.5	13.4
150	18.7	16.8	13.5	16.0	13.8	12.5
160	16.6	14.7	11.4	15.0	13.0	11.7
170	14.7	12.8	9.5	13.2	12.2	10.9
180	12.9	11.2	7.9	12.5	11.1	9.6
190		9.7	6.4	12.0	9.7	8.1
200		8.4	5.1	11.3	8.4	6.8
210		7.2	3.9	10.1	7.2	5.6
220		6.1	2.8	9.0	6.1	4.5
230		5.1	1.9	8.0	5.1	3.5
240			1.0		4.2	2.6
250					3.4	1.8
260					2.6	1.0
270					1.9	
280					1.0	

© 1999 Manitowoc 99-053

Component	Weight	Length	Width	Height
Base Carrier & Upperworks #1 — no Boom Butt, no Mast, no Boom Hoist Cylinders, no Outriggers	93,979 lbs.	40'0"	11'0"	13'6"
Base Carrier & Upperworks #2 — includes 9.5' Boom Butt Base Section, Mast, Boom Hoist Cylinders, no Outriggers	110,157 lbs.	61'3"	11'0"	13'6"

NOTE: Base configurations above include two standard speed full power hoist drums [1,120' of 1" rope on front drum, 900' of 1" rope on rear drum], full upper fuel tank (124 gal.), full carrier fuel tank (100 gal.), 14.00 x 24 Michelin radial tires.

Component	Weight	Length	Width	Height
13.0' Boom Butt Section with Wire Rope Guide and Boom Stop	4,290 lbs.	13'5"	8'9"	7'4"
17.5' Basic Boom Insert with Straps	4,670 lbs.	18'0"	7'6"	6'11"
Front Outriggers with Pads (2 required), each	4,035 lbs.	9'8"	3'0"	5'3"
Rear Outrigger Box Assembly with pads	13,970 lbs.	14'5"	4'0"	4'5"
High Speed Front Hoist Drum	250 lbs.	–	–	–
Upperworks Counterweight Tray	10,450 lbs.	11'0"	5'1"	2'3"
Upper Counterweight Boxes (5 total), each	16,990 lbs.	11'0"	3'4"	1'11"
Front Bumper Counterweight	25,000 lbs.	11'0"	4'0"	2'5"
10' Boom Insert with Straps	1,455 lbs.	10'6"	7'4"	6'2"
20' Boom Insert with Straps	2,530 lbs.	20'6"	7'4"	6'2"
50' Boom Insert with Straps	5,475 lbs.	50'6"	7'4"	6'2"
20' Heavy-Lift Boom Top with Straps and 6 Sheaves	8,175 lbs.	23'2"	7'6"	6'4"
50' Long-Reach Transition Insert with Straps	4,400 lbs.	50'6"	7'4"	6'2"
48' Long-Reach Tapered Top with Straps and 3 Sheaves	6,630 lbs.	48'6"	7'4"	6'0"
2-Axle Nitrogen Load-Transfer Assembly (approximate weight)	7,500 lbs.	126"	8'0"	5'4"
3-Axle Boom Dolly (approximate weight)	7,500 lbs.	14'6"	8'0"	5'0"
15-ton Hook & Weight Ball	1,250 lbs.	–	–	–
175-ton Hook Block	5,800 lbs.	–	–	–

NOTE: Consult MCC Sales Department for roading configurations & axle loadings for specific state(s).

Figure 2.8.5 (*By permission: Manitowoc Cranes, Inc., Manitowoc, Wisconsin*)

2.8.6 Manitowoc Model 2250—300-ton Crane

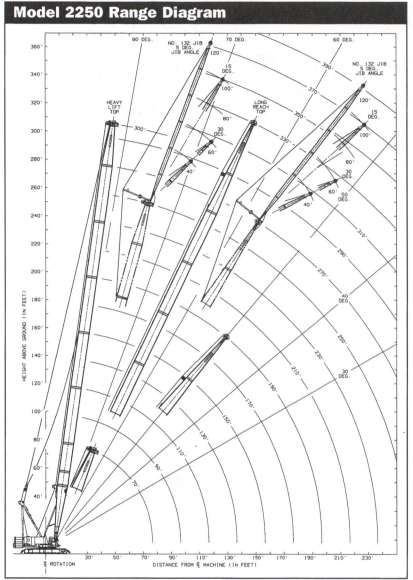

Model 2250 Range Diagram

SWING SPEED: 1.8 RPM   GRADEABILITY: 30%
TRAVEL SPEED: 1.0 MPH   BOOM HOIST SPEED: 300' boom, 0° to 82,° 2 minutes, 50 seconds

Reference Capacities
Model 2250 S-3 Liftcrane • Boom No. 44 • 360° Rating

Liftcrane Capacities (thousands of pounds)

Radius (feet)	Boom Length (feet)					
	70	110	170	210	270	300
18	600.0*					
20	560.0*					
22	541.5*					
24	506.3*	371.8*				
26	469.1*	367.0*				
28	436.9*	362.5*				
30	408.7*	358.2*	284.1*			
32	383.8*	354.3*	281.1*			
34	361.6*	350.6*	278.3*			
36	340.9*	339.8*	275.6*	225.3*		
38	315.4*	321.9*	273.1*	223.5*		
40	293.0*	303.9	270.7*	221.7*		
42	273.2*	282.9	263.0*	219.3*		
44	255.6*	264.5	249.0*	214.3*	153.9*	
46	239.8*	248.3	236.4*	209.4*	151.8*	123.3*
48	225.6*	233.8	225.0*	204.7*	148.9*	122.9*
50	212.6*	220.8	214.7*	200.1*	146.1*	122.6*
60	160.4*	171.8	169.8	163.3*	131.9*	116.0*
70	110.2*	139.5	137.4	135.6	118.3*	108.3*
80		116.6	114.5	112.6	105.8*	101.7*
90		97.6*	97.3	95.4	91.7*	88.0*
100		80.9*	84.0	82.0	79.9	76.6*
110			73.4	71.4	69.3	67.2*
120			64.7	62.7	60.5	59.4*
130			56.7*	55.5	53.3	52.4
140			49.1*	49.4	47.2	45.4*
150			42.4*	43.4*	41.9	38.6*
160			35.5*	37.6*	37.4	32.8*
170				32.5*	32.8*	27.8*
180				27.9*	28.3*	23.3*
190				23.9*	24.3*	19.5*
200					20.7*	16.1*
210					17.4*	13.1*
220					14.4*	10.4*
230					11.7*	8.1*
240					9.2*	6.0*
250					6.4*	4.2*

Lifting Capacities (thousands of pounds)

	2250 with MAX-ER™ 400		2250 with MAX-ER™ 225			
Radius (feet)	Boom Length (ft.)		Boom Length (ft.)			
	82	162	140	200	280	340
18	1,000.0*					
26	709.6*		500.0*			
28	660.3*	550.0*	499.0*			
34	545.1*	536.7*	482.1*	477.3*		
44	419.8*	411.7*	372.5*	367.8*	290.4*	
50	367.7*	359.7*	327.0*	322.3*	280.1*	163.1*
60	303.1*	295.3*	270.6*	265.9*	262.5*	153.8*
70	240.4*	233.3*	229.8*	225.2*	221.8*	144.4*
80	207.1*	200.2*	198.9*	194.3*	191.0*	135.1*
90		172.1	174.8*	170.1*	166.9*	126.1*
100		148.4	153.7	150.1	147.5*	117.4*
110		129.5	135.7	132.0	129.8	109.0*
120		114.0	120.9	117.2	115.0	101.0*
130		101.1	107.1*	104.9	102.6	93.3*
140		90.2		94.5	92.2	85.8*
150		80.8		85.6	83.2	78.7*
160				77.9	75.5	71.8*
170				71.1	68.7	65.2*
180				65.1	62.7	58.8*
190					57.3	52.6*
200					52.5	46.6*
210					48.2	40.8*
220					44.3	35.2*
230					40.7	29.8*
240					37.5	24.5*
250					34.5	19.4*
260					30.5*	14.4*
265						12.0*

*Capacity based on structural competence

Shipping Weights and Dimensions

Component	Weight	Length	Width	Height
Upperworks	78,000 lbs.	43'2"	10'4"	7'6"
Carbody, rotating module, and lower section of boom butt	67,000 lbs.	28'11"	9'11"	9'1"
Crawlers (2), each	58,000 lbs.	30'10"	5'5"	4'2"
Upper section of boom butt	10,000 lbs.	28'10"	8'6"	8'10"
Boom top	12,500 lbs.	33'0"	8'6"	10'1"
Upperworks center counterweight (on all Series)	37,000 lbs.	8'11"	7'0"	4'3"
Upperworks counterweight tray (on all Series)	39,000 lbs.	22'11"	7'2"	2'1"
Upperworks side counterweights (6, on all Series), each	15,500 lbs.	6'10"	6'6"	1'3"
Upperworks side counterweights (2 on S-2; 4 on S-3), each	20,000 lbs.	6'10"	6'6"	1'11"
Carbody center counterweights (2, only on S-2 and S-3), each	30,000 lbs.	11'0"	6'7"	3'0"
Carbody side counterweights (4, only on S-3), each	15,000 lbs.	7'2"	2'10"	3'0"
10' boom insert with straps	3,400 lbs.	10'7"	8'6"	8'6"
20' boom insert with straps	4,000 lbs.	20'7"	8'6"	8'6"
40' boom insert with straps	9,000 lbs.	40'7"	8'6"	8'6"

Figure 2.8.6 (*By permission: Manitowoc Cranes, Inc., Manitowoc, Wisconsin*)

2.8.7 Self-Assembly of the Model 2250

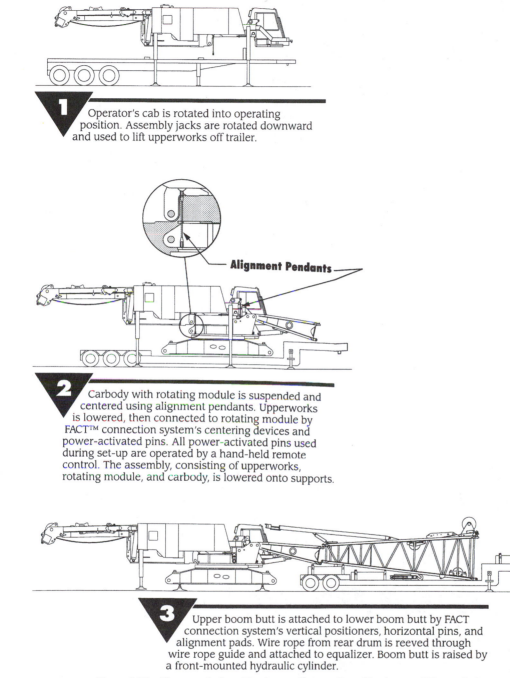

1 Operator's cab is rotated into operating position. Assembly jacks are rotated downward and used to lift upperworks off trailer.

Alignment Pendants

2 Carbody with rotating module is suspended and centered using alignment pendants. Upperworks is lowered, then connected to rotating module by FACT™ connection system's centering devices and power-activated pins. All power-activated pins used during set-up are operated by a hand-held remote control. The assembly, consisting of upperworks, rotating module, and carbody, is lowered onto supports.

3 Upper boom butt is attached to lower boom butt by FACT connection system's vertical positioners, horizontal pins, and alignment pads. Wire rope from rear drum is reeved through wire rope guide and attached to equalizer. Boom butt is raised by a front-mounted hydraulic cylinder.

Figure 2.8.7 *(By permission: Manitowoc Cranes, Inc., Manitowoc, Wisconsin.)*

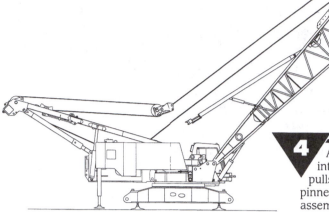

4 After hydraulic cylinders raise gantry to intermediate position, hoist line from rear drum pulls suspended equalizer to boom butt, where it is pinned. Wire rope is removed from wire rope guide, and assembly block is reeved using sheaves in boom butt.

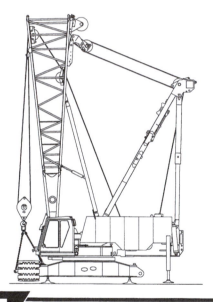

5 Gantry is pulled into upright position with boom hoist and secured by power-activated backhitch pins. Crawlers are removed from trailer and attached to carbody by FACT connection system's vertical positioners and horizontal power-activated pins. Crawler drive shafts are connected to motors on carbody.

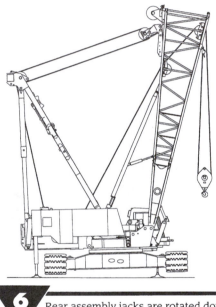

6 Rear assembly jacks are rotated downward and used to raise crane. Carbody supports are removed. Counterweight is removed from trailers and assembled.

Figure 2.8.7 *(cont.)*

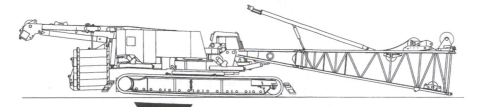

7 Crane is positioned near counterweight. Gantry is lowered. Counterweight is pendant-connected to gantry; raised into position using counterweight-raising cylinders; and attached to upperworks by power-activated pins.

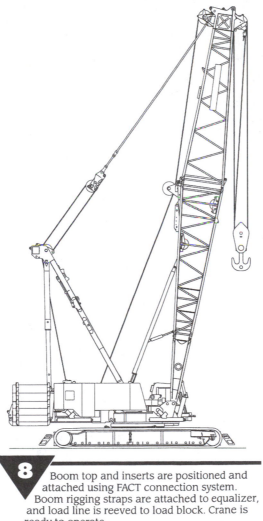

8 Boom top and inserts are positioned and attached using FACT connection system. Boom rigging straps are attached to equalizer, and load line is reeved to load block. Crane is ready to operate.

2.8.8 Manitowoc Full Range of Cranes, Draglines, and Clamshells

CRAWLER CRANES

	Model	Boom Length	Maximum Capacity on Minimum Boom (U.S. tons)	Lifting Capacities (lbs.) 80' Boom at 20' Radius	150' Boom at 40' Radius	200' Boom at 60' Radius	Boom + Jib Combinations[1] (Boom + Jib)	Jib Capacities (lbs.)	Maximum Rated Single-Line Pull (lbs.)	Crawlers Length	Stance	Pad Width	Counterweight (lbs.)	Operating Weight[2] (lbs.)	Model
EPIC®	111B	40' - 190'	80.0 at 10'	85,700	31,060	16,960[9]	180' + 50' 180' + 50'	20,000 20,000	20,000	19' 4"	16' 1" Ext 11' 10" Ret	35½"	57,700	145,050	111B
EPIC®	222B 222EXB	40' - 200'	100.0 at 10'	95,800	37,110	19,360	180' + 40' 190' + 60'	20,000 13,410	20,000	22' 4"	16' 9" Ext 12' 0" Ret	36"	77,300	184,650	222B 222EXB
EPIC®	222HDB	40' - 150'	100.0 at 11'	94,680	35,520	18,070[10]	150' + 50' 190' + 60'[10]	20,000 14,550	40,000	22' 4"	16' 9" Ext 12' 0" Ret	36"	77,300	190,680	222HDB
EPIC®	777 Series-1	60' -210'	169.6 at 13' R	224,900	78,900	42,100	190' + 30' 180' + 80'	59,000 20,700	29,500	24' 9"	19' 10¼"	38"[6]	105,000	261,590	777 Series-1
EPIC®	777 Series-2	60' - 270'	176.4 at 13' R	239,900	105,800	59,600	210' + 30' 220' + 80'	59,000 20,700	29,500	24' 9"	19' 10¼"	38"[6]	186,000	342,710	777 Series-2
EPIC®	888 Series-1	70' -280'[7]	217.0 at 15' R[3]	328,100[3]	124,100	66,500	230' + 30' 210' + 80'	59,000 20,700	29,500	28' 2½"	23' 2¾" Ext 19' 10¾" Ret	48"	144,100	339,685	888 Series-1
EPIC®	888 Series-2	70' - 310'[8]	230.0 at 15' R[3]	349,100[3]	157,000	86,200	260' + 30' 250' + 80'	59,000 20,700	29,500	28' 2½"	23' 2¾" Ext 19' 10¾" Ret	48"	223,100	418,835	888 Series-2
EPIC®	2250 Series-1	70' - 330'[4]	275.6 at 18' R	526,400	200,300	111,800[5]	190' + 40'[5] 260' + 120'[5]	100,000 32,900	30,000	30' 9⅜"	25' 11¼"	48"	168,350	449,965	2250 Series-1
EPIC®	2250 Series-2	70' - 330'[4]	300.0 at 18' R	526,400	250,200	141,300[5]	190' + 40'[5] 270' + 120'[5]	100,000 33,100	30,000	30'9⅜"	25' 11¼"	48"	268,350	550,485	2250 Series-2
EPIC®	2250 Series-3	70' - 330'[4]	300.0 at 18' R	526,400	282,300	166,900[5]	190' + 40'[5] 270' + 120'[5]	100,000 33,100	30,000	30'9⅜"	25' 11¼"	48"	368,350	650,485	2250 Series-3
EPIC®	21000	100' - 400'[11]	831.0 at 23' R	1,283,400[14]	954,200[12]	622,600[13]	-------------	-------------	48,800	46' 0"	47' 6" Std 30' 0" Opt	48"	730,700	1,586,500	21000
VICON®	3900	60' - 210'	100.0 at 16' R	137,000	48,600	25,900	190' + 30' 180' + 50'	40,000 20,000	22,500	20'4"	16' 8"	38"[6]	74,000	229,485	3900
VICON®	3900W Series-2	60' - 250'	140.0 at 15' R	216,600	75,300	41,800	240' + 30' 230' + 50'	40,000 20,000	25,800	24' 0"	19' 8" Ext 17' 2" Ret	48"	84,600	262,225	3900W Series-2
VICON®	3950W	70' - 260'	150.0 at 16' R	249,500	92,400	50,600	240' + 30' 230' + 50'	40,000 20,000	25,800	24' 0"	21' 1" Ext 18' 7" Ret	48"	102,400	301,670	3950W
VICON®	3950D	70' - 260'	150.0 at 16' R	250,800	96,400	52,900	240' + 30' 230' + 50'	40,000 20,000	25,800	24' 0"	21' 1" Ext 18' 7" Ret	48"	102,400	316,140	3950D
VICON®	4100W Series-1	70' - 260'	200.0 at 16' R	319,400	106,700	58,600	240' + 30' 230' + 60'	40,000 10,000	32,500	26' 5½"	21' 1" Ext 18' 7" Ret	48"	122,400	366,575	4100W Series-1
VICON®	4100W Series-2	70' - 260'	230.0 at 16.5' R	346,100	139,200	77,400	240' + 30' 230' + 60'	40,000 10,000	32,500	26' 5½"	21' 1" Ext 18' 7" Ret	48"	206,400	450,575	4100W Series-2
VICON®	4600 Series-3	80' - 260'	240.0 at 18' R	421,400	142,700	78,500	230' + 40' 230' + 50'	80,000 78,100	40,000	26' 1"	21' 0"	60"	113,500	482,015	4600 Series-3
VICON®	4600 Series-4	80' - 310'	350.0 at 20' R	700,000	215,200	118,400	270' + 60' 260' + 80'	88,000 80,000	44,000	30' 5"	25' 0"	60"	123,000	590,845	4600 Series-4
VICON®	4600 Series-5	80' - 310'	350.0 at 20' R	700,000	263,700	148,100	270' + 60' 260' + 80'	88,000 80,000	44,000	30' 5"	25' 0"	60"	274,400	742,245	4600 Series-5

[1]Top combination is longest that offers maximum capacity. Bottom combination offers maximum tip height. [2]Typically equipped crane with minimum boom. [3]On 22E boom. [4]Lengths over 300' require long-reach boom top. [5]With long-reach boom top. [6]48" pads optional. [7]Lengths over 250' require light-tapered boom top. [8]Lengths over 290' require light-tapered boom top. [9]190' boom [10]On 222HD/222 combination boom [11]Lengths over 340' require combination boom [12]160' boom [13]On combination boom [14]120' boom, 30' radius Key: R - Radius, Ext - Extended, Ret - Retracted, Bm - Boom.

TRUCK CRANES

	Model	Boom Length	Maximum Capacity[1] on Minimum Bm (U.S. tons)	Lifting Capacities[1] (lbs.) 80' Bm at 20' R	150' Bm at 40' R	200' Bm at 60'R	Boom + Jib Combinations (Boom + Jib)	Jib Cap.[1] (lbs.)	Outrigger Stance (Width x Length)	Operating Weight[2] (lbs.)	Counter-weight (lbs.)	Min. Road Weight (lbs.)	Min. Per-Axle Weight (lbs.)	Carrier Length (Road Ready)	Carrier Width (Road Ready)	Turning Radius (Outside of Bumper)	Travel Speed (mph)	Min. Travel Height	Gradeability for Road Travel	Model
EPIC®	777T	42½' - 298'[1,5]	220.0 at 11'R	252,000	112,000	72,500[4]	138' + 30'[1,4] 258' + 80'[1,4]	59,000 18,800	22'8" x 21'0"	252,500[6]	95,400	95,800	18,700	40' 10"	11' 1"	51' 10"	60	13'6"	30%	777T
EPIC®	M-250T	70' - 320'[3]	300.0 at 18' R	526,400	282,200	164,600[4]	190' + 40'[4] 260' + 120'[4]	100,000 32,900	30' 9" x 30' 9"	442,315	177,600	104,125	13,089	53' 1½"	9' 10"	51' 6"	50	11' 1"	30%	M-250T
EPIC®	M-250T with Pierce Carrier	70' - 320'[3]	300.0 at 18' R	526,400	282,200	164,600[4]	190' + 40'[4] 260' +120'[4]	100,000 32,900	30' 9" x 26' 4½"	466,265	177,600	131,015	29,944	54' 0½"	13' 0"	75' 6"	32	12' 0"	30%	M-250T with Pierce Carrier

[1]On fully extended outriggers . [2]Typically equipped crane with minimum boom. [3]Lengths over 300' require long-reach boom top. [4]With long-reach boom top. [5]Lengths over 270' require long reach top [6]Includes front bumper counterweight Key: Bm - Boom, R - Radius.

Figure 2.8.8 *(By permission: Manitowoc Cranes, Inc., Manitowoc, Wisconsin)*

	Model	Maximum Capacity (lbs.) Drag	Maximum Capacity (lbs.) Clam	Boom Lengths	Weight (lbs.) Drag	Weight (lbs.) Clam
DRAGLINES AND CLAMSHELLS						
EPIC®	111B	----------	17,500	40' - 70'	----------	----------
	222B 222EXB	----------	17,640	40' - 100'	----------	----------
	222HD	18,000	30,000	40' - 100'[4]	----------	----------
	777	20,000	29,500	60' - 120'[2]	----------	----------
	888	20,000	30,000	70' - 120'	----------	----------
	2250	----------	30,000	70' - 140'[1]	----------	387,035
VICON®	3900	20,000	28,000	60' - 120'[2]	198,505	198,070
	3900W Series-2	20,000	28,000	60' - 120'[2]	220,660	219,850
	3950W	20,000	32,000[3]	70' - 120'	232,700	232,700
	3950D	30,000	32,000	70' - 120'	282,430	282,430
	4100W	30,000	32,000[3]	70' - 120'	335,195	334,735
	4600 Series-1	42,000	36,000	100' - 140'	420,985	429,660
	4600 Series-3	42,000	50,000	80' - 140'	487,220	482,145
	4600 Series-4	---------	50,000	80' - 160'	----------	547,505

[1]Lengths over 120' require optional front drum. [2]120' clam, 100' drag. [3]With full-width tandem drums. [4]100' clam, 80' drag.

Figure 2.8.8 *(cont.)* *(By permission: Manitowoc Cranes, Inc., Manitowoc, Wisconsin)*

2.8.9 Capacity Enhancing Attachments

Model	Boom Length	Maximum Capacity (U.S. tons)	Lifting Capacities (lbs.) 260' Boom 80' Radius	Lifting Capacities (lbs.) 260' Boom 140' Radius	Boom + Jib Combinations[1] (Boom + Jib)	Boom + Jib Maximum Capacity (lbs.)	Auxiliary Counterweight (lbs.)	Tail Swing
CAPACITY-ENHANCING ATTACHMENTS								
4100W RINGER®	140' - 340'	300.0 at 45'	244,900	127,200	340' + 80'[2] / 340' + 140'[2]	65,000 / 31,300	275,000	29' 6¾"
36' PLATFORM-RINGER®	140' - 340'	300.0 at 45'	235,100	129,200	340' + 80'[2] / 340' + 140'[2]	65,000 / 31,300	550,000	28' 7¾"
888 RINGER	125' - 425'	661.4 at 40'	681,400 (250' Boom)	372,400 (250' Boom)	125' + 70' / 300' + 260'	400,000 / 78,500	1,395,100	29' 10"
4600 Series-4 RINGER	140' - 400'	750.0 at 70'	834,500	508,900	180' + 80' / 380' + 120' / 380' + 120' + 140'	631,600 / 220,700 / 18,700	978,700	36' 3"
60' PLATFORM-RINGER	140' - 400'	750.0 at 70'	834,500	509,200	180' + 80' / 380' + 120' / 380' + 120' + 140'	631,600 / 220,700 / 18,700	1,296,300	35' 7"
M-1200 RINGER (75A boom)	150' - 400'	900.0 at 65'	1,108,600 (250' Boom)	641,600 (250' Boom)	150' + 100' / 300' + 240'	508,700 / 194,800	1,577,600	36' 11"
M-1200 RINGER (72 or 72A boom)	153' - 403'	1,433.3 at 60' R	1,789,900 (253' Boom)	753,200 (253' Boom)	228' + 100' / 228' + 200'	1,744,600 / 1,070,100	2,061,600	36' 11"
M-1200 MAX-RINGER™	153' - 403'	1,433.3 at 60'	2,451,500 (253' Boom)	1,155,400 (253' Boom)	253' + 150' / 303' + 150'	1,442,600 / 1,339,200	2,843,600	51' 1½
2250 MAX-ER™ 225	140' - 340'	250.0 at 26'	191,800	92,700	340' + 40' / 340' + 120'	100,000 / 28,500	225,000	37' 2"
2250 MAX-ER™ 400	82' -162'	500.0 at 18'	200,200 (162' Boom)	90,200 (162' Boom)	--------------	--------------	400,000	37' 2¼"
2250 MAX-ER™ 2000	140' - 360'	400.0 at 28'	346,300	181,100	140' + 70' / 300' + 200' + 120'	--------------	460,000	57' 4"(max)
4100W Series-2 X-TENDER Series-2	150' - 350'	422.5 at 95'	712,300 at 100' R (250' Boom)	301,100 (250' Boom)	300' + 120' / 325' + 200'	195,000 / 108,000	--------------	75' 7"
21000 MAX-ER™	120' - 380'	1000.0 at 45'	1,141,900 (260' Boom)	616,100 (260' Boom)	--------------	--------------	1,102,300	84' 0"(max)

[1]Top combination is longest that offers maximum capacity. Bottom combination offers maximum tip height.   [2]25' strut required.

2.8.10 Tower and Luffing Jib Attachments

TOWER and LUFFING-JIB ATTACHMENTS

| | Model | Maximum Tower + Boom | | Maximum Capacity (lbs.) | Maximum Tip Height | Capacities (lbs.) Maximum Tower + Boom Raised Unassisted | | | Operating Weight[1] (lbs.) |
		Raised Unassisted	Raised With Assist			50' Radius	100' Radius	150' Radius	
TOWER	3900W Series-2	164' + 150'	204' + 150'	38,000	353' 7"	21,800	13,800	8,500	281,250
	4100W Series-1	183' + 170'	253' + 170'	61,300	421' 7"	32,200 at 55'	20,400	11,000	380,305
	4600 Series-4	194' + 180'	254' + 200'	143,100	454' 7"	77,600 at 60'	54,200	34,000	614,020
LUFFING-JIB	777 Series-1	140' + 150'	----------	104,600	294' 3"	39,500 at 55'	26,200	14,600	274,490
	777 Series-2	180' + 170'	----------	104,600	354' 0"	27,200 at 55'	23,200	14,900	355,610
	888 Series-1	160' + 170'	----------	105,500	332' 7"	36,200 at 55'	30,400	19,300	361,000
	888 Series-2	200' + 170'	----------	105,500	371' 7"	32,300 at 60'	29,200	21,300	440,000
	2250 Series-1	190' + 200'	----------	180,000	393' 7"	47,400 at 65'	48,500	30,800	481,765
	2250 Series-2	190' + 200'	250' + 200'	180,000	452' 10"	47,400 at 65'	48,500	38,400	582,285
	2250 Series-3	190' + 200'	250' + 200'	220,450	452' 10"	55,000 at 65'	55,000	41,800	650,485
	2250 MAX-ER™ 225	300' + 200'	----------	180,000	498' 6"	39,700 at 110'	37,200 at 160'	31,400 at 210'	871,633
	2250 MAX-ER™ 2000	300' + 200'	----------	380,700	498' 6"	----------	----------	----------	----------
	4100W Series-2 X-TENDER™ Series-2	325' + 200'	----------	195,000	530'	108,000 at 150'	103,900 at 200'	75,200 at 250'	865,300
	21000 MAX-ER™	300' + 300'	----------	500,000	596'	224,200 at 165'	134,800 at 260'	46,100 at 355'	----------
	888 RINGER®	300' + 300'	----------	551,200	600'	100,700 at 185'	94,200 at 235'	61,000 at 285'	2,146,600

[1]With minimum tower and boom, or minimum boom and luffing jib.

(By permission: Manitowoc Cranes, Inc., Manitowoc, Wisconsin)

2.8.11 Boom Top Configurations

BOOM TOPS TO MEET PROJECT REQUIREMENTS. . .PLUS high capacity jib that's adaptable to every boom top.

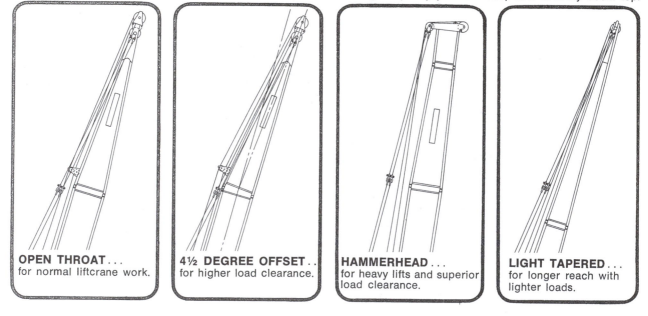

OPEN THROAT. . .
for normal liftcrane work.

4½ DEGREE OFFSET. .
for higher load clearance.

HAMMERHEAD. . .
for heavy lifts and superior load clearance.

LIGHT TAPERED. . .
for longer reach with lighter loads.

2.8.12 Hand Signals for Boom-Equipment Operators

USE LOAD LINE
Tap fist on head; then use regular signals.

USE WHIP LINE
Tap elbow with one hand; then use regular signals.

RAISE LOAD

LOWER LOAD

RAISE LOAD SLOWLY

LOWER LOAD SLOWLY

RAISE BOOM

LOWER BOOM

RAISE BOOM AND LOWER LOAD

Extend thumb upward and flex fingers in and out as long as load movement is desired.

LOWER BOOM AND RAISE LOAD

Extend thumb downward and flex fingers in and out as long as load movement is desired.

SWING—
Extend arm, with forefinger indicating direction of swing.

STOP—
Extend forearm and hand in horizontal position and make slicing motion.

Contents

3.0.0 Installation of Piping (General)

A great deal of construction activity involves the installation of rerouting of underground utilities (sanitary and storm sewers, domestic water lines and fire mains, electrical and telecommunications services, and natural gas lines). The nature and variety of these installations vary substantially from site to site, but the basic materials of construction generally do not.

Underground site utility work consists of the installation of conduits of various sizes and materials of construction to carry these utilities; the basic piping materials are either reinforced concrete pipe, thermoplastics, cast iron or ductile iron, lightweight aluminum or steel and corrugated metal pipe.

Installation of these types of pipes have several things in common:

1. *Excavation and pipe laying*   Depending upon the type of soil and the width and depth of excavation, either "open cut" will be used or a trench cut (utilizing sheet piling) or a trench box (to avoid collapse of the walls of the excavate).

2. *Bedding material*   Depending upon the type of pipe being installed and the nature of the subsoil, off-site bedding materials might be required, not only to place under the pipe, but for initial backfill.

3. Compaction of the soil above the pipe will also depend on the depth of the excavate, the soil conditions, and the percentage compaction required.

3.1.0 Bedding and Backfill Materials for Site Utility Work and the Pipe Zone

To discuss the backfill procedures for underground pipes, it is necessary to understand pipe zone terminology.

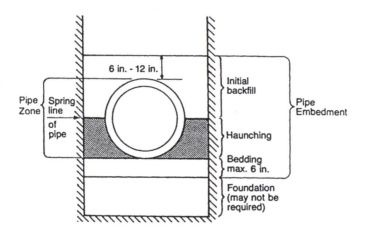

- *Foundation*   Might not be required if the trench bottom is stable and will support a rigid pipe without causing deviation in grade or such flexing of the pipe that will create flexural failure.

- *Bedding*   This material is required to bring the trench bottom up to grade and to provide uniform longitudinal support. Sand is often used for this purpose.

- *Haunching*   This material used in this zone will supply structural support for the pipe and prevent it from deflecting (if it is a flexible pipe) or having joint misalignment when further backfilling and compaction above occurs.

- *Initial backfill*   Material placed 6 to 12 inches above the spring line will only provide additional side support, most of the support coming from compaction of the soil in the haunching area.

3.1.1 Pipe-Zone Bedding Materials

- *Class I*   Angular stone, graded from ¼" (6.4 mm) to ½" (12.7 mm), including crushed stone, crushed shells, and cinders.

- *Class II*   Coarse sand with a maximum particle size of 1½″ (38.1 mm), including various graded sands and gravel containing small percentages of fines. Soil type GW, SP, SM, and C* (See the unified soil classification listing in Chapter 1).

- *Class III*   Fine sand and clayey gravel, including fine sand, sand-clay mixtures, and gravel-clay mixes. Soil types GM, GC, SM, and SC are included in this class.

- *Class IV*   Silt, silty clays (including inorganic clays), and silts of medium to high plasticity and liquid limits. Soil types MH, ML, CH, and CL are included in this class.

- *Class V*   Soils not recommended for bedding, haunching, or initial backfill consisting of organic silts, organic clays and peat, and other highly organic materials.

Common sense, experience, and OSHA regulations will dictate the precautions required during site utilities excavation. OSHA Handbook *Title 29 of the Code of Federal Regulations* (29 CFR Part 1926) is to be referred to for detailed regulations regarding excavation and trenching operations. OSHA *Construction Industry Digest* (OSHA 2202) is a pocket-sized digest of basic applicable standards, including excavation and trenching. This handy booklet can be obtained by calling the local U.S. Department of Labor office.

3.2.0 Setting Up Batter Boards and Determining Trench Widths and Depths

The figure shows how to set up a batter boards so as to establish a center line of underground pipe.

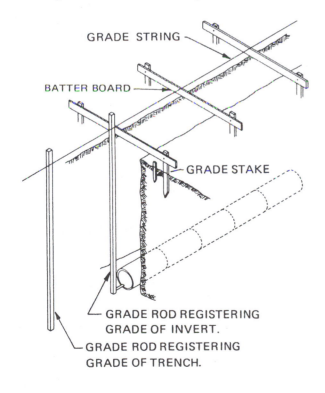

3.3.0 Laser System

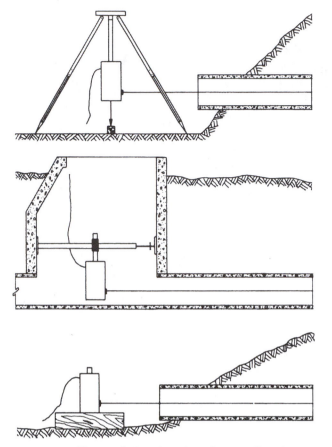

Figure 3.3.0   (*By permission: American Concrete Pipe Association, Irving, Texas*)

3.4.0 Excavation Limits

Excavation, pipe installation and backfill operations should succeed each other as rapidly as possible. Avoiding long stretches of open trench will:

- Reduce equipment requirements
- Reduce sheathing and shoring required at any one time
- Prevent trench flooding
- Reduce the need to control groundwater
- Minimize disruption to existing utilities
- Simplify traffic maintenance
- Reduce safety hazards
- Permit closer supervision and inspection of the work
- Permit better quality control
- Reduce adverse environmental impacts
- Assist in maintaining better public relations

3.5.0 Concrete Pipe Configurations

Concrete pipe is manufactured in five common shapes. Regional custom and demand usually determine availability.

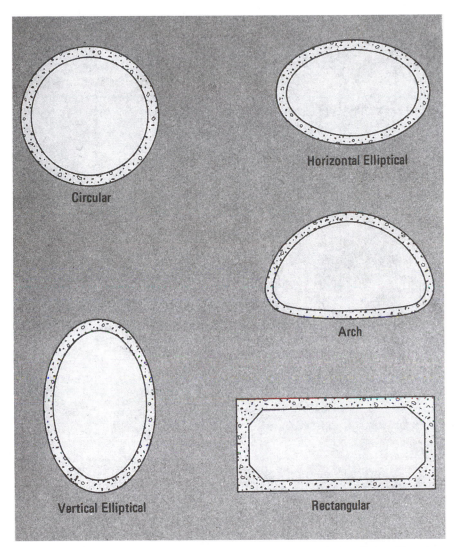

Figure 3.5.0 *(By permission: American Concrete Pipe Association, Irving, Texas)*

3.5.1 Nonreinforced Culvert Pipe—Sizes and Weights

Dimensions and Approximate Weights of Circular Concrete Pipe, Bell and Spigot Joint.

Internal Diameter, inches	ASTM C14—Nonreinforced Sewer and Culvert Pipe, Bell and Spigot Joint					
	Class 1		Class 2		Class 3	
	Minimum Wall Thickness, inches	Average Weight, pounds per foot	Minimum Wall Thickness, inches	Average Weight, pounds per foot	Minimum Wall Thickness, inches	Average Weight, pounds per foot
4	5/8	9.5	3/4	13	3/4	13
6	5/8	17	3/4	20	7/8	21
8	3/4	27	7/8	31	1-1/8	36
10	7/8	37	1	42	1-1/4	50
12	1	50	1-3/8	68	1-3/4	90
15	1-1/4	78	1-5/8	100	1-7/8	120
18	1-1/2	105	2	155	2-1/4	165
21	1-3/4	159	2-1/4	205	2-3/4	260
24	2-1/8	200	3	315	3-3/8	350
27	3-1/4	390	3-3/4	450	3-3/4	450
30	3-1/2	450	4-1/4	540	4-1/4	540
33	3-3/4	520	4-1/2	620	4-1/2	620
36	4	580	4-3/4	700	4-3/4	700

3.5.2 Reinforced Concrete Pipe—Sizes and Weights

Internal Diameter, inches	ASTM C76—Reinforced Concrete Culvert, Storm Drain and Sewer Pipe, Bell and Spigot Joint.			
	Wall A		Wall B	
	Minimum Wall Thickness, inches	Average Weight, pounds per foot	Minimum Wall Thickness, inches	Average Weight, pounds per foot
12	1-3/4	90	2	106
15	1-7/8	120	2-1/4	148
18	2	155	2-1/2	200
21	2-1/4	205	2-3/4	260
24	2-1/2	265	3	325
27	2-5/8	310	3-1/4	388
30	2-3/4	363	3-1/2	459

These tables are based on concrete weighing 150 pounds per cubic foot and will vary with heavier or lighter weight concrete.

Figure 3.5.2 (*By permission: American Concrete Pipe Association, Irving, Texas*)

3.5.3 SI and U.S. Customary Sizes for Circular Concrete Pipe

SUPPLEMENTAL DESIGN
CONSIDERATIONS AND PROCEDURES

SI and U.S. Customary Designated Sizes for
Circular Concrete Pipe.

SI			U.S. Customary	
DESIGNATED SIZE Diameter of Pipe, mm	Permissible Variation Internal Diameter of Pipe		Diameter of Pipe	
	Minimum, mm	Maximum, mm	Inches	Millimetres
100	100	110	4	101.6
150	150	160	6	152.4
200	200	210	8	203.2
250	250	260	10	254.0
300	300	310	12	304.8
375	375	390	15	381.0
450	450	465	18	457.2
525	525	545	21	533.4
600	600	620	24	609.6
675	675	695	27	685.8
750	750	775	30	762.0
825	825	850	33	838.2
900	900	925	36	914.4
1050	1050	1080	42	1066.8
1200	1200	1230	48	1219.2
1350	1350	1385	54	1371.6
1500	1500	1540	60	1524.0
1650	1650	1695	66	1676.4
1800	1800	1850	72	1828.8
1950	1950	2000	78	1981.2
2100	2100	2155	84	2133.6
2250	2250	2310	90	2286.0
2400	2400	2465	96	2438.4
2550	2550	2620	102	2590.8
2700	2700	2770	108	2743.2
2850	2850	2925	114	2895.6
3000	3000	3080	120	3048.0
3150	3150	3235	126	3200.4
3300	3300	3390	132	3352.8
3450	3450	3540	136	3505.2
3600	3600	3695	144	3657.6

NOTE: "SI" refers to the International System of Units (i.e., meter (m), kilogram (kg), second (s), ampere (a), kelvin (k), mole (mol), and candela (ca).

Figure 3.5.3   (*By permission: American Concrete Pipe Association, Irving, Texas*)

3.5.4 Reinforced Concrete Pipe (RCP) Specification ASTMC76 T&G Joints

ASTM C 76 Reinforced Concrete Culvert, Storm Drain and Sewer Pipe, Tongue and Groove Joints						
	WALL A		WALL B		WALL C	
Internal Diameter, inches	Minimum Wall Thickness, inches	Approximate Weight, pounds per foot	Minimum Wall Thickness, inches	Approximate Weight, pounds per foot	Minimum Wall Thickness, inches	Approximate Weight, pounds per foot
12	1¾	79	2	93	—	—
15	1⅞	103	2¼	127	—	—
18	2	131	2½	168	—	—
21	2¼	171	2¾	214	—	—
24	2½	217	3	264	3¾	366
27	2⅝	255	3¼	322	4	420
30	2¾	295	3½	384	4¼	476
33	2⅞	336	3¾	451	4½	552
36	3	383	4	524	4¾	654
42	3½	520	4½	686	5¼	811
48	4	683	5	867	5¾	1011
54	4½	864	5½	1068	6¼	1208
60	5	1064	6	1295	6¾	1473
66	5½	1287	6½	1542	7¼	1735
72	6	1532	7	1811	7¾	2015
78	6½	1797	7½	2100	8¼	2410
84	7	2085	8	2409	8¾	2660
90	7½	2395	8½	2740	9¼	3020
96	8	2710	9	3090	9¾	3355
102	8½	3078	9½	3480	10¼	3760
108	9	3446	10	3865	10¾	4160

Large Sizes of Pipe Tongue and Groove Joint			
Internal Diameter Inches	Internal Diameter Feet	Wall Thickness Inches	Approximate Weight, pounds per foot
114	9½	9½	3840
120	10	10	4263
126	10½	10½	4690
132	11	11	5148
138	11½	11½	5627
144	12	12	6126
150	12½	12½	6647
156	13	13	7190
162	13½	13½	7754
168	14	14	8339
174	14½	14½	8945
180	15	15	9572

These tables are based on concrete weighing 150 pounds per cubic foot and will vary with heavier or lighter weight concrete.

3.5.5 Reinforced Concrete Pipe Specifications ASTMC14, ASTMC76 (Bell and Spigot)

ASTM C 14—Nonreinforced Sewer and Culvert Pipe, Bell and Spigot Joint.						
	CLASS 1		CLASS 2		CLASS 3	
Internal Diameter, inches	Minimum Wall Thickness, Inches	Approx. Weight, pounds per foot	Minimum Wall Thickness, Inches	Approx. Weight, pounds per foot	Minimum Wall Thickness, inches	Approx. Weight, pounds per foot
4	⅝	9.5	¾	13	⅞	15
6	⅝	17	¾	20	1	24
8	¾	27	⅞	31	1⅛	36
10	⅞	37	1	42	1¼	50
12	1	50	1⅜	68	1¾	90
15	1¼	80	1⅝	100	1⅞	120
18	1½	110	2	160	2¼	170
21	1¾	160	2¼	210	2¾	260
24	2⅛	200	3	320	3⅜	350
27	3¼	390	3¾	450	3¾	450
30	3½	450	4¼	540	4¼	540
33	3¾	520	4½	620	4½	620
36	4	580	4¾	700	4¾	700

ASTM C 76—Reinforced Concrete Culvert, Storm Drain and Sewer Pipe, Bell and Spigot Joint.				
	WALL A		WALL B	
Internal Diameter, Inches	Minimum Wall Thickness Inches	Approximate Weight, pounds per foot	Minimum Wall Thickness, inches	Approximate Weight, pounds per foot
12	1¾	90	2	110
15	1⅞	120	2¼	150
18	2	160	2½	200
21	2¼	210	2¾	260
24	2½	270	3	330
27	2⅝	310	3¼	390
30	2¾	360	3½	450

3.5.6 Curved and Deflected Alignment of Concrete Pipe

CURVED ALIGNMENT

Change of direction in sewers is usually accomplished at manhole structures. Grade and alignment changes in concrete pipelines can be incorporated through the use of **deflected straight pipe, radius pipe** or **bends.**

When concrete pipe is to be installed in a curved alignment, local concrete pipe manufacturers should be consulted regarding manufacturing and installation feasibility. Many manufacturers have standardized joint configurations and deflections for specific radii and economies may be realized by using standard pipe.

DEFLECTED STRAIGHT PIPE

When concrete pipe in installed in a straight alignment and the joints are in a **home,** or normal, position, the joint space or distance between the ends of adjacent pipe sections is essentially uniform around the periphery of the pipe. Starting from this home position any joint may be opened the maximum permissible amount on one side while the other side remains in the home position. The difference between the home and opened joint space is generally designated as the **pull.** The maximum permissible pull must be limited to that opening which will provide satisfactory joint performance. This varies for different joint configurations and is best obtained from the pipe manufacturer.

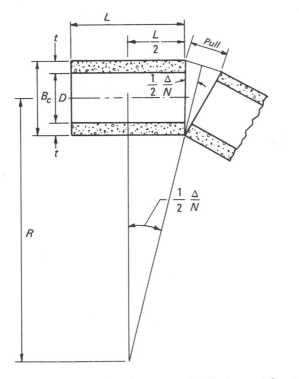

Deflection Angle Geometry for Deflected Straight Pipe.

Figure 3.5.6 (*By permission: American Concrete Pipe Association, Irving, Texas*)

3.5.7 Wyes for Concrete Pipe

Wyes are similar to tees except the centerline of the intersecting pipe intersects the centerline of the base pipe at an acute angle (Fig. 3.5.7). Wyes are also utilized to effect the junction of two pipelines without the necessity of a manhole or junction chamber. Wyes are commonly used to connect building sewers or house laterals to a sewer main.

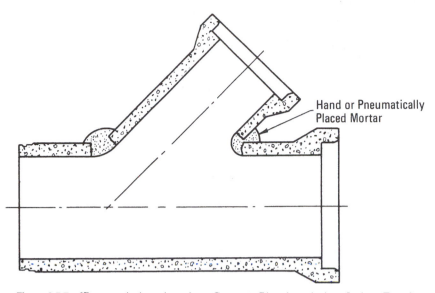

Figure 3.5.7 (*By permission: American Concrete Pipe Association, Irving, Texas*)

3.5.8 Concrete Pipe Tee

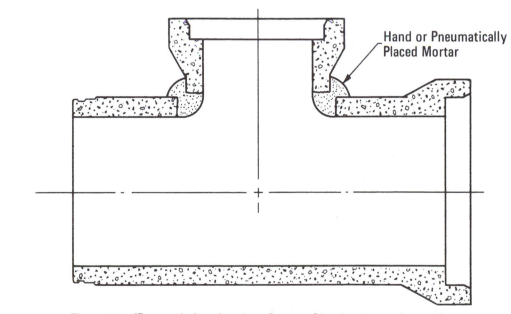

Figure 3.5.8 (*By permission: American Concrete Pipe Association, Irving, Texas*)

3.5.9 Offset Manhole Tee

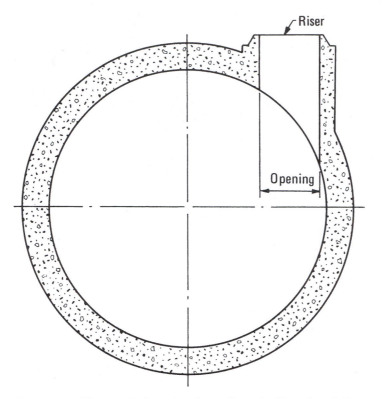

Figure 3.5.9 (*By permission: American Concrete Pipe Association, Irving, Texas*)

3.6.0 Trench Preparation for Concrete Pipe

Trench width required to install concrete pipe of varying pipe diameters (metric conversion).

Pipe, diameter, in (mm)	Trench width, ft	Pipe diameter, in (in)	Trench width, ft
4	1.6	60	8.5
6	1.8	66	9.2
8	2.0	72	10.0
10	2.3	78	10.7
12	2.5	84	11.4
15	3.0	90	12.1
18	3.4	96	12.9
21	3.8	102	13.6
24	4.1	108	14.3
27	4.5	114	14.9
33	5.2	120	15.6
36	5.6	126	16.4
42	6.3	132	17.1
48	7.0	138	17.8
54	7.8	144	18.5
100	0.47	150	25.0
150	0.54	165	28.0
200	0.60	180	30.0
250	0.68	195	32.0
300	0.80	210	34.0
375	0.91	225	36.0
450	1.02	240	39.0
525	1.10	255	41.0
600	1.20	270	43.0
675	13.0	285	45.0
825	16.0	300	48.0
900	17.0	315	50.0
1050	19.0	330	52.0
1200	21.0	345	54.0
1350	23.0	360	56.0

NOTE: Trench widths based on 1.25 Bc + 1 ft where Bc is the outside diameter of the pipe in inches, and + 300 where Bc is the outside diameter of the pipe in millimeters.

3.6.1 Trench Width Considerations

- *Cast-iron soil*   This is a rigid pipe that does not depend on sidefill stiffness; therefore, the trench can be as narrow as an installer requires in order to join the pipe sections together and complete the joint connections

- *Thermoplastic pipe*   This flexible pipe requires sidewall stiffness in the trench to limit deflections. ASTM D2321-89 recommends that the trench width be as wide as the outside diameter of the pipe being installed plus 16 inches (400 mm). An alternative to this formula is to multiply the outside diameter of the pipe by 1.25 and add 12 inches (300 mm). For example, a 6-inch (150 mm) pipe has an outside diameter of 6.625 inches (165.6 mm) and would require a 20-inch (500 mm) wide trench. The added width of the trench for "flexible" pipe is to allow for compaction equipment to operate in the "compaction zone" on each side of the flexible pipe to create sidewall stiffness.

3.6.2 Determining Trench Shield Size

If the company does not own a trench box, but plans to rent one, certain data, shown below, must be given to the rental company to ensure that the proper size box is ordered to fit the job at hand.

To size a trench shield

Depth of cut   _____

Soil Conditions*
    Type A (25#)   _____
    Type B (45#)   _____
    Type C (60#)   _____
    Hydrostatic   _____

Outside pipe diameter   _____
    (Shield 12 in wider than pipe OD)

Pipe length   _____
    (Shield 2 to 4 ft longer)

Bucket width   _____
    (Inside shield: 12 in less than shield)
    (Outside shield: 4 in more than shield)

Machine lift capacity   _____
    (1.5 times shield weight at 20-ft radius at grade)

* Soil conditions refer to OSHA classifications. (See Sec. 1.1.5.2 for a full explanation of Type A, B, and C soils.)

Figure 3.6.2 (*By permission: Efficiency Production, Lansing, Michigan*)

3.6.3 Essential Features of Various Types of Installations

The embankment group is further subdivided into **positive projection, negative projection** and **induced trench** subgroups, based upon the extent the pipe is exposed to direct embankment loading. The essential features of these are illustrated in the figure below. There are further subdivisions of the positive projection and negative projection embankment subgroups which are related to whether or not differential settlements may occur throughout the entire depth of backfill. Two additional classifications are **tunneled** and **jacked,** and **multiple pipe** installations.

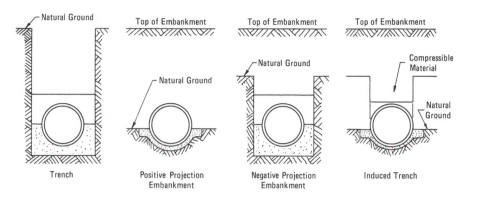

Trench · Positive Projection Embankment · Negative Projection Embankment · Induced Trench

3.6.4 Loads on the Pipe

Three types of loads must be considered:

1. **Earth loads**
2. **Live loads** from trucks, aircraft, and trains
3. **Surcharge loads** or loads from an additional earth fill or building over an installed pipe

The methods for determining the magnitude of these loads are discussed in the following section.

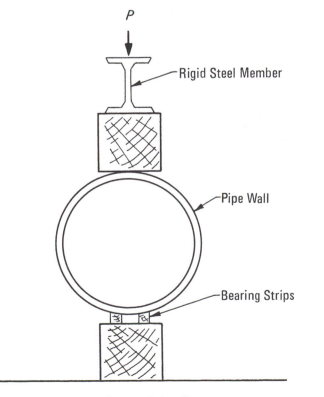

Three-Edge Bearing Test

3.6.5 Earth Loads

The **earth load** is the weight of the earth that must ultimately be carried by the pipe. This weight varies with the soil characteristics. More importantly, however, it varies with the installation conditions. The method for determining the earth loading is best approached by considering the two major classes of construction, trench and embankment.

3.6.6 Trench Installations

Trench installations are made in relatively narrow excavations and the pipeline covered with earth backfill which extends to the original ground surface. Sewers, drains and water mains are usually constructed in trenches.

3.6.7 Pipe Beddings and Classes

The type of bedding is one of the factors that determines the supporting strength of buried pipe. Four classes of beddings were proposed originally by Spangler and defined by a simplified and idealized distribution of the vertical reactive force acting on the bottom of the pipe. The distribution of the reactive force is described in terms of the central bedding angle (Figure 3.6.7)

- Class D   Impermissible, 0 Degree Central Bedding angle
- Class C   Ordinary, 60 Degree Central Bedding Angle
- Class B   First Class, 90 Degree Central Bedding

Impermissible, ordinary, and first class are the original descriptions used by Spangler. The class B, C, and D descriptions are current designations. The fourth bedding condition is designated as class A, in which the pipe is bedded in concrete.

To transform these mathematically defined loadings into real world conditions, Spangler proposed a series of installation methods. Improvements in construction equipment and methods have resulted in changes to the original installations proposed.

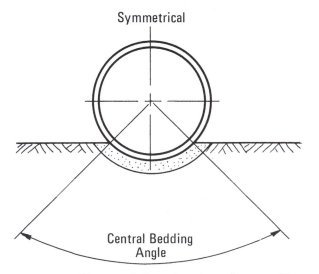

Figure 3.6.7 (*By permission: American Concrete Pipe Association, Irving, Texas*)

Classes of bedding for the trench condition:

- **Class A bedding** can be either a concrete cradle or a concrete arch.
- **Class B bedding** can be either a shaped subgrade with granular foundation or a granular foundation. A granular foundation without a shaped subgrade is used only with circular pipe.
- **Class C bedding** can be either a shaped subgrade or a granular foundation. A granular foundation without a shaped subgrade is used only with circular pipe.
- **Class D bedding** is a flat subgrade, and used only with circular pipe.

Classes of bedding for the embankment condition:

- **Class A bedding** is a concrete cradle.
- **Class B bedding** can be a shaped subgrade with granular foundation, granular foundation, or shaped subgrade. A granular foundation without a shaped subgrade is only used with circular pipe, and a shaped subgrade by itself is used only with arch and horizontal elliptical pipe.
- **Class C bedding** can be either a shaped subgrade or a granular foundation. A granular foundation without a shaped subgrade is used only with a circular pipe.
- **Class D bedding** is a flat subgrade and used only with circular pipe.

Figure 3.6.7 (*cont.*)

3.6.8 Trench Bedding—Circular Pipe (Illustrated)

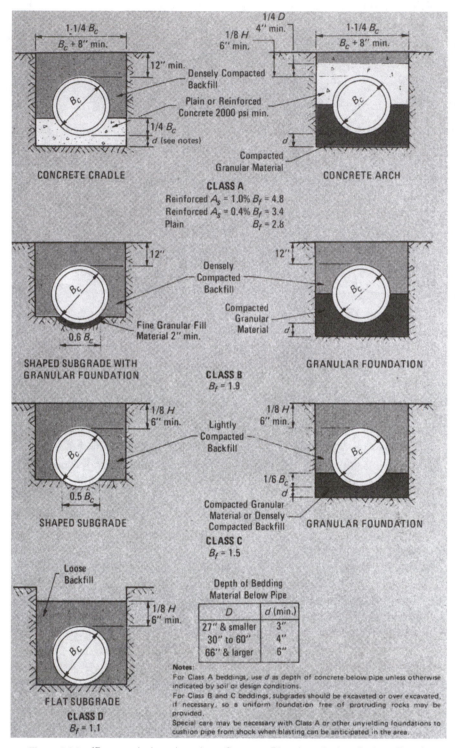

Figure 3.6.8 (*By permission: American Concrete Pipe Association, Irving, Texas*)

3.6.9 Trench Beddings—Elliptical and Arch Pipe (Illustrated)

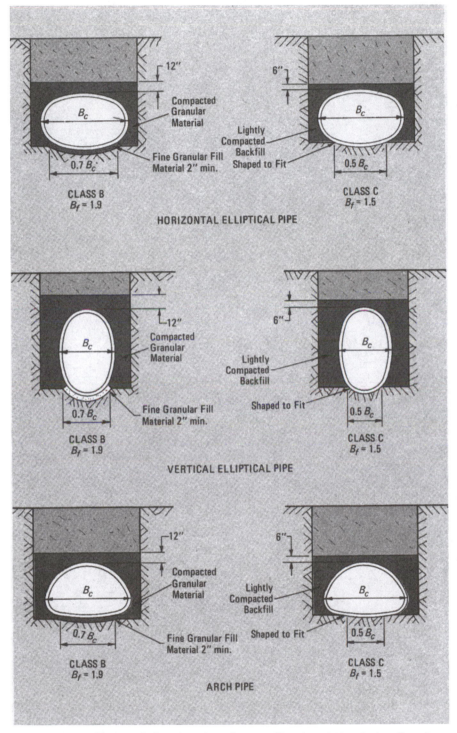

Figure 3.6.9 *(By permission: American Concrete Pipe Association, Irving, Texas)*

3.6.10 Embankment Bedding—Circular Pipe (Illustrated)

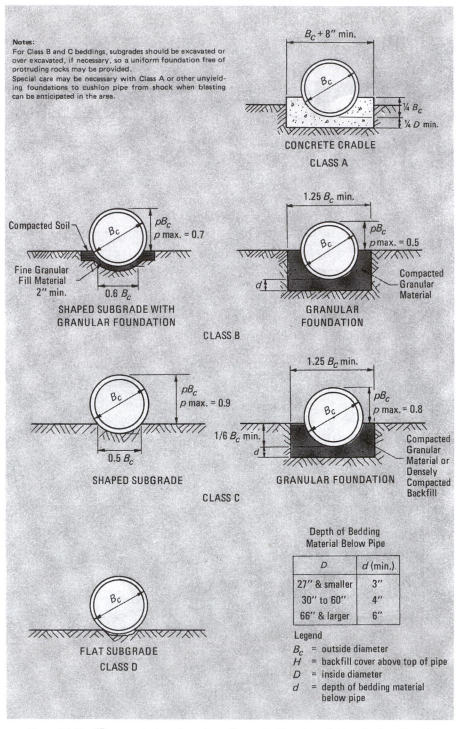

Figure 3.6.10   (*By permission: American Concrete Pipe Association, Irving, Texas*)

3.6.11 Embankment Beddings—Elliptical and Arch Pipe (Illustrated)

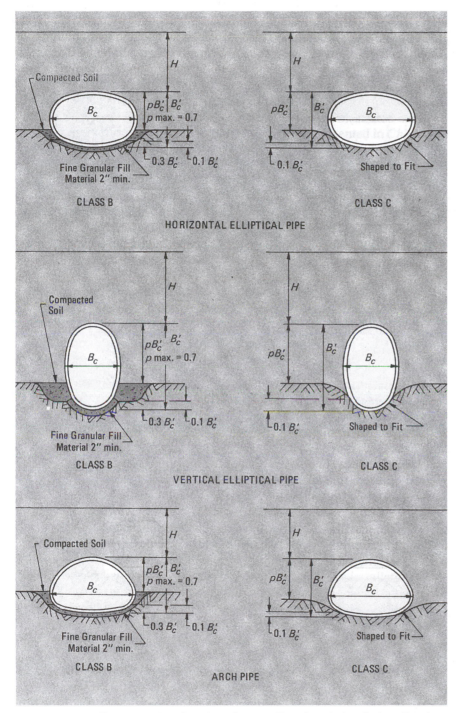

Figure 3.6.11   *(By permission: American Concrete Pipe Association, Irving, Texas)*

3.6.12 Installation Procedures for Bell and Spigot Concrete Pipe

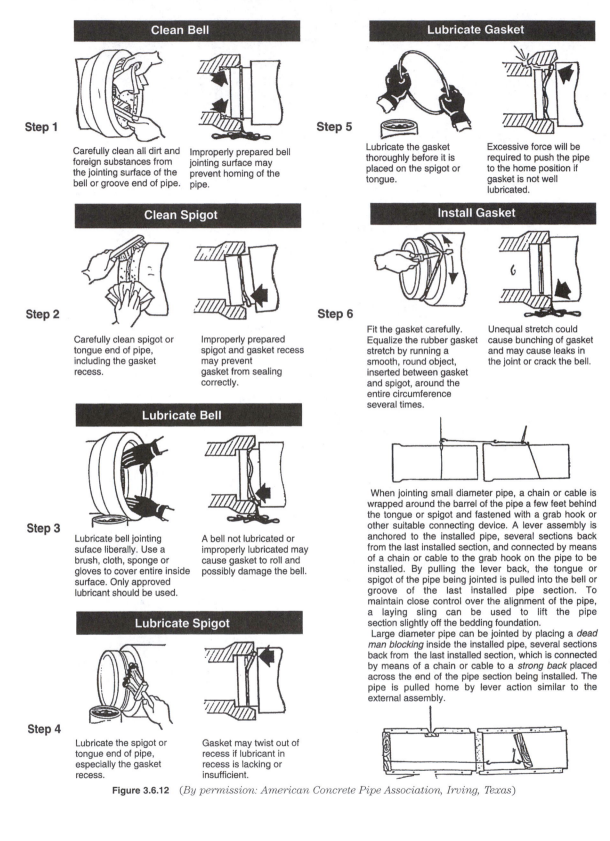

Clean Bell

Step 1

Carefully clean all dirt and foreign substances from the jointing surface of the bell or groove end of pipe.

Improperly prepared bell jointing surface may prevent homing of the pipe.

Clean Spigot

Step 2

Carefully clean spigot or tongue end of pipe, including the gasket recess.

Improperly prepared spigot and gasket recess may prevent gasket from sealing correctly.

Lubricate Bell

Step 3

Lubricate bell jointing suface liberally. Use a brush, cloth, sponge or gloves to cover entire inside surface. Only approved lubricant should be used.

A bell not lubricated or improperly lubricated may cause gasket to roll and possibly damage the bell.

Lubricate Spigot

Step 4

Lubricate the spigot or tongue end of pipe, especially the gasket recess.

Gasket may twist out of recess if lubricant in recess is lacking or insufficient.

Lubricate Gasket

Step 5

Lubricate the gasket thoroughly before it is placed on the spigot or tongue.

Excessive force will be required to push the pipe to the home position if gasket is not well lubricated.

Install Gasket

Step 6

Fit the gasket carefully. Equalize the rubber gasket stretch by running a smooth, round object, inserted between gasket and spigot, around the entire circumference several times.

Unequal stretch could cause bunching of gasket and may cause leaks in the joint or crack the bell.

When jointing small diameter pipe, a chain or cable is wrapped around the barrel of the pipe a few feet behind the tongue or spigot and fastened with a grab hook or other suitable connecting device. A lever assembly is anchored to the installed pipe, several sections back from the last installed section, and connected by means of a chain or cable to the grab hook on the pipe to be installed. By pulling the lever back, the tongue or spigot of the pipe being jointed is pulled into the bell or groove of the last installed pipe section. To maintain close control over the alignment of the pipe, a laying sling can be used to lift the pipe section slightly off the bedding foundation.

Large diameter pipe can be jointed by placing a *dead man blocking* inside the installed pipe, several sections back from the last installed section, which is connected by means of a chain or cable to a *strong back* placed across the end of the pipe section being installed. The pipe is pulled home by lever action similar to the external assembly.

Figure 3.6.12   *(By permission: American Concrete Pipe Association, Irving, Texas)*

3.6.13 Minimum Trench Width and Load Considerations

In sewer construction, the most important excavation limitations are trench width and depth. As excavation progresses, trench grades are continuously checked to obtain the elevations established on the sewer profile. Incorrect trench depths may adversely affect the hydraulic capacity of the sewer and require correction or additional maintenance after the line is completed.

The backfill load ultimately transmitted to the pipe is a function of trench width (Fig. 3.6.13.1). The designer assumes a certain trench width in determining the backfill load, and selects a pipe strength capable of withstanding that load. If the actual trench width exceeds the width assumed in design, the load on the pipe will be greater than estimated and structural distress may result. Therefore, trench widths should be as narrow as established in the plans or standard drawings. Side clearance must be adequate to permit proper compaction of backfill material at the sides of the pipe, and trenches are usually designed for a width of 1.25 times the outside diameter of the pipe plus one foot. Figure 3.6.13.2 illustrates the load carried by a pipe installed in a normal trench installation and if the width of the trench is increased, the load on the pipe is increased as shown in Fig. 3.6.13.3.

Figure 3.6.13.1 Concrete pipe being installed in a trench. (*By permission: American Concrete Pipe Association, Irving, Texas*)

If an excessively wide trench is excavated or the sides sloped back, the pipe can be installed in a narrow subtrench excavated at the bottom of the wider trench to avoid increase in the backfill load Fig. 3.6.13.4. The recommended depth of the subtrench is the vertical height of the pipe plus one foot.

For culverts installed under embankments, it may be possible to simulate a narrow subtrench by installing the pipe in the existing stream bed. When culverts are installed in a negative projection condition or by the induced trench method of construction the same excavation limits apply as for trench conditions.

For jacked or tunneled installations the excavation coincides as closely as possible to the outside dimensions and shape of the pipe. The usual procedure for jacking pipe is to equip the leading edge with a cutter, or shoe, to protect the lead pipe. As the pipe is jacked forward, soil is excavated and removed through the pipe. Materials should be trimmed approximately one or two inches larger than the outside diameter of the pipe and excavation should not precede pipe advancement more than necessary. This procedure results in minimum disturbance of the earth adjacent to the pipe.

Figure 3.6.13.1 *(cont.)*

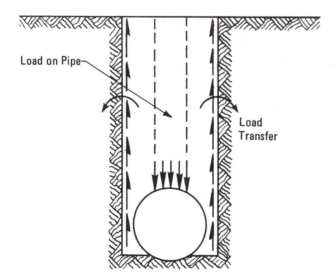

Figure 3.6.13.2 Trench installation. *(By permission: American Concrete Pipe Association, Irving, Texas)*

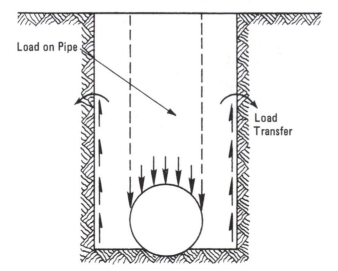

Figure 3.6.13.3 Wide trench installation. *(By permission: American Concrete Pipe Association, Irving, Texas)*

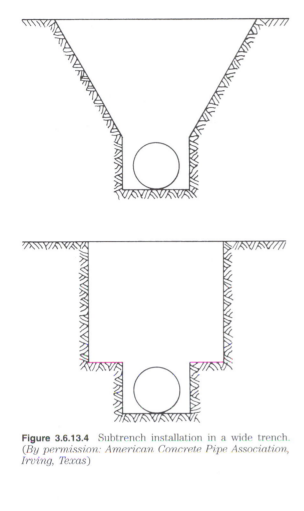

Figure 3.6.13.4 Subtrench installation in a wide trench. (*By permission: American Concrete Pipe Association, Irving, Texas*)

3.6.14 Types of Sheathing and Shoring

Accurate evaluation of all these factors is usually not possible, so the design and application of temporary bracing systems varies considerably. However, certain methods, materials, and terminology have evolved for stabilizing open trenches and serve as a general guide.

Shoring for trenches is accomplished by bracing one bank against the other, and the structural members which transfer the load between the trench sides are termed **struts.** Wood planks placed against the trench walls to retain the vertical banks are termed **sheathing.** The horizontal members of the bracing system which form the framework for the sheathing are termed **walers** or **stringers,** and the vertical members are termed **strongbacks.**

The four most common sheathing methods are (Fig. 3.6.14.1):

- Open or skeleton sheathing
- Close sheathing
- Tight sheathing
- Trench shields or boxes

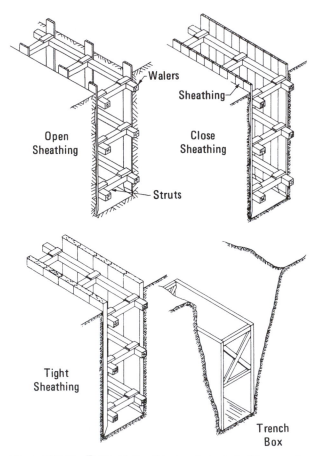

Figure 3.6.14.1 Types of sheathing and shoring. (*By permission: American Concrete Pipe Association, Irving, Texas*)

Open sheathing consists of a continuous frame with vertical sheathing planks placed at intervals along the open trench. This method of sheathing is used for cohesive, stable soils where groundwater is not a problem.

Close sheathing also uses a continuous frame, but the vertical sheathing planks are placed side by side so as to form a continuous retaining wall. This method of sheathing is used for noncohesive and unstable soils.

Tight sheathing is the same as close sheathing except the vertical sheathing planks are interlocked. This method of sheathing is used for saturated soils. Steel sheet piling is sometimes used instead of wood planking.

Trench shields (Fig. 3.6.14.2) are heavily braced steel or wood boxes which are moved along the trench bottom as excavation and pipe laying progress. Trench boxes are used to protect workmen installing pipe in stable soil where the trenches are deep and unsheathed. Trench shields are also used in lieu of other methods of sheathing and shoring for shallow excavations where the sides of the shield can extend from the trench bottom to the ground surface. When trench shields are used, care must be taken to avoid pulling the pipe apart or disrupting the bedding as the shield is moved.

Improper removal of sheathing can reduce soil friction along the trench wall and increase the backfill load on the pipe. Therefore, sheathing hould be removed in increments as the backfill is placed. Additional compaction of the backfill material may be necessary to fill any voids left by the sheathing.

Figure 3.6.14.2 Using trench box in unstable soil for concrete pipe installation. (*By permission: American Concrete Pipe Association, Irving, Texas*)

3.6.15 Foundation Preparation

A stable and uniform **foundation** is necessary for satisfactory performance of any pipe. The foundation must have sufficient load bearing capacity to maintain the pipe in proper alignment and sustain the loads imposed. The foundation should be checked for hard or soft spots. Where undesirable foundations exist, they should be stabilized by **ballasting** or **soil modification.**

Ballasting requires removal of undesirable foundation material and replacement with select materials such as sand, gravel, crushed rock, slag, or suitable earth backfill. The depth, gradation and size of the ballast depends on the specific material used and the amount of stabilization required. The ballast is usually well graded from coarse to fine, having a size not more than one inch per foot of pipe diameter with three inches maximum and placed to a minimum depth of four inches.

Soil modification involves the addition of select material to the native soil. Crushed rock, gravel, sand, slag or other durable inert materials with a maximum size of three inches is worked into the subsoil to accomplish the required stabilization. Soil modification can also be accomplished by the addition of lime, cement or chemicals to the soil.

Adequate foundation stability is difficult to evaluate by visual observation. However, when concrete pipe is set on the foundation with little or no care exercised to provide a bearing surface, the weight of the pipe exerts a pressure of approximately 1000 pounds per square foot. This pressure is about the same pressure a 200 pound man would exert when standing on one foot. If the foundation can support men working in the trench without sinking into the soil, the foundation should be stable enough to support the pipe and maintain it in proper alignment.

3.6.16 Pipe Bedding

Once a stable and uniform foundation is provided, it is necessary to prepare the **bedding** in accordance with the requirements of the plans, specifications or standard drawings. An important function of the bedding is to assure uniform support along the barrel of each pipe section. The bedding distributes the load reaction around the lower periphery of the pipe. The required supporting strength of the pipe is directly related to this load distribution, and several types of bedding have been established to enable specification of pipe strengths during the design phase of the project.

Pipe set on a flat foundation without bedding results in high load concentration at the bottom of the pipe (Fig. 3.6.16.1). Bedding the pipe so that the bottom reaction is distributed over 50 percent of the outside horizontal span of the pipe results in a 36 percent increase in supporting strength; a 60 percent distribution results in a 73 percent increase for the same amount of settlement; and a 100 percent distribution results in as much as a 150 percent increase depending on sidefill compaction.

If the pipe strength specified for a particular project is based on a design assumption that at least 60 percent of the outside horizontal span of the pipe is bedded, and the pipe is actually set on a flat foundation, a pipe strength significantly greater than specified would be required. The bedding being constructed needs to be continuously compared with the requirements in the plans or specifications.

Improved construction practices enable variations in the methods used to attain the required bearing surface at the bottom of the pipe. The general classifications of beddings are presented as a guideline of what is reasonably obtainable in the figures shown on pages 141–143. Based on current construction practices, it is generally more practical and economical to over excavate and bed the pipe on select materials, rather than shape the subgrade to conform to the shape of the pipe.

CLASS D BEDDING

Class D bedding is used only with circular pipe. Little or no care is exercised either to shape the foundation surface to fit the lower part of the pipe exterior or to fill all spaces under and around the pipe with granular materials. However, the gradient of the bed should be smooth and true to the established grade. This class of bedding also includes the case of pipe on rock foundations in which an earth cushion is provided under the pipe but is so shallow that the pipe, as it settles under the influence of vertical load, approaches contact with the rock.

Figure 3.6.16 (*By permission: American Concrete Pipe Association, Irving, Texas*)

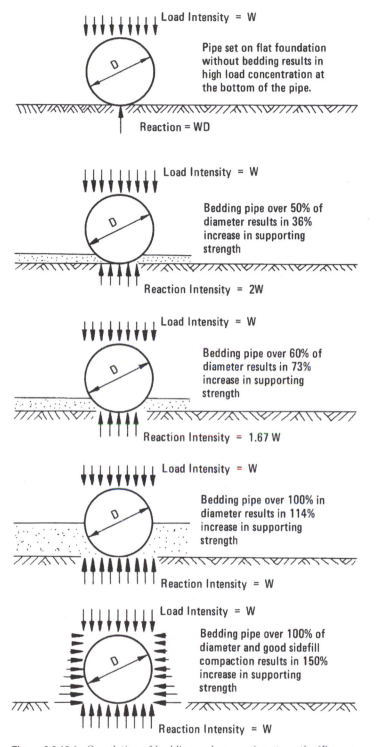

Load Intensity = W

Pipe set on flat foundation without bedding results in high load concentration at the bottom of the pipe.

Reaction = WD

Load Intensity = W

Bedding pipe over 50% of diameter results in 36% increase in supporting strength

Reaction Intensity = 2W

Load Intensity = W

Bedding pipe over 60% of diameter results in 73% increase in supporting strength

Reaction Intensity = 1.67 W

Load Intensity = W

Bedding pipe over 100% in diameter results in 114% increase in supporting strength

Reaction Intensity = W

Load Intensity = W

Bedding pipe over 100% of diameter and good sidefill compaction results in 150% increase in supporting strength

Reaction Intensity = W

Figure 3.6.16.1   Correlation of bedding and supporting strength. (*By permission: American Concrete Pipe Association, Irving, Texas*)

CLASS C BEDDING

With a **shaped subgrade** the pipe is bedded with ordinary care in a soil foundation, shaped to fit the lower part of the pipe exterior with reasonable closeness for a width of at least 50 percent of the outside diameter for a circular pipe, and one-tenth of the outside pipe rise for arch pipe, elliptical pipe and box sections. For trench installations the sides and area over the pipe are filled with lightly compacted backfill to a minimum depth of six inches above the top of the pipe. For embankment installations the pipe should not project more than 90 percent of the vertical height of the pipe above the bedding.

A **granular foundation** is used only with a circular pipe, and consists of a compacted granular material or densely compacted backfill placed on a flat bottom trench. The bedding material should extend up the sides for a height of at least one-sixth the outside diameter of the pipe.

CLASS B BEDDING

For a **shaped subgrade** with granular foundation the botom of the excavation is shaped to conform to the pipe surface but at least two inches greater than the outside dimensions of the pipe. The width should be sufficient to allow six-tenths of the outside pipe diameter for circular pipe and seven-tenths of the outside span for arch and elliptical pipe to be bedded in fine granular fill placed in the shaped excavation. Densely compacted backfill should be placed at the sides of the pipe to a depth of at least 12 inches above the top of the pipe.

A **granular foundation** without shaping is used only with circular pipe. The pipe is bedded in compacted granular material placed on the flat trench bottom. The granular bedding has a minimum thickness, and should extend at least halfway up the pipe at the sides. The remainder of the side fills, and a minimum depth of 12 inches over the top of the pipe, should be filled with densely compacted material.

CLASS A BEDDING

A **concrete craddle** bedding is used only with circular pipe. The pipe is bedded in nonreinforced or reinforced concrete having thickness, d, and extending up the sides for a height equal to one-fourth the outside diameter. The cradle should have a minimum width at least equal to the outside diameter of the pipe plus eight inches. The backfill above the cradle is densely compacted and extends 12 inches above the crown of the pipe. In rock, especially where blasting is likely in the adjacent vicinity, the concrete cradle should be cushioned from the shock of the blasting which can be transmitted through the rock.

The **concrete arch** is an alternate to the concrete cradle for trench installations. The pipe is bedded in carefully compacted granular material having the minimum thickness and extending halfway up the sides of the pipe. The top half of the pipe is covered with nonreinforced or reinforced concrete having a minimum thickness over the top of the pipe of one-fourth the inside pipe diameter. The arch should have a minimum width at least equal to the outside diameter of the pipe plus eight inches.

Figure 3.6.17 Compacting bedding under pipe haunches. (*By permission: American Concrete Pipe Association, Irving, Texas*)

BEDDING MATERIALS

Materials for bedding should be selected to intimate contact can be obtained between the bed and the pipe. Since most granular materials will shift to attain this contact as the pipe settles, an ideal load distribution can be realized. Granular materials are coarse sand, pea gravel or well graded crushed rock.

With the development of mechanical methods for subgrade preparation, pipe installation, backfilling and compaction, excellent results have been obtained with pipe installed on a flat bottom foundation and backfilled with well graded, job excavated soil. If this method of bedding is used, it is essential that the bedding material be uniformly compacted under the haunches of the pipe.

Where ledge rock, compacted rocky or gravel soil, or other unyielding foundation material is encountered, beddings should be modified as follows:

- For Class B and C beddings, subgrades should be excavated or over excavated, if necessary, so a uniform foundation free of protruding rocks is provided.

- Special care may be necessary with Class A beddings or other unyielding foundations to cushion pipe from shock when blasting can be anticipated in the area.

Figure 3.6.17   (*cont.*)

3.6.18 Joint Sealants

Rubber gaskets are either flat gaskets which may be cemented to the pipe tongue or spigot during manufacture, **O-ring** gaskets which are recessed in a groove on the pipe tongue or spigot and then confined by the bell or groove after the joint is completed, or **roll-on** gaskets which are placed around the tongue or spigot and then rolled into position as the tongue or spigot is inserted into the bell or groove.

When gaskets are used, dust, dirt and foreign matter must be removed from the joint surfaces. For flat and O-ring gaskets, the gasket and joint surfaces are lubricated with a lubricant recommended by the manufacturer. The lubricant can be applied with a brush, cloth pad, sponge or glove. For all gaskets not cemented to the pipe, a smooth round object should be inserted under the gasket and run around the circumference two or three times to equalize stretch in the gasket.

Mastic sealants consist of bitumen and inert mineral filler and are usually cold applied. The joint surfaces are thoroughly cleaned, dried and prepared in accordance with the manufacturer's recommendations. A sufficient amount of sealant is used to fill the annular joint space with some squeeze out. Better workability of the mastic sealant can be obtained during cold weather if the mastic and joint surfaces are heated.

Cement sealants consist of portland cement paste or mortar made with a mixture of portland cement, sand and water. The joint surface is thoroughly cleaned and soaked with water immediately before the joint is made. A layer of paste or mortar is placed in the lower portion of the bell or groove end of the installed pipe and on the upper portion of the tongue or spigot of the pipe section to be installed. The tongue or spigot is then inserted into the bell or groove of the installed pipe until the sealant material is squeezed out. Any annular joint space between the adjacent pipe ends is filled with mortar and the excess mortar on the inside of the pipes wiped and finished to a smooth surface.

Portland cement mortar bands are sometimes specified around the exterior of the pipe joint. A slight depression is excavated in the bedding material to enable mortar to be placed underneath the pipe. The entire external joint surface is then cleaned and soaked with water. Special canvas or cloth **diapers** can be used to hold the mortar as it is placed. Backfill material should be immediately placed around the pipe.

Rubber-mastic bands also can be used around the exterior of the pipe joint. The bands are stretched tightly around the barrel of the pipe and held firmly in place by the weight of the backfill material.

Regardless of the specific type of joint sealant used, each joint should be checked to be sure all pipe sections are in a **home** position. For joints sealed with ruber gaskets, it is important to follow the manufacturer's installation recommendations to assure that the gasket is properly positioned and under compression.

Figure 3.6.18 (*By permission: American Concrete Pipe Association, Irving, Texas*)

3.6.19 Jointing Procedures

Joints for pipe sizes up to 24 inches in diameter can usually be assembled by means of a bar. The axis of the pipe section to be installed should be aligned as closely as possible to the axis of the last installed pipe section, and the tongue or spigot end inserted slightly into the bell or groove. A bar is then driven into the bedding and wedged against the bottom bell or groove end of the pipe section being installed. A wood block is placed horizontally across the end of the pipe to act as a fulcrum point and to protect the joint end during assembly. By pushing the top of the vertical bar forward, lever action pushes the pipe into a home position.

When jointing larger diameter pipe, and when granular bedding is used, mechanical pipe pullers are required. Several types of pipe pullers or **come along** devices have been developed (Fig. 3.6.19.1), but the basic force principles are the same.

When jointing small diameter pipe, a chain or cable is wrapped around the barrel of the pipe a few feet behind the tongue or spigot and fastened with a grab hook or other suitable connecting device. A lever assembly is anchored to the installed pipe, several sections back from the last installed section, and connected by means of a chain or cable to the rab hook on the pipe to be installed. By pulling the lever back, the tongue or spigot of the pipe being jointed is pulled into the bell or groove of the last installed pipe section. To maintain close control over the alignment of the pipe, a laying sling can be used to lift the pipe section slightly off the bedding foundation.

Jointing Pipe with a Bar.

Figure 3.6.19 (*By permission: American Concrete Pipe Association, Irving, Texas*)

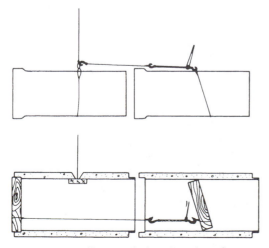

Figure 3.6.19.1 (*By permission: American Concrete Pipe Association, Irving, Texas*)

3.6.20 Jacking and Tunneling

In all jacking and tunneling operations, direction and distance are carefully established prior to beginning the operation. The first step is the excavation of **jacking pits** or shafts at each end of the proposed line. The pit must be of sufficient size to provide ample working space for the jacking head, jacks, jacking frame, reaction blocks, spoil removal and one or two sections of pipe (Fig. 3.6.20.1). Provisions are made for the erection of guide rails in the bottom of the pit. For large pipe it is desirable to set the rails in a concrete slab. If drainage is to be discharged from the jacking pit, a collection sump and drainage pump are required.

The number and capacity of jacks depend on the type of soil, size of the pipe and jacking distance. The jacks are placed on both sides of the pipe so the resultant jacking force is slightly below the springline. Use of a lubricant, such as bentonite, to coat the outside of the pipe is helpful in reducing frictional resistance and preventing the pipe from setting when forward movement is interrupted. Because of the tendency of soil friction to increase with time, it is usually desirable to continue jacking operations without interruption until completion.

Correct alignment of the guide frame, jacks and backstop is necessary for uniform distribution of the axial jacking force around the periphery of the pipe. By assuring that the pipe ends are parallel, and the jacking force properly distributed through the jacking frame to the pipe and parallel with the axis of the pipe, localized stress concentrations are avoided. A jacking head is often used to transfer the force from the jacks or jacking frame to the pipe. In addition to protecting the end of the pipe, a jacking head helps keep the pipe on proper line by maintaining equal force around the circumference of the pipe. Use of a cushion material, such as plywood, between adjacent pipe sections provides uniform load distribution.

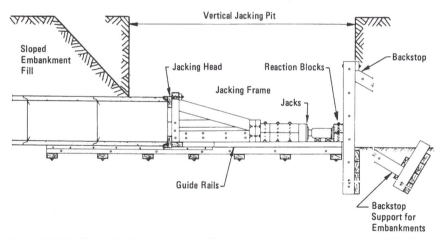

Figure 3.6.20.1   Typical jacking operation. (*By permission: American Concrete Pipe Association, Irving, Texas*)

Figure 3.6.20.1   Tunneling and jacking precast concrete pipe. (*cont.*)

3.6.21 Alignment for Radius Pipe

Curved alignment with deflected straight pipe.

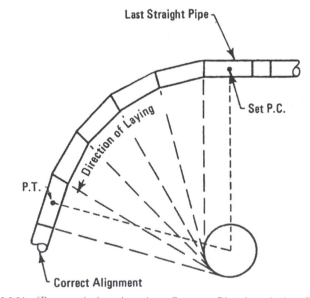

Figure 3.6.21 *(By permission: American Concrete Pipe Association, Irving, Texas)*

INFILTRATION AND EXFILTRATION

The United States Environmental Protection Agency, EPA, defines infiltration as the *volume of groundwater entering sewers and building sewer connections from the soil, through defective joints, broken or cracked pipe, improper connections, manhole walls, etc.* **Inflow** is defined as the *volume of any kind of water discharged into sewer lines from such sources as roof leaders, cellar and yard area drains, foundation drains, commercial and industrial so-called "clean water" discharges, drains from springs and swampy areas, etc. It does not include and is distinguished from, infiltration.* The most effective way to control infiltration and at the same time to assure the structural integrity and proper installation of the new sewer is to establish and enforce a maximum leakage limit as a condition of job acceptance. Limits may be stated in terms of water leakage for infiltration and exfiltration tests and should include a maximum allowable **test section rate** and a maximum allowable system **average rate.**

3.6.22 Infiltration Testing

The **infiltration test** (Fig. 3.6.22.1) is intended to measure the watertightness of a sewer to the infiltration of groundwaer and therefore, **is only applicable if the water table level is above the top of the pipe.** Although the test is a realistic method there are inherent difficulties in applying the test criteria because of seasonal fluctuations in the water table.

EPA states that *current information indicates that a maximum allowable test section rate of 200 gallons per inch of diameter per mile of pipe per day can normally be achieved with little or no effect on construction costs. This limit is appropriate when the average depth of the groundwater is between 2 feet and 6 feet over the crown of the pipe.*

The effect of soil permeability and increased depth of groundwater on infiltration allowances must be considered. EPA recommends *infiltration allowance should reflect a consideration of the permeability of the soil, particularly the envelope around the pipe, in additin to the depth of the groundwater over the pipe.* To adjust the infiltration allowance to reflect the effect of permeable soil, an average head of six feet of groundwater over the pipe is established as the base head. With heads of more than six feet, the infiltration limit is increased by the ratio of the square root of the actual average head to the square root of the base head. For example, with permeable soil and an average groundwater head of 12 feet, the 200 gallons per inch of diameter per mile of pipe per day infiltration limit should be increased by the ratio of the square root of the actual average head, 12 feet, to the square root of the base head, six feet, which results in an allowable infiltration limit of 282 gallons per inch of diameter per mile of pipe per day.

The American Society for Testing and Materials has developed Standard C 969, Standard Practice for Infiltration and Exfiltration Acceptance Testing of Installed Precast Concrete Pipe Sewer Lines. The criteria established in this recommended practice for infiltration testing follows the preceding recommendations of EPA. The practice includes procedures for preparing the sewer, conducting the infiltration test and evaluating the test results.

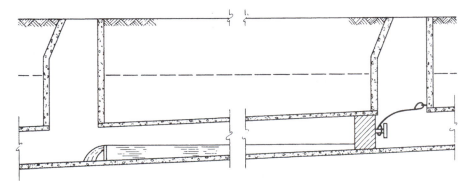

Figure 3.6.22.1 Typical infiltration test arrangement. (*By permission: American Concrete Pipe Association, Irving, Texas*)

The infiltration test is usually conducted between adjacent manholes with the upstream end of the sewer bulkheaded to isolate the test section. All service laterals, stubs and fittings are plugged or capped at the connection to the test section to prevent the entrance of groundwater. A V-notch weir or other suitable measuring device is installed in the pipe at the downstream manhole. When steady flow occurs over the weir, leakage is determined by direct reading from graduations on the weir or by converting the flow quantity to gallons per unit length of pipe per unit of time.

3.6.23 Exfiltration Testing

An **exfiltration test** (Fig. 3.6.23.1) may be specified if the groundwater level is below the top of the pipe. However, as cautioned by EPA, *Exfiltration limits, to achieve similar control of infiltration, should be set somewhat higher than the infiltration limits. Accordingly, the combined leakage from the pipe and manholes could be fixed at about 200 gallons per inch of diameter per mile of pipe per day when the average head on the test section is 3 feet.*

The effect of increased test head is accounted for in a manner similar to that for the infiltration test. The increased exfiltration limit is determined by multiplying 200 by the ratio of the square root of the actual average test head to the square root of the assumed base head, three feet. For example, if the actual average test head is eight feet, the exfiltration test allowance is 327 gallons per inch of diameter per mile of pipe per day, including manholes.

EPA states, *manholes may be tested separately and independently. An allowance for manholes of 0.1 gallon per hour per foot of diameter per foot of head would be appropriate.* To exclude both manholes from the test, it is necessary to bulkhead the outlet pipe of the upstream manhole. Provision must be made in the upstream bulkhead for a standpipe.

The criteria established in ASTM Standard C 969 for exfiltration testing follows the preceding recommendations of EPA. The practice includes procedures for preparing the sewer, conducting the exfiltration test and evaluating the test results.

The test is usually conducted between adjacent manholes. All service laterals, stubs and fittings within the test section are plugged or capped to withstand the test pressure. If manholes are included in the test, the inlet pipe to each manhole is bulkheaded and the test section filled with water through the upstream manhole. Water is added at a steady rate, to allow air to escape from the sewer, until the water is at the specified level above the crown of the pipe. After absorption into the pipe and the manhole has stabilized, the water in the upstream manhole is brought to the test level. At the end of the test period, drop in water elevation is measured and the loss of water calculated, or the water restored to the initial test level and the amount added used to determine leakage rate.

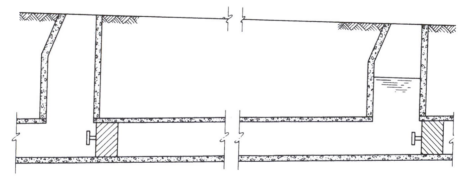

Figure 3.6.23.1   Typical exfiltration test arrangement. (*By permission: American Concrete Pipe Association, Irving, Texas*)

3.6.24 Low Pressure Air Testing

The **low pressure air test** (Fig 3.6.24.1) was developed to detect pipe that has been damaged or improperly jointed by measuring the rate at which air under pressure escapes from a section of the sewer. In applying low pressure air testing to sanitary sewers designed to carry fluid under gravity conditions, it is necessary to distinguish between air losses inherent in the type of pipe material used and those caused by damaged pipe or defective joints. Because of the physical difference between air and water, and the difference in behavior under pressure conditions, air loss does not necessarily mean there will be water infiltration. Depending on the porosity, moisture content and wall thickness of the pipe, a well constructed concrete sewer line which is impervious to water may still have some air loss through the pipe wall. EPA cautions, it should be understood that no direct mathematical correlation has been found applicable between air test limits and water exfiltration limits.

The American Society for Testing and Materials has developed Standard C 924, Standard Practice for Testing Concrete Pipe Sewer Lines by Low Pressure Air Test Method. The practice includes procedures for preparing the sewer, conducting the test and evaluating the test results. Based on field experience, the recommended practice establishes for each pipe size an appropriate allowable air loss which enables detection of any significant leak. The practice limits the maximum diameter of pipe to be tested to 24 inches for safety reasons. Larger pipe may be more conveniently accepted by visual inspection and individual joint testing.

Two air test methods are the **constant pressure method** and the **time pressure drop method.** With either method the section of pipe tested is plugged at each end. The ends of laterals, stubs and fittings to be included in the test section are plugged to prevent air leakage, and securely braced to prevent possible blowouts. Test equipment consists of valves and pressure gages to control air flow and to monitor pressure within the test section. The internal pressure is raised to a specified level and allowed to stabilize. For the constant pressure method, the air loss at a specified pressure is determined by use of an air flow measuring device. In the more commonly used pressure drop method, the air supply is disconnected and the time required for the pressure to drop to a certain level is determined. This time interval is then used to compute the rate of air loss.

In applying low pressure air testing to sanitary sewers intended to carry fluid under gravity conditions, several important factors should be understood and precautions followed:

- The air test is intended to detect defects in construction and pipe or joint damage and not to measure infiltration or exfiltration leakage under service conditions.

- Air test criteria are presently limited to pipe 24 inches in diameter and smaller.

- Plugs should be securely braced and not removed until all pressure has been released.

- For safety, no one should be allowed in the trench or manhole during the test.

- Testing apparatus should be equipped with a pressure relief device.

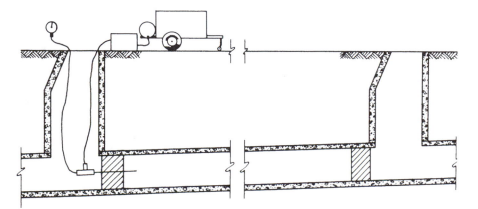

Figure 3.6.24.1 Typical low pressure air test arrangement. (*By permission: American Concrete Pipe Association, Irving, Texas*)

3.7.0 Design of Culverts

Culverts are constructed to convey water under a highway, railroad, canal, or other embankment.

Hydraulic design of a culvert is primarily influenced by **headwater depth.** As apparent in Fig. 3.7.0.1, if the flow does not pass through the culvert in sufficient quantity, water may go over the embankment or back up and cause flooding with damage and inconvenience upstream. The factors that affect the discharge from the culvert are headwater, pipe size, tailwater, roughness of the pipe, pipe slope, pipe length and inlet geometry.

Culvert flows are classified as being under either **inlet** or **outlet control,** that is, the discharge is controlled by either the outlet or inlet characteristics. The hydraulic capacity of a culvert is determined differently for the two types of control.

Under inlet control, the major factors are:

- Cross-sectional area of the culvert barrel

- Inlet geometry

- Headwater depth

Outlet control involves the additional consideration of:

- Tailwater depth

- Slope, roughness and length of the culvert barrel

Complex hydraulic computations can be used to determine the probable type of flow under which a culvert will operate for a given set of conditions. However, headwater depths for both inlet and outlet control can be easily calculated. The higher value determines the type of control and anticipated headwater depth.

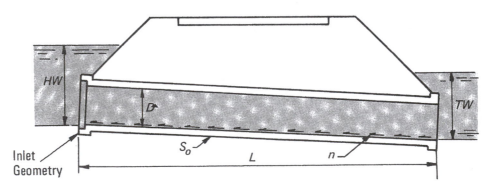

Figure 3.7.0.1 Factors affecting culvert discharge. (*By permission: American Concrete Pipe Association, Irving, Texas*)

3.7.1 Culverts Flowing with Inlet Control

Inlet control means that the discharge capacity is controlled at the culvert entrance, and the condition will exist as long as water can flow through the culvert at a greater rate than water can enter. Inlet conditions depend upon the depth of headwater, *HW*, entrance geometry, area, and shape and type of inlet edge. Types of inlet controlled flow are shown in Fig. 3.7.1.1. As indicated in Fig. 3.7.1c, a mitered or beveled entrance moves the control downstream to approximately the top of the miter.

For inlet control conditions, neither roughness, length of the culvert barrel, nor the outlet conditions are factors in determining culvert capacity. The barrel slope has some effect on discharge but a slope adjustment is considered minor and can be neglected for conventional culverts flowing with inlet control. Headwater-discharge relationships for various types of culverts flowing with inlet control have been developed in model studies and verified in some instances by prototype tests. The data were analyzed and nomographs for determining culvert capacity for inlet control were developed. These nomographs give headwater-discharge relationships for most conventional culverts flowing with inlet control through a range of headwater depths or discharges.

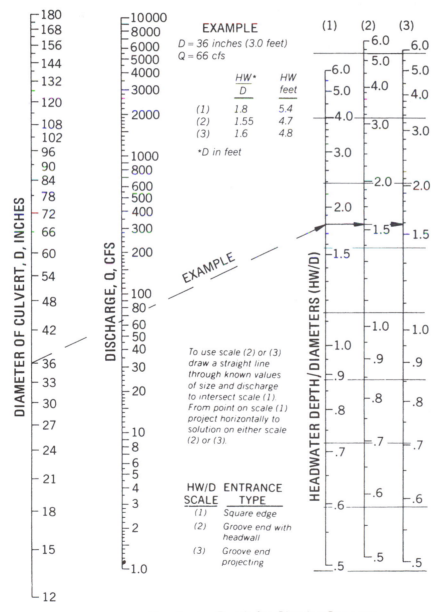

Headwater Depth for Circular Concrete
Pipe Culverts with Inlet Control.

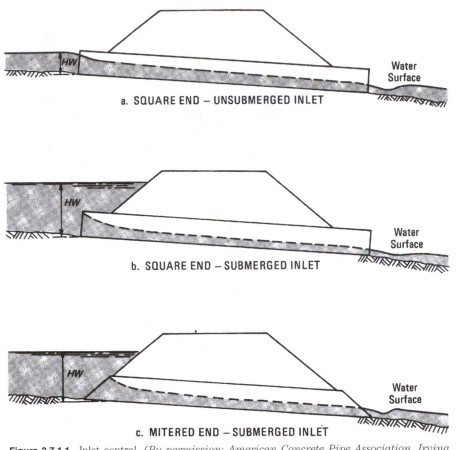

a. SQUARE END – UNSUBMERGED INLET

b. SQUARE END – SUBMERGED INLET

c. MITERED END – SUBMERGED INLET

Figure 3.7.1.1 Inlet control. (*By permission: American Concrete Pipe Association, Irving, Texas*)

3.7.2 Aluminum Box Culverts

The Solution for Small Bridges

CONTECH® Aluminum Box Culverts are a practical and cost-efficient solution for small bridge replacement. They have a lower installed cost because they are faster and easier to install than cast-in-place concrete structures. There are no forms to set and remove, no delays due to curing time, large installation crews are unnecessary and no special equipment is needed. Also, no heavy cranes are required, as with precast concrete structures.

These wide-span, low-rise structures are available in a large range of standard sizes (from 8'-9" span x 2'-6" rise to 25'-5" span x 10'-2" rise) that permit a minimum cover of only 17 inches for all spans.

Faster Installation Means Lower Installed Cost

Closing roads for bridge replacement causes extensive traffic detours, so minimizing installation time is critical. Aluminum Box Culverts may be quickly erected in place and are usually ready to be backfilled in a matter of hours. For faster installation, Aluminum Box Culverts can be completely assembled off site while the site is being prepared. Light equipment can then be used to set them in place.

Available Sizes for CONTECH Aluminum Box Culverts

Structure Number	Span "A" (Ft.-In.)	Rise "B" (Ft.-In.)	Area (Sq. Ft.)
1	8-9	2-6	18.4
2	9-2	3-3	25.4
3	9-7	4-1	32.6
4	10-0	4-10	40.2
5	10-6	5-7	48.1
6	10-11	6-4	56.4
7	11-4	7-2	65.0
8	10-2	2-8	23.0
9	10-7	3-5	31.1
10	10-11	4-3	39.5
11	11-4	5-0	48.2
12	11-8	5-9	57.2
13	12-1	6-7	66.4
14	12-5	7-4	76.0
15	11-7	2-10	28.1
16	11-11	3-7	37.4
17	12-3	4-5	46.9
18	12-7	5-2	56.6
19	12-11	6-0	66.6
20	13-3	6-9	76.9
21	13-0	3-0	33.8
22	13-4	3-10	44.2
23	13-7	4-7	54.8
24	13-10	5-5	65.6
25	14-1	6-2	76.6
26	14-5	3-3	40.0
27	14-8	4-1	51.5
28	14-10	4-10	63.2
29	15-1	5-8	75.1
30	15-4	6-5	87.2
31	15-6	7-3	99.4
32	15-9	8-0	111.8
33	15-10	3-6	46.8
34	16-0	4-3	59.5
35	16-2	5-1	72.3
36	16-4	5-11	85.2
37	16-6	6-8	98.3
38	16-8	7-6	111.5
39	16-10	8-3	124.8
40	17-9	3-10	54.4
41	18-2	4-7	68.3
42	18-7	5-4	82.5
43	19-0	6-1	97.1
44	19-5	6-11	111.9
45	19-10	7-8	127.1
46	20-3	8-5	142.6
47	19-1	4-2	63.3
48	19-5	4-11	78.3
49	19-9	5-8	93.6
50	20-1	6-6	109.2
51	20-6	7-3	125.0
52	20-10	8-1	141.2
53	21-2	8-10	157.6
54	20-4	4-6	73.1
55	20-7	5-3	89.2
56	20-11	6-1	105.5
57	21-3	6-10	122.1
58	21-6	7-8	139.0
59	21-10	8-5	156.0
60	22-1	9-3	173.3
61	21-7	4-11	83.8
62	21-10	5-8	101.0
63	22-1	6-6	118.4
64	22-3	7-3	135.9
65	22-6	8-1	153.7
66	22-9	8-10	171.6
67	23-0	9-8	189.8
68	22-9	5-4	95.5
69	23-0	6-1	113.7
70	23-2	6-11	132.1
71	23-4	7-8	150.6
72	23-6	8-6	169.3
73	23-8	9-3	188.1
74	23-10	10-1	207.0
75	24-0	5-9	108.2
76	24-1	6-6	127.5
77	24-3	7-4	146.8
78	24-4	8-2	166.2
79	24-5	8-11	185.7
80	24-7	9-9	205.3
81	24-8	10-6	225.0
82	25-2	6-2	122.0
83	25-2	7-0	142.2
84	25-3	7-9	162.4
85	25-4	8-7	182.6
86	25-4	9-5	202.9
87	25-5	10-2	223.3

Figure 3.7.2 *(By permission: CONTECH Construction Products Inc., Middletown, Ohio)*

3.8.0 U.S. Flood Frequency Data

The United States Geological Survey, **USGS,** has developed a nationwide series of water-supply papers. These reports contain tables of maximum known floods and charts for estimating the probable magnitude of floods of frequencies ranging from 1.1 to 50 years. Figure 3.8.0 shows the USGS regions, district and principal field offices, and the applicable water-supply paper numbers. Most states have adapted and consolidated those parts of the water-supply papers which pertain to specific hydrologic areas within their boundaries.

It is recommended that the culvert design flow be determined by methods based on USGS data. If such data are not available for a particular culvert location, flow quantities may be determined by the Rational Method or by statistical methods using records of flow and runoff.

Figure 3.8.0 Nationwide flood frequency projects. (*By permission: American Concrete Pipe Association, Irving, Texas*)

3.8.1 One Hour Expected Rainfall Maps

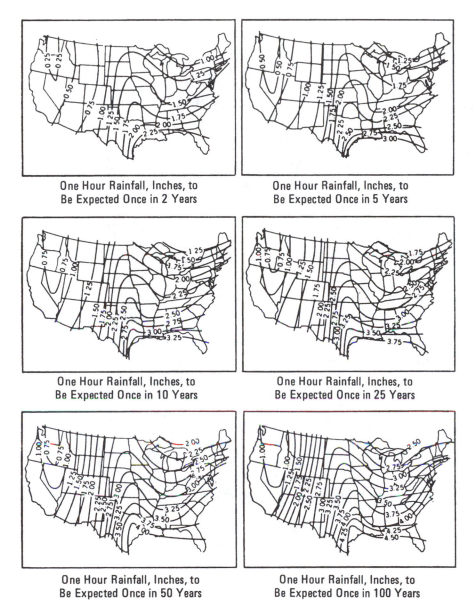

One Hour Rainfall, Inches, to
Be Expected Once in 2 Years

One Hour Rainfall, Inches, to
Be Expected Once in 5 Years

One Hour Rainfall, Inches, to
Be Expected Once in 10 Years

One Hour Rainfall, Inches, to
Be Expected Once in 25 Years

One Hour Rainfall, Inches, to
Be Expected Once in 50 Years

One Hour Rainfall, Inches, to
Be Expected Once in 100 Years

Figure 3.8.1 (*By permission: American Concrete Pipe Association, Irving, Texas*)

3.8.2 Dates When Cities Installed Concrete Sewer Pipes

Mohawk, New York	1842		Louisville, Kentucky	1889
Newark, New Jersey	1867		South Bend, Indiana	1889
Hudson, New York	1867		Wausau, Wisconsin	1890
Nashua, New Hampshire	1868		Superior, Wisconsin	1891
Chelsea, Massachusetts	1869		Salt Lake City, Utah	1893
New Haven, Connecticut	1869		Atlanta, Georgia	1895
Savannah, Georgia	1870		Augusta, Georgia	1895
San Francisco, California	1870		Springfield, Massachusetts	1895
Milwaukee, Wisconsin	1871		Bangor, Maine	1896
New London, Connecticut	1873		Eau Claire, Wisconsin	1899
Portland, Maine	1873		Oakland, California	1899
Kokomo, Indiana	1874		Kalamazoo, Michigan	1903
North Adams, Massachusetts	1874		Trenton, New Jersey	1903
Grand Rapids, Michigan	1875		Salem, Oregon	1903
Chicopee, Massachusetts	1875		Janesville, Wisconsin	1904
Indianapolis, Indiana	1878		Jackson, Michigan	1905
Lockport, New York	1879		Cleveland, Ohio	1906
Rutland, Vermont	1879		Springfield, Ohio	1907
Utica, New York	1879		Sacramento, California	1908
Bay City, Michigan	1880		Richmond, Indiana	1908
North Manchester, Indiana	1881		Albuquerque, New Mexico	1908
Racine, Wisconsin	1883		Spokane, Washington	1908
Appleton, Wisconsin	1883		Watertown, New York	1909
Oshkosh, Wisconsin	1884		Tacoma, Washington	1909
Los Angeles, California	1885		Lancaster Pennsylvania	1910
Minneapolis, Minnesota	1886		La Crosse, Wisconsin	1910
St. Paul, Minnesota	1886		Bakersfield, California	1911
Portland, Oregon	1888		Syracuse, New York	1911
Helena, Montana	1889		Lansing, Michigan	1912
Galveston, Texas	1889		Kansas City, Missouri	1912
Greeley, Colorado	1889		Everett, Washington	1913
Waukesha, Wisconsin	1889			

Figure 3.8.2 *(By permission: American Concrete Pipe Association, Irving, Texas)*

3.8.3 Dates When States Began Using Concrete Pipe Highway Culverts

State	Date	State	Date
Alabama	Prior to 1921	Montana	1922
Arizona	1927	Nebraska	1917
Arkansas	Prior to 1927	Nevada	1924
California	1925	New Hampshire	1927
Colorado	Prior to 1920	New Jersey	1920
Connecticut	1923	New Mexico	1939
Delaware	1920	New York	1923
District of Columbia	1935	North Carolina	1916
Florida	1933	North Dakota	1923
Georgia	1920	Ohio	1929
Hawaii	1925	Oklahoma	Prior to 1924
Idaho	1925	Oregon	1917
Illinois	1930	Pennsylvania	Prior to 1916
Indiana	1919	Rhode Island	1936
Iowa	1906	South Carolina	1918
Kansas	1917	South Dakota	1920
Kentucky	1923	Tennessee	1920
Louisiana	1921	Texas	Prior to 1930
Maine	1928	Utah	About 1920
Maryland	1908	Vermont	About 1921
Massachusetts	1924	Virginia	1923
Michigan	1918	Washington	About 1919
Minnesota	1921	West Virginia	1922
Mississippi	1926	Wisconsin	About 1930
Missouri	1930	Wyoming	1935

3.8.4 Dates When Concrete Pipe Railway Culverts Were First Installed

Railway	Date	Railway	Date
New York Central (Michigan Division)	1870	Elgin, Joliet & Eastern	1912
Louisville and Nashville	1870	Great Northern	1912
Boston and Maine	1903	Illinois Central	1912
Canadian Pacific	1903	Mississippi Central	1912
Delaware-Lackawanna & Western	1905	The Pennsylvania Railroad	1912
Southern Pacific Company	1905	The Western Pacific	1912
Chicago, Burlington & Quincy	1906	Atlantic Coast Line	1913
Duluth, Missabe & Iron Range	1908	Chicago, Milwaukee, St. Paul & Pacific R.R. Co.	1913
Northern Pacific	1909	Atchison, Topeka & Santa Fe	1914
Baltimore & Ohio	1910	Chicago & Illinois Midland	1914
Canadian National Railways	1910	Central of Georgia	1914
Kansas City Southern	1910	Southern Railway Co.	1914
Sand Springs Railway Co.	1911		

Figure 3.8.4 (*By permission: American Concrete Pipe Association, Irving, Texas*)

3.9.0 Reference Specifications for Corrugated Metal Pipe

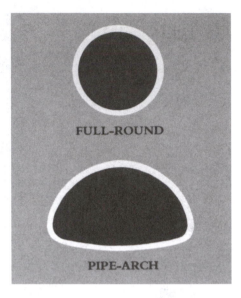

Reference Specifications

Material	Galvanized Steel	ASTM A 929 and AASHTO M218
	ALUMINIZED STEEL Type 2	ASTM A 929 and AASHTO M274
	FIBER-BONDED Steel	ASTM A 885
	Polymer-Coated Steel	ASTM A 742 and AASHTO M246
Pipe	Steel (Galvanized and ALUMINIZED STEEL Type 2)	ASTM A 760 and AASHTO M36
	Steel (Polymeric)	ASTM A 762 and AASHTO M245
Coating/Lining	Asphalt and Concrete	ASTM A 849 and AASHTO M190
Design	Steel	ASTM A 796 and AASHTO Standard Specification for Highway Bridges, Section 12
Installation	Steel	ASTM A 798 and AASHTO Standard Specification for Highway Bridges, Section 26

Figure 3.9.0   *(By permission: CONTECH Construction Products Inc., Middletown, Ohio)*

3.9.1 Approximate Weight per Foot/Corrugated Steel Pipe

(Estimated Average Weights—Not for Specification Use)

1½" x ¼" Corrugation

Inside Diameter, in.	Specified Thickness, in.	Galvanized & ALUMINIZED	Full Coated
6	0.052	4	5
	0.064	5	6
8	0.052	5	6
	0.064	6	7
10	0.052	6	7
	0.064	7	8

2⅝" x ½" Corrugation

Inside Diameter, in.	Specified Thickness, in.	Galvanized & ALUMINIZED*	Full Coated	Coated & PAVED-INVERT	SMOOTH-FLO	HEL-COR CL
12	0.052	8	10	13		
	0.064	10	12	15		
	0.079	12	14	17		
15	0.052	10	13	16	26	
	0.064	12	15	18	28	
	0.079	15	18	21	31	
18	0.052	12	16	19	31	
	0.064	15	19	22	34	
	0.079	18	22	25	37	
21	0.052	14	18	23	36	
	0.064	17	21	26	39	
	0.079	21	25	30	43	
24	0.052	15	20	26	41	
	0.064	19	24	30	45	65
	0.079	24	29	35	50	69
	0.109	33	38	44	59	77
30	0.052	20	26	32	51	
	0.064	24	30	36	55	82
	0.079	30	36	42	60	87
	0.109	41	47	53	72	96
36	0.052	24	31	39	50	
	0.064	29	36	44	65	98
	0.079	36	43	51	75	104
	0.109	49	56	64	90	116
	0.138	62	69	77	100	127
42	0.052	28	36	45	71	
	0.064	34	42	51	77	114
	0.079	42	50	59	85	121
	0.109	57	65	74	100	135
	0.138	72	80	89	115	149
48	0.064	38	48	57	85	128
	0.079	48	58	67	95	138
	0.109	65	75	84	112	154
	0.138	82	92	101	129	170
	0.168	100	110	119	147	186
54	0.079	54	65	76	105	156
	0.109	73	84	95	124	173
	0.138	92	103	114	143	191
	0.168	112	123	134	163	209
60	0.109	81	92	106	140	192
	0.138	103	114	128	162	212
	0.168	124	135	149	183	232
66	0.109	89	101	117	160	211
	0.138	113	125	141	180	233
	0.168	137	149	165	210	255
72	0.138	123	137	154	210	254
	0.168	149	163	180	236	278
78	0.168	161	177	194	260	302
84	0.168	173	190	208	270	325

3" x 1" or 5" x 1" Corrugation

Inside Diameter, in.	Specified Thickness, in.	Galvanized & ALUMINIZED*	Full Coated	Coated & PAVED-INVERT	SMOOTH-FLO	HEL-COR CL
54	0.064	50	66	84	138	197
	0.079	61	77	95	149	207
	0.109	83	100	118	171	226
	0.138	106	123	140	194	245
	0.168	129	146	163	217	264
60	0.064	55	73	93	153	218
	0.079	67	86	105	165	229
	0.109	92	110	130	190	251
	0.138	118	136	156	216	272
	0.168	143	161	181	241	293
66	0.064	60	80	102	168	240
	0.079	74	94	116	181	252
	0.109	101	121	143	208	276
	0.138	129	149	171	236	299
	0.168	157	177	199	264	322
72	0.064	66	88	111	183	262
	0.079	81	102	126	197	275
	0.109	110	132	156	227	301
	0.138	140	162	186	257	326
	0.168	171	193	217	288	351
78	0.064	71	95	121	198	
	0.079	87	111	137	214	298
	0.109	119	143	169	246	326
	0.138	152	176	202	279	353
	0.168	185	209	235	312	380
84	0.064	77	102	130	213	
	0.079	94	119	147	230	321
	0.109	128	154	182	264	351
	0.138	164	189	217	300	379
	0.168	199	224	253	335	409
90	0.064	82	109	140	228	
	0.079	100	127	158	246	
	0.109	137	164	195	283	376
	0.138	175	202	233	321	406
	0.168	213	240	271	359	438
96	0.064	87	116	149	242	
	0.079	107	136	169	262	
	0.109	147	176	209	302	401
	0.138	188	217	250	343	433
	0.168	228	257	290	383	467
102	0.064	93	124	158	258	
	0.079	114	145	179	279	
	0.109	155	186	220	320	426
	0.138	198	229	263	363	460
	0.168	241	272	306	406	496
108	0.079	120	153	188	295	
	0.109	165	198	233	340	
	0.138	211	244	279	386	487
	0.168	256	289	324	431	525
114	0.079	127	162	199	312	
	0.109	174	209	246	359	
	0.138	222	257	294	407	514
	0.168	271	306	343	456	554
120	0.109	183	220	259	378	
	0.138	234	271	310	429	541
	0.168	284	321	360	479	583
126	0.138	247	285	326	452	
132	0.138	259	299	342	474	
	0.168	314	354	397	529	
138	0.138	270	312	357	495	
	0.168	328	370	415	553	
144	0.168	344	388	435	579	

* Weights for polymer-coated pipe are 1% to 4% higher, varying by gage.

Figure 3.9.1   (*By permission: CONTECH Construction Products Inc., Middletown, Ohio*)

3.9.2 Corrugated Aluminum/Galvanized Steel Pipe Specifications—Aluminized Steel or Galvanized Steel Corrugated Metal Pipe

Diameter (Inches)	Weight (Pounds/Lineal Foot)		
	Specified Thickness and Gage		
	(.064") 16	(.079") 14	(.109") 12
18	14.9	18.3	
21	17.4	21.4	29.1
24	19.9	24.4	35.9
30	24.9	30.5	41.5
36	29.8	36.7	49.9
42	34.8	42.8	58.3
48	39.8	48.9	66.6
54	44.8	55.0	74.9
60	49.8	61.2	83.3
66		67.2	91.5
72		73.2	99.6
78		79.3	107.9
84			116.1
90			124.4
96			132.6
102			139.8

3.9.3 Aluminum Corrugated Pipe

Diameter (Inches)	Weight (Pounds/Lineal Foot)			
	Specified Thickness and Gage			
	(.060") 16	(.075") 14	(.105") 12	(.135") 10
18	5.2	6.4		
21	6.0	7.5	10.5	
24	6.9	8.6	12.0	
30	8.6	10.7	15.0	
36	10.3	12.8	18.0	22.4
42	12.1	15.0	21.0	26.2
48		17.1	24.0	29.9
54		19.3	27.0	33.7
60			30.0	37.4
66			33.0	41.1
72			36.0	44.8
78				48.5
84				52.2

Figure 3.9.3 (*By permission: CONTECH Construction Products Inc., Middletown, Ohio*)

3.9.4 Handling Weight of Aluminized Steel Corrugated Pipe

Handling Weight for **ALUMINIZED STEEL** Type 2 or Galvanized Steel
ULTRA FLO

Diameter (Inches)	Weight (Pounds/Lineal Foot) Specified Thickness and Gage		
	(0.064") 16	(0.079") 14	(0.109") 12
18	15		
21	18		
24	20		
30	25		
36	30	37	
42	35	43	59
48	40	49	67
54	45	55	75
60	50	61	83
66		67	92
72		73	100
78		80	108
84			116
90			125
96			133
102			140

3.9.5 Handling Weight of Aluminum Corrugated Pipe

Handling Weight for **ALUMINUM** ULTRA FLO

Diameter (Inches)	Weight (Pounds/Lineal Foot) Specified Thickness and Gage			
	(0.060") 16	(0.075") 14	(0.105") 12	(0.135") 10
18	5			
21	6			
24	7	9		
30	9	11	15	
36	11	13	18	23
42	12	15	21	26
48		17	24	30
54		19	27	34
60			30	37
66			33	41
72			36	45
78				49
84				52

Figures 3.9.4 and 3.9.5 *(By permission: CONTECH Construction Products Inc., Middletown, Ohio)*

3.9.6 Mannings "*n*"—Coefficient of Roughness

Values of Coefficient of Roughness (n)

For CONTECH Corrugated Steel Pipe (Manning's Formula)

	Annular Corrugations $2^2/_3$" x $^1/_2$" All Diameters	Helical* Corrugation									
		$1^1/_2$" x $^1/_4$" (11, 12)		Helical—$2^2/_3$" x $^1/_2$"							
		8 in.	10 in.	12 in.	15 in.	18 in.	24 in.	36 in.	48 in.	60 in. and Larger	
Unpaved	0.024	0.012	0.014	0.011	0.012	0.013	0.015	0.018	0.020	0.021	
PAVED-INVERT	0.021						0.014	0.017	0.020	0.019	
SMOOTH-FLO	0.012						0.012	0.012	0.012	0.012	
HEL-COR CL	0.012						0.012	0.012	0.012	0.012	

	Annular 3" x 1"	Helical—3" x 1"							
		36 in.	42 in.	48 in.	54 in.	60 in.	66 in.	72 in.	78 in. and Larger
Unpaved	0.027	0.022	0.022	0.023	0.023	0.024	0.025	0.026	0.027
PAVED-INVERT	0.023	0.019	0.019	0.020	0.020	0.021	0.022	0.022	0.023
SMOOTH-FLO	0.012			0.012	0.012	0.012	0.012	0.012	0.012
HEL-COR CL	0.012			0.012	0.012	0.012	0.012	0.012	0.012

	Annular 5" x 1"	Helical—5" x 1"						
		48 in.	54 in.	60 in.	66 in.	72 in.	78 in. and Larger	
Unpaved	0.025	0.022	0.022	0.023	0.024	0.024	0.025	
PAVED-INVERT	0.022	0.019	0.019	0.020	0.021	0.021	0.022	
SMOOTH-FLO	0.012			0.012	0.012	0.012	0.012	
HEL-COR CL	0.012			0.012	0.012	0.012	0.012	

* Tests on helically corrugated pipe demonstrate a lower coefficient of roughness than for annually corrugated steel pipe. Pipe-arches have the same roughness characteristics as their equivalent round pipes.

The values in the above table are based on standard helical pipe manufactured from a 24-inch net width strip of steel.

Figure 3.9.6 *(By permission: CONTECH Construction Products Inc., Middletown, Ohio)*

3.9.7 Heights of Cover for H20, H25, E80, Loads—1½–2⅔ inches

1½″ x ¼″ Height-of-Cover Limits for Corrugated Steel Pipe H 20, H 25 Load, and E 80 Live Loads

Diameter, Inches	Minimum Cover, Inches	Maximum Cover, Feet	
		Specified Thickness, Inches	
		0.052	0.064
6	12	388	486
8	12	291	365
10	12	233	292

2⅔″ x ½″ Height-of-Cover Limits for Corrugated Steel Pipe

H 20 and H 25 Live Loads

Diameter or Span, Inches	Minimum Cover, Inches	Maximum Cover, Feet					
		Specified Thickness, Inches					
		0.052	0.064	0.079	0.109	0.138	0.168
12	12	198	248	310			
15		158	199	248			
18		132	166	207			
21		113	142	178	249		
24		99	124	155	218		
30		79	99	124	174		
36		66	83	103	145	186	
42		56	71	88	124	160	195
48			62	77	109	140	171
54				66	93	122	150
60					79	104	128
66					68	88	109
72						75	93
78							79
84	12						66

E 80 Live Loads

Diameter or Span, Inches	Minimum Cover, Inches	Maximum Cover, Feet					
		Specified Thickness, Inches					
		0.052	0.064	0.079	0.109	0.138	0.168
12	12	198	248	310			
15		158	199	248			
18		132	166	207			
21		113	142	178	249		
24		99	124	155	218		
30		79	99	124	174		
36		66	83	103	145	186	
42		56	71	88	124	160	195
48	12		62	77	109	140	171
54	18			66	93	122	150
60					79	104	128
66					68	88	109
72	18					75	93
78	24						79
84	24						66

H 20 and H 25 Live Loads, Pipe-Arch

Size		Minimum Structural Thickness, Inches	Minimum Cover, Inches	Maximum Cover, Feet
Round Equivalent, Inches	Span x Rise, Inches			2 Tons/Ft.² Corner Bearing Pressure
15	17 x 13	0.064	12	16**
18	21 x 15	0.064		15**
21	24 x 18	0.064		
24	28 x 20	0.064		
30	35 x 24	0.064		
36	42 x 29	0.064		
42	49 x 33	0.064*		
48	57 x 38	0.064*		
54	64 x 43	0.079*		
60	71 x 47	0.109*		
66	77 x 52	0.109*		
72	83 x 57	0.138*	12	15**

E 80 Live Loads, Pipe-Arch

Size		Minimum Structural Thickness, Inches	Minimum Cover, Inches	Maximum Cover, Feet
Round Equivalent, Inches	Span x Rise, Inches			3 Tons/Ft.² Corner Bearing Pressure
15	17 x 13	0.079	24	22
18	21 x 15	0.079		
21	24 x 18	0.109		
24	28 x 20	0.109		
30	35 x 24	0.138		
36	42 x 29	0.138		
42	49 x 33	0.138*		
48	57 x 38	0.138*		
54	64 x 43	0.138*		
60	71 x 47	0.138*	24	22

* These values are based on the AISI Flexibility Factor limit (0.0433 x 1.5) for pipe-arch. Due to variations in arching equipment, thicker gages may be required to prevent crimping of the haunches.

** These values were calculated using K = 0.86 as adopted in the AISI Handbook, Fourth Edition, 1993.

Figure 3.9.7 *(By permission: CONTECH Construction Products Inc., Middletown, Ohio)*

3.9.8 Heights of Cover for H20, H25, E80 loads 5″ × 1″ and 3″ × 1″

5″ x 1″ or 3″ x 1″ Height-of-Cover Limits for Corrugated Steel Pipe

H 20 and H 25 Live Loads

Diameter or Span, Inches	Minimum Cover, Inches	Maximum Cover, Feet				
		Specified Thickness, Inches				
		0.064	0.079	0.109	0.138	0.168
54	12	56	70	98	126	155
60		50	63	88	114	139
66		46	57	80	103	126
72		42	52	73	95	116
78		39	48	68	87	107
84		36	45	63	81	99
90		33	42	59	76	93
96	12	31	39	55	71	87
102	18	29	37	52	67	82
108			35	49	63	77
114			32	45	58	71
120			30	41	54	66
126				39	50	62
132				36	47	57
138				33	43	53
144	18				39	49

To obtain maximum cover for 3″ x 1″, increase these values by 13%.

E 80 Live Loads

Diameter or Span, Inches	Minimum Cover, Inches	Maximum Cover, Feet				
		Specified Thickness, Inches				
		0.064	0.079	0.109	0.138	0.168
54	18	56	70	98	126	155
60		50	63	88	114	139
66		46	57	80	103	126
72	18	42	52	73	95	116
78	24	39	48	68	87	107
84		36	45	63	81	99
90		33[1]	42	59	76	93
96	24	31[1]	39	55	71	87
102	30	29[1]	37	52	67	82
108			35	49	63	77
114			32[1]	45	58	71
120	30		30[1]	41	54	66
126	36			39	50	62
132				36	47	57
138				33[1]	43	53
144	36				39	49

To obtain maximum cover for 3″ x 1″, increase these values by 13%.

[1] These diameters in these gages require additional minimum cover.

3″ x 1″ Pipe-Arch Height-of-Cover Limits for Corrugated Steel Pipe

H 20 and H 25 Live Loads

Size		Minimum Specified Thickness, Inches*	Minimum Cover, Inches	Maximum Cover, Feet
Equivalent Pipe Diameter	Span x Rise, Inches			2 Tons/Ft.² Corner Bearing Pressure
48	53 x 41	0.079	12	25
54	60 x 46	0.079	15	25
60	66 x 51	0.079	15	25
66	73 x 55	0.079	18	24
72	81 x 59	0.079	18	21
78	87 x 63	0.079	18	20
84	95 x 67	0.079	18	20
90	103 x 71	0.079	18	20
96	112 x 75	0.079	21	20
102	117 x 79	0.109	21	19
108	128 x 83	0.109	24	19
114	137 x 87	0.109	24	19
120	142 x 91	0.138	24	19

E 80 Live Loads

Size		Minimum Specified Thickness, Inches*	Minimum Cover, Inches	Maximum Cover, Feet
Equivalent Pipe Diameter	Span x Rise, Inches			2 Tons/Ft.² Corner Bearing Pressure
48	53 x 41	0.079	24	25
54	60 x 46	0.079	24	25
60	66 x 51	0.079	24	25
66	73 x 55	0.079	30	24
72	81 x 59	0.079	30	21
78	87 x 63	0.079	30	18
84	95 x 67	0.079	30	18
90	103 x 71	0.079	36	18
96	112 x 75	0.079	36	18
102	117 x 79	0.109	36	17
108	128 x 83	0.109	42	17
114	137 x 87	0.109	42	17
120	142 x 91	0.138	42	17

* Some 3″ x 1″ and 5″ x 1″ minimum gages shown for pipe-arch are due to manufacturing limitations.

Note: Sewer gage (trench conditions) tables for corrugated steel pipe can be found in the AISI book "Modern Sewer Design," 2nd Edition, 1990, pp. 202-203. These tables may reduce the minimum gage due to a higher flexibility factor allowed for a trench condition.

Figure 3.9.8   *(By permission: CONTECH Construction Products Inc., Middletown, Ohio)*

3.9.9 Pipe-Arch Height-of-Cover Limits

5" x 1" Pipe-Arch Height-of-Cover Limits for Corrugated Steel Pipe

H 20 and H 25 Live Loads

Equivalent Pipe Diameter	Span x Rise, Inches	Minimum Specified Thickness, Inches*	Minimum Cover, Inches	Maximum Cover, Feet 2 Tons/Ft.² Corner Bearing Pressure
72	81 x 59	0.109	18	21
78	87 x 63	0.109	18	20
84	95 x 67	0.109	18	20
90	103 x 71	0.109	18	20
96	112 x 75	0.109	21	20
102	117 x 79	0.109	21	19
108	128 x 83	0.109	24	19
114	137 x 87	0.109	24	19
120	142 x 91	0.138	24	19

E 80 Live Loads

Equivalent Pipe Diameter	Span x Rise, Inches	Minimum Specified Thickness, Inches*	Minimum Cover, Inches	Maximum Cover, Feet 2 Tons/Ft.² Corner Bearing Pressure
72	81 x 59	0.109	30	21
78	87 x 63	0.109	30	18
84	95 x 67	0.109	30	18
90	103 x 71	0.109	36	18
96	112 x 75	0.109	36	18
102	117 x 79	0.109	36	17
108	128 x 83	0.109	42	17
114	137 x 87	0.109	42	17
120	142 x 91	0.138	42	17

* Some 3" x 1" and 5" x 1" minimum gages shown for pipe-arch are due to manufacturing limitations.

3.9.10 Construction Loads for Temporary Construction Vehicles

Construction Loads

For temporary construction vehicle loads, an extra amount of **compacted cover** may be required over the top of the pipe. The height-of-cover shall meet the minimum requirements shown in the table below. The use of heavy construction equipment necessitates greater protection for the pipe than finished grade cover minimums for normal highway traffic.

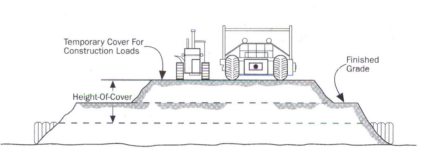

Pipe Span, Inches	Minimum Cover (feet) for Indicated Axle Loads (kips)			
	18-50	50-75	75-110	110-150
12-42	2.0	2.5	3.0	3.0
48-72	3.0	3.0	3.5	4.0
78-120	3.0	3.5	4.0	4.0
126-144	3.5	4.0	4.5	4.5

General Guidelines for Minimum Cover Required for Heavy Off-Road Construction Equipment

Minimum cover may vary, depending on local conditions. The contractor must provide the additional cover required to avoid damage to the pipe. Minimum cover is measured from the top of the pipe to the top of the maintained construction roadway surface.

Corrugated steel pipe is used extensively to rehabilitate failing reinforced concrete pipe.

Figure 3.9.10   *(By permission: CONTECH Construction Products Inc., Middletown, Ohio)*

3.9.11 Grate Inlet Hydraulics

Slotted drain can be used to intercept runoff in any one of the following ways:

1. Installed in a typical curb-and-gutter as a slot-on-grade to intercept flow from streets and highways.

2. Installed in a typical curb-and-gutter at a sag or low point in a grade to accommodate carryover from preceding slots on a grade and to intercept surface runoff sloped to the gutter.

3. Installed in wide, flat areas to intercept overland or sheet flow (as on a parking lot).

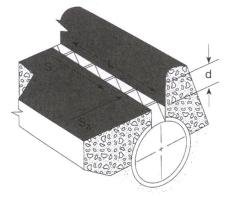

1. Slot-on-grade in typical curb and gutter

For any given discharge, Q, cross slope, S_x, and longitudinal gutter slope, S, the required slotted drain length can be determined from the nomograph (Figure A) on Page 7.

It is common practice in curb and gutter drainage design to carry over up to 35% of the total discharge, Q_d, to the next inlet. See Figure B on Page 7 for the carryover efficiency curve.

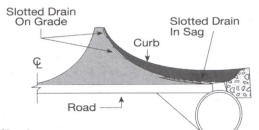

2. Slot-in-sag

When slotted drain is installed in a sag or at a low point in grade, the length of the slot is calculated from the formula:

$$L_r = \frac{1.4Q}{\sqrt{d}}$$

Normally a safety factor of two is used in a sag. $L_s = 2 \times L_r$.

3. Overland sheet flow

Slotted drain is used effectively to intercept runoff from wide, flat areas such as parking lots, highway medians—even tennis courts and airport taxiways. In these installations, the drain is placed transverse to the direction of flow, so that the open slot acts as a weir intercepting all of the flow uniformly along the entire length of the drain. The water is not collected and channeled against a berm (curb), as required by slot-on-grade installations.

Slotted drain has been tested for overland flow (sheet flow). These results are published in Report No. FHWA-RD-79-106 by the Federal Highway Administration.

The test system was designed to supply at least 0.025 cfs per foot, which corresponds to a rainstorm of 15 inches per hour over a 72-foot-wide roadway (six lanes).

At the design discharge of 0.025 cfs per foot, the total flow fell through the slot as a weir flow. The tests included flows up to 0.040 cfs per foot of slot.

Typical cross section of combination slot-on-grade and a slot-in-sag

Slopes ranged from a longitudinal slope of 9% and a Z of 16, to a longitudinal slope of 5% and a Z of 48.

The water ranged in depth from 0.38 inches to 0.56 inches. Velocity ranged from 1.263 ft/sec to 0.857 ft/sec.

Even at the maximum discharge of 0.04 cfs per foot and maximum slopes, nearly all the flow passed through the slot. Only some water hitting the spacer plates and splashing over was not intercepted.

Using:

$Q = CIA$, then $A = \dfrac{Q}{CI}$

Where:

Q given as 0.04 ft^3/sec/ft of slotted drain

C = 0.80 to 0.95 for asphalt pavement

After the engineer selects C and I (ft/sec), A can be calculated. Since Q is per foot of slot, A is ft^2/ft of slot. Since the units for A can be reduced to feet, the value of A is also the distance parallel to the flow intercepted by one foot of slot.

Example:

C = 0.85

I = 10 in./hr or 0.0002315 ft/sec

$A = \dfrac{0.04\ ft^3/sec/ft}{0.85 \times 0.0002315\ ft/sec}$

$A = 203.3\ ft^2$/ft

Therefore, at the selected C and I, one foot of slot will intercept flow from 203.3 linear feet upstream of the slot.

Figure 3.9.11   *(By permission: CONTECH Construction Products Inc., Middletown, Ohio)*

3.9.12 Grate Inlet Installation

One of CONTECH Slotted Drain's primary advantages is economical design and installation. Unlike typical parking lots that require grades to be sloped in four directions for *each grate*, a parking lot with slotted drain requires only one transverse and one longitudinal slope for the entire drainage area. That translates to a lower-cost installation for the contractor, and less stake-out for the engineer. And because of slotted drain's efficiency in removing surface water, fewer collectors—and fewer laterals under the roadway—are needed.

When properly installed, slotted drain provides a better-looking, more efficient drainage system at a lower cost. Photographs illustrate the basic steps for installing slotted drain as a curb inlet. The procedure is basically the same in other applications.

Experience has shown the best method for installing slotted drain is to place it in a contoured trench, level it to grade, backfill with high slump concrete, then pave with the desired surfacing material. The pipe must be placed so the slanted spacer plates are facing upstream, leaning against the direction of surface flow.

In long runs, construction joints should be placed perpendicular to the pipe runs.

Modified HUGGER Bands or the closure plate jointing system is used to join adjacent pipes.

Your CONTECH Sales Engineer can discuss various installation techniques with you.

Contoured trench

Installing slotted drain in a contoured trench reduces the amount of concrete required.

Leveling to grade

Contractors have developed many methods for positioning slotted drain in the trench prior to backfilling.

One popular method is to use positioning devices fastened through the slotted opening with a toggle bolt or similar device.

Another method involves leveling the pipe with granular material at selected points along the drain pipe. The remaining area is backfilled with high slump concrete.

Grate extensions

Grate extensions are available if the height needs to be raised at a future time.

Slotted drain is used often in interstate highway widening projects.

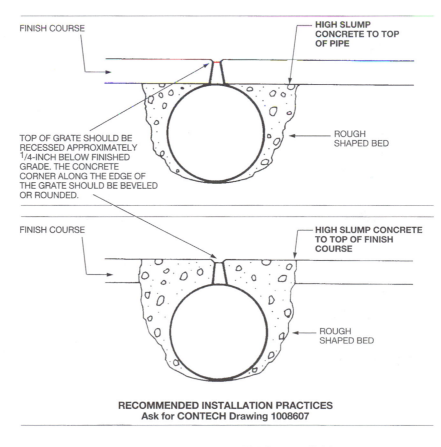

FINISH COURSE

HIGH SLUMP CONCRETE TO TOP OF PIPE

TOP OF GRATE SHOULD BE RECESSED APPROXIMATELY 1/4-INCH BELOW FINISHED GRADE. THE CONCRETE CORNER ALONG THE EDGE OF THE GRATE SHOULD BE BEVELED OR ROUNDED.

ROUGH SHAPED BED

FINISH COURSE

HIGH SLUMP CONCRETE TO TOP OF FINISH COURSE

ROUGH SHAPED BED

RECOMMENDED INSTALLATION PRACTICES
Ask for CONTECH Drawing 1008607

Figure 3.9.12 *(By permission: CONTECH Construction Products Inc., Middletown, Ohio)*

High slump concrete

After the slotted drain has been leveled to grade, it is important that a high slump concrete or lean grout (minimum 750 psi compressive strength) be used as backfill. The high slump concrete helps ensure a uniform foundation and side support, and transfers the live load to the surrounding earth. In non-live load areas, A-1-a AASHTO M145 backfill or cement stabilized sand is sufficient.

Surfacing

Once the slotted drain is backfilled with high slump concrete, cover the slotted opening before surfacing, and leave it covered until the paving operation is complete. Duct tape, metal strips, or lumber can be used to cover the slot.

See Variable Height Grate Drawing 1008732

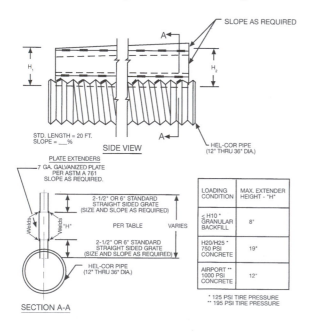

SLOPE AS REQUIRED

STD. LENGTH = 20 FT.
SLOPE = ___%

SIDE VIEW

HEL-COR PIPE
(12" THRU 36" DIA.)

PLATE EXTENDERS
7 GA. GALVANIZED PLATE
PER ASTM A 761
SLOPE AS REQUIRED.

2-1/2" OR 6" STANDARD
STRAIGHT SIDED GRATE
(SIZE AND SLOPE AS REQUIRED)

PER TABLE   VARIES

2-1/2" OR 6" STANDARD
STRAIGHT SIDED GRATE
(SIZE AND SLOPE AS REQUIRED)

HEL-COR PIPE
(12" THRU 36" DIA.)

SECTION A-A

LOADING CONDITION	MAX. EXTENDER HEIGHT - "H"
≤ H10 * GRANULAR BACKFILL	8"
H20/H25 * 750 PSI CONCRETE	19"
AIRPORT ** 1000 PSI CONCRETE	12"

* 125 PSI TIRE PRESSURE
** 195 PSI TIRE PRESSURE

Figure 3.9.12 *(cont.)*

3.9.13 Grate Welding and Hugger Band Details

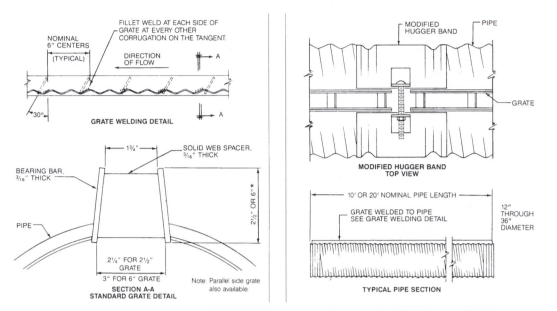

Figure 3.9.13 (*By permission: CONTECH Construction Products Inc., Middletown, Ohio*)

3.9.14 Heights of Cover

Table 3
ALUMINIZED STEEL Type 2 or Galvanized Steel ULTRA FLO HS 20 Live Load

Diameter (Inches)	Minimum/Maximum Cover (Feet) Specified Thickness and Gage		
	(0.064") 16	(0.079") 14	(0.109") 12
18	1.0/68		
21	1.0/58		
24	1.0/51		
30	1.0/41		
36	1.0/34	1.0/48	
42	1.0/29	1.0/41	1.0/69
48	1.0/25	1.0/36	1.0/60
54	1.25/22	1.25/32	1.0/53
60	1.25/20*	1.25/28	1.0/48
66		1.5/26	1.25/44
72		1.5/24*	1.25/40
78		1.75/22*	1.5/37
84			1.75/34
90			2.0/32*
96			2.0/30*
102			2.5/28*

Table 4
Aluminum ULTRA FLO HS 20 Live Load

Diameter (Inches)	Minimum/Maximum Cover (Feet) Specified Thickness and Gage			
	(0.060") 16	(0.075") 14	(0.105") 12	(0.135") 10
18	1.0/55			
21	1.0/47			
24	1.0/41	1.0/57		
30	1.25/33	1.0/45	1.0/73	
36	1.5/27	1.25/38	1.0/61	1.0/86
42	1.75/23*	1.5/32	1.25/52	1.0/74
48		2.0/28*	1.5/46	1.25/65
54		2.0/25*	1.75/40	1.25/57
60			2.0/36*	1.5/52
66			2.0/33*	1.75/47
72			2.25/30*	2.0/43
78				2.5/39*
84				2.5/34*

Table 5
Steel ULTRA FLO Pipe-Arch HS 20 Live Load

Equiv. Pipe Dia. (In.)	Span (In.)	Rise (In.)	Minimum/Maximum Cover (Feet) Specified Thickness and Gage		
			(0.064") 16	(0.075") 14	(0.109") 12
18	20	16	1.0/15		
21	23	19	1.0/15		
24	27	21	1.0/15		
30	33	26	1.0/15	1.0/15	
36	40	31	1.0/15	1.0/15	
42	46	36	M.L.[4]	M.L.[4]	1.0/15
48	53	41	M.L.[4]	M.L.[4]	1.0/15
54	60	46	M.L.[4]	M.L.[4]	1.0/15
60	66	51	M.L.[4]	M.L.[4]	1.25/15

Table 6
Aluminum ULTRA FLO Pipe-Arch HS 20 Live Load

Equiv. Pipe Dia (In.)	Span (In.)	Rise (In.)	Minimum/Maximum Cover (Feet) Specified Thickness and Gage			
			(0.060") 16	(0.075") 14	(0.105") 12	(0.135") 10
18	20	16	1.0/15			
21	23	19	1.0/15			
24	27	21	1.25/15	1.0/15		
30	33	26	1.50/15	1.25/15	1.0/15	
36	40	31	1.75/15	1.50/15	1.25/15	1.0/15
42	46	36			1.50/15	1.25/15
48	53	41			1.75/15	1.25/15
54	60	46			2.0/15	1.50/15
60	66	51			2.0/15	1.75/15

NOTES (Tables 3, 4, 5, and 6)

1. Allowable minimum cover is measured from top of pipe to bottom of flexible pavement or top of pipe to top of rigid pavement. Minimum cover in invade areas must be maintained.
2. All heights of cover are based on trench conditions. If embankment conditions exist, there may be restrictions on gages for the large diameters. Your CONTECH Sales Engineer can provide further guidance for a project in embankment conditions.
3. Tables 3, 4, 5 and 6 are for HS-20 loading only. For heavy construction loads, higher minimum compacted cover may be needed. See Table 7.
4. All steel ULTRA FLO is installed in accordance with ASTM A798 "Installing Factory-Made Corrugated Steel Pipe for Sewers and Other Applications."
5. Heights of cover are for 3/4" x 3/4" x 7-1/2" external rib corrugation.

   *These sizes and gage combinations are installed in accordance with ASTM A796 paragraphs 17.2.3 and ASTM A798. For aluminum ULTRA FLO refer to ASTM B790 and B788.

Table 7

Heavy Construction Loads Minimum Height of Cover Requirements for *Construction Loads* on ULTRA FLO Pipe

Diameter/Span (Inches)	Axle Load (Kips)			
	>32≤50	50≤75	75≤110	110≤150
	Steel 3/4" x 3/4" x 7-1/2"			
15-42	2.0 ft.	2.5 ft.	3.0 ft.	3.0 ft.
48-72	3.0 ft.	3.0 ft.	3.5 ft.	4.0 ft.
78-108	3.0 ft.	3.5 ft.	4.0 ft.	4.5 ft.
	Aluminum 3/4" x 3/4" x 7-1/2"			
15-42	2.5 ft.	3.0 ft.	3.5 ft.	3.5 ft.

NOTES (Tables 5 and 6 only)

1. The foundation in the corners should allow for 4,000 psf corner bearing pressure.
2. Maximum cover shown for all pipe arch is 15 feet.
3. Larger size pipe-arches may be available on special order.
4. M.L. (Heavier gage is required to prevent crimping at the haunches.)

Figure 3.9.14   (*By permission: CONTECH Construction Products Inc., Middletown, Ohio*)

3.9.15 Load Deflection Relationships

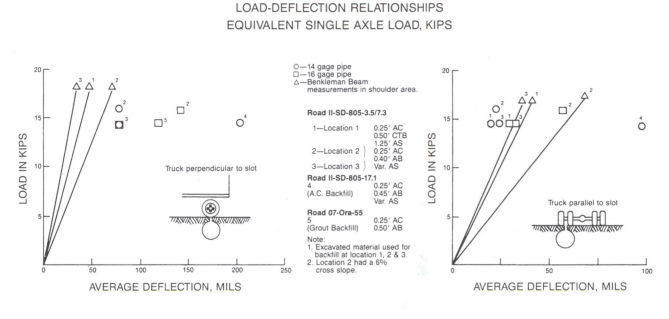

Figure 3.9.15   (*By permission: CONTECH Construction Products Inc., Middletown, Ohio*)

3.9.16 Nomograph—Slotted Drain on Grade in Curb and Gutter

Figure A: **NOMOGRAPH—SLOTTED DRAIN ON GRADE IN CURB AND GUTTER**

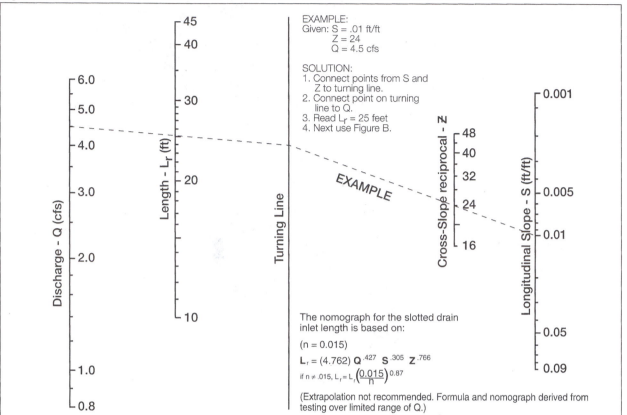

EXAMPLE:
Given: S = .01 ft/ft
       Z = 24
       Q = 4.5 cfs

SOLUTION:
1. Connect points from S and Z to turning line.
2. Connect point on turning line to Q.
3. Read L_r = 25 feet
4. Next use Figure B.

The nomograph for the slotted drain inlet length is based on:

(n = 0.015)

$$L_r = (4.762) \, Q^{.427} \, S^{.305} \, Z^{.766}$$

if n ≠ .015, $L_r = L_r \left(\dfrac{0.015}{n}\right)^{0.87}$

(Extrapolation not recommended. Formula and nomograph derived from testing over limited range of Q.)

Definitions

S — Longitudinal gutter or channel slope, ft/ft

S_x — Transverse slope, ft/ft

Z — Transverse slope reciprocal $\left(\dfrac{1}{S_x}\right)$, ft/ft

d — Depth of flow over the slot, ft

L — Length of slot, ft

L_r — Length of slot required for total interception, ft

L_s — A selected length of slot, ft

Q — Discharge into inlet, cfs

Q_d — Total discharge at an inlet, cfs

Q_a — An allowed discharge, cfs

C — Runoff coefficient

I — Rainfall intensity, ft/sec

A — Area drained

Figure B: **SLOTTED DRAIN CARRYOVER EFFICIENCY**

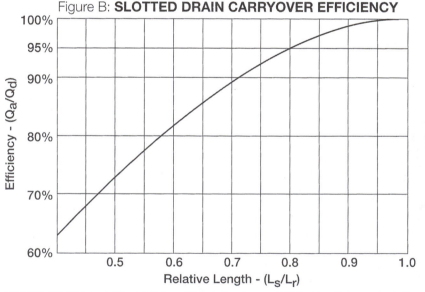

Example: Solution from Figure A is L_r = 25 feet. If a standard 20-foot length is used, relative length ratio L_s/L_r = 20 ft/25 ft = 0.8. From Figure B with a relative length ratio of 0.8, the efficiency is 95%. Ninety-five percent of the flow is intercepted by the 20-foot length, and 5' runs down the gutter to be intercepted by the next slot.

Figure 3.9.16 *(By permission: CONTECH Construction Products Inc., Middletown, Ohio)*

3.9.17 Groundwater Recharge Systems

RECHARGE SYSTEMS

The recharge retention method of storm water control is frequently a viable addition to detention facilities. In sites where soils drain well and the water table is low enough to accommodate a recharge system, such water management techniques may be the most economical means of managing runoff.

SUBSURFACE EFFECTS OF DEVELOPMENT

One potential effect of urbanization is water table reduction. Low water tables may weaken vegetation, undermine the soil structure and, in low-lying coastal areas, permit salt water intrusion.

Water table reductions can be minimized by a surface water collection system and a subsurface recharge network. These systems share many of the basic requirements for detention systems that regulate storm sewer discharge, except that much of the accumulated runoff water is allowed to percolate into the subsoil.

The most efficient underground recharge system is a perforated corrugated metal pipe surrounded by very porous materials such as uniformly graded stone. Typically, the same types of materials used around subdrainage pipes are excellent for recharge systems.

For protection against soil infiltration, the entire system is usually enclosed by a high-quality, soil-compatible geotextile to provide long-term filtration.

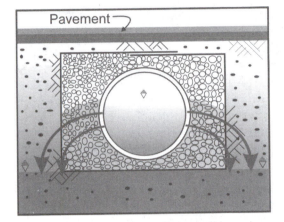

PRESERVING WATER TABLE LEVELS

Recharge systems may be designed to release detained storm water into the subsoil as needed to maintain original water table levels. If needed, the system can be designed to provide a higher water table than originally existed.

Protecting or improving the natural hydrologic environment is often essential to successful site development. Water quality is usually improved by the soil's natural filtering ability — important when runoff contains surface contaminants.

Recharge systems are a widely accepted, effective subsurface water disposal technique. In urbanized areas capable of supporting water table increases, these systems may be much less costly than larger-capacity storm sewer systems.

PERFORATED CMP SYSTEMS

Standard pipe-wall perforations (3/8" diameter holes meeting AASHTO M-36, Class 2. Reference CONTECH drawings 1008467C and 1008468B) provide approximately 2.5% open area. This provides adequate recharge flow for most soils. Perforated pipe-arch is a non-standard product with limited availability.

Your local CONTECH Sales Engineer can help guide you with specifications, details on bulkheads, reinforcement, configurations, and other items.

Figure 3.9.17 *(By permission: CONTECH Construction Products Inc., Middletown, Ohio)*

3.9.18 Run-off Detention Systems Using Corrugated Pipe

ENVIRONMENTAL PROTECTION

Natural hydrologic balances are often adversely affected by land development. The most significant change is increased storm water runoff caused by adding impervious surfaces such as roofs, streets, and parking areas.

Higher runoff volumes and greater peak flow rates following property development may inflict a number of hardships on the existing infrastructure and surrounding areas unless correctly managed.

Typical problems created by storm water runoff increases are:

❑ Overloads on existing storm sewers.

❑ Lower ground water tables.

❑ Soil erosion.

❑ Downstream flooding.

❑ Increased chemical pollution of streams and lakes.

❑ Excessive siltation of streams and lakes.

An analysis of the capacities of existing downstream storm water facilities to accommodate anticipated flow increases is an essential part of pre-development planning.

STORM WATER DETENTION FACILITIES

Whenever planning studies indicate that existing downstream facilities cannot accommodate the projected runoff increases, storm water detention facilities can be used to stabilize runoff rates. Such systems collect and temporarily store excess runoff, while discharging the water at rates not exceeding pre-development levels.

While both aboveground and underground detention systems limit storm water runoff from newly developed sites, the high maintenance costs of open ponds and a need to conserve land may be key considerations in using underground systems. Underground systems are also generally safer than open pond detention systems.

RUNOFF LEGISLATIONS

Increasingly, urban communities are imposing strict discharge and runoff regulations. Recent NPDES (National Pollution Discharge Elimination System) storm water regulations limit on site runoff to reduce damage to waterways and municipal sewers. Also, new developments are often required to limit runoff quantities to pre-development levels.

Traditional retention facilities, such as surface ponds and basins, frequently cannot be used for controlling runoff. Underground corrugated metal pipe or pipe-arch storm water detention systems are proven alternates to surface detention accommodations.

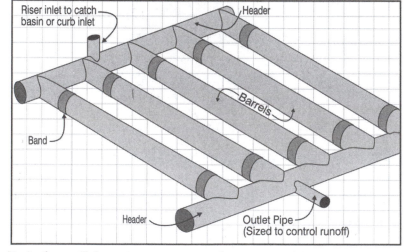

Typical runoff detention system. Number, sizes and lengths of barrels and headers can be varied according to the engineer's design.

Figure 3.9.18   *(By permission: CONTECH Construction Products Inc., Middletown, Ohio)*

3.9.19   Corrugated Metal Pipe Volumes for Detention/Recharge Systems

ROUND PIPE (CMP and Structural Plate Pipe Sizes)			PIPE ARCH (CMP Sizes) (1/2" Deep Corrugations)			PIPE ARCH (Structural Plate Pipe Sizes)		
Diameter (Inches)	Volume (Ft.3/Ft.)	Min. Cover Height	Shape* (Inches)	Volume (Ft.3/Ft.)	Min.Cover Height	Shape* (Feet-Inches)	Volume (Ft.3/Ft.)	Min. Cover Height
12	.78		17 x 13	1.1		18-Inch Corner Radius (Rc)		
15	1.22		21 x 15	1.6				
18	1.76		24 x 18	2.2		6-1 x 4-7	22	
21	2.40		28 x 20	2.9		6-4 x 4-9	24	
24	3.14		35 x 24	4.5		6-9 x 4-11	26	
30	4.9		42 x 29	6.5		7-0 x 5-1	29	12"
36	7.0		49 x 33	8.9	12"	7-3 x 5-3	31	
42	9.6		57 x 38	11.6		7-8 x 5-5	33	
48	12.5	12"	64 x 43	14.7		7-11 x 5-7	36	
54	15.9		71 x 47	18.1		8-2 x 5-9	38	
60	19.6		77 x 52	21.9		8-7 x 5-11	41	
66	23.7		83 x 57	26.0		8-10 x 6-1	43	
72	28.2					9-4 x 6-3	46	
78	33.1					9-6 x 6-5	49	
84	38.4					9-9 x 6-7	52	
90	44.1					10-3 x 6-9	55	18"
96	50.2		(1" Deep Corrugation)			10-8 x 6-11	58	
102	56.7					10-11 x 7-1	61	
108	63.6		Shape* (Inches)	Volume (Ft.3/Ft.)	Min. Cover Height	11-5 x 7-3	64	
114	70.8					11-7 x 7-5	68	
120	78.5					11-10 x 7-7	71	
126	86.5	18"	60 x 46	15.6	15"	12-4 x 7-9	74	
132	95.0		66 x 51	19.3		12-6 x 7-11	78	
138	103.8		73 x 55	23.2		12-8 x 8-4	85	
144	113.1		81 x 59	27.4		12-10 x 8-4	85	
150	122		87 x 63	32.1	18"	13-5 x 8-5	89	
156	132		95 x 67	37.0		13-11 x 8-7	93	
162	143		103 x 71	42.4		14-1 x 8-9	97	24"
168	153	24"	112 x 75	48.0	21"	14-3 x 8-11	101	
174	165		117 x 79	54.2		14-10 x 9-1	105	
180	176		128 x 83	60.5		15-4 x 9-3	109	
186	188		137 x 87	67.4	24"	15-6 x 9-5	114	
192	201		142 x 91	74.5		15-8 x 9-7	118	
198	213					15-10 x 9-10	122	
204	227					16-5 x 9-11	126	30"
210	240					16-7 x 10-1	131	
216	254		*Pipe Arch shape dimensions shown are for Span and Rise respectively.			31-Inch Corner Radius (Rc)		
222	268	30"				13-3 x 9-4	98	
228	283					13-6 x 9-6	102	
234	298					14-0 x 9-8	106	
240	314					14-2 x 9-10	111	

(Left margin label for rows 150–240: "Structural Plate Sizes")

MINIMUM PERMISSABLE SPACING FOR MULTIPLE INSTALLATIONS

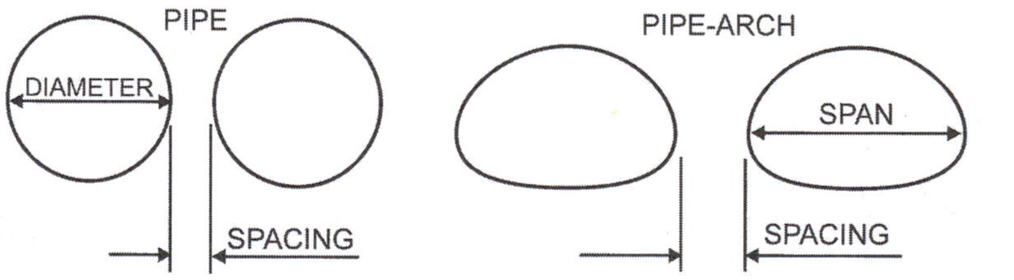

DIAMETER	SPACING	PIPE-ARCH SPAN	SPACING
Up to 24"	12"	Up to 36"	12"
24" to 72"	1/2 Dia. of Pipe	36" to 108"	1/3 Span of Pipe-Arch
72" and larger	36"	108" to 189"	36"

Spacings shown provide room for proper backfill to enable the structure to develop adequate side support.

31-Inch Corner Radius (Rc) (continued)

Shape* (Feet-Inches)	Volume (Ft.3/Ft.)	Min. Cover Height
14-5 x 10-0	115	24"
14-11 x 10-2	120	
15-4 x 10-4	124	
15-7 x 10-6	129	
15-10 x 10-8	134	
16-3 x 10-10	138	
16-6 x 11-0	143	
17-0 x 11-2	148	
17-2 x 11-4	153	
17-5 x 11-6	158	
17-11 x 11-8	163	
18-1 x 11-10	168	30"
18-7 x 12-0	174	
18-9 x 12-2	179	
19-3 x 12-4	185	
19-6 x 12-6	191	
19-8 x 12-8	196	
19-11 x 12-10	202	
20-5 x 13-0	208	36"
20-7 x 13-2	214	

Figure 3.9.19   (*By permission: CONTECH Construction Products Inc., Middletown, Ohio*)

3.9.20 Arch Pipe Installation (Illustrated)

When using a full invert,* it is strongly recommended that steps be taken to avoid undermining the invert.

For sites where the stream bed is non-erodible, footing pads are generally the most economical. Footing pads should be buried a minimum of 12 inches to resist the soil pressures resulting from backfilling operations.

Concrete footings

When concrete footings or concrete inverts are used, the Aluminum Box Culvert may be placed in a receiving channel or in a pre-formed slot.

Structures with .5N legs may require a wider slot to accommodate rib and shell.

Full Corrugated invert

The standard corrugated invert is .100" aluminum plate. (Supplemental plates under receiving channels serve as footing pads.) A standard 24" toewall is supplied for each end of the structure. Greater depths may be supplied when required. As an alternate, a concrete toewall can be used. Standard hook bolts are also available to attach the invert to this type of toewall.

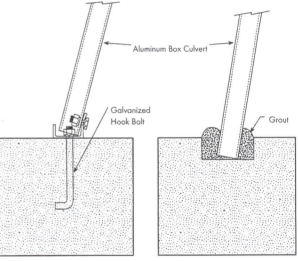

Concrete Footing with Receiving Channel

Slotted Concrete Footing

Geotextile is placed at the junction of shell and invert to prevent infiltration of backfill material.

*A minimum bearing capacity of 4,000 PSF is required when a full corrugated invert is used.

Figure 3.9.20 *(By permission: CONTECH Construction Products Inc., Middletown, Ohio)*

3.9.21 Concrete Lined Corrugated Pipe

Installation

Satisfactory site preparation, trench excavation, bedding and backfill operations are essential for proper soil interaction with pipe. The soil beneath, beside and above the pipe should meet the criteria for soil type and density specified in the section on Backfill, below.

Unsuitable materials within this zone should be excavated and replaced with suitable backfill. If the natural soil or pre-compacted fill at the site meets the criteria for soil type and density specified below, the pipe may be placed in a trench cut at least three feet wider than the pipe (18 inches on each side) to permit sufficient working area to compact the material below the haunches and adjacent to the sides.

For multiple structures one-half the pipe diameter or a three-foot spacing, whichever is smaller, should be left between barrels to allow room for proper compaction.

It should be noted that if the soils at the site appear to be unstable or otherwise unsuitable, a qualified soil engineer should be engaged to design the bedding and the backfill.

Bedding

The bedding preparation is important to both structure performance and service life.

The bed should be placed to uniform grade and line to ensure good vertical alignment and to avoid excessive stresses at pipe joints. The bedding should be free of rock formations, protruding stones, frozen lumps, roots and other foreign material. It is recommended that the bedding foundation be a stable, well-graded granular material. Any material that has inadequate bearing capability must be removed and replaced with a compacted, select fill approved by the engineer.

Placing HEL-COR CL on the bedding surface is generally accomplished by one of two methods to ensure satisfactory compaction beneath the haunches. One method is to shape the bedding surface to conform to the lower section of the pipe. The other method is to tamp a well-graded, granular or select material beneath the haunches to achieve a well-compacted condition.

Backfill

Satisfactory backfill material, proper placement and compaction are key factors in obtaining maximum strength and stability.

The fill material should be free of rocks, frozen lumps and foreign matter that could cause hard spots in backfill or that could decompose and create voids. Structural backfill material preferably should be a well-graded granular material.

Local site material is often adequate, provided that it is sufficiently compacted at controlled moisture content. Cohesive type materials generally do not have adequate shear strength unless moisture content is maintained near optimum during placement and compaction.

Highly plastic silts, highly plastic clays, organic silts, organic clays and peats should not be used as backfill materials.

Backfill should be placed symmetrically on each side of the structure in six-inch to eight-inch loose layers to one foot above the top of the pipe. Each layer is to be compacted to the specified density (minimum 90%) before placing the next layer. Reference ASTM A 798.

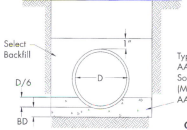

Select Backfill

Typical bedding composition, AASHTO M145*
Soil Classes A-1,A-3,A-2-4, or A-2-5 (Minimum 90% Proctor density per AASHTO T180)

Granular Foundation

*Typical Soil Composition
A-1: Stone fragments, gravel, and sand
A-3: Fine sand
A-2-4: Silty or clayey gravel and sand
A-2-5: Silty or clayey gravel and sand
For proper sieve analysis, see AASHTO M145

Depth of Bedding Material Below Pipe	
D	BD (Min.)
24"	3"
30" to 60"	4"
66" & larger	6"

Figure 3.9.21 *(By permission: CONTECH Construction Products Inc., Middletown, Ohio)*

3.9.22 Polymeric Coated Corrugated Pipe

Polymer coated corrugated steel pipe is produced in $2\frac{2}{3}$" x $\frac{1}{2}$", 3" x 1", and 125mm x 25mm corrugations to meet various project requirements. Sizes available in these corrugations range from 12" through 144" diameters and all the standard AASHTO pipe-arch sizes. In addition, hydraulically superior products—such as SmoothCor and ULTRA FLO—are available with polymer coatings in many sizes. Your local CONTECH Sales Engineer can provide details on the availability of various polymer coated corrugated steel pipe materials.

A polymeric corrugated steel pipe system also provides excellent performance in methane gas recovery applications in landfills. The CSP is structurally capable of handling the heavy and settling loads in a landfill. The polymeric coating provides added corrosion resistance to the pipe.

If you have a project with corrosive soil conditions such as bluish-gray or green clays, or other tough environmental conditions, polymeric corrugated steel pipe may be the answer to your drainage needs.

Call your local CONTECH Sales Engineer for more information about polymeric corrugated steel pipe.

Reference Specifications
for Polymer-Coated Corrugated Steel Pipe

	AASHTO	ASTM
Material	M246	A 742
Pipe	M245	A 762
Design	Section 12*	A 796
Installation	Section 26*	A 798

*Standard Specifications for Highway Bridges

This 66"-diameter polymeric corrugated steel pipe was installed in 1975 by the Arkansas State Highway Department as a test installation. After 22 years, this triple barrel installation exposed to pH 5.5 and moderate abrasion is in excellent condition. Other test sites in Michigan and Wisconsin under harsh conditions also demonstrate the excellent performance of TRENCHCOAT.

Figure 3.9.22   *(By permission: CONTECH Construction Products Inc., Middletown, Ohio)*

3.10.0 Physical Properties of Rigid PVC and CPVC Pipe

The following table lists typical physical properties of PVC and CPVC thermoplastic materials.
Variations may exist depending on specific compounds and product.

Mechanical

Properties	Unit	PVC	CPVC	Remarks	ASTM Test
Specific Gravity	g/cm³	1.40 ± .02	1.55 ± .02		D-792
Tensile Strength @ 73°F	PSI	7,200	8,000	Same in Circumferential Direction	D-638
Modules of Elasticity Tensile @ 73°F	PSI	430,000	360,000	Ratio of Stress on Bent Sample at Failure	D-638
Compressive Strength @ 73°F	PSI	9,500	10,100		D-695
Flexural Strength @ 73°F	PSI	13,000	15,100	Tensile Stress on Bent Sample at Failure	D-790
Izod Impact @ 73°F	Ft-Lbs/In of Notch	1.0	1.5	Impact Resistance of a Notched Sample to a Sharp Blow	D-256
Relative Hardness @ 73°F	Durometer "D" Rockwell "R"	80 ± 3 110-120	— 119	Equivalent to Aluminum —	D-2240 D-785

Thermodynamics

Properties	Unit	PVC	CPVC	Remarks	ASTM Test
Coefficient of Thermal Linear Expansion per °F	in/in/°F	2.8×10^{-5}	3.4×10^{-5}		D-696
Thermal Conductivity	BTU/hr/ft²/°F/in	1.3	0.95	Average Specific Heat of 0-100°C	C-177
Specific Heat	CAL/g/°C	0.20-0.28		Ratio of Thermal Capacity to that of Water at 15°C	
Maximum Operating Temperature	°F	140	210	Pressure Rating is Directly Related to Temperature	
Heat Distortion Temperature @ 264 PSI	°F	158	217	Thermal Vibration and Softening Occurs	D-648
Decomposition Point	°F	400+	400+	Scorching by Carbonization and Dehydrochloration	

Figure 3.10.0 *(Courtesy: George Fischer Sloane Inc., Little Rock, Arkansas)*

3.10.1 Standard Specifications for Schedule 80 PVC/CPVC Pipe

Standard Specifications: Schedule 80 PVC and CPVC Pipe

1. Scope

This specification covers requirements for Schedule 80 PVC and CPVC pressure pipe as described in ASTM D-1785 (PVC) and in ASTM F-441 (CPVC). All pipe shall be as manufactured by George Fischer Sloane, Inc., in Little Rock, Arkansas, or shall be in all respects the equal of such pipe. The purchaser reserves the right to require that the seller furnish proof of such equality where the pipe are of different manufacture. At the purchaser's discretion contract preference may be given those suppliers able to furnish all pipe required under the contract from a single manufacturer in order that responsibility will not be divided in warrantee claim situations.

2. Material

PVC used is of Type I, Grade 1 compound as stated in ASTM D-1784.

3. Dimensions

Dimensions and tolerances shall be as shown in the following tables when measured according to Method D-2122. Tolerances for out-of-roundness shall apply only to pipe prior to shipment.

4. Marking

Indicates manufacturer's name, material designation code, nominal pipe size, Schedule size with pressure rating in PSI for water at 73°F, ASTM designation, NSF seal for potable water and manufacturing date code.

Outside Diameters and Tolerances

Nominal Pipe Size	Outside Diameter	Tolerances	
		Average	For Maximum and Minimum Diameter (Out-of-Roundness)
			Schedule 80
$1/8$	0.405	±0.004	±0.008
$1/4$	0.540	±0.004	±0.008
$3/8$	0.675	±0.004	±0.008
$1/2$	0.840	±0.004	±0.008
$3/4$	1.050	±0.004	±0.010
1	1.315	±0.005	±0.010
$1^1/4$	1.660	±0.005	±0.012
$1^1/2$	1.900	±0.006	±0.012
2	2.375	±0.006	±0.012
$2^1/2$	2.875	±0.007	±0.015
3	3.500	±0.008	±0.015
$3^1/2$	4.000	±0.008	±0.015
4	4.500	±0.009	±0.015
5	5.563	±0.010	±0.030
6	6.625	±0.011	±0.035
8	8.625	±0.015	±0.075
10	10.750	±0.015	±0.075
12	12.750	±0.015	±0.075
14	14.000	±0.015	±0.100
16	16.000	±0.019	±0.160
18	18.000	±0.019	±0.180
20	20.000	±0.023	±0.200
24	24.000	±0.031	±0.240

Figure 3.10.1   (*Courtesy: George Fischer Sloane Inc., Little Rock, Arkansas*)

3.10.2 PVC Schedule 40/80 Socket Dimensions

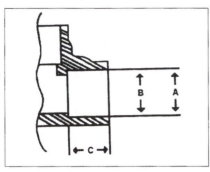

PVC IPS Schedule 40/80 Socket Dimensions

SIZE	PIPE O.D.	ENTRANCE (A)		BOTTOM (B)		MAX. OUT OF ROUND	SCHEDULE 40 SOCKET DEPTH (C) (MIN.)	SCHEDULE 80 SOCKET DEPTH (C) (MIN.)
		MAX.	MIN.	MAX.	MIN.			
1/4	.540	.556	.548	.540	.532	.016	.500	.625
3/8	.675	.691	.683	.675	.667	.016	.594	.750
1/2	.840	.852	.844	.840	.832	.016	.688	.875
3/4	1.050	1.062	1.054	1.050	1.042	.020	.719	1.000
1	1.315	1.330	1.320	1.315	1.305	.020	.875	1.125
1 1/4	1.660	1.675	1.665	1.660	1.650	.024	.938	1.250
1 1/2	1.900	1.918	1.906	1.900	1.888	.024	1.094	1.375
2	2.375	2.393	2.381	2.375	2.363	.024	1.156	1.500
2 1/2	2.875	2.896	2.882	2.875	2.861	.030	1.750	1.750
3	3.500	3.524	3.508	3.500	3.484	.030	1.875	1.875
3 1/2	4.000	4.024	4.008	4.000	3.984	.030	2.000	
4	4.500	4.527	4.509	4.500	4.482	.030	2.000	2.250
5	5.563	5.593	5.573	5.563	5.543	.060	3.000	
6	6.625	6.658	6.636	6.625	6.603	.060	3.000	3.000
8	8.625	8.670	8.640	8.625	8.595	.090	4.000	4.000

3.10.3 Wall Thickness and Tolerances—Schedule 80 PVC Pipe

Wall Thicknesses and Tolerances

Nominal Pipe Size	Wall Thickness	
	Schedule 80	
	Minimum	Tolerance
$1/8$	0.095	+0.020
$1/4$	0.119	+0.020
$3/8$	0.126	+0.020
$1/2$	0.147	+0.020
$3/4$	0.154	+0.020
1	0.179	+0.021
$1^1/4$	0.191	+0.023
$1^1/2$	0.200	+0.024
2	0.218	+0.026
$2^1/2$	0.276	+0.033
3	0.300	+0.036
$3^1/2$	0.318	+0.038
4	0.337	+0.040
5	0.375	+0.045
6	0.432	+0.052
8	0.500	+0.060
10	0.593	+0.071
12	0.687	+0.082
14	0.750	+0.090
16	0.843	+0.101
18	0.937	+0.112
20	1.031	+0.124
24	1.218	+0.146

Note: The minimum is the lowest wall thickness of the pipe at any cross section. The maximum permitted wall thickness, at any cross section, is the minimum wall thickness plus the stated tolerance. All tolerances are on the plus side of the minimum requirement. These dimensions conform to nominal IPS dimensions.

3.10.4 PVC and CPVC Pipe and Fittings

CHEMICAL:

Agricultural chemicals, bleach manufacturing, chemical laboratories, cosmetics, electroplating, fertilizers manufacturing, leather tanning, organic color industries, oil refineries, petrochemicals, pharmaceutical, photographic industry, soap manufacturing. For UV resistance of PVC Pipe and Fittings, paint with a white water-based latex paint. Service temperature of 33 degrees to 140 degrees F. Pipe is not recommended for use with compressed air or gases. PVC material is Type 1 according to the American Society for Testing Material (ASTM) D-1784. Fittings comform to ASTM D-2466. Temperature rating is 140 degrees F. INDUSTRIAL PLANTS: Air-conditioning, aircraft, aluminum industry, automotive, battery mfg, construction, fertilizer manufacturing, metals, glass, leather, marine, mining and smelting, paper and pulp, plumbing, shipping, steel industry, textile. PSI is based on water at 73 degrees F. Derate 50% at 110 degrees F and 78% at 140 degree F. MACHINABLE: Round, smooth inside and out. Resistant to cracking. Made from premium virgin Goodrich grey PVC resin. Superior to regular commercial grade used for regular plumbing.

PVC SCHEDULE 40 PIPE GRAY AND WHITE

40 PIPE SIZE	O.D.	AVE. I.D.	MIN. WALL THICK.	APPROX WT. PER 100 FT.	MAX. WORK PRESS PLAIN ENDS
1/4"	.540	.354	.088	7.88	780
3/8"	.675	.483	.091	10.53	620
1/2"	.840	.608	.109	15.63	600
3/4"	1.050	.810	.113	20.78	480
1"	1.315	1.033	.133	30.85	450
1 1/4"	1.660	1.364	.140	41.76	370
1 1/2"	1.900	1.592	.145	49.94	330
2"	2.375	2.049	.154	67.92	280
2 1/2"	2.875	2.445	.203	105.46	300
3"	3.500	3.042	.216	137.91	260
4"	4.500	3.998	.237	194.58	220
6"	6.625	6.031	.280	338.88	180

PVC SCHEDULE 80 PIPE GRAY AND WHITE

80 PIPE SIZE	O.D.	AVE. I.D.	MIN. WALL THICK.	APPROX WT. PER 100 FT.	WORK PRESS PLAIN ENDS	MAX. WORK PRESS. THRD. ENDS	GRAY	PRICE PER 100 FT.
1/4"	.540	.302	.119	9.4	1130	550	800-002	$54.52
3/8"	.675	.423	.126	13.0	920	450	800-003	73.75
1/2"	.840	.546	.147	19.2	850	420	800-005	32.77

3.10.5 Schedule 40/80 PVC Pipe Specifications

Schedule 40/80 pipe size: inside/outside dimensions, weight per foot for UL-rated PVC pipe.

J-M SCH. 40 CONDUIT
U.L. Listed

RIGID NON-METALIC CONDUIT FOR USE IN BOTH
ABOVE GROUND AND UNDERGROUND INSTALLATIONS

Schedule 40 Conduit					Rated for 90°C Conductors			
Size	Part Number	Avg. OD	Nom. ID	Min. Wall	Approx. Wt. 100/Ft	Ft. Per Bundle	Feet Per Lift	Price/ 100 Ft
1/2	40050	.840	.622	.109	18	100	6000	18.01
3/4	40075	1.050	.824	.113	24	100	4400	24.45
1	40100	1.315	1.049	.133	33	100	3600	35.32
1¼	40125	1.660	1.380	.140	45	50	3300	47.83
1½	40150	1.900	1.610	.145	56	50	2250	57.69
2	40200	2.375	2.067	.154	74	50	1400	76.35
2½	40250	2.875	2.469	.203	126	10	900	122.25
3	40300	3.500	3.068	.216	163	10	880	158.90
3½	40350	4.000	3.548	.226	197	10	630	190.17
4	40400	4.500	4.026	.237	234	10	480	224.77
5	40500	5.563	5.047	.258	319	10	230	319.35
6	40600	6.625	6.065	.280	411	10	220	410.11

Schedule 40 is furnished in standard 10' lengths with one bell end.
20 ft. lengths are available upon request.

J-M SCH. 80 CONDUIT
U.L. Listed

RIGID NON-METALIC CONDUIT FOR USE IN BOTH
ABOVE GROUND AND UNDERGROUND INSTALLATIONS

Schedule 80 Conduit					Rated for 90°C Conductors			
Size	Part Number	Avg. OD	Nom. ID	Min. Wall	Approx. Wt. 100/Ft	Ft. Per Bundle	Feet Per Lift	Price/ 100 Ft
1/2	80050	.840	.546	.147	22	100	6000	23.35
3/4	80075	1.050	.742	.154	30	100	4400	31.35
1	80100	1.315	.957	.179	42	100	3600	44.93
1¼	80125	1.660	1.278	.191	60	50	3300	62.72
1½	80150	1.900	1.500	.200	72	50	2250	74.57
2	80200	2.375	1.939	.218	98	10	1400	102.85
2½	80250	2.875	2.323	.276	151	10	900	157.30
3	80300	3.500	2.900	.300	213	10	880	209.90
4	80400	4.500	3.826	.337	310	10	480	305.50
5	80500	5.563	4.813	.375	430	10	230	440.42
6	80600	6.625	5.761	.432	590	10	220	583.00

Schedule 80 is furnished in standard 10' lengths with one bell end.
20 ft. lengths are available upon request.

3.10.6 PVC Expansion Loops

PVC Expansion Loops

PVC		Length of Run (feet)									
		10	20	30	40	50	60	70	80	90	100
Pipe Size (in.)	O.D. of Pipe (in.)	Minimum Deflected Pipe Length (DPL) (inches)									
1/2	0.840	11	15	19	22	24	27	29	31	32	34
3/4	1.050	12	17	21	24	27	30	32	34	36	38
1	1.315	14	19	23	27	30	33	36	38	41	43
1 1/4	1.660	15	22	26	30	34	37	40	43	46	48
1 1/2	1.900	16	23	28	33	36	40	43	46	49	51
2	2.375	18	26	32	36	41	45	48	51	55	58
3	3.500	22	31	38	44	49	54	58	62	66	70
4	4.500	25	35	43	50	56	61	66	71	75	79
6	6.625	30	43	53	61	68	74	80	86	91	96
8	8.625	35	49	60	69	78	85	92	98	104	110
10	10.750	39	55	67	77	87	95	102	110	116	122
12	12.750	42	60	73	84	94	103	112	119	127	133

3.10.7 PVC Offsets/Change of Directions

PVC Offsets and Change of Directions

PVC		Length of Run (feet)									
		10	20	30	40	50	60	70	80	90	100
Pipe Size (in.)	O.D. of Pipe (in.)	Minimum Deflected Pipe Length (DPL) (inches)									
1/2	0.840	15	22	27	31	34	37	41	43	46	48
3/4	1.050	17	24	30	34	38	42	45	48	51	54
1	1.315	19	27	33	38	43	47	51	54	57	61
1 1/4	1.660	22	30	37	43	48	53	57	61	65	68
1 1/2	1.900	23	33	40	46	51	56	61	65	69	73
2	2.375	26	36	45	51	58	63	68	73	77	81
3	3.500	31	44	54	62	70	77	83	88	94	99
4	4.500	35	50	61	71	79	87	94	100	106	112
6	6.625	43	61	74	86	96	105	114	122	129	136
8	8.625	49	69	85	98	110	120	130	139	147	155
10	10.750	55	77	95	110	122	134	145	155	164	173
12	12.750	60	84	103	119	133	146	158	169	179	189

Figure A: Guided Cantilever Beam

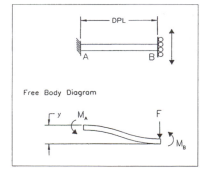

Figure B: Expansion Loop

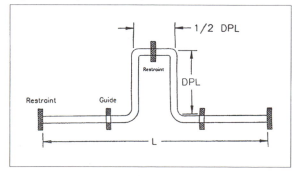

3.10.8 CPVC Expansion Loops

CPVC Expansion Loops

CPVC		Length of Run (feet)									
		10	20	30	40	50	60	70	80	90	100
Pipe Size (in.)	O.D. of Pipe (in.)	Minimum Deflected Pipe Length (DPL) (inches)									
1/2	0.840	15	21	26	30	33	36	39	42	44	47
3/4	1.050	17	23	29	33	37	40	44	47	50	52
1	1.315	18	26	32	37	41	45	49	52	55	58
1 1/4	1.660	21	29	36	42	46	51	55	59	62	66
1 1/2	1.900	22	31	39	44	50	54	59	63	67	70
2	2.375	25	35	43	50	56	61	66	70	75	79
3	3.500	30	43	52	60	67	71	80	85	91	95
4	4.500	34	4	59	68	77	84	91	97	103	108
6	6.625	42	59	72	83	93	102	110	117	125	131
8	8.625	47	67	82	95	106	116	125	134	142	150
10	10.750	53	75	92	106	118	130	140	150	159	167
12	12.750	58	81	100	115	129	141	152	163	173	182

3.10.9 CPVC Offsets/Change of Direction

CPVC Offsets and Change of Directions

CPVC		Length of Run (feet)									
		10	20	30	40	50	60	70	80	90	100
Pipe Size (in.)	O.D. of Pipe (in.)	Minimum Deflected Pipe Length (DPL) (inches)									
1/2	0.840	21	30	36	42	47	51	55	59	63	66
3/4	1.050	23	33	40	47	22	57	62	66	70	74
1	1.315	26	37	45	52	58	61	69	74	78	83
1 1/4	1.660	29	42	51	59	66	72	78	86	88	93
1 1/2	1.900	31	44	54	63	70	77	83	89	94	99
2	2.375	35	50	61	70	79	86	93	99	105	111
3	3.500	43	60	74	85	95	105	113	121	128	135
4	4.500	48	68	84	97	108	119	128	137	145	153
6	6.625	59	53	102	117	131	144	155	166	176	186
8	8.625	67	95	116	134	150	164	177	189	201	212
10	10.750	75	106	130	150	167	183	198	212	224	237
12	12.750	81	115	141	163	182	200	216	230	244	258

Figure C: Expansion Offset

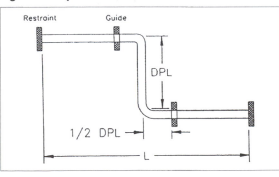

Figure D: Change of Direction

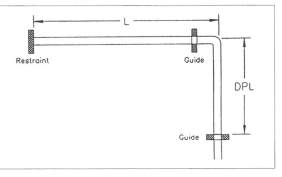

3.10.10 Friction Loss/Carrying Capacity—Schedule 40 PVC Pipe

Friction Loss — Schedule 40 Pipe

Carrying capacity, friction loss and flow data for Schedule 40 thermoplastic pipe are presented in tabular form in the table below. This table is applicable to pipe made of any of the thermoplastic piping materials as all have equally smooth interior surfaces.

Carrying Capacity and Friction Loss — Schedule 40 Thermoplastics Pipe

Independent variables: Gallons per minute and nominal pipe size O.D. (Min. I.D.)
Dependent variables: Velocity, friction head and pressure drop per 100 feet of pipe, interior smooth.

Each pipe-size group below has three sub-columns: **V** = Velocity Feet per Second, **H** = Friction Head Feet, **L** = Friction Loss Pounds per Square Inch. As gallons per minute increase, smaller pipe columns are replaced by larger sizes (4 in., 5 in., 6 in., 8 in., 10 in., 12 in.) in the same column positions, as labeled.

GALLONS PER MINUTE	1/2 in. → 4 in. V	H	L	3/4 in. → 5 in. V	H	L	1 in. → 6 in. V	H	L	1 1/4 in. → 8 in. V	H	L	1 1/2 in. → 10 in. V	H	L	2 in. → 12 in. V	H	L	3 in. V	H	L
1	1.13	2.08	0.90	0.63	0.51	0.22															
2	2.26	4.16	1.80	1.26	1.02	0.44	0.77	0.55	0.24	0.44	0.14	0.06	0.33	0.07	0.03						
5	5.64	23.44	10.15	3.16	5.73	2.48	1.93	1.72	0.75	1.11	0.44	0.19	0.81	0.22	0.09	0.49	0.066	0.029	0.30	0.015	0.007
7	7.90	43.06	18.64	4.43	10.52	4.56	2.72	3.17	1.37	1.55	0.81	0.35	1.13	0.38	0.17	0.69	0.11	0.048	0.49	0.021	0.009
10	11.28	82.02	35.51	6.32	20.04	8.68	3.86	6.02	2.61	2.21	1.55	0.67	1.62	0.72	0.31	0.98	0.21	0.091	0.68	0.03	0.013
15	**4 in.**			9.48	42.46	18.39	5.79	12.77	5.53	3.31	3.28	1.42	2.42	1.53	0.66	1.46	0.45	0.19	1.03	0.07	0.030
20	0.51	0.03	0.013	12.65	72.34	31.32	7.72	21.75	9.42	4.42	5.59	2.42	3.23	2.61	1.13	1.95	0.76	0.33	1.37	0.11	0.048
25	0.64	0.04	0.017	**5 in.**			9.65	32.88	14.22	5.52	8.45	3.66	4.04	3.95	1.71	2.44	1.15	0.50	1.71	0.17	0.074
30	0.77	0.06	0.026	0.49	0.02	0.009	11.58	46.08	19.95	6.63	11.85	5.13	4.85	5.53	2.39	2.93	1.62	0.70	2.05	0.23	0.10
35	0.89	0.08	0.035	0.57	0.03	0.013				7.73	15.76	6.82	5.66	7.36	3.19	3.41	2.15	0.93	2.39	0.31	0.13
40	1.02	0.11	0.048	0.65	0.03	0.013	**6 in.**			8.84	20.18	8.74	6.47	9.43	4.08	3.90	2.75	1.19	2.73	0.40	0.17
45	1.15	0.13	0.056	0.73	0.04	0.017				9.94	25.10	10.87	7.27	11.73	5.08	4.39	3.43	1.49	3.08	0.50	0.22
50	1.28	0.16	0.069	0.81	0.05	0.022	0.56	0.02	0.009	11.05	30.51	13.21	8.08	14.25	6.17	4.88	4.16	1.80	3.42	0.60	0.26
60	1.53	0.22	0.095	0.97	0.07	0.030	0.67	0.03	0.013				9.70	19.98	8.65	5.85	5.84	2.53	4.10	0.85	0.37
70	1.79	0.30	0.13	1.14	0.10	0.043	0.79	0.04	0.017							6.83	7.76	3.36	4.79	1.13	0.49
75	1.92	0.34	0.15	1.22	0.11	0.048	0.84	0.05	0.022							7.32	8.82	3.82	5.13	1.28	0.55
80	2.05	0.38	0.16	1.30	0.13	0.056	0.90	0.05	0.022	**8 in.**						7.80	9.94	4.30	5.47	1.44	0.62
90	2.30	0.47	0.20	1.46	0.16	0.069	1.01	0.06	0.026							8.78	12.37	5.36	6.15	1.80	0.78
100	2.56	0.58	0.25	1.62	0.19	0.082	1.12	0.08	0.035	0.65	0.03	0.012				9.75	15.03	6.51	6.84	2.18	0.94
125	3.20	0.88	0.38	2.03	0.29	0.125	1.41	0.12	0.052	0.81	0.035	0.015	**10 in.**						8.55	3.31	1.43
150	3.84	1.22	0.53	2.44	0.40	0.17	1.69	0.16	0.069	0.97	0.04	0.017							10.26	4.63	2.00
175	4.48	1.63	0.71	2.84	0.54	0.235	1.97	0.22	0.096	1.14	0.055	0.024								6.16	2.67
200	5.11	2.08	0.90	3.25	0.69	0.30	2.25	0.28	0.12	1.30	0.07	0.030	0.82	0.027	0.012	**12 in.**				7.88	3.41
250	6.40	3.15	1.36	4.06	1.05	0.45	2.81	0.43	0.19	1.63	0.11	0.048	1.03	0.035	0.015					11.93	5.17
300	7.67	4.41	1.91	4.87	1.46	0.63	3.37	0.60	0.26	1.94	0.16	0.069	1.23	0.05	0.022						
350	8.95	5.87	2.55	5.69	1.95	0.85	3.94	0.79	0.34	2.27	0.21	0.091	1.44	0.065	0.028	1.01	0.027	0.012			
400	10.23	7.52	3.26	6.50	2.49	1.08	4.49	1.01	0.44	2.59	0.27	0.12	1.64	0.09	0.039	1.16	0.04	0.017			
450				7.31	3.09	1.34	5.06	1.26	0.55	2.92	0.33	0.14	1.85	0.11	0.048	1.30	0.05	0.022			
500				8.12	3.76	1.63	5.62	1.53	0.66	3.24	0.40	0.17	2.05	0.13	0.056	1.45	0.06	0.026			
750							8.43	3.25	1.41	4.86	0.85	0.37	3.08	0.28	0.12	2.17	0.12	0.052			
1000							11.24	5.54	2.40	6.48	1.45	0.63	4.11	0.48	0.21	2.89	0.20	0.087			
1250										8.11	2.20	0.95	5.14	0.73	0.32	3.62	0.31	0.13			
1500										9.72	3.07	1.33	6.16	1.01	0.44	4.34	0.43	0.19			
2000													8.21	1.72	0.74	5.78	0.73	0.32			
2500													10.27	2.61	1.13	7.23	1.11	0.49			

3.10.11 Friction Loss/Carrying Capacity—Schedule 80 PVC Pipe

Friction Loss — Schedule 80 Pipe

Carrying capacity, friction loss and flow data for Schedule 80 thermoplastic pipe are presented in tabular form in the table below. This table is applicable to pipe made of any of the thermoplastic piping materials as all have equally smooth interior surfaces.

Friction Loss — Schedule 80 Fittings

The table "Friction Loss in Equivalent Feet of Pipe" (page 20) gives the estimated friction loss, in equivalent feet of pipe, through thermoplastic fittings of various sizes and configurations.

Carrying Capacity and Friction Loss — Schedule 80 Thermoplastics Pipe

Independent variables: Gallons per minute and nominal pipe size O.D. (Min. I.D.)
Dependent variables: Velocity, friction head and pressure drop per 100 feet of pipe, interior smooth.

Each pipe size below has three columns: **V** = Velocity Feet Per Second, **H** = Friction Head Feet, **L** = Friction Loss Pounds Per Square Inch.

GALLONS PER MINUTE	V	H	L	V	H	L	V	H	L	V	H	L	V	H	L	V	H	L	V	H	L	V	H	L
	1/2 in.			**3/4 in.**			**1 in.**			**1 1/4 in.**			**1 1/2 in.**			**2 in.**			**2 1/2 in.**			**3 in.**		
1	1.43	4.02	1.74	0.74	0.86	0.37																		
2	2.95	8.03	3.48	1.57	1.72	0.74	0.94	0.88	0.33	0.52	0.21	0.09	0.38	0.10	0.041									
5	7.89	45.23	19.59	3.92	9.67	4.19	2.34	2.78	1.19	1.30	0.66	0.29	0.94	0.30	0.126	0.56	0.10	0.040	0.39	0.05	0.022	0.25	0.02	0.009
7	10.34	83.09	35.97	5.49	17.76	7.59	3.23	5.04	2.19	1.82	1.21	0.53	1.32	0.55	0.24	0.78	0.15	0.088	0.54	0.07	0.032	0.35	0.023	0.013
10				7.84	33.84	14.65	4.68	9.61	4.16	2.60	2.30	1.00	1.88	1.04	0.45	1.12	0.29	0.13	0.78	0.12	0.052	0.50	0.04	0.017
15	**4 in.**			11.76	71.70	31.05	7.01	20.36	8.82	3.90	4.87	2.11	2.81	2.20	0.95	1.63	0.62	0.27	1.17	0.26	0.11	0.75	0.09	0.039
20	0.57	0.04	0.017				9.35	34.68	15.02	5.20	8.30	3.59	3.75	3.75	1.62	2.23	1.06	0.46	1.56	0.44	0.19	1.00	0.15	0.055
25	0.72	0.06	0.026	**5 in.**			11.69	52.43	22.70	6.50	12.55	5.43	4.69	5.67	2.46	2.79	1.60	0.69	1.95	0.67	0.29	1.25	0.22	0.095
30	0.86	0.08	0.035	0.54	0.03	0.013	14.03	73.48	31.62	7.80	17.59	7.62	5.63	7.95	3.44	3.35	2.25	0.97	2.34	0.94	0.41	1.49	0.31	0.13
35	1.00	0.11	0.048	0.63	0.04	0.017				9.10	23.40	10.13	6.57	10.58	4.58	3.91	2.99	1.29	2.73	1.25	0.64	1.74	0.42	0.13
40	1.15	0.14	0.061	0.72	0.04	0.017				10.40	29.97	12.98	7.50	13.55	5.87	4.47	3.86	1.66	3.12	1.60	0.89	1.99	0.54	0.23
45	1.29	0.17	0.074	0.81	0.05	0.020	**6 in.**			11.70	37.27	16.14	8.44	16.85	7.30	5.03	4.76	2.07	3.51	1.90	0.86	2.24	0.67	0.29
50	1.43	0.21	0.091	0.90	0.07	0.030	0.63	0.03	0.013	13.00	45.30	19.61	9.38	20.48	8.87	5.58	5.79	2.51	3.90	2.42	1.05	2.49	0.81	0.35
60	1.72	0.30	0.13	1.08	0.10	0.043	0.75	0.04	0.017				11.26	28.70	12.43	6.70	8.12	3.52	4.68	3.39	1.47	2.98	1.14	0.49
70	2.01	0.39	0.17	1.26	0.13	0.056	0.88	0.05	0.022							7.82	10.80	4.68	5.46	4.51	1.35	3.49	1.51	0.65
75	2.15	0.45	0.19	1.35	0.14	0.061	0.94	0.06	0.026							8.38	12.27	5.31	5.85	5.12	2.22	3.74	1.72	0.74
80	2.29	0.50	0.22	1.44	0.16	0.069	1.00	0.07	0.030	**8 in.**						8.93	13.83	5.99	6.24	6.77	2.50	3.99	1.94	0.84
90	2.58	0.63	0.27	1.62	0.20	0.087	1.13	0.08	0.035							10.05	17.20	7.45	7.02	7.18	3.11	4.48	2.41	1.04
100	2.87	0.76	0.33	1.80	0.24	0.10	1.25	0.10	0.043							11.17	20.90	9.05	7.80	8.72	3.78	4.98	2.93	1.27
125	3.59	1.16	0.50	2.25	0.37	0.16	1.57	0.16	0.068	0.90	0.045	0.019							9.75	13.21	5.72	6.23	4.43	1.92
150	4.30	1.61	0.70	2.70	0.52	0.23	1.88	0.22	0.095	1.07	0.05	0.022	**10 in.**						11.70	18.48	8.00	7.47	6.20	2.68
175	5.02	2.15	0.93	3.15	0.69	0.30	2.20	0.29	0.12	1.25	0.075	0.033										8.72	8.26	3.58
200	5.73	2.75	1.19	3.60	0.88	0.38	2.51	0.37	0.16	1.43	0.09	0.039										9.97	10.57	4.58
250	7.16	4.16	1.81	4.50	1.34	0.58	3.14	0.56	0.24	1.79	0.14	0.061				**12 in.**						12.46	16.00	8.93
300	8.60	5.33	2.52	5.40	1.87	0.81	3.76	0.78	0.34	2.14	0.20	0.087												
350	10.03	7.76	3.35	6.30	2.49	1.08	4.39	1.04	0.45	2.50	0.27	0.12	1.59	0.085	0.037	1.12	0.037	0.016						
400	11.47	9.93	4.30	7.19	3.19	1.38	5.02	1.33	0.68	2.86	0.34	0.15	1.81	0.11	0.048	1.28	0.05	0.022						
450				8.09	3.97	1.72	5.64	1.65	0.71	3.21	0.42	0.18	2.04	0.14	0.061	1.44	0.06	0.026						
500				8.99	4.82	2.09	6.27	2.00	0.87	3.57	0.51	0.22	2.27	0.17	0.074	1.60	0.07	0.030						
750							9.40	4.25	1.84	5.36	1.08	0.47	3.40	0.36	0.16	2.40	0.15	0.065						
1000							12.54	7.23	3.13	7.14	1.84	0.80	4.54	0.61	0.26	3.20	0.20	0.11						
1250										8.93	2.78	1.20	5.67	0.92	0.40	4.01	0.40	0.17						
1500										10.71	3.89	1.68	6.80	1.29	0.56	4.81	0.55	0.24						
2000													9.07	2.19	0.95	6.41	0.84	0.41						
2500													11.34	3.33	1.44	8.01	1.42	0.62						
3000																9.61	1.99	0.86						
3500																11.21	2.65	1.15						
4000																12.82	3.41	1.48						

3.10.12 Friction Loss/Fittings—Schedule 80 PVC Fittings

Friction Loss in Equivalent Feet of Pipe — Schedule 80 Thermoplastics Fittings

Nominal Pipe Size, In.	$3/8$	$1/2$	$3/4$	1	$1\frac{1}{4}$	$1\frac{1}{2}$	2	$2\frac{1}{2}$	3	$3\frac{1}{2}$	4	6	8
Tee, Side Outlet	3	4	5	6	7	8	12	15	16	20	22	32	38
90° Ell	$1\frac{1}{2}$	$1\frac{1}{2}$	2	$2\frac{3}{4}$	4	4	6	8	8	10	12	18	22
45° Ell	$3/4$	$3/4$	1	$1\frac{3}{8}$	$1\frac{3}{4}$	2	$2\frac{1}{2}$	3	4	$4\frac{1}{2}$	5	8	10
Insert Coupling	—	$1/2$	$3/4$	1	$1\frac{1}{4}$	$1\frac{1}{2}$	2	3	3	—	4	$6\frac{1}{4}$	—
Male-Female Adapters	—	1	$1\frac{1}{2}$	2	$2\frac{3}{4}$	$3\frac{1}{2}$	$4\frac{1}{2}$	—	$6\frac{1}{2}$	—	9	14	—

3.10.13 Head Loss of Water Flow Through Rigid Plastic Pipe

Head Loss Characteristics of Water Flow Thru Rigid Plastic Pipe

This nomograph provides approximate values for a wide range of plastic pipe sizes. More precise values should be calculated from the Williams and Hazen formula. Experimental test value of C (a constant for inside pipe roughness) ranges from 155 to 165 for various types of plastic pipe. Use of a value of 150 will ensure conservative friction loss values.

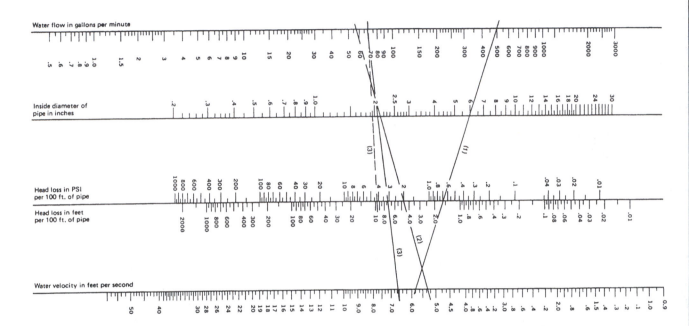

The values of this chart are based on the Williams & Hazen formula:

$$f = 0.2083 \left(\frac{100}{C} \right)^{1.852} \times \frac{g^{1.852}}{Di^{4.8655}}$$

$$= .0983 \frac{g^{1.852}}{Di^{4.8655}} \quad \text{for } C = 150$$

$$P = 4335f$$

Where:
- f   = Friction Head in ft. of Water per 100 ft. of Pipe
- P   = Pressure Loss in psi per 100 ft. of Pipe
- Di = Inside Pipe Diameter, in.
- g   = Flow Rate in U.S. gal./min.
- C   = Constant for Inside Roughness (C equals 150 for thermoplatics)

The nomograph is used by lining up values on the scales by means of a ruler or straight edge. Two independent variables must be set to obtain the other values. For example: line (1) indicates that 500 gallons per minute may be obtained with a 6-inch inside diameter pipe at a head loss of about 0.65 pounds per square inch at a velocity of 6.0 feet per second. Line (2) indicates that a pipe with 2.1 inch inside diameter will give a flow of about 60 gallons per minute at a loss in head of 2 pounds per square inch per 100 feet of pipe. Line (3) and lotted line (3) show that in going from a pipe 2.1 inch inside diameter to one of 2 inches inside diameter, the head loss goes from 3 to 4 pounds per square inch in obtaining a flow of 70 gallons per minute. Remember, velocities in excess of 5.0 feet per second are not recommended.

Nomograph courtesy of Plastics Pipe Institute, a division of The Society of The Plastics Industry.

3.10.14 Deflection in Thermoplastic Pipe

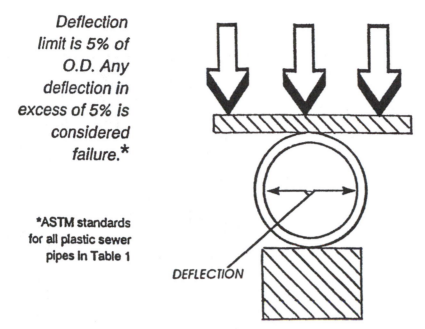

Deflection limit is 5% of O.D. Any deflection in excess of 5% is considered failure.*

*ASTM standards for all plastic sewer pipes in Table 1

DEFLECTION

Figure 3.10.14   *(Reprinted by permission of the Cast Iron Soil Pipe Institute)*

3.10.15 Expansion and Contraction of PVC Pipe

PVC non-metallic conduit will expand and contract with temperature variations. When it is necessary to allow for movement of PVC conduit because of temperature changes, the amount of movement can be determined from the chart below. The coefficient of thermal expansion of J-M PVC conduit is 3.0×10^{-5} in/in°F. If major temperature variations are expected, the use of expansion joints should be considered and should be installed in accordance with the engineer's design.

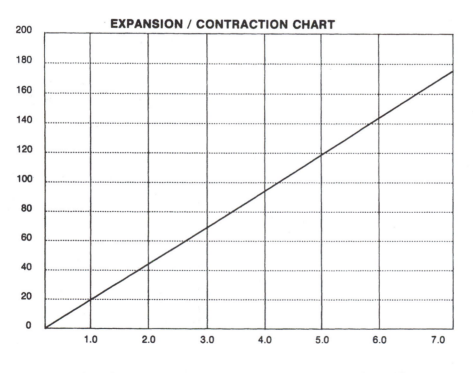

EXPANSION / CONTRACTION CHART

CHANGE IN LENGTH (INCHES) PER 100-FOOT LENGTH OF DUCT

Figure 3.10.15   *(By permission J-M Manufacturing Company, Inc., Livingston, N.J.)*

3.10.16 Expansion Characteristics of Metal and Plastic Pipe in Graph Form

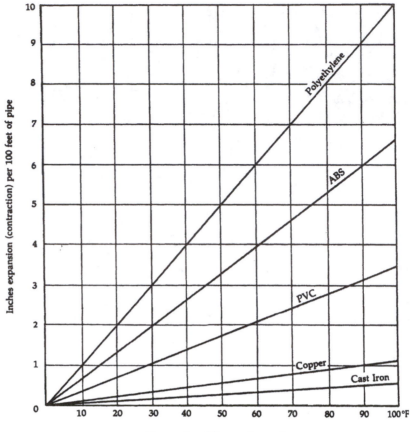

Temperature difference degrees F.

Example: Find the expansion allowance required for a 120 ft. run of ABS pipe in a concrete & masonry building and for a temperature difference of 90°F.

Answer: At a temperature difference of 90°F read from the chart, ABS expands 6″ and concrete expands ¼″.

$$(6 - ¼) \times \frac{120}{100} = 5¼ \times \frac{120}{100} = 6.3 \text{ inches}$$

Figure 3.10.16   *(By permission of Cast Iron Soil Pipe Institute, Chattanooga, Tn.)*

3.10.17 Expansion Characteristics of Various Metal/Thermoplastic Pipe

Expansion: Allowances for expansion and contraction of building materials are important design considerations. Material selection can create or prevent problems. Cast iron is in tune with building reactions to temperature. Its expansion is so close to that of steel and masonry that there is no need for costly expanison joints and special offsets. That is not always the case with other DWV materials.

Thermal expansion of various materials.			
Material	Inches per inch 10^{-6} X per °F	Inches per 100′ of pipe per 100°F.	Ratio-assuming cast iron equals 1.00
Cast iron	6.2	0.745	1.00
Concrete	5.5	0.66	.89
Steel (mild)	6.5	0.780	1.05
Steel (stainless)	7.8	0.940	1.26
Copper	9.2	1.11	1.49
PVC (high impact)	55.6	6.68	8.95
ABS (type 1A)	56.2	6.75	9.05
Polyethylene (type 1)	94.5	11.4	15.30
Polyethylene (type 2)	83.3	10.0	13.40

Here is the *actual* increase in length for 50 feet of pipe and 70° temperature rise.

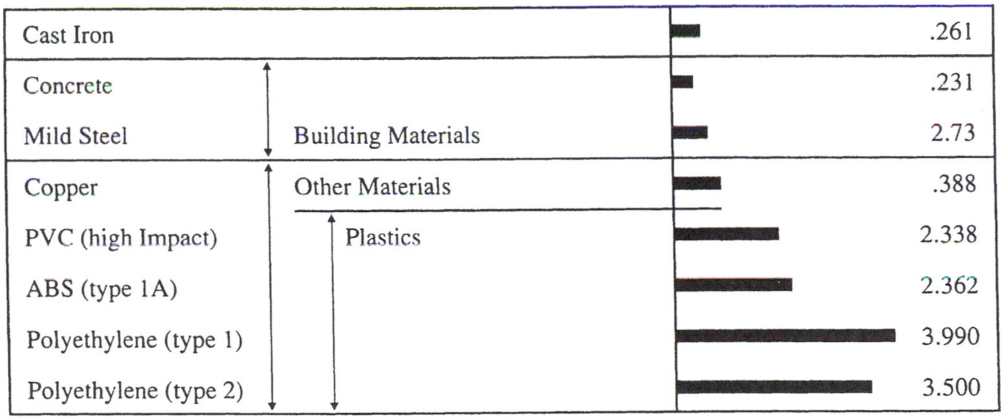

Cast Iron			.261
Concrete			.231
Mild Steel	Building Materials		2.73
Copper	Other Materials		.388
PVC (high Impact)	Plastics		2.338
ABS (type 1A)			2.362
Polyethylene (type 1)			3.990
Polyethylene (type 2)			3.500

Figure 3.10.17 *(By permission Cast Iron Soil Pipe Institute)*

3.10.18 Recommended Torque on Flange Bolts—PVC Pipe

Recommended Torque

Pipe Size (IPS)	No. Bolt Holes	Bolt Diameter	Approx. Bolt Length*	Recommended Torque ft/lbs
$1/2$	4	$1/2$	$2 1/2$	10-15
$3/4$	4	$1/2$	$2 1/2$	10-15
1	4	$1/2$	$2 1/2$	10-15
$1 1/4$	4	$1/2$	3	10-15
$1 1/2$	4	$1/2$	3	10-15
2	4	$5/8$	3	20-30
$2 1/2$	4	$5/8$	3	20-30
3	4	$5/8$	$3 1/2$	20-30
4	8	$5/8$	4	20-30
6	8	$3/4$	4	33-50
8	8	$7/8$	5	33-50
10	12	$7/8$	5	53-75
12	12	$7/8$	5	53-75

*Bolt lengths were calculated using two flanges. Additional accessories or different mating surfaces will alter these numbers.

3.10.19 Above-Ground Installation Procedures

Above-Ground Installation

Support Spacing

When thermoplastic piping systems are installed above-ground, they must be properly supported to avoid unnecessary stresses and possible sagging.

Horizontal runs require the use of hangers as described below, spaced approximately as indicated in tables in individual material sections. Note that additional support is required as temperatures increase. Continuous support can be accomplished by the use of smooth structural angle or channel.

Where the pipe is exposed to impact damage, protective shields should be installed.

Tables are based on the maximum deflection of a uniformly loaded, continuously supported beam calculated from:

$$y = .00541 \frac{wL^4}{EI}$$

Where:

y = Deflection or sag (in.)
w = Weight per unit length (lb./in.)
L = Support spacing (in.)
E = Modulus of elasticity at given temperature (lb./in.^{2})
I = Moment of inertia (in.^{4})

If 0.100 in. is chosen arbitrarily as the permissible sag (y) between supports, then:

$$L^4 = 18.48 \frac{EI}{w}$$

Where:

w = Weight of pipe + weight of liquid (lb./in.)

For a pipe $I = \frac{\pi}{64} (Do^4 - Di^4)$

Where:

Do = Outside diameter of the pipe (in.)
Di = Inside diameter of the pipe (in.)

Then:

$$L = \left(.907 \frac{E}{W} (Do^4 - Di^4)\right)^{1/4}$$

$$= .976 \left(\frac{E}{W} Do^4 - Di^4\right)^{1/4}$$

3.10.20 Making the Joint

1. Cutting: Pipe must be squarely cut to allow for the proper interfacing of the pipe end and the fitting socket bottom. This can be accomplished with a mitre box saw or wheel type cutter. Wheel type cutters are not generally recommended for larger diameters since they tend to flare the corner of the pipe end. If this type of cutter is used, the flare on the end must be completely removed.

Note: Power saws should be specifically designed to cut plastic pipe.

2. Deburring: Use a knife, plastic deburring tool, or file to remove burrs from the end of small diameter pipe. Be sure to remove all burrs from around the inside as well as the outside of the pipe. A slight chamfer (bevel) of about 10°-15° should be added to the end to permit easier insertion of the pipe into the fitting. Failure to chamfer the edge of the pipe may remove cement from the fitting socket, causing the joint to leak. For pressure pipe systems of 2" and above, the pipe must be end-treated with a 15° chamfer cut to a depth of approximately $3/_{32}$." Commercial power bevelers are recommended.

3. Test Dry Fit of the Joint: Tapered fitting sockets are designed so that an interference fit should occur when the pipe is inserted about $1/_3$ to $2/_3$ of the way into the socket. Occasionally, when pipe and fitting dimensions are at the tolerance extremes, it will be possible to fully insert dry pipe to the bottom of the fitting socket. When this happens, a sufficient quantity of cement must be applied to the joint to fill the gap between the pipe and fitting. The gap must be filled to obtain a strong, leak-free joint.

A 15° chamfer cut to a depth of approx. $3/_{32}$."

Step 1:

Step 2:

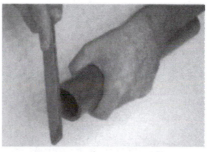

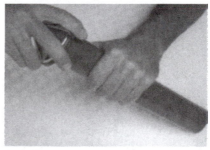

Step 3:

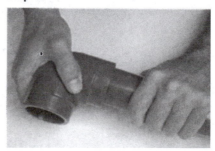

Figure 3.10.20

4. <u>Inspection, Cleaning, Priming:</u> Visually inspect the inside of the pipe and fitting sockets and remove all dirt, grease or moisture with a clean, dry rag or cloth. If wiping fails to clean the surfaces, a chemical cleaner must be used. Check for possible damage such as splits or cracks and replace if necessary.

<u>Depth-of-Entry Mark:</u> Marking the depth of entry is a way to check if the pipe has reached the bottom of the fitting socket in step #6. Measure the fitting socket depth and mark this distance on the pipe O.D. You may want to add several inches to the distance and make a second mark as the primer and cement will most likely destroy your first one.

Apply primer to the surface of the pipe and fitting socket with a natural bristle brush. This process softens and prepares the PVC or CPVC for the solvent cementing step. Move quickly without hesitation to the cementing procedure **while surfaces are still wet** with primer.

5. <u>Application of Solvent Cement:</u> Apply the solvent cement evenly and quickly around the outside of the *pipe* and at a width a little greater than the depth of the fitting socket while the primer is still wet.

Apply a light coat of cement evenly around the inside of the fitting socket. Avoid puddling.

Apply a second coat of cementing to the pipe end.

Note: When cementing bell-end pipe be careful not to apply an excessive amount of cement to the bell socket or spigot end. This will prevent solvent damage to the pipe. For buried pipe applications, do not throw empty primer or cement cans into the trench along side the pipe. Cans of cement and primer should be closed at all times when not in use to prevent evaporation of chemicals and hardening of cement.

Caution: Primers and cements are extremely flammable and must not be stored or used near heat or open flame. Read all warnings on primer and cement cans.

Step 4:

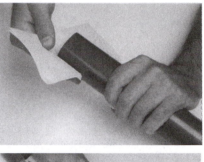

Step 5:

Figure 3.10.20 *(cont.)*

6. <u>Joint Assembly:</u> Working quickly, insert the pipe into the fitting socket bottom and give the pipe or fitting a $1/4''$ turn to evenly distribute the cement. Do not continue to rotate the pipe after it has hit the bottom of the fitting socket. A good joint will have sufficient cement to make a bead all the way around the outside of the fitting hub. The fitting will have a tendency to slide back on the pipe while the cement is wet, so hold the joint tightly together for about 15 seconds. For pipe sizes 4″ and above, greater axial forces are necessary for the assembly of interference fit joints. Mechanical forcing equipment may be needed to join the pipe and hold the joint until the cement "sets." The joint may have to be held together for up to 3 minutes. Consult the factory for specifics.

Note: Always wait at least 24 hours before pressure testing a piping system to allow cemented joints to cure properly. For colder temperatures, it may be necessary to wait a longer period of time.

7. <u>Clean-up & Joint Movement:</u> Remove all excess cement from around the pipe and fitting with a dry, cotton rag or cloth. This must be done while the cement is still soft.

The joint should not be disturbed immediately after the cementing procedure and sufficient time should be allowed for proper curing of the joint. Exact drying time is difficult to predict because it depends on variables such as temperature, humidity and cement integrity. For more specific information, contact your solvent cement manufacturer.

Step 6:

Step 7:

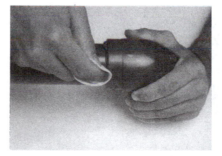

Figure 3.10.20 *(cont.)*

3.10.21 Joining Plastic Pipe in Hot Weather

There are many occasions when solvent cementing plastic pipe in 95°F temperatures and over cannot be avoided. If special precautions are taken, problems can be avoided.

Solvent cements for plastic pipe contain high-strength solvents which evaporate faster at elevated temperatures. This is especially true when there is a hot wind blowing. If the pipe is stored in direct sunlight, surface temperatures may be 20°F to 30°F above air temperature. Solvents attack these hot surfaces faster and deeper, especially inside a joint. Thus it is very important to avoid puddling inside socket and to wipe off excess cement outside.

By following our standard instructions and using a little extra care, as outlined below, successful solvent cemented joints can be made in even the most extreme hot weather conditions.

Tips to Follow When Solvent Cementing in High Temperatures

1. Store solvent cements and primers in a cool or shaded area prior to use.

2. If possible, store fitting and the pipe, or at least the ends to be solvent welded, in shady area before cementing.

3. Cool surfaces to be joined by wiping with a damp rag. Be sure that surface s dry prior to applying solvent cement.

4. Try to do the solvent cementing in cooler morning hours.

5. Make sure that both surfaces to be oined are still wet with cement when putting them together. With large size pipe more people on the crew may be necessary.

6. Use one of our heavier, high viscosity cements since they will provide a little more working time.

As you know, during hot weather there can be a greater expansion-contraction factor.

Good Joints Can Be Made at Sub-Zero Temperatures

By following our standard instructions and using a little extra care and patience, successful solvent cemented joints can be made at temperatures even as low as -15°F. In cold weather, solvents penetrate and soften the surfaces more slowly than in warm weather. Also the plastic is more resistant to solvent attack. Therefore, it becomes more important to presoften surfaces with a primer. And, because of slower evaporation, a longer cure time is necessary. Cure schedules already allow a wide margin for safety. For colder weather, simply allow more time.

Tips to Follow in Solvent Cementing During Cold Weather

1. Prefabricate as much of the system as possible in a heated working area.

2. Store cements and primers in a warmer area when not in use and make sure they remain fluid.

3. Take special care to remove moisture including ice and snow.

4. Use a primer to soften the joining surfaces before applying cement.

5. Allow a longer cure period before the system is used.

6. Read and follow all of our directions carefully before installation.

Regular cements are formulated to have well balanced drying characteristics and to have good stability in sub-freezing temperatures. Some manufacturers offer special cements for cold weather because their regular cements do not have that same stability.

For all practical purposes, good solvent cemented joints can be made in very cold conditions with our existing products provided proper care and a little common sense are used.

3.10.22 Joining Plastic Pipe in Cold Weather

Working in freezing temperatures is never easy. But sometimes the job is necessary. If that unavoidable job includes cementing plastic pipe . . . you can DO IT SUCCESSFULLY WITH REGULAR CEMENTS.

3.10.23 Recommended Spans for Various Size PVC Pipe

Recommended Support Spacing* (In Feet)

Nom. Pipe Size (In.)	PVC Pipe									CPVC Pipe					
	Schedule 40					Schedule 80					Schedule 80					
	Temp. °F					Temp. °F					Temp. °F					
	60	80	100	120	140	60	80	100	120	140	60	80	100	120	140	180
$1/2$	$4^1/2$	$4^1/2$	4	$2^1/2$	$2^1/2$	5	$4^1/2$	$4^1/2$	3	$2^1/2$	$5^1/2$	$5^1/2$	5	$4^1/2$	$4^1/2$	$2^1/2$
$3/4$	5	$4^1/2$	4	$2^1/2$	$2^1/2$	$5^1/2$	5	$4^1/2$	3	$2^1/2$	$5^1/2$	$5^1/2$	$5^1/2$	5	$4^1/2$	$2^1/2$
1	$5^1/2$	5	$4^1/2$	3	$2^1/2$	6	$5^1/2$	5	$3^1/2$	3	6	6	6	$5^1/2$	5	3
$1^1/4$	$5^1/2$	$5^1/2$	5	3	3	6	6	$5^1/2$	$3^1/2$	3	$6^1/2$	$6^1/2$	6	6	$5^1/2$	3
$1^1/2$	6	$5^1/2$	5	$3^1/2$	3	$6^1/2$	6	$5^1/2$	$3^1/2$	$3^1/2$	7	7	$6^1/2$	6	$5^1/2$	$3^1/2$
2	6	$5^1/2$	5	$3^1/2$	7	7	$6^1/2$	6	4	$3^1/2$	7	7	7	$6^1/2$	6	$3^1/2$
$2^1/2$	7	$6^1/2$	6	4	$3^1/2$	$7^1/2$	$7^1/2$	$6^1/2$	$4^1/2$	4	8	$7^1/2$	$7^1/2$	$7^1/2$	$6^1/2$	4
3	7	7	6	4	$3^1/2$	8	$7^1/2$	7	$4^1/2$	4	8	8	8	$7^1/2$	7	4
4	$7^1/2$	7	$6^1/2$	$4^1/2$	4	9	$8^1/2$	$7^1/2$	5	$4^1/2$	9	9	9	$8^1/2$	$7^1/2$	$4^1/2$
6	$8^1/2$	8	$7^1/2$	5	$4^1/2$	10	$9^1/2$	9	6	5	10	$10^1/2$	$9^1/2$	9	8	5
8	9	$8^1/2$	8	5	$4^1/2$	11	$10^1/2$	$9^1/2$	$6^1/2$	$5^1/2$	11	11	$10^1/2$	10	9	$5^1/2$
10	10	9	$8^1/2$	$5^1/2$	5	12	11	10	7	6	$11^1/2$	$11^1/2$	11	$10^1/2$	$9^1/2$	6
12	$11^1/2$	$10^1/2$	$9^1/2$	$6^1/2$	$5^1/2$	12	11	10	7	6	$12^1/2$	$12^1/2$	$12^1/2$	11	$10^1/2$	$6^1/2$
14	12	11	10	7	6	$13^1/2$	13	11	8	7						
16	$12^1/2$	$11^1/2$	$10^1/2$	$7^1/2$	$6^1/2$	14	$13^1/2$	$11^1/2$	$8^1/2$	$7^1/2$						

Note: This data is based on information supplied by the raw material manufacturers. It should be used as a general recommendation only and not as a guarantee of performance or longevity.

*Chart based on spacing for continuous spans and for uninsulated lines conveying fluids of specific gravity up to 1.00.

3.10.24 Recommended Hangers for Plastic Piping Systems

Hangers

A number of hangers designed for use with metal pipe are suitable for thermoplastic pipe as well. These include the shoe support, clamp, clevis, sling and other roller types. The hangers should, however, be modified to increase the bearing area. This is accomplished by inserting a protective sleeve of medium-gage sheet metal between the pipe and the hanger.

The pipe hangers, of whatever type, should be carefully aligned and there must be no rough or sharp edges in contact with the pipe. Plastic pipe must never be allowed to rub against any abrasive surface. If it rests on concrete piers, for example, wooden or thermoplastic pads should be used between the pipe and the concrete surface.

Vertical lines must also be supported at intervals so that the fittings at the lower end are not overloaded. The supports should be of the kind that do not exert a compressive strain on the pipe, such as the double-bolt type. Riser-type clamps that squeeze the pipe are not recommended. If possible, each clamp should be located just below a coupling or other fitting so that the shoulder of the coupling provides bearing support to the clamp.

Recommended Hangers for Plastic Piping Systems

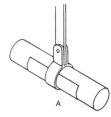

A

Band Hanger with Protective Sleeve

E

Roller Hanger

B

Clevis

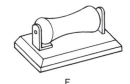

F

Pipe Roll and Plate

C

Adjustable Solid Ring Swivel Type

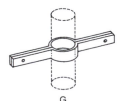

G

Riser Clamp

D

Single Pipe Roll

H

Double-Bolt Clamp

3.10.25 Anchorage Details

A Typical Method of Anchorage of a Change in Direction

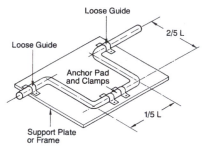

Typical Method of Anchorage

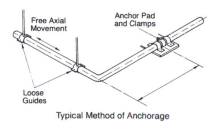

Typical Method of Anchorage

Typical Support Arrangements

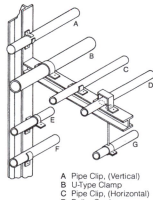

Note 1:
Pipes Must Be Free
to Move Axially

A Pipe Clip, (Vertical)
B U-Type Clamp
C Pipe Clip, (Horizontal)
D Roller Carrier
E Angle Bracket with U-Clamp
F Clamp (Vertical)
G Suspended Ring Clamp

Anchors and Guides

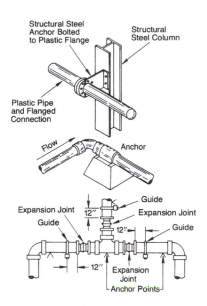

Anchors in a piping system direct movement of pipe within a defined reference frame. At the anchoring point, there is no axial or transverse movement. Guides are used to allow axial movement of pipe but prevent transverse movement. Anchoring and guides should be engineered to provide the required function without point loading the plastic pipe.

Guides and anchors are used whenever expansion joints are used and are also on long runs and directional changes in piping.

Continuous Support Arrangements

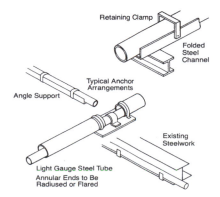

3.10.26 Storage and Handling of Plastic Pipe

For extended storage, the piping components should be covered with a light tarpaulin or kept under cover in a warehouse or shed that is well ventilated to prevent excessive temperature buildup and possible warpage. The storage area should not be located near steam lines or other heat sources.

To prevent sagging or "draping," particularly of the longer sections, pipe should be stored on racks that provide close or continuous support. Any sharp edges or burrs on the racks should be removed or covered. To prevent excessive deflection, loose stacks of pipe should not exceed a height of three feet. Bundled pipe can be stacked twice as high.

Fittings and flanges should be kept in their original packaging or in separate bins until they are needed. They should never be mixed in with metal piping components.

Since plastic pipe has lower impact strength and resistance to mechanical abuse than steel, it requires somewhat more care in handling. Pulling a length of pipe off a truck bed and letting the free end plummet to the ground should be avoided. So should dragging the pipe over rough ground, dropping heavy objects on it, or using any kind of chains. The resulting scratches, splits or gouges can reduce the pressure rating.

If damage from careless handling does occur, one of the advantages of plastic pipe is readily apparent. The damaged section can be quickly cut out and the pipe ends rejoined using the cutting and joining techniques described below.

Solvent Welding PVC and CPVC Pipe and Fittings

Basic Principles

The solvent cemented connection in thermoplastic pipe and fittings is the last vital link in a plastic pipe installation. It can mean the success or failure of the system as a whole. Accordingly, it requires the same professional care and attention that are given to other components of the system.

There are many solvent cementing techniques published covering step by step procedures on just how to make solvent cemented joints. However, we feel that if the basic principles involved are explained, known and understood, a better understanding would be gained, as to what techniques are necessary to suit particular applications, temperature conditions, and variations in sizes and fits of pipe and fittings.

To consistently make good joints the following should be clearly understood:

1. The joining surfaces must be dissolved and made semi-fluid.

2. Sufficient cement must be applied to fill the gap between pipe and fitting.

3. Assembly of pipe and fittings must be made while the surfaces are still wet and fluid.

4. Joint strength develops as the cement dries. In the tight part of the joint the surfaces will tend to fuse together, in the loose part the cement will bond to both surfaces.

Penetration and dissolving can be achieved by a suitable primer, or by the use of both primer and cement. A suitable primer will penetrate and dissolve the plastic more quickly than cement alone. The use of a primer provides a safety factor for the installer for he can know, under various temperature conditions, when he has achieved sufficient softening.

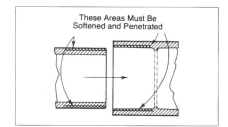
These Areas Must Be Softened and Penetrated

3.10.27 Basic Solvent Welding Principles

More than sufficient cement to fill the loose part of the joint must be applied. Besides filling the gap, adequate cement layers will penetrate the surface and also remain wet until the joint is assembled. Prove this for yourself. Apply on the top surface of a piece of pipe two separate layers of cement. First flow on a heavy layer of cement, then along side it a thin brushed out layer. Test the layers every 15 seconds or so by a gentle tap with your finger. You will note that the thin layer becomes tacky and then dries quickly (probably within 15 seconds). The heavy layer will remain wet much longer. Now check for penetration a few minutes after applying these layers. Scrape them with a knife. The thin layer will have achieved little or no penetration. The heavy one much more penetration.

As the solvent dissipates, the cement layer and the dissolved surfaces will harden with a corresponding increase in joint strength. A good joint will take the required working pressure long before the joint is fully dry and final strength will develop more quickly than in the looser (bonded) part of the joint.

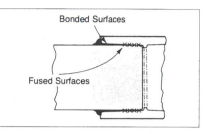

If the cement coatings on the pipe and fittings are wet and fluid when assembly takes place, they will tend to flow together and become one cement layer. Also, if the cement is wet the surfaces beneath them will still be soft, and these softened surfaces in the tight part of the joint will tend to fuse together.

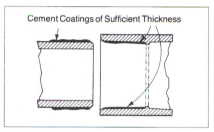

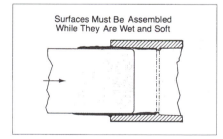

3.10.28 Basic Threading Techniques

Threading

While threaded thermoplastic systems are not recommended for high-pressure system, piping layouts where leaks would be dangerous, or for larger pipe sizes (more than two inches), they have two definite advantages. They quickly can be dismantled for temporary or take-down applications; and they can be used to join plastic to nonplastic materials.

The following recommendations for making threaded joints with thermoplastic pipe and fittings are adapted from PPI (Plastics Pipe Institute) Technical Note No. 8:

1. Thread only pipes that have wall thicknesses equal to or greater than those of Schedule 80 pipe.

2. For pressure-rated pipes of PVC and CPVC reduce the pressure rating of threaded pipe to one-half that of unthreaded pipe.

3. To cut the threads, use only pipe dies designed for plastic pipes. Keep the dies clean and sharp. Do not cut other materials with them.

4. Vises for holding the pipe during thread cutting and pipe wrenches should be designed and used in such a manner that the pipe is not damaged. Strap wrenches are recommended. Wooden plugs can be inserted into the end of the pipe, if needed to prevent distortion of the pipe walls and cutting of off-center threads.

5. The following general procedure for cutting threads may be used:

    A. Use a die stock with a proper guide so the die will start and go on square to the pipe axis. Any burrs or sharp edges on the guide that can scratch the pipe must be removed.

    B. Do not use cutting oil. However, a drop of oil may be rubbed onto the chasers occasionally. This prevents tearing and helps to promote clean, smooth threads.

6. Before assembly, the threads should be lubricated and sealed with a non-hardening pipe dope or wrapped with Teflon® tape.

7. In making up threaded joints, care must be taken not to overtighten the joints. Generally, $1/2$ to one thread past hand-tight is adequate if the male thread is wrapped with Teflon tape. Further tightening may split female threaded plastic parts.

8. In general, applications for threaded plastic pipe fittings fall into two categories:

    A. Fittings for use in an all plastic system where both the male and female parts are plastic.

    B. Fittings for use as transition fittings from plastic to metal.

Theoretically, it is possible to use any combination of threaded parts such as:

1. Metal male to plastic female.

2. Plastic male to plastic female.

3. Metal female to plastic male.

Practical experience, however, suggests that the METAL MALE TO PLASTIC FEMALE combination is more susceptible to premature failure than the other two applications.

The reason for this is due to the incompressability of metal. Standard instructions call for the male part to be run in hand tight and then tightened $1/2$ turn more. It has been our observation, however, that it is very common to find male metal parts screwed in for a total of 7 to 8 threads. This results in excessively high stress levels in the plastic female part.

The tensile strength of the Type I PVC is 7200 psi. However, all fittings have bondlines (where the melted material joins together after flowing around the core which forms the waterway) which are the weakest portions of the fitting. The tensile strength at the bondline is therefore lower than the minimum of 7200 psi. A metal nipple screwed in $7^1/2$ turns will generate a stress of approximately 6600 psi. This means that if the

3.10.29 American Standard Taper Thread Dimensions

fitting doesn't crack open immediately, there will probably be a small crack initiated on the inside which will ultimately cause failure. It is for this reason that George Fischer Sloane recommends that its threaded plastic pipe fittings be used only in the following two combinations:

1. PLASTIC MALE TO PLASTIC FEMALE

2. PLASTIC MALE TO METAL FEMALE

If it is absolutely necessary to use a plastic female thread for transition to metal nipple, then it is IMPERATIVE that the nipple not be turned more than $1/2$ turn past HANDTIGHT ("fingertight"for strong hands). To insure a leakproof joint, a good sealant is recommended (Teflon® tape or Teflon® pipe dope).

Note: If metal male to plastic female connections are used it is recommended that a steel ring or band clamp be used on the outside of the female adapter.

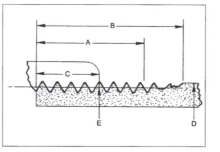

Teflon® is a registered trademark of E.I. DuPont de Nemours & Co.

Note: Angle between sides of thread is 60 degrees. Taper of thread, on diameter, is 3/4 inch per foot.

The basic thread is 0.8 x pitch of thread and the crest and root are truncated an amount equal to 0.033 x pitch, excepting 8 threads per inch which have a basic depth of 0.788 x pitch and are truncated 0.045 x pitch at the crest and 0.033 x pitch at the root.

American Standard Taper Pipe Thread Dimensions

Pipe		Thread					
Nominal Size (in.)	Outside Diameter (in.)	Number of Threads Per Inch	Normal Engagement by hand (in.)	Length of Effective Thread (in.)	Total Length End of Pipe to Vanish Point (in.)	Pitch Diameter at End of Internal Thread (in.)	Depth of Thread (Max.) (in.)
	D		C	A	B	E	
$1/8$	.405	27	.180	.2639	.3924	.39476	.02963
$1/4$	.540	18	.200	.4018	.5946	.48989	.04444
$3/8$	.675	18	.240	.4078	.6006	.62701	.04444
$1/2$	.840	14	.320	.5337	.7815	.77843	.05714
$3/4$	1.050	14	.339	.5457	.7935	.98887	.05714
1	1.315	$11 1/2$	.400	.6828	.9845	1.23863	.06957
$1 1/4$	1.660	$11 1/2$	.420	.7068	1.0085	1.58338	.06957
$1 1/2$	1.900	$11 1/2$	.420	.7235	1.0252	1.82234	.06957
2	2.375	$11 1/2$	.436	.7565	1.0582	2.29627	.06957
$2 1/2$	2.875	8	.682	1.1375	1.5712	2.76216	.10000
3	3.500	8	.766	1.2000	1.6337	3.38850	.10000
$3 1/2$	4.000	8	.821	1.2500	1.6837	3.88881	.10000
4	4.500	8	.844	1.3000	1.7337	4.38713	.10000
5	5.563	8	.937	1.4063	1.8400	5.44929	.10000
6	6.625	8	.958	1.5125	1.9472	6.50597	.10000
8	8.625	8	1.063	1.7125	2.1462	8.50003	.10000
10	10.750	8	1.210	1.9250	2.3587	10.62094	.10000
12	12.750	8	1.360	2.1250	2.5587	12.61781	.10000

3.10.30 Flanging

Flanged PVC and CPVC pipe has an advantage when used in a system where there is need to dismantle the pipe occasionally or when the system is temporary and mobility is required. Flanging can also be used when it is environmentally impossible to make solvent cemented joints on location.

Selection of Materials

1. <u>Gasket:</u> full-faced elastomeric (Durometer "A" scale of 55 to 80, usually 1/8" thick). Must be resistant to chemicals flowing through the line.

2. <u>Fasteners:</u> bolts, nuts and washers, also resistant to the chemical environment. (Threads should be well lubricated.)

3. <u>Torque Wrench:</u> a necessity for tightening bolts in a manner that guards against excessive torque.

Flange Assembly

1. <u>Join the flange to the pipe</u> as outlined in the solvent cementing section or in the threading section depending on the joining method desired.

2. <u>Align the flanges and gasket</u> by inserting all of the bolts through the matching bolt holes. Proper mating of flanges and gaskets is very important for a positive seal.

3. <u>Using a torque wrench, tighten each bolt</u> in a gradual sequence as outlined by the flange sketch. For final tightening of all bolts, find the recommended torque value in the chart on page 28.

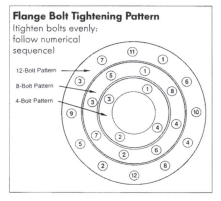

Flange Bolt Tightening Pattern
(tighten bolts evenly: follow numerical sequence)

12-Bolt Pattern
8-Bolt Pattern
4-Bolt Pattern

Note:
1. Do not over-torque flange bolts.
2. Use the proper bolt tightening sequence.
3. Make sure the system is in proper alignment.
4. Flanges should not be used to draw piping assemblies together.
5. Flat washers must be used under every nut and bolt head.

3.10.31 Below-Ground Installation

Trenching and Bedding

1. Depth: When installing underground piping systems, the depth of the trench is determined by the intended service and by local conditions (as well as by local, state and national codes that may require a greater trench depth and cover than are technically necessary).

   Underground pipes are subjected to external loads caused by the weight of the backfill material and by loads applied at the surface of the fill. These can range from static to dynamic loads.

   Static loads comprise the weight of the soil above the top of the pipe plus any additional material that might by stacked above ground. An important point is that the load on a flexible pipe will be less than on a rigid pipe buried in the same manner. This is because the flexible conduit transfers part of the load to the surrounding soil and not the reverse. Soil loads are minimal with narrow trenches until a pipe of 10 feet is attained.

   Dynamic loads are loads due to moving vehicles such as trucks, trains and other heavy equipment. For shallow burial conditions live loads should be considered and added to static loads, but at depths greater than 10 feet, live loads have very little effect.

   For static and dynamic soil loading tables, refer to specific materials sections, PVC and CPVC.

   Pipe intended for potable water service should be buried at least 12 inches below the maximum expected frost penetration.

2. Bedding: The bottom of the trench should provide a firm, continuous bearing surface along the entire length of the pipe run. It should be relatively smooth and free of rocks. Where hardpan, ledge rock or boulders are present, it is recommended that the trench bottom be cushioned with at least four (4) inches of sand or compacted fine-grained soils.

3. Snaking: To compensate for thermal contraction, the snaking technique of offsetting the pipe with relation to the trench centerline is recommended.

   Example: Snaking is particularly important when laying small diameter pipe in hot weather. For example, a 100-foot length of PVC Type I pipe will expand or contract about $3/4''$ fo reach 20°F temperature change. On a hot summer day, the direct rays of the sun on the pipe can drive the surface temperature up to 150°F. At night, the air temperature may drop to 70°F. In this hypothetical case, the pipe would undergo a temperature change of 80°F — and every 100 feet of pipe would contract 3". This degree of contraction would put such a strain on newly cemented pipe joints that a poorly made joint might pull apart.

   Installation: A practical and economical method is to cement the line together at the side of the trench during the normal working day. When the newly cemented joints have dried, the pipe is snaked from one side of the trench to the other in gentle, alternative curves. This added length will compensate for any contraction after the trench is back filled (see "Snaking of Pipe Within Trench illustration below).

The "Snaking Length" table below gives the required loop length, in feet, and offset in inches, for various temperature variations.

Snaking of Pipe Within Trench

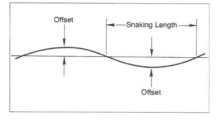

Snaking of thermoplastic pipe within trench to compensate for contraction.

Figure 3.10.31   *(cont.)*

3.10.32 Snaking Length vs. Offset (in.) to Compensate for Thermal Contraction

Snaking Length, (ft.)	Maximum Temperature Variation (°F) Between Time of Cementing and Final Backfilling									
	10°	20°	30°	40°	50°	60°	70°	80°	90°	100°
	Loop Offset, (in.)									
20	2.5	3.5	4.5	5.20	5.75	6.25	6.75	7.25	7.75	8.00
50	6.5	9.0	11.0	12.75	14.25	15.50	17.00	18.00	19.25	20.25
100	13.0	18.0	22.0	26.00	29.00	31.50	35.00	37.00	40.00	42.00

3.10.33 Backfilling Procedures

Backfilling

Before making the final connections and backfilling, the pipeline should be cooled to near the temperature of the soil. During hot weather, for example, backfilling should be done early in the morning, when the solvent-cemented joints are completely dried and the line is fully contracted.

Assuming that the pipe is uniformly and continuously supported over its entire length on firm, stable material, it should first be covered with 6 to 8 inches of soil that is free of debris and rocks larger than on-half inch in diameter. This initial layer should be compacted by hand or, preferably, by mechanical tamper so that it acts as a protective cushion against the final backfill. Any large, sharp rocks that could penetrate the tampered layer around the pipe should be removed from the final backfill.

Heavy Traffic: When plastic pipe is installed beneath streets, railroads or other surfaces that are subjected to heavy traffic and resulting shock and vibration, it should be run within a protective metal or concrete casing.

Locating Buried Pipe: The location of plastic pipelines should be accurately recorded at the time of installation. Since pipe is a non-conductor, it does not respond to the electronic devices normally used to locate metal pipelines. However, a copper or galvanized wire can be spiraled around, taped to or laid alongside or just above the pipe during installation to permit the use of a locating device.

Note: For additional information, see ASTM D-2774, "Underground Installation of Thermoplastic Piping."

3.10.34 Compaction of Backfill

The pipe, once installed and inspected, must be backfilled.

- **Cast iron soil pipe—Special compaction of the backfill is not necessary** except for meeting the requirements of normal compaction of the excavated area. Because cast iron is **"rigid,"** it does not depend on sidefill support.

- **Thermoplastic sewer pipe**—The **"flexible"** pipe design is dependent on sidefill support to gain **"stiffness"** to control deflections within acceptable limits (Fig. 3.10.34.1). **Compaction in six-inch maximum layers is required** to the springline of the pipe. Compaction around the pipe must be by hand. As noted earlier, trench width must be sufficient to allow this compaction. Depending on soil type, minimum density compaction can range from 85 to 95 percent. If the installation does not have suitable backfill material available, it must be imported.

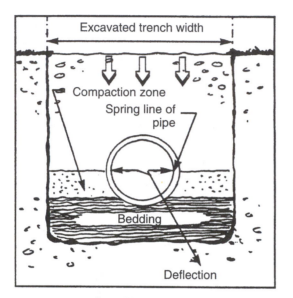

Figure 3.10.34.1 Special bedding requirements per ASTM D 2321-89.

3.10.35 Deflection

Deflection in all piping materials must be controlled in order to prevent obstruction to flow and to assure that the joints remain secure.

- **Cast iron soil pipe**—Because cast iron is rigid, deflection of the pipe wall is almost nonexistent.
- **Thermoplastic sewer pipe**—A **"flexible"** pipe (Fig. 3.10.35.1) is dependent on sidefill support to gain "stiffness" and some deflection of the pipe wall is both normal and expected. This deflection must be controlled within predetermined limits to assure clearance for inspection, cleaning, meeting flow requirements, and integrity of joint seals. The amount of allowed deflection must be determined before installation with a maximum of 5 percent deflection.

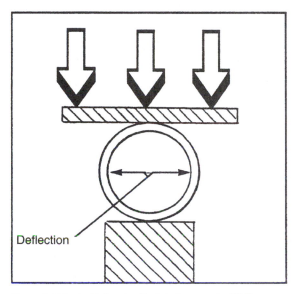

Figure 3.10.35.1 Deflection limit is 5 percent of O.D.; any deflection in excess is considered failure.

3.10.36 Soil Loads and Pipe Resistance for Flexible Thermoplastic Pipe

Trench Widths for PVC

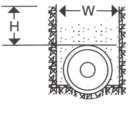

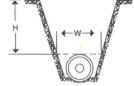

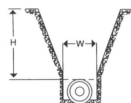

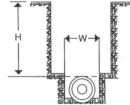

Note: W = Trench Width at Top of Pipe.

Soil Load and Pipe Resistance for Flexible Thermoplastic Pipe — PVC Schedule 80 Pipe

Nom. Size	Wc' = Load Resistance of Pipe (lb./ft.) Schedule 80 Pipe		H=Height of fill Above Pipe	Wc = Soil Loads at Various Trench Widths at Top of Pipe (lb./ft.)			
	E' = 200	E' = 700	(ft.)	2 ft	3 ft.	4 ft.	5ft.
1½	1375	1561	10	106	125	136	152
			20	138	182	212	233
			30	144	207	254	314
			40	—	214	269	318
2	1161	1400	10	132	156	170	190
			20	172	227	265	291
			30	180	259	317	392
			40	—	267	337	398
2½	1593	1879	10	160	191	210	230
			20	204	273	321	352
			30	216	306	377	474
			40		323	408	482
3	1416	1772	10	196	231	252	280
			20	256	336	392	429
			30	266	266	384	469
			40	—	394	497	586
3½	1318	1731	10	223	266	293	320
			20	284	380	446	490
			30	300	426	524	660
			40	—	450	568	670
4	1266	1735	10	252	297	324	360
			20	328	432	540	551
			30	342	493	603	743
			40	—	506	639	754
5	1206	1796	10	310	370	407	445
			20	395	529	621	681
			30	417	592	730	918
			40	—	625	790	932
6	1323	2028	10	371	437	477	530
			20	484	636	742	812
			30	503	725	888	1093
			40	—	745	941	1110
8	1319	2250	10	483	569	621	690
			20	630	828	966	1057
			30	656	945	1156	1423
			40	—	970	1225	1415
10	1481	2649	10	602	710	774	860
			20	785	1032	1204	1317
			30	817	1177	1405	1774
			40	—	1209	1527	1801
12	1676	3067	10	714	942	919	1020
			20	931	1225	1429	1562
			30	969	1397	1709	2104
			40	—	1434	1811	2136

Note 1: Figures are calculated from minimum soil resistance values (E' = 200 psi for uncompacted sandy clay foam) and compacted soil (E' = 700 for side-fill soil that is compacted to 90% or more of Proctor Density for distance of two pipe diameters on each side of the pipe). If Wc' is less than Wc at a given trench depth and width, then soil compaction will be necessary.

Note 2: These are soil loads only and do not include live loads.

3.10.37 Friction Loss in Thermoplastic Pipe

Friction-Loss Characteristics

Introduction

A major advantage of thermoplastic pipe is its exceptionally smooth inside surface area, which reduces friction loss compared to other materials.

Friction loss in plastic pipe remains constant over extended periods of time, in contrast to some other materials where the value of the Hazen and Williams C factor (constant for inside roughness) decreases with time. As a result, the flow capacity of thermoplastics is greater under fully turbulent flow conditions like those encountered in water service.

C Factors

Tests made both with new pipe and pipe that had been in service revealed C factor values for plastic pipe between 160 and 165. Thus, the factor of 150 recommended for water in the equation below is on the conservative side. On the other hand, the C factor for metallic pipe varies from 65 to 125, depending upon age and interior roughening. The obvious benefit is that with plastic systems it is often possible to use a smaller diameter pipe and still obtain the same or even lower friction losses.

Hazen and Williams Formula

The head losses resulting from various water flow rates in plastic piping may be calculated by means of the Hazen and Williams formula:

$$f = 0.2083 \left(\frac{100}{C}\right)^{1.852} \times \frac{q^{1.852}}{Di^{4.8655}}$$

$$= .0983 \frac{q^{1.852}}{Di^{4.8655}} \quad \text{for } C = 150$$

$$P = 4335f$$

Where:
- f = Friction Head in ft. of Water per 100 ft. of Pipe
- P = Pressure Loss in psi per 100 ft. of Pipe
- Di = Inside Pipe Diameter, in.
- g = Flow Rate in U.S. gal./min.
- C = Constant for Inside Roughness (C equals 150 for thermoplastics)

Figure 3.10.37

3.10.38 Glossary of Terms

Abrasion Resistance: Ability to withstand the effects of repeated wearing, rubbing, scraping, etc.

Acceptance Test: An investigation performed on an individual lot of a previously qualified product, by, or under the observation of, the purchaser to establish conformity with a purchase agreement.

Acetal Plastics: Plastics based on resins having a predominance of acetal linkages in the main chain.

Acids: One of a class of substances compounded of hydrogen and one or more other elements, capable of uniting with a base to form a salt, and in aqueous solution, turning blue litmus paper red.

Acrylate Resins: A class of thermoplastic resins produced by polymerization of acrylic and acid derivatives.

Acrylonitrile-Butadiene-Sytrene (ABS) Pipe and Fitting Plastics: Plastics containing polymers and/or blends of polymers, in which the minimum butadiene content is 6 percent, the minimum acrylonitrile content is 15 percent, the minimum syrene and/or substituted styrene content is 15 percent, and the maximum content of all othe monomers is not more than 5 percent, and lubricants, stabilizers and colorants.

Adhesive: A substance capable of holding materials together by surface attachment.

Adhesive, solvent: An adhesive having a volatile organic liquid as a vehicle.

Aging: The effect of time on plastics exposed indoors at ordinary conditions of temperature and relatively clean air.

Alkalies: Compounds capable of neutralizing acids and usually characterized by an acrid taste. Can be mild like baking soda or highly caustic like lye.

Aliphatic: Derived from or related to fats and other derivatives of the parrafin hydrocarbons, including unsaturated compounds of the ethylene and acetylene series.

Alkyd Resins: A class of resins produced by condensation of a polybasic acid or anhydride and a polyhydric alcohol.

Allyl Resins: A class of resins produced from an ester or other derivative of allyl alcohol by polymerization.

Anneal: To prevent the formation of or remove stresses in plastic parts by controlled cooling from a suitable temperature.

Antioxidant: A compounding ingredient added to a plastic composition to retard possible degradation from contact with oxygen (air), particularly in processing at or exposures to high temperatures.

Aromatic: A large class of cyclic organic compounds derived from, or characterized by the presence of the benezene ring and its homologs.

Artificial Weathering: The exposure of plastics to cyclic laboratory conditions involving changes in temperature, relative humidity, and ultraviolet radiant energy, with or without direct water spray, in an attempt to produce changes in the material similar to those observed after long-term continuous outdoor exposure.

Bell End: The enlarged portion of a pipe that resembles the socket portion of a fitting and that is intended to be used to make a joint by inserting a piece of pipe into it. Joining may be accomplished by solvent cements, adhesives, or mechanical techniques.

Beam Loading: The application of a load to a pipe between two pints of support, usually expressed in pounds and the distance between the centers of the supports.

Blister: Undesirable rounded elevation of the surface of a plastic, whose boundaries may be either more or less sharply defined, somewhat resembling in shape or blister on the human skin. A blister may burst and become flattened.

Bond: To attach by means of an adhesive.

Burned: Showing evidence of thermal decomposition through some discoloration, distortion, or destruction of the surface of the plastic.

Burst Strength: The internal pressure required to break a pipe or fitting. This pressure will vary with the rate of build-up of the pressure and the time during which the pressure is held.

Figure 3.10.38

Butylene Plastics: Plastics based on resins made by the polymerization of butene or copolymerization of butene with one or more unsaturated compounds, the butene being in in greatest amount by weight.

Calendering: A process by which a heated rubber plastic product is squeezed between heavy rollers into a thin sheet or film. the film may be frictioned into the interstices of cloth, or it may be coated onto cloth or paper.

Cast Resin: A resinous product prepared by pouring liquid resins into a mold and heat-treating the mass to harden it.

Catalysis: The acceleration (or retardation) of the speed of a chemical reaction by the presence of a comparatively small amount of a foreign substance called a catalyst.

Cellulose: Inert substance, chemically a carbohydrate, which is the chief component of the solid structure of plants, wood, cotton, linen, etc.

Cellulose Acetate: A class of resins made from a cellulose base, either cotton linters or purified wood pulp, by the action of acetic anhydride and acetic acid.

Cement: A dispersion of "solution" of unvulcanized rubber or a plastic in a volatile solvent. This meaning is peculiar to the plastics and rubber industries and may or may not be an adhesive composition.

Chemical Resistance: (1) The effect of specific chemicals on the properties of plastic piping with respect to concentration, temperature and time exposure. (2) The ability of a specific plastic pipe to render service for a useful period in the transport of a specific chemical at a specified concentraiton and temperature.

Coalescence: The union or fusing together of fluid globules or particles to form larger drops or a continuous mass.

Cold Flow: Change in dimensions or shape of some materials when subjected to external weight or pressure at room temperature.

Compound: A combination of ingredients before being processed or made into a finished product. Sometimes used as a synonym for material, formulation.

Condensation: A chemical reaction in which two or more molecules combine, usually with the separation of water or some other substance.

Copolymer: The product of simultaneous polymerization of two or more polymerizeable chemicals, commonly known as monomers.

Crazing: Fine cracks at or under the surface of a plastic.

Creep: The unit elongation of a particular dimension under load for a specific time following the initial elastic elongation caused by load application. It is expressed usually in inches per inch per unit of time.

Cure: To change the properties of a polymeric system into a final, more stable, usable condition by the use of heat, radiation, or reaction with chemical additives.

Deflection Temperature: The temperature at which a specimen will deflect a given distance at a given load under prescribed conditions of test.

Degradation: A deleterious change in the chemical structure of a plastic.

Delamination: The separation of the layers of material in a laminate.

Deterioration: A permanent change in the physical properties of a plastic evidenced by impairment of these properties.

Dielectric Constant: Specific inductive capacity. The dielectric constant of a material is the ratio of the capacitance of a condenser having that material as dielectric to the capacity of the same condenser having a vacuum as dielectric.

Dielectric Strength: This is the force required to drive an electric current through a definite thickness of the material; the voltage required to break down a specified thickness of insulation.

Diffusion: The migration or wandering of the particles or molecules of a body of fluid matter away from the main body through a medium or into another medium.

Dimension Ratio: The diameter of a pipe divided by the wall thickness. Each pipe can have two dimension ratios depending on whether the outside or inside diameter is used. In practice, the outside

Figure 3.10.38 *(cont.)*

diameter is used if the standards requirement and manufacturing control are based on this diameter. The inside diameter is used when this measurement is the controlling one.

Dimensional Stability: Ability of a plastic part to maintain its original proportions under conditions of use.

Dry-Blend: A free-flowing dry compound prepared without fluxing or addition of solvent.

Durometer: Trade name of the Shore Instrument Company for the instrument that measures hardness. The rubber or plastic durometer determines the "hardness" of rubber or plastics by measuring the depth of penetration (without puncturing) of blunt needles compressed on the surface for a short period of time.

Elastic Limit: The load at which a material will no longer return to its original form when the load is released.

Elastomer: The name applied to substances having rubberlike properties.

Electrical Properties: Primarily the resistance of a plastic to the passage of electricity, e.g. dielectric strength.

Elevated Temperature Testing: Tests on plastic pipe above 23°C (73°F).

Elongation: The capacity to take deformation before failure in tension and is expressed as a percentage of the original length.

Emulsion: A dispersion of one liquid in another—possibly only when they are mutually insoluble.

Environmental Stress Cracking: Cracks that develop when the material is subjected to stress in the presence of specific chemicals.

Ester: A compound formed by the elimination of waste during the reaction between an alcohol and an acid; many esters are liquids. They are frequently used as plasticizers in rubber and plastic compounds.

Ethyl Cellulose: A thermoplastic material prepared by the ethylation of cellulose by diethyl sulfate halides and alkali.

Ethylene Plastics: Plastics based on resins made by the polymerization of ethylene or copolymerization of ethylene with one or more unsaturated compounds, the ethylene being in greatest amount by weight.

Extrusion: Method of processing plastic in a continuous or extended form by forcing heat-softened plastic through an opening shaped like the cross-section of the finished project.

Extender: A material added to a plastic composition to reduce its cost.

Fabricate: Method of forming a plastic into a finished article by machining, drawing and similar operations.

Failure, adhesive: Rupture of an adhesive bond, such that the place of separation appears to be at the adhesive-adherence interface.

Fiber Stress: The unit stress, usually in pounds per square inch (psi), in a piece of material that is subjected to external load.

Filler: A material added to a plastic composition to impart certain qualities in the finished article.

Flexural Strength: The outer fiber stress which must be attained in order to produce a given deformation under a beam load.

Formulation: A combination of ingredients before being processed or made into a finished product. Sometimes used as a synonym for material, compound.

Fungi Resistance: The ability of plastic pipe to withstand fungi growth and/or their metabolic products under normal conditions of service or laboratory tests simulating such conditions.

Fuse: To join two plastic parts by softening the material by heat or solvents.

Generic: Common names for types of plastic materials. They may be either chemical terms or coined names. they contrast with trademarks which are the property of one company.

Hardness: A comparative gauge of resistance to indentation, not of surface hardness or abrasion resistance.

Heat Joining: Making a pipe joint by heating the edges of the parts to be joined so that they fuse and become essentially one piece with or without the use of additional material.

Figure 3.10.38 *(cont.)*

Heat Resistance: The ability to withstand the effects of exposure to high temperature. Care must be exercised in defining precisely what is meant when this term is used. Descriptions pertaining to heat resistance properties include: boilable, washable, cigaretteproof, sterilizable, etc.

Hoop Stress: The circumferential stress imposed on a cylindrical wall by internal pressure loading.

Hydrostatic Design Stress: The estimated maximum tensile stress in the wall of the pipe in the circumferential orientation due to internal hydrostatic pressure that can be applied continuously with a high degree of certainty that failure of the pipe will not occur.

Hydrostatic Strength (quick): The hoop stress calculated by means of the ISO equation at which the pipe breaks due to an internal pressure build-up, usually within 60 to 90 seconds.

Impact, Izod: A specific type of impact test made with a pendulum type machine. The specimens are molded or extruded with a machined notch in the center.

Impact Strength: Resistance or mechanical energy absorbed by a plastic part to such shocks as dropping and hard blows.

Impact, Tup: A failing weight (tup) impact test developed specifically for pipe and fittings. There are several variables that can be selected.

Impermeability: Permitting no passage into or thorugh a material.

Injection Molding: Method of forming a plastic into a relatively cool cavity where it rapidly solidifies.

ISO Equation: An equation showing the interrelations between stress, pressure and dimensions of pipe, namely

$$S = \frac{P(ID) + t}{2t} \text{ or } \frac{P(OD) - t}{2t}$$

where:

S = stress
P = pressure
ID = average inside diameter
OD = average outside diameter
t = minimum wall thickness

Ketones: Compounds containing the carbonyl group (CO) to which is attached two alkyl groups. Ketones, such as methyl ethyl ketone, are commonly used as solvents for resins and plastics.

Light Stability: Ability of a plastic to retain its original color and physical properties upon exposure to sun or artificial light.

Light Transmission: The amount of light that a plastic will pass.

Longitudinal Stress: The stress imposed on the long axis of any shape. It can be either a compressive or tensile stress.

Long-Term Hydrostatic Strength: The estimated tensile stress in the wall of the pipe in the circumferential orientation (hoop stress) that when applied continuously will cause failure of the pipe at 100,000 hours (11.43 years). These strengths are usually obtained by extrapolation of log-log regression equations or plots.

Lubricant: A substance used to decrease the friction between solid faces and sometimes used to improve processing characteristics of plastic compositions.

Modulus: The load in pounds per square inch or kilos per square centimeter of initial cross-sectional area necessary to produce a stated percentage-elongation which is used in the physical testing of plastics.

Moisture Resistance: Ability to resist absorption of water.

Molding, Compression: A method of forming objects from plastics by placing the material in a confining mold cavity and applying pressure and usually heat.

Molding, Injection: A method of forming plastics from granular or powdered plastics by the fusing of plastic in a chamber with heat and pressure and then forcing part of the mass into a cooler chamber where it solidifies.

Note: This method is commonly used to form objects from thermoplastics.

Monomer: A relatively simple chemical which can react to form a polymer.

Non-Flammable: Will not support combustion.

Nonrigid Plastic: A plastic which has a stiffness or apparent modulus of elastic-

Figure 3.10.38 *(cont.)*

ity of not over 10,000 psi at 23°C which determined in accordance with the Standard Method of Test for Stiffness in Flexure of Plastics.

Non-Toxic: Non-poisonous.

Nylon Plastics: Plastics based on resins composed principally of a long-chain synthetic polymeric amide which has recurring amide groups as an integral part of the main polymer chain.

Olefin Plastics: Plastics based on resins made by the polymerization of olefins or copolymerization of olefins with other unsaturated compounds, the olefins being in greatest amount by weight. Polyethylene, polypropylene and polybutylene are the most common olefin plastics encountered in pipe.

Orange-Peel: Uneven surface somewhat resembling an orange peel.

Organic Chemical: Originally applied to chemicals derived from living organisms, as distinguished from "inorganic" chemicals found in minerals and inanimate substances; modern chemists define organic chemicals more exactly as those which contain the element carbon.

Outdoor Exposure: Plastic pipe placed in service or stored so that it is not protected from the elements of normal weather conditions, i.e., the sun's rays, rain, air and wind. Exposure to industrial and waste gases, chemicals, engine exhausts, etc. are not considered normal "outdoor exposure."

Permanence: The property of a plastic which describes its resistance to appreciable changes in characteristics with time and environment.

Phenolic Resins: Resins made by reaction of a phenolic compound or tar acid with an aldehyde; more commonly applied to thermosetting resins made from pure phenol.

Plasticity: A property of plastics and resins which allows the material to be deformed continuously and permanently without rupture upon the application of a force that exceeds the yield value of the material.

Plasticizer: A liquid or solid incorporated in natural and synthetic resins and related substances to develop such properties as resiliency, elasticity and flexibility.

Plastics Conduit: Plastic pipe or tubing used as an enclosure for electrical wiring.

Plastics Pipe: A hollow cylinder of a plastic material in which the wall thicknesses are usually small when compared to the diameter and in which the inside and outside walls are essentially concentric.

Plastics Tubing: A particular size of plastics pipe in which the outside diameter is essentially the same as that of copper tubing.

Polybutylene: A polymer prepared by the polynmerization of butene-1 as the sole monomer.

Polybutylene Plastics: Plastics based on polymers made with butene-1 as essentially the sole monomer.

Polyethylenes: A class of resins formed by polymerizing ethylene, a gas obtained from petroleum hydrocarbons.

Polymer: A product resulting from a chemical change involving the successive addition of a large number of relatively small molecules (monomer) to form the polymer, and whose molecular weight is usually a multiple of that of the original substance.

Polymerization: Chemical change resulting in the formation of a new compound whose molecular weight is usually a multiple of that of the original substance.

Polyolefin: A polymer prepared by the polymerization of an olefins(s) as the sole monomer(s).

Polyolefin Plastics: Plastics based on polymers made with an olefin(s) as essentially the sole monomer(s).

Polypropylene: A polymer prepared by the polymerization of propylene as the sole monomer.

Polypropylene Plastics: Plastics based on polymers nmade with propylene as the sole monomer.

Polystyrene: A plastic based on a resin made by polymerization of styrene as the sole monomer.

Note: Pollystyrene may contain minor proportions of lubricants, stabilizers, fillers, pigments and dyes.

Figure 3.10.38 (*cont.*)

Polyvinyl Chloride: Polymerized vinyl chloride, a synthetic resin, which when plasticized or softened with other chemicals has some rubber-like properties. It is derived from acetylene and anhydrous hydrochloric acid.

Polyvinyl Chloride Plastics: Plastics made by combining polyvinyl chloride with colorants, fillers, plasticizers, stabilizers, lubricants, other polymers and other compounding ingredients. Not all of these modifiers are used in pipe compounds.

Porosity: Presence of numerous visible voids.

Power Factor: The ratio of the power in watts delivered in an alternating current circuit (real power) to the voltampere input (apparent power). The power factor of an insulation indicates the amount of the power input which is consumed as a result of the impressed voltage forcing a small leakage current through the material.

Pressure: When expressed with reference to pipe the force per unit area exerted by the medium in the pipe.

Pressure Rating: The estimated maximum pressure that the medium in the pipe can exert continuously with a high degree of certainty that failure of the pipe will not occur.

Propylene Plastics: Plastics based on resins made by the polymerization of propylene or copolymerization of propylene with one or more other unsaturated compounds, the propylene being in greatest amount by weight.

Qualification Test: An investigation, independent of a procurement action, performed on a product to determine whether or not the product conforms to all requirements of the applicable specification.

Note: The examination is usually conducted by the agency responsible for the specification, the purchaser, or by a facility approved by the purchaser, at the request of the supplier seeking inclusion of his product on a qualified products list.

Quick Burst: The internal pressure required to burst a pipe or fitting due to an internal pressure build-up, usually within 60 to 70 seconds.

Resilience: Usually regarded as another name for elasticity. While both terms are fundamentally related, there is a distinction in meaning. Elasticity is a general term used to describe the property of recovering original shape after a deformation. Resilience refers more to the energy of recovery; that is, a body may be elastic but not highly resilient.

Resin: An organic substance, generally synthetic, which is used as a base material for the manufacture of some plastics.

Reworked Material: A plastic material that has been reprocessed, after having been previously processed by molding, extrusions, etc., in a fabricator's plant.

Rigid Plastic: A plastic which has a stiffness or apparent modulus of elasticity greater than 100,000 psi at 23°C when determined in accordance with the Standard Method of Test for Stiffness in Flexure of Plastics.

Rubber: A material that is capable of recovering from large deformations quickly and forcibly.

Sample: A small part or portion that is capable of recovering from large deformations quickly and forcibly,

Saran Plastics: Plastics based on resins made by the polymerization of vinylidene chloride or copolymerization of vinylidene chloride with other unsaturated compounds, the vinylidene chloride being in greatest amount of weight.

Schedule: A pipe size system (outside diameters and wall thicknesses) originated by the iron pipe industry.

Self-Extinguishing: The ability of a plastic to resist burning when the source of heat or flame that ignited it is removed.

Service Factor: A factor which is used to reduce a strength value to obtain an engineering design stress. The factor may vary depending on the service conditions, the hazard, the length of service desired and the properties of the pipe.

Set: To convert an adhesive into a fixed or hardened state by chemical or physical action, such as condensation, polymerization, oxidation, vulcanization, gelatin, hydration or evaporation of volatile constituents.

Simulated Weathering: The exposure of plastics to cyclic laboratory conditions of high and low temperatures, high and low relative humidities and ultraviolet ra-

Figure 3.10.38 (*cont.*)

diant energy in an attempt to produce changes in their properties similar to those observed on long-term continuous exposure outdoors. The laboratory exposure conditions are usually intensified beyond those encountered in actual outdoor exposure in an attempt to achieve an accelerated effect.

Simulated Aging: The exposure of plastics to cyclic laboratory conditions of high and low temperatues, and high and low relative humidities in an attempt to produce changes in their properties similar to those observed on long-time continuous exposure to conditions of temperature and relative humidity commonly encountered indoors or to obtain an acceleration of the effects of ordinary indoor exposure. The laboratory exposure conditions are usually intensified beyond those actually encountered in an attempt to achieve an accelerated effect.

Softening Range: The range of temperature in which a plastic changes from a rigid to a soft nature.

Note: Actual values will depend on the method of test. Sometimes referred to as softening point.

Pressure: When expressed with reference to pipe the force per unit area exerted by the medium in the pipe.

Solvent: The medium within which a substance is dissolved; most commonly applied to liquids used to bring particular solids into solution, e.g., acetone is a solvent for PVC.

Solvent Cement: In the plastic piping field, a solvent adhesive that contains a solvent that dissolves or softens the surfaces being bonded so that the bonded assembly becomes essentially one piece of the same type of plastic.

Solvent Cementing: Making a pipe joint with a solvent cement.

Specific Gravity: Ratio of the mass of a body to the mass of an equal body of volume of water at 4°C, or some other specified temperature.

Specific Heat: Ratio of the thermal capacity of a substance to that of water at 15°C.

Specimen: An individual piece or portion of a sample used to make a specific test. Specific tests usually require specimens of specific shape and dimensions.

Stabilizer: A chemical substance which is frequently added to plastic compounds to inhibit undesirable changes in the material, such as discoloration due to heat or light.

Standard Dimension Ratio: A selected series of numbers in which the dimension ratios are constants for all sizes of pipe for each standard dimension, ratio and which are the USASI Preferred Number Series 10 modified by +1 or −1. If the outside diameter (OD) is used the modifier is +1. If the outside diameter (ID) is used the modifier is −1.

Standard Thermoplastic Pipe Materials Designation Code: A means for easily identifying a thermoplastic pipe material by means of three elements. The first element is the abbreviation for the chemical type of the plastic in accordance with ASTM D-1600. The second is the type and grade (based on properties in accordance with the ASTM materials specification): in the case of ASTM specifications which have no types and grades or those in the cell structure system, two digit numbers are assigned by the PI that are used in place of the larger numbers. The third is the recommended hydrostatic design stress (RHDS) for water at 23°C (73°F) in pounds per square inch divided by 100 and with decimals dropped, e.g., PVC 1120 indicates that the plastic in polyvinyl chloride, Type I, Grade 1 according to ASTM D-1748 with a RHDS of 2000 psi for water at 73°F. PE 3306 indicates that the plastic is polyethylene. Type III Grade 3 according to ASTM D-1248 with a RHDS of 630 psi for water at 73°F. PP 1208 is polypropylene. Class I-19509 in accordance with ASTM D-2146 with a RHDS of 800 psi for water at 73°F; the designation of PP 12 for polypropylene Class I-19509 will be covered in the ASTM and Product Standards for polyproptylene pipe when they are issued.

Stiffness Factor: A physical property of plastic pipe that indicates the degree of flexibility of the pipe when subjected to external loads.

Strain: The ratio of the amount of deformation to the length being deformed caused by the application of a load on a piece of material.

Strength: The mechanical properties of a plastic, such as a load or weight-carrying ability, and ability to withstand sharp blows. Strength properties include tensile, flexural and tear strength, toughness, flexibility, etc.

Figure 3.10.38 *(cont.)*

Stress: When expressed with reference to pipe the force per unit area in the wall of the pipe in the circumferential orientation due to internal hydrostatic pressure.

Stress-Crack: External or internal cracks in a plastic caused by tensile stresses less than that of its short-term mechanical strength.

Note: The development of such cracks is frequently accelerated by the environment to which the plastic is exposed. The stresses which cause cracking may be present internally or externally or may be combinations of these stresses. The appearance of a network of fine cracks is called crazing.

Stress Relaxation: The decrease of stress with respect to time in a piece of plastic that is subject to an external load.

Styrene Plastics: Plastics based on resins made by the polymerization of styrene or copolymerization of styrene with other unsaturated compounds, the styrene being in greatest amount by weight.

Styrene-Rubber (SR) Pipe and Fittings Plastics: Plastics containing at least 50 percent styrene plastics combined with rubbers and other compounding materials, but not more than 15 percent acrylonitrile.

Styrene-Rubber Plastics: Compositions based on rubbers and styrene plastics, the styrene plastics being in greatest amount by weight.

Sustained Presure Test: A constant internal pressure test for 1000 hours.

Tear Strength: Resistance of a material to tearing (strength).

Tensile Strength: The capacity of a material to resist a force tending to stretch it. Ordinarily the term is used to denote the force required to stretch a material to rupture, and is known variously as "breaking load," "breaking stress," "ultimate tensile strength," and sometimes erroneously as "breaking strain." In plastics testing, it is the load in pounds per square inch or kilos per square centimeter of original cross-sectional area, supported at the moment of rupture by a piece of test sample on being elongated.

Thermal Conductivity: Capacity of a plastic material to conduct heat.

Thermal Expansion: The increase in length of a dimension under the influence of a change in temperature.

Thermoforming: Forming with the aid of heat.

Thermoplastic Materials: Materials which soften when heated to normal processing temperatures without the occurrence of appreciable chemical change, but are quickly hardened by cooling. Unlike the thermosetting materials they can be reheated to soften, and recooled to "set," almost indefinitely; they may be formed and reformed many times by heat and pressure.

Thermoset: A plastic which, when cured by application of heat or chemical means, changes into a substantially infusible and insoluble product.

Thermosetting: Plastic materials which undergo a chemical change and harden permanently when heated in processing. Further heating will not soften these materials.

Translucent: Permitting the passage of light, but diffusing it so that objects beyond cannot be clearly distinguished.

Vinyl Chloride Plastics: Plastics based on resins made by the polymerization of vinyl chloride or copolymerization of vinyl chloride with minor amounts (not over 50 percent) of other unsaturated compounds.)

Vinyl Plastics: Plastics based on resins made from vinyl monomers, except those specifically covered by other classifications, such as acrylic and styrene plastics. Typical vinyl plastics are polyvinyl chloride, polyvinyl acetate, polyvinyl alcohol, and polyvinyl butyral, and copolymers of vinyl monomers with unsaturated compounds.

Virgin Material: A plastic material in the form of pellets, granules, powder, floc or liquid that has not been subjected to use or processing other than that required for its original manufacture.

Viscosity: Inernal friction of a liquid because of its resistance to shear, agitation or flow.

Volatile: Property of liquids to pass away by evaporation.

Volume Resistivity: The electrical resistance of a 1-centimeter cube of

Figure 3.10.38 *(cont.)*

the material expressed in ohm-centimeters.

Water Absorption: The percentage by weight of water absorbed by a sample immersed in water. Dependent upon area exposed.

Water Vapor Transmission: The penetration of a plastic by moisture in the air.

Weather Resistance: Ability of a plastic to retain its original physical properties and appearance upon prolonged exposure to outdoor weather.

Weld-or-Knitline: A mark on a molded plastic formed by the union of two or more streams of plastic flowing together.

Welding: The joining of two or more pieces at adjoining or nearby areas either with or without the addition of plastic from another source.

Yield Point: The point at which a material will continue to elongate at no substantial increase in load during a short test period.

Yield Stress: The force which must be applied to a plastic to initiate flow.

Figure 3.10.38 *(cont.)*

3.11.0 Ductile Iron Pipe Dimensions/Weights for Push-on Mechanical Joint Pipe

Standards Applicable to Ductile Iron Pipe and Fittings

Thickness Design of Ductile Iron Pipe	ANSI/AWWA C150/A21,50
Ductile Iron Pipe for Water and Other Liquids	ANSI/AWWA C151/A21, 51 FEDERAL WWP421D, Grade C
Ductile Iron Pipe for Gravity Flow Service	ANSI/ASTM A746
Ductile Iron Fittings for Water and Other Liquids 30 in through 36 in	ANSI/AWWA C110/A21.10
Ductile Iron Compact Fittings 3 in through 24 in	ANSI/AWWA C153/A21.53
Flanged Fittings	ANSI/AWWA C110/A21.10 ANSI B16-1
Ductile Iron Pipe with Threaded Flanges	ANSI/AWWA C115/21.15
Castings and Linings: Asphaltic	ANSI/AWWA C151/A21.51 ANSI/AWWA C110/A21.10 ANSI/AWWA C153/A21.53
Cement Lining	ANSI/AWWA C104/A21.4
Various Epoxy Linings and Casings	MANUFACTURER'S STANDARD
Exterior Polyethylene Encasement	ANSI/AWWA C105/A21.5
Joints—Pipe and Fittings	
Push-on and Mechanical Rubber-Gasket joints	ANSI/AWWA C111/A21.11 FEDERAL WWP421D
Flanged	ANSI/AWWA C115/A21.15 ANSI B16.1
Grooved and Shouldered	ANSI/AWWA C606
Pipe Threads	ANSI B2.1
Installation	ANSI/AWWA C600

3.11.1 Laying Conditions

Laying Conditions

Type 1†*	Type 2	Type 3	Type 4	Type 5
Flat-bottom trench.† Loose backfill.	Flat-bottom trench.† Backfill lightly consolidated to centerline of pipe.	Pipe bedded in 4-in. minimum loose soil.‡ Backfill lightly consolidated to top of pipe.	Pipe bedded in sand, gravel or crushed stone to depth of ⅛ pipe diameter, 4-in. minimum. Backfill compacted to top of pipe. (Approximately 80% Standard Proctor, AASHTO§ T-99).	Pipe bedded in compacted granular material to centerline of pipe. Compacted granular or select** material to top of pipe. (Approximately 90% Standard Proctor, AASHTO§-T-99)

Notes: Consideration of the pipe-zone embedment conditions included in this figure may be influenced by factors other than pipe strength. For additional information on pipe bedding and backfill, see ANSI/AWWA C600.

* For nominal pipe sizes 14 in and larger, consideration should be given to the use of laying conditions other than Type 1.

† Flat bottom is defined as undisturbed earth.

‡ Loose soil or select material is defined as native soil excavated from the trench, free of rocks, foreign materials, and frozen earth.

§ American Association of State Highway and Transportation Officials, 444 N. Capitol St. N.W., Suite 225, Washington, DC 20001.

Figure 3.11.1

The success of a pipe product is due not only to its material composition but also to its proper design. The design must consider and respond to many adverse influences. The primary design considerations for ductile iron pipe are as follows:

Trench Load

This consideration includes the basic static earth load, the dynamic truck load together with the force of the resulting impact upon pipe structure. To resist this influence, a design ring bending stress of 48,000 psi is utilized.

Internal Pressure

Here the concern is with the intended working pressure of the pipeline together with an additional surge allowance of 100 psi. These values combine to provide the required design standard. In response to this requirement, a yield strength of 42,000 psi is utilized.

Laying Conditions

Once the above criteria has been determined, it is necessary to decide on how the pipeline shall be bedded and backfilled in the field. There are five standard methods, each offering its particular design value.

The above is intended to provide a general understanding of the design environment of an underground pressure pipe product.

Detailed information is available by referring to the latest AMERICAN NATIONAL STANDARD for the THICKNESS DESIGN of DUCTILE IRON PIPE (ANSI/AWWA C150/A21.50). Griffin Regional Product Engineers are available for consultation regarding unusual design or installation problems.

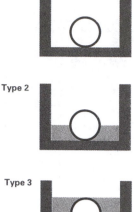

Type 1 Flat bottom trench with loose backfill.[†]

Type 2 Flat bottom trench.[†] Backfill lightly consolidated to centerline of pipe.

Type 3 Pipe bedded in 4 in. minimum loose soil.[††] Backfill lightly consolidated to top of pipe.

Type 4 Pipe bedded in sand, gravel or crushed stone to depth of 1/8 pipe diameter, 4-in. minimum. Backfill compacted to top of pipe. (Approximately 80% Standard Proctor, AASHO T-99.)

Type 5 Pipe bedded in compacted granular material to centerline of pipe. Compacted granular or select[††] material to top of pipe. (Approximately 90% Standard Proctor, AASHO T-99.)

Figure 3.11.1 *(cont.)*

3.11.2 Maximum Deflection—Push-on Pipe Full Length

Push-on Joint Pipe,
Maximum Deflection Full Length Pipe

Size of pipe, in	Max. joint deflection, degrees	Deflection, in		Approximate radius feet of curve, produced by succession of joints, 3 in. same as 4 in	
		18 ft length	20 ft length	18 ft length	20 ft length
3	5		21		230
4	5		21		230
6	5	19	21	206	230
8	5	19		206	
10	5	19		206	
12	5	19		206	
14	5	19		206	
16	5	19		206	
18	5	19		206	
20	5	19		206	
24	5	19		206	
30	5	19		206	
36	5	19		206	

*20-ft length.

3.11.3 Maximum Deflection—Mechanical Joint

Mechanical Joint Pipe,
Maximum Allowable Joint Deflection

Size of Pipe, in	Y, max. Joint Deflection	X, Deflection Inches per 18-ft. Length	Approx. Radius, Feet of Curve, Produced by Succession of Joints 18 ft. Length
3	8° 18'	35*	140*
4	8° 18'	35*	140*
6	7° 7'	30*	160*
48	5° 21'	20	195
10	5° 21'	20	195
12	5° 21'	20	195
14	3° 35'	13.5	285
16	3° 35'	13.5	285
18	3° 0'	11	340
20	3° 0'	11	340
24	2° 23'	9	450

* 20-ft length.

3.11.4 Ductile Iron Specifications, 3- to 36-Inch Standard Pressure Classes of Pipe

Nominal Thickness for Standard
Pressure Classes of Ductile-Iron Pipe

Size, in	Outside diameter, in	Pressure class*				
		150	200	250	300	350
		Nominal thickness, in				
3	3.96	—	—	—	—	0.25†
4	4.80	—	—	—	—	0.25†
6	6.90	—	—	—	—	0.25†
8	9.05	—	—	—	—	0.25†
10	11.10	—	—	—	—	0.26
12	13.20	—	—	—	—	0.28
14	15.30	—	—	0.28	0.30	0.31
16	17.40	—	—	0.30	0.32	0.34
18	19.50	—	—	0.31	0.34	0.36
20	21.60	—	—	0.33	0.36	0.38
24	25.80	—	0.33	0.37	0.40	0.43
30	32.00	0.34	0.38	0.42	0.45	0.49
36	38.30	0.38	0.42	0.47	0.51	0.56

*Pressure classes are defined as the rated water pressure of the pipe in psi. The thicknesses shown are adequate for the rated water working pressure plus a surge allowance of 100 psi. Calculations are based on a minimum yield strength of 42,000 and a 2.0 safety factor times the sum of the working pressure and 100 psi surge allowance.

†Calculated thicknesses for these sizes and pressure ratings are less than those shown above. Presently these are the lowest nominal thicknesses available in these sizes.

Note: Per ANSI/AWWA C150/A21.50 the thicknesses above include the 0.06" service allowance and the casting tolerance listed below by size ranges:

Size, in	Casting tolerances, in
3–8	−0.05
10–12	−0.06
14–36	−0.07

3.11.5 Ductile Iron Specifications, 3- to 36-Inch Push-on Joint Pipe

Standard Dimensions and Weights of
3 in through 36 in Push-on-Joint Ductile Iron Pipe

Size, in	Pressure class	Thickness, in	Outside diameter,* in	18-ft laying length	
				Weight per length,† lb	Ave. weight lb/ft
3§	350	0.25	3.96	185	9.2
4§	350	0.25	4.80	225	11.3
6§	350	0.25	6.90	300	16.6
8	350	0.25	9.05	395	22.0
10	350	0.26	11.10	510	28.4
12	350	0.28	13.20	655	36.4
14	250	0.28	15.30	770	42.9
	300	0.30	15.30	825	45.8
	350	0.31	15.30	850	47.2
16	250	0.30	17.40	940	52.3
	300	0.32	17.40	1000	55.5
	350	0.34	17.40	1060	58.8
18	250	0.31	19.50	1090	60.5
	300	0.34	19.50	1185	65.9
	350	0.36	19.50	1250	65.9
20	250	0.33	21.60	1290	71.6
	300	0.36	21.60	1395	77.6
	350	0.38	21.60	1470	81.6
24	200	0.33	25.80	1550	86.1
	250	0.37	25.80	1725	95.8
	300	0.40	25.80	1855	103.0
	350	0.43	25.80	1985	110.2
30	150	0.34	32.00	2000	111.2
	200	0.38	32.00	2220	123.2
	250	0.42	32.00	2435	135.2
	300	0.45	32.00	2595	144.2
	350	0.49	32.00	2810	156.1
36	150	0.38	38.30	2675	148.7
	200	0.42	38.30	2935	163.1
	250	0.47	38.30	3260	181.1
	300	0.51	38.30	3520	195.5
	350	0.56	38.30	3840	213.4

*Tolerance of OD of spigot end: 3–12 in, ±0.06 in, 14–24 in, +0.05-in, −0.08 in, 30–36 in, +0.08 in, −0.06 in.

†Including bell; calculated weight of pipe rounded off to nearest 5 lbs

‡Including bell; average weight, per foot, based on calculated weight of pipe before rounding.

§Available in 20-ft lengths.

3.11.6 Types of Push-on Restrained Joint (Diagrams)

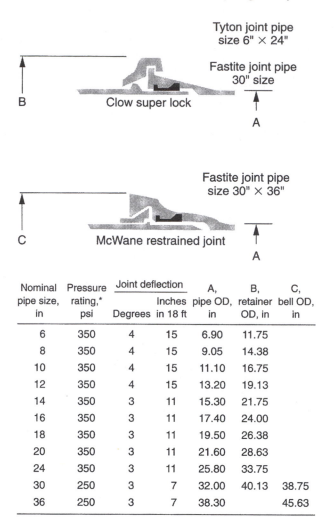

Nominal pipe size, in	Pressure rating,* psi	Joint deflection Degrees	Joint deflection Inches in 18 ft	A, pipe OD, in	B, retainer OD, in	C, bell OD, in
6	350	4	15	6.90	11.75	
8	350	4	15	9.05	14.38	
10	350	4	15	11.10	16.75	
12	350	4	15	13.20	19.13	
14	350	3	11	15.30	21.75	
16	350	3	11	17.40	24.00	
18	350	3	11	19.50	26.38	
20	350	3	11	21.60	28.63	
24	350	3	11	25.80	33.75	
30	250	3	7	32.00	40.13	38.75
36	250	3	7	38.30		45.63

*In the 14-in and larger sizes pressure rating limited to the rating of the pipe barrel thickness selected.

All BND Socket Joint Pipe

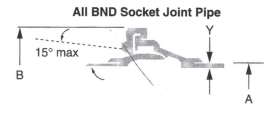

Figure 3.11.6 *(Reprinted by permission from Atlantic State Cast Iron Pipe Company, Phillipsburg, New Jersey)*

3.11.7 Assembly of Field-Cut Type Pipe

When pipe are cut in the field, the cut end may be readily conditioned so that it can be used to make up the next joint. The outside of the cut end should be beveled about ¼-in at an angle of about 30° (Figure 1). This can be quite easily done, with a coarse file or a portable grinder. The operation removes any sharp, rough edges which otherwise might injure the gasket.

Figure 1

When ductile iron pipe 14 in and larger is to be cut in the field, the material should be ordered as "*gauged full length*." Pipe that is gauged full "length" is specially marked to avoid confusion. The ANSI/AWWA standard for ductile iron pipe requires factory gauging of the spigot end. Accordingly, pipe selected for field cutting should also be field-gauged in the location of the cut and found to be within the tolerances shown in Table 1. In the field a mechanical joint gland can be used a gauging device.

Table 1. Suitable Pipe Diameters, for Field Cuts
and Restrained Joint Field Fabrication

Nominal pipe size, in	Min. pipe diameter, in	Max. pipe diameter, in	Min. pipe circumference, in	Max. pipe circumference, in
3	3.90	4.02	12¼	12⅝
4	4.74	4.86	14²⁹⁄₃₂	15⁵⁄₃₂
6	6.84	6.96	21½	21⅞
8	8.99	9.11	28¼	28⅝
10	11.04	11.16	34¹¹⁄₁₆	35¹⁄₁₆
12	13.14	13.26	41⁹⁄₃₂	41²¹⁄₃₂
14	15.22	15.35	47¹³⁄₁₆	48⁷⁄₃₂
16	17.32	17.45	54¹³⁄₃₂	54¹⁵⁄₁₆
18	19.42	19.55	61	61¹³⁄₃₂
20	21.52	21.65	67¹⁹⁄₃₂	68
24	25.72	25.85	80¹³⁄₁₆	81⁷⁄₃₂
30	31.94	32.08	100¹¹⁄₃₂	100²⁵⁄₃₂
36	38.24	38.38	120⅛	120⁹⁄₁₆

Table based on ANSI/AWWA C151/A21.51 guidelines for push-on joints.

THE BACKHOE METHOD OF ASSEMBLY

A backhoe may be used to assemble pipe of intermediate and larger sizes. The plain end of the pipe should be carefully guided by hand into the bell of the previously assembled pipe. The bucket of the backhoe may then be used to push the pipe until fully seated. A timber header should be used between the pipe and backhoe bucket to avoid damage to the pipe.

Figures 3.11.7 (*Reprinted by permission from Atlantic State Cast Iron Pipe Company, Phillipsburg, New Jersey*)

3.11.8 Ductile Iron-Pipe Specifications

Size, in	Thickness Class (A21.51)	Thickness in	A, pipe OD in	B, retainer OD in	Full-length weight,* lb As shipped	Under water Full of air	Under water Full of water	Safe end pull (lb)
6	55	0.40	6.90	13⅞	545	240	465	50,000
8	55	0.42	9.05	16⅞	770	240	655	70,000
10	55	0.44	11.10	19⅛	1005	200	860	95,000
12	55	0.46	13.20	22	1270	155	1080	120,000
14	56	0.51	15.30	24½	1655	160	1410	145,000
16	56	0.52	17.40	27	1990	45	1685	165,000
18	56	0.53	19.50	30	2375	−70	2015	195,000
18	58⁺	0.59			2560	110	2170	
20	56	0.54	21.60	32¾	2810	−200	2375	210,000
20	59⁺	0.63			3110	100	2635	
24	56	0.56	25.80	38¼	3700	−620	3100	260,000
24	62⁺	0.74			4415	95	3715	
30	58	0.71	32.00	46¼	5855	−900	4920	330,000
30	61⁺	0.83			6435	−180	5360	
36	57	0.78	38.30	54½	8145	−1300	6880	400,000
36	59⁺	0.88			8725	−725	7330	

*Weights are for 18 ft 0 in laying lengths. Nominal full lengths vary by size.

Pipe, bell, ball, and retainer are ductile iron.

Dimensions and weights subject to manufacturing tolerances.

6–24-in pressure rating: 350 psi.

30–36-in pressure rating: 250 psi.

⁺Thickness required to overcome buoyancy.

3.11.9 Ductile Iron-Pipe Specifications for 3- to 36-Inch Pipe

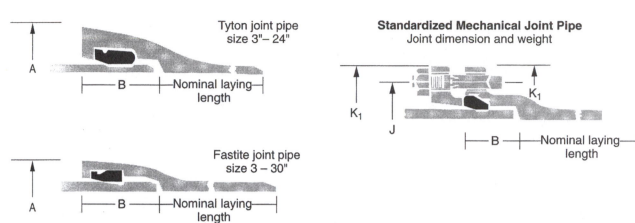

Tyton joint pipe
size 3"– 24"

Fastite joint pipe
size 3 – 30"

Standardized Mechanical Joint Pipe
Joint dimension and weight

Size, in	Pipe thickness,* in		Outside diameter, in	Dimensions, in	
	From	To		A	B
3	0.25	0.40	3.96	5.80	3.00
4	0.25	0.41	4.80	6.86	3.15
6	0.25	0.43	6.90	8.75	3.38
8	0.25	0.45	9.05	11.05	3.69
10	0.26	0.47	11.10	13.15	3.75
12	0.28	0.49	13.20	15.30	3.75
14	0.28	0.51	15.30	17.85	5.00
16	0.30	0.52	17.40	20.00	5.00
18	0.31	0.53	19.50	22.10	5.00
20	0.33	0.54	21.60	24.25	5.00
24	0.33	0.56	25.80	28.50	5.00
30	0.34	0.63	32.00	34.95	6.50
36	0.38	0.73	38.30	41.37	6.50

*3–4-in nominal 20-ft laying length; 6-in nominal 18 or 20 ft; 8–36-in nominal 18-ft laying length.

Dimensions subject to manufacturing tolerances.

3.11.10 Ductile Iron-Pipe with Bolted Glands

Size inches	Pipe thickness inches From	Pipe thickness inches To	Out-side dia-meter inches	B	J	K*	K*	Bolts No	Bolts Size inches	Bolts Length inches	Bell weight pounds	Gland bolts gasket weight pounds
3	0.25	0.40	3.96	2.50	6.19	7.62	7.69	4	⅝	3	11	7
4	0.26	0.41	4.80	2.50	7.50	9.06	9.12	4	¾	3½	16	10
6	0.25	0.43	6.90	2.50	9.50	11.06	11.12	6	¾	3½	18	16
8	0.27	0.45	9.05	2.50	11.75	13.31	13.37	6	¾	4	24	25
10	0.29	0.47	11.10	2.50	14.00	15.62	15.62	8	¾	4	31	30
12	0.31	0.49	13.20	2.50	16.25	17.88	17.88	8	¾	4	37	40
14	0.33	0.51	15.30	3.50	18.75	20.25	20.25	10	¾	4½	61	45
16	0.34	0.52	17.40	3.50	21.00	22.50	22.50	12	¾	4½	74	55
18	0.35	0.53	19.50	3.50	23.25	24.75	24.75	12	¾	4½	85	65
20	0.36	0.54	21.60	3.50	25.50	27.00	27.00	14	¾	4½	98	85
24	0.38	0.56	25.80	3.50	30.00	31.50	31.50	16	¾	5	123	105

*3"–4" nominal 20' laying length. –6" nominal 18' or 20'
8"–24" nominal 18' laying length.
Dimensions subject to manufacturing tolerances.

Figure 3.11.10   (*Reprinted by permission from Atlantic State Cast Iron Pipe Company, Phillipsburg, New Jersey*)

3.11.11 Ductile Iron-Pipe—Rated Working Pressure/Maximum Depth of Cover

Rated Working Pressure & Maximum Depth of Cover

Size in.	Pressure † Class psi	Nominal Thickness in.	Type 1 Trench	Type 2 Trench	Type 3 Trench	Type 4 Trench	Type 5 Trench
			Maximum Depth of cover‡–ft.				
3	350	0.25	78	88	99	100§	100§
4	350	0.25	53	61	69	85	100§
6	350	0.25	26	31	37	47	65
8	350	0.25	16	20	25	34	50
10	350	0.26	11**	15	19	28	45
12	350	0.28	10**	15	19	28	44
14	250	0.28	††	11**	15	23	36
	300	0.30	††	13	17	26	42
	350	0.31	††	14	19	27	44
16	250	0.30	††	11**	15	24	34
	300	0.32	††	13	17	26	39
	350	0.34	††	15	20	28	44
18	250	0.31	††	10**	14	22	31
	300	0.34	††	13	17	26	36
	350	0.36	††	15	19	28	41
20	250	0.33	††	10	14	22	30
	300	0.36	††	13	17	26	35
	350	0.38	††	15	19	28	38
24	200	0.33	††	8**	12	17	25
	250	0.37	††	11	15	20	29
	300	0.40	††	13	17	24	32
	350	0.43	††	15	19	28	37

† Ductile iron pipe is adequate for the working pressure indicated for each nominal size plus a surge allowance of 100 psi (689 kPa). Calculations are based on a 2.0 safety factor times the sum of working pressure and 100 psi (689 kPa) surge allowance. (See ANSI/AWWA C150/A21.50 for design formulae.) Ductile-iron pipe for working pressures higher than 350 psi (2413 kPa) is available.
‡ An allowance for a single H-20 truck with 1.5 impact factor is included for all depths of cover.
§ Calculated maximum depth of cover exceeds 100 ft. (30.5 m).
** Minimum allowable depth cover is 3 ft.(0.9 m).
†† For pipe 14 in. (350 mm) and larger, consideration should be given to the use of laying conditions other than Type 1.

3.11.12 Standard Laying Conditions

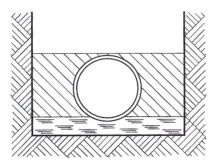

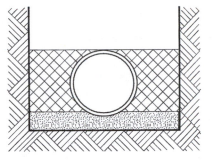

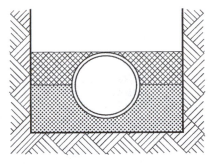

Type 3 - Pipe bedded in 4 in. minimum loose soil.[††] Backfill lightly consolidated to top of pipe.

Type 4 - Pipe bedded in sand, gravel or crushed stone to depth of 1/8 pipe diameter, 4-in. minimum. Backfill compacted to top of pipe. (Approximately 80% Standard Proctor, AASHO T-99.)

Type 5 - Pipe bedded in compacted granular material to centerline of pipe. Compacted granular or select[††] material to top of pipe. (Approximately 90% Standard Proctor, AASHO T-99.)

[††]"Loose soil" or "select material" is defined as native soil excavated from the trench, free of rocks, foreign materials and frozen earth.

3.11.13 Push-on Fitting Assembly Instructions, 4- to 16-Inch Pipe

Assembling AMERICAN Fastite, Lok-Ring, and Flex-Ring fittings is simple. It is very similar to the assembly of Fastite pipe shown in Section 2. (For instructions on complete assembly of Fast-Grip, Flex-Ring, Field Flex-Ring and Lok-Ring joints, check Section 9.) Fast-Grip gaskets may be used in lieu of standard Fastite gaskets in the bells of same size (4"-24") Fastite and Flex-Ring joint pipe and fittings where easy, field-adaptable restraint is desired in pipelines with working pressure up to 250 psi. No Flex-Ring restraining mechanism is necessary when using Fast-Grip gaskets in Flex-Ring bells.

Push-on fittings may be assembled on individual pipe aboveground, or assembled onto the pipe line below ground. Many installers, however, do prefer to pull restrained joint fittings in particular Fast-Grip, Flex-Ring, etc. onto a piece of pipe above ground. This is sometimes accomplished by simply bracing one end of the pipe against a heavy piece of equipment (e.g. backhoe) and pulling the fitting onto the far end of the pipe using the method shown below. Then the pipe and fitting can be lowered as a single unit into the trench. Fastite fittings with Fastite gaskets may be assembled above ground if a situation arises which would make trench assembly difficult. However, in this case the assembly yokes and pulling sling must be kept safely in place and the sling must remain taut while the Fastite assembly is lowered into the ground. This should prevent the Fastite fitting from slipping off the end of the pipe.

1. CLEANING OF SOCKET AND SPIGOT

Clean the socket and plain end of the pipe thoroughly, removing mud, sand, gravel, ice, frozen material, or other matter which could prevent a proper joint seal. Material in the gasket grooves may cause the gasket to protrude into the path of the entering spigot. Therefore, **it is important that all joint recesses be kept clean during insertion of the gasket and assembly of the joint to prevent gasket dislodgment and/or subsequent leakage.**

2. PLACEMENT OF GASKET

Wipe the gasket clean. After flexing one or more "loops" in the gasket, insert the gasket in the gasket recess of the socket with the large sealing end of the gasket toward the rear of the socket (Photo 1). If Fast-Grip gaskets are used, the center of the gasket loops should be positioned between tooth locations. Press the gasket into the mating socket recesses so the metal-carrying retainer end of the gasket is seated completely and uniformly in the socket groove. Take care that no gasket loops or bulges protrude into the path of the entering pipe spigot. In extremely cold weather conditions, gaskets should be warmed before installing.

3. LUBRICATION OF THE JOINT

With a clean brush apply a liberal amount of regular Fastite lubricant completely over the exposed inner surface of the gasket after it is placed in the socket (Photo 2). Also, apply lubricant completely over the plain end of the pipe, the spigot radius, and the outer surface of the pipe up to the assembly stripe (Photo 3).

NOTES:

*Larger diameter fittings which cannot be assembled with equipment shown may be assembled using a similar procedure with heavier or stronger field rigging and more powerful equipment

Figure 3.11.13 *(By permission: American Cast Iron Pipe Company, Birmingham, Alabama)*

Use only lubricant provided by AMERICAN. For underwater or very wet conditions, special AMERICAN

underwater lubricant is recommended and is available upon request. This special lubricant for underwater service is relatively insoluble in water immersion or exposure to flowing water.

4. INITIAL PLACEMENT OF BEVEL END INTO SOCKET †

The spigot end of the pipe should be in reasonably straight alignment when entered in the socket. Center the spigot in the installed gasket, so it makes firm and even contact with the inner surface of the gasket. Do not place pipe spigot in socket while in a substantially deflected position.

5. ASSEMBLY OF LUGGED FITTINGS AND OPTIONAL USE OF YOKES

4"-12" Fastite fittings may normally then be pulled readily onto pipe ends by first firmly supporting the fitting in aligned assembly position, and then pulling on the center of a wire rope or chain sling attached to the fitting lugs with any suitable pulling mechanism (such as a backhoe, come-along, pry bar, etc.) as in Photo 4 until the spigot is fully inserted into the socket. However, if any difficulties in assembly with available rigging are encountered with

any 4"-16" push-on joint fittings, special steel assembly yokes, custom wire rope slings, and come-alongs in all 4"-16" sizes are available from AMERICAN. The simple placement of one the semi-circular steel yokes under the spigot end about 3 or 4 inches back from the spigot stripe and the placement of a wire rope or chain sling through the yoke and attached to the lugs of 4"-12" Fastite fittings as in Photo 5 assures that the pulling force is applied where it is most effective and straight at the springline of the joint.

6. ASSEMBLY OF YOKES

Special steel assembly yokes are particularly helpful in the assembly of 4"-16" Flex-Ring and 14"-16" Fastite fittings, which are standardly furnished without

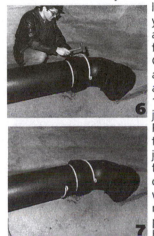

lugs. Such yokes are available from AMERICAN for assembly of all sizes 4"-16" push-on joint fittings. Place one of these yokes just beyond the bell end of the fitting which is connected to the pipe. (It may be necessary to tap the yoke with a hammer.) The yoke should be on the top side of the fitting (Photo 6). Place a second yoke underneath the spigot pipe about 3 or 4 inches from the spigot stripe, 180 degrees from the first yoke (Photo 7). Put a wire rope sling (or other similar device) underneath the fitting, thread the sling through the hooks of the second yoke, and connect the ends of the sling to the pulling mechanism (e.g. backhoe, prybar, come-along with choker, etc.) as shown in Photos 8 & 9. It is likewise possible to assemble even larger diameter fittings with field made rigging similar in

† Joints using Fast-Grip gaskets should be assembled in proper orientation so as to avoid rotation of the fitting after assembly. See Fast-Grip Gasket Assembly Instructions in Secton 9.

Figure 3.11.13 (cont.)

function to the smaller custom steel yokes, but instead using a chain or wire rope sling, etc. looped under the spigot to in similar effect direct a single point pulling force into two pulling sling legs straight at the springline (see Photo 10).

7. COMPLETE ASSEMBLY OF PLAIN END INTO SOCKET

Pull the sling with smooth, steady force until the fitting is "pulled up" and the plain end uniformly contacts the rear of the socket. It is best practice to make assemblies smoothly and progressively in one motion, without repeated "wobbling" (or joint deflection) back and forth. Desired joint deflection may then be set. Any abnormal joint assembly loads or behavior, such as unexplained exposure of the assembly stripe outside the bell, may indicate improper cleaning, gasket insertion, spigot placement, or lubrication. In any such case, it may be advisable to feel for correct gasket positioning by passing a thin (automotive or other) feeler gage between the bell and spigot all around the assembled joint. Any joint with apparent problems (pushed gasket locations found by the probe, etc.) should be disassembled and corrected. (See Section 9 for disassem-

bly involving Fast-Grip gaskets or Field Flex-Rings.)

DOUBLE FITTING ASSEMBLY

In areas involving short lengths of pipe between two fittings, it may be desirable to simultaneously assemble the two fittings onto a single piece of pipe. This may be accomplished with AMERICAN push-on joint fittings due to the self-centering nature of the joints.

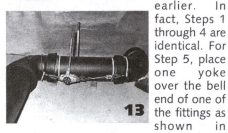

This "double assembly" procedure is similar to the single fitting assembly procedure mentioned earlier. In fact, Steps 1 through 4 are identical. For Step 5, place one yoke over the bell end of one of the fittings as shown in Photo 4. Then, place the second yoke under the bell of the other fitting (Photo 11). Attach a wire rope sling (or other similar device) to the yokes in a manner similar to Step 5 (Photo 12). Note that both yokes can also be placed on the bottoms of the fittings for the same effect, if desired. Complete the assembly by simply pulling the ends of the sling with a pulling mechanism as in Step 5.

Another method for the "double assembly" may be more convenient for pipes of sufficient length and any in-trench applications where there may not be sufficient bracing for the fitting. This

Figure 3.11.13 *(cont.)*

method involves the use of two come-alongs (one on each side of the pipe) in order to provide a reasonably steady, evenly distributed assembly force. The assembly procedure is similar to the previously mentioned procedure. Place yokes on the fittings in the same manner as the other "double assembly" procedure (Photo 9). Then, place one short assembly sling under each of the two fittings and attach the sling to the yokes. Position the come-alongs on each side of the assembly and attach the sling thimbles or loops to both sides of the come-alongs (Photo 13). Work the come-alongs simultaneously to make a smooth, even assembly.

FIELD-CUT PIPE

When pipe are cut in the field, the cut end must be properly prepared prior to assembly into the push-on socket. Using a portable grinder, place an approximately 1/4" to 1/2" long smooth assembly chamfer or bevel on the outside end of the pipe. This bevel should make an angle of 30-45° with the axis of the pipe. Care should be taken to insure that all corners are rounded and no sharp or rough edges remain that might damage or dislodge the gasket. Finally, it is good practice to mark at least a rough assembly stripe on the newly beveled pipe. The distance from the beveled end of the pipe to the opposite edge of the stripe should be about 1/8" less than the socket depth (See Table No. 2-2 or 4-1 for typical socket depths of pipe or fittings, respectively, or measure same in the field). This stripe is helpful in confirming proper joint insertion and also as an indication of joint deflection.

FIELD RIGGING EXAMPLE

Assembly of a larger push-on joint fitting.

Figure 3.11.13 (*cont.*)

3.11.14 Lok-Ring Plug Assembly

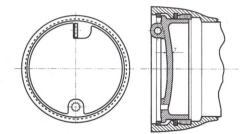

Typical Lok-Ring Plug

NOTE: A Lok-Ring plug in effect simulates the spigot end of a Lok-Ring pipe or fitting, with the outside face of the plug "bulkhead" plate roughly even with the bearing face of the plug abutment (simulating the weld ring on a pipe end). The Lok-Ring plug is normally shipped with the lok-ring bolted down (in effect "backwards") on the beveled plain end barrel of the plug. This lok-ring is completely removed prior to insertion of the plug completely up inside a socket, then the lok-ring is inserted and spread completely into the socket locking groove (in effect "behind" the plug) to restrain it in the socket.

Assembly Instructions:

1. Remove (unbolt) the lok-ring completely off the plug by wrenching or manipulating only the outside closure nut opposite the "locked" (nut) side of the closure mechanism.

2. Prepare spigot and sockets, insert gasket, and lubricate the plug spigot end and gasket in accordance with basic Fastite (and Lok-Ring) joint assembly instructions. See page 2-10.

3. Push or pull the plug completely up inside the socket. Due to the "short" nature of the push-on joint plug, some means is normally necessary to stabilize or brace the plug so that it does not pivot or "buck sideways" in joint assembly. A longer pipe, the end of which can be placed against the outside face of the plug, or a (large bearing face) timber braced between the plug face and the flat lower face of a backhoe bucket are normally quite effective for this purpose. Of course, if the end of a pipe is used to push the plug in, any conventional pipe assembly means could be used to pull or push the pipe and the plug into the socket. Sufficient socket locking groove "width" should be clearly visible after pushing the plug up inside the socket to allow insertion of the lok-ring.

4. Compress the ends of the loose lok-ring together and push it **completely** into the socket locking groove to restrain the plug in the socket.

5. Wrenching only the inside spreader nut opposite the locked side of the closure mechanism, mechanically spread the lok-ring into firm contact with the inner socket surface of the socket locking groove.

6. Inspect the installed lok-ring <u>making sure that the ring is completely inserted in the socket locking groove and completely restrained by the socket restraint lip from one end of the ring to the other end all around the joint</u>. If the ring is out of the groove at any point, correct this condition prior to applying any pressure load to the plug.

Figure 3.11.14 *(By permission: American Cast Iron Pipe Company, Birmingham, Alabama)*

3.11.15 Specifications/Diagrams of Tees and Crosses, 4 to 30 Inches

ANSI/AWWA C153/A21.53* and AMERICAN Standard

Tees and Crosses

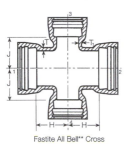

Fastite All Bell** Tee

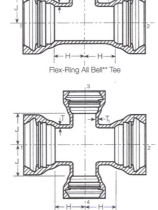

Flex-Ring All Bell** Tee

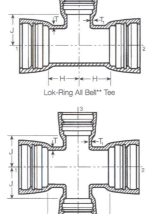

Lok-Ring All Bell** Tee

Fastite All Bell** Cross

Flex-Ring All Bell** Cross

Lok-Ring All Bell** Cross

Table No. 4-10

Size in.		Pressure Rating psi †	Dimensions in Inches				Weight in Pounds					
							Tee			Cross		
Run	Branch		T	T₁	H	J	Fastite All Bell **	**Flex-Ring All Bell	**Lok-Ring All Bell	Fastite All Bell **	**Flex-Ring All Bell	**Lok-Ring All Bell
4	4	350	.34	.34	4.0	4.0	47	80	-	60	-	-
6	4	350	.36	.34	4.0	5.0	64	105	-	-	-	-
6	6	350	.36	.36	5.0	5.0	70	115	-	88	-	-
8	4	350	.38	.34	4.0	6.5	89	125	-	-	-	-
8	6	350	.38	.36	5.0	6.5	94	140	-	-	-	-
8	8	350	.38	.38	6.5	6.5	103	160	-	127	-	-
10	4	350	.40	.34	4.0	7.5	131	185	-	-	-	-
10	6	350	.40	.36	5.0	7.5	135	200	-	-	-	-
10	8	350	.40	.38	6.5	7.5	142	220	-	-	-	-
10	10	350	.40	.40	7.5	7.5	158	255	-	-	-	-
12	4	350	.42	.34	4.0	8.75	167	225	-	-	-	-
12	6	350	.42	.36	5.0	8.75	170	240	-	-	-	-
12	8	350	.42	.38	6.5	8.75	176	265	-	-	-	-
12	10	350	.42	.40	7.5	8.75	191	300	-	-	-	-
12	12	350	.42	.42	8.8	8.75	202	330	-	244	-	-
14	4	350	.47††	.34††	14.0	14.0	-	535	-	-	570	-
14	6	350	.47††	.36††	14.0	14.0	249	545	-	-	595	-
14	8	350	.47††	.38††	14.0	14.0	-	560	-	-	625	-
14	10	350	.47††	.40††	14.0	14.0	266	595	-	-	695	-
14	12	350	.47††	.42††	14.0	14.0	276	625	-	-	750	-
14	14	350	.47††	.47††	14.0	14.0	304	670	-	368	840	-

*4"-16" and 54"-64" Fastite fittings are per AWWA C153.

**18" and larger Fastite and 14" and larger Flex-Ring and Lok-Ring Tees and Crosses may be furnished on the runs with Fastite Plain Ends, Flex-Ring Ends, or Lok-Ring Ends. See Table No. 4-2.

† Higher pressure ratings are available on special applications. Check AMERICAN.

Note: Tees and Crosses with smaller reductions may be available; however, welded-on outlets are normally preferable in these cases from a layout, installation, and economical standpoint. See Section 7.

†† 14" and 16" Flex-Ring fittings may have greater "T" and "T₁" dimensions than those shown.

Figure 3.11.15 *(By permission: American Cast Iron Pipe Company, Birmingham, Alabama)*

ANSI/AWWA C153/A21.53* and AMERICAN Standard

Tees and Crosses

Table No. 4-10 —Continued

Size in.		Pressure Rating psi †	Dimensions in Inches				Weight in Pounds					
							Tee			Cross		
Run	Branch		T	T₁	H	J	Fastite All Bell **	**Flex-Ring All Bell	**Lok-Ring All Bell	Fastite All Bell **	**Flex-Ring All Bell	**Lok-Ring All Bell
16	4	350	.50††	.34††	15.0	15.0	-	645	-	-	675	-
16	6	350	.50††	.36††	15.0	15.0	298	655	-	-	705	-
16	8	350	.50††	.38††	15.0	15.0	302	670	-	-	730	-
16	10	350	.50††	.40††	15.0	15.0	313	705	-	-	800	-
16	12	350	.50††	.42††	15.0	15.0	320	730	-	-	850	-
16	14	350	.50††	.47††	15.0	15.0	347	775	-	-	940	-
16	16	350	.50††	.50††	15.0	15.0	362	805	-	432	1000	-
18	6	350	.75	.55	13.0	15.5	635	810	-	670	850	-
18	8	350	.75	.60	13.0	15.5	650	820	-	695	880	-
18	10	350	.75	.68	13.0	15.5	670	855	-	740	940	-
18	12	350	.75	.75	13.0	15.5	690	875	-	780	990	-
18	14	350	.75	.66	16.5	16.5	825	1015	-	965	1185	-
18	16	350	.75	.70	16.5	16.5	855	1045	-	1025	1240	-
18	18	350	.75	.75	16.5	16.5	880	1125	-	1075	1405	-
20	6	350	.80	.55	14.0	17.0	750	905	-	790	950	-
20	8	350	.80	.60	14.0	17.0	765	920	-	815	980	-
20	10	350	.80	.68	14.0	17.0	790	950	-	860	1040	-
20	12	350	.80	.75	14.0	17.0	805	975	-	900	1090	-
20	14	350	.80	.66	14.0	17.0	845	1020	-	980	1175	-
20	16	350	.80	.70	18.0	18.0	1000	1170	-	1170	1365	-
20	18	350	.80	.75	18.0	18.0	1025	1255	-	1220	1530	-
20	20	350	.80	.80	18.0	18.0	1055	1270	-	1275	1565	-
24	6	350	.89	.55	15.0	19.0	1070	1290	-	1105	1335	-
24	8	350	.89	.60	15.0	19.0	1080	1305	-	1130	1360	-
24	10	350	.89	.68	15.0	19.0	1105	1335	-	1170	1425	-
24	12	350	.89	.75	15.0	19.0	1120	1360	-	1205	1470	-
24	14	350	.89	.66	15.0	19.0	1160	1400	-	1285	1550	-
24	16	350	.89	.70	15.0	19.0	1185	1425	-	1335	1600	-
24	18	350	.89	.75	22.0	22.0	1515	1810	-	1720	2100	-
24	20	350	.89	.80	22.0	22.0	1540	1830	-	1775	2135	-
24	24	350	.89	.89	22.0	22.0	1650	1970	-	1995	2420	-
30	10	250	1.03	.68	18.0	23.0	1757	1952	-	1828	2046	-
30	12	250	1.03	.75	18.0	23.0	1776	1977	-	1867	2097	-
30	14	250	1.03	.66	18.0	23.0	1802	1999	-	1918	2140	-
30	16	250	1.03	.70	18.0	23.0	1824	2019	-	1962	2180	-
30	18	250	1.03	.75	18.0	23.0	1840	2088	-	1994	2318	-
30	20	250	1.03	.80	18.0	23.0	1858	2097	-	2030	2336	-
30	24	250	1.03	.89	25.0	25.0	2369	2640	-	2658	3028	-
30	30	250	1.03	1.03	25.0	25.0	2566	2824	-	3051	3395	-

*4"-16" and 54"-64" Fastite fittings are per AWWA C153.

**18" and larger Fastite and 14" and larger Flex-Ring and Lok-Ring Tees and Crosses may be furnished on the runs with Fastite Plain Ends, Flex-Ring Ends, or Lok-Ring Ends. See Table No. 4-2.

† Higher pressure ratings are available on special applications. Check AMERICAN.

Note: Tees and Crosses with smaller reductions may be available; however, welded-on outlets are normally preferable in these cases from a layout, installation, and economical standpoint. See Section 7.

†† 14" and 16" Flex-Ring fittings may have greater "T" and "T₁" dimensions than those shown.

Figure 3.11.15 (*cont.*)

3.11.16 Specifications/Diagrams of 22½-degree Bends, 4 to 64 Inches

ANSI /AWWA C153/A21.53* and AMERICAN Standard

22 1/2° Bends (1/16th)

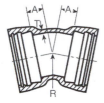
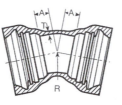
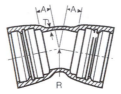

Fastite Bell-Bell**   **Flex-Ring Bell-Bell****   **Lok-Ring Bell-Bell****

Table No. 4-7

Size in.	Pressure† Rating psi	Dimensions in Inches			Weight in Pounds		
		T	A	R	Fastite	Flex-Ring	Lok-Ring
4	350	.34	1.5	2.51	26	45	-
6	350	.36	2.0	5.03	37	65	-
8	350	.38	2.5	6.94	54	90	-
10	350	.40	3.0	8.80	78	140	-
12	350	.42	3.5	10.05	100	180	-
14	350	.47††	7.5	25.12	158	360	-
16	350	.50††	8.0	27.62	185	430	-
18	350	.75	8.5	30.19	450	600	-
20	350	.80	9.5	35.19	540	670	-
24	350	.89	11.0	37.69	810	1005	-
30	250	1.03	15.0	57.81	1460	1635	-
36	250	1.15	18.0	72.88	2220	2345	-
42	250	1.28	21.0	88.00	2625	-	2815
48	250	1.42	24.0	103.06	3695	-	3890
54	250	.90	10.24	37.65	2125	-	2305
60	250	.94	10.63	38.36	2425	-	2650
64	250	.99	11.02	39.06	2785	-	3000

*4"-16" and 54"-64" Fastite fittings are per AWWA C153.

**18" and larger Fastite, and 14" and larger Flex-Ring and Lok-Ring 22½° Bends may be furnished with Fastite Plain Ends, Flex-Ring Ends, or Lok-Ring Ends. See Table No. 4-2.

† Higher pressure ratings are available on special applications. Check AMERICAN.

†† 14" and 16" Flex-Ring fittings may have greater "T" dimensions than those shown.

Figure 3.11.16   *(By permission: American Cast Iron Pipe Company, Birmingham, Alabama)*

3.11.17 Specifications/Diagrams of 30-Degree Bends

ANSI /AWWA C153/A21.53* and AMERICAN Standard

30° Bends (1/12th)

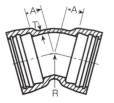
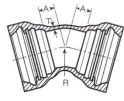
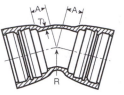

Fastite Bell-Bell**   **Flex-Ring Bell-Bell****   **Lok-Ring Bell-Bell****

Table No. 4-6

Size*** in.	Pressure† Rating psi	Dimensions in Inches			Weight in Pounds		
		T	A	R	Fastite	Flex-Ring	Lok-Ring
14	350	.66	7.5	18.7	310	360	-
16	350	.70	8.0	20.5	380	430	-
18	350	.75	8.5	22.4	450	600	-
20	350	.80	9.5	26.1	535	670	-
24	350	.89	11.0	28.0	805	1005	-
30	250	1.03	15.0	42.9	1455	1630	-
36	250	1.15	18.0	54.1	2210	2335	-
42	250	1.28	21.0	65.3	2610	-	2800
48	250	1.42	24.0	76.5	3670	-	3865
54	250	.90	16.0	49.4	2585	-	2765
60	250	.94	19.0	59.7	3180	-	3405
64	250	.99	18.5	56.9	3545	-	3760

*54"-64" sizes are per AWWA C153.
**Fastite, Flex-Ring, and Lok-Ring Bends may be furnished with Fastite Plain Ends, Flex-Ring Ends, or Lok-Ring Ends. See Table No. 4-2.
***Smaller sizes are available on special applications. Check AMERICAN.
† Higher pressure ratings are available on special applications. Check AMERICAN.

Figure 3.11.17   *(By permission: American Cast Iron Pipe Company, Birmingham, Alabama)*

3.11.18 Specifications/Diagrams of 45-Degree Bends, 4 to 64 Inches

ANSI /AWWA C153/A21.53* and AMERICAN Standard
45° Bends (1/8th)

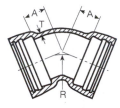
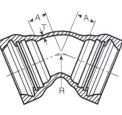
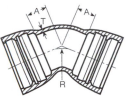

Fastite Bell-Bell**   **Flex-Ring Bell-Bell****   **Lok-Ring Bell-Bell****

Table No. 4-5

Size in.	Pressure† Rating psi	Dimensions in Inches			Weight in Pounds		
		T	A	R	Fastite	Flex-Ring	Lok-Ring
4	350	.34	2.0	2.41	27	45	-
6	350	.36	3.0	4.83	40	70	-
8	350	.38	3.5	5.75	59	95	-
10	350	.40	4.5	7.85	88	150	-
12	350	.42	5.5	9.66	117	195	-
14	350	.47††	7.5	12.06	177	355	-
16	350	.50††	8.0	13.25	210	425	-
18	350	.75	8.5	14.50	445	595	-
20	350	.80	9.5	16.88	530	665	-
24	350	.89	11.0	18.12	800	995	-
30	250	1.03	15.0	27.75	1440	1610	-
36	250	1.15	18.0	35.00	2180	2305	-
42	250	1.28	21.0	42.25	2560	-	2755
48	250	1.42	24.0	49.50	3605	-	3795
54	250	.90	20.28	42.32	2890	-	3070
60	250	.94	21.26	44.08	3335	-	3560
64	250	.99	22.24	45.85	3865	-	4085

*4"-16" and 54"-64" Fastite fittings are per AWWA C153.

**18" and larger Fastite, and 14" and larger Flex-Ring and Lok-Ring 45° Bends may be furnished with Fastite Plain Ends, Flex-Ring Ends, or Lok-Ring Ends. See Table No. 4-2.

† Higher pressure ratings are available on special applications. Check AMERICAN.

†† 14" and 16" Flex-Ring fittings may have greater "T" dimensions than those shown.

Figure 3.11.18   *(By permission: American Cast Iron Pipe Company, Birmingham, Alabama)*

3.11.19 Specifications/Diagrams of 60-Degree Bends, 14 to 64 Inches

ANSI /AWWA C153/A21.53* and AMERICAN Standard

60° Bends (1/6th)

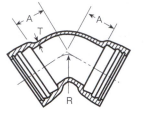
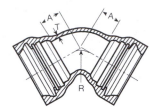
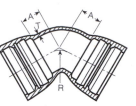

Fastite Bell-Bell**   **Flex-Ring Bell-Bell****   **Lok-Ring Bell-Bell****

Table No. 4-4

Size*** in.	Pressure† Rating psi	Dimensions in Inches			Weight in Pounds		
		T	A	R	Fastite	Flex-Ring	Lok-Ring
14	350	.66	9.5	12.1	330	380	-
16	350	.70	9.0	11.3	390	440	-
18	350	.75	9.5	12.1	460	610	-
20	350	.80	9.5	12.1	525	655	-
24	350	.89	12.0	14.7	820	1020	-
30	250	1.03	15.0	19.9	1415	1585	-
36	250	1.15	28.0	42.4	2780	2895	-
42	250	1.28	31.0	47.6	3315	-	3510
48	250	1.42	24.0	35.5	3505	-	3700
54	250	.90	24.0	36.8	3115	-	3295
60	250	.94	24.0	36.4	3495	-	3720
64	250	.99	30.0	46.3	4520	-	4740

*54"-64" sizes are per AWWA C153.

**Fastite, Flex-Ring, and Lok-Ring Bends may be furnished with Fastite Plain Ends, Flex-Ring Ends, or Lok-Ring Ends. See Table No. 4-2.

***Smaller sizes are available on special applications. Check AMERICAN.

† Higher pressure ratings are available on special applications. Check AMERICAN.

Figure 3.11.19   *(By permission: American Cast Iron Pipe Company, Birmingham, Alabama)*

3.11.20 Specifications/Diagrams of 90-Degree Bends, 4 to 64 Inches

ANSI /AWWA C153/A21.53* and AMERICAN Standard
90° Bends (1/4th)

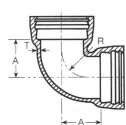

Fastite Bell-Bell**

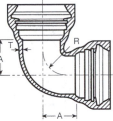

Flex-Ring Bell-Bell**

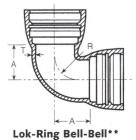

Lok-Ring Bell-Bell**

Table No. 4-3

Size in.	Pressure† Rating psi	Dimensions in Inches			Weight in Pounds		
		T	A	R	Fastite	Flex-Ring	Lok-Ring
4	350	.34	4.0	3.00	32	50	-
6	350	.36	5.0	4.00	48	75	-
8	350	.38	6.5	5.38	72	105	-
10	350	.40	7.5	6.25	109	165	-
12	350	.42	9.0	7.50	142	215	-
14	350	.47††	14.0	11.50	231	425	-
16	350	.50††	15.0	12.50	277	515	-
18	350	.75	16.5	14.00	565	720	-
20	350	.80	18.0	15.50	680	815	-
24	350	.89	22.0	18.50	1065	1265	-
30	250	1.03	25.0	21.50	1750	1920	-
36	250	1.15	28.0	24.50	2565	2690	-
42	250	1.28	31.0	27.50	3015	-	3210
48	250	1.42	34.0	30.50	4120	-	4315
54	250	.90	37.0	34.25	3750	-	3930
60	250	.94	39.5	36.50	4390	-	4620
64	250	.99	42.0	38.75	5170	-	5385

*4"-16" and 54"-64" Fastite fittings are per AWWA C153.
**18" and larger Fastite, and 14" and larger Flex-Ring and Lok-Ring 90° Bends may be furnished with Fastite Plain Ends, Flex-Ring Ends, or Lok-Ring Ends. See Table No. 4-2.
† Higher pressure ratings are available on special applications. Check AMERICAN.

†† 14" and 16" Flex-Ring fittings may have greater "T" dimensions than those shown.

Figure 3.11.20 *(By permission: American Cast Iron Pipe Company, Birmingham, Alabama)*

3.11.21 Specifications/Diagrams of Lok-Ring/Flex-Ring/Fastite Fittings

AMERICAN Ductile Iron Fastite, Flex-Ring, and Lok-Ring Fittings
ANSI/AWWA C153/A21.53 and AMERICAN Standard
Fastite, Flex-Ring, and Lok-Ring Joint Dimensions

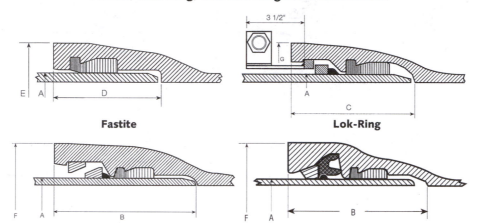

Fastite   Lok-Ring

4"-12" Flex-Ring   14"-36" Flex-Ring

Table No 4-1

Size in.	Dimensions in Inches						
	A Outside Diameter	B Socket Depth Flex-Ring	C Socket Depth Lok-Ring	D Socket Depth Fastite	E* Bell O.D. Fastite	F* Bell O.D. Flex-Ring	G* Bell O.D. Lok-Ring
4	4.80	5.53	-	3.75	7.44	7.44	-
6	6.90	5.53	-	3.75	9.54	9.54	-
8	9.05	5.65	-	4.00	11.78	11.78	-
10	11.10	6.62	-	4.50	14.05	14.05	-
12	13.20	6.62	-	4.50	16.34	16.34	-
14	15.30	7.38	-	4.50	18.42	19.37	-
16	17.40	7.38	-	4.50	20.72	21.49	-
18	19.50	8.20	?	4.50	22.92	23.71	-
20	21.60	8.20	-	4.50	25.50	25.83	-
24	25.80	8.96	-	4.75	29.87	30.70	-
30	32.00	9.38	-	5.25	36.74	37.04	-
36	38.30	9.38	-	5.25	43.64	43.54	-
42	44.50	-	10.07	7.50	48.64	-	48.64
48	50.80	-	10.07	8.00	55.14	-	55.14
54	57.56	-	10.07	8.50	62.14	-	62.14
60	61.61	-	10.57	8.75	66.27	-	66.27
64	65.67	-	10.57	9.00	70.45	-	70.45

*Dimensions subject to change at our option.
For Fastite pipe dimensions, see Section 2.
For Flex-Ring and Lok-Ring pipe dimensions, see Section 9.

Figure 3.11.21   *(By permission: American Cast Iron Pipe Company, Birmingham, Alabama)*

3.11.22 Specifications/Diagrams Base Flange Details, 4 to 64 Inches

ANSI /AWWA C110/A21.10* and AMERICAN Standard

Base Flange Details

Table No. 6-21

Fitting Size in.	Dimensions in Inches		No. of Bolts
	B.C. Bolt Circle	Bolt Hole Diameter	
4	4.75	¾	4
6	5.50	¾	4
8	7.50	¾	4
10	7.50	¾	4
12	9.50	⅞	4
14	9.50	⅞	4
16	9.50	⅞	4
18	11.75	⅞	4
20	11.75	⅞	4
24	11.75	⅞	4
30	14.25	1	4
36	17.00	1	4
42	21.25	1⅛	4
48	22.75	1¼	4
*54	25.00	1¼	4
*60	29.50	1⅜	4
*64	36.00	1⅜	4

*These sizes are not included in AWWA C110 or C153.

BASE FACED AND DRILLED: Bases are not faced or drilled unless so specified on the purchase order. When a base fitting is ordered with base faced and drilled, base will be plain faced and drilled as shown in the table.

For supporting pipe sizes see Table No. 6-22. See Table Nos. 6-19 and 6-20 for additional base dimensions.

Size of Supporting Flanged Pipe for Base Fittings

Table No. 6-22

Fitting Size-Inches	4	5	6	8	10	12	14	16	18	20	24	30	36	42	48	54	60	64
Supporting Pipe Size-Inches	2	2½	2½	4	4	6	6	6	8	8	8	10	12	16	18	20	24	30

Figure 3.11.22 (*By permission: American Cast Iron Pipe Company, Birmingham, Alabama*)

3.11.23 Specifications/Diagrams of Flanged Fittings, 1 to 96 Inches

ANSI/AWWA C110/A21.10, C111/A21.11 or C153/A21.53

Flange Details

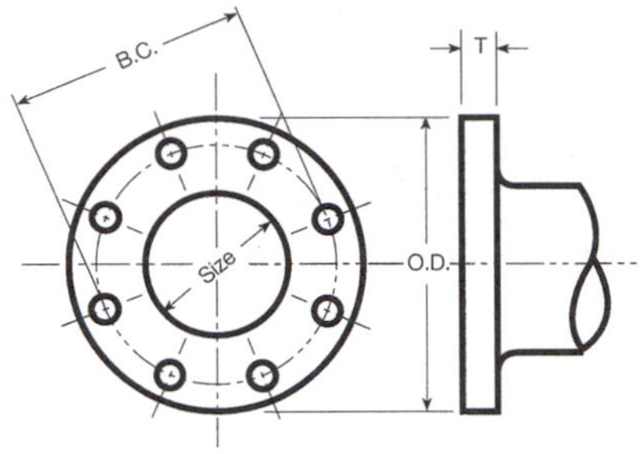

Table No. 6-1

Size in.	O.D. in.	B.C. in.	T in.	Bolt Hole Diameter in.	Bolts		Ring Gasket in.
					No. Per Joint	Size in.	
*1	4.25	3.12	.44	⅝	4	½ x 2	1 x 2⅝
*1¼	4.62	3.50	.50	⅝	4	½ x 2	1¼ x 3
*1½	5.00	3.88	.56	⅝	4	½ x 2	1½ x 3⅜
*2	6.00	4.75	.62	¾	4	⅝ x 2½	2 x 4⅛
*2½	7.00	5.50	.69	¾	4	⅝ x 2½	2½ x 4⅞
3	7.50	6.00	.75	¾	4	⅝ x 2½	3 x 5⅜
*3½	8.50	7.00	.81	¾	8	⅝ x 3	3½ x 6⅜
4	9.00	7.50	.94	¾	8	⅝ x 3	4 x 6⅞
*5	10.00	8.50	.94	⅞	8	¾ x 3	5 x 7¾
6	11.00	9.50	1.00	⅞	8	¾ x 3½	6 x 8¾
8	13.50	11.75	1.12	⅞	8	¾ x 3½	8 x 11
10	16.00	14.25	1.19	1	12	⅞ x 4	10 x 13⅜
12	19.00	17.00	1.25	1	12	⅞ x 4	12 x 16⅛
14	21.00	18.75	1.38	1⅛	12	1 x 4½	14 x 17¾
16	23.50	21.25	1.44	1⅛	16	1 x 4½	16 x 20¼
18	25.00	22.75	1.56	1¼	16	1⅛ x 5	18 x 21⅝
20	27.50	25.00	1.69	1¼	20	1⅛ x 5	20 x 23⅞
24	32.00	29.50	1.88	1⅜	20	1¼ x 5½	24 x 28¼
30	38.75	36.00	2.12	1⅜	28	1¼ x 6½	30 x 34¾
36	46.00	42.75	2.38	1⅝	32	1½ x 7	36 x 41¼
42	53.00	49.50	2.62	1⅝	36	1½ x 7½	42 x 48
48	59.50	56.00	2.75	1⅝	44	1½ x 8	48 x 54½
54	66.25	62.75	3.00	2	44	1¾ x 8½	54 x 61
60	73.00	69.25	3.12	2	52	1¾ x 9	60 x 67½
64	80.00	76.00	3.38	2	52	1¾ x 9	64 x 74¼
*66	80.00	76.00	3.38	2	52	1¾ x 9	66 x 74¼
*72	86.50	82.50	3.50	2	60	1¾ x 9½	72 x 80¾
*84	99.75	95.50	3.88	2¼	64	2 x 10½	84 x 93½
*96	113.25	108.50	4.25	2½	68	2¼ x 11½	96 x 106¼

* These sizes, listed for information only, are not included in AWWA C110, C111 or C153.

FACING: Flanges are plain faced and are finished smooth or with shallow serrations (AMERICAN's option).

BACK FACING: Flanges may be back faced or spot faced, AMERICAN's option, for compliance with the flange thickness tolerance.

FLANGES: The flanges shown above are adequate for water service of 250 psi working pressure and should not be confused with Class 250 flanges per ANSI B16.1. The bolt circle and the bolt holes match those of ANSI B16.1 Class 125. If flanges are required to be made in accordance with other ratings or other standards, this must be specified on the purchase order. 12" and smaller flanges are adequate for water service of 350 psi with the use of AMERICAN's Toruseal® gaskets.

Drilling of flanges can be rotated when required; for those sizes with an even number of bolt holes in each quadrant, fitting can be rotated 45° with standard drilling.

As listed in the Appendix of AWWA C110, gaskets are rubber—either ring or full face—and are 1/8" thick unless otherwise specified. **AMERICAN recommends AMERICAN Toruseal® gaskets shown on page 6-26 or ⅛" thick ring gaskets for normal water service.**

See Section 8, Table No. 8-3 for information on bolts and studs.

Figure 3.11.23 *(By permission: American Cast Iron Pipe Company, Birmingham, Alabama)*

3.11.24 Specifications/Diagrams of Companion/Blind Flanges, 4 to 64 Inches

Threaded Companion Flanges and Blind Flanges
F&D ANSI/AWWA C110/A21.10 or C115/A21.15

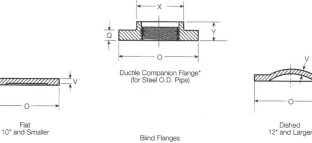

Ductile Companion Flange*
(for Steel O.D. Pipe)

Flat
10" and Smaller

Blind Flanges

Dished
12" and Larger

Table No. 6-16

Size in.	O Diameter of Flange in.	Q Thickness of Flange in.	V Thickness in.	X Hub Diameter in.	Y Overall Thickness in.	Weight in Pounds Blind Flange	Weight in Pounds Companion Flange*
4	9.00	.94	.88	5.31	1.31	14	12
6	11.00	1.00	.94	7.56	1.56	25	17
8	13.50	1.12	1.06	9.69	1.75	40	25
10	16.00	1.19	1.12	11.94	1.94	60	35
12	19.00	1.25	.81	14.06	2.19	80	55
14	21.00	1.38	.88	-	-	110	-
16	23.50	1.44	1.00	-	-	145	-
18	25.00	1.56	1.06	-	-	175	-
20	27.50	1.69	1.12	-	-	225	-
24	32.00	1.88	1.25	-	-	328	-
30	38.75	2.12	2.32	-	-	703	-
36	46.00	2.38	2.99	-	-	1232	-
42	53.00	2.62	1.81	-	-	1230	-
48	59.50	2.75	2.00	-	-	1657	-
54	66.25	3.00	2.25	-	-	2249	-
60	73.00	3.12	2.38	-	-	2863	-
64	80.00	3.38	2.56	-	-	3761	-

*Ductile Companion Flanges are threaded for fabrication on pipe of standard steel pipe outside diameter unless specified otherwise. No gray iron Companion Flanges are produced. Check AMERICAN if larger Companion Flanges are desired.

See Section 8 for Companion Flanges for ductile iron pipe.
Reducing Blind and Companion Flanges can be furnished. Specify flange outside diameter and the size of pipe.
Dished Blind Flanges 24" and larger are furnished with a lifting eye. The "V" thicknesses for 30" and 36" Blind Flanges are special manufacturing thicknesses. Flat Blind Flanges and Blind Flanges smaller than 20" may be ordered with a special, fabricated lifting eye.

Table No. 6-17

Blind Flange Design

Size	Standard Design	Pressure Rating psi
4"-10"	Flat	250
12"-24"	Dished	250
30"-48"	Dished	250
54"-64"	Dished	150
Special Design		
12"-24"	Flat	200
30"-48"	Flat	150

Figure 3.11.24 *(By permission: American Cast Iron Pipe Company, Birmingham, Alabama)*

3.11.25 Specifications/Diagrams of Flanged Fittings

ANSI/AWWA C110/A21.10 or AMERICAN Standard

Flanged Fitting Dimensions and Designation of Outlets

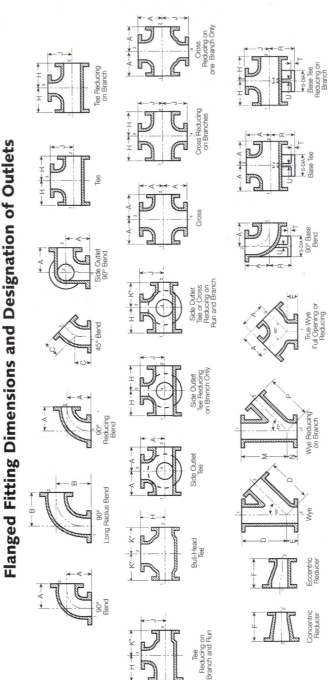

Dimensions for above fittings are given in Table No. 6-2.

Some of above fittings are not listed in AWWA C110, but do meet applicable requirements of AWWA C110.

The largest opening establishes the basic size of a reducing fitting. The largest opening is named first, except for bull-head tees and for double branch bends where both branches are reducing; in these two cases the largest opening is named last.

In the designation of the openings of reducing fittings, they should be read in the order indicated by the numbers 1, 2, 3 and 4 as shown. In designating the outlets of side outlet reducing fittings, the side outlet is named last, and in the case of a cross, which is not shown, the side outlet is designated by the number 5.

In describing tees reducing on the branch only, the first size applies to both run outlets and the same size for both branch outlets, it is necessary to indicate only two sizes. For example: for an 8 x 6 Tee or Wye—it is not necessary to show 8 x 8 x 6; for an 8 x 6 Cross—it is not necessary to show 8 x 8 x 6 x 6.

"K" dimensions are shown in Table Nos. 6-8 and 6-9.

Figure 3.11.25 (*By permission: American Cast Iron Pipe Company, Birmingham, Alabama*)

252

Flanged Fittings Dimensions

Table No. 6-2

Size in.	A	B	C	D**	E**	F	R	S (Dia.)	T	U	Reducing Tees and Crosses — Size of Branches and Smaller***	H	J	Reducing Wyes — Size of Branches and Smaller	M	N	P
3	5.5	7.75	3.0	10.0	3.0	6	4.88	5.00	.56	.50	-	†	†	-	††	††	††
4	6.5	9.00	4.0	12.0	3.0	7	5.50	6.00	.62	.50	-	†	†	-	††	††	††
6	8.0	11.50	5.0	14.5	3.5	9	7.00	7.00	.69	.62	-	†	†	-	††	††	††
8	9.0	14.00	5.5	17.5	4.5	11	8.38	9.00	.94	.88	-	†	†	-	††	††	††
10	11.0	16.50	6.5	20.5	5.0	12	9.75	9.00	.94	.88	-	†	†	-	††	††	††
12	12.0	19.00	7.5	24.5	5.5	14	11.25	11.00	1.00	1.00	-	†	†	-	††	††	††
14	14.0	21.50	7.5	27.0	6.0	16	12.50	11.00	1.00	1.00	-	†	†	-	††	††	††
16	15.0	24.00	8.0	30.0	6.5	18	13.75	11.00	1.00	1.00	-	†	†	-	††	††	††
18	16.5	26.50	8.5	32.0	7.0	19	15.00	13.50	1.12	1.12	12	13	15.5	8	25.0	1.00	27.5
20	18.0	29.00	9.5	35.0	8.0	20	16.00	13.50	1.12	1.12	14	14	17.0	10	27.0	1.00	29.5
24	22.0	34.00	11.0	40.5	9.0	24	18.50	13.50	1.12	1.12	16	15	19.0	12	31.5	0.50	34.5
30	25.0	41.50	15.0	See Table No. 6-13		30	23.00	16.00	1.19	1.15	20	18	23.0	-	See Table No. 6-13		
36	28.0	49.00	18.0			36	26.00	19.00	1.25	1.15	24	20	26.0	-			
42	31.0	56.50	21.0			42	30.00	23.50	1.44	1.28	24	23	30.0	-			
48	34.0	64.00	24.0			48	34.00	25.00	1.56	1.42	30	26	34.0	-			
54	39.0	-	20.5	-	-	§	38.00	27.50	1.69	1.55	§§	§§	§§	-	-	-	-
60	43.0	-	23.5	-	-	§	42.00	32.00	1.88	1.75	§§	§§	§§	-	-	-	-
64	48.0	-	25.0	-	-	§	44.00	38.75	2.12	1.75	§§	§§	§§	-	-	-	-

*These sizes are not included in AWWA C110 or C153; also AWWA C110 and C153 do not include data on Wyes of any size.
**See footnote to Table No. 6-13.
***For larger branches, use "A" dimensions.
†For these smaller diameter reducing Tees and Crosses the "A" dimension applies as shown in "cut" of standard Tee and Cross.
††For these smaller diameter reducing Wyes the "D" and "E" dimensions apply as shown in "cut" of standard Wye.

§"F" dimensions for 54"-64" reducers vary with diameters of reduction.
§§"H" and "J" dimensions for 54"-64" tees and crosses vary with diameters of reduction.

Figure 3.11.25 (cont.) (By permission: American Cast Iron Pipe Company, Birmingham, Alabama)

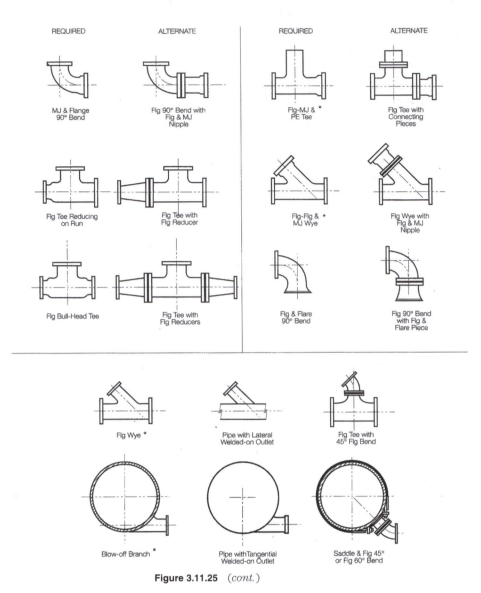

REQUIRED   ALTERNATE   REQUIRED   ALTERNATE

MJ & Flange
90° Bend

Flg 90° Bend with
Flg & MJ
Nipple

Flg-MJ & *
PE Tee

Flg Tee with
Connecting
Pieces

Flg Tee Reducing
on Run

Flg Tee with
Flg Reducer

Flg-Flg & *
MJ Wye

Flg Wye with
Flg & MJ
Nipple

Flg Bull-Head Tee

Flg Tee with
Flg Reducers

Flg & Flare
90° Bend

Flg 90° Bend
with Flg &
Flare Piece

Flg Wye *

Pipe with Lateral
Welded-on Outlet

Flg Tee with
45° Flg Bend

Blow-off Branch *

Pipe with Tangential
Welded-on Outlet

Saddle & Flg 45°
or Flg 60° Bend

Figure 3.11.25 *(cont.)*

3.11.26 Method of Designating Location of Tapped Holes for Flanges

Method of Designating Location of Tapped Holes and Sequence of Openings

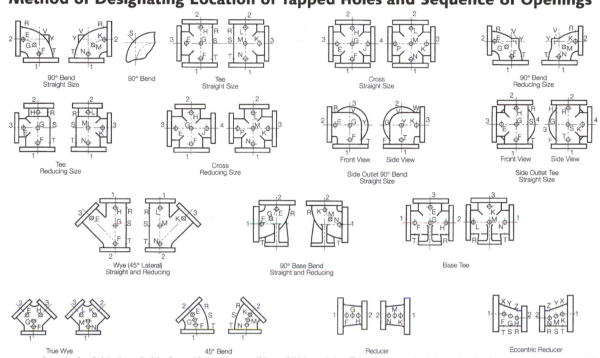

Taps are usually made directly into the wall of the flanged fitting. However, if the wall thickness is insufficient for the required size and angle of tap, a tapping boss is provided on the fitting.

In the designation of the openings of reducing fittings, they should be read in the order indicated by the sequence of the numbers 1, 2, 3 and 4 as shown. In designating the outlets of side outlet reducing fittings, the side outlet is named last, and in the case of a cross, which is not shown, the side outlet is designated by the number 5.

Note that numbering sequence for wye ends differs from that shown in ANSI B16.1.

Figure 3.11.26 *(By permission: American Cast Iron Pipe Company, Birmingham, Alabama)*

3.11.27 Specifications/Diagrams of Flanged Base Tees, 4 to 64 Inches

Flanged Base Tees

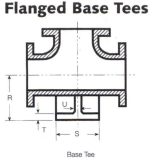

Base Tee

Table No. 6-20

Size in.	Pressure Rating psi	Dimensions in Inches				Weight in Pounds	
		R	S Dia.	T	U	Base Fitting	Base Only
4	250	5.50	6.00	.62	.50	75	10
6	250	7.00	7.00	.69	.62	110	15
8	250	8.38	9.00	.94	.88	185	30
10	250	9.75	9.00	.94	.88	300	30
12	250	11.25	11.00	1.00	1.00	430	45
14	250	12.50	11.00	1.00	1.00	485	50
16	250	13.75	11.00	1.00	1.00	600	50
18	250	15.00	13.50	1.12	1.12	740	75
20	250	16.00	13.50	1.12	1.12	930	75
24	250	18.50	13.50	1.12	1.12	1410	80
30	250	23.00	16.00	1.19	1.15	2270	120
36	250	26.00	19.00	1.25	1.15	3320	160
42	150	30.00	23.50	1.44	1.28	4740	270
42	250	30.00	23.50	1.44	1.28	5850	270
48	150	34.00	25.00	1.56	1.42	6235	335
48	250	34.00	25.00	1.56	1.42	7720	335
*54	250	38.00	27.50	1.69	1.55	6590	565
*60	250	42.00	32.00	1.88	1.75	8905	855
*64	250	44.00	38.75	2.12	1.75	12275	1275

*Not included in AWWA C110 or C153.
   Bases are faced and drilled only when specified. Dimension "R" is finished dimension; unfinished bases will be slightly longer. For supporting pipe sizes see Table No. 6-22.
   For base drilling see Table No. 6-21.
   See tables on preceding pages for dimensions and weights of all tees. Base dimensions and weights are the same for full opening tees and for reducing tees. Proper base is determined by largest opening. To compute total weight of reducing base tee, add weight of base only (shown above) to weight of reducing tee of size and class as selected from preceding tables. To order base tees reducing on the branch, specify sizes in proper order and give figure number shown above. Bases other than shown above—such as on the side of fitting—are special and may be available on 30" and larger bends. Some available base locations are shown on page 6-5. Check AMERICAN for special base locations.
   Bases of these fittings are intended for support in compression and are not to be used for thrust anchors or supports in tension or shear.

Figure 3.11.27 (*By permission: American Cast Iron Pipe Company, Birmingham, Alabama*)

3.11.28 Polyethylene Encasement of Ductile Iron Pipe

Polyethylene Encasement for Ductile Iron Pipe in Corrosive Soils

Most soils encountered in the construction of underground ductile iron pipe lines do not cause corrosion of any magnitude. There are, however, some soils that are corrosive and can cause problems. Fortunately, remedies do exist. Soil surveys can be conducted to determine whether or not corrosive soils are to be expected in the area of excavation. Polyethylene encasement is easily accomplished if required.

Polyethylene encasement is a proven method of protecting cast iron and ductile iron pipe in areas of severely corrosive soil. The protection is provided by isolating the pipe from the corrosive environment. A completely air and water-tight enclosure is not necessary.

The dielectric capability of polyethylene also provides shielding against stray direct current at most levels encountered in the field.

Optionally, a four (4) mil thick high density cross laminated polyethylene tube or an eight (8) mil thick low density polyethylene tube is furnished in the flat tube widths listed at the right.

More detailed information on polyethylene encasement is available upon request. Both material and installation procedures are standardized in ANSI/AWWA C105/A21.5.

Recommended Polyethylene Flat Tube Width by Pipe Size

Nominal Pipe Size Inches	Minimum Flat Tube Width–Inches (Layflat Size)
4	14
6	16
8	20
10	24
12	27
14	30
16	34
18	37
20	41
24	54

Figure 3.11.28 (*By permission: American Cast Iron Pipe Company, Birmingham, Alabama*)

3.12.0 Cast Iron Soil Pipe Equivalents

	1½	2	3	4	5	6	8	10	12	15
1½	1	1.8	4	7.1	10.8	15.7	28.	44.4	63.4	100
2		1	2.3	4	6.1	8.8	15.8	25	35.6	56.3
3			1	1.8	2.7	3.9	7	11.1	15.8	25
4				1	1.5	2.2	3.9	6.3	8.9	14.1
5					1	1.4	2.6	4.1	5.8	9.2
6						1	1.8	2.8	4.	6.4
8							1	1.6	2.3	3.6
10								1	1.4	2.3
12									1	1.6
15										1

EXAMPLE: A 4″ cast iron soil pipe is equivalent to how many 2″ cast iron soil pipe? In the vertical column under 4″, and opposite 2″, read the equivalent which is 4: This means that four 2″ cast iron soil pipe are the equivalent of one 4″ cast iron soil pipe in inside cross-sectional area.

Figure 3.12.0 (*By permission Cast Iron Soil Pipe Institute*)

3.12.1 Ring Test Crushing Loads on Cast Iron Soil Pipe

NO-HUB				SERVICE WEIGHT				EXTRA HEAVY			
Pipe Size In.	Nominal O.D. (D_o)	Nominal Thickness (1)	Ring Crushing Load* (w)	Pipe Size In.	Nominal O.D. (D_o)	Nominal Thickness (1)	Ring Crushing Load* (w)	Pipe Size In.	Nominal O.D. (D_o)	Nominal Thickness (1)	Ring Crushing Load* (w)
1½	1.90	.16	8328	—	—	—	—	—	—	—	—
2	2.35	.16	6617	2	2.30	.17	7680	2	2.38	.19	9331
3	3.35	.16	4542	3	3.30	.17	5226	3	3.50	.25	10885
4	4.38	.19	4877	4	4.30	.18	4451	4	4.50	.25	8324
5	5.30	.19	3999	5	5.30	.18	3582	5	5.50	.25	6739
6	6.30	.19	3344	6	6.30	.18	2997	6	6.50	.25	5660
8	8.38	.23	3674	8	8.38	.23	3674	8	8.62	.31	6546
10	10.56	.28	4317	10	10.50	.28	4342	10	10.75	.37	7465
				12	12.50	.28	3632	12	12.75	.37	6259
				15	15.88	.36	4727	15	15.88	.44	7097

*Pounds per linear foot

Figure 3.12.1 (*By permission Cast Iron Soil Pipe Institute*)

3.12.2 Equivalent Length of Pipe for 90-Degree Elbows (in Feet)

Velocity, ft/s	1/2	3/4	1	1–1/4	1–1/2	2	2–1/2	3	3–1/2	4	5	6	8	10	12
1	1.2	1.7	2.2	3.0	3.5	4.5	5.4	6.7	7.7	8.6	10.5	12.2	15.4	18.7	22.2
2	1.4	1.9	2.5	3.3	3.9	5.1	6.0	7.5	8.6	9.5	11.7	13.7	17.3	20.8	24.8
3	1.5	2.0	2.7	3.6	4.2	5.4	6.4	8.0	9.2	10.2	12.5	14.6	18.4	22.3	26.5
4	1.5	2.1	2.8	3.7	4.4	5.6	6.7	8.3	9.6	10.6	13.1	15.2	19.2	23.2	27.6
5	1.6	2.2	2.9	3.9	4.5	5.9	7.0	8.7	10.0	11.1	13.6	15.8	19.8	24.2	28.8
6	1.7	2.3	3.0	4.0	4.7	6.0	7.2	8.9	10.3	11.4	14.0	16.3	20.5	24.9	29.6
7	1.7	2.3	3.0	4.1	4.8	6.2	7.4	9.1	10.5	11.7	14.3	16.7	21.0	25.5	30.3
8	1.7	2.4	3.1	4.2	4.9	6.3	7.5	9.3	10.8	11.9	14.6	17.1	21.5	26.1	31.0
9	1.8	2.4	3.2	4.3	5.0	6.4	7.7	9.5	11.0	12.2	14.9	17.4	21.9	26.6	31.6
10	1.8	2.5	3.2	4.3	5.1	6.5	7.8	9.7	11.2	12.4	15.2	17.7	22.2	27.0	32.0

3.12.3 Equivalent Length (Pipe, Elbows, Tees, and Valves)

Find the nominal pipe size being used in the left-most column. For each fitting, read the value under the appropriate heading and add this to the length of piping. This allows total system pressure drop to be calculated. (This is valid for any fluid.)

PIPE SIZE	EQUIVALENT LENGTH OF STRAIGHT PIPE (FEET)				
	STANDARD ELBOW	STANDARD TEE	GATE VALVE FULL OPEN	GLOBE VALVE FULL OPEN	ANGLE VALVE FULL OPEN
1-1/2	4	9	0.9	41	21
2	5	11	1.2	54	27
2-/1/2	6	13	1.4	64	32
3	8	16	1.6	80	40
3-1/2	9	18	2.0	91	45
4	11	21	2.2	110	55
5	13	26	2.8	140	70
6	16	32	3.4	155	81
8	20	42	4.5	210	110
10	25	55	5.5	270	140
12	30	65	6.5	320	160
14	35	75	8.0	370	190

This table contains the number of feet of straight pipe usually allowed for standard fittings and valves.

3.12.4 Maximum Capacity of Gas Pipe (in Cubic Feet Per Hour)

Nominal Iron Pipe Size, in.	Internal Diameter, in.	Length of Pipe, ft													
		10	20	30	40	50	60	70	80	90	100	125	150	175	200
1/4	0.364	32	22	18	15	14	12	11	11	10	9	8	8	7	6
3/8	0.493	72	49	40	34	30	27	25	23	22	21	18	17	15	14
1/2	0.622	132	92	73	63	56	50	46	43	40	38	34	31	28	26
3/4	0.824	278	190	152	130	115	105	96	90	84	79	72	64	59	55
1	1.049	520	350	285	245	215	195	180	170	160	150	130	120	110	100
1-1/4	1.380	1050	730	590	500	440	400	370	350	320	305	275	250	225	210
1-1/2	1.610	1600	1100	890	760	670	610	560	530	490	460	410	380	350	320
2	2.067	3050	2100	1650	1450	1270	1150	1050	990	930	870	780	710	650	610
2-1/2	2.469	4800	3300	2700	2300	2000	1850	1700	1600	1500	1400	1250	1130	1050	980
3	3.068	8500	5900	4700	4100	3600	3250	3000	2800	2600	2500	2200	2000	1850	1700
4	4.026	17,500	12,000	9700	8300	7400	6800	6200	5800	5400	5100	4500	4100	3800	3500

Notes: 1. Capacity is in cubic feet per hour at gas pressures of 0.5 psig or less and a pressure drop of 0.5 in. of water; Specific gravity = 0.60.   2. Copyright by the American Gas Association and the National Fire Protection Association. Used by permission of the copyright holder.

Figure 3.12.4   (*By permission of American Society of Heating, Refrigerating and Air-Conditioning Engineers, Inc. Atlanta, Georgia, from their* 1993 ASHRAE Fundamentals Handbook)

3.12.5 Metric Equivalent of NPS, ASHRAE, AWWA, NFPA, and ASTM Tube and Pipe Sizes

Nominal pipe size (NPS), in IP	ASHRAE std. wt. size, mm	AWWA pipe size, mm	NFPA pipe size, mm	ASTM copper tube size, mm	Nominal pipe size DN, mm
1/8	—	—	—	6	6
3/16	—	—	—	8	8
1/4	8	—	—	10	10
3/8	10	—	—	12	12
1/2	15	12.7 & 13	12	15	15
5/8	—	—	—	18	18
3/4	20	—	—	22	20
1	25	25	25 & 25.4	28	25
1 1/4	32	—	33	35	32
1 1/2	40	45	38 & 38.1	42	40
2	50	50 & 50.8	51	54	50
2 1/2	65	63 & 63.5	63.5 & 64	67	65
3	80	75	76 & 80	79	80
3 1/2	—	—	89	—	90
4	100	100	102	105	100
4 1/2	—	114.3			115
5	—	—	127	130	125
6	150	150	152	156	150
8	200	200	203	206	200
10	250	250	—	257	250
12	300	300	305	308	300
14	—	350	—		350
18	—	400	—		400
18	—	—	—		450
20	—	500	—		500
24	—	600	—		600
28					700
30					750
32					800
36					900
40					1000
44					1100
48					1200
52					1300
56					1400
60					1500

Figure 3.12.5 (*By permission: McGraw-Hill Inc.* Plumber's and Pipefitter's Calculations Manual. *R. Dodge Woodson.*)

3.12.6 Dimensions of Cast Iron, Steel, and Bronze Flanges

125 lb. CAST IRON **ASA B16.1**

Pipe Size	½	¾	1	1¼	1½	2	2½	3	3½	4	5	6	8	10	12
Diameter of Flange			4¼	4⅝	5	6	7	7½	8½	9	10	11	13½	16	19
Thickness of Flange (min)[1]			7/16	½	9/16	⅝	11/16	¾	13/16	15/16	15/16	1	1⅛	1 3/16	1¼
Diameter of Bolt Circle			3⅛	3½	3⅞	4¾	5½	6	7	7½	8½	9½	11¾	14¼	17
Number of Bolts			4	4	4	4	4	4	8	8	8	8	8	12	12
Diameter of Bolts			½	½	½	⅝	⅝	⅝	⅝	⅝	¾	¾	¾	⅞	⅞

[1] 125 lb. flanges have plain faces.

250 lb CAST IRON **ASA B16.2**

Pipe Size	½	¾	1	1¼	1½	2	2½	3	3½	4	5	6	8	10	12
Diameter of Flange			4⅞	5¼	6⅛	6½	7½	8¼	9	10	11	12½	15	17½	20½
Thickness of Flange (min)[2]			11/16	¾	13/16	⅞	1	1⅛	1 3/16	1¼	1⅜	1 7/16	1⅝	1⅞	2
Diameter of Raised Face			2 11/16	3 1/16	3 9/16	4 3/16	4 15/16	6 5/16	6 5/16	6 15/16	8 5/16	9 11/16	11 15/16	14 1/16	16 7/16
Diameter of Bolt Circle			3½	3⅞	4½	5	5⅞	6⅝	7¼	7⅞	9¼	10⅝	13	15¼	17¾
Number of Bolts			4	4	4	8	8	8	8	8	8	12	12	16	16
Diameter of Bolts			⅝	⅝	¾	⅝	¾	¾	¾	¾	¾	¾	⅞	1	1⅛

[2] 250 lb. flanges have a 1/16″ raised face which is included in the flange thickness dimensions.

150 lb BRONZE **ASA B16.24**

Pipe Size	½	¾	1	1¼	1½	2	2½	3	3½	4	5	6	8	10	12
Diameter of Flange	3½	3⅞	4¼	4⅝	5	6	7	7½	8½	9	10	11	13½	16	19
Thickness of Flange (min)[3]	5/16	11/32	⅜	13/32	7/16	½	9/16	⅝	11/16	11/16	¾	13/16	15/16	1	1 1/16
Diameter of Bolt Circle	2⅜	2¾	3⅛	3½	3⅞	4¾	5½	6	7	7½	8½	9½	11¾	14¼	17
Number of Bolts	4	4	4	4	4	4	4	4	8	8	8	8	8	12	12
Diameter of Bolts	½	½	½	½	½	⅝	⅝	⅝	⅝	⅝	¾	¾	¾	⅞	⅞

[3] 150 lb. bronze flanges have plain faces with two concentric gasket-retaining grooves between the port and the bolt holes.

300 lb. BRONZE **ASA B16.24**

Pipe Size	½	¾	1	1¼	1½	2	2½	3	3½	4	5	6	8	10	12
Diameter of Flange	3¾	4⅝	4⅞	5¼	6⅛	6½	7½	8¼	9	10	11	12½	15		
Thickness of Flange (min)[4]	½	17/32	19/32	⅝	11/16	¾	13/16	29/32	31/32	1 1/16	1⅛	1 3/16	1⅜		
Diameter of Bolt Circle	2⅝	3¼	3½	3⅞	4½	5	5⅞	6⅝	7¼	7⅞	9¼	10⅝	13		
Number of Bolts	4	4	4	4	4	8	8	8	8	8	8	12	12		
Diameter of Bolts	½	⅝	⅝	⅝	¾	⅝	¾	¾	¾	¾	¾	¾	⅞		

[4] 300 lb. bronze flanges have plain faces with two concentic gasket-retaining grooves between the port and the bolt holes.

150 lb. STEEL **ASA B16.5**

Pipe Size	½	¾	1	1¼	1½	2	2½	3	3½	4	5	6	8	10	12
			4¼	4⅝	5	6	7	7½	8½	9	10	11	13½	16	19
Thickness of Flange (min)[5]			7/16	½	9/16	⅝	13/16	¾	13/16	15/16	15/16	1	1⅛	1 3/16	1½
Diameter of Raised Face			2	2½	2⅞	3⅝	4⅛	5	5½	6 3/16	7 5/16	8½	10⅝	12¾	15
Diameter of Bolt Circle			3⅛	3½	3⅞	4¾	5½	6	7	7½	8½	9½	11¾	14¼	17
Number of Bolts			4	4	4	4	4	4	8	8	8	8	8	12	12
Diameter of Bolts			½	½	½	⅝	⅝	⅝	⅝	⅝	¾	¾	¾	⅞	⅞

[5] 150 lb. steel flanges have a 1/16″ raised face which is included in the flange thickness dimensions.

300 lb. STEEL **ASA B16.5**

Pipe Size	½	¾	1	1¼	1½	2	2½	3	3½	4	5	6	8	10	12
Diameter of Flange			4⅞	5¼	6⅛	6½	7½	8¼	9	10	11	12½	15	17½	20½
Thickness of Flange (min)[6]			11/16	¾	13/16	⅞	1	1⅛	1 3/16	1¼	1⅜	1 7/16	1⅝	1⅞	2
Diameter of Raised Face			2	2½	2⅞	3⅝	4⅛	5	5½	6 3/16	7 5/16	8½	10⅝	12¾	15
Diameter of Bolt Circle			3½	3⅞	4½	5	5⅞	6⅝	7¼	7⅞	9¼	10⅝	13	15¼	17¾
Number of Bolts			4	4	4	8	8	8	8	8	8	12	12	16	16
Diameter of Bolts			⅝	⅝	¾	⅝	¾	¾	¾	¾	¾	¾	⅞	1	1⅛

[6] 300 lb. steel flanges have a 1/16″ raised face which is included in the flange thickness dimensions.

	400 lb. STEEL									ASA B16.5						
Pipe Size	½	¾	1	1¼	1½	2	2½	3	3½	4	5	6	8	10	12
Diameter of Flange	3¾	4⅝	4⅞	5¼	6⅛	6½	7½	8¼	9	10	11	12½	15	17½	20½
Thickness of Flange (min)⁷	9⁄16	⅝	11⁄16	13⁄16	⅞	1	1⅛	1¼	1⅜	1⅜	1½	1⅝	1⅞	2⅛	2¼
Diameter of Raised Face	1⅜	1 11⁄16	2	2½	2⅞	3⅝	4⅛	5	5½	6 3⁄16	7 5⁄16	8½	10⅝	12¾	15
Diameter of Bolt Circle	2⅝	3¼	3½	3⅞	4½	5	5⅞	6⅝	7¼	7⅞	9¼	10⅝	13	15¼	17¾
Number of Bolts	4	4	4	4	4	8	8	8	8	8	8	12	12	16	16
Diameter of Bolts	½	⅝	⅝	⅝	¾	⅝	¾	¾	⅞	⅞	⅞	⅞	1	1⅛	1¼

⁷ 400 lb. steel flanges have a ¼″ raised face which is NOT included in the flange thickness dimensions.

	600 lb. STEEL									ASA B16.5						
Pipe Size	½	¾	1	1¼	1½	2	2½	3	3½	4					
Diameter of Flange	3¾	4⅝	4⅞	5¼	6⅛	6½	7½	8¼	9	10¾	13	14	16½	10	12
Thickness of Flange (min)⁸	9⁄16	⅝	11⁄16	13⁄16	⅞	1	1⅛	1¼	1⅜	1½	1¾	1⅞	2 3⁄16	2½	2⅝
Diameter of Raised Face	1⅜	1 11⁄16	2	2½	2⅞	3⅝	4⅛	5	5½	6 3⁄16	7 5⁄16	8½	10⅝	12¾	15
Diameter of Bolt Circle	2⅝	3¼	3½	3⅞	4½	5	5⅞	6⅞	7¼	8½	10½	11½	13¼	17	19½
Number of Bolts	4	4	4	4	4	8	8	8	8	8	8	12	12	16	20
Diameter of Bolts	½	⅝	⅝	⅝	¾	⅝	¾	¾	⅞	⅞	1	1	1⅛	1¼	1¼

⁸ 600 lb. steel flanges have a ¼″ raised face which is NOT included in the flange thickness dimensions.

Figure 3.12.6 *(cont.)*

3.12.7 Slopes Required for Self-Cleaning Cast Iron Pipe

Slopes required to obtain self-cleaning velocities of 2.0 and 2.5 ft/sec. (based on Mannings Formula with $N = 0.012$)

Pipe Size (In.)	Velocity (Ft./Sec.)	¼ FULL		½ FULL		¾ FULL		FULL	
		Slope (Ft./Ft.)	Flow (Gal./Min.)	Slope (Ft./Ft.)	Flow (Gal./Min.)	Slope (Ft./Ft.)	Flow (Gal./Min.)	Slope (Ft./Ft.)	Flow (Gal./Min.)
2.0	2.0	0.0313	4.67	0.0186	9.34	0.0148	14.09	0.0186	18.76
	2.5	0.0489	5.84	0.0291	11.67	0.0231	17.62	0.0291	23.45
3.0	2.0	0.0178	10.77	0.0107	21.46	0.0085	32.23	0.0107	42.91
	2.5	0.0278	13.47	0.0167	26.82	0.0133	40.29	0.0167	53.64
4.0	2.0	0.0122	19.03	0.0073	38.06	0.0058	57.01	0.0073	76.04
	2.5	0.0191	23.79	0.0114	47.58	0.0091	71.26	0.0114	95.05
5.0	2.0	0.0090	29.89	0.0054	59.79	0.0043	89.59	0.0054	119.49
	2.5	0.0141	37.37	0.0085	74.74	0.0067	11.99	0.0085	149.36
6.0	2.0	0.0071	43.18	0.0042	86.36	0.0034	129.54	0.0042	172.72
	2.5	0.0111	53.98	0.0066	107.95	0.0053	161.93	0.0066	214.90
8.0	2.0	0.0048	77.20	0.0029	154.32	0.0023	231.52	0.0029	308.64
	2.5	0.0075	96.50	0.0045	192.90	0.0036	289.40	0.0045	385.79
10.0	2.0	0.0036	120.92	0.0021	241.85	0.0017	362.77	0.0021	483.69
	2.5	0.0056	151.15	0.0033	302.31	0.0026	453.46	0.0033	604.61
12.0	2.0	0.0028	174.52	0.0017	349.03	0.0013	523.55	0.0017	698.07
	2.5	0.0044	218.15	0.0026	436.29	0.0021	654.44	0.0026	872.58
15.0	2.0	0.0021	275.42	0.0012	550.84	0.0010	826.26	0.0012	1101.68
	2.5	0.0032	344.28	0.0019	688.55	0.0015	1032.83	0.0019	1377.10

Figure 3.12.7 *(By permission Cast Iron Soil Pipe Institute)*

3.12.8 Typical Pipe Joining Methods for Cast Iron Pipe

Note: lead and oakum will be found on older piping installations only.

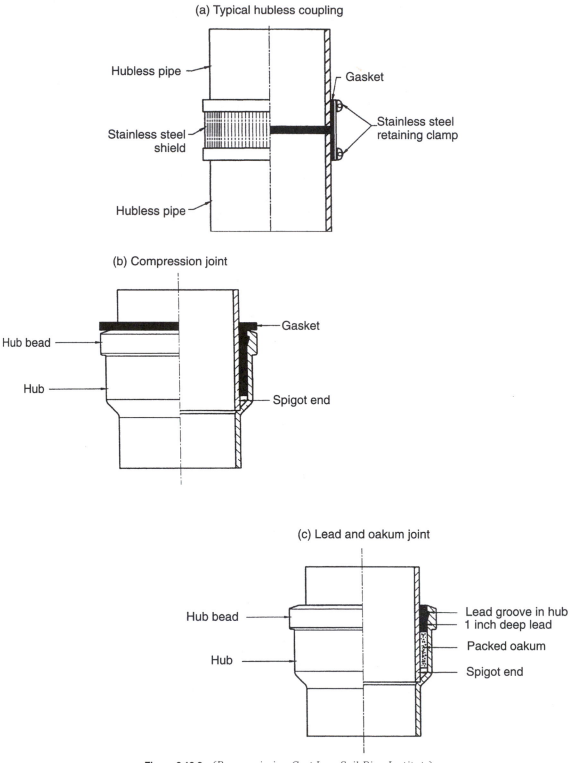

Figure 3.12.8 (*By permission Cast Iron Soil Pipe Institute*)

3.12.9 Iron and Copper Elbow Size Equivalents

Fitting	Iron Pipe	Copper Tubing
Elbow, 90°	1.0	1.0
Elbow, 45°	0.7	0.7
Elbow, 90° long turn	0.5	0.5
Elbow, welded, 90°	0.5	0.5
Reduced coupling	0.4	0.4
Open return bend	1.0	1.0
Angle radiator valve	2.0	3.0
Radiator or convector	3.0	4.0
Boiler or heater	3.0	4.0
Open gate valve	0.5	0.7
Open globe valve	12.0	17.0

[a]See Table 4 for equivalent length of one elbow.
Source: Giesecke (1926) and Giesecke and Badgett (1931, 1932).

Figure 3.12.9 (*By permission of American Society of Heating, Refrigerating and Air-Conditioning Engineers, Inc. Atlanta, Georgia, from their* 1993 ASHRAE Fundamentals Handbook)

3.12.10 Water Velocities (Types of Service)

Type of Service	Velocity, ft/s	Reference
General service	4 to 10	a, b, c
City water	3 to 7	a, b
	2 to 5	c
Boiler feed	6 to 15	a, c
Pump suction and drain lines	4 to 7	a, b

[a]Crane Co. 1976. Flow of fluids through valves, fittings, and pipe. Technical Paper 410.
[b]*System Design Manual.* 1960. Carrier Air Conditioning Co., Syracuse, NY.
[c]*Piping Design and Engineering.* 1951. Grinnell Company, Inc., Cranston, RI.

Maximum Water Velocity to Minimize Erosion

Normal Operation, h/yr	Water Velocity, ft/s
1500	15
2000	14
3000	13
4000	12
6000	10

Source: *System Design Manual,* Carrier Air Conditioning Co., 1960.

Figure 3.12.10 (*By permission of American Society of Heating, Refrigerating and Air-Conditioning Engineers, Inc. Atlanta, Georgia, from their* 1993 ASHRAE Fundamentals Handbook)

3.12.11 Flow Rates/Demand for Various Plumbing Fixtures

Proper Flow and Pressure Required during Flow for Different Fixtures

Fixture	Flow Pressure[a]	Flow, gpm
Ordinary basin faucet	8	3.0
Self-closing basin faucet	12	2.5
Sink faucet—3/8 in.	10	4.5
Sink faucet—1/2 in.	5	4.5
Dishwasher	15–25	—[b]
Bathtub faucet	5	6.0
Laundry tube cock—1/4 in.	5	5.0
Shower	12	3–10
Ball cock for closet	15	3.0
Flush valve for closet	10–20	15–40[c]
Flush valve for urinal	15	15.0
Garden hose, 50 ft, and sill cock	30	5.0

[a]Flow pressure is the pressure (psig) in the pipe at the entrance to the particular fixture considered.
[b]Varies; see manufacturers' data.
[c]Wide range due to variation in design and type of flush valve closets.

3.12.12 Demand Weights of Fixtures in Fixture Units

Fixture or Group[b]	Occupancy	Type of Supply Control	Weight in Fixture Units[c]
Water closet	Public	Flush valve	10
Water closet	Public	Flush tank	5
Pedestal urinal	Public	Flush valve	10
Stall or wall urinal	Public	Flush valve	5
Stall or wall urinal	Public	Flush tank	3
Lavatory	Public	Faucet	2
Bathtub	Public	Faucet	4
Shower head	Public	Mixing valve	4
Service sink	Office, etc	Faucet	3
Kitchen sink	Hotel or restaurant	Faucet	4
Water closet	Private	Flush valve	6
Water closet	Private	Flush tank	3
Lavatory	Private	Faucet	1
Bathtub	Private	Faucet	2
Shower head	Private	Mixing valve	2
Bathroom group	Private	Flush valve for closet	8
Bathroom group	Private	Flush tank for closet	6
Separate shower	Private	Mixing valve	2
Kitchen sink	Private	Faucet	2
Laundry trays (1 to 3)	Private	Faucet	3
Combination fixture	Private	Faucet	3

Note: See Hunter (1941).
[a]For supply outlets likely to impose continuous demands, estimate continuous supply separately, and add to total demand for fixtures.
[b]For fixtures not listed, weights may be assumed by comparing the fixture to a listed one using water in similar quantities and at similar rates.
[c]The given weights are for total demand. For fixtures with both hot and cold water supplies, the weights for maximum separate demands can be assumed to be 75% of the listed demand for the supply.

Figure 3.12.12 *(By permission of American Society of Heating, Refrigerating and Air-Conditioning Engineers, Inc. Atlanta, Georgia, from their* 1993 ASHRAE Fundamentals Handbook)

3.12.13 Head-of-Water Equivalents (in PSI)

Head Ft.	0	1	2	3	4	5	6	7	8	9
0	....	0.433	0.866	1.299	1.732	2.165	2.598	3.031	3.464	3.987
10	4.330	4.763	5.196	5.629	6.062	6.495	6.928	7.361	7.794	8.277
20	8.660	9.093	9.526	9.959	10.392	10.825	11.258	11.691	12.124	12.557
30	12.990	13.423	13.856	14.289	14.722	15.155	15.588	16.021	16.454	16.887
40	17.320	17.753	18.186	18.619	19.052	19.485	19.918	20.351	20.784	21.217
50	21.650	22.083	22.516	22.949	23.382	23.815	24.248	24.681	25.114	25.547
60	25.980	26.413	26.846	27.279	27.712	28.145	28.578	29.011	29.444	29.877
70	30.310	30.743	31.176	31.609	32.042	32.475	32.908	33.341	33.774	34.207
80	34.640	35.073	35.506	35.939	36.372	36.805	37.238	37.671	38.104	38.537
90	38.970	39.403	39.836	40.269	40.702	41.135	41.568	42.001	42.436	42.867

3.12.14 Pipe Sizes for Horizontal Rainwater Piping

Size of Pipe in Inches 1/8" Slope	Maximum Rainfall in Inches per Hour				
	2	3	4	5	6
3	1644	1096	822	657	548
4	3760	2506	1880	1504	1253
5	6680	4453	3340	2672	2227
6	10700	7133	5350	4280	3566
8	23000	15330	11500	9200	7600
10	41400	27600	20700	16580	13800
11	66600	44400	33300	26650	22200
15	109000	72800	59500	47600	39650

Size of Pipe in Inches 1/4" Slope	Maximum Rainfall in Inches per Hour				
	2	3	4	5	6
3	2320	1546	1160	928	773
4	5300	3533	2650	2120	1766
5	9440	6293	4720	3776	3146
6	15100	10066	7550	6040	5033
8	32600	21733	16300	13040	10866
10	58400	38950	29200	23350	19450
11	94000	62600	47000	37600	31350
15	168000	112000	84000	67250	56000

Size of Pipe in Inches 1/2" Slope	Maximum Rainfall in Inches per Hour				
	2	3	4	5	6
3	3288	2295	1644	1310	1096
4	7520	5010	3760	3010	2500
5	13660	8900	6680	5320	4450
6	21400	13700	10700	8580	7140
8	46000	30650	23000	18400	15320
10	82800	55200	41400	33150	27600
11	133200	88800	66600	53200	44400
15	238000	158800	119000	95300	79250

Figure 3.12.14 *(By permission of Cast Iron Soil Pipe Institute)*

3.12.15 Size of Roof Drains for Varying Amounts of Rainfall (in Square Feet)

Rain Fall in Inches	Size of Drain or Leader in Inches*					
	2	3	4	5	6	8
1	2880	8800	18400	34600	54000	116000
2	1440	4400	9200	17300	27000	58000
3	960	2930	6130	11530	17995	38660
4	720	2200	4600	8650	13500	29000
5	575	1760	3680	6920	10800	23200
6	480	1470	3070	5765	9000	19315
7	410	1260	2630	4945	7715	16570
8	360	1100	2300	4325	6750	14500
9	320	980	2045	3845	6000	12890
10	290	880	1840	3460	5400	11600
11	260	800	1675	3145	4910	10545
12	240	730	1530	2880	4500	9660

*Round, square or rectangular rainwater pipe may be used and are considered equivalent when closing a scribed circle quivalent to the leader diameter.

Source: Uniform Plumbing Code (IAPMO) 1985 Edition

Figure 3.12.15 *(By permission of Cast Iron Soil Pipe Institute)*

3.12.16 Velocity/Flow in Cast Iron Sewer Pipe of 2 Inches (5.08 cm) to 15 Inches (38.1 cm)

Pipe Size (In.)	SLOPE (In./Ft.)	(Ft./Ft.)	¼ FULL Velocity (Ft./Sec.)	Flow (Gal./Min.)	½ FULL Velocity (Ft./Sec.)	Flow (Gal./Min.)	¾ FULL Velocity (Ft./Sec.)	Flow (Gal./Min.)	FULL Velocity (Ft./Sec.)	Flow (Gal./Min.)
2.0	0.0120	0.0010	0.36	0.83	0.46	2.16	0.52	3.67	0.46	4.35
	0.0240	0.0020	0.51	1.18	0.66	3.06	0.74	5.18	0.66	6.15
	0.0360	0.0030	0.62	1.45	0.80	3.75	0.90	6.35	0.80	7.53
	0.0480	0.0040	0.72	1.67	0.93	4.33	1.04	7.33	0.93	8.69
	0.0600	0.0050	0.80	1.87	1.04	4.84	1.16	8.20	1.04	9.72
	0.0720	0.0060	0.88	2.04	1.13	5.30	1.27	8.98	1.13	10.65
	0.0840	0.0070	0.95	2.21	1.23	5.72	1.38	9.70	1.23	11.50
	0.0960	0.0080	1.01	2.36	1.31	6.12	1.47	10.37	1.31	12.29
	0.1080	0.0090	1.07	2.50	1.39	6.49	1.56	11.00	1.39	13.04
	0.1200	0.0100	1.13	2.64	1.47	6.84	1.64	11.59	1.47	13.75
	0.2400	0.0200	1.60	3.73	2.07	9.67	2.33	16.39	2.07	19.44
	0.3600	0.0300	1.96	4.57	2.54	11.85	2.85	20.07	2.54	23.81
	0.4800	0.0400	2.26	5.28	2.93	13.68	3.29	23.18	2.93	27.49
	0.6000	0.0500	2.53	5.90	3.28	15.29	3.68	25.92	3.28	30.74
	0.7200	0.0600	2.77	6.47	3.59	16.75	4.03	28.39	3.59	33.67
	0.8400	0.0700	2.99	6.98	3.88	18.10	4.35	30.66	3.88	36.37
	0.9600	0.0800	3.20	7.47	4.14	19.35	4.65	32.78	4.14	38.88
	1.0800	0.0900	3.39	7.92	4.40	20.52	4.93	34.77	4.40	41.24
	1.2000	0.1000	3.58	8.35	4.63	21.63	5.20	36.65	4.63	43.47
3.0	0.0120	0.0010	0.47	2.55	0.61	6.56	0.69	11.05	0.61	13.12
	0.0240	0.0020	0.67	3.61	0.86	9.28	0.97	15.63	0.86	18.55
	0.0360	0.0030	0.82	4.42	1.06	11.36	1.19	19.14	1.06	22.72
	0.0480	0.0040	0.95	5.11	1.22	13.12	1.37	22.10	1.22	26.24
	0.0600	0.0050	1.06	5.71	1.37	14.67	1.53	24.71	1.37	29.33
	0.0720	0.0060	1.16	6.25	1.50	16.07	1.68	27.07	1.50	32.13
	0.0840	0.0070	1.25	6.75	1.62	17.35	1.81	29.24	1.62	34.71
	0.0960	0.0080	1.34	7.22	1.73	18.55	1.94	31.26	1.73	37.11
	0.1080	0.0090	1.42	7.66	1.83	19.68	2.06	33.16	1.83	39.36
	0.1200	0.0100	1.50	8.07	1.93	20.74	2.17	34.95	1.93	41.49
	0.2400	0.0200	2.21	11.42	2.73	29.33	3.07	49.43	2.73	58.67
	0.3600	0.0300	2.60	13.98	3.35	35.93	3.76	60.53	3.35	71.86
	0.4800	0.0400	3.00	16.14	3.87	41.49	4.34	69.90	3.87	82.97
	0.6000	0.0500	3.35	18.05	4.32	46.38	4.85	78.15	4.32	92.77
	0.7200	0.0600	3.67	19.77	4.74	50.81	5.31	85.61	4.74	101.62
	0.8400	0.0700	3.96	21.36	5.12	54.88	5.74	92.47	5.12	109.76
	0.9600	0.0800	4.24	22.83	5.47	58.67	6.13	98.85	5.47	117.34
	1.0800	0.0900	4.50	24.22	5.80	62.23	6.51	104.85	5.80	124.46
	1.2000	0.1000	4.74	25.53	6.11	65.29	6.86	110.52	6.11	131.19

Figure 3.12.16 *(By permission of Cast Iron Soil Pipe Institute)*

Velocity and Flow in Cast Iron Soil Pipe Sewers and Drains
(Based on Mannings Formula with N = .012)

Pipe Size (In.)	SLOPE (In./Ft.)	SLOPE (Ft./Ft.)	¼ FULL Velocity (Ft./Sec.)	¼ FULL Flow (Gal./Min.)	½ FULL Velocity (Ft./Sec.)	½ FULL Flow (Gal./Min.)	¾ FULL Velocity (Ft./Sec.)	¾ FULL Flow (Gal./Min.)	FULL Velocity (Ft./Sec.)	FULL Flow (Gal./Min.)
4.0	0.0120	0.0010	0.57	5.45	0.74	14.08	0.83	23.63	0.74	28.12
	0.0240	0.0020	0.81	7.70	1.05	19.91	1.17	33.42	1.05	39.77
	0.0360	0.0030	0.99	9.44	1.28	24.38	1.44	40.92	1.28	48.71
	0.0480	0.0040	1.15	10.90	1.48	28.16	1.66	47.26	1.48	56.25
	0.0600	0.0050	1.28	12.18	1.65	31.48	1.85	52.83	1.65	62.88
	0.0720	0.0060	1.40	13.34	1.81	34.48	2.03	57.88	1.81	68.89
	0.0840	0.0070	1.51	14.41	1.96	37.25	2.19	62.51	1.96	74.41
	0.0960	0.0080	1.62	15.41	2.09	39.82	2.34	66.83	2.09	79.54
	0.1080	0.0090	1.72	16.34	2.22	42.23	2.49	70.88	2.22	84.37
	0.1200	0.0100	1.81	17.23	2.34	44.52	2.62	74.72	2.34	88.93
	0.2400	0.0200	2.56	24.36	3.31	62.96	3.71	105.67	3.31	125.77
	0.3600	0.0300	3.14	29.84	4.05	77.11	4.54	129.42	4.05	154.04
	0.4800	0.0400	3.62	34.46	4.68	89.04	5.24	149.44	4.68	177.86
	0.6000	0.0500	4.05	38.52	5.23	99.55	5.86	167.08	5.23	198.86
	0.7200	0.0600	4.43	42.20	5.73	109.05	6.42	183.02	5.73	217.84
	0.8400	0.0700	4.79	45.58	6.19	117.79	6.94	197.69	6.19	235.29
	0.9600	0.0800	5.12	48.73	6.62	125.92	7.41	211.34	6.62	251.54
	1.0800	0.0900	5.43	51.68	7.02	133.56	7.86	224.15	7.02	266.80
	1.2000	0.1000	5.73	54.48	7.40	140.78	8.29	236.28	7.40	281.23
5.0	0.0120	0.0010	0.67	9.94	0.86	25.71	0.96	43.15	0.86	51.37
	0.0240	0.0020	0.94	14.06	1.22	36.35	1.36	61.02	1.22	72.65
	0.0360	0.0030	1.15	17.22	1.49	44.52	1.67	74.74	1.49	88.98
	0.0480	0.0040	1.33	19.88	1.72	51.41	1.93	86.30	1.72	102.75
	0.0600	0.0050	1.49	22.23	1.92	57.48	2.15	96.49	1.92	114.87
	0.0720	0.0060	1.63	24.35	2.11	62.97	2.36	105.70	2.11	125.84
	0.0840	0.0070	1.76	26.30	2.28	68.01	2.55	114.17	2.28	135.92
	0.0960	0.0080	1.88	28.12	2.43	72.71	2.72	122.05	2.43	145.31
	0.1080	0.0090	2.00	29.82	2.58	77.12	2.89	129.45	2.58	154.12
	0.1200	0.0100	2.10	31.44	2.72	81.29	3.05	136.45	2.72	162.46
	0.2400	0.0200	2.97	44.46	3.85	114.96	4.31	192.97	3.85	229.75
	0.3600	0.0300	3.64	54.45	4.71	140.80	5.28	236.34	4.71	281.38
	0.4800	0.0400	4.21	62.88	5.44	162.58	6.09	272.91	5.44	324.91
	0.6000	0.0500	4.70	70.30	6.08	181.77	6.81	305.12	6.08	363.26
	0.7200	0.0600	5.15	77.01	6.66	199.12	7.46	334.24	6.66	397.94
	0.8400	0.0700	5.56	83.18	7.19	215.07	8.06	361.02	7.19	429.82
	0.9600	0.0800	5.95	88.92	7.69	229.92	8.62	385.95	7.69	459.50
	1.0800	0.0900	6.31	94.31	8.16	243.92	9.14	409.36	8.16	487.37
	1.2000	0.1000	6.65	99.42	8.60	257.06	9.63	431.50	8.60	513.73

Figure 3.12.16 *(cont.)*

Velocity and Flow in Cast Iron Soil Pipe Sewers and Drains
(Based on Mannings Formula with N = .012)

Pipe Size (In.)	SLOPE (In./Ft.)	SLOPE (Ft./Ft.)	¼ FULL Velocity (Ft./Sec.)	¼ FULL Flow (Gal./Min.)	½ FULL Velocity (Ft./Sec.)	½ FULL Flow (Gal./Min.)	¾ FULL Velocity (Ft./Sec.)	¾ FULL Flow (Gal./Min.)	FULL Velocity (Ft./Sec.)	FULL Flow (Gal./Min.)
6.0	0.0120	0.0010	0.75	16.23	0.97	41.98	1.09	70.55	0.97	83.96
	0.0240	0.0020	1.06	22.95	1.37	59.37	1.54	99.77	1.37	118.74
	0.0360	0.0030	1.30	28.11	1.68	72.71	1.89	122.20	1.68	145.42
	0.0480	0.0040	1.50	32.46	1.94	83.96	2.18	141.10	1.94	167.92
	0.0600	0.0050	1.68	36.29	2.17	93.87	2.44	157.76	2.17	187.74
	0.0720	0.0060	1.84	39.75	2.38	102.83	2.67	172.81	2.38	205.66
	0.0840	0.0070	1.99	42.94	2.57	111.07	2.88	186.66	2.57	222.13
	0.0960	0.0080	2.13	45.90	2.75	118.74	3.08	199.55	2.75	237.47
	0.1080	0.0090	2.26	48.69	2.92	125.94	3.27	211.65	2.92	251.88
	0.1200	0.0100	2.38	51.32	3.07	132.75	3.44	223.10	3.07	265.50
	0.2400	0.0200	3.36	72.58	4.35	187.74	4.87	315.51	4.35	375.47
	0.3600	0.0300	4.12	88.89	5.32	229.93	5.97	386.42	5.32	459.86
	0.4800	0.0400	4.75	102.64	6.15	265.50	6.89	446.20	6.15	531.00
	0.6000	0.0500	5.32	114.76	6.87	296.84	7.70	498.87	6.87	593.68
	0.7200	0.0600	5.82	125.71	7.53	325.17	8.44	546.27	7.53	650.34
	0.8400	0.0700	6.29	135.78	8.13	351.22	9.11	590.27	8.13	702.45
	0.9600	0.0800	6.72	145.16	8.70	375.47	9.74	631.02	8.70	750.95
	1.0800	0.0900	7.13	153.96	9.22	398.25	10.33	669.30	9.22	796.50
	1.2000	0.1000	7.52	162.29	9.72	419.79	10.89	705.51	9.72	839.59
8.0	0.0120	0.0010	0.91	35.25	1.18	91.04	1.32	153.06	1.18	182.09
	0.0240	0.0020	1.29	49.85	1.67	128.75	1.87	216.46	1.67	257.51
	0.0360	0.0030	1.58	61.05	2.04	157.69	2.29	265.11	2.04	315.38
	0.0480	0.0040	1.83	70.50	2.36	182.09	2.64	306.12	2.36	364.17
	0.0600	0.0050	2.04	78.82	2.64	203.58	2.96	342.26	2.64	407.16
	0.0720	0.0060	2.24	86.34	2.89	223.01	3.24	374.92	2.89	446.02
	0.0840	0.0070	2.42	93.26	3.12	240.88	3.50	404.96	3.12	481.75
	0.0960	0.0080	2.58	99.70	3.34	257.51	3.74	432.92	3.34	515.02
	0.1080	0.0090	2.74	105.75	3.54	273.13	3.97	459.18	3.54	546.26
	0.1200	0.0100	2.89	111.47	3.73	287.90	4.18	484.02	3.73	575.81
	0.2400	0.0200	4.08	157.64	5.28	407.16	5.91	684.51	5.28	814.32
	0.3600	0.0300	5.00	193.06	6.46	498.66	7.24	838.35	6.46	997.33
	0.4800	0.0400	5.78	222.93	7.46	575.81	8.36	968.05	7.46	1151.62
	0.6000	0.0500	6.46	249.24	8.34	643.77	9.35	1082.31	8.34	1287.55
	0.7200	0.0600	7.07	273.03	9.14	705.22	10.24	1185.61	9.14	1410.44
	0.8400	0.0700	7.64	294.91	9.87	761.72	11.06	1280.60	9.87	1523.45
	0.9600	0.0800	8.17	315.27	10.55	814.31	11.83	1369.02	10.55	1628.63
	1.0800	0.0900	8.66	334.40	11.19	863.71	12.54	1452.07	11.19	1727.42
	1.2000	0.1000	9.13	352.48	11.80	910.43	13.22	1530.61	11.80	1820.86

Figure 3.12.16   *(cont.)*

Velocity and Flow in Cast Iron Soil Pipe Sewers and Drains
(Based on Mannings Formula with N = .012)

Pipe Size (In.)	SLOPE (In./Ft.)	SLOPE (Ft./Ft.)	¼ FULL Velocity (Ft./Sec.)	¼ FULL Flow (Gal./Min.)	½ FULL Velocity (Ft./Sec.)	½ FULL Flow (Gal./Min.)	¾ FULL Velocity (Ft./Sec.)	¾ FULL Flow (Gal./Min.)	FULL Velocity (Ft./Sec.)	FULL Flow (Gal./Min.)
10.0	0.0120	0.0010	1.06	64.08	1.37	165.75	1.54	278.56	1.37	331.51
	0.0240	0.0020	1.50	90.62	1.94	234.41	2.17	393.95	1.94	468.83
	0.0360	0.0030	1.84	110.99	2.37	287.10	2.66	482.48	2.37	574.19
	0.0480	0.0040	2.12	128.16	2.74	331.51	3.07	557.12	2.74	663.02
	0.0600	0.0050	2.37	143.29	3.07	370.64	3.43	622.88	3.07	741.28
	0.0720	0.0060	2.60	156.96	3.36	406.01	3.76	682.33	3.36	812.03
	0.0840	0.0070	2.80	169.54	3.63	438.55	4.06	737.01	3.63	877.09
	0.0960	0.0080	3.00	181.24	3.88	468.82	4.34	787.89	3.88	937.65
	0.1080	0.0090	3.18	192.24	4.11	497.26	4.61	835.69	4.11	994.53
	0.1200	0.0100	3.35	202.64	4.33	524.16	4.86	880.89	4.33	1048.32
	0.2400	0.0200	4.74	286.57	6.13	741.28	6.87	1245.77	6.13	1482.55
	0.3600	0.0300	5.80	350.98	7.51	907.88	8.41	1525.75	7.51	1815.75
	0.4800	0.0400	6.70	405.27	8.67	1048.32	9.71	1761.78	8.67	2096.65
	0.6000	0.0500	7.49	453.11	9.69	1172.06	10.86	1969.73	9.69	2344.13
	0.7200	0.0600	8.21	496.36	10.62	1283.93	11.90	2157.74	10.62	2567.86
	0.8400	0.0700	8.87	536.12	11.47	1386.80	12.85	2330.62	11.47	2773.61
	0.9600	0.0800	9.48	573.14	12.26	1482.55	13.74	2491.54	12.26	2965.11
	1.0800	0.0900	10.05	607.91	13.00	1572.49	14.57	2642.67	13.00	3144.97
	1.2000	0.1000	10.60	640.79	13.71	1657.55	15.36	2785.62	13.71	3315.09
12.0	0.0120	0.0010	1.20	104.53	1.55	270.34	1.74	454.27	1.55	540.68
	0.0240	0.0020	1.69	147.83	2.19	382.32	2.45	642.43	2.19	764.63
	0.0360	0.0030	2.07	181.05	2.68	468.24	3.01	786.82	2.68	936.48
	0.0480	0.0040	2.40	209.06	3.10	540.68	3.47	908.54	3.10	1081.35
	0.0600	0.0050	2.68	233.74	3.46	604.49	3.88	1015.78	3.46	1208.99
	0.0720	0.0060	2.93	256.05	3.79	662.19	4.25	1112.73	3.79	1324.38
	0.0840	0.0070	3.17	276.56	4.10	715.25	4.59	1201.88	4.10	1430.50
	0.0960	0.0080	3.39	295.66	4.38	764.63	4.91	1284.87	4.38	1529.27
	0.1080	0.0090	3.59	313.59	4.65	811.01	5.21	1362.81	4.65	1622.03
	0.1200	0.0100	3.79	330.56	4.90	854.88	5.49	1436.53	4.90	1709.77
	0.2400	0.0200	5.36	467.48	6.93	1208.99	7.76	2031.55	6.93	2417.98
	0.3600	0.0300	6.56	572.54	8.48	1480.71	9.50	2488.14	8.48	2961.41
	0.4800	0.0400	7.58	661.11	9.80	1709.77	10.98	2873.05	9.80	3419.54
	0.6000	0.0500	8.47	739.14	10.95	1911.58	12.27	3212.17	10.95	3823.17
	0.7200	0.0600	9.28	809.69	12.00	2094.03	13.44	3518.76	12.00	4188.07
	0.8400	0.0700	10.02	874.57	12.96	2261.81	14.52	3800.69	12.96	4523.63
	0.9600	0.0800	10.71	934.95	13.86	2417.98	15.52	4063.11	13.86	4835.96
	1.0800	0.0900	11.36	991.67	14.70	2564.65	16.46	4309.57	14.70	5129.30
	1.2000	0.1000	11.98	1045.31	15.49	2703.38	17.35	4542.69	15.49	5406.76

Figure 3.12.16 *(cont.)*

Velocity and Flow in Cast Iron Soil Pipe Sewers and Drains
(Based on Mannings Formula with N = .012)

Pipe Size (In.)	SLOPE (In./Ft.)	(Ft./Ft.)	¼ FULL Velocity (Ft./Sec.)	Flow (Gal./Min.)	½ FULL Velocity (Ft./Sec.)	Flow (Gal./Min.)	¾ FULL Velocity (Ft./Sec.)	Flow (Gal./Min.)	FULL Velocity (Ft./Sec.)	Flow (Gal./Min.)
15.0	0.0120	0.0010	1.39	192.03	1.80	496.67	2.02	834.85	1.80	993.34
	0.0240	0.0020	1.97	271.58	2.55	702.40	2.86	1180.65	2.55	1404.79
	0.0360	0.0030	2.42	332.61	3.12	860.25	3.50	1445.99	3.12	1720.51
	0.0480	0.0040	2.79	384.07	3.61	993.34	4.04	1669.69	3.61	1986.67
	0.0600	0.0050	3.12	429.40	4.03	1110.58	4.52	1866.77	4.03	2221.17
	0.0720	0.0060	3.42	470.38	4.42	1216.58	4.95	2044.95	4.42	2433.17
	0.0840	0.0070	3.69	508.07	4.77	1314.06	5.35	2208.79	4.77	2628.12
	0.0960	0.0080	3.94	543.15	5.10	1404.79	5.72	2361.30	5.10	2809.58
	0.1080	0.0090	4.18	576.10	5.41	1490.01	6.06	2504.54	5.41	2980.01
	0.1200	0.0100	4.41	607.26	5.70	1570.60	6.39	2640.01	5.70	3141.21
	0.2400	0.0200	6.24	858.80	8.06	2221.17	9.04	3733.54	8.06	4442.34
	0.3600	0.0300	7.64	1051.81	9.88	2720.37	11.07	4572.64	9.88	5440.73
	0.4800	0.0400	8.82	1214.52	11.41	3141.21	12.78	5280.03	11.41	6282.41
	0.6000	0.0500	9.86	1357.88	12.75	3511.98	14.29	5903.25	12.75	7023.95
	0.7200	0.0600	10.80	1487.48	13.97	3847.18	15.65	6466.69	13.97	7694.35
	0.8400	0.0700	11.67	1606.66	15.09	4155.43	16.91	6984.82	15.09	8310.85
	0.9600	0.0800	12.47	1717.60	16.13	4442.33	18.07	7467.07	16.13	8884.66
	1.0800	0.0900	13.23	1821.78	17.11	4711.80	19.17	7920.03	17.11	9423.61
	1.2000	0.1000	13.94	1920.33	18.03	4966.68	20.21	8348.44	18.03	9933.35

Figure 3.12.16 *(cont.)*

3.12.17 Expansion Characteristics of Metal and Plastic Pipe

Expansion: Allowances for expansion and contraction of building materials are important design considerations. Material selection can create or prevent problems. Cast iron is in tune with building reactions to temperature. Its expansion is so close to that of steel and masonry that there is no need for costly expansion joints and special offsets. That is not always the case with other DWV materials.

Thermal expansion of various materials.

Material	Inches per inch 10^{-6} X per °F	Inches per 100' of pipe per 100°F.	Ratio-assuming cast iron equals 1.00
Cast iron	6.2	0.745	1.00
Concrete	5.5	0.66	.89
Steel (mild)	6.5	0.780	1.05
Steel (stainless)	7.8	0.940	1.26
Copper	9.2	1.11	1.49
PVC (high impact)	55.6	6.68	8.95
ABS (type 1A)	56.2	6.75	9.05
Polyethylene (type 1)	94.5	11.4	15.30
Polyethylene (type 2)	83.3	10.0	13.40

Here is the *actual* increase in length for 50 feet of pipe and 70° temperature rise.

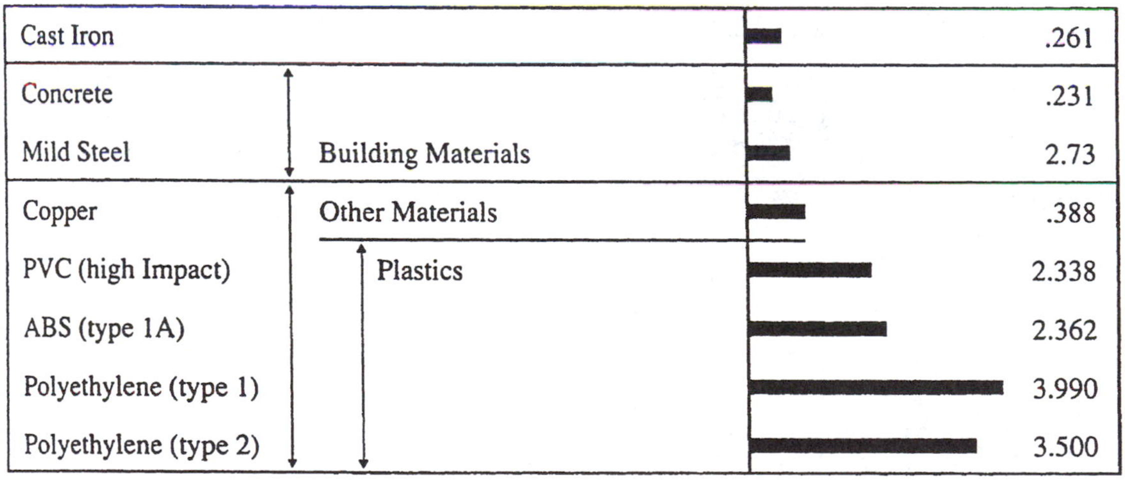

Material			Value
Cast Iron			.261
Concrete			.231
Mild Steel	Building Materials		2.73
Copper	Other Materials		.388
PVC (high Impact)		Plastics	2.338
ABS (type 1A)			2.362
Polyethylene (type 1)			3.990
Polyethylene (type 2)			3.500

Figure 3.12.17   *(By permission of Cast Iron Soil Pipe Institute)*

3.12.18 Pipe Diameters and Trench Widths (U.S. and Metric Sizes)

Pipe Diameter (millimeters)	Trench Width (millimeters)	Pipe Diameter (millimeters)	Trench Width (millimeters)
100	470	1500	2500
150	540	1650	2800
200	600	1800	3000
250	680	1950	3200
300	800	2100	3400
375	910	2250	3600
450	1020	2400	3900
525	1100	2550	4100
600	1200	2700	4300
675	1300	2850	4500
825	1600	3000	4800
900	1700	3150	5000
1050	1900	3300	5200
1200	2100	3450	5400
1350	2300	3600	5600

NOTE: Trench widths based on 1.25 Bc + 300 where Bc is the outside diameter of the pipe in millimeters.

Pipe Diameter (inches)	Trench Width (feet)	Pipe Diameter (inches)	Trench Width (feet)
4	1.6	60	8.5
6	1.8	66	9.2
8	2.0	72	10.0
10	2.3	78	10.7
12	2.5	84	11.4
15	3.0	90	12.1
18	3.4	96	12.9
21	3.8	102	13.6
24	4.1	108	14.3
27	4.5	114	14.9
33	5.2	120	15.6
36	5.6	126	16.4
42	6.3	132	17.1
48	7.0	138	17.8
54	7.8	144	18.5

NOTE: Trench widths based on 1.25 Bc + 1 ft where Bc is the outside diameter of the pipe in inches.

3.12.19 Pipe Test Plugs (Illustrated)

Typical test plugs used for air/water tests.

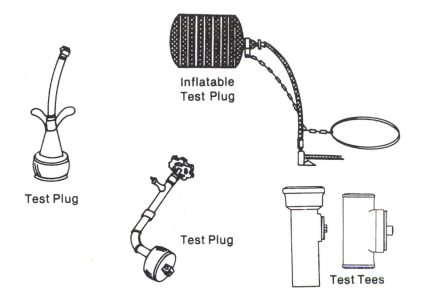

Inflatable Test Plug

Test Plug

Test Plug

Test Tees

3.12.20 Thrust Pressures When Hydrostatically Testing Soil Pipe

PIPE SIZE		1½″	2″	3″	4″	5″	6″	8″	10″
HEAD, Feet of Water	PRESSURE PSI	THRUST lb.	THRUST lb.	THRUST lb.	THRUST lb.	THRUST lb.	THRUST lb.	THRUST lb.	THRUST lb.
10	4.3	12	19	38	65	95	134	237	377
20	8.7	25	38	77	131	192	271	480	762
30	13.0	37	56	115	196	287	405	717	1139
40	17.3	49	75	152	261	382	539	954	1515
50	21.7	62	94	191	327	479	676	1197	1900
60	26.0	74	113	229	392	574	810	1434	2277
70	30.3	86	132	267	457	668	944	1671	2654
80	34.7	99	151	306	523	765	1082	1914	3039
90	39.0	111	169	344	588	860	1216	2151	3416
100	43.4	123	188	382	654	957	1353	2394	3801
110	47.7	135	208	420	719	1052	1487	2631	4178
120	52.0	147	226	458	784	1147	1621	2868	4554
AREA, OD. in.²		2.84	4.34	8.81	15.07	22.06	31.17	55.15	87.58

Thrust = Pressure × Area

Figure 3.12.20   *(By permission of Cast Iron Soil Pipe Institute)*

3.12.21 Pressure Drop in Flow of Water through Schedule 40 Steel Pipe

Flow of water through Schedule 40 steel pipe

Discharge Gals. per Min.	Vel. Ft. per Sec.	Pressure Drop	Vel. Ft. per Sec.	Pressure Drop	Vel. Ft. per Sec.	Pressure Drop	Vel. Ft. per Sec.	Pressure Drop	Vel. Ft. per Sec.	Pressure Drop	Vel. Ft. per Sec.	Pres. sure Drop	Vel. Ft. per Sec.	Pressure Drop	Vel. Ft. per Sec.	Pressure Drop	Vel. Ft. per Sec.	Pres. sure Drop
	1″		**1¼″**		**1½″**		**2″**		**2½″**		**3″**		**3½″**		**4″**		**5″**	
1	.37	0.49																
2	.74	1.70	0.43	0.45														
3	1.12	3.53	0.64	0.94	0.47	0.44												
4	1.49	5.94	0.86	1.55	0.63	0.74												
5	1.86	9.02	1.07	2.36	0.79	1.12												
6	2.24	12.25	1.28	3.30	0.95	1.53	.57	0.46										
8	2.98	21.1	1.72	5.52	1.26	2.63	.76	0.75										
10	3.72	30.8	2.14	8.34	1.57	3.86	.96	1.14	.67	0.48								
15	5.60	64.6	3.21	17.6	2.36	8.13	1.43	2.33	1.00	0.99								
20	7.44	110.5	4.29	29.1	3.15	13.5	1.91	3.86	1.34	1.64	.87	0.59						
25			5.36	43.7	3.94	20.2	2.39	5.81	1.68	2.48	1.08	0.67	.81	0.42				
30			6.43	62.9	4.72	29.1	2.87	8.04	2.01	3.43	1.30	1.21	.97	0.60				
35			7.51	82.5	5.51	38.2	3.35	10.95	2.35	4.49	1.52	1.58	1.14	0.79	.88	0.42		
40					6.30	47.8	3.82	13.7	2.68	5.88	1.74	2.06	1.30	1.00	1.01	0.53		
45					7.08	60.6	4.30	17.4	3.00	7.14	1.95	2.51	1.46	1.21	1.13	0.67		
50					7.87	74.7	4.78	20.6	3.35	8.82	2.17	3.10	1.62	1.44	1.26	0.80		
60							5.74	29.6	4.02	12.2	2.60	4.29	1.95	2.07	1.51	1.10		
70							6.69	38.6	4.69	15.3	3.04	5.84	2.27	2.71	1.76	1.50	1.12	0.48
80							7.65	50.3	5.37	21.7	3.48	7.62	2.59	3.53	2.01	1.87	1.28	0.63
	6″																	
90							8.60	63.6	6.04	26.1	3.91	9.22	2.92	4.46	2.26	2.37	1.44	0.80
100	1.11	0.39					9.56	75.1	6.71	32.3	4.34	11.4	3.24	5.27	2.52	2.81	1.60	0.95
125	1.39	0.56							8.38	48.2	5.45	17.1	4.05	7.86	3.15	4.38	2.00	1.48
150	1.67	0.78							10.06	60.4	6.51	23.3	4.86	11.3	3.78	6.02	2.41	2.04
175	1.94	1.06	**8″**						11.73	90.0	7.59	32.0	5.67	14.7	4.41	8.20	2.81	2.78
200	2.22	1.32									8.68	39.7	6.48	19.2	5.04	10.2	3.21	3.46
225	2.50	1.66	1.44	0.44							9.77	50.2	7.29	23.1	5.67	12.9	3.61	4.37
250	2.78	2.05	1.60	0.55							10.85	61.9	8.10	28.5	6.30	15.9	4.01	5.14
275	3.06	2.36	1.76	0.63							11.94	75.0	8.91	34.4	6.93	18.3	4.41	6.22
300	3.33	2.80	1.92	0.75							13.02	84.7	9.72	40.9	7.56	21.8	4.81	7.41

Pressure Drop per 1000 Feet of Schedue 40 Steel Pipe, in Pounds per Square Inch.

Figure 3.12.21 *Pressure Drop in Flow of Water through Schedule 40 Steel Pipe*

3.12.22 ANSI, ASTM, and Other Standard Designations for Ferrous and Nonferrous Plastic Pipe and Fittings

Materials and products	ANSI	ASTM	FS	IAPMO	Other standards	Footnote remarks
FERROUS PIPE AND FITTINGS:						
Cast Iron Screwed Fittings (125 & 250 lbs) (56.8 & 113.5 Kg)	B16.4-1963	A 126-66				
Cast Iron Soil Pipe and Fittings		A 74-82				Note 4
Cast Iron Soil Pipe and Fittings for Hubless Cast Iron Sanitary Systems					CISPI 301-85	Note 4
Cast Iron Threaded Drainage Fittings	B16.12-1971					Note 4
Gray Iron and Ductile Iron Pressure Pipe		A 377-66				
Hubless Cast Iron Sanitary And Rainwater Systems (Installation)				IS 6-89		
Malleable Iron Threaded Fittings (150 & 300 lb) (68.1 & 136.2 Kg)	B16.3-1977					
Neoprene Rubber Gaskets for Hub and Spigot Cast Iron Soil Pipe and Fittings					CISPI HSN-85	
Pipe, Steel, Black and Hot-Dipped, Zinc-Coated Welded and Seamless		A 53-83				
Pipe, Steel, Black and Hot-Dipped, Zinc-Coated (Galvanized) Welded and Seamless, For Ordinary Uses		A 120-82				
Pipe Threads, General Purpose (Inch)	B1.20.1-83					
Roof Drains	A112.21.2M-1983					
Shielded Couplings for Use with Hubless Cast Iron Soil Pipe and Fittings				PS 35-89		

Figure 3.12.22 (*By permission: McGraw-Hill Inc. Plumber's and Pipefitter's Calculations Manual, R. Dodge Woodson.*)

Special Cast Iron Fittings.................	PS 5-84
Subdrains For Built-up Shower Pans..........	PS16-90
Threaded Cast Iron Pipe For Drainage, Vent and Waste Services...................	A40.5-1943
Welded and Seamless Carbon Steel and Austenitic Stainless Steel Pipe Nipples...........	A 733-76

3.12.23 Nonferrous Pipe and Fittings

Brass-, Copper-, and Chromium-Plated Pipe Nipples.................................	B 687-81
Bronze Pipe Flanges and Flanged Fittings (Class 150 & 300)......................	B16.24-1979
Cast Brass and Tubing P-Traps..........	PS 2-89
Cast Copper Alloy Fittings for Flared Copper Tubes............................	B16.26-1987
Cast Bronze Threaded Fittings (Classes 125 & 250)................................	B16.15-1985
Cast Copper Alloy Solder-Joint Drainage Fittings-DWV.........................	B16.23-1984
Cast Copper Alloy Solder-Joint Pressure Fittings..............................	B16.18-1984 Note 4

Figure 3.12.22 *(cont.)*

Materials and products	ANSI	ASTM	FS	IAPMO	Other standards	Footnote remarks
NONFERROUS PIPE FITTINGS:						
Copper Drainage Tube (DWV)................		B 306-86				
Copper Plumbing Tube, Pipe and Fittings (Installation)................				IS 3-89		
Diversion Tees and Twin Waste Elbow........				PS 9-84		
Drains for Prefabricated and Precast Showers..				PS 4-90		
Flexible Metallic Water Connectors..........				PS 14-89		
General Requirements for Wrought Seamless Copper and Copper-Alloy Tube............		B 251-87				
Seamless Brass Tube................		B 135-86(a)				
Seamless Copper Pipe, Standard Sizes.........		B 42-87				
Seamless Copper Tube................		B 75-86				
Seamless Copper Water Tube............		B 88-86				
Seamless Red Brass Pipe, Standard Sizes......		B 43-87				
Seamless and Welded Copper Distribution Tube (Type D)................		B 641-86				
Threadless Copper Pipe................		B 302-87				
Tubing Trap Wall Adapters..............				PS 7-84		
Welded Brass Tube................		B 587-88				
Welded Copper-Alloy UNS No. C21000 Water Tube................		B 642-86				
Welded Copper and Copper Alloy Water Tube (Installation)................				IS 21-89		

Figure 3.12.23 *(cont.)*

281

Welded Copper Tube	B 447-86	
Welded Copper Water Tube	B 716-86	
Wrought Copper and Copper Alloy Solder-Joint Pressure Fittings	B16.22-1986	
Wrought Copper and Wrought Copper Alloy Solder-Joint Drainage Fittings-DWV	B16.29-1986	Note 4

Figure 3.12.23 *(cont.)*

3.12.24 Nonmetallic Pipe

Acrylonitrile-Butadiene-Styrene (ABS) Building Drain, Waste and Vent Pipe and Fittings (Installation)	IS 5-89	Note 4
Acrylonitrile-Butadiene-Styrene (ABS) Schedule 40 Plastic Drain Waste and Vent Pipe	D 2661-87a	Note 4
Acrylonitrile-Butadiene-Styrene (ABS) Schedule 40 Plastic Drain Waste and Vent Pipe With a Cellular Core	F 628-88	Note 4
Acrylonitrile-Butadiene-Styrene (ABS) Sewer Pipe and Fittings	D 2751-88	Note 4

Figure 3.12.24

Materials and products	ANSI	ASTM	FS	IAPMO	Other standards	Footnote remarks
NON-METALLIC PIPE:						
Acrylonitrile-Butadiene-Styrene (ABS) Sewer Pipe and Fittings (Installation)				IS 11-87		Notes 1 & 3
Asbestos-Cement Nonpressure Sewer Pipe		C 428-74				
Asbestos Cement Pressure Pipe		C 296-73				
Asbestos-Cement Pressure Pipe For Water and other Liquids					AWWA C400-72	
Asbestos Cement Pressure Pipe For Water Service and Yard Piping (Installation)				IS 15-82		
Borosilicate Glass Pipe and Fittings for Drain, Waste and Vent (DWV) Applications		C 1053-85				
Chlorinated Poly (Vinyl Chloride) (CPVC) Plastic Pipe, Schedules 40 and 80		F 441-88				
Chlorinated Poly (Vinyl Chloride) (CPVC) Plastic Hot and Cold Water Distribution Systems		D 2846-89e1				
Chlorinated Poly (Vinyl Chloride) (CPVC) Solvent Cemented Hot and Cold Water Distribution Systems (Installation)				IS 20-89		
Coextruded Poly (Vinyl Chloride) Plastic Pipe with a Cellular Core		F891-86e1				
Concrete Drain Tile		C 412-80				Note 3
Concrete Sewer, Storm Drain and Culvert Pipe		C 14-80				

Figure 3.12.24 (cont.)

Materials and products	ANSI	ASTM	FS	IAPMO	Other standards	Footnote remarks
NON-METALLIC PIPE:						
Polybutylene Hot and Cold Water Distribution Pipe, Tubing and Fitting Systems Using Heat Fusion (Installation)				IS 23-90		
Polybutylene Hot and Cold Water Distribution Pipe, Tubing and Fitting systems Using Pressure-Lock Fittings (Installation)				IS 24-90		
Polybutylene (PB) Plastic Hot- & Cold-Water Distribution Systems		D 3309-88a				
Polybutylene (PB) Plastic Pipe (SIDR-PR) Based on Controlled Inside Diameter		D 2662-88				
Polybutylene (PB) Plastic Tubing		D 2666-88				
Polyethylene (PE) Cold Water Building Supply and Yard Piping (Installation)				IS 7-90		
Polyethylene (PE) For Gas Yard Piping (Installation)				IS 12-90		
Polyethylene (PE) Plastic Pip (SIDR-PR) Based on Controlled Inside Diameter		D 2239-88				
Poly (Vinyl Chloride) (PVC) Building Drain, Waste and Vent Pipe and Fittings (Installation)				IS 9-90		
Poly (Vinyl Chloride) (PVC) Cold Water Building Supply and Yard Piping (Installation)				IS 8-89		

Figure 3.12.24 (*cont.*)

Materials and products	ANSI	ASTM	FS	IAPMO	Other standards	Footnote remarks
NON-METALLIC PIPE:						
Socket-Type Chlorinated Poly (Vinyl Chloride) (CPVC) Plastic Pipe Fittings, Schedule 40		F 438-88				
Socket-Type Chlorinated Poly (Vinyl Chloride) (CPVC) Plastic Pipe Fittings, Schedule 80		F 439-88				
Socket-Type Poly (Vinyl Chloride) (PVC) Plastic Pipe Fittings Schedule 80		D 2467-88				Note 4
Solvent Cement For Acrylonitrile-Butadiene-Styrene (ABS) Plastic Pipe and Fittings		D 2235-88				
Solvent Cements For Chlorinated Poly (Vinyl Chloride) (CPVC) Plastic Pipe and Fittings		F 493-88				
Solvent Cements For Poly (Vinyl Chloride) (PVC) Plastic Pipe and Fittings		D 2564-88				
Thermoplastic Accessible and Replaceable Plastic Tube and Tubular Fittings		F 409-88				Note 4
Thermoplastic Gas Pressure Pipe, Tubing and Fittings		D 2513-88b				
Type PS-46 Poly (Vinyl Chloride) (PVC) Plastic Gravity Flow Sewer Pipe and Fittings		F 789-85				
Type PSM Poly (Vinyl Chloride) (PVC) Sewer Pipe and Fittings		D 3034-88				
Threaded Poly (Vinyl Chloride) (PVC) Plastic Pipe Fittings Schedule 80		D 2464-88				Note 4
Vitrified Clay Pipe, Extra Strength, Standard Strength and Perforated		C 700-78				

Figure 3.12.24 (*cont.*)

Poly (Vinyl Chloride) (PVC) Corrugated Sewer Pipe with a Smooth Interior and Fittings	F 949-86a	Note 4
Poly (Vinyl Chloride) (PVC) Natural Gas Yard Piping (Installation)	IS 10-90	
Poly (Vinyl Chloride) (PVC) Plastic Drain, Waste and Vent Pipe and Fittings	D 2665-88	
Poly (Vinyl Chloride) (PVC) Pressure-Rated Pipe (SDR Series)	D 2241-88	
Poly (Vinyl Chloride) (PVC) Plastic Pipe, Schedules 40, 80 and 120	D 1785-88	
Poly (Vinyl Chloride) (PVC) Plastic Pipe Fittings (Schedule 40)	D 2466-88	Note 4
Primers For Use in Solvent Cement Joints of Poly (Vinyl Chloride) (PVC) Plastic Pipe and Fittings	F 656-88	
Rubber Rings for Asbestos-Cement Pipe	D 1869-79	
Safe Handling Of Solvent Cements, Primers, and Cleaners Used For Joining Thermoplastic Pipe and Fittings	F 402-88	
Smoothwall Polyethylene (PE) Pipe for Use in Drainage and Waste Disposal Absorption Fields	F810-85	

Figure 3.12.24 (*cont.*)

Description	Standard	Note
Drain, Waste and Vent (DWV) Plastic Fittings Patterns	D 3311-86	Note 4
Extra Strength Vitrified Clay Pipe in Building Drains (Installation)	IS 18-85	
Fittings for Joining Polyethylene Pipe for Water Service and Yard Piping	PS 25-84	
Joints for IPS PVC Pipe Using Solvent Cement	D 2672-88	
Non-Metallic Building Sewers (Installation)	IS 1-90	
Plastic Insert Fittings For Polybutylene (PB) Tubing	F 845-88	Note 4
Plastic Insert Fittings for Polyethylene (PE) Plastic Pipe	D 2609-88	Note 6
Polybutylene (PB) Cold Water Building Supply and Yard Piping and Tubing (Installation)	IS 17-90	
Polybutylene Hot and Cold Water Distribution Tubing Systems Using Insert Fittings (Installation)	IS 22-90	
Polybutylene Hot and Cold Water Distribution Tubing Systems Using Compression Joints (Installation)	IS 25-90	

Figure 3.12.24 (*cont.*)

3.12.25 Various City Water Temperatures and Hardness Figures

City Water Data

State and City	Source of Supply	Maximum Water Temp. F	Hardness PPM
Alabama			
Anniston	W	70	104
Birmingham	S	85	43
Alaska			
Fairbanks	W	46	120
Ketchikan	S	44	4
Arizona			
Phoenix	W	81	210
Tucson	W	80	222
Arkansas			
Little Rock	WS	89	26
California			
Fresno	W	72	87
Los Angeles	WS	79	195
Sacramento	S	83	76
San Francisco	S	66	181
Colorado			
Denver	S	74	123
Pueblo	S	77	279
Connecticut			
Hartford	S	73	12
New Haven	S	76	46
Delaware			
Wilmington	S	83	48
District of Columbia			
Washington	S	84	162
Florida			
Jacksonville	WS	90	305
Miami	W	82	78
Georgia			
Atlanta	S	87	14
Savannah	W	85	120
Hawaii			
Honolulu	S	70	57
Idaho			
Boise	WS	65	71
Illinois			
Chicago	5	73	125
Peoria	W	67	386
Springfield	84	164	
Indiana			
Evansville	S	87	140
Fort Wayne	S	84	95
Indianapolis	WS	85	279
Iowa			
Des Moines	S	77	340
Dubuque	W	60	324
Sioux City	W	62	548
Kansas			
Kansas City	S	92	230
Kentucky			
Ashland	S	85	93
Louisville	S	85	104
Louisiana			
New Orleans	S	93	150
Shreveport	S	90	36
Maine			
Portland	S	70	12

City Water Data

State and City	Source of Supply	Maximum Water Temp. F	Hardness PPM
Maryland			
Baltimore	S	75	50
Massachusetts			
Cambridge	S	74	46
Holyoke	S	77	23
Michigan			
Detroit	S	78	100
Muskegon	S	71	153
Minnesota			
Duluth	S	58	54
Minneapolis	S	83	172
Mississippi			
Jackson	S	85	38
Meridian	WS	89	7
Missouri			
Springfield	WS	80	187
St. Louis	S	88	83
Montana			
Butte	WS	54	63
Helena	WS	57	96
Nebraska			
Lincoln	W	63	188
Omaha	S	85	135
New Hampshire			
Berlin	S	69	10
Nashua	W	70	25
Nevada			
Reno	S	63	114
New Jersey			
Atlantic City	WS	73	12
Newark	S	75	29
New Mexico			
Albuquerque	W	72	155
New York			
Albany	S	70	42
Buffalo	S	76	118
New York	WS	73	30
North Carolina			
Asheville	S	79	4
Wilmington	S	89	34
North Dakota			
Bismarck	S	80	172
Ohio			
Cincinnati	S	85	120
Cleveland	S	77	121
Oklahoma			
Oklahoma City	S	83	100
Tulsa	S	85	80
Oregon			
Portland	S	65	10
Pennsylvania			
Philadelphia	S	83	98
Pittsburgh	84	95	
Rhode Island			
Providence	S	71	26
South Carolina			
Charleston	S	85	18
Greenville	S	79	4

Figure 3.12.25 *(By permission of The Trane Company, LaCrosse, Wisconsin)*

3.12.26 Abbreviations, Definitions, and Symbols that Appear on Plumbing Drawings

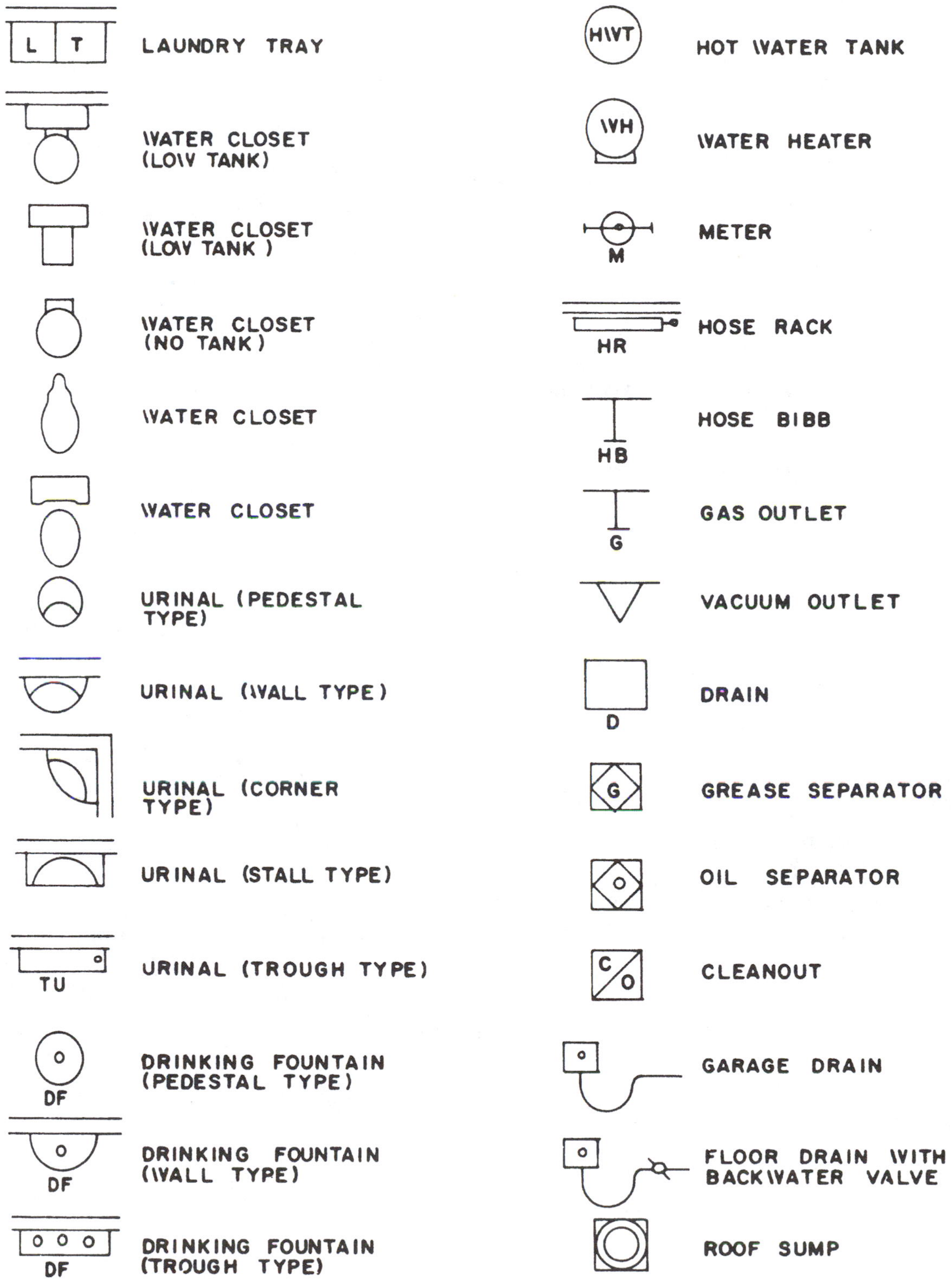

Symbol	Description	Symbol	Description
	LAUNDRY TRAY		HOT WATER TANK
	WATER CLOSET (LOW TANK)		WATER HEATER
	WATER CLOSET (LOW TANK)		METER
	WATER CLOSET (NO TANK)		HOSE RACK
	WATER CLOSET		HOSE BIBB
	WATER CLOSET		GAS OUTLET
	URINAL (PEDESTAL TYPE)		VACUUM OUTLET
	URINAL (WALL TYPE)		DRAIN
	URINAL (CORNER TYPE)		GREASE SEPARATOR
	URINAL (STALL TYPE)		OIL SEPARATOR
	URINAL (TROUGH TYPE)		CLEANOUT
	DRINKING FOUNTAIN (PEDESTAL TYPE)		GARAGE DRAIN
	DRINKING FOUNTAIN (WALL TYPE)		FLOOR DRAIN WITH BACKWATER VALVE
	DRINKING FOUNTAIN (TROUGH TYPE)		ROOF SUMP

Figure 3.12.26 (*By permission of Cast Iron Soil Pipe Institute*)

3.12.27 Recommended Symbols for Plumbing on Plumbing Drawings

Symbols for Fixtures.[1]

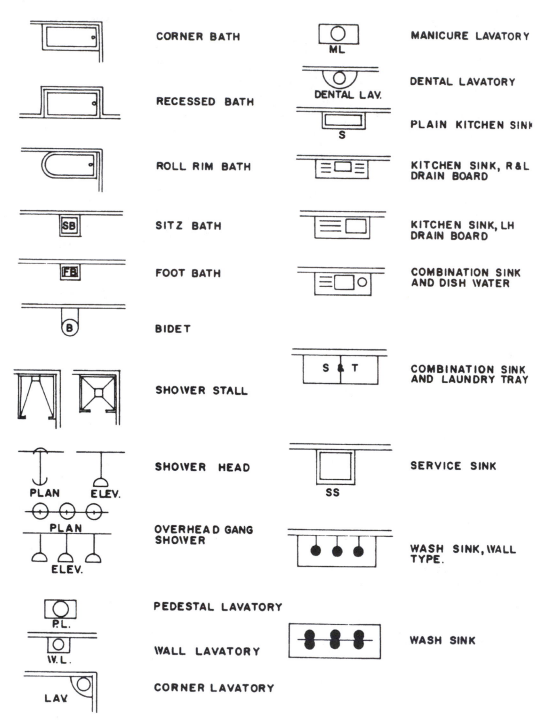

[1]Symbols adopted by the American National Standards Association (ANSI).

Figure 3.12.27 (*By permission of Cast Iron Soil Pipe Institute*)

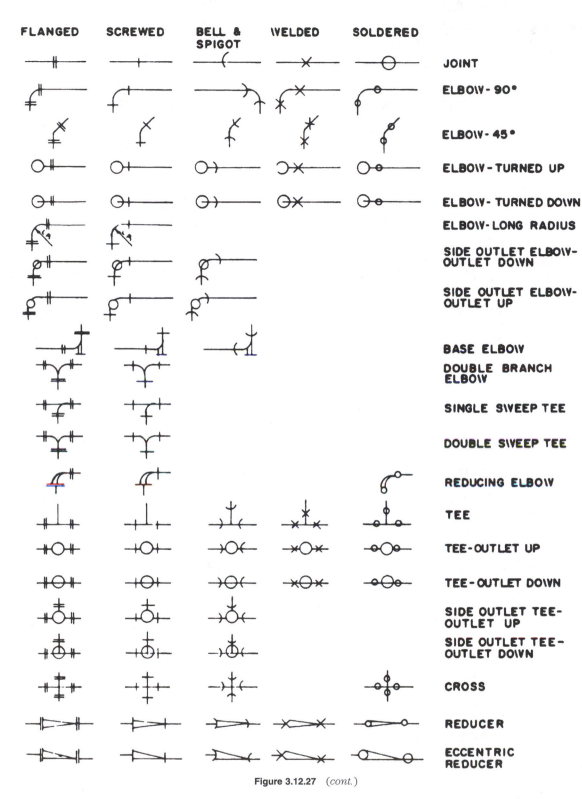

Figure 3.12.27 (*cont.*)

3.12.28 Symbols for Pipe Fittings and Valves

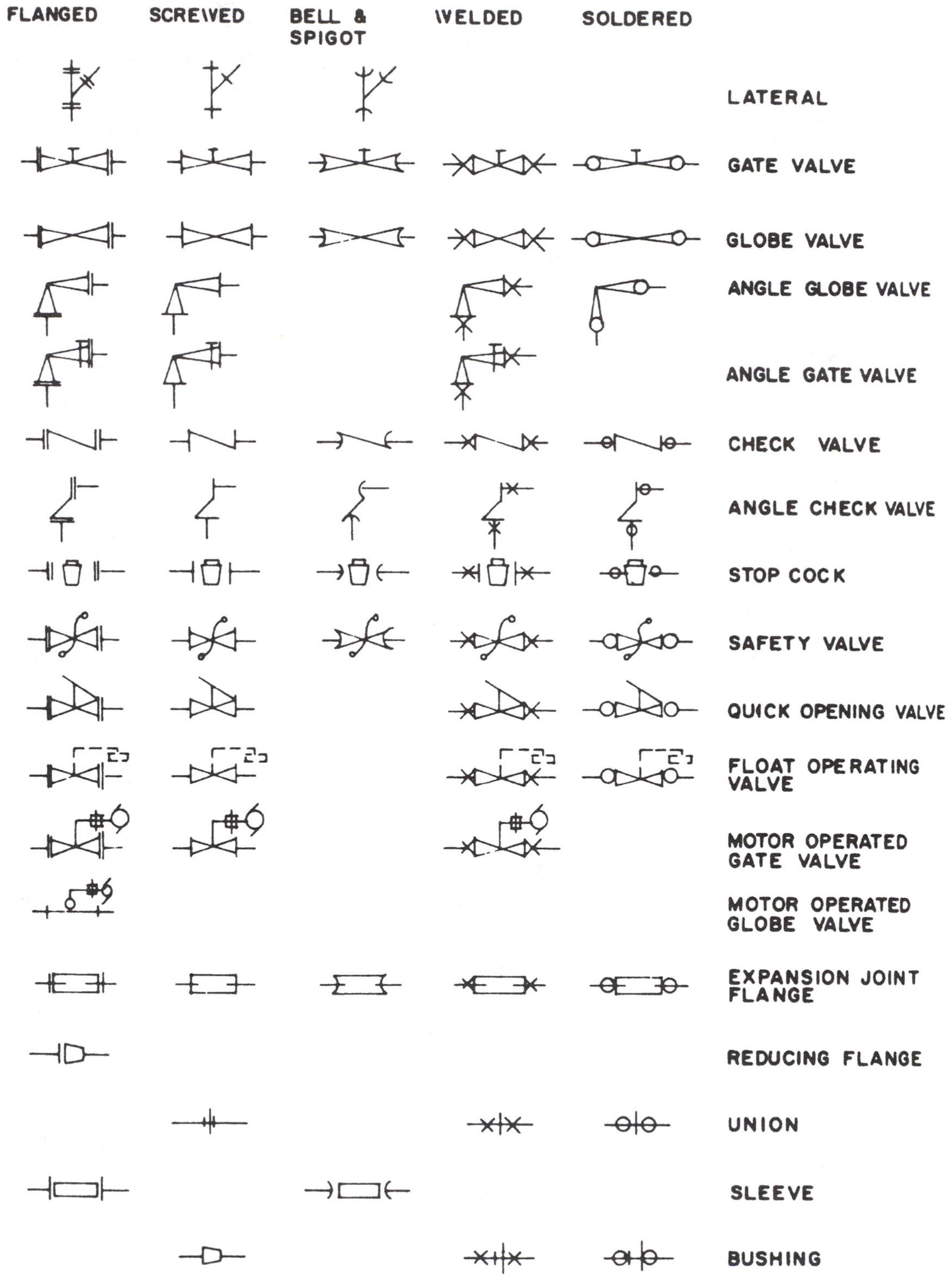

FLANGED	SCREWED	BELL & SPIGOT	WELDED	SOLDERED	
					LATERAL
					GATE VALVE
					GLOBE VALVE
					ANGLE GLOBE VALVE
					ANGLE GATE VALVE
					CHECK VALVE
					ANGLE CHECK VALVE
					STOP COCK
					SAFETY VALVE
					QUICK OPENING VALVE
					FLOAT OPERATING VALVE
					MOTOR OPERATED GATE VALVE
					MOTOR OPERATED GLOBE VALVE
					EXPANSION JOINT FLANGE
					REDUCING FLANGE
					UNION
					SLEEVE
					BUSHING

Figure 3.12.28 *(By permission of Cast Iron Soil Pipe Institute)*

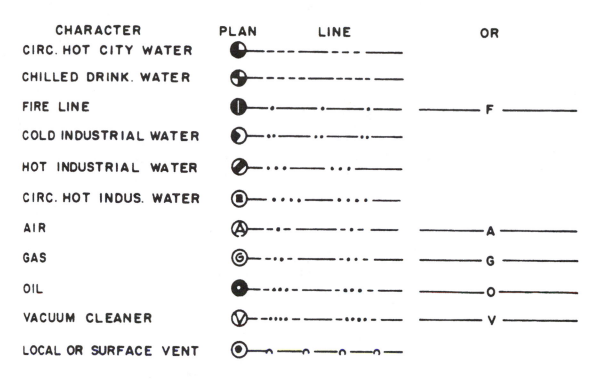

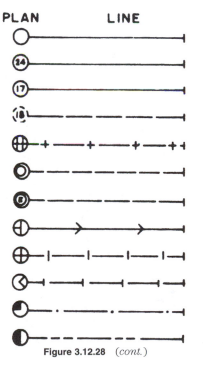

Figure 3.12.28 (*cont.*)

Concrete and Masonry Basics

Contents

4.1.0 History

Concrete is an ancient material of construction, first used during the Roman Empire, which extended from about 20 B.C. to 200 A.D. The word concrete is derived from the Roman *concretus*, meaning to grow together. Although this early mixture was made with lime, cement, and a volcanic ash material called *pozzolana,* concrete today is a sophisticated material to which exotic constitutents can be added and, with computer-controlled batching, can produce a product capable of achieving 50,000 psi compressive strength.

The factors contributing to a successful batch of concrete are

- Precise measurement of water content;
- Type, size, and amount of cement and aggregate;
- Type, size, and location of reinforcement within the concrete pour to compensate for the lack of tensile strength basic in concrete;
- Proper curing procedures during normal hot or cold weather conditions.

4.1.1 General Properties

With some exceptions, the two most widely used concrete mixtures are

- Normal-weight (stone) concrete with a dry weight of 145 psf (6.93 kPa);
- Lightweight concrete (LWC) with a weight of approximately 120 psf (5.74 kPa). Extra light concrete, with weights as low as 80 psf (3.82 kPa), an be achieved with the use of special aggregates.

Other Types of Concrete

- *Lightweight Insulating* Containing perlite, vermiculite, and expanded polystyrene, which is used as fill over metal roof decks, in partitions, and in panel walls.
- *Cellular* Contains air or gas bubbles suspended in mortar and either no coarse aggregates or very limited quantities are included in the mixture. Use where high insulating properties are required.
- *Shot-crete or Gunite* The method of placement characterizes this type of concrete, which is applied via pneumatic equipment. Typical uses are swimming pools, shells, or domes, where formwork would be complicated because of the shape of the structure.
- *Ferrocement* Basically a mortar mixture with large amounts of light-gauge wire reinforcing. Typical uses include bins, boat hulls, and other thin, complex shapes.

4.2.0 Portland Cement as a Major Component

Different types of portland cement are manufactured to meet specific purposes and job conditions.

- Type I is a general-purpose cement used in pavements, slabs, and miscellaneous concrete pads and structures.
- Type IA is used for normal concrete, to which an air-entraining admixture is added.
- Type II creates a moderate sulfur-resistant product that is used where concrete might be exposed to groundwater that contains sulfates.
- Type IIA is the same as Type II, but is suited for an air-entrainment admixture.
- Type III is known as *high early strength* and generates high strength in a week or less.
- Type IIIA is high early, to which an air-entrainment admixture is added.
- Type IV cement produces low heat of hydration and is often used in mass pours, such as dam construction or thick mat slabs.
- Type V is a high sulfate-resistant cement that finds application in concrete structures exposed to high sulfate-containing soils or groundwater.

- White Portland cement is generally available in Type I or Type III only and gains its white color from the selection of raw materials containing negligible amounts of iron and magnesium oxide. White cement is mainly used as a constituent in architectural concrete.

4.2.1 High Early Cement

High early cement does exactly what its name implies: it provides higher compressive strength at an earlier age. Although Type III or Type IIIA cement can produce high early strength, there are other ways to achieve the same end result:

- Add more cement to the mixture [600 lb (272 kg) to 1000 lb (454 kg)];
- Lower the water content (0.2 to 0.45) by weight;
- Raise the curing temperature after consultation with the design engineer;
- Introduce an admixture into the design mix;
- Introduce microsilica, also known as *silica fume*, to the design mix;
- Cure the cast-in-place concrete by autoclaving (steam curing);
- Provide insulation around the formed, cast-in-place concrete to retain heat of hydration.

4.2.2 How Cement Content Affects Shrinkage

When low slumps, created in conjunction with minimum water requirements, are used with correct placement procedures, the shrinkage of concrete will be held to a minimum. Conversely, high water content and high slumps will increase shrinkage. A study at the Massachusetts Institute of Technology, as reported by the Portland Cement Association, indicated that for every 1% increase in mixing water, shrinkage of concrete increased by 2%. This study produced the following chart, showing the correlation of water and cement content to shrinkage.

4.2.3 Effect of Cement/Water Content on Shrinkage

Cement Content Bags/cubic yard	Concrete composition			Aggregate	Water + air	Water cement ratio by weight	Slump (inches)	Shrinkage (av. 3 × 3 × 10* prism)
	Cement	Water	Air					
4.99	0.089	0.202	0.017	0.692	0.219	0.72	3.3	0.0330
5.99	0.107	0.207	0.016	0.670	0.223	0.62	3.6	0.330
6.98	0.124	0.210	0.014	0.652	0.224	0.54	3.8	0.0289
8.02	0.143	0.207	0.015	0.635	0.223	0.46	3.8	0.0300

4.3.0 Control Joints

Thermal shrinkage will occur and the object of control joints, sometimes referred to as construction joints is to avoid the *random cracking* that often comes about when a concrete slab dries and produces excess tensile stress. Control joint spacing depends upon the slab thickness, aggregate size, and water content, as reported by the Portland Cement Association in their articles "Concrete Floors on Concrete," second edition, 1983.

4.3.1 Maximum Spacing of Control Joints

Slab Thickness	Slump of 4–6 inches (101.6 mm–152.4 mm)		Slump less than 4 inches (101.6 mm)
	Max. size aggregate less than ¾ inches (19.05 mm)	Max. size aggregate larger than ¾ inches	
4" (101.6 mm)	8' (2.4 m)	10' (3.05 m)	12' (3.66 m)
5" (126.9 mm)	10' (3.05 m)	13' (3.96 m)	15' (4.57 m)
6" (152.4 mm)	12' (3.66 m)	15' (4.57 m)	18' (5.49 m)
7" (177.8 mm)	14' (4.27 m)	18' (5.49 m)	21' (6.4 m)
8" (203.1 mm)	16' (4.88 m)	20' (6.1 m)	24' (7.32 m)
9" (228.6 mm)	18' (5.49 m)	23' (7.01 m)	27' (8.23 m)
10" (253.9 mm)	20' (6.1 mm)	25' (7.62 m)	30' (9.14 m)

The term *control joint* is often used as being synonymous with *construction joint,* however, there is a difference between the two. A *control joint* is created to provide for movement in the slab and induce cracking at that point, whereas a *construction joint* (Fig. 4.3.2.1) is a bulkhead that ends that day's slab pour. When control joints are created by bulkheading off a slab pour, rather than saw-cutting after the slab has been poured, steel dowels are often inserted in the bulkhead to increase load transfer at this joint.

4.3.2 Dowel Spacing

Slab Depth in. (mm)	Diameter (bar number)	Total length in. (mm)	Spacing in. (mm) center to center
5" (126.9 mm)	#5	12 in. (304.8 mm)	12 in. (304.8 mm)
6" (152.4 mm)	#6	14 in. (355.6 mm)	12 in. (304.8 mm)
7" (177.8 mm)	#7	14 in. (355.6 mm)	12 in. (304.8 mm)
8" (203.1 mm)	#8	14 in. (355.6 mm)	12 in. (304.8 mm)
9" (228.6 mm)	#9	16 in. (406.4 mm)	12 in. (304.8 mm)
10" (253.9 mm)	#10	16 in. (406.4 mm)	12 in. (304.8 mm)

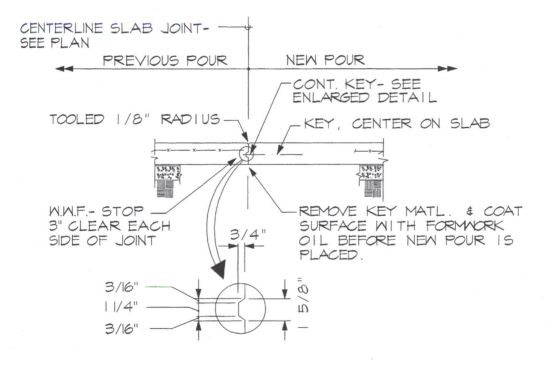

ENLARGED KEY DETAIL

Figure 4.3.2.1   Keyed construction joint detail #2 (not to scale; detail T2-SKCJ2). (*By permission from the McGraw-Hill Co.,* Structural Details Manual, *David R. Williams.*)

4.4.0 Admixtures

Although concrete is an extremely durable product, it faces deterioration from various sources: chemical attack, permeation by water and/or gases from external sources, cracking because of the chemical reaction (known as *heat of hydration*), corrosion of steel reinforcement, freeze/thaw cycles, and abrasion. Much of the deterioration caused by these internal and external factors can be drastically delayed by the addition of a chemical admixture to the ready-mix concrete.

Admixtures are chemicals developed to make it easier for a contractor to produce a high-quality concrete product. Some admixtures retard curing, some accelerate it; some create millions of microscopic bubbles in the mixture; others allow a substantial reduction in water content, but still permit the concrete to flow like thick pea soup.

- *Water-reducing admixtures*   Improve strength, durability, workability of concrete. Available in normal range and high range.

- *High-range water-reducing admixture*   Also known as superplasticizer, it allows up to 30% reduction in water content with no loss of ultimate strength, but it creates increased flowability. It is often required where reinforcing steel is placed very close together in intricate forms.

- *Accelerating admixtures*   They accelerate the set time of concrete, thereby reducing the protection time in cold weather, allowing for earlier stripping of forms. Accelerating admixtures are available in both chloride- and nonchloride-containing forms. Nonchloride is required if concrete is to be in contact with metal and corrosion is to be avoided.

- *Retarder admixtures*   Retards the setting time, a desirable quality during very hot weather.

- *Air-entraining admixtures*   Creates millions of microscopic bubbles in the cured concrete, allowing for expansion of permeated water, which freezes and is allowed to expand into these tiny bubbles, thereby resisting hydraulic pressures caused by the formation of ice.

- *Fly ash*   When added to the concrete mixture, it creates a more dense end product, making the concrete extremely impermeable to water, which affords more protection to steel reinforcement contained in the pour. The addition of fly ash can increase ultimate strength to as much as 6500 psi (44.8 MPa), in the process, making the concrete more resistant to abrasion.

- *Silica fume*   Also known as microsilica, it consists of 90 to 97% silicon dioxide, containing various amounts of carbon that are spherical in size and average about 0.15 microns in size. These extremely fine particles disperse into the spaces around the cement grains and create a uniform dense microstructure that produces concrete with ultra-high compressive strengths, in the nature of 12,000 (82.73 MPa) to 17,000 psi (117.20 MPa).

- *Multifilament or fibrillated fibers*   This material is not a chemical admixture per se, but several manufacturers of concrete chemical additives also sell containers of finely chopped synthetic fibers, generally polypropylene, which, when added to the ready-mix concrete, serve as secondary reinforcement and prevent cracks.

4.4.1 Chloride Content in the Mixing Water

Excessive chloride ions in mixing water can contribute to accelerated reinforcing-steel corrosion and should be a concern when evaluating a mix design. Maximum water-soluble chloride ions, in various forms of concrete (as a percentage), should not exceed the following:

- Prestressed concrete   0.06%

- Reinforced concrete exposed to chloride in service (e.g., garbage slab)   0.15%

- Reinforced concrete that will be dry and/or protected from moisture infiltration   1.00%

- Other reinforced concrete   0.30%

4.5.0 Guidelines for Mixing Small Batches of Concrete (by Weight)

Max. size aggregate	Cement (lb/kg)	Wet-fine aggregate (lb/kg)	Wet-coarse aggregate (lb/kg)	Water (lb/kg)
⅜" (9.52 mm)	29 lb (13.15 kg)	59 lb (26.76 kg)	46 lb (20.87 kg)	11 lb (4.99 kg)
½" (12.6 mm)	27 lb (12.25 kg)	53 lb (24.04 kg)	55 lb (24.95 kg)	11 lb (4.99 kg)
¾" (19.05 mm)	25 lb (11.34 kg)	47 lb (21.32 kg)	65 lb (29.66 kg)	10 lb (4.54 kg)
1" (25.39 mm)	24 lb (10.89 kg)	45 lb (20.41 kg)	70 lb (31.75 kg)	10 lb (4.54 kg)
1½" (37.99 mm)	23 lb (10.43 kg)	43 lb (19.50 kg)	75 lb (34.02 kg)	9 lb (4.08 kg)

4.5.1 Guidelines for Mixing Small Batches of Concrete (by Volume)

Max. size aggregate	Cement	Wet-fine aggregate	Wet-coarse aggregate	Water
⅜" (9.52 mm)	1	2½	1½	½
½" (12.6 mm)	1	2½	2	½
¾" (19.05 mm)	1	2½	2½	½
1" (25.39 mm)	1	2½	2¾	½
1½" (37.99 mm)	1	2½	3	½

4.5.2 Recommended Slumps

The Portland Cement Association recommends the following slumps:

Component	Max. slump (inches	Min. slump (inches)
Footings (reinforced or not)	3	1
Foundation walls	3	1
Substructure walls	3	1
Caissons	3	1
Beams and reinforced walls	4	1
Building columns	4	1
Pavements and slabs	3	1
Mass concrete	2	1

4.5.3 The Slump Test

Slump, as it relates to concrete, is a measure of consistency equal to the decrease in height, measured to the nearest ¼ inch (6 mm) of the molded mass immediately after it has been removed from this molded mass created by the "slump cone" (Fig. 4.5.3.1).

The mold is in the form of a frustum (part of a solid cone intersected by the use of parallel lines) 12 inches (2.5 cm) high, with a base diameter of 8 inches (2 cm) and a top diameter of 4 inches (1 cm).

This mold (slump cone) is filled with freshly mixed concrete in three layers, each being rodded with a ⅝-inch (15.9 mm) bullet-shaped rod 25 times. When the mold has been filled, the top is struck off and the mold is lifted. The amount by which the mass settles after mold removal is referred to as "slump." A small slump is an indication of a very stiff mix, and a very large slump is indicative of a very wet consistency.

Recommended slumps are:

Type of construction	Maximum slump (inches)	Minimum slump (inches)
Reinforced walls/footings	3 (76.2 mm)	1 (25.4 mm)
Caissons, substructure walls	3	1
Beams, reinforced walls	4 (102 mm)	1
Building columns	4	1
Pavements, slabs	3	1
Mass concrete	2 (50.8 mm)	1

Rule of thumb: To raise the slump 1 inch (25.4 mm), add 10 punds of water for each cubic yard of concrete. (One gallon of water equals 8.33 pounds.)

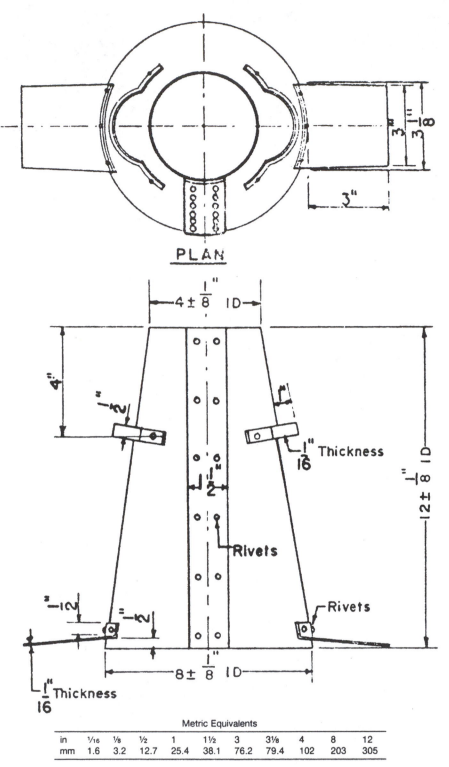

Metric Equivalents									
in	$\frac{1}{16}$	$\frac{1}{8}$	$\frac{1}{2}$	1	$1\frac{1}{2}$	3	$3\frac{1}{8}$	4	8	12
mm	1.6	3.2	12.7	25.4	38.1	76.2	79.4	102	203	305

Figure 4.5.3.1 The slump cone.

4.6.0 Forms for Cast-in-Place Concrete

Many different types of forms are on the market: wood, steel, aluminum, and fiberglass. Each has its advantage and disadvantage; however, some items (form ties and form-release materials) are common to all forms. Also, numerous types and configurations of form liners are available, primarily for architectural concrete use.

4.6.1 Maximum Allowable Tolerances for Form Work

The American Concrete Institute (ACI), in their ACI 347 Manual, include recommended maximum allowable tolerances for various types of cast-in-place and precast concrete, for example:

- *Maximum variations from plumb*   In column and wall surfaces in any 10 feet (3.05 m) of length: ¼ inch (6.35 mm)

- *Maximum for entire length*
  ½ inch (12.7 mm)

- *Maximum variations from established position in plan shown in drawings—walls*
  ¾ inch (19.05 mm)

- *Variations in cross-sectional dimensions of beams/slab-wall thickness*
  Minus: ⅛ inch (3.175 mm)
  Plus: ¼ inch (6.35 mm)

4.6.2 Release Agents for Forms

A number of commercially available form release agents are on the market and some contractors use their own formula, but precautions (as seen below) are necessary, in some instances, to protect the form material.

Form face material   Release agent comments and precautions.

Wood forms   Oils penetrate wood and extend its life.

Unsealed plywood   Apply a liberal amount of release agent several days before using, then wipe off, so only a thin layer remains prior to placing concrete.

Sealed/overlaid plywood   Do not use diesel oil or motor oil on HDO/MDO plywood. Products containing castor oil can discolor concrete.

Steel   Use a product with a rust inhibitor.

Aluminum   Avoid products that contain wax or paraffin.

Glass-fiber reinforced   Follow the form manufacturer's recommendations to avoid damage to forms.

Rigid plastic forms   Follow the form manufacturer's recommendations to avoid damage to forms.

Elastomeric liners   These often do not require release agents, but using the proper agent can prolong life. When deep textures are required, release agents should be used. Follow the manufacturer's recommendations to avoid damage to forms.

Foam expanded plastic liners   Petroleum-based agents can dissolve the foam. These liners are generally "one-time" use only.

Rubber liners/molds   Do not use petroleum, mineral oil, or solvent-based form oils to avoid damage to liner.

Concrete molds   Avoid chemically active release agents and avoid match-cast or slab-on-slab work when the casting surface used as the form is only a few days old.

Controlled-permeability forms   No release agent required.

Plaster waste molds   Pretreat the mold with shellac or some other type of waterproof coating. Yellow cup grease (thinned) is an effective release agent.

4.6.3 Principal Types of Commercially Available Form Ties

	Type of Tie	Typical Working Loads In Tension* (LB)	Notes/Comments
One-Piece Ties	**LOOP TIE** (Breakback point; Hardware that connects adjacent panels also secures tie through loop)	Standard: 2,250 Heavy: 3,000	Shown with manufactured panel; also used with combination lock and bearing-plate hardware in job-built forms.
	FLAT TIE (Notched for breakback; Hardware that connects adjacent panels also secures tie through loop)	Standard: 2,250 Heavy: 3,000	Also available for 1,500-pound loads.
	SNAP TIE (Waterstop (optional); With cone spreaders)	Standard: 2,250 Heavy: 3,000-3,200	Shown with cone spreaders; also available with washer spreaders.
	FIBERGLASS TIE (Spreaders and waterstop available)	3,000; 7,500; and 25,000, with diameters of 0.3, 0.5, and 1 inch, respectively	Available in 10- and 12-foot pieces for cutting to any desired length. Spreaders available.
	TAPER TIE (Taper permits easy pull out)	7,500-64,000, depending on diameter and grade of steel	Completely reusable; grease before installation to facilitate removal. No spreaders included.
	THREADED BAR TIE (Plastic tube and cones prevent bar from bonding to concrete)	7,000-69,000, with diameters from ½ inch to 1½ inches	Stock up to 50 feet long can be cut to required length. Plastic sleeve makes it removable.
Internally Disconnecting Ties	**SHE-BOLT TIE** (Threaded hole in tapered end of the she-bolt screws onto inner tie rod; Inner tie rod; She-bolt)	5,000-64,000	No internal spreader. External spreader bracket available.
	COIL TIE WITH BOLTS (Coil bolt; Cone spreaders; 2-strut coil tie)	Two-strut: 4,500-64,000 Four-strut: 18,000-27,000	Shown with cone spreader, but can be used as combination tie/spreader where it is not necessary to keep the tie ends at the back of the wall face.

*Based on manufacturers' data, using a 2-1 factor of safety. Wide working-load ranges indicate a range of form-tie diameters and grades of steel.

Figure 4.5.3.1 (*By permission Aberdeen's Concrete Construction*)

4.7.0 Curing of Concrete

To attain design strength, curing is a crucial part of the cast-in-place concrete process in order that the proper amount of moisture content and ambient temperature is maintained immediately following the placement of the concrete. The optimum curing cycle will take into account the prevention or replenishment of moisture content from the concrete and the maintenance of a favorable temperature for a specific period of time. During winter months, temporary protection and heat is required in conjunction with the curing process, and, during summer months, moisture replenishment becomes an integral part of the curing process.

4.7.1 Curing Procedures

1. Apply a membrane-curing compound—either by spraying or rolling on the surface immediately after the troweling process on slabs has ceased, or on walls, columns, beams, after the forms have been removed.

2. Curing by water in other than cold-weather conditions is acceptable, as long as it is continuous.

3. Waterproof paper, applied directly over the concrete surface after it has received a spray of water, is often effective.

4. Damp burlap, free of foreign substances that could leach out and stain the concrete, is also a proven curing procedure, as long a the burlap is kept moist.

5. Polyethylene sheets can be used as a blanket in much the same manner as waterproof paper, as long as its edges are lapped and sealed properly.

6. Damp sand or straw is also used on occasion, when nothing else is available. These materials must also be sprayed from time to time to maintain the moisture content.

The length of curing depends upon a number of factors, including the type of cement used and ambient temperatures. The following can be used as a guideline to determine the length of curing time.

4.7.2 Curing Times

At 50°F (10°C)

Percentage design strength required	Type cement used in mix		
	I	II	III
50%	6	9	3
65%	11	14	5
85%	21	28	16
95%	29	35	26

At 70°F (21°C) Days

Percentage design strength required	Type cement used in mix		
	I	II	III
50%	6	9	3
65%	11	14	5
85%	21	28	16
95%	29	35	26

4.8.0 Concrete Reinforcing Bar Size/Weight Charts

Because of concrete's low resistance to shear and tensile strength, the type configuration and placement of reinforcement is crucial to achieve the project's design criteria. The most common form of concrete reinforcement is the deformed reinforcing bar and welded wire fabric. The most commonly used reinforcing bars are set forth in the following chart (Fig. 4.8.0.1).

BAR SIZE DESIGNATION	WEIGHT POUNDS PER FOOT	NOMINAL DIMENSIONS–ROUND SECTIONS		
		DIAMETER INCHES	CROSS-SECTIONAL AREA-SQ INCHES	PERIMETER INCHES
#3	.376	.375	.11	1.178
#4	.668	.500	.20	1.571
#5	1.043	.625	.31	1.963
#6	1.502	.750	.44	2.356
#7	2.044	.875	.60	2.749
#8	2.670	1.000	.79	3.142
#9	3.400	1.128	1.00	3.544
#10	4.303	1.270	1.27	3.990
#11	5.313	1.410	1.56	4.430
#14	7.650	1.693	2.25	5.320
#18	13.600	2.257	4.00	7.090

Figure 4.8.0.1   Concrete reinforcing bar size/weight chart.

4.8.1 ASTM Standards, Including Soft Metric

Soft metric size	Nom diam mm	Area mm^2	Weight factors		Imperial size	Nom diam inches	Area in^2	Weight factors	
			kg/m	kg/ft				lb/ft	lb/m
10	9.5	71	.560	.171	3	.375	.11	.376	1.234
13	12.7	129	.994	.303	4	.500	.20	.668	2.192
16	15.9	199	1.552	.473	5	.625	.31	1.043	3.422
19	19.1	284	2.235	.681	6	.750	.44	1.502	4.928
22	22.2	387	3.042	.927	7	.875	.60	2.044	6.706
25	25.4	510	3.973	1.211	8	1.000	.79	2.670	8.760
29	28.7	645	5.060	1.542	9	1.128	1.00	3.400	11.155
32	32.3	819	6.404	1.952	10	1.270	1.27	4.303	14.117
36	35.8	1006	7.907	2.410	11	1.410	1.56	5.313	17.431
43	43.0	1452	11.384	3.470	14	1.693	2.25	7.650	25.098
57	57.3	2581	20.239	6.169	18	2.257	4.00	13.600	44.619

Comparison of Steel Grades

Soft metric			Imperial		
Grade	mPa	psi	Grade	mPa	psi
300	300	43,511	40	257.79	40,000
420	420	60,716	60	413.69	60,000
520	520	75,420	75	517.11	75,000

Figure 4.8.1   *(By permission, Concrete Reinforcing Steel Institute, Schramsburg, Illinois)*

4.8.2 Recommended End Hooks—All Grades

Soft metric size	mm				Imperial size	Inches			
	D	90° Hooks	180° Hooks			D	90° Hooks	180° Hooks	
		A or G	A or G	J			A or G	A or G	J
10	60	150	125	80	3	2.25	6	5	3
13	80	200	150	105	4	3.0	8	6	4
16	95	250	175	130	5	3.75	10	7	5
19	115	300	200	155	6	4.50	12	8	6
22	135	375	250	180	7	5.25	14	10	7
25	155	425	275	205	8	6.0	16	11	8
29	240	475	375	300	9	9.50	19	15	11.75
32	275	550	425	335	10	10.75	22	17	13.25
36	305	600	475	375	11	12.0	24	19	14.75
43	465	775	675	550	14	18.25	31	27	21.75
57	610	1050	925	725	18	24.0	41	36	28.50

4.8.3 Stirrup and Tie Hooks—All Grades (General)

Soft size	mm				Imperial size	Inches			
	D	90° Hooks	135° Hooks			D	90° Hooks	135° Hooks	
		A or G	A or G	H approx			A or G	A or G	H approx
10	40	105	105	65	3	1.50	4	4	2.5
13	50	115	115	80	4	2.00	4.5	4.5	3
16	65	155	140	95	5	2.50	6	5.5	3.75
19	115	305	205	115	6	4.50	12	8	4.5
22	135	355	230	135	7	5.25	14	9	5.25
25	155	410	270	155	8	6.00	16	10.5	6

4.8.4 Stirrup and Tie Hooks—All Grades (Seismic)

Soft metric size	mm			Imperial size	Inches		
	D	135° Seismic			D	135° Seismic	
		A or G	H approx			A or G	H approx
10	40	110	80	3	1.50	4.25	3
13	50	115	80	4	2.00	4.5	3
16	65	140	95	5	2.50	5.5	3.75
19	115	205	115	6	4.50	8	4.5
22	135	230	135	7	5.25	9	5.25
25	155	270	155	8	6.00	10.5	6

4.8.5 Welded Wire Fabric (WWF)

Wire size number		Nominal diameter, in.	Nominal weight, lb/ft	Area per width (in.^{2}/ft) for various spacings (in)						
Plain	Deformed			2	3	4	6	8	12	16
W45	D45	0.757	1.53	2.70	1.80	1.35	0.90	0.68	0.45	0.34
W31	D31	0.628	1.05	1.86	1.24	0.93	0.62	0.47	0.31	0.23
W20	D20	0.505	0.680	1.2	0.80	0.60	0.40	0.30	0.20	0.15
W18	D18	0.479	0.612	1.1	0.72	0.54	0.36	0.27	0.18	0.14
W16	D16	0.451	0.544	0.96	0.64	0.48	0.32	0.24	0.16	0.12
W14	D14	0.422	0.476	0.84	0.56	0.42	0.28	0.21	0.14	0.11
W12	D12	0.391	0.408	0.72	0.48	0.36	0.24	0.18	0.12	0.09
W11	D11	0.374	0.374	0.66	0.44	0.33	0.22	0.17	0.11	0.08
W10.5		0.366	0.357	0.63	0.42	0.32	0.21	0.16	0.11	0.08
W10	D10	0.357	0.340	0.60	0.40	0.30	0.20	0.15	0.10	0.08
W9.5		0.348	0.323	0.57	0.38	0.29	0.19	0.14	0.095	0.07
W9	D9	0.338	0.306	0.54	0.36	0.27	0.18	0.14	0.090	0.07
W8.5		0.329	0.289	0.51	0.34	0.26	0.17	0.13	0.085	0.06
W8	D8	0.319	0.272	0.48	0.32	0.24	0.16	0.12	0.080	0.06
W7.5		0.309	0.255	0.45	0.30	0.23	0.15	0.11	0.075	0.06
W7	D7	0.299	0.238	0.42	0.28	0.21	0.14	0.11	0.070	0.05
W6.5		0.288	0.221	0.39	0.26	0.20	0.13	0.097	0.065	0.05
W6	D6	0.276	0.204	0.36	0.24	0.18	0.12	0.090	0.060	0.05
W5.5		0.265	0.187	0.33	0.22	0.17	0.11	0.082	0.055	0.04
W5	D5	0.252	0.170	0.30	0.20	0.15	0.10	0.075	0.050	0.04
W4.5		0.239	0.153	0.27	0.18	0.14	0.090	0.067	0.045	0.03
W4	D4	0.226	0.136	0.24	0.16	0.12	0.080	0.060	0.040	0.03
W3.5		0.211	0.119	0.21	0.14	0.11	0.070	0.052	0.035	0.03
W3		0.195	0.102	0.18	0.12	0.090	0.060	0.045	0.030	0.02
W2.9		0.192	0.099	0.17	0.12	0.087	0.058	0.043	0.029	0.02
W2.5		0.178	0.085	0.15	0.10	0.075	0.050	0.037	0.025	0.02
W2.1		0.162	0.070	0.13	0.84	0.063	0.042	0.031	0.021	0.02
W2		0.160	0.068	0.12	0.080	0.060	0.040	0.030	0.020	0.02
W1.5		0.138	0.051	0.090	0.060	0.045	0.030	0.022	0.015	0.01
W1.4		0.134	0.048	0.084	0.056	0.042	0.028	0.021	0.014	0.01

Figure 4.8.5 (*By permission, Concrete Reinforcing Steel Institute, Schramsburg, Illinois*)

4.8.6 Common Types of Welded Wire Fabric

Style designation (W = Plain, D = Deformed)	Steel area (in²/ft)		Approximate weight (lb per 100 sq ft)
	Longitudinal	Transverse	
4 x 4-W1.4 x W1.4	0.042	0.042	31
4 x 4-W2.0 x W2.0	0.060	0.060	43
4 x 4-W2.9 x W2.9	0.087	0.087	62
4 x 4-W/D4 x W/D4	0.120	0.120	86
6 x 6-W1.4 x W1.4	0.028	0.028	21
6 x 6-W2.0 x W2.0	0.040	0.040	29
6 x 6-W2.9 x W2.9	0.058	0.058	42
6 x 6-W/D4 x W/D4	0.080	0.080	58
6 x 6-W/D4.7 x W/D4.7	0.094	0.094	68
6 x 6-W/D7.4 x W/D7.4	0.148	0.148	107
6 x 6-W/D7.5 x W/D7.5	0.150	0.150	109
6 x 6-W/D7.8 x W/D7.8	0.156	0.156	113
6 x 6-W/D8 x W/D8	0.160	0.160	116
6 x 6-W/D8.1 x W/D8.1	0.162	0.162	118
6 x 6-W/D8.3 x W/D8.3	0.166	0.166	120
12 x 12-W/D8.3 x W/D8.3	0.083	0.083	63
12 x 12-W/D8.8 x W/D8.8	0.088	0.088	67
12 x 12-W/D9.1 x W/D9.1	0.091	0.091	69
12 x 12-W/D9.4 x W/D9.4	0.094	0.094	71
12 x 12-W/D16 x W/D16	0.160	0.160	121
12 x 12-W/D16.6 x W/D16.6	0.166	0.166	126

*Many styles may be obtained in rolls.

4.9.0 Protection of Residential Concrete Exposed to Freeze–Thaw Cycles

SECTION 1928 — GENERAL

1928.1 Purpose. The purpose of this appendix is to provide minimum standards for the protection of residential concrete exposed to freezing and thawing conditions.

1928.2 Scope. The provisions of this appendix apply to concrete used in buildings of Groups R and U Occupancies that are three stories or less in height.

1928.3 Special Provisions. Normal-weight aggregate concrete used in buildings of Groups R and U Occupancies three stories or less in height which are subject to de-icer chemicals or freezing and thawing conditions as determined shall comply with the requirements of Table 4.9.0.1.

TABLE 4.9.0.1   Minimum Specified Compressive Strength of Concrete[1]

TYPE OR LOCATION OF CONCRETE CONSTRUCTION	MINIMUM SPECIFIED COMPRESSIVE STRENGTH[2] (f'_c)		
	× 6.89 for kPa		
	Weathering Potential		
	Negligible	Moderate	Severe
Basement walls and foundations not exposed to the weather	2,500	2,500	2,500[3]
Basement slabs and interior slabs on grade, except garage floor slabs	2,500	2,500	2,500[3]
Basement walls, foundation walls, exterior walls and other vertical concrete work exposed to the weather	2,500	3,000[4]	3,000[4]
Porches, carport slabs and steps exposed to the weather, and garage floor slabs	2,500	3,000[4]	3,500[4]

[1]Increases in compressive strength above those used in the design shall not cause implementation of the special inspection provisions of Section 1701.5, Item 1.

[2]At 28 days, pounds per square inch (kPa).

[3]Concrete in these locations which may be subject to freezing and thawing during construction shall be air-entrained concrete in accordance with Footnote 5.

[4]Concrete shall be air entrained. Total air content (percentage by volume of concrete) shall not be less than 5 percent or more than 7 percent.

(Reproduced from the 1997 Edition of the Uniform Building Code, Volumes 1, 2, 3, copyright 1997, with the permission of the publisher, the International Conference of Building Officials (ICBO). ICBO assumes no responsibility for the accuracy or the completion of summaries provided therein.)

4.9.1 Total Air Content for Frost-Resistant Concrete

NOMINAL MAXIMUM AGGREGATE SIZE (inches)	AIR CONTENT, PERCENTAGE	
× 25.4 for mm	Severe Exposure	Moderate Exposure
$3/8$	$7^1/2$	6
$1/2$	7	$5^1/2$
$3/4$	6	5
1	6	$4^1/2$
$1^1/2$	$5^1/2$	$4^1/4$
2^1	5	4
3^1	$4^1/2$	$3^1/2$

[1]These air contents apply to total mix, as for the preceding aggregate sizes. When testing this concrete, however, aggregate larger than $1^1/2$ inches (38 mm) is removed by hand picking or sieving, and air content is determined on the minus $1^1/2$-inch (38 mm) fraction.

Figure 4.9.1 *(Reproduced from the 1997 Edition of the Uniform Building Code, Volumes 1, 2, 3, copyright 1997, with the permission of the publisher, the International Conference of Building Officials (ICBO). ICBO assumes no responsibility for the accuracy or the completion of summaries provided therein.)*

4.9.2 Requirements for Special Exposure Conditions

EXPOSURE CONDITION	MAXIMUM WATER-CEMENTITIOUS MATERIALS RATIO, BY WEIGHT, NORMAL-WEIGHT AGGREGATE CONCRETE	MINIMUM f'_c, NORMAL-WEIGHT AND LIGHTWEIGHT AGGREGATE CONCRETE, psi × 0.00689 for MPa
Concrete intended to have low permeability when exposed to water	0.50	4,000
Concrete exposed to freezing and thawing in a moist condition or to deicing chemicals	0.45	4,500
For corrosion protection for reinforced concrete exposed to chlorides from deicing chemicals, salt, saltwater, brackish water, seawater or spray from these sources	0.40	5,000

Figure 4.9.2 *(Reproduced from the 1997 Edition of the Uniform Building Code, Volumes 1, 2, 3, copyright 1997, with the permission of the publisher, the International Conference of Building Officials (ICBO). ICBO assumes no responsibility for the accuracy or the completion of summaries provided therein.)*

4.9.3 Requirements for Concrete Exposed to Deicing Chemicals

CEMENTITIOUS MATERIALS	MAXIMUM PERCENT OF TOTAL CEMENTITIOUS MATERIALS BY WEIGHT[1]
Fly ash or other pozzolans conforming to ASTM C 618	25
Slag conforming to ASTM C 989	50
Silica fume conforming to ASTM C 1240	10
Total of fly ash or other pozzolans, slag and silica fume	50[2]
Total of fly ash or other pozzolans and silica fume	35[2]

[1]The total cementitious materials also includes ASTM C 150, C 595 and C 845 cement.

[2]Fly ash or other pozzolans and silica fume shall constitute no more than 25 and 10 percent, respectively, of the total weight of the cementitious materials.

The maximum percentages above shall include:

1. Fly ash or other pozzolans present in Type IP or I(PM) blended cement in accordance with ASTM C 595.
2. Slag used in the manufacture of a IS or I(SM) blended cement in accordance with ASTM C 595.
3. Silica fume, ASTM C 1240, present in a blended cement.

Figure 4.9.3 *(Reproduced from the 1997 Edition of the Uniform Building Code, Volumes 1, 2, 3, copyright 1997, with the permission of the publisher, the International Conference of Building Officials (ICBO). ICBO assumes no responsibility for the accuracy or the completion of summaries provided therein.)*

4.9.4 Requirements for Concete Exposed to Sulfate-Containing Solutions

SULFATE EXPOSURE	WATER-SOLUBLE SULFATE (SO_4) IN SOIL, PERCENTAGE BY WEIGHT	SULFATE (SO_4) IN WATER, ppm	CEMENT TYPE	MAXIMUM WATER-CEMENTITIOUS MATERIALS RATIO, BY WEIGHT, NORMAL-WEIGHT AGGREGATE CONCRETE[1]	MINIMUM f'_c, NORMAL-WEIGHT AND LIGHTWEIGHT AGGREGATE CONCRETE, psi × 0.00689 for MPa
Negligible	0.00-0.10	0-150	—	—	—
Moderate[2]	0.10-0.20	150-1,500	II, IP(MS), IS (MS)	0.50	4,000
Severe	0.20-2.00	1,500-10,000	V	0.45	4,500
Very severe	Over 2.00	Over 10,000	V plus pozzolan[3]	0.45	4,500

[1]A lower water-cementitious materials ratio or higher strength may be required for low permeability or for protection against corrosion of embedded items or freezing and thawing (Table 19-A-2).

[2]Seawater.

[3]Pozzolan that has been determined by test or service record to improve sulfate resistance when used in concrete containing Type V cement.

Figure 4.9.4 *(Reproduced from the 1997 Edition of the Uniform Building Code, Volumes 1, 2, 3, copyright 1997, with the permission of the publisher, the International Conference of Building Officials (ICBO). ICBO assumes no responsibility for the accuracy or the completion of summaries provided therein.)*

4.10.0 Minimum Cover for Reinforcement in Cast-in-Place Concrete

Cast-in-place concrete (nonprestressed). The following minimum concrete cover shall be provided for reinforcement:

		MINIMUM COVER, inches (mm)
1.	Concrete cast against and permanently exposed to earth	3 (76)
2.	Concrete exposed to earth or weather:	
	No. 6 through No. 18 bar	2 (51)
	No. 5 bar, W31 or D31 wire, and smaller	$1^{1}/_{2}$ (38)
3.	Concrete not exposed to weather or in contact with ground:	
	Slabs, walls, joists:	
	No. 14 and No. 18 bar	$1^{1}/_{2}$ (38)
	No. 11 bar and smaller	$^{3}/_{4}$ (19)
	Beams, columns:	
	Primary reinforcement, ties, stirrups, spirals	$1^{1}/_{2}$ (38)
	Shells, folded plate members:	
	No. 6 bar and larger	$^{3}/_{4}$ (19)
	No. 5 bar, W31 or D31 wire, and smaller	$^{1}/_{2}$ (12.7)
4.	*Concrete tilt-up panels cast against a rigid horizontal surface, such as a concrete slab, exposed to the weather:*	
	No. 8 and smaller	*1 (25)*
	No. 9 through No. 18	*2 (51)*

Precast concrete (Manufactured under plant control conditions). The following minimum concrete cover shall be provided for reinforcement:

Figure 4.10.0 (*Reproduced from the 1997 Edition of the Uniform Building Code, Volumes 1, 2, 3, copyright 1997, with the permission of the publisher, the International Conference of Building Officials (ICBO). ICBO assumes no responsibility for the accuracy or the completion of summaries provided therein.*)

4.10.1 Minimum Cover for Reinforcement in Precast Concrete

		MINIMUM COVER, inches (mm)
1.	Concrete exposed to earth or weather:	
	Wall panels:	
	No. 14 and No. 18 bar	$1^{1}/_{2}$ (38)
	No. 11 bar and smaller	$^{3}/_{4}$ (19)
	Other members:	
	No. 14 and No. 18 bar	2 (51)
	No. 6 through No. 11 bar	$1^{1}/_{2}$ (38)
	No. 5 bar W31 or D31 wire, and smaller	$1^{1}/_{4}$ (32)
2.	Concrete not exposed to weather or in contact with ground:	
	Slabs, walls, joists:	
	No. 14 and No. 18 bar	$1^{1}/_{4}$ (32)
	No. 11 bar and smaller	$^{5}/_{8}$ (16)
	Beams, columns:	
	Primary reinforcement	d_b but not less than $^{5}/_{8}$ (16) and need not exceed $1^{1}/_{2}$ (38)
	Ties, stirrups, spirals	$^{3}/_{8}$ (9.5)
	Shells, folded plate members:	
	No. 6 bar and larger	$^{5}/_{8}$ (16)
	No. 5 bar, W31 or D31 wire, and smaller	$^{3}/_{8}$ (9.5)

Figure 4.10.1 (*Reproduced from the 1997 Edition of the Uniform Building Code, Volumes 1, 2, 3, copyright 1997, with the permission of the publisher, the International Conference of Building Officials (ICBO). ICBO assumes no responsibility for the accuracy or the completion of summaries provided therein.*)

4.11.0 Weathering Regions (Weathering Index)

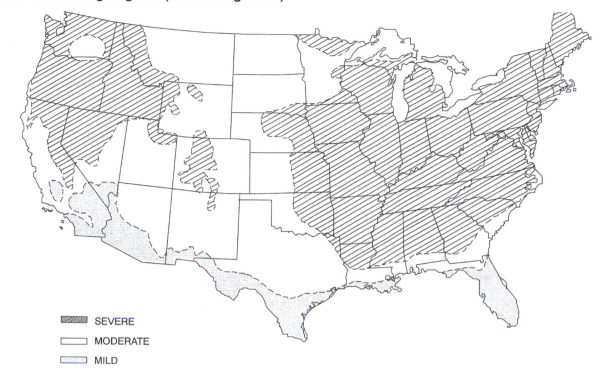

SEVERE

MODERATE

MILD

NOTES:

[1]The three exposures are:

A. Severe—Outdoor exposure in a cold climate where concrete may be exposed to the use of de-icing salts or where there may be a continuous presence of moisture during frequent cycles of freezing and thawing. Examples are pavements, driveways, walks, curbs, steps, porches and slabs in unheated garages. Destructive action from de-icing salts may occur either from direct application or from being carried onto an unsalted area from a salted area, such as on the undercarriage of a car traveling on a salted street but parked on an unsalted driveway or garage slab.

B. Moderate—Outdoor exposure in a climate where concrete will not be exposed to the application of de-icing salts but will occasionally be exposed to freezing and thawing.

C. Mild—Any exposure where freezing and thawing in the presence of moisture is rare or totally absent.

[2]Data needed to determine the weathering index for any locality may be found or estimated from the tables of Local Climatological Data, published by the Weather Bureau, U.S. Department of Commerce.

[3]The weathering regions map provides the location of severe, moderate and mild winter weathering areas as they occur in the United States (Alaska and Hawaii are classified as severe and mild, respectively). The map cannot be precise. This is especially true in mountainous areas where conditions change dramatically within very short distances. It is intended to classify as severe any area in which weathering conditions may cause de-icing salt to be used, either by individuals or for street or highway maintenance. These conditions are significant snowfall combined with extended periods during which there is little or no natural thawing. If there is any doubt about which of two regions is applicable, the more severe exposure should be selected.

[4]The Weathering Index:

Severe—As a guideline, the number of days during which the temperature does not rise above 32°F (0°C) is multiplied by the inches of snowfall. An index of 150 or more is classified as severe. Cold, humid climates may be more severe than cold, dry climates for a given index.

Moderate, Mild—Multiply the inches of precipitation times the number of days the temperature registers below 32°F (0°C) Use the occurrence between the first day in the fall and the last day in the spring that the temperature registers below 32°F (0°C) An index above 200 is moderate. An index below 200 is mild.

Figure 4.11.0 *(Reproduced from the 1997 Edition of the Uniform Building Code, Volumes 1, 2, 3, copyright 1997, with the permission of the publisher, the International Conference of Building Officials (ICBO). ICBO assumes no responsibility for the accuracy or the completion of summaries provided therein.)*

4.12.0 Typical Concrete Wall Form

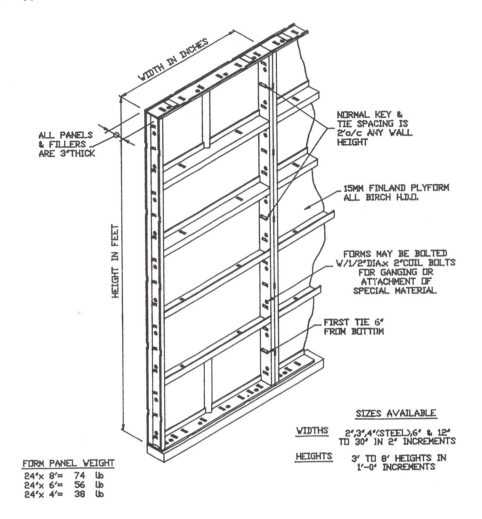

ALL PANELS & FILLERS ARE 3" THICK

WIDTH IN INCHES

HEIGHT IN FEET

NORMAL KEY & TIE SPACING IS 2'o/c ANY WALL HEIGHT

15MM FINLAND PLYFORM ALL BIRCH H.D.O.

FORMS MAY BE BOLTED W/1/2" DIA.× 2" COIL BOLTS FOR GANGING OR ATTACHMENT OF SPECIAL MATERIAL

FIRST TIE 6" FROM BOTTOM

SIZES AVAILABLE

WIDTHS   2",3",4"(STEEL),6" & 12" TO 30" IN 2" INCREMENTS

HEIGHTS   3' TO 8' HEIGHTS IN 1'-0" INCREMENTS

FORM PANEL WEIGHT
24"× 8'= 74 lb
24"× 6'= 56 lb
24"× 4'= 38 lb

4.12.1 Typical Concrete Wall Form Schematic—One Side in Place

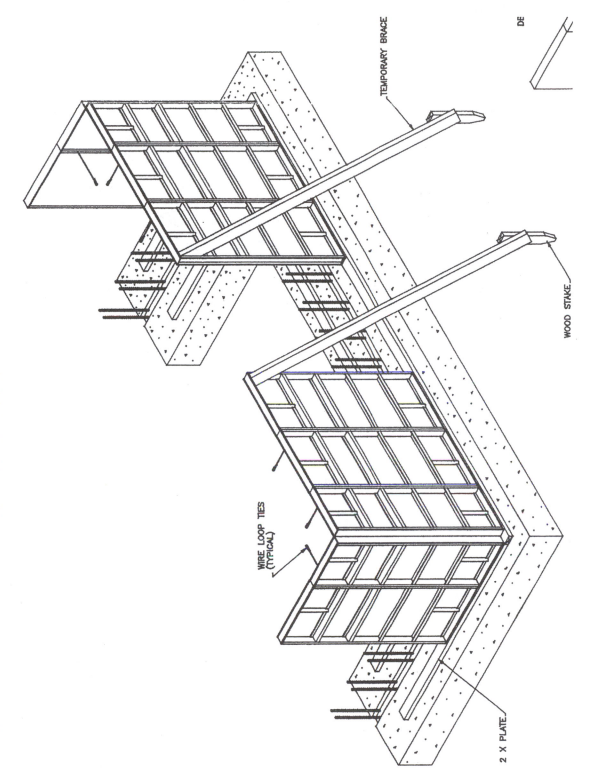

TEMPORARY BRACE

WOOD STAKE

WIRE LOOP TIES
(TYPICAL)

2 X PLATE

4.12.2 Typical Concrete Wall From Schematic with Walkway Bracket Installed—One Side in Place

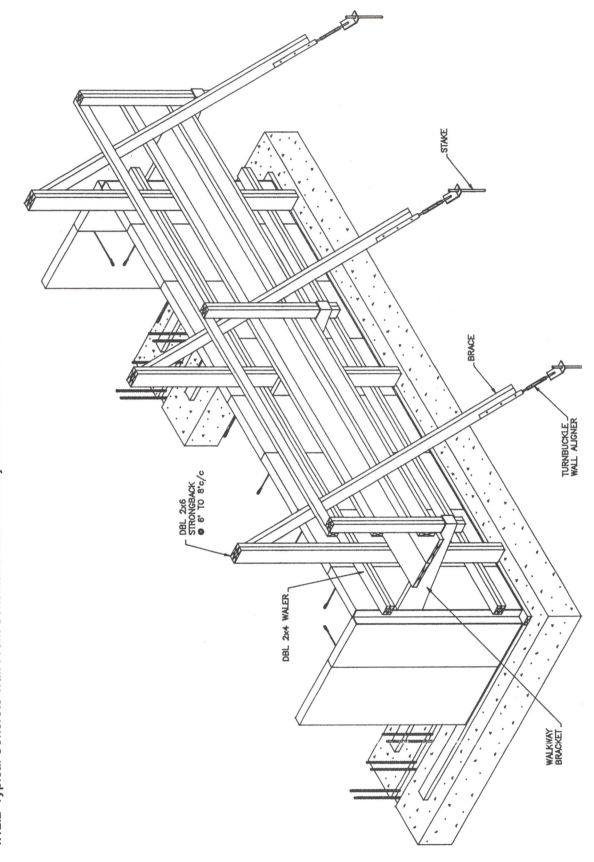

STAKE

BRACE

TURNBUCKLE
WALL ALIGNER

DBL 2x6
STRONGBACK
@ 6' TO 8'c/c

DBL 2x4 WALER

WALKWAY
BRACKET

4.12.3 Typical Plaster Form

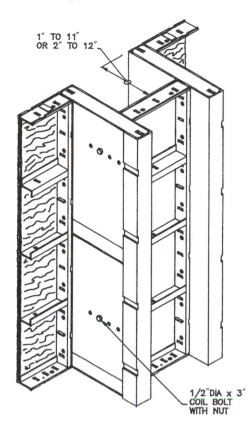

1" TO 11"
OR 2" TO 12"

1/2"DIA x 3"
COIL BOLT
WITH NUT

4.12.4 Typical Tie Connection

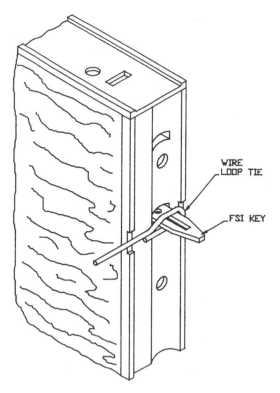

WIRE
LOOP TIE

FSI KEY

4.12.5 Typical 45-Degree Corner (as shown for 12" wall)

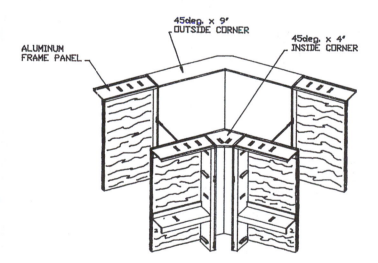

45deg. x 9"
OUTSIDE CORNER

45deg. x 4"
INSIDE CORNER

ALUMINUM
FRAME PANEL

4.12.6 Typical Waler and Walkway Bracket Attachment

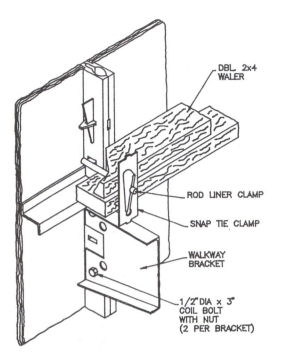

DBL. 2x4
WALER

ROD LINER CLAMP

SNAP TIE CLAMP

WALKWAY
BRACKET

1/2" DIA x 3"
COIL BOLT
WITH NUT
(2 PER BRACKET)

4.12.7 Typical 90-Degree Outside Corner

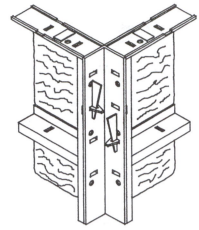

4.12.8 Typical 90-Degree Inside Corner

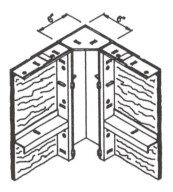

4.12.9 Typical Wood Filler

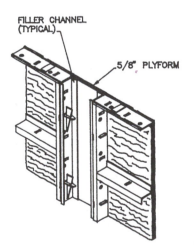

FILLER CHANNEL
(TYPICAL)

5/8" PLYFORM

4.12.10 Attachment of Form to Plate Using Double-Headed Nails

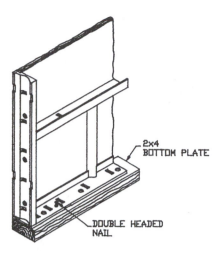

2x4
BOTTOM PLATE

DOUBLE HEADED
NAIL

4.12.11 Long Key Installation

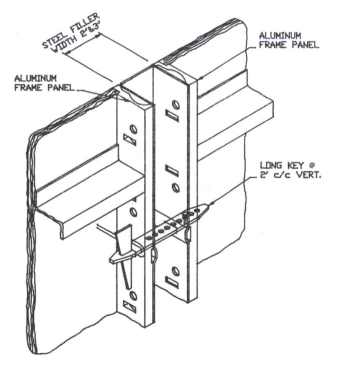

STEEL FILLER
WIDTH 2"&3"

ALUMINUM
FRAME PANEL

ALUMINUM
FRAME PANEL

LONG KEY @
2' c/c VERT.

4.12.12 Proper Key and Wedge Connections and Installation Diagrams

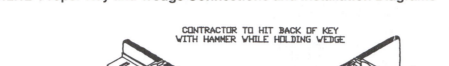

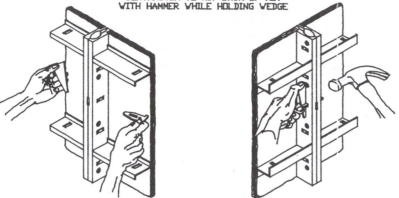

PROPER KEY & WEDGE INSTALLATION

SIDE RAIL TO END
RAIL CONNECTION

END RAIL TO END
RAIL CONNECTION

TYPICAL KEY & WEDGE CONNECTION

4.12.13 Various Clamps, Ties, Keys, and Wedges (Illustrated)

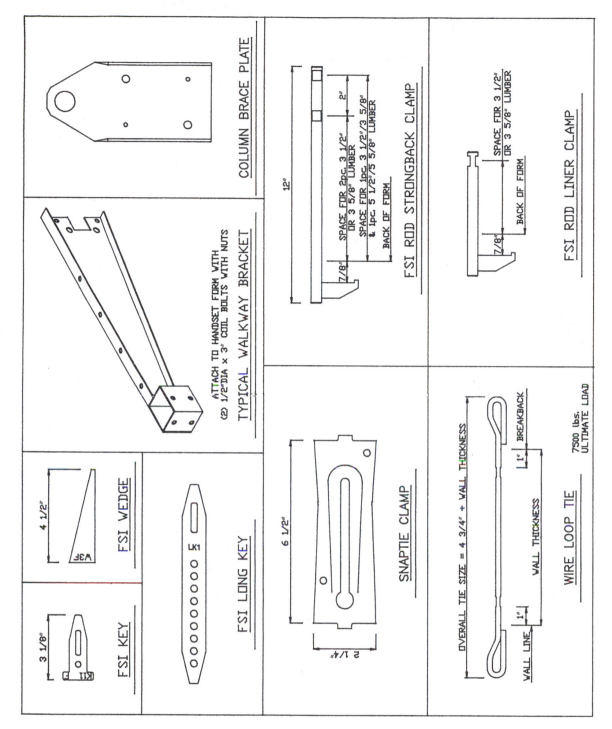

COLUMN BRACE PLATE

TYPICAL WALKWAY BRACKET

ATTACH TO HANDSET FORM WITH
(2) 1/2"DIA x 3" COIL BOLTS WITH NUTS

FSI ROD STRONGBACK CLAMP

12"
2"
7/8"
SPACE FOR 2pc. 3 1/2"
OR 3 5/8" LUMBER
SPACE FOR 1pc. 3 1/2"/3 5/8"
& 1pc 5 1/2"/5 5/8" LUMBER
BACK OF FORM

FSI ROD LINER CLAMP

SPACE FOR 3 1/2"
OR 3 5/8" LUMBER
BACK OF FORM
7/8"

FSI WEDGE

4 1/2"
W3F

FSI KEY

3 1/8"

FSI LONG KEY

LK1

SNAPTIE CLAMP

6 1/2"
2 1/4"

WIRE LOOP TIE

7500 lbs.
ULTIMATE LOAD

OVERALL TIE SIZE = 4 3/4" + WALL THICKNESS
BREAKBACK
1"
WALL THICKNESS
WALL LINE
1"

4.13.0 Concrete Form Design Using Plywood Panels

FORM DESIGN

INTRODUCTION

This section presents tables and shows how to use them to choose the right Plyform thickness for most applications. It also includes tables for choosing the proper size and spacing of joists, studs, and wales.

Though many combinations of frame spacing and plywood thicknesses will meet the structural requirements, it is probably better to use only one thickness of plywood and then vary the frame spacing for different pressures. Plyform can be manufactured in various thicknesses, but it is good practice to base designs on 19/32", 5/8", 23/32" and 3/4" Plyform Class I, as they are most commonly available. Plywood thickness should be compatible with form tie dimensions. For large jobs or those having special requirements, other thicknesses may be preferable, but could require a special order.

CONCRETE PRESSURES

The required plywood thickness, as well as size and spacing of framing, will depend on the maximum load. So the first step in form design is to determine maximum concrete pressure. It will depend on such things as pour rate, concrete temperature, concrete slump, cement type, concrete density, method of vibration, and height of form.

PRESSURES ON COLUMN AND WALL FORMS

Table 1 shows the lateral pressure for newly placed concrete that should be used for the design of column and wall formwork. This pressure is based on the recommendations of the American Concrete Institute (ACI). When form work is to be designed for exterior vibration or to be used in conjunction with pumped concrete placement systems, the design pressures listed should increase in accordance with accepted concrete industry standards.

TABLE 1
Concrete Pressures for Column and Wall Forms

Pour Rate (ft–hr)	Pressures of Vibrated Concrete (psf)[a][b]			
	50°F		70°F	
	Columns	Walls	Columns	Walls
1	330	330	280	280
2	510	510	410	410
3	690	690	540	540
4	870	870	660	660
5	1050	1050	790	790
6	1230	1230	920	920
7	1410	1410	1050	1050
8	1590	1470	1180	1090
9	1770	1520	1310	1130
10	1950	1580	1440	1170

(a) Maximum pressure need not exceed 150h, where h is maximum height of pour.
(b) Based on concrete with density of 150 pcf and 4 in. slump.

TABLE 2
Design Loads for Slab Forms

Slab Thickness (in.)	Design Load (psf)	
	Nonmotorized Buggies[a]	Motorized Buggies[b]
4	100[c]	125[c]
5	113	138
6	125	150
7	138	163
8	150	175
9	163	188
10	175	200

(a) Includes 50 psf load for workers, equipment, impact, etc.
(b) Includes 75 psf load for workers, equipment, impact, etc.
(c) Minimum design load regardless of concrete weight.

Figure 4.13.0   *(Courtesy: APA, The Engineered Wood Assoc., Takoma, Washington)*

4.13.1 Recommended Pressures on Plyform

RECOMMENDED PRESSURES ON PLYFORM

Recommended pressures on the more common thicknesses of Plyform Class I are shown in Tables 3 and 4. Tables 5 and 6 show pressures for Structural I Plyform. Use unshaded columns for design of architectural concrete forms where appearance is important. Calculations for these pressures were based on deflection limitations of 1/360th or 1/270 of the span, or shear or bending strength: whichever provided the most conservative (lowest load) value.

Though not manufactured specifically for concrete forming, grades of plywood other than Plyform have been used for forming when thin panels are needed for curved forms. The recommended pressures shown in the following tables give a good estimate of performance for sanded grades such as APA A-C Exterior and APA B-C Exterior, and unsanded grades such as APA Rated Sheathing Exterior and Exposure 1 (CDX) (marked PS 1), provided face grain is across supports. For Group 1 sanded grades, use the tables for Plyform Class I. For unsanded grades (Span Rated PS 1 panels) use the Plyform Class I tables

assuming 15/32″ Plyform for 32/16 panels, 19/32″ for 40/20 and 23/32″ for 48/24.

Textured plywood has recently been used to obtain various patterns for architectural concrete. Many of these panels have some of the face ply removed due to texturing. Consequently, strength and stiffness will be reduced. As textured plywood is available in a variety of patterns and wood species, it is impossible to give exact factors for strength and stiffness reductions. For approximately equivalent strength, specify the desired grade in Group 1 species and determine the thickness assuming Plyform Class I. When

TABLE 3
Recommended Maximum Pressures on Plyform Class I (psf)[a]
Face Grain Across Supports[b]

Support Spacing (in.)	Plywood Thickness (in.)													
	15/32		1/2		19/32		5/8		23/32		3/4		1-1/8	
4	2715	2715	2945	2945	3110	3110	3270	3270	4010	4010	4110	4110	5965	5965
8	885	885	970	970	1195	1195	1260	1260	1540	1540	1580	1580	2295	2295
12	335	395	405	430	540	540	575	575	695	695	730	730	1370	1370
16	150	200	175	230	245	305	265	325	345	390	370	410	740	770
20	–	115	100	135	145	190	160	210	210	270	225	285	485	535
24	–	–	–	–	–	100	–	110	110	145	120	160	275	340
32	–	–	–	–	–	–	–	–	–	–	–	–	130	170

(a) Deflection limited to 1/360th of the span, 1/270th where shaded.
(b) Plywood continuous across two or more spans.

TABLE 4
Recommended Maximum Pressures on Plyform Class I (psf)[a]
Face Grain Parallel To Supports[b]

Support Spacing (in.)	Plywood Thickness (in.)													
	15/32		1/2		19/32		5/8		23/32		3/4		1-1/8	
4	1385	1385	1565	1565	1620	1620	1770	1770	2170	2170	2325	2325	4815	4815
8	390	390	470	470	530	530	635	635	835	835	895	895	1850	1850
12	110	150	145	195	165	225	210	280	375	400	460	490	1145	1145
16	–	–	–	–	–	–	–	120	160	215	200	270	710	725
20	–	–	–	–	–	–	–	–	115	125	145	155	400	400
24	–	–	–	–	–	–	–	–	–	–	–	100	255	255

(a) Deflection limited to 1/360th of the span, 1/270th where shaded.
(b) Plywood continuous across two or more spans.

Figure 4.13.1   *(Courtesy: APA, The Engineered Wood Assoc., Takoma, Washington)*

TABLE 5
Recommended Maximum Pressures on Structural 1 Plyform (psf)[a]
Face Grain Across Supports[b]

Support Spacing (in.)	15/32		1/2		19/32		5/8		23/32		3/4		1-1/8	
4	3560	3560	3925	3925	4110	4110	4305	4305	5005	5005	5070	5070	7240	7240
8	890	890	980	980	1225	1225	1310	1310	1590	1590	1680	1680	2785	2785
12	360	395	410	435	545	545	580	580	705	705	745	745	1540	1540
16	155	205	175	235	245	305	270	330	350	400	375	420	835	865
20	–	115	100	135	145	190	160	215	210	275	230	290	545	600
24	–	–	–	–	–	100	–	110	110	150	120	160	310	385
32	–	–	–	–	–	–	–	–	–	–	–	–	145	190

(a) Deflection limited to 1/360th of the span, 1/270th where shaded.
(b) Plywood continuous across two or more spans.

TABLE 6
Recommended Maximum Pressures on Structural 1 Plyform (psf)[a]
Face Grain Parallel To Supports[b]

Support Spacing (in.)	15/32		1/2		19/32		5/8		23/32		3/4		1-1/8	
4	1970	1970	2230	2230	2300	2300	2515	2515	3095	3095	3315	3315	6860	6860
8	470	530	605	645	640	720	800	865	1190	1190	1275	1275	2640	2640
12	130	175	175	230	195	260	250	330	440	545	545	675	1635	1635
16	–	–	–	–	–	110	105	140	190	255	240	315	850	995
20	–	–	–	–	–	–	–	100	135	170	170	210	555	555
24	–	–	–	–	–	–	–	–	–	–	–	115	340	355

(a) Deflection limited to 1/360th of the span, 1/270th where shaded.
(b) Plywood continuous across two or more spans.

3/8″ textured plywood is used as a form liner, assume that the plywood backing must carry the entire load.

In some cases, it may be desirable to use two layers of plywood. The recommended pressures shown in Tables 3 through 6 are additive for more than one layer.

Tables 3 through 6 are based on the plywood acting as a continuous beam which spans between joists or studs. No blocking is assumed at the unsupported panel edges. Under conditions of high moisture or sustained load to the panel however, edges may have greater deflection than the center of the panel and may exceed the calculated deflection unless panel edges are supported. For this reason, and to minimize differential deflection between adjacent panels, some form designers specify blocking at the unsupported edge, particularly when face grain is parallel to supports.

EXAMPLE 1:
SELECTION OF PLYFORM CLASS I FOR WALL FORMS

Internally vibrated concrete will be placed in wall forms at the rate of 3 feet per hour; concrete temperature is 70°. What is the maximum support spacing for 5/8″ Plyform Class I for architectural concrete if the wall is 9 feet high?

The concrete to be used is made with Type I cement, weighs approximately 150 lbs per cubic foot, contains no pozzolans or admixtures, has a 4-inch slump and is internally vibrated to a depth of 4 feet or less.

Find Maximum Concrete Pressure: Table 1 shows 540 psf pressure for 70° and a pour rate of 3 feet per hour. This is less than 150h (150 x 9 ft. = 1350 psf), therefore, use 540 psf maximum pressure.

Select Table Giving Recommended Pressures: Assume the plywood will be placed with its face grain across supports. Therefore, see Table 3.

Determine Maximum Support Spacing: Look down the column for 5/8″ Plyform. It shows 575 psf for supports at 12 inches on center. In this case, 12 inches is the maximum support spacing recommended.

Figure 4.13.1 *(cont.)*

4.13.2 Section Properties of Class I and Class II Plyform

TABLE 9
Section Properties for Plyform Class I and Class II, and Structural I Plyform[a]

Thickness (inches)	Approx. Weight (psf)	Properties for Stress Applied Parallel with Face Grain			Perpendicular to Face Grain Properties for Stress Applied		
		Moment of Inertia I (in.⁴/ft.)	Effective Section Modulus KS (in.³/ft.)	Rolling Shear Constant Ib/Q (in.²/ft.)	Moment of Inertia I (in.⁴/ft.)	Effective Section Modulus KS (in.³/ft.)	Rolling Shear Constant Ib/Q (in.²/ft.)
CLASS I							
15/32	1.4	0.066	0.244	4.743	0.018	0.107	2.419
1/2	1.5	0.077	0.268	5.153	0.024	0.130	2.739
19/32	1.7	0.115	0.335	5.438	0.029	0.146	2.834
5/8	1.8	0.130	0.358	5.717	0.038	0.175	3.094
23/32	2.1	0.180	0.430	7.009	0.072	0.247	3.798
3/4	2.2	0.199	0.455	7.187	0.092	0.306	4.063
7/8	2.6	0.296	0.584	8.555	0.151	0.422	6.028
1	3.0	0.427	0.737	9.374	0.270	0.634	7.014
1-1/8	3.3	0.554	0.849	10.430	0.398	0.799	8.419
CLASS II							
15/32	1.4	0.063	0.243	4.499	0.015	0.138	2.434
1/2	1.5	0.075	0.267	4.891	0.020	0.167	2.727
19/32	1.7	0.115	0.334	5.326	0.025	0.188	2.812
5/8	1.8	0.130	0.357	5.593	0.032	0.225	3.074
23/32	2.1	0.180	0.430	6.504	0.060	0.317	3.781
3/4	2.2	0.198	0.454	6.631	0.075	0.392	4.049
7/8	2.6	0.300	0.591	7.990	0.123	0.542	5.997
1	3.0	0.421	0.754	8.614	0.220	0.812	6.987
1-1/8	3.3	0.566	0.869	9.571	0.323	1.023	8.388
STRUCTURAL I							
15/32	1.4	0.067	0.246	4.503	0.021	0.147	2.405
1/2	1.5	0.078	0.271	4.908	0.029	0.178	2.725
19/32	1.7	0.116	0.338	5.018	0.034	0.199	2.811
5/8	1.8	0.131	0.361	5.258	0.045	0.238	3.073
23/32	2.1	0.183	0.439	6.109	0.085	0.338	3.780
3/4	2.2	0.202	0.464	6.189	0.108	0.418	4.047
7/8	2.6	0.317	0.626	7.539	0.179	0.579	5.991
1	3.0	0.479	0.827	7.978	0.321	0.870	6.981
1-1/8	3.3	0.623	0.955	8.841	0.474	1.098	8.377

a) The section properties presented here are specifically for Plyform, with its special layup restrictions.

Figure 4.13.2   *(Courtesy: APA, The Engineered Wood Assoc., Takoma, Washington)*

4.13.3 Selection of Framing Supports for Plywood Forms

TABLE 8
Hem-Fir No. 2

Equivalent Uniform Load (lb/ft)	Continuous Over 2 or 3 Supports (1 or 2 Spans)								Continuous Over 4 or More Supports (3 or More Spans)							
	Nominal Size								Nominal Size							
	2x4	2x6	2x8	2x10	2x12	4x4	4x6	4x8	2x4	2x6	2x8	2x10	2x12	4x4	4x6	4x8
200	48	71	90	110	127	64	96	117	52	77	97	118	137	78	112	137
400	34	50	63	77	90	51	76	99	37	54	69	84	97	56	83	109
600	28	41	52	63	73	43	62	82	30	44	56	68	79	46	67	89
800	24	35	45	55	64	37	54	71	26	38	48	59	69	40	58	77
1000	22	32	40	49	57	33	48	64	23	34	43	53	61	36	52	69
1200	20	29	37	45	52	30	44	58	21	31	40	48	56	33	48	63
1400	18	27	34	41	48	28	41	54	20	29	37	45	52	30	44	58
1600	17	25	32	39	45	26	38	50	18	27	34	42	49	28	41	54
1800	16	24	30	37	42	25	36	48	17	26	32	39	46	27	39	51
2000	15	22	28	35	40	23	34	45	17	24	31	37	43	25	37	49
2200	15	21	27	33	38	22	33	43	16	23	29	36	41	24	35	46
2400	14	20	26	32	37	21	31	41	15	22	28	34	40	23	34	44
2600	13	20	25	30	35	21	30	40	15	21	27	33	38	22	32	42
2800	13	19	24	29	34	20	29	38	14	20	26	32	37	21	31	40
3000	12	18	23	28	33	19	28	37	14	20	25	31	35	21	30	39
3200	12	18	22	27	32	18	27	36	13	19	24	30	34	20	29	38
3400	12	17	22	27	31	18	26	35	13	19	23	29	33	19	28	36
3600	11	17	21	26	30	17	25	34	12	18	23	28	32	19	28	35
3800	11	16	21	25	29	17	25	33	12	18	22	27	31	18	27	35
4000	11	16	20	24	28	17	24	32	12	17	22	26	31	18	26	34
4500	10	15	19	23	27	16	23	30	11	16	20	25	29	17	25	32
5000	10	14	18	22	25	15	22	29	10	15	19	24	27	16	23	31

Example 2
Selection of Framing For Wall Forms

Design the lumber studs and double wales for the Plyform selected in Example 1. Maximum concrete pressure is 540 psf.

Design Studs: Since the plywood must be supported at 12″ on center, space studs 12″ on center. The load carried by each stud equals the concrete pressure multiplied by the stud spacing in feet:*

$$540 \text{ psf} \times \frac{12}{12} \text{ ft} = 540 \text{ lb per ft}$$

*This method is applicable to most framing systems. It assumes the maximum concrete pressure is constant over the entire form. Actual distribution is more nearly "trapezoidal" or "triangular". Design methods for these distributions are covered in the American Concrete Institute's FORMWORK FOR CONCRETE.

Assuming No. 2 Douglas-Fir or Southern Pine 2 x 6 studs continuous over 3 supports (2 spans), Table 7 shows a 51″ span for 400 lb per ft and a 41″ span for 600 lb per ft. Interpolate between these spans for a 540 lb per ft load:

$$\frac{540 - 400}{600 - 400} \times (51 - 41) = \frac{140}{200} \times 10 = 7″$$

For 540 lb per ft, span = 51″ − 7″ = 44″

The 2 x 6 studs must be supported at 44″ on center. Assume this support is provided by double 2 x 6 wales spaced 44″ on center.

Design Double Wales: Load carried by the double wales equals the maximum concrete pressure multiplied by the wale spacing in feet, or

$$540 \text{ psf} \times \frac{44}{12} \text{ ft} = 1980 \text{ lb per ft}$$

Since the wales are doubled, each wale carries 990 lb per ft (1980 ÷ 2 = 990). Assuming 2 x 6 wales continuous over 4 or more supports, Table 7 shows a 35″ span for 1000 lb per ft and 39″ span for 800 lb per ft.** Interpolation shows that 2 x 6's can span 35″ for 990 lb per ft. Support 2 x 6's at 35″ on center with form ties. (Place bottom wale about 10″ from bottom of form).

**Tables 7 and 8 are for uniform loads but the wales actually receive point loads from the studs. This method of approximating the capacity of the wales is probably adequate when there are three or more studs between the ties. A point load analysis should be performed when there are only one or two studs between the ties.

Figure 4.13.3 *(Courtesy: APA, The Engineered Wood Assoc., Takoma, Washington)*

4.13.4 Maximum Spans for Framing Lumber

SELECTING SIZE OF JOISTS, STUDS, AND WALES

The loads carried by slab joists, and by wall studs and wales are proportional to their spacings as well as to the maximum concrete pressure. Tables 7 and 8 give design information for lumber framing directly supporting the plywood. Note that tables show spans for two conditions: members over 2 or 3 supports (1 or 2 spans) and over 4 or more supports (3 or more spans). Some forming systems use doubled framing members. Even though Tables 7 and 8 are for single members, these tables can be adapted for use with multiple members.

MAXIMUM SPANS FOR LUMBER FRAMING, INCHES

Spans are based on the 1991 NDS allowable stress values.
Spans are based on dry, single-member allowable stresses multiplied by a 1.25 duration-of-load factor for 7-day loads.
Deflection is limited to 1/360th of the span with 1/4″ maximum. Spans are measured center-to-center on the supports.
Lumber with no end splits or checks is assumed.
Width of supporting members (e.g. wales) assumed to be 3-1/2″ net (double 2x lumber plus 1/2″ for tie).

TABLE 7
Douglas-Fir Larch No. 2 or Southern Pine No. 2

Equivalent Uniform Load (lb/ft)	Continuous Over 2 or 3 Supports (1 or 2 Spans) Nominal Size								Continuous Over 4 or More Supports (3 or More Spans) Nominal Size							
	2x4	2x6	2x8	2x10	2x12	4x4	4x6	4x8	2x4	2x6	2x8	2x10	2x12	4x4	4x6	4x8
200	49	72	91	111	129	68	101	123	53	78	98	120	139	81	118	144
400	35	51	64	79	91	53	78	102	38	55	70	85	98	57	84	111
600	28	41	53	64	74	43	63	84	31	45	57	69	80	47	68	90
800	25	36	45	56	64	38	55	72	27	39	49	60	70	41	59	78
1000	22	32	41	50	58	34	49	65	24	35	44	54	62	36	53	70
1200	20	29	37	45	53	31	45	59	22	32	40	49	57	33	48	64
1400	19	27	34	42	49	28	41	55	20	29	37	45	53	31	45	59
1600	17	25	32	39	46	27	39	51	19	27	35	42	49	29	42	55
1800	16	24	30	37	43	25	37	48	18	26	33	40	46	27	40	52
2000	16	23	29	35	41	24	35	46	17	25	31	38	44	26	37	49
2200	15	22	27	34	39	23	33	44	16	23	30	36	42	24	36	47
2400	14	21	26	32	37	22	32	42	15	22	28	35	40	23	34	45
2600	14	20	25	31	36	21	30	40	15	22	27	33	39	22	33	43
2800	13	19	24	30	34	20	29	39	14	21	26	32	37	22	32	42
3000	13	19	23	29	33	19	28	37	14	20	25	31	36	21	31	40
3200	12	18	23	28	32	19	27	36	13	19	25	30	35	20	30	39
3400	12	17	22	27	31	18	27	35	13	19	24	29	34	20	29	38
3600	12	17	21	26	30	18	26	34	13	18	23	28	33	19	28	37
3800	11	16	21	25	30	17	25	33	12	18	23	28	32	19	27	36
4000	11	16	20	25	39	17	25	32	12	17	22	27	31	18	27	35
4500	10	15	19	23	27	16	23	30	11	16	21	25	29	17	25	33
5000	10	14	18	22	26	15	22	29	11	16	20	24	28	16	24	31

Figure 4.13.4   *(Courtesy: APA, The Engineered Wood Assoc., Takoma, Washington)*

4.14.0 Typical Sleeve Through Concrete Footing Detail

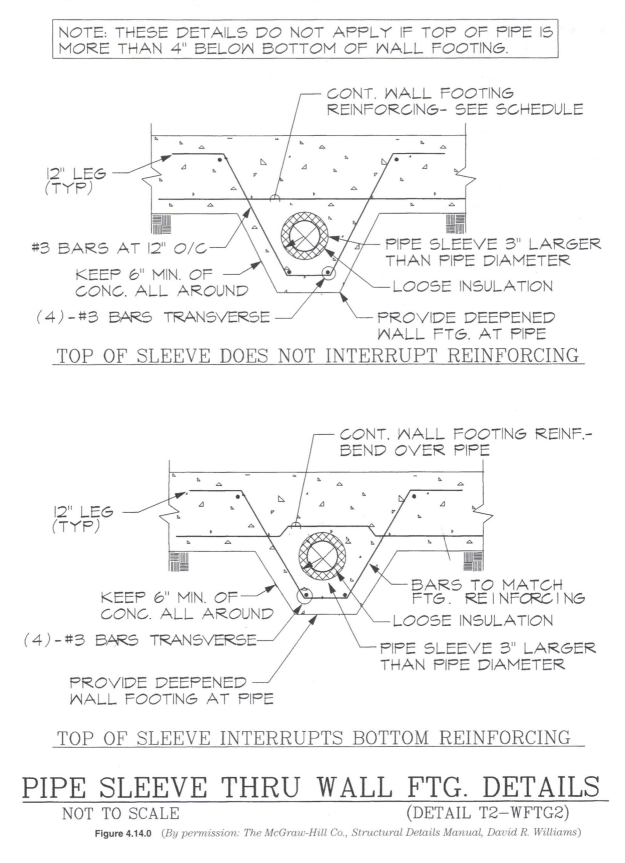

NOTE: THESE DETAILS DO NOT APPLY IF TOP OF PIPE IS
MORE THAN 4" BELOW BOTTOM OF WALL FOOTING.

CONT. WALL FOOTING
REINFORCING- SEE SCHEDULE

12" LEG
(TYP)

#3 BARS AT 12" O/C

KEEP 6" MIN. OF
CONC. ALL AROUND

(4)-#3 BARS TRANSVERSE

PIPE SLEEVE 3" LARGER
THAN PIPE DIAMETER

LOOSE INSULATION

PROVIDE DEEPENED
WALL FTG. AT PIPE

TOP OF SLEEVE DOES NOT INTERRUPT REINFORCING

CONT. WALL FOOTING REINF.-
BEND OVER PIPE

12" LEG
(TYP)

KEEP 6" MIN. OF
CONC. ALL AROUND

(4)-#3 BARS TRANSVERSE

PROVIDE DEEPENED
WALL FOOTING AT PIPE

BARS TO MATCH
FTG. REINFORCING

LOOSE INSULATION

PIPE SLEEVE 3" LARGER
THAN PIPE DIAMETER

TOP OF SLEEVE INTERRUPTS BOTTOM REINFORCING

PIPE SLEEVE THRU WALL FTG. DETAILS
NOT TO SCALE   (DETAIL T2-WFTG2)

Figure 4.14.0 *(By permission: The McGraw-Hill Co., Structural Details Manual, David R. Williams)*

4.14.1 Typical Sleeve Through Concrete Grade Beam Detail

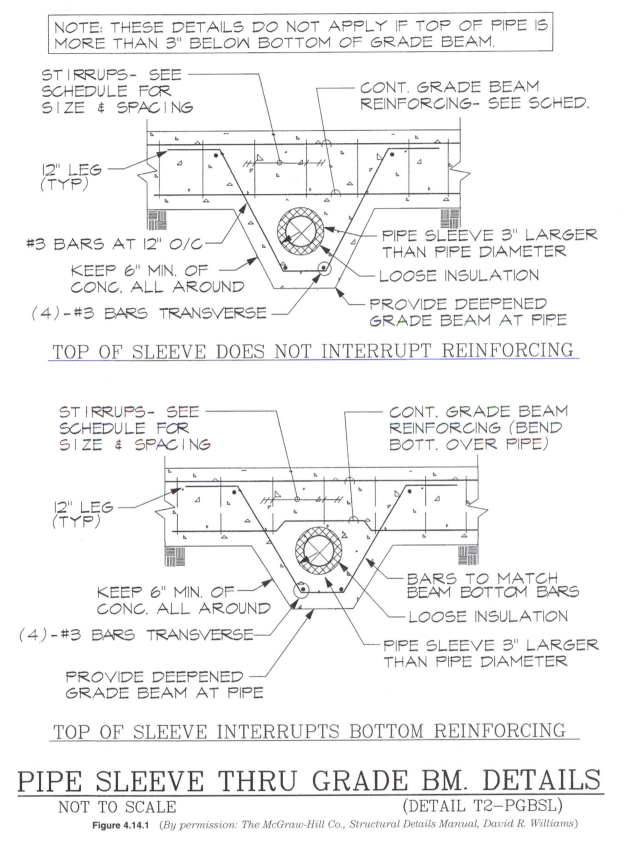

NOTE: THESE DETAILS DO NOT APPLY IF TOP OF PIPE IS MORE THAN 3" BELOW BOTTOM OF GRADE BEAM.

STIRRUPS- SEE SCHEDULE FOR SIZE & SPACING

CONT. GRADE BEAM REINFORCING- SEE SCHED.

12" LEG (TYP)

#3 BARS AT 12" O/C

KEEP 6" MIN. OF CONC. ALL AROUND

(4)-#3 BARS TRANSVERSE

PIPE SLEEVE 3" LARGER THAN PIPE DIAMETER

LOOSE INSULATION

PROVIDE DEEPENED GRADE BEAM AT PIPE

TOP OF SLEEVE DOES NOT INTERRUPT REINFORCING

STIRRUPS- SEE SCHEDULE FOR SIZE & SPACING

CONT. GRADE BEAM REINFORCING (BEND BOTT. OVER PIPE)

12" LEG (TYP)

KEEP 6" MIN. OF CONC. ALL AROUND

(4)-#3 BARS TRANSVERSE

BARS TO MATCH BEAM BOTTOM BARS

LOOSE INSULATION

PIPE SLEEVE 3" LARGER THAN PIPE DIAMETER

PROVIDE DEEPENED GRADE BEAM AT PIPE

TOP OF SLEEVE INTERRUPTS BOTTOM REINFORCING

PIPE SLEEVE THRU GRADE BM. DETAILS

NOT TO SCALE   (DETAIL T2-PGBSL)

Figure 4.14.1 *(By permission: The McGraw-Hill Co., Structural Details Manual, David R. Williams)*

4.14.2 Typical Wood Post/Concrete Footing Detail

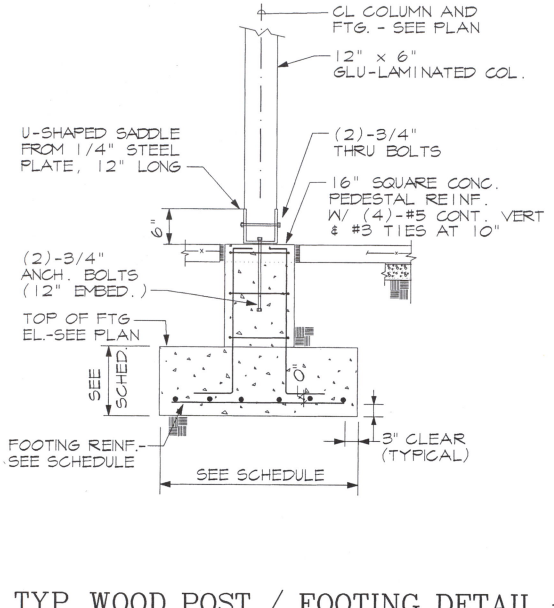

CL COLUMN AND
FTG. - SEE PLAN

12" x 6"
GLU-LAMINATED COL.

U-SHAPED SADDLE
FROM 1/4" STEEL
PLATE, 12" LONG

(2)-3/4"
THRU BOLTS

16" SQUARE CONC.
PEDESTAL REINF.
W/ (4)-#5 CONT. VERT
& #3 TIES AT 10"

6"

(2)-3/4"
ANCH. BOLTS
(12" EMBED.)

TOP OF FTG
EL.-SEE PLAN

SEE SCHED.

FOOTING REINF.-
SEE SCHEDULE

3" CLEAR
(TYPICAL)

SEE SCHEDULE

TYP. WOOD POST / FOOTING DETAIL #2
NOT TO SCALE   (DETAIL T2-CFTG5)

Figure 4.14.2 *(By permission: The McGraw-Hill Co., Structural Details Manual, David R. Williams)*

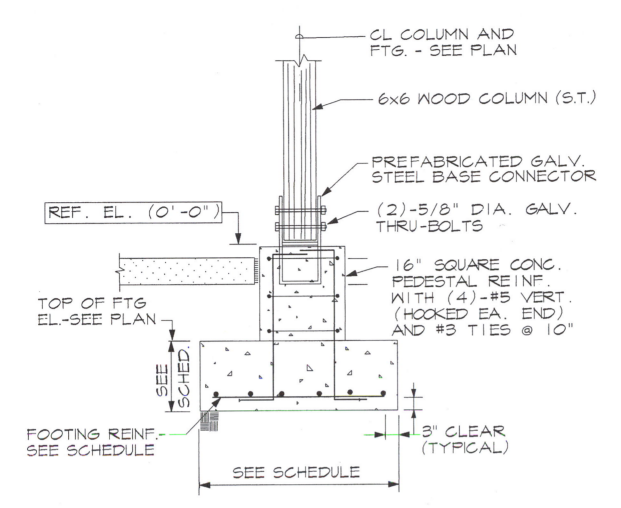

CL COLUMN AND
FTG. - SEE PLAN

6x6 WOOD COLUMN (S.T.)

PREFABRICATED GALV.
STEEL BASE CONNECTOR

REF. EL. (0'-0")

(2)-5/8" DIA. GALV.
THRU-BOLTS

16" SQUARE CONC.
PEDESTAL REINF.
WITH (4)-#5 VERT.
(HOOKED EA. END)
AND #3 TIES @ 10"

TOP OF FTG
EL.-SEE PLAN

SEE SCHED.

FOOTING REINF.-
SEE SCHEDULE

3" CLEAR
(TYPICAL)

SEE SCHEDULE

TYP. WOOD POST / FOOTING DETAIL #1
NOT TO SCALE   (DETAIL T2–CFTG4)

Figure 4.14.2 *(cont.)*

4.14.3 Typical Detail—Continuous Concrete Reinforcing at Corners

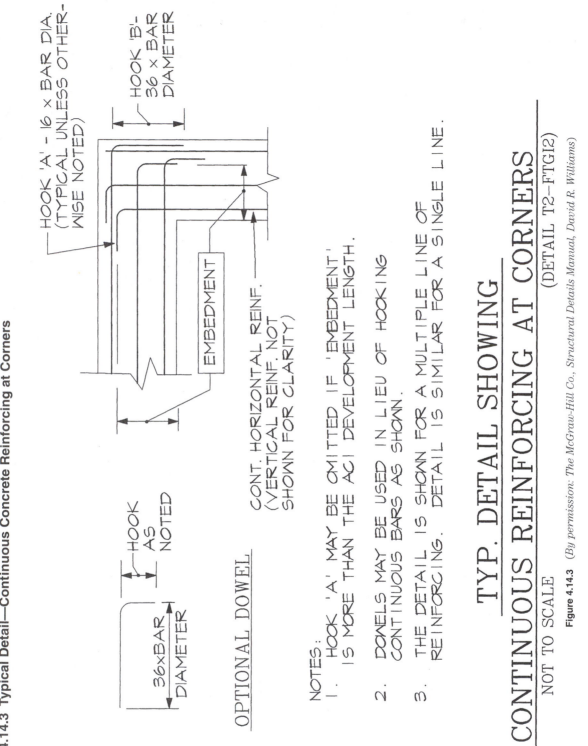

HOOK 'A' - 16 × BAR DIA. (TYPICAL UNLESS OTHER-WISE NOTED)

HOOK 'B'- 36 × BAR DIAMETER

HOOK AS NOTED

36×BAR DIAMETER

OPTIONAL DOWEL

EMBEDMENT

CONT. HORIZONTAL REINF. (VERTICAL REINF. NOT SHOWN FOR CLARITY)

NOTES:

1. HOOK 'A' MAY BE OMITTED IF 'EMBEDMENT' IS MORE THAN THE ACI DEVELOPMENT LENGTH.

2. DOWELS MAY BE USED IN LIEU OF HOOKING CONTINUOUS BARS AS SHOWN.

3. THE DETAIL IS SHOWN FOR A MULTIPLE LINE OF REINFORCING. DETAIL IS SIMILAR FOR A SINGLE LINE.

TYP. DETAIL SHOWING

CONTINUOUS REINFORCING AT CORNERS

(DETAIL T2-FTGI2)

NOT TO SCALE

Figure 4.14.3 (*By permission: The McGraw-Hill Co., Structural Details Manual, David R. Williams*)

4.14.4 Typical Perimeter Foundation Deatil

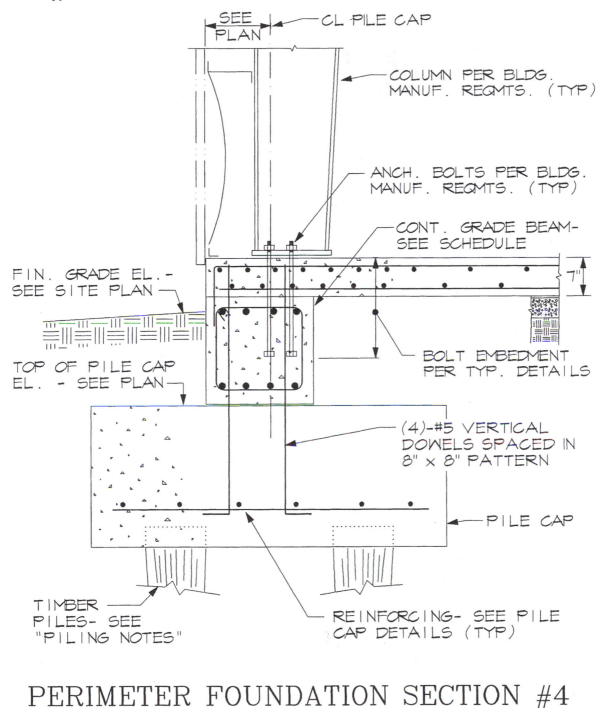

SEE PLAN

CL PILE CAP

COLUMN PER BLDG. MANUF. REQMTS. (TYP)

ANCH. BOLTS PER BLDG. MANUF. REQMTS. (TYP)

CONT. GRADE BEAM- SEE SCHEDULE

FIN. GRADE EL.- SEE SITE PLAN

7"

TOP OF PILE CAP EL. - SEE PLAN

BOLT EMBEDMENT PER TYP. DETAILS

(4)-#5 VERTICAL DOWELS SPACED IN 8" × 8" PATTERN

PILE CAP

TIMBER PILES- SEE "PILING NOTES"

REINFORCING- SEE PILE CAP DETAILS (TYP)

PERIMETER FOUNDATION SECTION #4
NOT TO SCALE   (DETAIL T2–PERM4)

Figure 4.14.4 *(By permission: The McGraw-Hill Co., Structural Details Manual, David R. Williams)*

4.14.5 Typical Anchor Bolt Detail

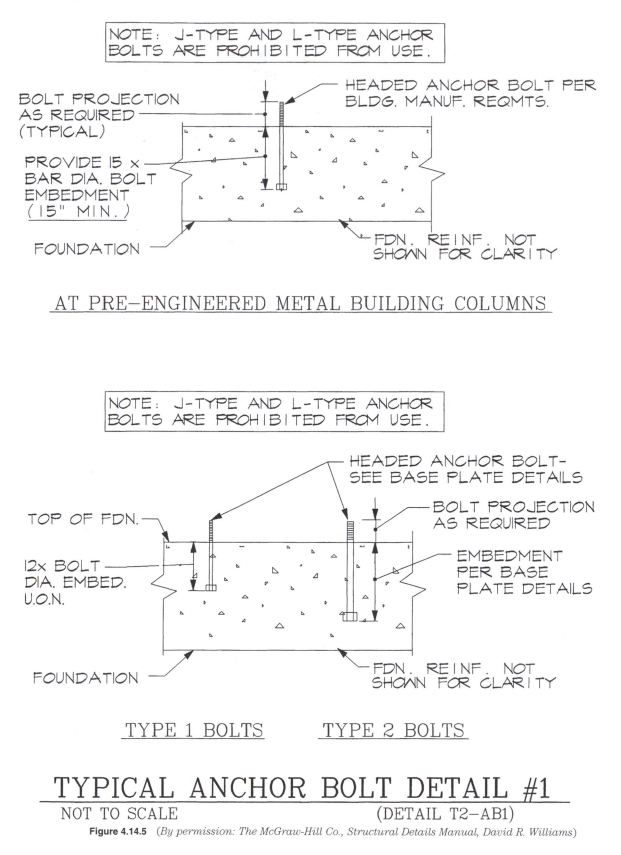

AT PRE-ENGINEERED METAL BUILDING COLUMNS

TYPE 1 BOLTS   TYPE 2 BOLTS

TYPICAL ANCHOR BOLT DETAIL #1

NOT TO SCALE   (DETAIL T2-AB1)

Figure 4.14.5 *(By permission: The McGraw-Hill Co., Structural Details Manual, David R. Williams)*

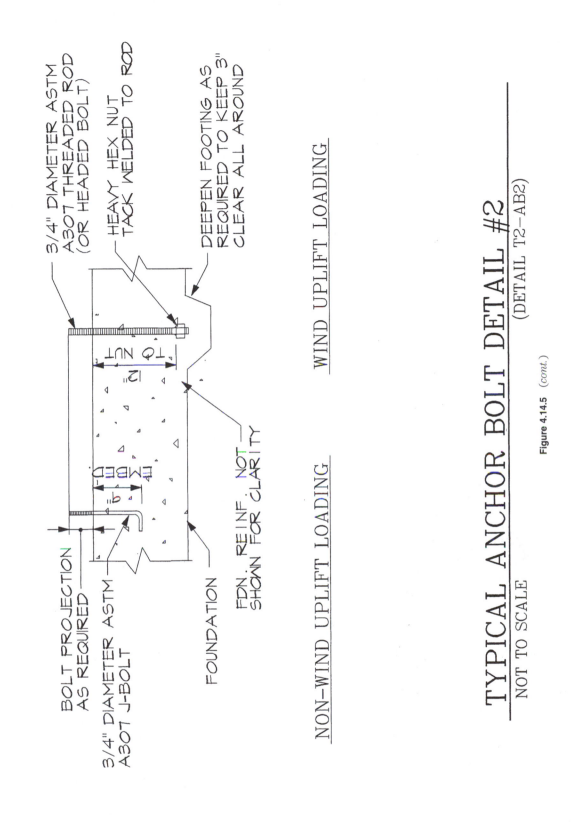

3/4" DIAMETER ASTM A307 THREADED ROD (OR HEADED BOLT)

HEAVY HEX NUT TACK WELDED TO ROD

DEEPEN FOOTING AS REQUIRED TO KEEP 3" CLEAR ALL AROUND

BOLT PROJECTION AS REQUIRED

3/4" DIAMETER ASTM A307 J-BOLT

FOUNDATION

FDN. REINF. NOT SHOWN FOR CLARITY

12" MIN. TO NUT

9" MIN. EMBED

WIND UPLIFT LOADING

NON-WIND UPLIFT LOADING

TYPICAL ANCHOR BOLT DETAIL #2
(DETAIL T2-AB2)
NOT TO SCALE

Figure 4.14.5 *(cont.)*

4.14.6 Strengthening Existing Footing Detail

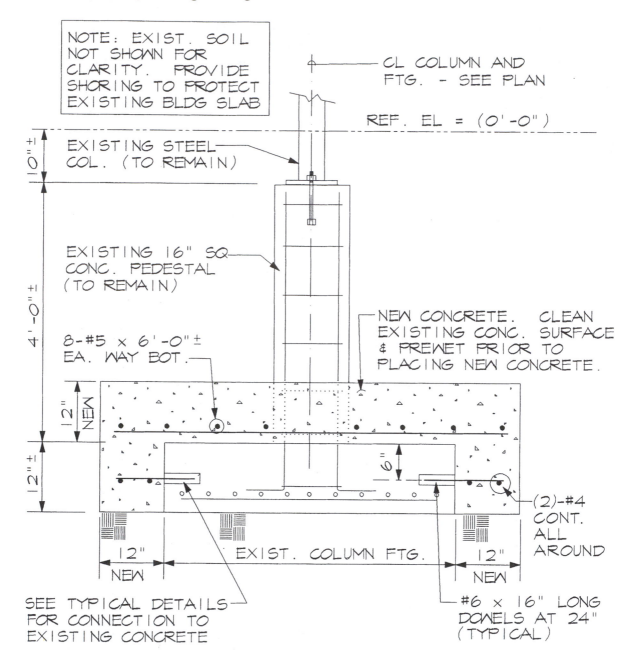

NOTE: EXIST. SOIL NOT SHOWN FOR CLARITY. PROVIDE SHORING TO PROTECT EXISTING BLDG SLAB

CL COLUMN AND FTG. - SEE PLAN

REF. EL = (0'-0")

EXISTING STEEL COL. (TO REMAIN)

EXISTING 16" SQ. CONC. PEDESTAL (TO REMAIN)

8-#5 × 6'-0"± EA. WAY BOT.

NEW CONCRETE. CLEAN EXISTING CONC. SURFACE & PREWET PRIOR TO PLACING NEW CONCRETE.

(2)-#4 CONT. ALL AROUND

SEE TYPICAL DETAILS FOR CONNECTION TO EXISTING CONCRETE

#6 × 16" LONG DOWELS AT 24" (TYPICAL)

EXIST. COLUMN FTG.

12" NEW

12" NEW

EXIST. FOOTING STRENGTHENING DETAIL
NOT TO SCALE   (DETAIL T2-MREN1)

Figure 4.14.6 (*By permission: The McGraw-Hill Co., Structural Details Manual, David R. Williams*)

4.14.7 Typical Wall Construction Joint Detail

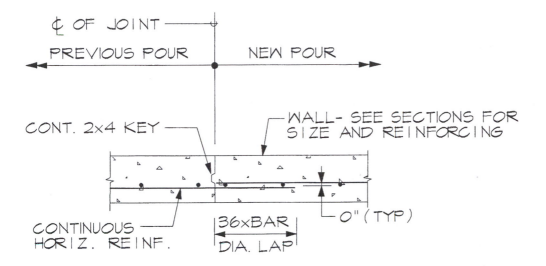

¢ OF JOINT

PREVIOUS POUR   NEW POUR

CONT. 2x4 KEY

WALL- SEE SECTIONS FOR SIZE AND REINFORCING

CONTINUOUS HORIZ. REINF.

36xBAR DIA. LAP

0" (TYP)

AT SINGLE LAYER OF REINFORCING

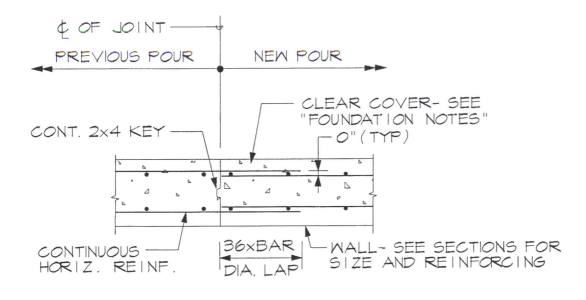

¢ OF JOINT

PREVIOUS POUR   NEW POUR

CONT. 2x4 KEY

CLEAR COVER- SEE "FOUNDATION NOTES"

0" (TYP)

CONTINUOUS HORIZ. REINF.

36xBAR DIA. LAP

WALL- SEE SECTIONS FOR SIZE AND REINFORCING

AT DOUBLE LAYER OF REINFORCING

WALL CONSTRUCTION JOINT DETAILS

NOT TO SCALE   (DETAIL T2–RWJT2)

Figure 4.14.7 *(By permission: The McGraw-Hill Co., Structural Details Manual, David R. Williams)*

4.14.8 Typical Elevator Pit Detail

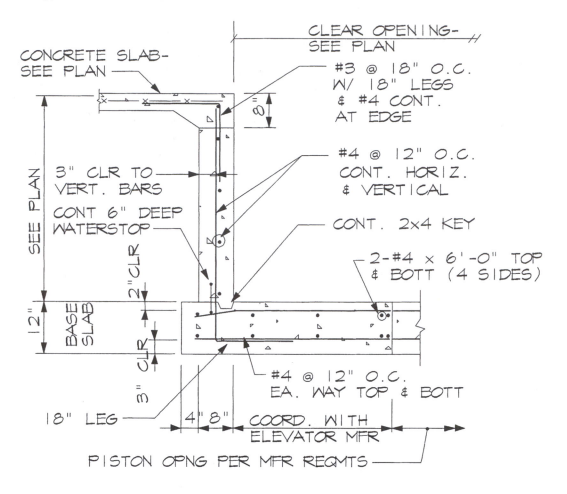

CLEAR OPENING-
SEE PLAN

CONCRETE SLAB-
SEE PLAN

#3 @ 18" O.C.
W/ 18" LEGS
& #4 CONT.
AT EDGE

8"

3" CLR TO
VERT. BARS

#4 @ 12" O.C.
CONT. HORIZ.
& VERTICAL

CONT 6" DEEP
WATERSTOP

CONT. 2x4 KEY

2-#4 x 6'-0" TOP
& BOTT (4 SIDES)

SEE PLAN

2"CLR

12"

BASE SLAB

3" CLR

18" LEG

4" 8"

#4 @ 12" O.C.
EA. WAY TOP & BOTT

COORD. WITH
ELEVATOR MFR

PISTON OPNG PER MFR REQMTS

SHALLOW FOUNDATION SYSTEM

TYPICAL ELEVATOR PIT DETAIL #1

NOT TO SCALE   (DETAIL T2-PIT3)

Figure 4.14.8 *(By permission: The McGraw-Hill Co., Structural Details Manual, David R. Williams*

4.15.0 Concrete—Quality Control Checklist

Quality
Control
Checklist

	Project no.
Section	No.
Concrete	**03301**
Accepted By	Date

1. Shop drawings are approved and on site.
2. Verify approval of forms and rebar prior to pour.
3. Requirements for tests, mix design, ingredients, etc.
4. Test lab notified prior to pour and tests are performed (cylinder, sump, etc.)
5. Testing is arranged at plant and site.
6. Previously placed conc. properly prepared for new.
7. Vibrators are used during pour.
8. Temporary form openings, tremies, chutes, provided.
9. Sub-base and capillary fills compacted.
10. Arrange for specified curing and sawed joints.
11. Arrange for cold weather protection if needed.
12. Bolts and loose items properly located and installed.   Verify contractor has coordinated.
13. Date and location of pours is recorded.
14. Verify finishes: smooth, exposed aggregate, colored.
15. No troweling while bleed water is on surface.
16. Curing compounds compatible with finishes have been verified.
17. Slopes are provided for proper drainage.
18. Wet spray or moist curing is adequately performed.
19. Loading and traffic are controlled over surfaces.
20. Methods of repairing provided as soon as possible.

USE REVERSE SIDE FOR ADDITIONAL REMARKS AND COMMENTS

Accepted By _____

4.15.1 Concrete Reinforcement—Quality Control Checklist

Quality
Control
Checklist

	Project no.
Section	No.
Concrete Reinforcement	03200
	Date

1. Shop drawings are approved and on site.

2. Grade of steel delivered as required.

3. Spacing coordinated to suit masonry units.

4. Required clearance of steel from forms provided.

5. Length of splices, and staggering splices as required.

6. Bends within radii and tolerance are uniformly made.

7. Additional bars at intersects, openings, and corners.

8. Bars cleaned of material that might reduce bond.

9. Dowels for marginal bars at openings.

10. Bars tied and supported to prevent displacement.

11. Spacers, tie wires, chairs as required.

12. Conduits are separated 3 conduit diameters minimum.

13. No conduit or pipe place below rebar mat except where approved.

14. No contact of bars is made with dissimilar metals.

15. Bars not near surfaces that allow rusting.

16. Adequate clearance provided for deposit of conc.

17. Verify that contractor has coordinated and reviewed drawings for anchors, piping, sleeves, boxes.

18. Verify that contractor has resolved conflicts between embedded items and reinf.

19. Special coating as required.

20. Cantilever – proper placement.

21. No bent bars and tension members are installed except where approved.

22. No conduit or piping is placed below rebar mat in suspended slabs unless approved by three conduit diameters minimum.

23. Unless approved. boxing-out is not allowed for subsequent grouting-in.

24. Rules of thumb for bar splices:
    For 24d lap: multiply bar size by 3 = lap in inches.
    For 32d lap: multiply bar size by 4 = lap in inches.
    For 40d lap: multiply bar size by 5 = lap in inches.

25. Avoid bending rebar excessively ("hickeying").   Max slope = 1:6.   Field bending of partially embedded bar should be done with consultants approval.

26. Agency inspection is performed if required.

USE REVERSE SIDE FOR ADDITIONAL REMARKS AND COMMENTS

Accepted By

4.15.2 Concrete Form Removal—Quality Control Checklist

Quality
Control
Checklist

	Project no.
Section	No.
Concrete Form Removal	03801
	Date

1. Method of patching approved and applied early.
2. No troweling while bleed water is on surface.
3. Overtroweling is to be avoided.
4. Finishing method provides surface within tolerance.
5. Slopes provide proper drainage.
6. Check edges are finished and match.
7. Moist curing is adequately performed.
8. Protective covers have sufficient lap and seal.
9. Methods to repair defects are applied early.

USE REVERSE SIDE FOR ADDITIONAL REMARKS AND COMMENTS

Accepted By _____

4.15.3 Quality Control Checklist—Concrete and Asphalt Paving

Quality
Control
Check List

Section	No.
	Project no.
Pavement and Walks	**02500**
	Date

SUBGRADE AND BASES

1. Observe that contractor has installed subgrade to proper elevation and crosssection.

2. Subgrade is dense and properly compacted.

3. All drains, utilities, and other underground construction are in place and contractor has coordinated the work.

4. Trench backfilling is performed as required.

5. Control testing of subgrade and subgrade materials is being performed and recorded if required.

6. Sub-base and base courses are of source, type, thickness and material specified.

7. Source of material is sampled and approved by testing laboratory.

8. Materials delivered are of uniform quality.

9. Hauling equipment does not produce ruts in subgrade.

10. Location of all manholes, outlets, and surface features is known and coordinated with other site features.

ASPHALT PAVING

11. Certificates or other requirements are submitted, approved and on site.

12. Approved samples are on site as required.

13. Existing and adjacent work tie-ins are suitable.

14. Arrangement has been made for Testing labs, Agencies, Mfg. Reps.

15. Record location and time of streches of batch spread.

16. Collect, review, and file all copies of tests.

17. Prime applied to waterfree surface.

18. Prime coat seals surface voids.

19. Tack coat provides proper binding.

20. Observe prime coat curing time.

21. Adjacent concrete is to be at least 7 days old.

22. Certificates which indicate asphalt is of proper mix.

23. Plant inspection has been made if required.

24. Size and type of roller is as specified.

25. Headers and screeds installed for thickness control.

26. Weather limitations are observed.

27. Temperature of delivered mix within limits.

28. Materials are properly spread and raked.

29. Raking to be kept to a minimum.

30. Rollers are operated within speed range.

31. Temperature after rolling within limits.

32. Damage to adjacent surfaces is corrected.

33. Drainage tests are made.

34. Check seal coat curing period.

35. Check special topping surfaces, coatings, striping.

continued on next page

Pavement and Walks continued

CONCRETE PAVING

36. Contractor submits proof that requirements for mixes, batching plants, and mixing plant are met and approved if required. Admixtures are as approved.

37. Base course is maintained in a firm, moist condition and is as required.

38. Observe that all forms, headers, outlets, boxes, and equipment are in place before pouring.

39. Observe that all embedded items, sleeves, dowels, and reinforcement are as required.

40. Joint methods and materials are provided and observed.

41. Grade, slope, pitch, and thickness control is provided as required.

42. Concrete is deposited, rodded, and vibrated to suit conditions. Observe that reinforcement is maintained at elevation required.

43. Time interval between pours allows for continuous working.

44. Controls joints, construction joints, and expansion joints are provided as required.

45. Color, if required, is of proper type, tone, and amount.

46. Finishing treatment and texture is as required.

47. Curing provisions are as required and work is properly protected.

48. Overtroweling is avoided.

49. Jointing of old concrete to new work is properly performed.

50. Forms are not removed until minimum required time after placement has elapsed.

51. Sawed joints are made at proper time and are properly aligned.

52. Sawed joints are of proper width and depth.

53. Joints are cleaned and cured as required.

54. Joints are sealed properly as required.

55. Concrete is protected from damage during backfilling.

56. Tests for drainage and surface variation are made.

USE REVERSE SIDE FOR ADDITIONAL REMARKS AND COMMENTS

Accepted By _____

Figure 4.15.3 *(cont.)*

4.16.0 History of Masonry

The first recorded brick masonry units were made by the Egyptians in 10,000 B.C. and the Romans used brick in many of their structures 2000 years go. The Great Pyramid of Giza in Egypt is the first recorded use of mortar. Brick manufacture and use occurred in the mid-1600s and was patterned on English methods and practices. It was not until 1930, however, that cavity wall construction (as we know it today) was introduced into the United States from Europe as a means of controlling moisture. This method provides a physical separation between the inner and outer wythes to serve as a drainage cavity for water, which would be expelled through weep holes in the outer wythe.

Masonry today is primarily devoted to the construction of brick, block, structural clay products, and natural and cast stone. Walls can be basically categorized as load-bearing or non-load-bearing walls, cavity walls, veneer walls, and solid walls. No matter the type of material used or the method by which the masonry wall is constructed, two components remain crucial: mortar and wall reinforcement.

4.17.0 Mortar

Mortar is the bonding agent that holds all of the masonry units together. Bond strength is the crucial element that differs from its close relative concrete, where compressive strength is the most important physical property.

Mortar serves four functions:

1. It bonds the masonry units together and seals the space between them.

2. It allows for dimensional variations in the masonry units while still maintaining a high degree of levelness.

3. It bonds to the reinforcing steel in the wall.

4. It provides an added decorative effect to the wall inasmuch as various colors or tooled joints can be introduced.

4.17.1 Mortar Types

- *Type M*   High compressive strength (2500 psi average), containing greater durability than other types. Therefore, it is generally recommended for unreinforced masonry walls below grade.

- *Type S*   Reasonable high compressive strength (1800 psi average) and having great tensile bond strength. It is usually recommended for reinforced masonry walls, where maximum flexural strength is required.

- *Type N*   Midrange compressive strength (750 psi average) and suitable for general above-grade masonry construction for parapets and chimneys.

- *Type O*   Load compressive strength (350 psi average) and suitable for interior non-load-bearing masonry walls.

- *Type K*   Very low compressive strength (75 psi average) and occasionally used for interior non-load-bearing walls, where permitted by local building codes.

Workability or plasticity of the mortar is an essential characteristic of proper mortar mixes. The mortar must have both cohesive and adhesive qualities when it makes contact with the masonry units. Hardness or high strength is not necessarily a measure of durability. Mortar that is stronger than the masonry units to which it is applied might not "give," thereby causing stress to be relieved by the masonry units. This could result in these units cracking or spalling.

4.17.2 Mortar Additives

Like concrete, mortar admixtures can be added for many reasons:

- *Accelerators*   To speed up the setting time by 30 to 40% and increase the 24-hour strength. Some accelerators contain calcium chloride and are not acceptable to the architect/engineer.

- *Retarders* Extends the board life of the mortar by as much as 4 to 5 hours. it slows down the set time of mortar when temperatures exceed 70°F.

- *Integral water repellents* It reduces water absorption and is useful when a single wythe wall will be exposed to the elements.

- *Bond modifiers* Improves adhesion to block. It is particularly useful when glass block walls are being built.

- *Corrosion inhibitors* Used in marine environments where salt air could penetrate the mortar and begin to corrode any wall reinforcement.

4.17.3 Mortar Testing

Mortar testing is performed by the "prism" test method, in accordance with ASTM E 447, Method B. The compressive strength is the average strength of three prisms.

Net area compressive strength of concrete masonry units, psi (MPa)		Net area compressive strength of masonry, psi[1] (MPa)
Type M or S mortar	Type N mortar	
1250 (8.6)	1300 (9.0)	1000 (6.9)
1900 (13.1)	2150 (14.8)	1500 (10.3)
2800 (19.3)	3050 (21.0)	2000 (13.8)
3750 (25.8)	4050 (27.9)	2500 (17.2)
4800 (33.1)	5250 (36.2)	3000 (20.1)

[1]For units of less than 4 in. (102 mm) height, 85 percent of the values listed.

4.17.4 Compressive Strength of Mortars Made with Various Types of Cement

Type of cement	Minimum compressive strength, psi				ASTM designation
	1 day	3 days	7 days	28 days	
Portland cements					C150-85
I	—	1800	2800	4000*	
IA	—	1450	2250	3200*	
II	—	1500	2500	4000*	
	—	1000†	1700†	3200*†	
IIA	—	1200	2000	3200*	
	—	800†	1350†	2560*†	
III	1800	3500	—	—	
IIIA	1450	2800	—	—	
IV	—	—	1000	2500	
V	—	1200	2200	3000	
Blended cements					C595-85
I(SM), IS,					
I(PM), IP	—	1800	2800	3500	
I(SM)-A, IS-A					
I(PM)-A, IP-A	—	1450	2250	2800	
IS(MS), IP(MS)	—	1500	2500	3500	
IS-A(MS), IP-A(MS)	—	1200	2000	2800	
S	—	—	600	1500	
SA	—	—	500	1250	
P	—	—	1500	3000	
PA	—	—	1250	2500	
Expansive cement					C845-80
E-1	—	—	2100	3500	
Masonry cements					C91-83a
N	—	—	500	900	
S	—	—	1300	2100	
M	—	—	1800	2900	

*Optional requirement.

†Applicable when the optional heat of hydration or chemical limit on the sum of C2S and C3A is specified.

Note: When low or moderate heat of hydration is specified for blended cements (ASTM C595), the strength requirements is 80% of the value shown.

(By permission from the Masonry Society, ACI, ASCE from their manual Building Code Requirements for Masonry Structures.)

4.17.5 Proportions by Volume of M, S, N, and O Mortars

MORTAR	TYPE	Portland Cement or Blended Cement	Masonry Cement[1] M	S	N	Mortar Cement[2] M	S	N	Hydrated Lime or Lime Putty	AGGREGATE MEASURED IN A DAMP, LOOSE CONDITION
Cement-lime	M	1	—	—	—	—	—	—	$1/4$	
	S	1	—	—	—	—	—	—	over $1/4$ to $1/2$	
	N	1	—	—	—	—	—	—	over $1/2$ to $1\,1/4$	
	O	1	—	—	—	—	—	—	over $1\,1/4$ to $2\,1/2$	
Mortar cement	M	1	—	—	—	—	—	1	—	Not less than $2\,1/4$ and not more than 3 times the sum of the separate volumes of cementitious materials.
	M	—	—	—	—	1	—	—	—	
	S	$1/2$	—	—	—	—	—	1	—	
	S	—	—	—	—	—	1	—	—	
	N	—	—	—	—	—	—	1	—	
Masonry cement	M	1	—	—	1	—	—	—	—	
	M	—	1	—	—	—	—	—	—	
	S	$1/2$	—	—	1	—	—	—	—	
	S	—	—	1	—	—	—	—	—	
	N	—	—	—	1	—	—	—	—	
	O	—	—	—	1	—	—	—	—	

[1]Masonry cement conforming to the requirements of UBC Standard 21-11.
[2]Mortar cement conforming to the requirements of UBC Standard 21-14.

Figure 4.17.5 (*Reproduced from the 1997 Edition of the Uniform Building Code, Volumes 1, 2, 3, copyright 1997, with the permission of the publisher, the International Conference of Building Officials (ICBO). ICBO assumes no responsibility for the accuracy or the completion of summaries provided therein.*)

4.17.6 Compressive Strength of Masonry with M, S, and N Mortars

COMPRESSIVE STRENGTH OF CLAY MASONRY UNITS[1,2] (psi)	SPECIFIED COMPRESSIVE STRENGTH OF MASONRY, f'_m Type M or S Mortar[3] (psi)	Type N Mortar[3] (psi)
	× 6.89 for kPa	
14,000 or more	5,300	4,400
12,000	4,700	3,800
10,000	4,000	3,300
8,000	3,350	2,700
6,000	2,700	2,200
4,000	2,000	1,600

COMPRESSIVE STRENGTH OF CONCRETE MASONRY UNITS[2,4] (psi)	SPECIFIED COMPRESSIVE STRENGTH OF MASONRY, f'_m Type M or S Mortar[3] (psi)	Type N Mortar[3] (psi)
	× 6.89 for kPa	
4,800 or more	3,000	2,800
3,750	2,500	2,350
2,800	2,000	1,850
1,900	1,500	1,350
1,250	1,000	950

[1]Compressive strength of solid clay masonry units is based on gross area. Compressive strength of hollow clay masonry units is based on minimum net area. Values may be interpolated. When hollow clay masonry units are grouted, the grout shall conform to the proportions in Fig. 2.1.10.
[2]Assumed assemblage. The specified compressive strength of masonry f'_m is based on gross area strength when using solid units or solid grouted masonry and net area strength when using ungrouted hollow units.
[3]Mortar for unit masonry, proportion specification, as specified in Fig. 2.1.7. These values apply to portland cement-lime mortars without added air-entraining materials.
[4]Values may be interpolated. In grouted concrete masonry, the compressive strength of grout shall be equal to or greater than the compressive strength of the concrete masonry units.

Figure 4.17.6 (*Reproduced from the 1997 Edition of the Uniform Building Code, Volumes 1, 2, 3, copyright 1997, with the permission of the publisher, the International Conference of Building Officials (ICBO). ICBO assumes no responsibility for the accuracy or the completion of summaries provided therein.*)

4.18.0 Allowable Compressive Stresses for Masonry

Construction; compressive strength of unit, gross area, psi (MPa)	Allowable compressive stresses[1] gross cross-sectional area, psi (MPa)	
	Type M or S mortar	Type N mortar
Solid masonry of brick and other solid units of clay or shale; sand-lime or concrete brick:		
8000 (55.1) or greater	350 (2.4)	300 (2.1)
4500 (31.0)	225 (1.6)	200 (1.4)
2500 (17.2)	160 (1.1)	140 (0.97)
1500 (10.3)	115 (0.79)	100 (0.69)
Grouted masonry, of clay or shale; sand-lime or concrete:		
4500 (31.0) or greater	225 (1.6)	200 (1.4)
2500 (17.2)	160 (1.1)	140 (0.97)
1500 (8.3)	115 (0.79)	100 (0.69)
Solid masonry of solid concrete masonry units:		
3000 (20.7) or greater	225 (1.6)	200 (1.4)
2000 (13.8)	160 (1.1)	140 (0.97)
1200 (8.3)	115 (0.79)	100 (0.69)
Masonry of hollow load bearing units:		
2000 (13.8) or greater	140 (0.97)	120 (0.83)
1500 (10.3)	115 (0.79)	100 (0.69)
1000 (6.9)	75 (0.52)	70 (0.48)
700 (4.8)	60 (0.41)	55 (0.38)
Hollow walls (noncomposite masonry bonded) Solid units:		
2500 (17.2) or greater	160 (1.1)	140 (0.97)
1500 (10.3)	115 (0.79)	100 (0.69)
Hollow units	75 (0.52)	70 (0.48)
Stone ashlar masonry:		
Granite	720 (5.0)	640 (4.4)
Limestone or marble	450 (3.1)	400 (2.8)
Sandstone or cast stone	360 (2.5)	320 (2.2)
Rubble stone masonry Coursed, rough, or random	120 (0.83)	100 (0.69)

Net area compressive strength of units, psi (MPa)	Moduli of elasticity[1] E, psi $\times 10^6$ (MPa $\times 10^3$)	
	Type N mortar	Type M or S mortar
6000 (41.3) and greater	—	3.5 (24)
5000 (34.5)	2.8 (19)	3.2 (22)
4000 (27.6)	2.6 (18)	2.9 (20)
3000 (20.7)	2.3 (16)	2.5 (17)
2500 (17.2)	2.2 (16)	2.4 (17)
2000 (13.8)	1.8 (12)	2.2 (15)
1500 (10.3)	1.5 (10)	1.6 (11)

[1]Linear interpolation permitted.

Figure 4.18.0 *(By permission from the Masonry Society, ACI, ASCE from their manual Building Code Requirements for Masonry Structures.)*

4.19.0 Grouting Masonry

SCOPE
This Construction Guide presents information on grouting masonry walls. The intent of this guide is to provide general information on grouting methods and procedures. Knowledge of applicable codes and standards, in conjunction with acceptable field practices, is required to assure successful grouting results.

INTRODUCTION
Masonry can be grouted and reinforced to produce efficient load-resistant structures. Reinforced masonry is used to produce taller, thinner and more economical walls. Most grouted masonry walls include reinforcing steel to provide additional lateral strength. Walls constructed in certain seismic zones are required to be reinforced and grouted to resist the dynamic forces of an earthquake.

Placing Grout
Grout is placed in lifts either between two wythes of masonry or into the cells of masonry units. Lifts should be poured in increments not to exceed 5 feet. If it is demonstrated that the grout space can be properly filled, it might be allowable to place grout in increments greater than 5 feet. One or more lifts constitute a grout pour, which is the total height of grout placed in a masonry wall prior to the construction of additional masonry. For example, if a wall is constructed to a height of 10 feet, the total grout pour would be 10 feet, with the grout being placed in two 5-foot lifts. The

FIGURE 1: Vertical Grout Barriers.

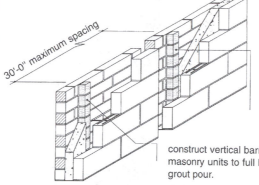

construct vertical barrier from partial Masonry units to full height of the grout pour

construct vertical barrier from partial masonry units to full height of the grout pour.

maximum height of a grout pour is restricted by the size and type of grout space and the type of grout mix (fine or coarse) listed in Table 1.

GROUTING METHODS
There are two methods of placing grout: low-lift grouting and high-lift grouting. In low-lift grouting, 5 feet or less of wall height is grouted in one day. The grout is usually placed with buckets. In high-lift grouting, grout can be placed up to a full story height in one day. Grout is usually placed with a grout pump.

The method for placing grout is usually determined by the contractor. However, the specifications can require a particular grouting method. Each of the two basic methods for placing grout—low lift grouting and high lift grouting—has its advantages. The method ultimately select-

ed depends upon the type of masonry wall, the size of the project, the equipment available, and the experience of the contractor.

Low-lift Grouting
The primary benefit of low lift grouting is that cleanouts are not required. Since the total grout pour cannot exceed 5 feet in one day, all visual inspections of the grout space can be made from the top of the wall. Also, low-lift grouting is better suited to smaller projects and multi-wythe construction and when construction sequencing prevents the use of high-lift grouting.

There are two procedures for low-lift grouting:

Pours 12 inches or less - This method involves grouting the masonry as the wall is being constructed. The grout is placed in the grout space in lifts not to exceed six times the width of the space or a maximum of 8 inches. The grout lift should be terminated approximately 1 inch below the top of the uppermost units. When grout is placed between multi-wythe walls, vertical barriers must be constructed to contain grout flow. Designed to prevent excessive flowage, which can cause segregation of grout materials, vertical barriers should be constructed in grout spaces at a maximum spacing of 30 feet. These barriers can be comprised of partial masonry units constructed to the full height of the wall (See Figure 1). Consolidate the grout shortly after it is placed. Masonry units must not be displaced or dislodged while consolidating grout (See Figure 2).

TABLE 1: Grout Pour Heights and Space Requirements

Specified Grout Type	Maximum Grout Pour Height (ft)	Minimum Width of Grout Space Between Wythes (in.)	Minimum Grout Space Dimensions for Grouting Cells of Hollow Units (in. x in.)	Cleanout Requirement
Fine	1	3/4	1 1/2x2	No
Fine	5	2	2 x 3	No
Fine	12	2 1/2	2 1/2x3	Yes
Fine	24	3	3x3	Yes
Coarse	1	1 1/2	1 1/2x3	No
Coarse	5	2	2 1/2 x 3	No
Coarse	12	2 1/2	3x3	Yes
Coarse	24	3	3x4	Yes

NOTES: The minimum grout space dimension is the distance between any masonry protrusion and shall be increased by the width of horizontal bars installed within the space.

Figure 4.19.0 *(By permission from the Brick Industry Association, Reston, Virginia.)*

Pours greater than 12 inches and up to 5 feet - For multi-wythe walls, first construct the masonry to a height of 4 or 5 feet. The wythes must be bonded together with wire ties or joint reinforcement to prevent bulging or blowouts; and the masonry must be allowed to cure for approximately 12 to 18 hours prior to grouting to withstand hydrostatic grout pressure. A minimum 3/4 inch grout space is required between the wythes. Vertical barriers must be constructed to contain grout flow. Next, install vertical reinforcement (if required); then place grout in two or three lifts, evenly distributing the grout throughout the space in each lift. Consolidate the grout shortly after it is placed and reconsolidate after initial water loss and settlement have occurred (See Figure 2).

For single-wythe walls, first construct the masonry to height of 4 or 5 feet with vertical cells sufficiently aligned and clear of debris and mortar obstructions. Lay units with cross webs bedded with mortar to contain grout. Next install vertical reinforcement (if required); then pour grout into the cells of units, terminating the grout approximately 1 to 2 inches below the top of the upper most unit. Consolidate the grout shortly after it is placed and reconsolidate after initial water loss and settlement has occurred (See Figure 2).

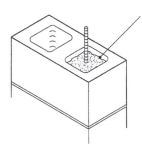

Terminate grout 1 inch to 2 inch below top of masonry unit

FIGURE 3D: Grout shear key

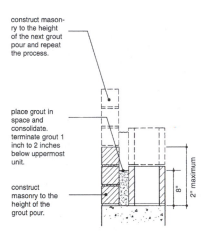

construct masonry to the height of the next grout pour and repeat the process.

place grout in space and consolidate. terminate grout 1 inch to 2 inches below uppermost unit.

construct masonry to the height of the grout pour.

8"

2" maximum

FIGURE 2A: Grout pours 12 inch or less for multi-wythe wall

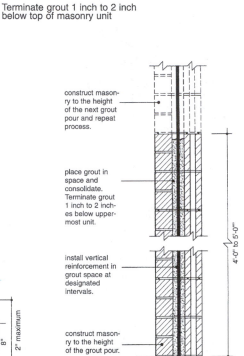

construct masonry to the height of the next grout pour and repeat process.

place grout in space and consolidate. Terminate grout 1 inch to 2 inches below uppermost unit.

install vertical reinforcement in grout space at designated intervals.

construct masonry to the height of the grout pour.

4'-0" to 5'-0"

FIGURE 2B: Grout pours up to 5 feet for multi-wythe wall

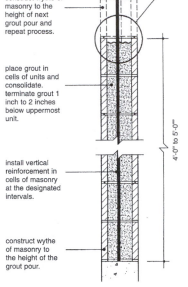

FIGURE 3D

construct wythe of masonry to the height of next grout pour and repeat process.

place grout in cells of units and consolidate. terminate grout 1 inch to 2 inches below uppermost unit.

install vertical reinforcement in cells of masonry at the designated intervals.

construct wythe of masonry to the height of the grout pour.

4'-0" to 5'-0"

FIGURE 2C: Grout pours up to 5 feet for single-wythe wall

FIGURE 2: Methods of low-lift grouting.

High-lift Grouting

Grouting masonry walls that have been constructed to a full story height has several advantages. Reinforcement bars are placed after the masonry wall has been constructed. Productivity is increased because the mason does not have to lift and place the unit over the reinforcement bar for single-wythe grouting. Large amounts of grout can be placed at one time, which also increases productivity and produces more consistent workmanship.

Cleanout openings are required at the base of the wall for high-lift grouting. Used to remove mortar droppings and debris from the grout space, cleanouts also can be used to inspect the placement of reinforcing steel. These openings are formed by removing face shells from units, cubing holes in face shells or by deleting entire units, and should be a minimum of 3 inches long by 3 inches high. Cleanouts should be located at the base of the wall every 32 inches or less in a multi-wythe wall and at each vertical bar location when grouting cells of masonry units. Cleanouts must be covered prior to grouting - with a face shell or a form board that is braced or anchored to the wall (See Figure 3).

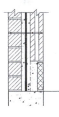

FIGURE 3A. Reinstall face shell or CMU soap

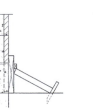

FIGURE 3B. Install 2x10 braced against masonry

FIGURE 3C. Install plywood form board mechanically fastened to masonry

FIGURE 3: Methods for sealing cleanout openings.

Figure 4.19.0 *(cont.)*

PRECAUTIONARY MEASURES

Certain precautions must be taken to assure successful grouting of masonry.

Keeping Grout Space Clean

Certain provisions must be met to keep the grout space clear of mortar while the masonry is being constructed. In multi-wythe walls, beveling the mortar bed joints back and upward slightly from the grout space eliminates most mortar extrusions. If the grout space is wide enough, the mason can pick out excess mortar extrusions with his trowel. When the cells of masonry units are to be grouted, mortar extrusions (mortar fins) in the cells should be removed with a trowel while the masonry is being constructed, or the mortar fins should be knocked off with a piece of wood or rebar down to the base of the cell shortly after the mortar begins to set. Also, it is good practice to clean the grout space with high-pressure air or water to remove mortar build-up.

Preventing Blowouts

Blowouts of mortar joints occur when the hydrostatic grout pressure exceeds the strength of the mortar joint. These blowouts can be prevented by providing proper curing time for the masonry prior to placing grout—at least 24 hours when grouting cells of masonry units and at least 72 hours when grouting multi-wythe walls.

Additional precautions against blowouts should be taken for grouting multi-wythe walls.

- Bond wythes of masonry together with a 9-gauge rectangular wall tie and a 3/16 inch diameter wire tie, or joint reinforcement. Place a minimum of one rectangular tie, one-tie or one cross-wire of joint reinforcement every 2 square feet of wall.
- Sufficiently embed masonry ties to ensure proper bonding of the wythes. Masonry ties and joint reinforcement should be at least 2 inches less in width than the actual thickness of the wall.
- Consolidate grout properly to help prevent build-up of highpressure. Do not continue to grout at one location, forcing the grout to flow throughout the space. Shift grout placement to other locations.

INSTALLING REINFORCEMENT

Methods for installing rebar are dictated by code requirements. Some codes require reinforcement to be installed prior to units being laid.

Grout Consolidation

Grout must be consolidated (vibrated) as it is being placed to minimize the voids that are left when water is absorbed by the masonry units. Grout consolidation, which usually takes place 5 to 10 minutes after placement, can be accomplished by puddling the grout with a piece of rebar or a 1 x 2 inch board if grout lifts do not exceed 12 inches. Lifts greater than 12 inches must be consolidated with a mechanical vibrator fitted with a small diameter vibrator head (3/4 inch to 1 inch diameter). Mechanically vibrate grout placed in cells of masonry units for only several seconds at a given location. Leaving the vibrator head stationary at one location may cause mortar or face-shell blowouts.

The grout should be reconsolidated within 1/2 hour after it has been consolidated to assure proper bond. Reconsolidation prevents separations from developing between the grout and masonry by eliminating water build-up between the two materials.

Given this requirement, the only productive way to construct masonry would be with open-ended units (A- or H-shaped), so that the units would not have to be lifted and placed over the reinforcement.

If the code allows for reinforcement after wall construction, there are two installation methods available to the contractor. One method is to install full-length reinforcement bars into masonry after the units have been laid. Rebar positioners can be installed into the masonry units during construction to assure proper bar location.

Reinforcement bars also can be installed in shorter lengths as the wall is being constructed. In this method, masonry can be constructed to a height of 4 feet and allowed to cure properly. Then, a 6-foot reinforcement bar can be inserted into the masonry and grouted to an approximate height of 4 feet, leaving an adequate length of reinforcement bar exposed to provide a lap splice. This process is repeated until the wall is complete.

PARTIAL GROUTING

Single-wythe masonry walls may be partially grouted, confining grout to areas of the wall containing vertical or horizontal reinforcement. Place hardware cloth or plastic mesh material below and sometimes above bond beam units containing horizontal reinforcement to confine grout to the units or cells that form the bond beam. Bed cross-webs with mortar to confine grout only to those areas that contain vertical reinforcement.

WEATHER PROTECTION

When constructing and grouting masonry under adverse conditions, follow recommendations and procedures stated within the applicable masonry code. However, consider additional means of protection when grouting masonry under the following conditions:

- When the climatic conditions are extremely hot and arid, moisten the exterior of the masonry with water prior to grouting. This will cool down the wall and help prevent the grout from setting prematurely.
- When climatic conditions are extremely cold, construct and grout the masonry within an adequately heated enclosure to assure that all excessive water is extracted from the grout. Keeping the newly constructed masonry warm will allow proper curing and prevent the masonry from freezing. Heat the masonry until it has thoroughly dried and cured.

DISCLAIMER

This document is intended to assist the industry in avoiding design and construction problems sometimes associated with masonry construction. It is intended for mason contractors, field personnel, architects, engineers, building officials, general contractors, construction managers, students, suppliers, manufacturers and other industry representatives. It is not the intent of this report to cover every aspect of masonry construction, but to focus on issues that may lead to problems. This document should not be used as the sole guide for designing and constructing masonry. It is imperative to refer to relevant codes and standards and other industry-related documents. As such, the IMI assumes no liability for any consequences that may follow from the use of this document.

Figure 4.19.0   *(cont.)*

4.18.1 Embedment Lengths for Anchor Bolts

f'_m (psi)	EMBEDMENT LENGTH, l_b, or EDGE DISTANCE, l_{be} (inches)						
	2	3	4	5	6	8	10
× 6.89 for kPa	× 25.4 for mm × 4.45 for N						
1,500	240	550	970	1,520	2,190	3,890	6,080
1,800	270	600	1,070	1,670	2,400	4,260	6,660
2,000	280	630	1,120	1,760	2,520	4,500	7,020
2,500	310	710	1,260	1,960	2,830	5,030	7,850
3,000	340	770	1,380	2,150	3,100	5,510	8,600
4,000	400	890	1,590	2,480	3,580	6,360	9,930
5,000	440	1,000	1,780	2,780	4,000	7,110	11,100
6,000	480	1,090	1,950	3,040	4,380	7,790	12,200

[1]The allowable tension values shown here are based on compressive strength of masonry assemblages.
[2]Values are for bolts of at least A 307 quality. Bolts shall be those specified in Section 2106.2.14.1.
[3]Values shown are for work with or without special inspection.

Figure 4.18.1 *(Reproduced from the 1997 Edition of the Uniform Building Code, Volumes 1, 2, 3, copyright 1997, with the permission of the publisher, the International Conference of Building Officials (ICBO). ICBO assumes no responsibility for the accuracy or the completion of summaries provided therein.)*

4.19.1 Grout Proportions by Volume

TYPE[1]	PARTS BY VOLUME OF PORTLAND CEMENT OR BLENDED CEMENT	PARTS BY VOLUME OF HYDRATED LIME OR LIME PUTTY	AGGREGATE MEASURED IN A DAMP, LOOSE CONDITION	
			Fine	Coarse
Fine grout	1	0 to $^1/_{10}$	$2^1/_4$ to 3 times the sum of the volumes of the cementitious materials	
Coarse grout	1	0 to $^1/_{10}$	$2^1/_4$ to 3 times the sum of the volumes of the cementitious materials	1 to 2 times the sum of the volumes of the cementitious materials

[1]Grout shall attain a minimum compressive strength at 28 days of 2,000 psi (13.8 MPa). The building official may require a compressive field strength test of grout made in accordance with UBC Standard 21-18.

Figure 4.19.1 *(Reproduced from the 1997 Edition of the Uniform Building Code, Volumes 1, 2, 3, copyright 1997, with the permission of the publisher, the International Conference of Building Officials (ICBO). ICBO assumes no responsibility for the accuracy or the completion of summaries provided therein.)*

4.19.2 Grouting Limitations

GROUT TYPE	GROUT POUR MAXIMUM HEIGHT (feet)[1]	MINIMUM DIMENSIONS OF THE TOTAL CLEAR AREAS WITHIN GROUT SPACES AND CELLS[2,3]	
		× 25.4 for mm	
	× 304.8 for mm	Multiwythe Masonry	Hollow-unit Masonry
Fine	1	$^3/_4$	$1^1/_2 \times 2$
Fine	5	$1^1/_2$	$1^1/_2 \times 2$
Fine	8	$1^1/_2$	$1^1/_2 \times 3$
Fine	12	$1^1/_2$	$1^3/_4 \times 3$
Fine	24	2	3×3
Coarse	1	$1^1/_2$	$1^1/_2 \times 3$
Coarse	5	2	$2^1/_2 \times 3$
Coarse	8	2	3×3
Coarse	12	$2^1/_2$	3×3
Coarse	24	3	3×4

[1]See also Section 2104.6.
[2]The actual grout space or grout cell dimensions must be larger than the sum of the following items: (1) The required minimum dimensions of total clear areas in this figure; (2) The width of any mortar projections within the space; and (3) The horizontal projections of the diameters of the horizontal reinforcing bars within a cross section of the grout space or cell.
[3]The minimum dimensions of the total clear areas shall be made up of one or more open areas, with at least one area being $^3/_4$ inch (19 mm) or greater in width.

Figure 4.19.2 *(Reproduced from the 1997 Edition of the Uniform Building Code, Volumes 1, 2, 3, copyright 1997, with the permission of the publisher, the International Conference of Building Officials (ICBO). ICBO assumes no responsibility for the accuracy or the completion of summaries provided therein.)*

4.20.0 Foundation Wall Construction (Depth of Unbalanced Backfill)

Wall construction	Nominal wall thickness, in. (mm)	Maximum depth of unbalanced backfill, ft (m)
Hollow unit masonry	8 (203) 10 (254) 12 (305)	5 (1.53) 6 (1.83) 7 (2.14)
Solid unit masonry	8 (203) 10 (254) 12 (305)	5 (1.53) 7 (2.14) 7 (2.14)
Fully grouted masonry	8 (203) 10 (254) 12 (305)	7 (2.14) 8 (2.44) 8 (2.44)

Figure 4.20.0 (*By permission from the Masonry Society, ACI, ASCE from their manual Building Code Requirements for Masonry Structures.*)

4.20.1 Exterior Foundation Requirements—6- and 8-inch-Thick Walls

Wood or Steel Framing
Width of Footings in Inches[1,2]

WALL HEIGHT (feet)	SPAN TO BEARING WALLS (feet)	ONE-STORY BUILDINGS Roof Live Load × 0.0479 for kN/m²			TWO-STORY BUILDINGS Roof Live Load (psf) × 0.0479 for kN/m² Plus Floor Live Load (psf) × 0.0479 for kN/m²					
					20		30		40	
		20 psf (inches)	30 psf (inches)	40 psf (inches)	50	100	50	100	50	100
× 304.8 for mm		× 25.4 for mm Minimum Width of Footing (inches)								
8	8	12			12	12	12	12	12	12
	16				12	14	12	14	12	14
	24				14	18	14	18	16	18
	32				16	20	18	20	18	20
10	8	12			12	12	12	12	12	12
	16				14	16	14	16	14	16
	24				16	20	16	18	16	20
	32				20	24	20	22	20	24
12	8	12	12	12	12	14	12	14	12	14
	16	12	12	12	16	18	16	16	14	16
	24	12	12	14	18	20	18	20	18	20
	32	12	14	16	20	22	22	22	22	24

[1]For buildings with under-floor space or basements, footing thickness is to be a minimum of 12 inches (305 mm). It shall be reinforced with No. 4 bars at 24 inches (610 mm) on center when its width is required to be 18 inches (457 mm) or larger and it supports more than the roof and one floor.

[2]Footings are a minimum of 10 inches (254 mm) thick for a one-story building and 12 inches (305 mm) thick for a two-story building. Bottom of footing to be 18 inches (457 mm) below grade or the frost depth, whichever is deeper. Footing to be reinforced with No. 4 bars at 24 inches (610 mm) on center when supporting more than the roof and one floor.

Figure 4.20.1 (*Reproduced from the 1997 Edition of the Uniform Building Code, Volumes 1, 2, 3, copyright 1997, with the permission of the publisher, the International Conference of Building Officials (ICBO). ICBO assumes no responsibility for the accuracy or the completion of summaries provided therein.*)

4.20.2 Interior Foundation Requirements—6- and 8-inch-Thick Walls

Wood or Steel Framing
Width of Footings in Inches[1,2,3]

WALL HEIGHT (feet)	SPAN TO BEARING WALLS (feet)	ONE-STORY BUILDINGS Roof Live Load[4] × 0.0479 for kN/m²			TWO-STORY BUILDINGS Roof Live Load[4] (psf) × 0.0479 for kN/m² 20 Plus Floor Live Load[5] (psf) × 0.0479 for kN/m²		30		40	
		20 psf (inches)	30 psf (inches)	40 psf (inches)	50	100	50	100	50	100
× 304.8 for mm					Minimum Width of Footing (inches) × 25.4 for mm					
8	8	12	12	12	12	14	12	14	12	14
	16	12	12	12	16	20	18	20	18	22
	24	12	12	14	20	26	22	28	22	28
	32	14	14	16	24	28	26	32	28	34
10	8	12	12	12	14	16	14	16	14	16
	16	12	12	12	20	24	20	22	20	22
	24	12	14	14	22	28	22	28	22	28
	32	14	14	16	26	34	26	32	28	34
12	8	12	12	12	14	16	16	18	16	18
	16	12	14	16	20	24	20	22	20	22
	24	14	14	16	24	28	22	28	24	28
	32	16	16	18	28	30	28	32	28	34

[1]For buildings with under-floor space or basements, footing thickness is to be a minimum of 12 inches (305 mm). It shall be reinforced with No. 4 bars at 24 inches (610 mm) on center when its width is required to be 18 inches (457 mm) or larger and it supports more than the roof and one floor.

[2]Footings are 10 inches (254 mm) thick for up to 24 inches (610 mm) wide and 12 inches (305 mm) thick for up to 34 inches (864 mm) wide. Footings shall be reinforced with No. 4 bars at 24 inches (610 mm) on center when supporting more than the roof and one floor.

[3]These interior footings support roof-ceiling or floors or both for a distance on each side equal to the span length shown. A tributary width equal to the span length may be used.

[4]From local snow load tables. For areas without snow loads use 20 pounds per square foot (0.96 kN/m²).

[5]For intermediate floor loads go to next higher value.

Figure 4.20.2 (*Reproduced from the 1997 Edition of the Uniform Building Code, Volumes 1, 2, 3, copyright 1997, with the permission of the publisher, the International Conference of Building Officials (ICBO). ICBO assumes no responsibility for the accuracy or the completion of summaries provided therein.*)

4.20.3 Empirical Design—Wall Lateral Support Requirements

CONSTRUCTION	MAXIMUM *l/t* or *h/t*

Bearing walls
    Solid or solid grouted
    All other
	20
	18

Nonbearing walls
    Exterior
    Interior
	18
	36

Figure 4.20.3 (*Reproduced from the 1997 Edition of the Uniform Building Code, Volumes 1, 2, 3, copyright 1997, with the permission of the publisher, the International Conference of Building Officials (ICBO). ICBO assumes no responsibility for the accuracy or the completion of summaries provided therein.*)

4.20.4 Empirical Design—Thickness of Foundation Walls

FOUNDATION WALL CONSTRUCTION	NOMINAL THICKNESS (inches)	MAXIMUM DEPTH OF UNBALANCED FILL (feet)
	× 25.4 for mm	× 304.8 for mm
Masonry of hollow units, ungrouted	8 10 12	4 5 6
Masonry of solid units	8 10 12	5 6 7
Masonry of hollow or solid units, fully grouted	8 10 12	7 8 8

Masonry of hollow units
  reinforced vertically with No. 4 bars and grout at 24″ o.c. Bars located not
  less than 4$\frac{1}{2}$″ from pressure side of wall.
	8	7

Figure 4.20.4 *(Reproduced from the 1997 Edition of the Uniform Building Code, Volumes 1, 2, 3, copyright 1997, with the permission of the publisher, the International Conference of Building Officials (ICBO). ICBO assumes no responsibility for the accuracy or the completion of summaries provided therein.)*

4.20.5 Empirical Design—Allowable Shear on Bolts in Unburned Units

DIAMETER OF BOLTS (inches)	EMBEDMENTS (inches)	SHEAR (pounds)
× 25.4 for mm		× 4.45 for N
$\frac{1}{2}$	—	—
$\frac{5}{8}$	12	200
$\frac{3}{4}$	15	300
$\frac{7}{8}$	18	400
1	21	500
$1\frac{1}{8}$	24	600

Figure 4.20.5 *(Reproduced from the 1997 Edition of the Uniform Building Code, Volumes 1, 2, 3, copyright 1997, with the permission of the publisher, the International Conference of Building Officials (ICBO). ICBO assumes no responsibility for the accuracy or the completion of summaries provided therein.)*

4.20.6 Empirical Design—Allowable Shear on Bolts for All Masonry Except Unburned Clay Units

DIAMETER BOLT (inches)	EMBEDMENT[1] (inches)	SOLID MASONRY (shear in pounds)	GROUTED MASONRY (shear in pounds)
× 25.4 for mm		× 4.45 for N	
$\frac{1}{2}$	4	350	550
$\frac{5}{8}$	4	500	750
$\frac{3}{4}$	5	750	1,100
$\frac{7}{8}$	6	1,000	1,500
1	7	1,250	1,850[2]
$1\frac{1}{8}$	8	1,500	2,250[2]

[1]An additional 2 inches of embedment shall be provided for anchor bolts located in the top of columns for buildings located in Seismic Zones 2, 3 and 4.
[2]Permitted only with not less than 2,500 pounds per square inch (17.24 MPa) units.

Figure 4.20.6 *(Reproduced from the 1997 Edition of the Uniform Building Code, Volumes 1, 2, 3, copyright 1997, with the permission of the publisher, the International Conference of Building Officials (ICBO). ICBO assumes no responsibility for the accuracy or the completion of summaries provided therein.)*

4.20.7 Empirical Design—Allowable Compressive Stresses for Masonry

CONSTRUCTION: COMPRESSIVE STRENGTH OF UNIT, GROSS AREA	ALLOWABLE COMPRESSIVE STRESSES[1] GROSS CROSS-SECTIONAL AREA (psi)	
× 6.89 for kPa	× 6.89 for kPa	
	Type M or S Mortar	Type N Mortar
Solid masonry of brick and other solid units of clay or shale; sand-lime or concrete brick:		
8,000 plus, psi	350	300
4,500 psi	225	200
2,500 psi	160	140
1,500 psi	115	100
Grouted masonry, of clay or shale; sand-lime or concrete:		
4,500 plus, psi	275	200
2,500 psi	215	140
1,500 psi	175	100
Solid masonry of solid concrete masonry units:		
3,000 plus, psi	225	200
2,000 psi	160	140
1,200 psi	115	100
Masonry of hollow load-bearing units:		
2,000 plus, psi	140	120
1,500 psi	115	100
1,000 psi	75	70
700 psi	60	55
Hollow walls (cavity or masonry bonded)[2] solid units:		
2,500 plus, psi	160	140
1,500 psi	115	100
Hollow units	75	70
Stone ashlar masonry:		
Granite	720	640
Limestone or marble	450	400
Sandstone or cast stone	360	320
Rubble stone masonry Coarse, rough or random	120	100
Unburned clay masonry	30	—

[1]Linear interpolation may be used for determining allowable stresses for masonry units having compressive strengths which are intermediate between those given in the table.

[2]Where floor and roof loads are carried upon one wythe, the gross cross-sectional area is that of the wythe under load. If both wythes are loaded, the gross cross-sectional area is that of the wall minus the area of the cavity between the wythes.

Figure 4.20.7 (*Reproduced from the 1997 Edition of the Uniform Building Code, Volumes 1, 2, 3, copyright 1997, with the permission of the publisher, the International Conference of Building Officials (ICBO). ICBO assumes no responsibility for the accuracy or the completion of summaries provided therein.*)

4.21.0 Brick Sizes (Nomenclature)

MODULAR BRICK SIZES

Unit Designation	Nominal Dimensions, in.			Joint Thickness[2], in.	Specified Dimensions[3], in.			Vertical Coursing
	w	h	l		w	h	i	
Modular	4	2⅔	8	⅜	3⅜	2¼	7⅝	3C = 8 in.
				½	3½	2¼	7½	
Engineer Modular	4	3⅓	8	⅜	3⅜	2¾	7⅝	5C = 16 in.
				½	3½	2¹³⁄₁₆	7½	
Closure Modular	4	4	8	⅜	3⅝	3⅝	7⅝	1C = 4 in.
				½	3½	3½	7½	
Roman	4	2	12	⅜	3⅝	1⅝	11⅝	2C = 4 in.
				½	3½	1½	11½	
Norman	4	2⅔	12	⅜	3⅝	2¼	11⅝	3C = 8 in.
				½	3½	2¼	11½	
Engineer Norman	4	3⅓	12	⅜	3⅝	2¾	11⅝	5C = 16 in.
				½	3½	2¹³⁄₁₆	11½	
Utility	4	4	12	⅝	3⅝	3⅝	11⅝	1C = 4 in.
				½	3½	3½	11½	

NONMODULAR BRICK SIZES

Unit Designation	Nominal Dimensions, in.			Joint Thickness[2], in.	Specified Dimensions[3], in.			Vertical Coursing
	w	h	l		w	h	l	
Standard				⅜	3⅝	2¼	8	3C = 8 in.
				½	3½	2¼	8	
Engineer Standard				⅝	3⅝	2¾	8	5C = 16 in.
				½	3½	2¹³⁄₁₆	8	
Closure Standard				⅜	3⅝	3⅝	8	1C = 4 in.
				½	3½	3½	8	
King				⅜	3	2¾	9⅝	5C = 16 in.
					3	3⅝	9¾	
Queen				⅜	3	2¾	8	5C = 16 in.

[1] 1 in. = 25.4 mm; 1 ft = 0.3m

[2] Common joint sizes used with length and width dimensions. Joint thicknesses of bed joints vary based on vertical coursing and specified unit height.

[3] Specified dimensions may vary within this range from manufacturer to manufacturer.

(Reprinted by permission from the Brick Institute of America, Reston, Virginia.)

4.22.0 Estimating Concrete Masonry

NOMINAL LENGTH OF CONCRETE MASONRY WALLS BY STRETCHERS

(Based on units 15⅝" long and half units 7⅝" long with ⅜" thick head joints)

LENGTH OF WALL	NO. OF UNITS	LENGTH OF WALL	NO. OF UNITS	LENGTH OF WALL	NO. OF UNITS	LENGTH OF WALL	NO. OF UNITS	LENGTH OF WALL	NO. OF UNITS	LENGTH OF WALL	NO. OF UNITS
0'-8"	½	20'-8"	15½	40'-8"	30½	60'-8"	45½	80'-8"	60½	100'-8"	75½
1'-4"	1	21'-4"	16	41'-4"	31	61'-4"	46	81'-4"	61	101'-4"	76
2'-0"	1½	22'-0"	16½	42'-0"	31½	62'-0"	46½	82'-0"	61½	102'-0"	76½
2'-8"	2	22'-8"	17	42'-8"	32	62'-8"	47	82'-8"	62	102'-8"	77
3'-4"	2½	23'-4"	17½	43'-4"	32½	63'-4"	47½	83'-4"	62½	103'-4"	77½
4'-0"	3	24'-0"	18	44'-0"	33	64'-0"	48	84'-0"	63	104'-0"	78
4'-8"	3½	24'-8"	18½	44'-8"	33½	64'-8"	48½	84'-8"	63½	104'-8"	78½
5'-4"	4	25'-4"	19	45'-4"	34	65'-4"	49	85'-4"	64	105'-4"	79
6'-0"	4½	26'-0"	19½	46'-0"	34½	66'-0"	49½	86'-0"	64½	106'-0"	79½
6'-8"	5	26'-8"	20	46'-8"	35	66'-8"	50	86'-8"	65	106'-8"	80
7'-4"	5½	27'-4"	20½	47'-4"	35½	67'-4"	50½	87'-4"	65½	107'-4"	80½
8'-0"	6	28'-0"	21	48'-0"	36	68'-0"	51	88'-0"	66	108'-0"	81
8'-8"	6½	28'-8"	21½	48'-8"	36½	68'-8"	51½	88'-8"	66½	108'-8"	81½
9'-4"	7	29'-4"	22	49'-4"	37	69'-4"	52	89'-4"	67	109'-4"	82
10'-0"	7½	30'-0"	22½	50'-0"	37½	70'-0"	52½	90'-0"	67½	110'-0"	82½
10'-8"	8	30'-8"	23	50'-8"	38	70'-8"	53	90'-8"	68	110'-8"	83
11'-4"	8½	31'-4"	23½	51'-4"	38½	71'-4"	53½	91'-4"	68½	111'-4"	83½
12'-0"	9	32'-0"	24	52'-0"	39	72'-0"	54	92'-0"	69	112'-0"	84
12'-8"	9½	32'-8"	24½	52'-8"	39½	72'-8"	54½	92'-8"	69½	112'-8"	84½
13'-4"	10	33'-4"	25	53'-4"	40	73'-4"	55	93'-4"	70	113'-4"	85
14'-0"	10½	34'-0"	25½	54'-0"	40½	74'-0"	55½	94'-0"	70½	114'-0"	85½
14'-8"	11	34'-8"	26	54'-8"	41	74'-8"	56	94'-8"	71	114'-8"	86
15'-4"	11½	35'-4"	26½	55'-4"	41½	75'-4"	56½	95'-4"	71½	115'-4"	86½
16'-0"	12	36'-0"	27	56'-0"	42	76'-0"	57	96'-0"	72	116'-0"	87
16'-8"	12½	36'-8"	27½	56'-8"	42½	76'-8"	57½	96'-8"	72½	116'-8"	87½
17'-4"	13	37'-4"	28	57'-4"	43	77'-4"	58	97'-4"	73	117'-4"	88
18'-0"	13½	38'-0"	28½	58'-0"	43½	78'-0"	58½	98'-0"	73½	118'-0"	88½
18'-8"	14	38'-8"	29	58'-8"	44	78'-8"	59	98'-8"	74	118'-8"	89
19'-4"	14½	39'-4"	29½	59'-4"	44½	79'-4"	59½	99'-4"	74½	119'-4"	89½
20'-0"	15	40'-0"	30	60'-0"	45	80'-0"	60	100'-0"	75	120'-0"	90

NOMINAL HEIGHT OF CONCRETE MASONRY WALLS BY COURSES

(Based on units 7⅝" high ⅜" thick mortar joints)

HEIGHT OF WALL	NO. OF UNITS	HEIGHT OF WALL	NO. OF UNITS	HEIGHT OF WALL	NO. OF UNITS	HEIGHT OF WALL	NO. OF UNITS
0'-8"	1	8'-8"	13	16'-8"	25	24'-8"	37
1'-4"	2	9'-4"	14	17'-4"	26	25'-4"	38
2'-0"	3	10'-0"	15	18'-0"	27	26'-0"	39
2'-8"	4	10'-8"	16	18'-8"	28	26'-8"	40
3'-4"	5	11'-4"	17	19'-4"	29	27'-4"	41
4'-0"	6	12'-0"	18	20'-0"	30	28'-0"	42
4'-8"	7	12'-8"	19	20'-8"	31	28'-8"	43
5'-4"	8	13'-4"	20	21'-4"	32	29'-4"	44
6'-0"	9	14'-0"	21	22'-0"	33	30'-0"	45
6'-8"	10	14'-8"	22	22'-8"	34	30'-8"	46
7'-4"	11	15'-4"	23	23'-4"	35	31'-4"	47
8'-0"	12	16'-0"	24	24'-0"	36	30'-0"	48

HOW TO USE THESE TABLES

The tables on this page are an aid to estimating and designing with standard concrete masonry units. The following are examples of how they can be used to advantage.

Example:

Estimate the number of units required for a wall 76' long and 12' high.

From table: 76' = 57 units
                12' = 18 courses
57 × 18 = 1026 = No. masonry units required

Example:

Estimate the number of units required for a foundation 24' × 30' = 11 courses high.

2 (24 + 30) = 108' = distance for a foundation
From table: 108' = 81 units
81 × 11 = 891 = No. masonry units required.

This table can also be useful in the layout of a building on a modular basis to eliminate cutting of units. Example: If design calls for a wall 41' long it can be found from the table that making wall 41'-4", will eliminate cutting units and consequent waste. Example: If the distance between two openings has been tentatively established at 2'-9", consulting the table will show that 2'-8" dimension would eliminate cutting of units.

4.23.0 Nominal Height of Brick and Block Walls by Coursing

COURSES	REGULAR 4 2¼" bricks + 4 equal joints =					MODULAR 3 bricks + 3 joints =	CONCRETE BLOCKS	
	10" ¼" joints	10½" ⅜" joints	11" ½" joints	11½" ⅝" joints	12" ¾" joints	8"	3⅝" blocks ⅜" joints	7⅝" blocks ⅜" joints
1	2½"	2⅝"	2¾"	2⅞"	3"	2¹¹⁄₁₆"	4"	8"
2	5"	5¼"	5½"	5¾"	6"	5⁵⁄₁₆"	8"	1'4"
3	7½"	7⅞"	8¼"	8⅝"	9"	8"	1'0"	2'0"
4	10"	10½"	11"	11½"	1'0"	10¹¹⁄₁₆"	1'4"	2'8"
5	1'0½"	1'1⅛"	1'1¾"	1'2⅜"	1'3"	1'1⁵⁄₁₆"	1'8"	3'4"
6	1'3"	1'3¾"	1'4½"	1'5¼"	1'6"	1'4"	2'0"	4'0"
7	1'5½"	1'6⅜"	1'7¼"	1'8⅛"	1'9"	1'6¹¹⁄₁₆"	2'4"	4'8"
8	1'8"	1'9"	1'10"	1'11"	2'0"	1'9⁵⁄₁₆"	2'8"	5'4"
9	1'10½"	1'11⅝"	2'0¾"	2'1⅞"	2'3"	2'0"	3'0"	6'0"
10	2'1"	2'2¼"	2'3½"	2'4¾"	2'6"	2'2¹¹⁄₁₆"	3'4"	6'8"
11	2'3½"	2'4⅞"	2'6¼"	2'7⅝"	2'9"	2'5⁵⁄₁₆"	3'8"	7'4"
12	2'6"	2'7½"	2'9"	2'10½"	3'0"	2'8"	4'0"	8'0"
13	2'8½"	2'10⅛"	2'11¾"	3'1⅜"	3'3"	2'10¹¹⁄₁₆"	4'4"	8'8"
14	2'11"	3'0¾"	3'2½"	3'4¼"	3'6"	3'1⁵⁄₁₆"	4'8"	9'4"
15	3'1½"	3'3⅜"	3'5¼"	3'7⅛"	3'9"	3'4"	5'0"	10'0"
16	3'4"	3'6"	3'8"	3'10"	4'0"	3'6¹¹⁄₁₆"	5'4"	10'8"
17	3'6½"	3'8⅝"	3'10¾"	4'0⅞"	4'3"	3'9⁵⁄₁₆"	5'8"	11'4"
18	3'9"	3'11¼"	4'1½"	4'3¾"	4'6"	4'0"	6'0"	12'0"
19	3'11½"	4'1⅞"	4'4¼"	4'6⅝"	4'9"	4'2¹¹⁄₁₆"	6'4"	12'8"
20	4'2"	4'4½"	4'7"	4'9½"	5'0"	4'5⁵⁄₁₆"	6'8"	13'4"
21	4'4½"	4'7⅛"	4'9¾"	5'0⅜"	5'3"	4'8"	7'0"	14'0"
22	4'7"	4'9¾"	5'0½"	5'3¼"	5'6"	4'10¹¹⁄₁₆"	7'4"	14'8"
23	4'9½"	5'0⅜"	5'3¼"	5'6⅛"	5'9"	5'1⁵⁄₁₆"	7'8"	15'4"
24	5'0"	5'3"	5'6"	5'9"	6'0"	5'4"	8'0"	16'0"
25	5'2½"	5'5⅝"	5'8¾"	5'11⅞"	6'3"	5'6¹¹⁄₁₆"	8'4"	16'8"
26	5'5"	5'8¼"	5'11½"	6'2¾"	6'6"	5'9⁵⁄₁₆"	8'8"	17'4"
27	5'7½"	5'10⅞"	6'2¼"	6'5⅝"	6'9"	6'0"	9'0"	18'0"
28	5'10"	6'1½"	6'5"	6'8½"	7'0"	6'2¹¹⁄₁₆"	9'4"	18'8"
29	6'0½"	6'4⅛"	6'7¾"	6'11⅜"	7'3"	6'5⁵⁄₁₆"	9'8"	19'4"
30	6'3"	6'6¾"	6'10½"	7'2¼"	7'6"	6'8"	10'0"	20'0"
31	6'5½"	6'9⅜"	7'1¼"	7'5⅛"	7'9"	6'10¹¹⁄₁₆"	10'4"	20'8"
32	6'8"	7'0"	7'4"	7'8"	8'0"	7'1⁵⁄₁₆"	10'8"	21'4"
33	6'10½"	7'2⅝"	7'6¾"	7'10⅞"	8'3"	7'4"	11'0"	22'0"
34	7'1"	7'5¼"	7'9½"	8'1¾"	8'6"	7'6¹¹⁄₁₆"	11'4"	22'8"
35	7'3½"	7'7⅞"	8'0¼"	8'4⅝"	8'9"	7'9⁵⁄₁₆"	11'8"	23'4"
36	7'6"	7'10½"	8'3"	8'7½"	9'0"	8'0"	12'0"	24'0"
37	7'8½"	8'1⅛"	8'5¾"	8'10⅜"	9'3"	8'2¹¹⁄₁₆"	12'4"	24'8"
38	7'11"	8'3¾"	8'8½"	9'1¼"	9'6"	8'5⁵⁄₁₆"	12'8"	25'4"
39	8'1½"	8'6⅜"	8'11¼"	9'4⅛"	9'9"	8'8"	13'0"	26'0"
40	8'4"	8'9"	9'2"	9'7"	10'0"	8'10¹¹⁄₁₆"	13'4"	26'8"
41	8'6½"	8'11⅝"	9'4¾"	9'9⅞"	10'3"	9'1⁵⁄₁₆"	13'8"	27'4"
42	8'9"	9'2¼"	9'7½"	10'0¾"	10'6"	9'4"	14'0"	28'0"
43	8'11½"	9'4⅞"	9'10¼"	10'3⅝"	10'9"	9'6¹¹⁄₁₆"	14'4"	28'8"
44	9'2"	9'7½"	10'1"	10'6½"	11'0"	9'9⁵⁄₁₆"	14'8"	29'4"
45	9'4½"	9'10⅛"	10'3¾"	10'9⅜"	11'3"	10'0"	15'0"	30'0"
46	9'7"	10'0¾"	10'6½"	11'0¼"	11'6"	10'2¹¹⁄₁₆"	15'4"	30'8"
47	9'9½"	10'3⅜"	10'9¼"	11'3⅛"	11'9"	10'5⁵⁄₁₆"	15'8"	31'4"
48	10'0"	10'6"	11'0"	11'6"	12'0"	10'8"	16'0"	32'0"
49	10'2½"	10'8⅝"	11'2¾"	11'8⅞"	12'3"	10'10¹¹⁄₁₆"	16'4"	32'8"
50	10'5"	10'11¼"	11'5½"	11'11¾"	12'6"	11'1⁵⁄₁₆"	16'8"	33'4"

4.24.0 Corner, Beam, and Jamb Details

CORNER DETAILS

W (Wall Width)	L (Corner Unit Length)
3½"	11½"
5½"	13½"
7½"	15½" (Reg. Stretcher Unit, No Knifed Corner Req'd.)

W (Wall Width)	A	B	C
3½"	1¼"	9¼"	7½"
5½"	2¹/₁₆"	10¹/₁₆"	7½"
7½"	2⅞"	10⅞"	7½"

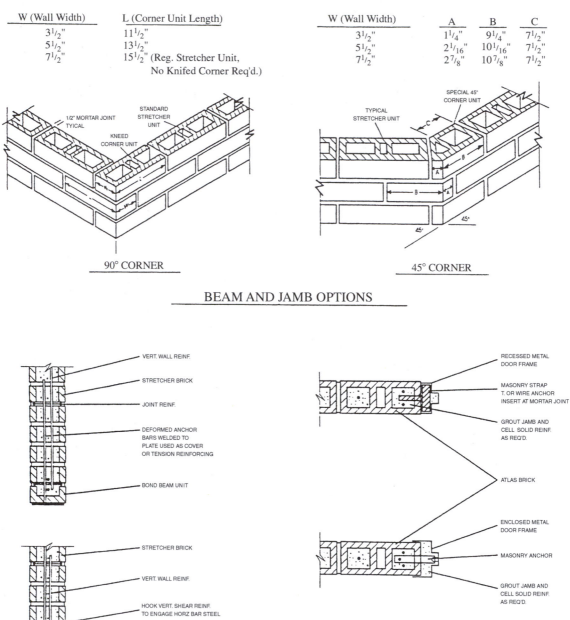

90° CORNER

45° CORNER

BEAM AND JAMB OPTIONS

LINTEL BEAM OPTIONS

DOOR JAMB OPTIONS

Figure 4.24.0   (*Reprinted with permission from Interstate Brick, West Jordan, Utah.*)

4.24.1 Pilaster Details

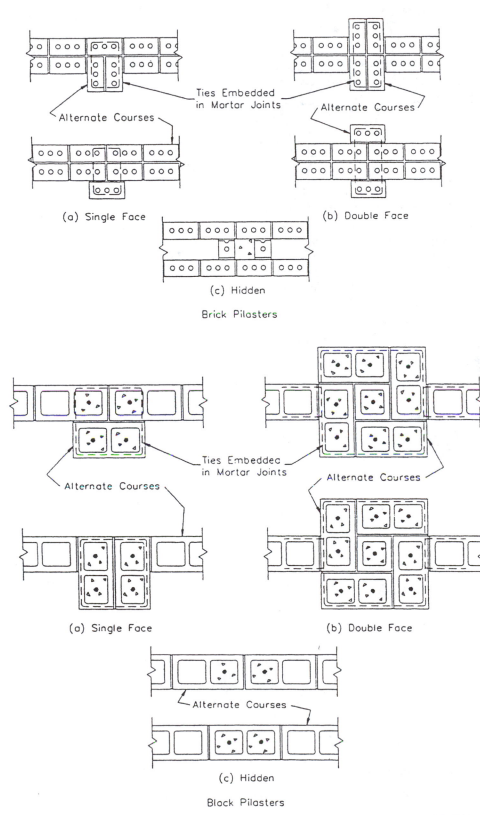

Figure 4.24.1 *(By permission from the Masonry Society, ACI, ASCE from their manual Building Code Requirements for Masonry Structures.)*

4.25.0 Masonry Reinforcement—Types of Ties

Whenever a double wythe wall is constructed or a cavity wall containing a masonry veneer is built anchors, ties, or reinforcement is required to stablize the two components. Seismic requirements add other components to the conventional masonry wall reinforcement to stabilize the structure in case of a seismic event.

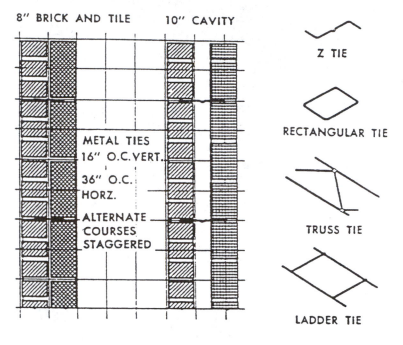

Metal-Tied Masonry Walls

Figure 4.25.0   *(By permission from the Brick Institute of America, Reston, Virginia.)*

4.25.1 Masonry Reinforcement (Materials and Physical Properties of Bars/Wire)

Reinforcement and metal accessories

ASTM specification	Material	Use	Yield strength, ksi (MPa)	ASTM yield stress, MPa
A 36	Structural steel	Connectors	36 (248)	250
A 82	Steel wire	Joint reinforcement, ties	70 (483)	485
A 167	Stainless steel	Bolts, reinforcement, ties	30 (207)	205
A 185	Steel wire	Wire fabric, ties	75 (517)	485
A 307	Carbon steel	Connectors	60 (414)	
A 366	Carbon steel	Connectors	—	
A 496	Steel wire	Reinforcement	75 (517)	485
A 497	Steel wire fabric	Reinforcement, wire fabric	70 (483)	485
A 615	Billet steel	Reinforcement	40,60 (276, 414)	300,400
A 616	Rail steel	Reinforcement	50,60 (345, 414)	350,400
A 617	Axle steel	Reinforcement	40,60 (276, 414)	300,400
A 706	Low alloy steel	Reinforcement	60 (414)	

Physical properties of steel reinforcing wire and bars

Designation		Diameter, in (mm)	Area, in.² (mm²)	Perimeter, in (mm)
Wire				
W1.1 (11 gage)		0.121 (3.07)	0.011 (7.10)	0.330 (9.65)
W1.7 (9 gage)		0.148 (3.76)	0.017 (11.0)	0.465 (11.8)
W2.1 (8 gage)		0.162 (4.12)	0.020 (12.9)	0.509 (12.9)
W2.8 (3/16 wire)		0.187 (4.75)	0.027 (17.4)	0.587 (14.9)
W4.9 (¼ wire)		0.250 (6.35)	0.049 (31.6)	0.785 (19.9)
Bars	Metric			
#3		0.375 (9.53)	0.11 (71.0)	1.178 (29.92)
	10	0.445 (11.3)	0.16 (100)	1.398 (35.5)
#4		0.500 (12.7)	0.20 (129)	1.571 (39.90)
#5	15	0.625 (15.9)	0.31 (200)	1.963 (49.86)
#6		0.750 (19.1)	0.44 (284)	2.456 (62.38)
	20	0.768 (19.5)	0.47 (300)	2.413 (61.3)
#7		0.875 (22.2)	0.60 (387)	2.749 (69.83)
	25	0.992 (25.2)	0.76 (500)	3.118 (79.2)
#8		1.000 (25.4)	0.79 (510)	3.142 (79.81)
#9		1.128 (28.7)	1.00 (645)	3.544 (90.02)
	30	1.177 (29.9)	1.09 (700)	3.697 (93.9)
#10		1.270 (32.2)	1.27 (819)	3.990 (101.3)
	35	1.406 (35.7)	1.55 (1000)	4.417 (112.2)
#11		1.410 (35.8)	1.56 (1006)	4.430 (112.5)

Wire size	Minimum number of ties required
W1.7	one wall tie per 2²/₃ ft² (0.25 m²) of wall
W2.8	one wall tie per 4¹/₂ ft² (0.42 m²) of wall

Figure 4.25.1 (By permission from the Brick Institute of America, Reston, Virginia.)

4.25.2 Wall Anchorage Details

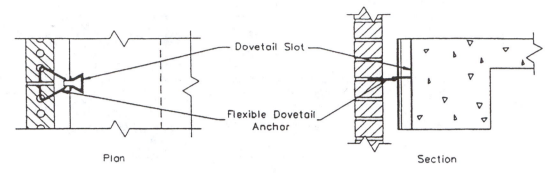

(a) Wall Anchorage to Concrete Beams

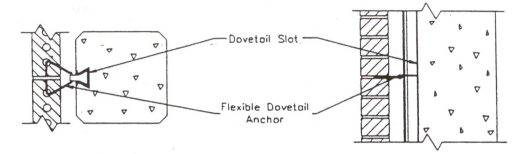

(b) Wall Anchorage to Concrete Columns

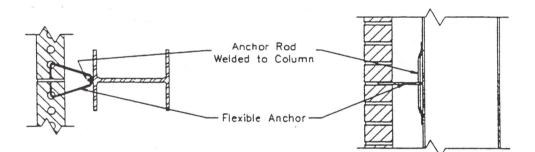

(c) Wall Anchorage to Steel Column

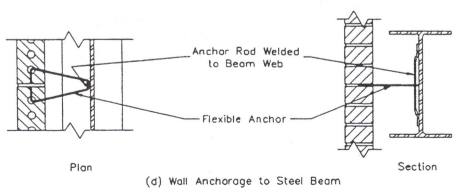

(d) Wall Anchorage to Steel Beam

Figure 4.25.2   *(By permission from the Masonry Society, ACI, ASCE from their manual Building Code Requirements for Masonry Structures.)*

4.25.3 Truss and Ladur Reinforcement

DUR-O-WAR® TRUSS

D/A 310 TRUSS   D/A 310 TR TRI-ROD   D/A 310 DSR DOUBLE SIDE ROD

LADUR TYPE®

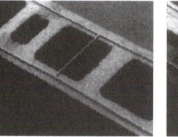

D/A 320 LADUR   D/A 320 TR TRI-ROD   D/A 320 DSR DOUBLE SIDE ROD

INSTALLATION – TRUSS AND LADUR

Use at least one longitudinal side rod for each bed joint. Out-to-out spacing of the side rods is approximately 2" (50mm) less than the nominal thickness of the wall or wythe in which the reinforcement is placed.

Splices

Side rods should be lapped 6" (150mm) at splices in order to provide adequate continuity of the reinforcement when subjected to normal shrinkage stresses.

Centering and Placement

Place joint reinforcement directly on masonry and place mortar over wire to form bed joint. This applies to both truss type (shown) and ladur type.

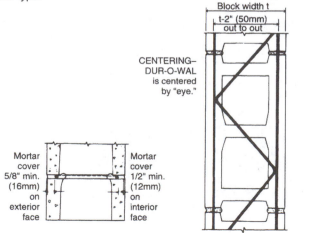

Figure 4.25.3 *(By permission from Dur-O-Wall, Inc., Arlington Heights, Illinois.)*

4.25.4 Masonry Wall Ties

D/A 5801

Recommended for non-insulated cavity/walls. The channel base plate is secured to the back-up and has a 1-1/4" (30mm) slot for coursing adjustability. The 3/16" (5mm) triangular wire tie is mortared in the veneer. Hot dipped galvanized and stainless steel finishes are available.

D/A 5431

Recommended for reconstructing brick wythes of composite walls. The 14 gauge (1.9 mm) corrugated strap has a 1-1/4" (30mm) of adjustability. The tie is mortared in place with the new brick wythe. Shear lugs accommodate seismic ladur or pencil rod. Hot dipped galvanized and stainless steel finishes are available.

D/A 5213S with Seismic Ladur

Figure 4.25.4   *(By permission from Dur-O-Wall, Inc., Arlington Heights, Illinois.)*

4.25.5 Masonry Veneer Anchors

D/A 213   D/A 207 WITH D/A 701   D/A 709 WITH D/A 701

Dovetail Slots and Anchors

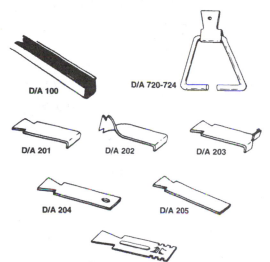

D/A 100

D/A 720-724

D/A 201   D/A 202   D/A 203

D/A 204   D/A 205

D/A 131 (PATENT PENDING)

Channel Slots and Anchors

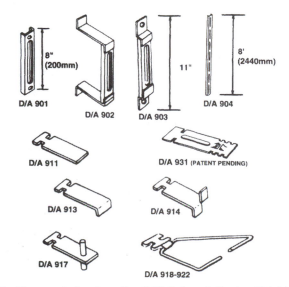

D/A 901

8" (200mm)

D/A 902

D/A 903

11"

D/A 904

8' (2440mm)

D/A 911   D/A 931 (PATENT PENDING)

D/A 913   D/A 914

D/A 917   D/A 918-922

Figure 4.25.5 *(By permission from Dur-O-Wall, Inc., Arlington Heights, Illinois.)*

4.25.6 Seismic Masonry Veneer Anchors

Seismic Veneer Anchoring Application

DUR-O-WAL's seismic veneer anchors are designed to meet performance criteria as defined by building codes. These anchors can be used for tieing brick veneers to wood stud, steel studs, steel framing, masonry, brick and concrete, They are fabricated with shear lugs that accommodate 9 gauge veneer reinforcement. The connectors are individually mounted and are easily installed.

Seismic Veneer Anchors (patented)

This anchor has the same plate and pintle design as Seismic Dur-O-Eye. The plate is engineered to be attached to the face of a CMU or concrete (D/A 5213) steel stud, wood stud or steel frame (D/A 213S) rather than embedded in mortar. The pintle shear lugs hold pencil rod or Seismic Ladur in place for greater pull out stress resistance and ductility. Adjusts 1-1/4" (30mm) up or down to allow for different course heights and allows at least 1/2" (13mm) horizontal in-plane movement to accommodate expansion and contraction. A hot dipped galvanized finish (1.5 oz. zinc per sq. ft.) (458g/m2) is standard, and 304 stainless steel is available. DUR-O-WAL recommends the use of two screws for stud applications, either the D/A 807 for steel, D/A 808 for wood, or D/A 995, or a special 1/4" (6mm) expansion bolt for concrete or masonry retrofit applications (D/A 5213).

D/A 931 Seismic Channel Slot Anchor Assembly (patent pending)

Engineered for use with standard channel slots. Pencil rod or Seismic Ladur fits inside shear lug for positive placement without the need for special clips.

D/A 431 Seismic Strap Anchor
(patent pending)

A special 14 ga. (1.9mm) adjustable seismic corrugated veneer anchor with two shear lugs, which is engineered for use with pencil rod or Seismic Ladur to resist out of plane movement and afford greater ductility in seismic zones 3 and 4 or Seismic Performance Categories D and E can be nailed or screwed to wood stud backup (D/A 808).

D/A 131 Seismic Dovetail Anchor Assembly (patent pending)

Specially designed tie with shear lug locks for pencil rod or Seismic Ladur to assure positive positioning and reinforcement without the need for special clips. Engineered to fit standard dovetail slots with 5/8" (16mm) throat opening.

Figure 4.25.6 *(By permission from Dur-O-Wall, Inc., Arlington Heights, Illinois.)*

4.25.7 Seismic Masonry Ladur and Comb Reinforcement

D/A 360 S SEISMIC LADUR-EYE

D/A 370 S SEISMIC DUR -O-EYE

D/A 320 S SEISMIC LADUR

D/A 5213/Seismic 5213S

Recommended for brick cavity walls with or without insulation. Dual leg $3/16$" (5mm) pintle adjusts vertically $1-1/4$" (30mm), up or down. The plate projects off the back-up wall to accommodate insulation, or bridge cavities. Hot dip galvanized, and stainless steel finishes available.

Seismic Comb (patent pending)

Masonry confinement reinforcement located in horizontal mortar joint to improve seismic performance of shear walls. Provides the Vertical Rebar confinement requirements in Section 2108.2.5.6 (1994). Made with $3/16$" diameter wire conforming to ASTM A82. A hot dipped galvanized finish (1.5 oz., 458 g/m2, zinc per square foot), per ASTM A153, is standard. Available for 6" (150mm), 8" (200mm), 10" (250mm) and 12" (300mm) hollow masonry units.

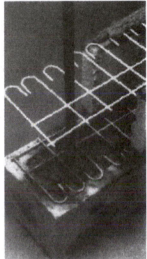

Figure 4.25.7 *(By permission from Dur-O-Wall, Inc., Arlington Heights, Illinois.)*

4.26.0 Investigating Unstable Masonry Conditions to Prevent Failures

Although masonry walls are extremely durable, "old age" and neglect can take its toll on even the most durable structure. When inspecting a masonry facade for potential problems and restoration, a number of contributing factors must be considered. Often, it is necessary to cut out a small section of wall in the area/areas where failures are suspected.

The following checklist will aid in this investigation:

1. When initially built, were all ties and anchors installed as required?
2. Were the ties properly installed (e.g., embedded adequately in the bed joint and connected to the backup correctly)
3. Does there appear to be excessive differential wall movement caused by thermal movement, settlement, or freeze/thaw conditions?
4. Were the proper size and type of ties/anchors used to avoid stresses that exceed the facade materials' capacity?
5. Were the proper type of expansion and control joints installed at the proper distances?
6. Have the ties, anchors, fasteners, relieving angles, and lintels corroded because of moisture being trapped? Is there accelerated corrosion from chlorides or has galvanic action taken place because of a combination of carbon steel anchors in contact with dissimilar materials?
7. Has excessive water penetrated the wall system from any poorly maintained parapet flashings or roof-coping flashings?
8. Have the caulk joints been allowed to deteriorate?
9. Have the weep holes been caulked when maintenance caulking was performed and have the lintels been caulked at the point where brick is bearing on them?
10. Have the mortar joints deteriorated and not been tuckpointed during routine maintenance inspections?

4.26.1 Restabilizing and Reanchoring a Masonry Veneer Wall System

At times both new and old brick veneer walls may require reanchoring to provide structural stability and ensure watertight integrity.

In new construction, after the veneer wall has been built it may be determined that the desired quality level of construction had not been achieved and additional anchors may be required. Corrosion of the existing wall ties over the years may also call for restabilization of the brick veneer wall assembly. In order to establish the proper method to achieve stabilization in either new or older brick veneer wall assemblies, a structural engineer will most likely consider the following:

- Relative stiffness of the existing veneer;
- The construction of the veneer wall, i.e., CMU back-up, steel stud assembly;
- Whether the anchoring device should be a friction fit, mechanically activated, or an adhesive device;
- Load versus deflection characteristics of the anchoring device.

Changing seismic code requirements have made it necessary to reinforce existing brick veneer wall assemblies and stabilization anchors afford the contractor a relatively easy way to fulfill these code requirements.

Also, renewed interest in preserving old, historic, or architecturally important structures has been another impetus behind product development to aid restoration contractors.

The various types of veneer anchors displayed on the following pages have been developed by the Dur-O-Wall Company, and several other leading masonry accessories manufacturers have also developed masonry restabilization products that basically function in a similar manner.

Substructures—Piles, Caissons, Slurry Walls, and Rock and Soil Anchors

Contents

5.0.0 Site Preparation

Generally, specifications will require that excavation be carried out to at least pile cutoff grade before piles are driven. There may be conditions such as extreme variations in pile cutoff grades, unusual subsoil conditions, or closely spaced pile groups which make excavation to below pile cutoff impractical before piles are driven. If piles are to be driven through overburden (soils which will subsequently be removed) the inspector should be aware of the possible effects of driving piles through overburden soils on such things as pile butt location, pile payment, the cutting off of piles, or the use of followers. There should be agreements as to the responsibilities of the parties concerned with these matters.

Excavation banks should be properly sloped or adequately sheeted and braced to take all possible surcharge loads during and after pile driving and to protect the piles from detrimental ground movements. Groundwater and surface water should be controlled to provide reasonably dry work areas.

The various types of concrete piles are shown in Fig. 5.0.0.1. Other pile types are discussed under Special-Type Piles. These include bored piles, auger grout piles, cast-in-place piles, compacted concrete piles, enlarged base piles, and minipiles.

Subsoil Information.   Pile foundations should not be designed or installed without adequate subsoil information. This is obtained by conventional test borings from which soil and rock samples are recovered for analyses and tests. To be meaningful, the depth to which test borings are taken must be below the depth to which piles may be installed. Various types of soil tests can also be made in the borehole to determine soil properties. The inspector should have access to the subsoil information and be familiar with existing subsurface conditions which may affect pile installation.

It should be understood that test boring data show only conditions existing at the location of the boring and that conditions can vary widely between boreholes. A general knowledge of the geological history of the area and its previous development is useful in anticipating possible variations in subsoil conditions.

Underground Structures.   It is not the responsibility of the inspector to determine the existence, location, or condition of underground structures including pipelines. However, the inspector should be advised of such, should be alert to potential damage during pile installation, and should call the contractor's attention to any dangerous situation which seems to be developing.

Pile Material

Inspection of pile material involves various items to ensure that material meets the requirements of the specifications and conforms with material on which the design was based. The basic pile materials are timber, concrete, and steel, and each has its unique inspection requirements.

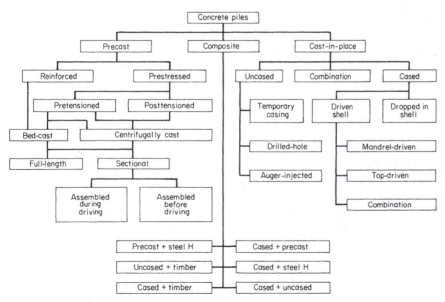

Figure 5.0.0.1   Various types of concrete piles. (*Courtesy: McGraw-Hill: Field Inspection Handbook; Brock, Levy, Sutcliffe*)

5.1.0 Timber Piles

Species. The species will often be specified since strength and other properties vary with species of timber. In some cases the subspecies may also be specified. The species should be noted on inspection reports received from the point of origin of each pile shipment.

Conditioning and Preservative Treatment. Adequate inspection certificates should accompany each shipment of piles treated with a preservative. Certificates should show the type of conditioning used, such as air-dried, kiln-dried, steam, Bolton, or heating, in the preservative, and the type and retention of the preservative. These should be as specified. The type of conditioning as well as the temperature and duration of the conditioning process could seriously affect the strength of the timber. It should be noted that the American Wood-Preservers' Association (AWPA) specifications prohibit steam conditioning of certain species such as Pacific coast Douglas fir and oak. For all piles, the temperature and duration of the conditioning process are limited by AWPA specifications.

Peeled or Unpeeled. Sometimes specifications will permit unpeeled piles, but in general, and especially for all treated piles, piles will be clean-peeled, which requires the removal of all outer bark and at least 80 percent of the inner bark. Unpeeled piles require no bark removal, and rough-peeled piles require complete removal of the outer bark only.

Lengths. Timber piles will be shipped to the jobsite according to approved ordered lengths. Standard brands on each pile should indicate the pile length. Piles shorter than those ordered are not acceptable, with the exception of piles specified according to American Society for Testing and Materials (ASTM) D25, Standard Specifications for Round Timber Piles, which may be 6 in shorter or, depending on the ordered length, may be from 1 to 2 ft (0.3 to 0.6 m) longer. Payment for timber piles is frequently based upon the full furnished lengths, and thus delivered lengths must agree with ordered lengths. Ordered lengths may be somewhat longer than the anticipated driven lengths to allow for extra lengths which may be required owing to subsoil conditions that vary from those indicated by test borings.

Dimensions. Typically the minimum diameter at the pile tip or butt, or both, is specified. ASTM D25 is essentially developed by the timber industry and reflects current practice as set by that industry. According to ASTM D25-88, timber piles are classified as either friction piles or end-bearing piles. For friction piles, a minimum butt circumference is specified, and the minimum tip circumference results from the pile length. The pile inspector should understand that, regardless of current ASTM specifications, the engineer is at liberty to specify any wanted dimensions or other properties for timber piles and the owner is entitled to receive piles as specified, regardless of industry standards.

Taper and Straightness. Timber piles should have a continuous taper from tip to butt, and the piles should be sufficiently straight so that a line from the center of the butt to the center of the tip lies entirely within the body of each pile. Short crooks or reverse bends could cause problems during driving, and piles with these defects should be carefully checked to see that they comply with specifications on straightness.

Knots. The pile inspector should be familiar with ASTM D25-88. It requires that sound knots be no larger than one-sixth the circumference of the pile located where the knot occurs; cluster knots are considered as a single knot; and the entire cluster cannot be greater in size than permitted for a single knot; the sum of knot diameters in any 1-ft (304-mm) length of pile should not exceed one-third of the circumference at the point where they occur.

Piles may have unsound knots not exceeding half the permitted size of a sound knot, provided that the unsoundness extends to not more than $1\frac{1}{2}$-in (38-mm) depth and that the adjacent areas of the trunk are not affected.

Checks, Shakes, and Splits. The extent of checks, shakes, and splits is limited by ASTM D25. However, they are permitted by D25 to some degree, and for timber piles that are not required to conform to ASTM D25 severe limitation on these types of defects should be imposed or the defects prohibited. Such initial defects in a pile may be made more severe during driving and, in the case of treated timber piles, could render the preservative treatment ineffective.

Soundness. There should be no evidence of decay or insect attack. The inspection certificate obtained from the point of origin or the treatment plant should cover this point, but the inspector should spot-check.

Special Fittings. Specifications may require that pile tips be fitted with steel drive shoes or that during driving pile butts be protected with steel bands. All special fittings should comply with the specifications.

(Courtesy McGraw-Hill: Field Inspection Handbook; Brock, Levy, Sutcliffe)

5.1.1 Precast Concrete Piles

Design Requirements.   Adequate plant inspection reports should accompany each pile shipment, identifying the piles and certifying that they meet the design specifications including such things as the amount of reinforcing steel, 28-day concrete strength, and effective pre-stress. Piles should be marked or stamped with the date of manufacture. Inspection reports should come from an independent testing agency and not from the manufacturer.

Lengths.   Precast piles will be shipped to the jobsite according to specified or approved ordered lengths. Each pile should be of the full ordered length except when sectional-type piles are permitted. Sometimes piles are ordered with sufficient extra length to permit stripping back the concrete and exposing the reinforcing steel for the pile-to-cap connection (see under Pile Installation). Ordered lengths may be somewhat larger than anticipated driven lengths to allow for variations in subsoil conditions.

Dimensions.   Piles should be of the shape and size specified.

Tolerances.   Piles should be straight within specified tolerances. Butt ends should be square to the longitudinal axis and free of any major surface irregularities.

Chamfers.   All corners or edges of square piles should be chamfered. The width of the chamfer should be limited to about 1.5 in (38.1 mm) so that the reduction in any side dimension due to chamfer is not more than about 2 in (50.8 mm).

Damage.   Check piles for detrimental cracks, spalling, slabbing, or other damage. Hairline cracks are normal but should not be too numerous.

5.1.2 Concrete for Cast-In-Place Piles

Design Mix.   A design mix with results of tests on standard cylinders should be furnished by the contractor. Copies of these data should be made available to the inspector at the start of pile installation.

Concrete Production Facilities.   Concrete may be mixed in portable mixers brought to the pile locations, but generally it will be ready-mixed. Ready-mix concrete may be (1) batched and mixed at a central plant and delivered to the pile locations in agitating or nonagitating trucks (central-mixed), (2) batched at a central plant and mixed in a truck mixer in transit to or after reaching the jobsite (truck-mixed), or (3) partially centrally mixed with mixing completed in a truck mixer in route to the job or on the jobsite (shrink-mixed). The central plant may be located on the jobsite. The concrete batch and mixing plant should be inspected for adequacy of storage facilities for materials, accuracy, and reliability of batching equipment, condition of mixing equipment, and proper operational procedures.

Storage Facilities.   Cement must be kept dry whether it is stored in bulk containers or in bags. To avoid contamination, stockpiles of aggregates that have been cleaned, graded, and prepared for batching should be on a hard, clean base with the area around the stockpiles spread with a bedding material of sand, gravel, or rock. Side slopes of stockpiles should not exceed 7 in/ft (583 mm or less per meter) to prevent segregation. Coarse aggregate should be separated by type and size gradation. Overlapping of stockpiles should be prevented and suitable drainage should be provided. All reasonable precautions should be taken to keep the moisture content of aggregates as nearly uniform as possible.

Batching Equipment.   Concrete is usually batched by weight. Batching scales should have a recent calibration and certificate of inspection and must be clean and free of interference by other objects. Separate weight-batching facilities should be provided for cement. Batch-weight recording and cutoff devices must operate accurately. The bottom of batch bins must be fully sloped in all directions. Water-metering devices, whether at a central mixing plant or mounted on a truck mixer, must be accurate and equipped with indicating dials and totalizers.

Mixing Equipment.   All mixing equipment, whether stationary or truck-mounted, must be in good operating condition. The interior of drums should be clean, and mixing blades should not show signs of wear in excess of 1 in (25.4 mm). Truck mixers must be equipped with a reliable revolution counter.

Operations.   All materials must be accurately batched, and batching should be by weight. Admixtures, if required, must also be accurately measured. Mixing drums must be cleaned after each use to prevent an accumulation of hardened concrete on the blades. All washwater must be removed from the mixing drum prior to batching. Cement should be used on the basis of first in-first out. The free-water content of the aggregates should be included as part of the total mix water. Aggregates should be allowed sufficient time to drain, and it may be necessary to have a moisture meter in the sand batcher to monitor moisture content. Proper equipment and methods must be used for handling aggregates to avoid segregation and breakage. Segregation of coarse aggregate can be reduced by separating it into several size fractions and batching them separately. Finished screening of aggregates at the batcher is recommended to avoid problems of segregation and contamination.

Concrete Materials.   Materials including cement, sand, coarse aggregate, and water should be inspected for compliance with specifications and accepted practice.

Cement.   Cement must be of the type specified or permitted with the approval of the engineer. Mill certificates should be furnished to show that cement conforms with the requirements of the specifications and ASTM C150, Standard Specifications for Portland Cement. Type IV cement should not be used for pile concrete. Type III, or high-early, cement may be permitted for cast-in-place concrete test piles to get a fast gain in strength. Type II or Type V cement may be specified for sulfate exposure.

Cement remaining in bulk storage for more than 6 months or cement stored in bags longer than 3 months should be retested before use to ensure that it meets the requirements of ASTM C150. Cement should not be used directly from the mill if it is still hot. The cement should be allowed to cool before using to reduce the possible occurrence of false sets.

Cement should be inspected for lumps caused by moisture. Cement bags should be inspected for rips, punctures, or other defects. If cement is to be batched by bag, the weights of bags should be spot-checked and should not vary by more than 3 percent.

Sand.   Sand should be clean, sharp, well graded, and free of silt, clay, or organic material. The specific gravity and/or fineness modulus may be specified for special mixes such as reduced coarse aggregate concrete.

Coarse Aggregate.   Specifications may permit gravel or crushed stone. The use of crushed rock aggregate requires more cement and sand for comparable workability. Air entrainment also improves workability. Lightweight aggregates are not recommended, and slag aggregates are not generally used. Alkali-reactive aggregates or aggregates from shales, friable sandstone, chert, and clayey or micaceous rock should not be permitted. Aggregates should be uncoated and free of silt, clay, organic material, and chemical salts. The specific gravity of the coarse aggregate may be specified. Aggregates should be well graded with a maximum size of $3/4$ in (19.05 mm) and with the amounts of aggregates less than $3/16$ in (4.762 mm undersize) held uniform and within 3 percent.

Water.   As a general rule, mix water should be potable. It should contain no impurities which would affect the quality of the concrete. It should not have a sweet, saline, or brackish taste or contain silt or suspended solids. Very hard water may contain high concentrations of sulfate. Well water from arid regions may contain harmful dissolved mineral salts. If questionable, the water can be chemically analyzed. The quality of the water can be checked by comparing the strength of concrete reached at various ages for a mix using the water of unknown quality with the results of similar age tests on a mix made with water which is known to be acceptable. Impurities in mix water may affect both the compressive strength of the concrete and its setting time.

Admixtures.   The authorized or mandatory use of admixtures will be noted on the mix design report. Special admixtures such as retarders and fluidizers may be required for pumped concrete.

Cold-Weather Operations.   The minimum temperature of fresh concrete as mixed should be about 45°F (7.2°C) for air temperatures above 30°F (−1.1°C), 50°F (10°C) for air temperatures from 0 to 30°F (−17.2 to −1.1°C), and 55°F (12.7°C) for air temperatures below 0°F (−17.2°C). Frozen aggregate or aggregates containing lumps of ice should be thawed before being used. It may be necessary to preheat the mix water and/or the aggregate. For air temperatures between 30 and 40°F (−1.1 and 7.2°C), it is usually necessary only to heat the water to a maximum of about 140°F (60°C). For air temperatures below 30°F (−1.1°C), the water can be heated to 140 to 212°F (60 to 100°C) and the aggregate to about 45 to 55°F (7.2 to 12.7°C). Overheating should be avoided. If both the mix water and the aggregates are preheated, it is recommended that the water be mixed with the aggregates before adding the cement to avoid a flash set. The temperature of the water-aggregate mixture should not be higher than 80°F (26.6°C) and preferably about 60°F (15.5°C).

Hot-Weather Operations.   If the temperature of the concrete during mixing is above 80°F (26.6°C), it could result in increased water demand (slump loss) or an accelerated set. The easiest way to control and reduce the concrete temperature is by using cold mix water, which can be achieved by mechanical refrigeration or by using crushed ice as part or all of the mix water. Mixing time should be kept to a minimum, and mixing drums, water tanks, and pipe should be painted white.

Mixing Time.   Mixing time starts when the water is added to the mix and should be adequate but not excessive. Minimum mixing times vary with the size and type of the mixer and range from 1 to 3 minutes. Maximum mixing times can range from 3 to 10 minutes. For stationary mixers, minimum mixing time can be established by tests on mixer performance. For truck-mixed concrete, complete mixing requires from 50 to 100 revolutions of the drum at mixing speed. Check the manufacturer's plate on the mixer. If, after mixing, drum speed is reduced to agitation speed or stopped, the drum should be rotated at mixing speed for from 10 to 15 revolutions just before concrete is discharged.

Elapsed Time.   For normal temperatures, the total time from start of mixing to discharge should not exceed about $1\frac{1}{2}$ hours and should be reduced as temperatures increase. The mix should be discharged before 300 revolutions of the drum.

Slump.   Slump tests should be made periodically in accordance with ASTM C143, Standard Method of Test for Slump of Portland Cement Concrete, to ensure that concrete has the specified slump for proper placement in pile casings, shells, or holes. The slump for concrete as delivered to the top of the pile casing or hole should be 5 in (127 mm) for conventional concrete or 4 in (101.6 mm) for reduced coarse aggregate concrete, both with a tolerance of $+2$ in, -1 in ($+50.8$ mm, -25.4 mm). Special-type piles may require concrete having different slumps. See Special-Type Piles. Sometimes it is advisable to check the slump just before adding the final water at the jobsite to avoid too high a slump or a wet mix.

Slump Loss.   Slump loss can be caused by overmixing, hot weather, pumping through long lines, or delays in delivery and placement of concrete. Overmixing can and should be avoided. If necessary, all the mix water can be added and all mixing done upon delivery at the jobsite. This could prevent overmixing and may help in eliminating slump loss due to hot weather. If concrete is to be pumped to the pile locations, the slump should be increased without changing the water-cement ratio or concrete strength to compensate for slump loss during pumping. All preparations should be made for depositing concrete upon delivery, and delivery schedules should be arranged to eliminate delays in placing concrete.

Retempering.   The addition of water to the concrete mix to compensate for slump loss resulting from delays in delivery or placing is permissible provided the design water-cement ratio is not exceeded and the concrete has not attained its initial set. Initial set is not to be confused with a false set, when the concrete appears to stiffen but can be made workable with agitation.

Delivery Tickets.   A delivery ticket must accompany each load or batch of concrete. The delivery ticket is for the purchaser, but the inspector should be furnished a copy. It should include sufficient data to identify the producer, project, contractor (purchaser), truck mixer used, and specified concrete mix or strength. Other information which should be on the delivery ticket is the date of delivery, type and brand of cement, maximum aggregate size, weights of cement, sand, and coarse aggregate, type and amount of admixtures, quantity of water, time batched, reading of revolution counter and time when water was first added, volume of batch, and amount of water added by the receiver. The inspector should note the times of delivery and placement and the air temperature.

Concrete Strength.   Standard cylinders for compression tests should be made periodically or as specified in accordance with ASTM C31, Standard Method for Making and Curing Concrete Test Specimens in the Field, to ensure that concrete of required strength is being furnished. The frequency for making test cylinders will vary with the job size and other factors, but generally a test set (minimum of two cylinders) should be made for each daily pour or for every 50 yd^3 (38.2 m^3) placed. Also a test set should be made for each age at which compression tests are to be run. The inspector should ensure that cylinders are properly cast, handled, stored, sealed, packaged for shipment, and shipped so as not to invalidate test results. For strict concrete control, test specimens should be cast in cast-iron or tin-can molds. Although widely used, cardboard or paper molds are not recommended for molding test cylinders for strict concrete control. If cardboard molds are used, they should conform with ASTM C470, Specifications for Molds for Forming Concrete Test Cylinders Vertically. Jobsite curing or the use of cardboard molds may contribute to low strength-test results. Grout strengths for special-type piles are determined by standard cube tests in accordance with ASTM C91, Specification for Masonry Cement. (See Auger-Grout Pile, Cast-in-Place Pile, and Minipiles under Special-Type Piles.)

Results of Tests.   The pile inspector should be furnished with copies of the results of all concrete compression tests as called for in the specifications. It is advisable to obtain 3- and 7-day results at the beginning of the job in order to detect trends in concrete strengths. The results of 7-day tests are also valuable in monitoring concrete strength trends as the job progresses so that, if necessary, remedial measures can be taken before too much concrete is placed.

Strength Variations.   Variations in concrete strength as determined by standard cylinder tests are normal. Several criteria are used to determine the acceptability of variations. For example, the concrete is considered satisfactory if the average of three consecutive tests is equal to or greater than the required 28-day strength and no test falls below the required 28-day strength by more than 500 psi (3447 kPa). Another acceptance criterion is that 80 percent of the tests show strengths greater than the design strength and that not more than 1 test in 10 is less than the required 28-day strength. A third is that the average strength from consecutive tests is greater than the required 28-day strength. ACI 214.3-88 provides recommendations on evaluating concrete strength-test results. If the test results show that concrete strength is below that specified, the cause of low-strength concrete should be investigated. Low strength could be caused by unsatisfactory materials, by improper batching and mixing, or by the use of excess water in the mix. Low cylinder breaks could also result from improperly preparing, curing, handling, or testing cylinder specimens.

Verification of Concrete Strengths.   If the results of standard cylinder tests are low, cores can be removed from piles for testing. Core tests are considered satisfactory if the average of three cores is equal to or greater than 85 percent of the required 28-day strength and if no core strength is less than 75 percent of the specified 28-day strength. The results of tests on cores are normally lower than those on standard cylinders owing to microfracturing of the concrete. It should be noted that pile concrete in a long steel shell embedded in the ground will cure at a rate slower than that for test cylinders or exposed concrete. Curing conditions are ideal, but the rate of strength gain is lower than normal. Concrete strength in completed piles can also be checked by various nondestructive methods such as penetration-resistance tests. See ASTM C803, Method of Tests for Penetration Resistance of Hardened Concrete.

5.1.3 Steel Pipe and Tube Piles

Material Specifications.   Mill certificates or laboratory test reports should be furnished to show that the pipe or tube conforms to the required material specifications covering yield strength. Pipe may be specified by grade with reference to ASTM A252, Standard Specification for Welded and Seamless Steel Pipe Piles.

Lengths.   Pipe pile material may be delivered in random or double random lengths or in ordered lengths. The lengths of tapered portions of tube piles will be determined by the type and size specified. The lengths of extension pieces for tube piles will be as ordered in 10- to 40-ft (3- to 12-m) lengths in 5-ft (1.5-m) increments. Generally payment for pipe and tube pile material is made for the quantity incorporated in the finished foundation. However, if pile material is to be paid for on the basis of approved ordered or furnished lengths, the total delivered lengths of pipe or tube piles should be checked. Extra pieces of pipe and extra lengths or pieces of tube piles may be ordered to provide for piles that may drive longer than anticipated.

Dimensions.   Diameter and wall thickness or gauge should meet the minimum specified. Although a minimum wall thickness or gauge may be specified, this does not relieve the contractor from responsibility to furnish thicker walls or heavier gauges if necessary to install the piles properly without damage or to achieve required pile penetration. For tapered tube piles the type of lower pile section (designating the degree of taper) should be as specified.

Pile Fittings.   Special closure plates or points may be required by the specifications. When pipe piles are closed with a flat steel plate, the diameter of the closure plate should not be more than $1/4$ in (19.05 mm) larger than the outside diameter of the pipe. The specifications may permit splicing pipe piles with steel sleeves with or without welding. Splicing sleeves should conform with the specifications.

Welding.   The welding on of boot plates at the tips of pipe piles will be done on a horizontal rack in advance of pile driving, or pipe may be shipped to the job with closure plates attached. Tube piles will be furnished with closure points welded on by the manufacturer. Splice welding of pipe or tube piles may be done on a horizontal rack or "in the leaders." See Pile Splicing. The qualifications of welders and all welding should conform with American Welding Society (AWS) D1.1, Structural Welding Code.

5.1.4 Steel H Piles

Material Specifications.   Mill certificates or laboratory test reports showing that the material meets the specifications as to type of steel and properties including yield strength should be furnished. If the required yield strength is higher than 36 ksi (248.2 MPa), the identity of all steel must be continuously maintained from the point of manufacture to the pile location, or coupons must be taken and tested from each piece to see that it meets the specifications.

Lengths.   If pile material is to be paid for on the basis of approved ordered or furnished lengths, the total delivered lengths of H piles should be checked. Piles will generally be shipped in mill lengths. Extra lengths of H piles may be ordered to provide for driving piles longer than anticipated.

Dimensions.   The pile size, weight per foot, and thickness should meet the minimum requirements of the specifications. Piles should be of H section conforming with American Institute of Steel Construction (AISC) HP shapes.

Types.   Specifications may permit either rolled or built-up shapes.

5.2.0 Material Handling and Storage—Timber Piles

Unloading.   Timber piles may be unloaded by controlled roll-off. Dumping should not be permitted.

Handling.   Generally treated timber piles should not be handled with timber tongs, cant hooks, peaveys, or pile chains. Piles should be handled so as to avoid puncturing or breaking through their outer treated portion. AWPA standard M4-80 permits the use of pointed tools provided that the side surfaces of the pile are not penetrated more than $1/2$ in (12.7 mm). This may be difficult to control. Treated timber piles should not be dragged along the ground.

Storage.   Timber piles in storage for any length of time should be on adequate blocking and supported to avoid permanent bends. Piles should be stacked on treated or nondecaying material and with an air space beneath them. Storage areas should be free of debris, decayed wood, and dry vegetation (this presents a fire hazard) and should have sufficient drainage to prevent the piles from lying in water.

5.2.1 Precast Piles

Unloading.   Precast piles should be unloaded by lifting them in a horizontal position. Dumping or rolling off the precast piles should not be permitted.

Handling.   Precast piles should be handled with proper slings attached to designated pickup points or inserts. Impact loads should be avoided.

Storage.   If precast piles are stored on blocking, it should be placed at designated support points to avoid overstressing and cracking the piles.

5.2.2 Steel Pipe and Tube Piles

Unloading.   Controlled dumping or roll-off unloading of pipe or tube piles may be permitted.

Handling.   Sufficient pickup points should be used to avoid bends in pipe or tube piles. A closed-end pile should not be dragged along the ground with the open end first.

Steel H Piles

Unloading.   H piles should be unloaded by lifting them in a horizontal position. Dumping piles should be prohibited.

Handling.   H piles lifted in a horizontal position should have their webs vertical to avoid bending. Coated H piles must be carefully handled so as to avoid damage to the coating.

Storage.   H piles should be stored on adequate blocking. Nesting of piles with their flanges vertical is recommended.

Pile Shells

Unloading.   Dumping of pile shells should not be permitted, but they may be roll-off unloaded.

Handling.   Pile shells should be handled at all times so as to avoid permanent deformations. A closed-end shell should not be dragged along the ground with the open end first.

Storage.   Pile shells should be stored out of mud or standing water. If in storage for a long period of time, shells should be protected from the elements.

5.2.3 Handling Cement and Concrete for Pilings

Cement

Storage.   Bag cement must be stored off the ground on adequate racks and protected from the elements, especially moisture.

Concrete Aggregates

Handling.   Aggregates should be handled so as to avoid breakage, segregation, and contamination. The required gradation must be maintained.

Storage.   See Concrete Production Facilities: Storage Facilities under Pile Material.

5.2.4 Handling Reinforcement

Reinforcement

Handling.   Reinforcing steel should be handled in bundles with appropriate lifting slings located at sufficient pickup points to avoid permanent bending. Bundles should not be broken until the steel is to be used. All necessary precautions must be taken to maintain the identification of the steel after the bundles have been broken. This can be done by keeping the steel separated according to type, size, and length with a tagged piece in each stack.

Storage.   Reinforcing steel should be stored off the ground on suitable racks or blocking so as to avoid permanent bends. The steel should be stored so as to prevent excessive rusting and contamination by dirt, grease, or other bond-breaking coatings.

5.3.0 Pile Equipment

Figure 5.3.0.1 illustrates the basic parts of a typical pile-driving rig.

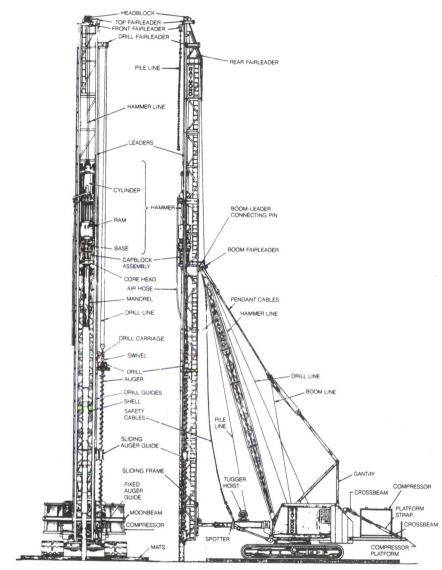

Figure 5.3.0.1   Basic parts of typical pile-driving rig. (*Courtesy: McGraw-Hill: Field Inspection Handbook; Brock, Levy, Sutcliffe*)

Leaders

Leaders hold the pile and hammer in proper alignment to avoid eccentric hammer blows, which could damage the pile, drive it off location, or reduce the driving energy. They also serve to hold the pile in the specified alignment and provide means for supporting the pile during driving. The leaders, in combination with the boom, spotter, moonbeam, and sliding frame, also permit driving the pile on any required angle or batter. Leaders should be of sturdy construction and fixed at two points, normally the boom tip and the spotter. The use of swinging leaders or spud-type leaders is generally not recommended except possibly for small jobs. Leaders should be equipped with some type of extension which will permit the hammer to travel below the bottom of the leaders during driving. Leaders also serve as a mounting for ancillary equipment such as a jet or a drill.

Preexcavation Equipment

The jet or drill should be mounted on the leaders on the same driving radius as the hammer and pile to assure proper positioning for pile alignment and location. If necessary to avoid whipping, the jet pipe or drill stem should be supported by some type of traveling guide. Drill bits or augers should be of the specified size to avoid overdrilling. Drill stems should be of adequate size to handle the necessary water pressures and volumes for effective preexcavation and removal of soil cuttings for the size of hole drilled. Drill stems, kelly bars, and jet pipes must be long enough to reach the necessary preexcavation depths. Jet pumps should be of sufficient size and capacity to provide the required volume of water at the necessary rates and pressures.

Power Source

The power source used to operate the pile-driving hammer such as a boiler or a compressor should be of adequate capacity. Boilers can be sized by horsepower, which can be based on the pounds of steam consumed per hour, or the ASME rating or on the square footage of heating area. Compressors will be rated according to the standard volume of compressed air in cubic feet produced per minute (1 ft^3/minute = 0.000472 m^3/second). The hammer manufacturer's data can be used as a guide for determining the required boiler or compressor capacity. However, the actual required capacity will depend upon such things as the length and size of piping and hose connecting the hammer to the power source, the condition of the hammer (wear of piston rings, etc.), and leaks in the system. For air-operated hammers, the consumption figure normally reported is based upon adiabatic compression and the required pressure at the hammer. However, as the air is compressed, it is cooled down for practical reasons. Further heat losses occur before it reaches the hammer. As the air cools, the volume decreases and the pressure drops. Therefore, unless the air is reheated after normal compression, the output of the compressor will be from 30 to 50 percent greater than the adiabatic consumption figure normally reported. The condition of the boiler and the quality of water used can affect the volume of steam produced and therefore the effective capacity of the boiler.

The manufacturer's data will generally specify the minimum pressure required at the hammer for operation at normal efficiency. Adequate pressure at the hammer is important for double-acting, differential, or compound hammers. It is impractical to measure the pressure at the hammer, and therefore it is customary to mount the pressure gauge at the boiler or compressor discharge. The pressure as measured at the boiler or compressor should be higher to provide for line losses. For steam, the amount of excess pressure would depend upon the length and sizes of piping and hose. With the boiler mounted on the rig and with recommended pipe and hose sizes used, the normal minimum required increase in pressure is about 15 psi (10.34 kPa). For air, friction losses in the line tend to be compensated for by pressure gains due to the friction heating the air. There are no energy losses and no reduction in ability to do work resulting from friction losses. Hydraulic pile hammers operate on electrical power supplied by either a carrier unit or a separate power pack unit.

5.3.1 Pile-Driving Hammers

Hammer Types. Typical pile-driving hammers include drop, single-acting, double-acting, differential, compound, diesel, and vibratory hammers. In addition, there are special-type hammers such as air-gun, vibratory-impact, and electrohydraulic hammers.

Drop Hammer.   Drop hammers are rarely used to install foundation piles except for compacted-concrete piles as described in Compacted-Concrete Pile under Special-Type Piles.

Single-Acting Hammers.   Single-acting hammers are powered with compressed air or steam pressure, which is used to raise the hammer ram for each stroke. The delivered energy results from the kinetic energy developed by the gravity fall of the ram.

Double-Acting Hammers.   Double-acting hammers can be powered by steam but usually are powered with compressed air, which is used both to raise the ram and to accelerate its fall. This type of hammer exhausts at both the upstroke and the downstroke, and the operating pressure on the downstroke is applied to the full top area of the piston. Double-acting hammers have light rams and high speed and are not as effective in driving piles as hammers with heavy rams.

Differential Hammer.   Generally, differential hammers are powered by either steam or compressed air. However, this type of hammer has been designed and built to operate by high-pressure hydraulic fluid. The steam, air, or hydraulic fluid is used both to raise the ram and to accelerate its fall. Differential hammers exhaust during the upstroke, but those powered by hydraulics do not exhaust to the atmosphere, thus reducing noise and pollution. During the downstroke, the cylinder both above and below the piston is open to the operating fluid and thus under equal pressure. The accelerating downward force results from the operating-fluid pressure acting on the top of the piston, which is of larger area than the bottom of the piston (difference equals area of piston rod). These hammers have shorter strokes than comparable single-acting hammers and combine the advantages of the heavy ram of the single-acting hammer with the higher operating speed of the double-acting hammer.

Compound Hammer.   The compound hammer is a variation of the differential hammer utilizing compressed air or steam both to raise the ram and to accelerate its fall. The hammer exhausts during the upstroke, at the top of which both exhaust and inlet valves close and no additional motive fluid is introduced into the cylinder. The cylinder spaces both above and below the piston are interconnected, permitting the motive fluid to raise the ram and also to enter the top of the cylinder as it expands. After the piston or ram reaches the top of its stroke, the continued expansion of the motive fluid acting on the differential area between the top and the bottom of the piston accelerates the piston on the downstroke.

Diesel Hammer.   Diesel hammers are self-contained power units using the explosion of diesel fuel under the ram or piston both to raise the ram for the next stroke and to exert some push on the pile. The hammer is started by raising with a line the ram or piston to the top of its stroke and releasing it, permitting it to fall by gravity inside the cylinder. During the downstroke, the ram activates a fuel pump, which injects diesel fuel into the combustion chamber at the base of the cylinder just before the ram reaches the end of its stroke. Some hammers are of the impact-atomization type, whereas others have an atomized fuel-injection system. The ram compresses the air in the cylinder and the air-fuel mixture ignites under pressure and heat. The resulting explosion raises the ram for its next stroke. As long as sufficient pile penetration resistance is encountered, the hammer will continue to operate. If there is little or no resistance to pile penetration, the hammer will stop. The hammer can be stopped at any time by cutting off the fuel flow.

There are two basic types, single-acting and double-acting. For the single-acting type, the cylinder in which the ram operates is open at the top. For the double-acting type, the cylinder is closed at the top, and entrapped air above the ram is compressed as the ram rises. This compressed air may help to accelerate the fall of the ram but also shortens the stroke.

Hydraulic Hammer.   Figure 5.3.1.1 shows a Junttan HHK-6 hydraulic hammer. Hydraulic hammers operate by high-pressure hydraulic fluid used to raise the ram. Typically, the hydraulic ram free-falls with no hydraulic force acting. However, other hydraulic hammers, such as the Junttan HHKA series or the IHC hydrohammer, provide an additional force at the top of the stroke. The Junttan HHKA is accelerated with hydraulic pressure whereas the IHC utilizes nitrogen gas. In some hydraulic hammers the hydraulic lines create a small resistance to the free fall of the hammer. Many hydraulic hammers have adjustable weights and drop heights. It is important for the inspector to understand the operation of the hammer, and dynamic testing can be valuable in determining the actual energy transfer or efficiency of the hammer.

Vibratory Hammer.   Vibratory hammers or drivers apply a dynamic force to the pile from paired rotating weights set eccentrically from their centers of rotation and positioned so that when rotated the horizontal forces are canceled and the vertical forces are added. The oscillator containing the ro-

Figure 5.3.1.1 Junttan HHK-6 hydraulic hammer.
(*Courtesy: Norwalk Marine Corp., Norwalk, Conn.*)

tating weights must be rigidly connected to the pile to transmit the longitudinal vibrations effectively. Vibratory drivers are more effective for installing nondisplacement-type piles (or steel sheeting) in granular or cohesionless soils. The driving effectiveness of the vibratory hammer can be increased by applying a bias weight or force to the nonvibrating portion of the driver. Vibratory drivers of the low-frequency type have operating frequencies ranging from 0 to about 2000 vibrations per minute, and the oscillator is powered with an electric or a hydraulic motor. The high-frequency type operates at frequencies ranging from 0 to about 8000 vibrations per minute, which is within the resonant frequency range of most piles. The oscillators for these drivers are powered by either gasoline or diesel engines.

5.3.2 Hammer Energy

Pile-driving hammers are available in various sizes, ranging from a few hundred ft·lb (1 ft·lb = 1.356 J) per blow to over 800,000 ft·lb (1084.7 kJ). Normally foundation piles are installed with hammers ranging from about 10,000 to 50,000 ft·lb (13.56 to 67.79 kJ). Hammers smaller than about 10,000 ft·lb are used principally for underpinning work or for installing sheeting; hammers larger than about 50,000 ft·lb have been developed to drive long, heavy piles for offshore and other marine structures.

The required type and size (rated energy) of the pile-driving hammer will generally be specified. For vibratory drivers the required size may be stated in terms of dynamic force available at a specific frequency. Hammers somewhat larger than specified may be used, except that the hammer energy used to drive timber piles should be limited to about 15,000 ft·lb (20.34 kJ) per blow.

The energy delivered by an impact hammer depends upon the mass of the ram and its velocity at impact. Note that hammer energy is not a function of operating speed. Under field conditions it is very difficult to measure the terminal velocity of the ram.

Special equipment has been developed to measure the amount of hammer energy being delivered to the top of the pile. This equipment, known as a pile-driving analyzer (PDA), is sometimes used to monitor and control pile installation. It requires that transducers be attached near the top of the pile to measure the longitudinal strain and acceleration. The PDA then performs computations to predict transferred energy, driving stresses, and pile capacity. A more thorough description of the PDA is provided later in this chapter in the discussion on pile-load tests.

Drop Hammer.   The theoretical energy delivered by the drop hammer can be determined from the weight of the hammer and its fall. However, friction losses will occur, and other energy reductions could result from such things as the force necessary to overhaul the hammer line or the snubbing of the hammer just prior to impact.

Single-Acting Hammer.   The rated energy of a single-acting hammer can be checked by determining the weight of the ram from its material and volume and measuring the stroke. In addition to normal energy reduction due to mechanical losses in the hammer, the delivered energy could be affected by improper positioning of the valve trip wedges on the slide bar. Sufficient steam or air pressure must be furnished to raise the ram to the full height of its stroke, and the opening of the inlet valve must be timed so as to avoid back pressure. The energy of the single-acting hammer can be varied by several methods such as changing hammer rams, changing slide bars and the hammer stroke, adding or removing weight from the ram, or using a split trip combined with a multiple-wedge slide bar. The inspector should check to see that the hammer is set up to deliver the required rated energy.

Double-Acting Hammer.   For double-acting hammers, delivered energy depends upon the ram weight and stroke plus the force developed by the operating pressure. Although the correct pressure must be maintained at the hammers, it is very impractical to try to measure this pressure. However, gauges could be mounted at the power source or somewhere in the line with sufficient excess pressure furnished to compensate for line losses. The kinetic energy developed by any double-acting hammer cannot exceed the total weight of the hammer multiplied by the length of the ram stroke.

Differential Hammer.   As with the double-acting hammer, the delivered energy of the differential hammer is a function of operating pressure. To assure sufficient operating pressure, gauges can be mounted at the power source. The energy delivered by the differential hammer can be varied by changing slide bars and adjusting the stroke. Valve trips and slide-bar wedges must be correctly positioned to ensure a full stroke and avoid back pressure. Differential hammers are delivering their full available energy when the hammer base rises slightly during the downstroke.

Diesel Hammer.   The method used to determine the rated energy of diesel hammers varies within the industry. Many methods are based on the weight of the ram or piston and its stroke, but there is no uniformity on the lengths of strokes used in rating the hammers. Some manufacturers use the maximum stroke attainable, while others use a more conservative and realistic normal stroke. The actual stroke will depend upon the type of pile being driven and the driving resistance.

Theoretically, three basic forces result from the diesel cycle: (1) compression, (2) impact, and (3) combustion. To these might be added a fourth force resulting from the compression and subsequent expansion of the air at the top of the cylinder in a double-acting diesel hammer. During the compression cycle on the downstroke the impact of the ram is cushioned, resulting in some energy loss. This loss could be considered as being compensated for by the force gained from the explosion. Therefore, overall energy would be equal to the ram weight times its stroke. During the compression cycle on the upstroke for double-acting hammers, the ram slows down and the stroke shortens. The accelerating force resulting from the expansion of the compressed air could be considered to compensate for the shortened stroke.

The energy resulting from the explosion of the diesel fuel will depend upon the type and amount of fuel injected and the efficiency of combustion. The type and quantity of fuel used should be in accordance with the manufacturer's data. The fuel pump can be checked to see if it delivers the correct amount of fuel by disconnecting the pump and activating it for several strokes while measuring the quantity of fuel dispensed. When the driving resistance is low, when the hammer is driving on lightweight springy piles, or when the piston rings are badly worn, the full compression and explosive force cannot be achieved. Evidence of poor combustion is a black exhaust, whereas a light-blue exhaust indicates good combustion. Preignition which could cushion the hammer blow or prevent actual impact could occur if the wrong type of fuel is used. Fuel with too low a flash point could lead to preignition with reduced hammer performance. Preignition could also occur with impact atomization hammers owing to overheating.

Unless the rated energy is verified by testing, such as the pile-driving analyzer, it is recommended that the theoretical energy delivered by the hammer is computed from the weight of the ram and its stroke. All diesel hammers should be equipped with a stroke indicator. When the energy being delivered by a diesel hammer can be controlled by a throttle adjustment, the throttle must be set to deliver the required energy. Some diesel hammers are equipped with interchangeable pistons or rams. Others are convertible to either the double-acting or the single-acting mode. The inspector should check the piston weight being used and the mode of hammer operation.

Vibratory Driver.   The amount of dynamic force delivered by a vibratory driver will vary with the steady-state frequency. This frequency will depend upon the type and weight of the pile being driven, the type of soil into which the pile is being driven, and the available horsepower. A limiting factor may be the design of the oscillator itself. For electric-powered drivers the operating frequency can be adjusted by changing the sprocket ratios. The inspector should learn which sprocket ratio is being used. The available dynamic force is also a function of the eccentric moment being developed by the rotating weights, which in turn depends upon the mass of the weights and their eccentricity. For the high-frequency resonant driver, the available dynamic force will depend upon the type and size of rollers (eccentric weights) being used, which will determine not only the eccentric moment developed but also the available frequency. Check the type and size of the rollers in the oscillator.

5.3.3 Hammer Cushion (Capblock)

The hammer cushion, or capblock, is inserted between the striking part of the hammer and the pile or drive cap to protect the hammer from damaging impact stresses. However, the hammer cushion must be stiff enough to transmit the driving energy adequately to the pile. A commonly used capblock is a hardwood block with its grain parallel to the pile axis and enclosed in a tight-fitting steel sleeve. This type of capblock becomes crushed and burnt during pile driving, resulting in variations in elastic properties, and requires frequent changing. Many hammer cushions in use today are of laminated construction with alternating layers of aluminum and micarta disks or similar material. These capblocks are generally stiffer than the wood capblock and more efficiently transmit hammer energy to the pile. Also, these hammer cushions retain fairly constant elastic properties and are relatively long-lived.

The minimum type of capblock should be as specified. Capblock materials with low elastic moduli such as wood chips, pieces of wire rope, etc., are generally prohibited. The inspector should record the type and description of the hammer cushion used.

Pile Cushion

When precast concrete piles are being driven, pile cushions will be required to protect the heads of the piles from sharp impact stresses and to distribute the hammer blows uniformly over the tops of the piles. A pile cushion is placed between the drive cap and the head of the pile. A new pile cushion should be used at the start of driving each pile. The type of wood and thickness of the cushion should be as specified or with properties as used in the wave equation analysis used to establish driving criteria. The inspector should record the type and description of the pile cushion used.

5.3.4 Drive Cap (Drive Head, Bonnet)

Drive caps are steel castings or forgings which are attached to the hammer base and into which the head of the pile is inserted for driving. Drive caps used for timber, steel H, or tube piles should be of the correct size and provide full bearing over the entire cross section of the pile. Drive caps for precast-concrete piles should be sufficiently loose so as not to restrain the pile from its tendency to rotate during driving.

5.3.5 The Follower

The follower is a structural member used as an extension of the pile to drive the head of the pile below the ground or water surface or below the level to which the hammer can reach. The follower may also serve to accommodate reinforcing steel or prestressing strand which may project beyond the head of a precast pile for the pile-to-cap connection.

Followers should be of steel and sufficiently stiff to assure adequate transmission of hammer energy. The bottom of the follower should be formed like, or fitted with, a drive cap suitable for the type of pile being driven. A pile cushion may be required between the pile head and the bottom end of the follower. The use of a follower may be subject to the approval of the engineer of record. If a follower is used, it should be described as to material, size, length, and weight.

(Courtesy: McGraw-Hill: Field Inspection Handbook; Brock, Levy, Sutcliffe)

5.3.6 Impact Pile-Hammer Specifications, 4000 to 100,000 ft·lb[a]

Rated energy, ft·lb[b]	Model or size	Make or manufacturer[c]	Type[d]	Ram or piston weight, lb[e]	Total operating weight, lb[f]	Normal stroke, in[f]	Boiler capacity, hp	Air consumption, scfm[g]	Operating pressure at hammer, psi[h]	Required hose size, in[i]
100,000	40X	Raymond	S-A	40,000	62,000	30	250[j]	...[k]	135	4
93,220	MRBS-850	Menck	S-A	18,960	27,800	59	185	1950	142	4
91,100	K-45	Kobe	S-A-D	9,920	25,600	36	247	1833	150	3
90,000[l]	030	Vulcan	S-A	30,000	55,410	36	247	1903	150	3
90,000[l]	300	Conmaco	S-A	30,000	55,390					
87,000	D-44	Delmag	S-A-D	9,500	22,300					
84,000	M-43	Mitsubishi	S-A-D	9,460	22,660					
83,180	D-36	Delmag	S-A-D	7,940	17,700					
83,100	D36-02	Delmag	S-A-D	7,900	17,700					
81,250	8/0	Raymond	S-A	25,000	34,000	39	140[j]	...[k]	135	3
80,000	B45	BSP	D-A-D	10,000	27,500					
79,500	J-44	IHI	S-A-D	9,720	21,500					
79,000	K-42	Kobe	S-A-D	9,260	23,300					
75,000	30X	Raymond	S-A	30,000	52,000	30	200[j]	...[k]	135	3
75,000	B-500	Berming-hammer	S-A-D	6,900	16,500					
70,800	K-35	Kobe	S-A-D	7,720	18,700					
67,825	MRBS-750	Menck	S-A	16,500	25,520	49	185	1590	114	3
66,100	D30-02	Delmag	D-A-D	6,600	13,150					
63,900	B35	BSP	D-A-D	7,700	12,200					
63,500	J-35	IHI	S-A-D	7,730	16,900					
62,000	M-33	Mitsubishi	S-A-D	7,260	16,940					
60,100	K-32	Kobe	S-A-D	7,050	16,500					
60,000	S-20	MKT	S-A	20,000	38,650	36	190	NA	150	3
60,000[l]	020	Vulcan	S-A	20,000	42,020	36	217	1634	120	3
60,000[l]	200	Conmaco	S-A	20,000	44,560	36	217	1634	120	3
59,500	DE-70B	MKT	S-A-D	7,000	14,600					
59,500[m]	DE70B/50B	MKT	S-A-D	7,000	14,600					
56,900	22X	Raymond	S-A	22,050	31,750	31	100[j]	...[k]	135	3
56,875	5/0	Raymond	S-A	17,500	26,450	39	100[j]	...[k]	135	3
54,250	D-30	Delmag	S-A-D	6,600	12,300					
50,700	K-25	Kobe	S-A-D	5,510	13,100					

387

5.3.6 Impact Pile-Hammer Specifications, 4000 to 100,000 ft·lb[a] (Continued)

	Model	Manufacturer	Type							
50,200	200C	Vulcan	Diff.	20,000	39,050	15.5	260	1746	142	4
48,750[f]	016	Vulcan	S-A	16,250	33,340	36	210	1275	120	3
48,750[f]	160	Conmaco	S-A	16,250	33,200	36	198	1275	120	3
48,750	4/0	Raymond	S-A	15,000	23,800	39	85[j]	...[k]	120	2.5
48,750	150C	Raymond	Diff.	15,000	32,500	18	150[j]	...[k]	120	3
48,500	J-22	IHI	S-A-D	4,850	10,800					
48,500	D22-02	Delmag	S-A-D	4,850	11,400					
45,700	B25	BSP	D-A-D	5,510	15,200					
45,200	MRBS-500	Menck	S-A	11,000	15,200	49	80	1060	110	3
45,000	M-23	Mitsubishi	S-A-D	5,060	11,220					
43,400	N-60	VN-Vulcan	S-A-D	5,280	12,760					
43,400	DA-55-B	MKT	S-A-D	5,000	17,000					
42,500	DE-50B	MKT	S-A-D	5,000	12,000					
42,500[m]	DE70B/50B	MKT	S-A-D	5,000	12,600					
42,000[f]	014	Vulcan	S-A	14,000	27,500	36	200	1282	110	3
42,000[f]	140	Conmaco	S-A	14,000	30,750	36	179	1164	110	3
41,300	K-22	Kobe	S-A-D	4,850	11,700					
41,280	160D	Conmaco	Diff.	16,000	35,400	15.5	237[i]	1550	160	3
40,625	3/0	Raymond	S-A	12,500	21,225	39	70[j]	...[i]	120	2.5
40,625	125CX	Raymond	Diff.	15,000	32,800	15	150[j]	...[k]	120	3
40,625	125	Conmaco	S-A	12,500	21,430	39	119	940	125	2.5
39,700	D-22	Delmag	S-A-D	4,850	11,200					
37,500	S-14	MKT	S-A	14,000	31,700	32	155	NA	100	3
37,375[f]	115(K)	Conmaco	S-A	11,500	20,250	39	161	1066	120	2.5
37,375[f]	115(C)	Conmaco	S-A	11,500	20,780	39	99	910	120	2.5
36,000	140C	Vulcan	Diff.	14,000	27,984	15.5	211	1425	140	3
36,000	140D	Conmaco	Diff.	14,000	31,200	15.5	211	1425	140	3
32,549	N-46	VN-Vulcan	S-A-D	3,960	9,845					
32,500	2/0	Raymond	S-A	10,000	18,550	39	55[j]	...[k]	110	2
32,500	S-10	MKT	S-A	10,000	22,380	39	130	1000	80	2.5
32,500	010	Vulcan	S-A	10,000	18,750	39	157	1002	105	2.5
32,500[l]	100(K)	Conmaco	S-A	10,000	18,700	39	145	1002	100	2.5
32,500[l]	100(C)	Conmaco	S-A	10,000	19,280	39	85	820	100	2.5
32,000	DE-40	MKT	S-A-D	4,000	9,825					
30,225	OR	Vulcan	S-A	9,300	18,050	39	140	1020	100	2.5
30,000	520	Link Belt	D-A-D	5,070	12,545					

Rated energy, ft·lb	Model or size	Make or manufacturer[c]	Type[d]	Ram or piston weight, lb	Total operating weight, lb[e]	Normal stroke, in[f]	Boiler capacity, hp	Air consumption, scfm[g]	Operating pressure at hammer, psi[h]	Required hose size, in[i]
30,000	CPD-15	Bolt	S-A-pneu.	14,200	30,000	...	...	600	250	2
27,100	D-15	Delmag	S-A-D	3,300	6,600					
26,200	B15	BSP	D-A-D	3,300	9,000					
26,000	08	Vulcan	S-A	8,000	16,750	39	127	880	83	2.5
26,000[l]	80(K)	Conmaco	S-A	8,000	16,700	39	121	880	85	2.5
26,000[l]	80(C)	Conmaco	S-A	8,000	17,280	39	75	730	85	2.5
26,000	S-8	MKT	S-A	8,000	18,300	39	120	850	80	2.5
26,000	M-14S	Mitsubishi	S-A-D	2,970	7,260					
25,200	K-13	Kobe	S-A-D	2,870	8,000					
25,000	B-225	Berning-hammer	S-A-D	2,850	6,800					
24,600	N-33	VN-Vulcan	S-A-D	3,000	7,645	16.5	180	1245	120	2.5
24,450	80C	Vulcan	Diff.	8,000	17,885	16.5	...	...[k]	5100	
24,450	80CH	Raymond	Diff.-hyd	8,000	17,780	16.5	80[j]	...[k]	120	2.5
24,450	80C	Raymond	Diff.	8,000	17,780	39	128	841	80	2.5
24,375	0	Vulcan	S-A	7,500	16,250	39	50[j]	...[k]	110	2
24,374	1/0	Raymond	S-A	7,500	16,100	18	120	875	125	2.5
24,000	C-826	MKT	Comp.	8,000	17,750					
23,800	DE-30B	MKT	S-A-D	2,800	7,250					
23,800[m]	DE30B/20B	MKT	S-A-D	2,800	7,250					
23,800[n]	DA-35-B	MKT	S-A-D	2,800	10,000					
23,150[o]	MS300	MKT	S-A	6,614	9,800	420	50	750	115	2.5
22,500	D-12	Delmag	S-A-D	2,750	6,050					
22,400	DE-30	MKT	S-A-D	2,800	8,125					
21,000[n]	DA-35-B	MKT	D-A-D	2,800	10,000					
19,500	1-S	Raymond	S-A	6,500	12,500	36	40[j]	...[k]	104	1.5
19,500	65C	Raymond	Diff.	6,500	14,675	16	70[j]	...[k]	120	2
19,500	65CH	Raymond	Diff.-hyd	6,500	14,615	16	...	...	5000	
19,500[p]	06 & 106	Vulcan	S-A	6,500	11,200	36	94	625	100	2
19,500[l]	65(K)	Conmaco	S-A	6,500	11,200	36	94	650	100	2
19,500[l]	65(C)	Conmaco	S-A	6,500	12,100	36	94	650	100	2
19,200	65C	Vulcan	Diff.	6,500	14,886	15.5	152	991	150	2
19,150	11B3	MKT	D-A	5,000	14,000	19	126	900	100	2.5

5.3.6 Impact Pile-Hammer Specifications, 4000 to 100,000 ft·lb[a] (Continued)

Energy (ft·lb)	Mfr.	Model	Type							
18,200	Link Belt	440	D-A-D	4,000	10,300					
18,000	Link Belt	312	D-A-D	3,855	10,375					
17,000	MKT	DE-20B	S-A-D	2,000	6,450					
17,000[m]	MKT	DE30B/20B	S-A-D	2,000	6,450					
16,250	MKT	S-5	S-A	5,000	12,460	39	85	600	80	2
16,000	MKT	DE-20	S-A-D	2,000	5,375					
16,000	MKT	C-5	Comp.	5,000	11,880	18	80	585	100	2.5
15,100	Vulcan	50C	Diff.	5,000	11,782	15.5	125	880	120	2
15,000[p]	Vulcan	1 & 106	S-A	5,000	9,700	36	81	565	80	2
15,000	Raymond	1	S-A	5,000	11,000	36	40[j]	..[k]	80	1.5
15,000[q]	Raymond	15M	Diff.	5,000	10,300	18	60[j]	...	120	2
15,000[r]	Conmaco	50(K)	S-A	5,000	9,700	36	81	565	80	2
15,000	Conmaco	50(C)	S-A	5,000	10,600	36	81	565	80	2
15,000	Link Belt	312	S-A-D	3,855	10,375					
13,100	MKT	10B3	D-A	3,000	10,850	19	104	750	100	2.5
9,100	Delmag	D-5	S-A-D	1,100	2,750					
9,000	MKT	C-3	Comp.	3,000	8,500	16	60	450	100	2
9,000	MKT	S-3	S-A	3,000	9,030	36	50	400	80	1.5
8,800	MKT	DE-10	S-A-D	1,100	3,100					
8,750	MKT	9B3	D-A	1,600	7,000	17	85	600	100	2
8,300	MKT	DA-15	D-A-D	1,100	5,000					
8,100	Link Belt	180	D-A-D	1,725	4,550					
7,500	Link Belt	105	D-A-D	1,445	3,885					
7,260	Vulcan	30C	Diff.	3,000	7,036	12.5	85	488	120	1.5
7,260	Vulcan	2	S-A	3,000	6,700	29	49	336	80	1.5
6,500	Link Belt	105	S-A-D	1,445	3,885					
4,150	MKT	7	D-A	800	5,100	9.5	65	450	100	1.5
4,000	Vulcan	DGH-900	Diff.	900	5,000	10	75	580	78	1.5

[d] S-A = single-acting steam-air; D-A = double-acting steam-air; S-A-D = single-acting diesel; D-A-D = double-acting diesel; diff. = differential; comp. = compound; hyd. = hydraulic; pneu. = pneumatic.

[e] Total weights for diesel hammers are without drive caps. For other hammers, weight includes base.

[f] Strokes for diesel hammers vary with driving conditions and other factors.

[g] Air consumption is based on adiabatic compression. Unless the air is reheated before entering the hammer, the actual consumption will be greater. The compressor capacity should be from 30 to 50 percent higher than the air consumption indicated.

[h] Pressure at boiler should be from 15 to 25 psi higher to provide for line losses.

[i] Piping should be the next larger size.

[j] Based on 10 ft^2 of heating surface per horsepower.

[k] Required air to operate any hammer at normal efficiency depends upon many variables. The manufacturer prefers to consider each case rather than assign fixed scfm numbers.

[l] Adjustable energy. Check on trip setting.

[m] Interchangeable pistons. Check on the piston used.

[n] Convertible to either double-acting or single-acting mode.

[p] For Model 106 the ram weight can be either 6500 or 5000 lb. Weight is changed by adding or removing two 750-lb weights which fit inside cylinders formed in the ram. Check on the ram weight used.

[q] Variable energy due to variable stroke. Check on the stroke setting by operator.

[r] Internal hammer used with special Raymond mandrel.

NOTE: 1 ft^2 = 0.929 m^2; 1 ft·lb = 135.6 J; 1 hp = 746 W; 1 in = 25.4 mm; 1 lb = 0.4536 kg; 1 psi = 6.894 kPa.

5.4.0 Pile Spacing

Normal pile center-to-center spacing is about twice the average diameter or diagonal dimension. Special-type piles may require increased spacing (see Special-Type Piles). The required spacing for friction piles may be greater than that for end-bearing piles. If there is any evidence of pile interference during driving such as driving a pile into a pile already driven, it may be necessary to increase pile spacing. The possibility of pile interference increases with increasing pile lengths and flexibility. The use of a stiffer pile or increasing pile spacing beforehand is recommended for very long piles.

Identification of Piles

Each pile on the job should have a unique designation which should be noted on the foundation plan and used on all pile-driving logs, pile reports, and other job records.

Preexcavation

The use of preexcavation methods such as predrilling or prejetting may be permitted by the specifications or be subject to the approval of the engineer. Preexcavation may be required to assist in pile penetration of dense upper strata, to reduce pile heave, or to assist in driving piles to the required penetration. Sometimes the depth of preexcavation will be limited by the specifications, and in most cases the specifications will require driving the pile below the depth preexcavated. The inspector should observe the discharge of soil cuttings in order to estimate the volume of soil removed. Prolonged jetting or drilling without advancing the jet or drill should be prohibited. Jetting should not be permitted in soils containing large gravel, cobbles, or boulders. These could collect at the bottom of the jetted hole, making subsequent pile driving impossible.

Pile Spotting

The tip of the pile (or auger or drill bit) should be accurately spotted over the pile-location stake (Fig. 5.4.1.1). To avoid errors due to parallax, the positioning of the pile can be checked by measuring to two reference stakes 90° apart. Sometimes a template can be used to assure accurate positioning.

Figure 5.4.1.1   Accurate placement of pile tip over pile location.

5.4.1 Pile Alignment

Piles are installed either vertically (plumb) or on a batter. The degree or rate of batter and the direction of batter will be indicated on the plans. Both the leaders and the pile or casing should be positioned as accurately as possible to the alignment required by the plans. For drilled piles or when piles are preexcavated, the drill stem or jet pipe must also be properly lined up. If batter piles are installed, the pile or holes must be oriented in the direction shown on the plans.

Before driving starts and after the pile point has been properly spotted, the alignment of the pile and equipment should be checked by using a mason's level (Fig. 5.4.1.2); for batter piles the level can be held on a simple wood template cut to the correct batter angle and held against the leaders, pile, or drill stem. The pile or drill rig should be on a stable and level hard standing (preferably timber mats) so that alignment of the pile and equipment will be maintained. Alignment of piles should be checked periodically during driving, and such checks should be made on exposed lengths of not less than 5 ft (1.5 m).

If a pile is to be driven through soils which are to be subsequently removed or through water, the initial alignment of the pile is critical. Any misalignment could result in the pile butt being off design location at cutoff; the amount of mislocation would increase with increasing depths of overburden or water.

If pile alignment below the ground surface is considered critical enough to specify an axial-alignment tolerance for the full length of the pile, only those piles for which axial alignment after installation can be determined throughout the pile length should be used. This includes pipe, tube, or shell-type piles plus those precast-concrete and steel H piles which are fitted with a full-length inspection duct down which an inclinometer could be lowered to measure axial deviations.

Any type of pile can be deflected off required axial alignment during driving by subsurface obstructions, sloping rock surfaces, densification of soil during driving, or certain subsoil conditions such as cavernous limestone. Under soft soil conditions, the forces of gravity could cause flexible piles driven on a batter to deflect downward as they are driven. Piles that are less stiff, for example, steel H piles, are more readily deflected off line. Also, flexible drill stems used for preexcavation or for installing drilled piles could easily be deflected off line. The chances of bending increase with increasing pile lengths or flexibility.

Excessive deviations from the specified axial alignment can sometimes be controlled by using a stiffer pile, by predrilling, or by the removal of obstructions. Under some conditions such as driving piles in cavernous limestone, there is no practical method for eliminating pile bending.

Figure 5.4.1.2 Checking placement with mason's level. (*Courtesy of Norwalk Marine Corp., Norwalk, Conn.*)

5.4.2 Batter Piles

If the actual driven lengths of vertical piles, batter piles, or opposing batter piles are longer than anticipated, the possible interference of piles should be investigated. It may be necessary to relocate the batter piles or change the degree or direction of the batter to avoid pile interference.

Pile Length in Leaders

If payment for piles is based on the length of pile raised in the leaders, the inspector should record such length for each pile. It should be noted that additional lengths of pile may have to be added during installation. If the lengths of piles cut off are used, pay lengths of piles should be adjusted so that double payment is not made for pile material.

5.5.0 Driving Formulas

Dynamic pile formulas have been in use for over 100 years. The formulas attempt to relate pile load-carrying capacity to the potential energy of he hammer and the pile set per blow. Driving formulas range from the relatively simple Engineering News formula

$$R = \frac{2WH}{}$$

to the complicated Hiley formula

$$R_u = \frac{e_f W_r h}{s + \frac{1}{2}(C_1 + C_2 + C_3)} \times \frac{W_r + e^2 W_p}{W_r + W_p}$$

See *Pile Foundations* by R. D. Chellis (in References) for more details on driving formulas. There is no indication that the more complicated of these empirical formulas are any more accurate than the simple ones. All can show a wide scatter when compared with load test results. R. D. Chellis is attributed with saying, "There are many dynamic formulas, they give widely varying results and obviously they can not all be right."

The wave equation was first applied to pile driving in 1931 by D. V. Isaacs. Although the analysis method is relatively simple, it involves a large number of steps for different masses, springs, and time increments. Therefore, the method did not come into use until the late 1960s when computer programs were available for mainframe computers. In the mid-1980s microcomputer versions became available. The wave-equation analysis requires the following input data:

Ram—weight (and stiffness for diesel hammers)

Hammer—efficiency (and combustion force for diesel hammers)

Hammer cushion—stiffness and coefficient of restitution

Drive head—weight and coefficient of restitution

Pile cushion—stiffness and coefficient of restitution

Pile—length, weight, area, and modulus of elasticity

Quakes at the side and tip of the pile

Damping at the side and tip of the pile

Distribution of frictional capacity along the side of the pile

Percentage of load carried by end bearing

Quake is the magnitude of pile displacement at which there is a transition from elastic to plastic behavior. Quake is typically 0.1 in/ft (3.05 cm/30.48 cm) width of pile tip, although for piles end bearing on bedrock values will be less and for piles terminating in spongy glacial tills the quake may be much higher.

Output from the wave-equation analysis provides:

Required sets for specified ultimate capacity

Maximum compressive and tensile driving stresses

Energy transferred to pile

Estimated driving time (this requires the input of soil strengths)

It must be understood that wave-equation analyses provide ultimate pile capacities, as compared to dynamic formulas which provide working-load capacities. Typically, wave-equation-derived ultimate capacities are divided by a factor of safety of 2.0 to obtain working-load capacity. When used in conjunction with dynamic pile testing, the driving efficiency can be checked by comparing the wave-equation-computed transferred energies with the transferred energies derived from dynamic measurements.

5.5.1 Engineering News Formula

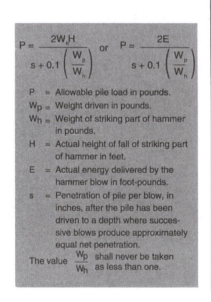

$$P = \frac{2W_h H}{s + 0.1 \left(\dfrac{W_p}{W_h} \right)} \quad \text{or} \quad P = \frac{2E}{s + 0.1 \left(\dfrac{W_p}{W_h} \right)}$$

P = Allowable pile load in pounds.

W_p = Weight driven in pounds.

W_h = Weight of striking part of hammer in pounds.

H = Actual height of fall of striking part of hammer in feet.

E = Actual energy delivered by the hammer blow in foot-pounds.

s = Penetration of pile per blow, in inches, after the pile has been driven to a depth where successive blows produce approximately equal net penetration.

The value $\dfrac{W_p}{W_h}$ shall never be taken as less than one.

Many dynamic pile driving formulas have been developed as an aid in determining pile capacity. Such formulas serve a useful purpose particularly on small projects, or where previous experience and static testing in a given area have confirmed their reliability.

The modified Engineering News Formula, which takes into account the relationship between the relative weights of the pile and hammer, is commonly used.

This formula has proven to be most reliable when uniformly tapered piles are embedded in a cohesionless bearing stratum.

Foundation Engineers recognize the varying results with any formula, or static analysis approach, in the determination of pile capacity and/or length. The reliability of soil information, its variability within the site, the selection of appropriate soil values, pile shape, hammer efficiency, and the effect of pile driving forces on in-situ soil conditions are among the variables which influence pile performance and require sound engineering judgement.

On large projects involving high capacity piles, predesign or prebid programs incorporating static load tests, or dynamic methods coordinated with static tests can provide rational driving criteria resulting in substantial cost savings.

5.5.2 Pile Bearing Capacity

MATERIAL
Suggested Specifications:

Piles: All piles shall be fluted steel tubes, having a uniformly tapered lower section which may be combined with extensions of essentially constant diameter, of a type wherein the tubes are driven to a required penetration, left permanently in place, and filled with concrete.

Material Specifications: The material in the pile shall comply with SAE-1010 steel specifications having the following chemical ranges and limits:

Carbon............0.08 to 0.13 (%)

Manganese.....0.30 to 0.60 (%)

Sulfur.................0.05 (% max.)

Phosphorous.....0.04 (% max.)

Minimum tensile yield strength of the steel in fabricated pile, as determined by standard tests, shall be 50,000 psi.

Steel piles shall be strong enough to withstand safely the forces of installation, but in no case shall the wall thickness of the pile be less than 1/10 of an inch.

Point diameter shall be not less than 8 inches. Piles shall taper not less than one inch in eight feet, excepting that when such taper has developed a diameter of *____ inches, the piles may be extended with sections having a diameter which is virtually constant. All splices shall consist of a continuous weld to form an integral pile.

NOTE: *Insert appropriate butt diameter; i.e., 12″, 14″, 16″, 18″.

DESIGN COMPUTATION

Allowable structural capacity in accordance with provisions of the Standard Building Code — Sections 1303.2, 1304.5, 1306 and the BOCA National Building Code — Sections 1214.5, 1216.2.

EXAMPLE

Allowable load
$$P_a = 0.35 F_y A_s + 0.33 F'_c A_c$$

For 7 ga. 12 in. dia. Monotube:
$A_s = 6.97$ in^2 (p. 8)
$A_c = 101$ in^2 (p. 8)

Concrete $F'_c = 4000$ p.s.i.

Steel $F'_y = 50000$ p.s.i. (p. 11)
$P_a = 0.35 \times 50,000 \times 6.97$
$+ 0.33 \times 4000 \times 101$
$= 255295\#$ or 128 tons

5.5.3 Installation Criteria

Piles can be installed by dead weight, dynamic impact, drilling, jacking, vibratory driving, or combinations of these methods.

Dead Weight.   For the dead-weight method and in the absence of soil freeze the ultimate capacity of the pile as installed is limited to the amount of weight applied. This is not a practical way to install piles. If used, the amount of weight applied must be recorded.

Dynamic Impact.   The required set of piles installed by dynamic impact should be established by static or dynamic load test or wave-equation analysis. The use of arbitrary rule-of-thumb sets is strongly discouraged even for lightly loaded piles where they may seem to be conservative. While a high blow count may seem safe, it can lead to overdriving and damaging piles. Specifications may require that the piles be driven to a minimum required length or tip elevation. The introduction of fresh capblock material just prior to the measurements of final pile set should not be allowed, since it will result in an artificially high blow count.

Adherence to an established driving resistance permits each pile to see its own required capacity regardless of normal variations in depth, density, and quality of the bearing strata or variations in pile length. This characteristic is not found in piles installed by drilling.

Drilling.   Drilled piles will generally be installed to a required depth. The drilling operation and soil cuttings should be observed to assist in evaluating the subsoil conditions at each pile location and to correlate with those at test pile locations. If conditions such as a hole of adequate size, absence of groundwater, and insertion of a temporary steel liner permit, the hole may be entered and the bottom inspected or tested to ensure that an adequate bearing stratum has been reached. (See Bored Pile under Special-Type Piles.)

Jacking.   For piles installed by jacking, ultimate capacity is generally limited by the reaction available to resist the jacking force. If this method is used, the jacking force must be recorded.

Vibratory Driving.   If piles are to be installed with a vibratory driver, the specifications may require that the piles be driven to a required length or tip elevation. Additional subsoil information based upon closely spaced borings may be necessary to establish required pile lengths. Driving criteria may involve a minimum rate of penetration for a specified dynamic force or frequency. The specifications may require that installation criteria be verified by means of conventional impact hammer and/or load test. The specifications may also require that the final penetration resistance of all piles be checked with an impact hammer.

5.6.0 Damage to Nearby Structures

Possible causes of damage to nearby structures by pile driving include excessive levels of vibration, vibration-induced settlements of loose granular materials, and loss of ground. Vibrations are typically monitored with portable seismographs, most of which can be set to trigger at critical particle velocities. Peak particle velocities of up to 2 in/second (0.05 m/s) are typically considered to be safe for new construction, and values of 0.5 in/second (0.012 m/s) are considered safe for structures which are old or are more massive and have lower natural frequencies. Normally measurements taken on the structure will be less than those taken on the ground surface adjacent to a structure.

In loose granular soils including poorly compacted fills and uniform sands, vibrations from pile driving can cause the soil grains to rearrange and find a more compact soil structure leading to settlement. Vibration levels as low as 0.2 in/second may trigger this type of settlement. Generally, impact hammers cause less damage than vibratory hammers even though the resultant particle velocities may be higher. This is due to the vibratory hammers imparting many more blows per foot of penetration and their operating frequency being near the resonant frequency of most soils. To minimize settlements caused by loss of ground the inspector should compare the quantities of soil removed with theoretical values and also, when conditions warrant, take settlement measurements on adjacent structures.

Driving Log

Probably the most common duty of the pile inspector is the recording of the driving log. This involves more than just counting blows. The inspector should observe, record, and evaluate occurrences during pile installation which may affect the integrity or bearing capacity of the pile, prevent it from meeting the specifications, or affect existing piles or structures.

Impact Driving. The specifications normally require that the number of hammer blows be recorded for each foot of pile penetration and that the final driving resistance be recorded in blows per inch (1 in = 25.4 mm). The basic driving log can be obtained by marking the pile off at even 1-ft (0.3-m) intervals and counting the number of hammer blows delivered as adjacent 1-ft marks pass a fixed reference point. If the pile is penetrating rapidly, it may be impossible to record blows per foot, and under such conditions the total number of blows for a given length of penetration may be recorded. A convenient way of observing final driving resistance in blows per inch is to put marks on the pile at 12-in intervals and count the number of blows delivered as one of the marks passes an adjacent stationary plank. For piles that are inserted in predrilled or jetted holes, or for very heavy piles or soft soil conditions, the pile will run into the ground under its weight and the weight of the hammer. The depth of run should be recorded.

Vibratory Driving. If piles are to be installed with a vibratory driver, the inspector should record, in addition to pile length and other data, the final penetration rate in feet per minute and the steady-rate frequency in vibrations per minute during final pile penetration.

5.7.0 Soil Freeze, Relaxation, and Changes in Driving Resistance

Soil Freeze

Some fine-grain soils exhibit a sharp decrease in shear strength when being remolded and disturbed by pile driving but regain their strength with time after pile driving stops. This is known as soil freeze or setup. During driving, penetration resistance is relatively low and dynamic methods will under-predict capacity. The occurrence of soil freeze can be checked by retapping piles some time after final driving. The amount of soil freeze and the rate at which it occurs can vary over a wide range. Some soils show substantial soil freeze in a matter of hours, but others may require days or even weeks to regain full strength. Under such conditions the application of dynamic testing or wave-equation analyses should be based on retap data.

Relaxation

Relaxation is a term applied to a decrease with time of final pile penetration resistance. It could result from driving friction piles into dense, fine submerged sand, inorganic silt or stiff, fissured clay, or from driving end-bearing piles into a friable shale or a claystone. Where soil conditions indicate that pile relaxation could occur, some piles should be retapped several hours after they and adjacent piles have been driven. If upon retapping it is observed that the original final driving resistance has decreased, all piles should be retapped and further such checking should be done until penetration resistance and/or depths of penetration are satisfactory.

Interruptions of Pile Driving

Generally, specifications will require that piles be driven their full lengths without interruption. If the driving of a pile is interrupted, the cause and duration should be noted. Where soil freeze conditions exist, any prolonged interruption in pile driving will result in an increase in pile penetration resistance when driving is resumed. In extreme cases, it may not be possible to achieve further pile penetration. In such cases, if the pile is not long enough or has not reached the required bearing strata, it may be necessary to abandon and replace the pile. If piles are to be driven to a specified penetration resistance, driving after an interruption should continue long enough to break the soil freeze.

Abrupt Changes in Driving Resistance

Sudden increases in driving resistance, either real or apparent, could be caused by the pile hitting underground obstructions or hard soil strata, by the introduction of fresh capblock material, by the brooming of the butts of timber piles, by the cracking or spalling of the butts of concrete piles, or by elastic yielding of steel piles. Sudden decreases in pile-driving resistance could be caused by the breaking below ground surface of timber or precast piles, by the plastic yielding of a steel pile, or by any pile encountering unexpected soft soil strata.

Abrupt changes in pile-driving resistance should be investigated to determine the cause. If they are not investigated or if pile breakage is found to be the cause, the pile should be rejected. The investigation may indicate that some remedial measures, such as changing pile-installation equipment or methods or using a different type of pile, must be taken.

5.7.1 Obstructions Encountered While Driving

Obstructions may be buried timber, logs or tree stumps, boulders, old foundations, slag or rock fill, or other objects which may prevent the pile from reaching the necessary penetration, cause the pile to bend or drift off location, or damage the pile. Obstructions near the ground surface will be evident by the pile tending to drift off location at the start of driving or by the refusal of the pile to penetrate. The presence of deep obstructions may be difficult to detect and can seriously damage the pile or prevent the pile from reaching the required bearing layer. Deep obstructions may be indicated by an unexpected refusal of the pile to penetrate or by substantial differences in tip elevations of adjacent piles. Such conditions should be investigated by test borings. It may be necessary to remove or dislodge obstructions to install piles properly. Sometimes obstructions can be pushed aside, broken up, or pierced with a spud. Shallow obstructions can often be readily removed, but for deep obstructions special methods may be required to break up or bypass the obstructions, or the pile may have to be relocated.

If obstructions such as boulder formations are known to exist, the inspector should observe the pile driving or installation carefully to detect any unusual occurrence which may indicate pile damage or misalignment.

The tips of steel H piles are very susceptible to overstressing and deformation. The use of reinforced tips for steel H piles is recommended if obstructions are anticipated. The tips of timber piles also can be easily damaged by obstructions. The use of steel drive shoes may help prevent tip damage, but if too much reliance is placed on the reinforced tip and the pile is driven hard, it may be damaged elsewhere. The tips of precast-concrete piles can also be damaged by obstructions, but the use of tip protection is unusual. The most effective remedy for steel, timber, or concrete piles is to avoid overdriving. (See Overdriving below.)

Piles such as pipe, tube, or the shell type can be visually inspected for damage after the casting has been driven. If other types of piles are driven in ground which is seriously obstructed, some of the driven piles should be pulled for inspection as the work progresses. In extreme cases, a change in pile type and/or method of installation may be required.

5.7.2 Overdriving

Overdriving of piles should be avoided. Whether or not piles are being overdriven is often a matter of judgment based upon the pile material involved, the subsoil conditions, the driving equipment used, and the installation method. A wave-equation analysis is valuable in that it will provide the inspector with maximum compressive and tensile stresses over a wide range of blow counts. Maximum permissible driving stresses are typically taken as $0.9\,F_y$ for steel, 3000 psi compressive stress for timber, and $0.85\,f_c'$ compressive stress and zero tensile stress for concrete. The concrete stresses must include any prestress.

Pile damage can also occur by a repetition of hammer blows that result in fatigue failure. This is especially true for timber piles. Overdriving of timber piles can cause brooming at the pile tip or butt, or breakage. If a sudden high resistance to penetration is encountered when driving timber piles, driving should cease immediately.

Steel piles deform by plastic yielding, rupture, or both. Damage to precast-concrete piles can be in the form of cracks, spalling, or breakage.

5.7.3 Pile Spacing

The type of splice to be used may be specified. If the splice is to withstand prolonged driving, it should be as strong in all respects as the pile material. The splice must be capable of transferring all required stresses including compression, tension, and bending. Whether pile sections are spliced together on a horizontal rack or in the leaders, the individual sections must be in accurate axial alignment. Specifications may prohibit the splice being located in the upper portion of the pile as installed.

Where soil conditions are such that soil setup or freeze can occur rapidly, the use of pile splicing during driving should be limited. Under extreme conditions it may be impossible to resume further penetration of the pile after the splice has been made. If splicing is necessary under such conditions, the splice should be a type which can be completed in the shortest length of time.

Precast Piles.   There are several types of precast pile splices or joints. These include various types of mechanical joints, welded joints, sleeve joints, and dowel joints. Some are specifically designated for installing sectional precast piles. Some of these joints cannot take bending or tension.

Pipe Piles.   Pipe piles can be spliced by using either inside or outside drive sleeves, with or without welding. Splices can also be made by butt welding.

Tube Piles.   Tube piles may require special preparation for heavy-duty welded splices. The manufacturer's recommendations should be followed.

H Piles.   H piles are spliced by simple butt welding, using special splice fittings or welding on fishplates. If special splice fittings are used, they should at least be tack-welded.

Pile Shells.   *Uniform Section.*   Shells are spliced with butt welds.
Step-Tapered Sections.   Joints are generally screw-connected and made watertight by the application of a waterproofing compound or the use of an O-ring gasket. Joints can also be welded. Note that step-tapered piles may be made of different section lengths giving different rates of taper.

5.7.4 Pile Heave

When driving displacement piles in highly incompressible soils, ground heave may occur. Under such conditions the possibility of pile heave should be checked. This can be done by taking level readings on the tops of driven piles before and after adjacent piles have been driven. However, for corrugated pile shells, such check level readings should be made on a telltale pipe resting on the closure plate at the pile point. Upward movements of the tops of such shells are not harmful when they are due to shell stretch.

When pile heave is observed, piles should be redriven to their original depth and set. Pile heave is more critical for end-bearing piles and might not affect the capacity of friction piles. Cast-in-place concrete should not be placed until pile driving has progressed beyond the pile heave range.

If piles are not suitable for redriving (for example, uncased cast-in-place concrete piles) and heave occurs, the driving sequence and/or methods of installation must be modified to eliminate heave and all heaved piles should be replaced. If ground or pile heave occurs, the effects on existing structures should be monitored. Pile heave can be controlled or eliminated by proper preexcavation methods.

5.7.5 Retapping

When it is necessary to redrive or retap piles to a specified penetration resistance, it should be done with the same or equivalent type and size of hammer used for initial driving. It should be noted that the first few hammer blows may not be indicative of actual driving resistance. Soil freeze may have occurred. Furthermore, the hammer may not be delivering its full stroke until it has warmed up.

Retapping or redriving of uncased cast-in-place piles should not be permitted. Redriving on cast-in-place concrete encased in a steel shell or pipe is permissible provided the proper hammer-pile-cushion system has been established by prior use or designed by a wave-equation analysis. For pile shells such as pipe or tube not originally driven with an internal mandrel, the concrete filling produces a much stiffer pile and results in more effective transmission of hammer energy. Therefore, for such piles the actual retap resistance will be less than the original final resistance for the same formula capacity. A wave-equation analysis will show the comparable final resistances for the load involved.

Variations in Pile Lengths

If possible, extreme variations in pile lengths in any one group should be avoided. The cause for extreme variations in pile tip elevations for adjacent piles or piles within a group should be investigated. Length variations could be caused by obstructions or by unusual subsoil conditions such as cavernous-limestone formations or variable glacial deposits. There may be no practical remedy for extreme variations in pile lengths. Additional piles may be required. The gradual densification of the subsoil as driving progresses may result in a shortening of pile depth. To minimize this, pile driving for large groups should start at the center of the group and continue outward uniformly.

5.7.6 Pile Cutoffs and Waste Pilings

Pile Cutoff

Pile cutoffs should be made perpendicular to the longitudinal axis within 1 in (24.5 mm) of required elevation or below all damaged portions. By checking against the pile plans, the inspector can visually verify that the piles have been cut off at approximately the correct elevation. Such visual checking should detect major discrepancies in pile cutoff elevations.

Waste Piling

Pile lengths which are cut off will be the property of the owner if pile material is paid for on the basis of furnished lengths or become the property of the contractor if piles are paid for from tip to cutoff grade. In either case, the specifications may require the contractor to dispose of waste piling.

Treatment of Timber-Pile Butts

The butts of treated timber piles should be treated with preservatives in accordance with the specifications. Generally three coats of hot creosote oil are required, followed by sealing the end surface of the pile with a heavy application of coal-tar pitch.

5.7.7 Cap Plates

Steel H piles which are installed to carry very high design loads and are capped with concrete may require steel cap plates to be welded to the tops of all piles for adequate load transfer by bearing between the piles and the concrete caps. Alternatively the piles may be embedded a sufficient length into the concrete cap to transfer the load through bond. Cap plates should at least be tack-welded to the piles.

5.7.8 Pile Buildup

Pile butts that are too low in elevation can be built up to the proper grade. This is readily accomplished for cast-in-place concrete piles by adding a section of shell or pipe before concrete is placed. For other types of piles, especially when the specifications prohibit splices along the upper portion of the pile, it may be necessary to verify with the engineer the acceptability of a pile buildup. In some cases, the bottom of the pile cap can be lowered to accommodate such a pile, the top of which is below the required cutoff grade.

5.8.0 Cleaning Out Piles

Open-End Pipe Piles.   The specifications may require that pipe piles driven open-ended be cleaned out and filled with concrete. The length of pile to be cleaned out may vary from a few feet at the top to the full length. Subsoil conditions may dictate that open-end pipe piles be cleaned out periodically to achieve the necessary penetration. It should be noted that an open-end pipe pile which picks up an immovable plug during driving will behave as a displacement-type pile.

Soil inside such piles can be removed by drilling, by washing with a jet with or without an airlift, or with various types of grab buckets. It may be necessary to break up large boulders with some type of churn-drill chopping bit, or cable tool or a hammer grab. Care must be taken to avoid removing soil from beneath the final pile tip elevation or from alongside the pile.

Pile Shells and Casings.   Unless pile shells, casings, and drilled holes are covered until concrete is placed, various types of debris can fall in. Soil in pile shells or closed-end casings can usually be removed by jetting either without or in combination with air or steam pressure. If soil enters the shell through a tear in the side, the pile can be salvaged by cleaning out provided the hole is small and plugs. It may be necessary to keep the shell filled with water and place concrete by tremie (see Tremie Placement, p. 401).

Foreign objects can often be removed with some type of grab or hook. Pieces of wood can sometimes be speared with a pointed length of reinforcing steel or floated to the top by filling the shell with water. Water can be removed with a bailing bucket, steam siphon, or small submersible pump. If the lift is not too high, conventional pumps can be used.

5.9.0 Pile Alignment/Deformation after Installation

Pile Alignment after Installation

The major factor in pile alignment is the alignment of the upper portion of the pile as it enters the pile cap or structure; this should be in accordance with the design and within the specified axial-alignment tolerance. The alignment tolerance may be specified in degrees or as a rate such as 1 in in 4 ft (25.4 mm in 1.2 m) or as a percentage of length. The alignment along the top portion of the pile can be checked with a mason's level over a minimum 5-ft (1.5-m) length of pile. For batter or tapered piles a template should be used to compensate for the batter or taper.

Of secondary importance is the alignment of the pile below ground surface. If the alignment tolerance relates to the full length of the pile, there should be an understanding by the parties concerned as to how the tolerance is to be applied, for example, to each incremental length of the pile or to the overall length of the pile. Alignment tolerances applied throughout the pile length are not considered practical or warranted.

Misalignment of piles below ground surfaces can take the shape of (1) the axis of the pile being straight but not on required alignment, (2) the pile being bent, or (3) a combination of these. The pile bend can be either sharp (called a dogleg) or smooth with various radii.

The approximate axial alignment of pipe, tube, or shell-type piles or drilled holes can be checked visually by using a mirror and reflected sunlight or a droplight. Alignment could also be checked by lowering a plumb bob down the pile and noting the offset for a given depth. For pipe piles, the alignment can be checked by lowering a proving "plug," the diameter and length of which are designed to prevent the plug from passing a bend in the pile having a minimum specified radius of curvature.

Deviation from axial alignment could be measured with some type of inclinometer lowered down the pile or through a special tube fitted to or cast into the pile. Except for research types of projects, the use of such instruments is considered impractical.

Generally, long, sweeping bends in friction or end-bearing piles are not considered detrimental. The capacity of piles with sharp bends or doglegs can be checked by an analysis involving structural and soil mechanics principles. Except possibly for very sharp doglegs in extremely soft soil, soil resistance is generally more than sufficient to restrain the pile from buckling under its design load. If the capacity of a bent pile is in doubt, it can be checked by a proof load test.

Pile Deformation

Piles can be deformed at the tip, along the shaft, or at the butt. Tip deformation for timber, precast concrete, or steel H piles cannot be detected unless the piles are pulled for visual inspection. See Obstructions (p. 411). Pile deformation along the pile shaft could be in the form of a reduction in the cross-sectional area, a break in the pile, or in severe cases of overdriving, a folded bend in the pile. Except for pile shells or casings which can be visually inspected internally after being driven, such deformations cannot be observed unless the pile is pulled or excavated.

Reductions in cross-sectional area could be caused by obstructions, by soil pressures, or by driving pipe or tube piles into rock fissures or crevices. For shell, pipe, or tube piles the cross section of the pile can be checked visually during internal inspection or by lowering a proving ball down the pile.

Generally any reduction of 10 percent or less is considered satisfactory. For reductions in cross-sectional area for uncased cast-in-place-concrete piles, see Special-Type Piles.

The occurrence of detrimental deformation may be controlled or eliminated for shell, pipe, or tube piles by the use of heavier gauges or wall thickness, preexcavation, filling a section of driven pipe with concrete before adding and driving additional pipe lengths, inserting dummy cores in pile shells, or filling them with water until such time as shells are concreted.

5.9.1 Placing Reinforcing Steel

The number and arrangement of longitudinal bars and the diameter of the reinforcing cage should be as specified. Longitudinal steel to resist bending should be tied with hoops or a spiral with the specified pitch or spacing and should have sufficient spacers to center the cage in the pile casing or hole. Longitudinal steel to resist uplift loads may be bundled and placed in the center of the pile. Except for short dowels, reinforcing steel for cast-in-place piles should be placed before concrete is cast. All steel must be free from dirt, rust scale, loose mill scale, oil, grease, and any other coating which would affect the steel-to-concrete bond. The concrete cover should be at least 3 in (76.2 mm) for uncased cast-in-place-concrete piles and $1\frac{1}{2}$ in (38.1 mm) for cased piles.

5.9.2 Placing Concrete for Cast-in-Place Piles

A minimum distance between pile driving and concrete placement may be specified. Tests have indicated that vibrations from pile driving have no detrimental effect on fresh concrete, and a criterion of one open pile between the driving and concreting operations is considered satisfactory. However, for practical reasons a minimum distance of 10 ft (3 m) is often specified. Concrete should not be placed, however, until pile driving has progressed beyond the heave range and all heaved piles have been properly reseated. Also, concrete should not be placed as long as relaxation is occurring.

The shell, tube, or pipe should be inspected just prior to filling it with concrete and should be free of all foreign matter and contain not more than about 4 in (101.6 mm) of water unless tremie placing of concrete is permitted. Concrete should be placed in each pile shell, casing, or hole without interruption. If it is necessary to interrupt the concreting process long enough so that the concrete could take its initial set or harden, steel dowels should be inserted in the top of the concreted portion of the pile. When concreting is resumed, all laitance should be removed and the concrete surface should be flushed with neat fluid grout.

When discharging concrete from the mixer, the concrete flow must not be restricted by a partially opened gate. Concrete chutes must be steep enough so that the concrete flows freely and does not have to be pushed or shoveled. If a hopper or concrete bucket is being used, concrete should not be discharged into a funnel-type downpipe centered over the hopper or bucket. When discharged from the hopper or bucket, the concrete should be drawn off from the center.

Concrete containing a reduced amount of coarse aggregate and a compensating increased quantity of sand and cement is recommended for filling long pile shells or those driven on a batter or containing heavy reinforcement. Such concrete, if properly designed and mixed, is very workable and cohesive. The top 6 to 10 ft (1.8 to 3 m) of the concreted pile should be rodded. Vibration of concrete is unnecessary and may promote bleeding.

Conventional Placement. The concrete mix design and the equipment and techniques used in placing the concrete must be selected to prevent separation of the coarse aggregate. Concrete should be of proper slump (see Slump under Pile Material) and deposited in a rapid continuous pour through a steep-sided funnel centered at the top of the pile. The diameter of the discharge end of the funnel should not be larger than 10 in (254 mm). When conventional concrete is used, about a pail full of flowable grout should be poured into the pile shell before placing concrete. Pregrouting is not necessary when reduced coarse-aggregate concrete is used.

Tremie Placement. The tremie method should be used for placing concrete through water. Concrete can be poured into the tremie tube or pumped through a tremie pipe. The bottom of the tremie should be closed with a detachable plug or hinged-flap valve before it is lowered through the water, or a movable plug can be inserted in the top of the tremie just before concrete is placed to prevent the concrete from coming in contact with water in the tremie. The tremie tube must be resting on the bottom of the casing or hole before concreting begins. It should then be raised only a few inches (1 in = 25.4 mm) to start the flow of concrete and to ensure good contact between the concrete and the bottom of the casing or hole.

As the tremie is raised during concreting, it must be kept below the surface of the concrete in the pile. Before withdrawing the tremie completely, sufficient concrete should be placed to displace all free water and watery concrete.

Bleeding.   Bleeding is evidenced by a collection of water and cement on top of the concrete after placement. It could be caused by excessive water in the mix or by poorly graded aggregates. The amount of mixing water could be reduced by a water-reducing admixture. If the quality of aggregate is the problem, corrective measures should be taken to improve the gradation.

Cold Weather.   For concrete placed in pile shells, casings, or holes, cold weather is not as critical as it is for concrete placed in aboveground forms. However, suitable precautions should be taken. See Cold-Weather Operations under Pile Material. The temperature of concrete as placed should not be less than about 40°F (4.4°C). Concrete should be discharged promptly upon delivery. The tops of freshly poured cast-in-place concrete piles should be adequately protected from freezing. This can be done by covering the pile butts with salt hay or soil, or in extreme cases, by using tarpaulins and heaters.

Hot Weather.   If the weather is extremely hot, special measures may be required to keep the concrete temperature down to acceptable limits during mixing and placing. See Hot-Weather Operations under Pile Material. The time between mixing and placing should be reduced to a minimum, and concrete must be placed promptly upon delivery. The temperature of the concrete as placed should be below 80°F (26.6°C) to avoid flash sets.

Concrete for Special-Type Piles.   See Special-Type Piles for other concreting requirements for special types of piles.

Piles Overlooked

Before the pile rig is moved away, the inspector should check that all piles in the immediate area have been installed.

5.10.0 Pile-to-Cap Connections

The plans or specifications may require that the tops of piles be anchored into the pile caps. For steel piles this can be done by welding short dowels to the pile. For cast-in-place concrete without internal reinforcement, dowels can be embedded in the concrete at the pile butt; internal reinforcement, if used, can be extended into the cap. However, if such pile-to-cap connections are required for uplift loads, reinforcing steel should extend the full length of the concrete piles.

For timber piles special fittings consisting of steel bars and/or straps can be bolted or otherwise fastened to the top of the pile. All exposed steel including bolts must be galvanized, and all holes or other cuts must be treated with several applications of hot creosote oil followed by a seal coat of coal-tar pitch.

For connections involving precast piles, the piles could be made with the longitudinal or pre-stressing steel projecting the required length beyond the pile butt. For such piles, a special drive cap or follower is required to accommodate the steel extension. In other cases it may be necessary to strip the concrete back from the top of the pile to expose a sufficient length of reinforcing steel or prestressing strand for the pile-to-cap connection or for attaching additional lengths of steel. The pile top has to be left a sufficient height above cutoff grade, and stripping must be done carefully to avoid damage to the pile. To make the cap connection, the precast pile may also be cast with dowel holes at the top into which steel dowels are grouted after the pile has been driven.

Pile-Butt Locations

Final pile positions should be checked as the work progresses so that, if necessary, additional piles can be installed before the equipment is moved off or cannot reach the required locations of the additional piles. If piles are found to be installed beyond the permissible tolerance for butt location, the resulting loads on the piles should be checked, and if any have more than 10 percent overload, additional piles may be required. The number of additional piles required in any one group may be determined by the number necessary to balance the group to maintain the design center of gravity. Added piles must be installed at the proper locations. If a pile group to be corrected is adjacent to a property line, the locations of additional piles should be checked to ensure that they do not encroach on other property.

Piles Off Location

After piles have been installed and butt locations checked, piles could be pushed off location by various types of construction activity or by general ground movements. Under soft soil conditions, heavy construction equipment should not be permitted to operate immediately adjacent to piles. Excavations along a single side of a pile should not be made without bracing the piles or the excavation bank. General ground movements could result from unstable slopes or from inadequately braced or tied sheeting. Heavy rains could cause ground loss or movement.

Pile shells not yet filled with concrete can sometimes be pulled back into position. If concrete has already been placed in pile shells or for uncased cast-in-place concrete piles, piles that have been pushed off location should be inspected for breaks. If the piles cannot be inspected properly or repaired satisfactorily, they should be rejected and replaced.

If timber, precast-concrete, or steel piles are pushed off location, the piles should be inspected for breaks or severe bends. It may be necessary to excavate alongside piles. If the pile can be pulled back into position, the force applied to the pile should be limited to avoid breaking the pile. This is especially true when there is a long moment arm as in marine construction.

Unsupported Pile Lengths

If, after installation and cutoff, a portion of the pile extends above the adjacent ground surface, the unsupported lengths may have to be braced until backfill operations have been completed or the piles incorporated into the structure. Backfilling around high cutoff piles must be done carefully and preferably by hand. Construction equipment must not be allowed to hit the pile, and the backfill should be placed in uniform lifts around the pile.

Excavating Overburden

If piles have been installed with tops below ground surface, the excavation to pile-cutoff grade must be carried out carefully so as not to damage the piles. Hand excavation is recommended.

Pile Lengths

The inspector should record the length of each pile installed as measured from tip to butt. If pile tip and butt elevations are recorded, they should be referenced to an established permanent datum which is identified.

For pipe, tube, or shell piles the lengths can be measured by lowering the weighted end of a steel tape down the pile after it has been driven and cut off to grade and before concrete is placed. Because of stretching, a cloth tape should not be used.

For timber, precast-concrete, and steel H piles the installed length of pile can be determined by measuring the length driven and subtracting the cutoff length. The driven length will include all lengths added in the leaders.

For measurement of installed lengths of special-type piles see Special-Type Piles.

Pay Quantities

The inspector may have to record or verify quantities of pay items. Payment for the pile foundation could be by lump sum, principal sum with adjustment prices, unit price, or a combination of these methods.

Usually no extra payment is made for withdrawn, rejected, or abandoned piles, for pile splicing, or for cutting off piles. Pile buildups may be included in the total pile footage installed. If pile materials are paid for on the basis of full ordered and furnished lengths, waste piling actually becomes the property of the owner. The contract documents may require that the contractor dispose of this material, and if it has any salvage value, the approximate quantities should be recorded and proper credits given to the owner.

Lump-Sum Method.   For contracts on a lump-sum basis it is necessary only to verify that all work is completed in accordance with the plans and specifications or subsequent revisions. Revisions should be covered by change orders. If additional work is performed under a work order, the satisfactory completion of such work should be verified.

Principal-Sum Method.   For the principal-sum method a stipulated payment is made for a specified total aggregate length of piling or for a specified number of piles having a stated base individual length. For final payment, the principal sum is adjusted on the basis of added or omitted piles or increases or decreases in pile lengths. The method of payment, and therefore the method for adjusting the principal sum on the basis of actual quantities of piles installed, should be understood so that the information necessary to make the computations can be properly recorded and certified.

Payment is usually made for actual pile lengths installed, measured from tip to cutoff elevation. However, when the pile-cutoff grade is below the ground or water surface existing at the time of pile driving and the pile is installed to that ground or water surface (the pile is driven through overburden), payment may be made from pile tip to ground or water surface. This does not apply to any type of uncased cast-in-place concrete pile. For cased concrete piles, the concrete is generally placed to cutoff grade, and if full payment is made for overburden footage, a credit should be given for omitted concrete. If, after final pile cutoff, the lengths cut off are reused, double payment for these pile lengths should be avoided.

The actual installed length of each pile as well as the total number of piles installed should be recorded with added or omitted piles identified. Adjustments to the principal sum will be either plus or minus, depending upon whether they represent added or omitted work. Adjustments will be made on the basis of adjustment unit prices in the contract.

Unit-Price Method.   All piling may be paid for on the basis of unit prices; sometimes the furnishing of pile materials and the driving of piles are paid for under separate unit prices. In such cases, the furnished pile lengths as well as the installed lengths must be recorded accurately. Payment may be made for the full lengths delivered to the jobsite or for the lengths of each pile raised in the leaders. The pay quantity for installing piles will generally be that measured from pile tip to cutoff grade. The exception may be when piles are installed through overburden; payment for overburden footage may be made as described under Principal-Sum Method. This should not apply when a follower is used to drive the pile head to cutoff grade below the ground or water surface or for uncased cast-in-place concrete piles, especially auger-grout piles. See Auger Grout Piles: Forming Pile Butts under Special-Type Piles.

Combination Method.   Sometimes the lump-sum method is combined with either the principal-sum or the unit-price method. For example, mobilization and demobilization of the pile installation equipment and contractor's spread may be paid for on a lump-sum basis with payment for piles on either a principal-sum or a unit-price basis.

Other Items.   Other items of work relating to pile installation are frequently paid for on a unit basis. Typically these are items which may or may not be required or, if required, the extent or quantity is not known in advance. Examples are:

Load testing, per test (for stated total load)

Reinforcing steel, per pound (1 lb = 0.4536 kg)

Preexcavation (drill or jet), per foot or per pile (1 ft = 0.3048 m)

Overburden footage, per foot

Retapping piles, per pile or per rig-hour

Use of casing (bored piles), per foot

Extra grout (auger-grout piles), per cubic yard (1 yd = 0.7646 m^3)

Some items of work, such as extra moving of the pile-driving equipment, cribbing, spudding, removal of obstructions, and delays caused by others, may be paid for on an hourly or a daily rate.

Force-Account Work.   Sometimes the contractor will be authorized to perform work on a force-account basis. For such work, the contractor is usually reimbursed for all labor, material, and equipment involved with appropriate factors for tools, payroll costs, overhead, and profit. For force-account work, the inspector should verify that the work was performed together with the quantity and type of labor including supervision, the type of equipment, the quantity and type of materials, and the time involved in such work.

Pile Caps

Pile installation will generally be considered complete when piles are capped or otherwise incorporated into the structure. The proper design and construction of pile caps are necessary for the satisfactory performance of the pile foundation.

Cap Excavation.   Pile cap excavations must be of the required depth and lateral dimensions to provide for the proper embedment of pile butts and minimum edge distances. The bottom of the excavation should be level and clear of all loose material. If the soil at the bottom is very soft, a working mat should be installed. This base course could consist of about 4 in (101.6 mm) of lean concrete or about 6 in (152.4 mm) of well-graded gravel or crushed rock. The base course is not considered a part of the pile cap, and if it is installed, the excavation level should be lowered accordingly. Note that the size or shape of pile caps may be revised from that shown on the plans owing to redesigns required because of added piles or other field conditions.

Pile Butts.   Pile butts must extend the required embedment length above the bottom of the excavation or form (or top of the base course). The top and sides of all pile butts must be clean and free of all soil and other foreign matter.

Forms.   Forms must be sufficiently tight to prevent leakage of mortar and constructed and braced so as to retain the wet concrete without distortion. Shapes, positions, dimensions, and edge distances to piles should be checked for conformance with the plans or subsequent revisions. Subject to the approval of the engineer or if permitted by the specifications, side forms may be of soil, provided it will stand without caving in and the sides of the bank are cut neatly to the minimum required dimensions. Clearances to reinforcement greater than shown on the plans may be required. Proper provisions must be made for accurate positioning and support of anchor bolts.

Reinforcement.   Reinforcement should be of the specified type and size. It should be positioned and spaced according to the plans and subsequent redesigns and supported on metal or plastic chairs or small concrete blocks. All steel must be clean and free from dirt, rust scale, loose mill scale, oil, grease, and any other coating which could affect the bond. Reinforcing should have a minimum concrete cover of 3 in (76.2 mm).

Concrete.   All concrete must be in accordance with the specifications regarding type, strength, ingredients, and slump. See Pile Material: Concrete for Cast-in-Place Piles; Material Handling and Storage: Cement; Concrete Aggregates; and Pile Installation: Placing Concrete for Cast-in-Place Piles for applicable control provisions. For example, the maximum size of the coarse aggregate and the required slump will differ from those used for concrete for cast-in-place piles. Forms or excavations should not contain free water and concrete should be spread and compacted with vibrators. If soil is used for the sides of the form, all precautions must be taken to prevent caving of the soil during concrete placement. Anchor bolts or other embedments as called for on the drawings should be placed in accurate positions as concrete is poured or immediately thereafter.

Curing.   Until it is cured, concrete should be protected from extreme heat or cold. Precautions must be taken to prevent the concrete from a rapid loss of moisture during the curing period. Completed caps should be kept wet for at least 10 days after the placement of concrete. Note that wind can have a substantial drying effect; concrete could be protected by windbreaks or covers if necessary.

Payment.   Generally, pile caps are not constructed by the pile contractor. However, if such work is included in the pile contract, the inspector should verify that all pile caps were completed according to plans and specifications or subsequent redesigns. Pile caps may be paid for on a unit basis per cubic yard, or separate unit prices may apply to such things as excavation, formwork, reinforcement, and concrete. Depending upon contract payment provisions, the inspector should keep accurate records of appropriate quantities involved. If pile caps are redesigned, the required additional quantities should be recorded separately.

5.11.0 Special-Type Piles

Special-type piles include the bored pile, the auger-grout pile, the cast-in-place pile, the compacted-concrete pile, the enlarged base pile, and the minipile. Each of these requires special installation and inspection considerations. There are other types of special piles, but these are not commonly used.

For some of the following described special-type piles, there are no means such as a measure of penetration resistance to reveal variations in subsoil conditions during installation. Therefore, more closely spaced test borings may be required to properly define the elevation, thickness, and adequacy of the bearing strata.

5.11.1 Bored Pile

This type of pile includes those uncased cast-in-place-concrete piles installed by drilling a hole to the required depth and filling the hole with concrete. There are three basic installation methods: (1) the dry method, (2) the casting method (often combined with a slurry), and (3) the slurry-displacement method.

The dry method is applicable in those soils in which the drilled hole will remain open for placement of concrete. Where soil conditions, such as soft soils of low shear strength or granular soils with no cohesion, can cause caving or sloughing of the soil, the casing method is used. A temporary casing or liner is installed through the layer of unstable soil and sealed off in a stratum of impermeable soil. If the layer of unstable soil is relatively thick, it may be necessary to use a drilling mud slurry to maintain the hole until the casing can be set. After the casing has been set, normal drilling proceeds and the hole is cleaned out before concrete is placed. The casing is generally removed during concrete placement. If the temporary casing cannot be sealed off or if the soils are predominantly unstable, the slurry-displacement method can be used. During drilling, the hole is kept filled with a slurry which is displaced during the concreting operation. Generally these piles are uncased, but sometimes ground conditions necessitate leaving a steel casing or liner in place.

Required Lengths.   Required pile lengths may be specified for each area of similar subsurface conditions. However, additional subsoil information may be required. The inspector should be alert to any changes in the drilling operation which may indicate variations in subsoil conditions, such as changes in the rate of advance of the drill or changes in the type or consistency of the soil cuttings. Any observed changes should be investigated in case pile lengths may have to be adjusted. If the drilled hole is large enough and a temporary steel liner is inserted, the soil at the bottom of the hole can be inspected or tested for adequacy. In other cases a test boring can be made within the drilled hole and soil or rock samples recovered. Such test borings can be used to determine the adequacy of the bearing stratum.

Unstable Soils.   If the hole is to be drilled through soft clay, soft silt, peat, or loose granular soil, it will be necessary to install a steel casing through the zones of unstable soils or to use a drilling mud slurry to maintain the hole. If the soil is extremely soft, the steel casing should not be removed.

Groundwater Control.   If groundwater is permitted to enter the hole either through the side or from the bottom, the rate of flow may be such as to cause soil erosion, resulting in contamination of the concrete, collection of unsatisfactory bearing material in the bottom of the hole, or loss of bearing capacity of the soil at the bottom of the hole. In extreme cases, such conditions could preclude drilling the hole to the required depth. Temporary or permanent steel liners should be inserted and, if necessary, the hole should be filled with water or slurry to balance the hydrostatic head. If an artisan aquifer is encountered, the water pressure may be such as to prevent the drilling of the holes or the proper placement of concrete. Under such conditions the drilling of relief wells may provide a solution.

Use of Slurry.   The slurry must be of sufficient density to offset hydrostatic and lateral soil pressures and maintain the hole. The shear strength characteristics of the slurry as determined by its viscosity and the density of the slurry should be controlled to obtain optimum conditions for displacement of the slurry during the concreting process. Tests to determine density, viscosity, and shear strength should be carried out initially to establish a suitable mix.

All reasonable steps should be taken to prevent contamination of the slurry. Discarded slurry which has been pumped or displaced from a drilled hole should be removed from the site. If slurry is to be reused, its quality should be checked periodically.

Cleaning the Bottom of the Holes.   After drilling to the required depth, the bottom of the holes should be cleaned of loose materials by using a cleanout bucket. If the hole is underreamed, the underreaming tool should not be permitted to ride up while underreaming is in process.

Inspection of Bottom.   If it is necessary to enter the hole for direct inspection of the bearing material or conditions at the bottom of the hole, the hole must be a minimum of 24 in (609.6 mm) in diameter and a full-length steel casing must be inserted. All safety precautions, such as ensuring that the hole is free of noxious gas, groundwater is controlled, methods of communication are adequate, lifelines are furnished, and support personnel are in attendance at all times, must be taken.

Placing Concrete.   Concrete may be placed conventionally for both the dry and casing methods, provided the drilled hole can be dewatered. See Placing Concrete for Cast-in-Place Piles: Conventional Placement under Pile Installation. Concrete must be deposited vertically in the center of the hole through a funnel hopper. The chuting of concrete directly into the hole, allowing it to hit the side of the hole, will cause the mix to segregate and must be prohibited. For the slurry displacement method or when concrete is to be placed through water, the tremie method should be used. See Placing Concrete for Cast-in-Place Piles: Tremie Placement under Pile Installation. Concrete should be of sufficiently high slump to prevent arching in the steel casing. See Placing Concrete for Cast-in-Place Piles for other concreting requirements.

To prevent contamination of concrete or an accumulation of unsatisfactory material at the bottom of holes, all holes deeper than about 20 ft (6 m) or which extend through soils which could slough off during concrete placement should have a steel liner or casing inserted before concrete is poured. A steel liner may also be required to seal off the flow of groundwater, which could affect the quality of the concrete.

Withdrawal of Casing.   If the steel casing or liner is to be withdrawn as concrete is placed, it is recommended that withdrawal be by the vibratory method to reduce the possibility of concrete arching in the casing. As the casing is withdrawn, observations should be made to ensure that the concrete is not being raised up with the casing and that the concrete level is always above the bottom of the casing. The height of concrete above the bottom of the casing should be sufficient to resist lateral soil pressures. This is to prevent the necking down of the concrete shaft or complete discontinuity in the shaft as the soil tends to squeeze in. It also ensures positive displacement of all slurry. After each use, the temporary steel casing or liner should be thoroughly cleaned of all concrete to prevent a buildup of hardened concrete, which could contribute to arching and cause the fresh concrete to be lifted with the casing.

Casing Left in Place.   Steel casings or liners should be left in place through zones of extremely soft soils or to protect the concrete from groundwater movements.

Drop in Concrete Level.   If the level of concrete in the pile shaft drops after completion, this may be caused by the weight of concrete pushing into soft surrounding soils. This can be prevented by leaving a steel casing in place. Before additional concrete is placed to bring the pile to the required cutoff grade, reinforcing steel dowels should be inserted in the top of the pile and all laitance and contaminated concrete removed.

Installation Sequence.   When piles are to be installed through soft soils and especially if permanent steel liners or casings are not used, it may be necessary to install piles in a staggered sequence to permit the concrete in completed piles to set up before adjacent piles are drilled. Otherwise, the weight of unset concrete may be sufficient to break through the wall of soft soil between the completed pile and the pile being drilled.

Installed Length.   The installed lengths of all bored piles should be recorded. If possible, the elevations of the tops of piles should be recorded and referenced to a permanent fixed datum. Pile lengths can be determined by measuring with a steel tape the depth of the hole from a fixed reference and deducting the distance down to the cutoff grade.

5.11.2 Auger-Grout Pile

This type of pile is installed by drilling a hole with a continuous-flight hollow-stem auger and pumping grout down through the hollow stem, filling the hole as the auger is withdrawn. It is also called an *auger cast pile.*

Required Lengths.   Required pile lengths may be specified for each area of similar subsoil conditions. However, additional subsoil data may be required. The inspector should be alert to any changes in the drilling operation which may indicate different subsoil conditions, such as torque required to drill, the rate of advance of the auger, or changes in type or consistency of soil cuttings. Any observed changes should be investigated as to cause in case pile lengths may have to be adjusted.

Equipment.   The auger flight should be continuous from top to bottom with no gaps or other breaks. The discharge hole at the bottom of the auger should be below the bar containing the cutting teeth. Augers over 40 ft (12.2 m) long should be laterally supported by intermediate movable guides spaced a maximum of 20 ft (6.1 m) apart. The mortar pump should be a positive-displacement piston pump capable of developing pressures at the auger tip during grout pumping in excess of any hydrostatic or lateral soil pressures. A pressure gauge should be mounted in the grout line as close to the auger head as is practical. The grout pump should be equipped at the discharge end with a metering device capable of accurately measuring the volume of grout pumped.

Grout.   The grout should be of proper consistency to be pumped yet not too fluid. If the water-cement ratio is less than 0.45, the grout mix is too dry. If the grout splashes when poured into a container of grout or if, when flowing down a chute, the mix gets a glossy sheen (sand grains disappear below water and cement surface), the grout is too wet. Strengths will be determined from cube tests, ASTM C91. Admixtures are used to improve the pumpability of the grout and to retard its set.

Drilling.   When drilling to the required depth, the auger should be advancing continuously at a rate which would prevent removing excess soil. If the auger is advancing too slowly or rotation is continued without advancement, the removal of excess soil could have a detrimental effect on adjacent structures or piles. After reaching the required depth, rotation of the auger should stop. Auger flights full of soil assist in maintaining the hole and subsequent pumping pressures, and help prevent removal of excess soil.

Pumping Grout.   At the start of pumping grout, the auger should be raised from 6 to 12 in (152.4 to 304.8 mm), and after grout pressure has built up, indicating discharge of grout, the auger should be redrilled to the original depth before forming the pile. A positive pressure must be maintained during grout pumping. To ensure a properly filled hole, the pumping pressure, considering all line losses, should be greater than any hydrostatic or lateral soil pressures. However, when installing such piles in very soft soils, excessively high pumping pressures should be avoided. Such pressures could cause upward or lateral movement of unset adjacent piles. Under such conditions it is advisable to install piles in a staggered sequence. Pumping pressures are normally reduced during final withdrawal of the auger.

Withdrawing Auger.   During pumping of grout to form the pile the auger should be withdrawn in a smooth, continuous motion and not in jerks or lifts. The auger may turn very slowly during withdrawal, but in no case should a counterclockwise rotation be permitted. The rate of withdrawal should be such that a positive pressure is maintained in the grout line at all times. Unless these precautions are taken, necking down or discontinuities of the grout shaft could occur, or the grout column could contain soil inclusions.

Interruptions and Pressure Drops. The formation of each pile should be a continuous, uninterrupted operation once grouting has started. If the grouting process is interrupted or grouting pressures drop below acceptable levels, the auger should be redrilled to the original tip elevation and the pile re-formed.

Volume of Grout. The volume of grout placed should be greater by about 15 percent than the theoretical volume of the hole created by the auger. The normal wobble of the auger will create a hole larger than theoretical. If a considerable excess of grout is pumped, the cause should be investigated. The excess may be due to excessive pumping pressures in very soft soils or to the grout flowing into fissures, crevices, or solution channels and cavities in limestone formations. Grout may also be lost in underground structures such as sewer lines. The injection of excess grout may be totally wasteful or damaging to underground or adjacent structures.

Forming Pile Butts. To form and protect the pile butt properly, a steel sleeve should be placed at the top of the pile before grouting that portion of the pile and removing the auger. The steel sleeve should extend from pile cutoff grade or ground surface, whichever is higher, to a point not less than 1 ft (0.3048 m) below pile cutoff grade or the ground surface, whichever is lower, and should be left in place. If a steel sleeve is used, the ground surface adjacent to the pile should be at least 1 ft higher than pile cutoff grade and the hole should be filled with grout to the ground surface; the pile butt is then trimmed to grade. Excess grout should be pumped to displace as much potential laitance as possible.

Placing Reinforcing Steel. Reinforcing steel commonly consists of single bars or cages conforming to ASTM A615, Grade 60, or high-strength threaded bars with yield strengths of 150 or 160 ksi conforming to ASTM A722. The advantage of threaded bars is ease of splicing and installation into long piles as compared with steel cages. Reinforcing steel is normally placed after the pile is grouted to grade.

Drop in Grout Level. Completed piles which are still unset should be checked periodically during and after installation of adjacent piles to see if the grout level is maintained. If any drop in grout level occurs in a completed pile during the drilling of adjacent piles, the completed pile should be rejected and replaced. If the grout level subsides when no adjacent piles are being installed, the subsidence may be caused by the fluid grout squeezing out into soft soils. The cause of subsidence and the extent of pile damage should be investigated by a test boring and core drilling of the pile shaft. If the grout level drops and the pile is salvageable, precautions should be taken to prevent contamination of grout at the top of the pile resulting from sloughing soils or surface water. Before additional grout or concrete is placed to bring the butts of satisfactory piles to grade, all laitance and contaminated grout should be removed and steel dowels should be placed at the joint. If the grout has set, it would be preferable to increase the depth of the pile cap rather than build up the pile.

Installed Length. The installed length of each auger-grout pile can be determined by knowing the total length of the auger used and observing the depth or elevation to which the auger tip is drilled before grouting starts. It may be convenient to have the leaders marked off in 1-ft (0.3-m) increments and an index mark on the top drill guide to measure auger penetration. Depths or elevations should be referenced to a fixed datum. Pile lengths can be calculated from tip and specified butt elevations or distances to the tip and butt from a fixed reference.

Payment Quantities. Auger-grout piles should be paid for in accordance with Pile Lengths under Pile Installation. If grout is placed above cutoff grade, no extra payment should be made for pile lengths above cutoff or for trimming pile tops back to cutoff grade.

The contract may provide for payment per cubic yard for the volume of grout pumped in excess of the theoretical volume of the piles. In such cases grout metering devices must be accurate, and the inspector must keep records of the actual volume of grout pumped. Waste grout resulting from spillage or excess pumping once the pile hole has been filled should not be included in this pay item.

5.11.3 Cast-in-Place Pile

This type of pile includes those uncased cast-in-place concrete piles which are installed by driving a casing with a sacrificial closure plate or a removable internal mandrel, filling the casing with concrete or grout during or after driving, and withdrawing the casing.

Required Lengths.   Required pile lengths may be specified for each area of similar subsoil conditions. The required driving criteria may also include a minimum penetration resistance. Penetration resistance during driving may include frictional resistance, which could be altered during pile installation by withdrawal of the casing. In some cases the pile-driving logs may give an indication of varying subsoil conditions. If substantial variations are indicated, the cause should be investigated by means of a test boring.

Spacing.   Normal pile spacing is from 4 to 5 times the pile diameter. This spacing may have to be increased if piles are installed in very soft soil or boulder formations unless the piles are driven in a staggered sequence.

Driving Sequence.   If piles have to be driven in a staggered sequence, the concrete or grout in a completed pile should be at least 3 days old before the adjacent piles are driven.

Placing Reinforcing Steel   *Steel to Resist Bending.*   Reinforcing steel required to resist bending should be fabricated into cages of specified diameter, and if the embedment length is longer than about 5 ft (1.5 m), it should be installed centered in the casing before concrete or grout is placed. The steel cage must be prevented from coming up as the casing is withdrawn. If the embedded length of the cage is not greater than about 5 ft, it may be pushed into the fluid concrete or grout column. Such steel may be carefully centered in the pile and accurately aligned with the pile's longitudinal axis. See Placing Reinforcing Steel under Pile Installation.

Steel to Resist Uplift.   Reinforcing required to resist uplift should be full-length. It may be placed in the center of the concrete or grout column, provided the control provisions for placing steel in this manner are in accordance with Placing Reinforcing Steel above.

Placing Concrete or Grout.   After the casing has been driven to the required depth, concrete may be placed in accordance with Placing Concrete for Cast-in-Place Piles under Pile Installation. Concrete must be of sufficiently high slump to prevent arching in the casing as it is withdrawn. If concrete or grout is placed during the driving operation, it must be of very high slump, 10 in or more (254 mm or more). The normal slump test is not applicable. The suitable viscosity of the mix is determined by a flow-cone test. A water-reducing and -retarding admixture must be used to give at least a 4-hour retardation. Grout strengths will be determined by cube tests, ASTM C91.

Filling Pile Hole to Cutoff Grade.   To ensure that the pile hole is filled to cutoff grade, it may be necessary to have a sufficient extra length of casing filled with concrete or grout to compensate for the concrete or grout which will fill in the space occupied by the steel casing as it is withdrawn. Alternatively, extra grout or concrete could be added during withdrawal of the casing. Also, the upper portion of the pile hole could be cased off with a steel liner to prevent soil from caving in on top of the concrete or grout column after the casing is withdrawn and to permit placing additional concrete or grout to grade. If the liner is withdrawn, the concrete or grout must be sufficiently fluid to prevent it from arching and coming up with the liner. Withdrawal by vibratory means should help prevent this.

When installing this type of pile, it is often difficult to fill the hole precisely to the required cutoff grade. In most cases excess concrete or grout will be placed, and pile butts must be trimmed back to the required elevation.

Withdrawing Casing.   The driven casing should be withdrawn by vibratory methods to help prevent arching of the concrete or grout in the casing. The casing should be thoroughly cleaned periodically to prevent a buildup of hardened concrete or grout which may cause arching. This should be done at each interruption in the installation schedule no matter how short.

Protection of Butt.   A short steel sleeve should be placed at the top of the pile to protect the unset concrete or grout at the butt and to ensure proper filling without contamination. This is especially necessary if the cutoff grade is below the ground surface. The sleeve should not be removed.

Pile Heave.   If ground heave occurs during driving of adjacent piles, all completed piles should be checked for evidence of pile heave. If pile heave is detected, all heaved piles should be abandoned and replaced.

Obstructions.   If subsurface obstructions including boulders are known to exist, the piles should be installed in a staggered sequence, permitting the concrete or grout to harden before driving adjacent piles. If the ground is heavily obstructed, completed piles should be carefully observed for signs of movement or damage when adjacent piles are driven. If movement or damage is evident, further investigation may be required to determine whether or not the pile is satisfactory, or the pile may have to be rejected and replaced.

Concrete or Grout Subsidence.   When installing piles in extremely soft soils, the concrete or grout level of completed piles should be checked for subsidence. If the concrete or grout level in the pile has subsided, the cause and possible damage to the pile should be investigated. This can be done by making a test boring alongside the pile, by excavation, or by coring the pile. Before additional concrete or grout is placed to bring the tops of satisfactory piles to the required grade, all laitance and contaminated concrete or grout should be removed and steel dowels inserted in the tops of the piles.

Lateral Support.   If, after withdrawing the casing, a space is left between the pile and the soil, the space should be filled with fluid grout or by washing in sand to reestablish the lateral support of the soil.

Installed Length.   The installed lengths of cast-in-place piles can be determined by knowing the length of drive casing used and observing the depth or elevation to which the bottom end is driven. The casing should be marked off in 5-ft (1.5-m) increments with incremental marks numbered according to distance from the bottom. Depths or elevations should be referenced to a fixed datum. Pile lengths can be calculated from pile tip and specified butt elevations or distances to the tip and butt from a fixed reference.

5.11.4 Compacted-Concrete Pile

Compacted-concrete piles are also called pressure-injected footings and Franki piles. This type of pile is installed by (1) driving to the required depth a steel casing or drive tube closed at the bottom with a gravel or zero-slump concrete plug using a drop weight operating inside the casing, (2) restraining the casing from further penetration while driving out the closure plug, (3) placing small batches of zero-slump concrete in the bottom of the casing and ramming out each batch to form an enlarged base, and (4) forming either an uncased or cased concrete shaft to the required pile cutoff grade. For uncased shafts, small batches of zero-slump concrete are placed in the casing and rammed out with the drop weight as the casing is withdrawn. For cased shafts, a steel shell or pipe is placed inside the drive casing and, with a small amount of zero-slump concrete at the bottom, is tapped into the enlarged base. The drive tube is then withdrawn and the shell filled with conventional concrete.

Required Lengths.   The bases for compacted-concrete piles should be formed in granular materials, which should be of adequate thickness to contain the compaction resulting from forming the base and to spread the load so as not to overstress underlying soils. The elevations or depths at which pile bases are to be formed will be specified. It is necessary that the vertical location and thickness of the bearing stratum as well as the character and properties of the underlying soils be clearly defined by test borings.

Spacing.   The minimum spacing for compacted-concrete piles is 4.5 ft (1.35 m). Greater spacing may be required for piles installed with uncased shafts in soft soils or boulder formations unless the piles are installed in a staggered sequence.

Driving the Casing.   The bottom of the drive casing or tube should be closed with a gravel or zero-slump concrete plug of sufficient thickness so that it will arch in the drive tube and not be driven out under the blows of the drop weight. If the plug is driven out, the tube must be withdrawn and redriven with an adequate plug. The drive tube should be driven to the depth at which the enlarged base is to be formed. The inspector should observe the driving of the casing in comparison with that of a test pile or a pile installed at a test-boring location. If a radical change is noted, the cause should be investigated by means of a test boring to determine if pile lengths need to be adjusted.

Concrete.   All concrete should conform with specifications for cast-in-place piles, except that concrete for forming the bases and uncased shafts of compacted-concrete piles should be of zero slump. However, the concrete should contain enough water to ensure hydration of the cement.

Hammer Energy.   The weight of the drop hammer must be known to calculate the fall required to deliver the specified energy. To check on the hammer fall and thus the delivered energy, a mark should be placed on the hammer line which will show above the top of the casing when the hammer is raised to its full required height. The drive casing should be at least as long as the required fall plus the length of the drop hammer. During the fall, the hammer should not be restrained by the operator, nor should the hammer line be snubbed just at impact. As the hammer falls, it must overhaul the hammer line, and therefore all sheaves, drums, etc., must turn freely and easily. The inspector must carefully observe the actions of the rig operator and the operations of the hammer to determine whether or not the hammer energy is being seriously affected.

Formation of Base.   After being driven to the required depth, the drive tube should be raised not more than about 6 in (152.4 mm) and restrained from further penetration. If, in driving out the closure plug, the entire plug is expelled, creating a void below the casing or allowing soil and/or water to enter, the casing should be withdrawn and redriven with an adequate plug.

In forming the enlarged base, batches of zero-slump concrete of known volume should be placed in the bottom of the drive tube and rammed out with the drop weight, delivering a specified energy. For the design load involved, the specifications will indicate the minimum number of hammer blows of a stated energy required to drive out the last batch of concrete of a stated volume. The minimum volume of the base may also be specified.

During base forming, a sufficient height of dry concrete must be kept in the drive tube at all times to maintain a seal and exclude all water and soil. If water or soil enters the tube, the pile must be abandoned and redriven. The inspector should keep a constant record of the quantities of concrete placed in the tube, the elevation of the bottom of the tube, and the elevation of the top of the compacted concrete. The intrusion of soil can be monitored by observing the position of a mark on the hammer line relative to the top of the drive tube as each charge of concrete is placed. After each batch of concrete has been placed in the tube, the distance to which the mark rises must not exceed the additional depth of concrete in the tube. If the mark rises above the calculated increase in depth of concrete, it must be assumed that foreign matter has entered the tube, and the pile should be abandoned. The entry of water will be indicated by a softening of the hammer blows.

If, in attempting to form the base for any pile, the total volume of concrete placed is more than 50 percent greater than that for a test-pile base, the cause should be investigated. A test boring may reveal a change in subsoil conditions requiring an adjustment in pile lengths. Intrusion of groundwater may soften the concrete, precluding the formation of a normal base. Possible interference with adjacent pile bases should be considered, and if necessary, pile spacing may have to be increased.

Formation of Shaft   *Uncased Shaft.*   The number of hammer blows and the energy required for driving out each batch of concrete in forming the shaft will be specified; the volume of each batch of zero-slump concrete will also be specified. In forming the shaft, the concrete must not be driven out below the bottom of the drive casing. The concrete level in the drive tube should be kept at a sufficient height above the bottom of the tube at all times to form an adequate seal. This can be monitored in a manner similar to that described under Formation of Base. The casing should be withdrawn as the concrete is rammed out and not in lifts. If an adequate seal is not maintained, water and soil could enter the tube and contaminate the concrete. If the concrete is driven out below the bottom of the drive tube or if the tube is raised in lifts, water and soil may fill the void created, producing a reduction (necking) of the shaft cross section or even a complete interruption of the concrete shaft. If soil or water enters the drive tube or the concrete shaft, the pile should be rejected. If the shaft is to be formed through unstable or organic or extremely soft soils, a cased shaft is recommended.

Cased Shafts.   The steel shell or pipe inserted in the drive tube should be adequately connected to the pile base before the drive tube is withdrawn to ensure a positive seal and a good, sound joint between base and shaft. This is usually done by placing a small batch of zero-slump concrete in the shell and tapping it into the base with the drop weight. The inner shell or pipe should be restrained from moving upward as the drive tube is withdrawn. After withdrawal of the drive tube and before the shell is filled with concrete, the bottom of the shell of pipe should be inspected to determine whether or not any soil or water has entered. If soil has intruded, the pile should be rejected. If only water has entered, it should be removed before concrete is placed. Concrete should be placed in the shell or pipe in accordance with Placing Concrete for Cast-in-Place Piles under Pile Installation. The annular space between the cased shaft and the soil should be filled with grout or clean sand, washed down to reestablish lateral support.

Reinforcing Steel *Uncased shaft.*   The formation of zero-slump concrete shafts containing reinforcing cages can easily result in damage to the cage or inadequate concrete cover. Cased shafts are much preferred. The reinforcing steel is made up in cages and placed inside the drive tube after the base has been formed or just before the last batch of base concrete is compacted. If the steel cage is placed after the base has been formed, a small batch of zero-slump concrete should be placed in the tube and tapped down with the drop weight operating inside the cage to anchor the cage to the base. The formation of the shaft should be in accordance with Formation of Shaft: Uncased Shaft, with the drop hammer operating inside the cage. Care must be exercised to ensure that the cage is not lifted when the drop weight is raised or damaged as the weight falls. If the cage consists of a number of large-diameter longitudinal bars enclosed in a spiral steel on a small pitch, it may be difficult to ram zero-slump concrete out through the steel cage without deforming the cage; the use of a more workable concrete mix or a cased shaft may be necessary. Extra caution must be exercised in forming the upper portions of the pile shaft to ensure that the drop hammer is not raised out of the cage and allowed to damage or dislodge the reinforcement as it drops.

Cased Shafts.   For cased shafts, the reinforcing steel should be placed inside the steel shell or pipe before concrete is poured and in accordance with Placing Reinforcing Steel under Pile Installation.

Heave.   If pile heave occurs (see Pile Heave under Pile Installation), all compacted-concrete piles that have heaved must be rejected and replaced. If only ground heave occurs, piles with uncased shafts should be rejected.

Pile Butts.   When forming uncased shafts, it is difficult to form the shaft precisely to cutoff grade. The top of the pile shaft should be at or above the required elevation and trimmed back if necessary.

Records.   In addition to the normal records of installed lengths and payment quantities, the inspector should keep an accurate record of the volume of concrete in each pile base, the number of hammer blows and energy required to compact the last batch of concrete [generally 5 ft^3 (0.1416 m^3)] in forming the base, and the elevation at which each base was formed.

Installed Length.   The installed length of a compacted-concrete pile can be determined from the elevation or depth to which the bottom of the drive tube is driven with reference to a fixed datum and the cutoff elevation or depth referenced to the same datum. To facilitate taking measurements, the drive tube should be marked off in 1-ft (0.3-m) increments with each 5-ft (1.5-m) mark numbered with the distance from the bottom.

Pay Length.   The pay length of a compacted-concrete pile should be that measured from the depth to which the drive tube is driven to the cutoff grade. No extra payment should be made for bases regardless of size or for trimming off pile butts.

Load Tests.   The actual capacity of a compacted-concrete pile is a function of many variables such as the density, thickness, and physical properties of the granular stratum in which the base is formed, the type and bearing capacity of the underlying soils, the proper shape and formation of the base, the degree of compaction of the surrounding soils, the dynamic energy applied in forming the base, and the integrity of the pile shaft and its connection with the base. The method of installation and the final product are sensitive to the actions of the rig operator.

For these reasons the actual capacities of the piles may require periodic proof load tests. This is especially important if the results of closely spaced reliable test borings are not available. Piles to be tested should be selected at random by the engineer.

5.11.5 Enlarged-Base Pile

This type of pile is generally installed by driving with or without a mandrel, a steel shell, or pipe to which is attached an enlarged precast-concrete base at the tip. After the base is driven to the required depth, the mandrel (if used) is removed and the shell or pipe is filled with concrete to the required cutoff grade. The enlarged base could also be attached to a steel H pile or formed with a precast concrete pile. The enlarged-base pile is similar in function to the compacted-concrete pile and requires many of the same controls.

General.   The applicable control provisions under Pile Material, Material Handling and Storage, and Pile Installation for the type of pile shaft used must be enforced. The following are special provisions for the enlarged-base pile.

Required Lengths.   The required pile lengths or the elevation to which the enlarged base is to be driven may be specified. The development of high capacity for this type of pile requires that the base be driven into a stratum of granular soil which can be compacted by the displacement action of the base. The bearing stratum must be of adequate thickness and underlain by satisfactory soils. More closely spaced test borings may be required to define the type, thickness, and vertical position of the bearing stratum and the adequacy of the underlying soils. If sufficient and reliable subsoil information is available to establish the required elevations of the pile bases, final penetration resistance may be correlated with that of test piles.

Spacing.   The center-to-center spacing of enlarged-base piles should be at least 4.5 ft.

Base Construction.   Precast-concrete bases should be of proper design and construction to withstand driving stresses without damage and must conform with the requirements of the plans and specifications. For inspection details see Pile Material: Precast-Concrete Piles for plant-produced bases or Concrete for Cast-in-Place Piles for bases made on the jobsite.

Base-to-Shaft Connection.   The pipe or shell should be attached to the base to prevent the entry of soil or water. If the pile is to resist uplift loads, the joint between the base and the casing or the concrete filling must be designed and installed to transfer the tension load into the base. Joints between bases and any type of pile shaft must be capable of resisting all driving and service load stresses. For short piles the joint may be subjected to bending from lateral loads resulting from ground movements or construction activity. See Piles Off Location under Pile Installation.

Uncased Shafts.   The use of uncased cast-in-place-concrete shafts with enlarged precast-concrete bases should not be permitted. The chances of installing an adequate foundation are quite remote. If such shafts are used, all the applicable control provisions under Cast-in-Place Pile, Compacted-Concrete Pile, and Enlarged-Base Pile must be strictly enforced.

Heave.   If ground or pile heave occurs (see Pile Heave under Pile Installation), check levels should be taken on the tops of all driven piles until driving progresses beyond heave range. For cased cast-in-place concrete shafts, check levels should also be taken on telltales bearing on the bases.

Heaved piles may be redriven as necessary provided the joints between base and casing have not separated. If the joint cannot take the tension resulting from heave and if check levels indicate possible joint separation, the pile should be rejected.

Relaxation.   If relaxation occurs (see Relaxation under Pile Installation) and if pile bases are not driven to hard material, consideration should be given to changing the pile type or assigning a lower-design capacity to each pile.

Reinforcing Steel.   If internal reinforcing steel is required for cased concrete shafts, it should be placed in accordance with Placing Reinforcing Steel under Pile Installation. If such steel is required to resist uplift loads, it must be positively connected to the precast base to transfer the tensile forces.

Concrete.   Concrete for cased cast-in-place shafts should conform with the control provision of Concrete for Cast-in-Place Piles under Pile Material. Placing Concrete for Cast-in-Place Piles under Pile Installation covers the placement of such concrete.

Lateral Support.   The annular space between the soil and the pile shaft created by the enlarged base must be backfilled to provide lateral support to the pile. The natural sloughing or caving of soils may not be sufficient to restore lateral support. Grout can be pumped into the annular space or clean sand washed down if the space remains open. If the space closes in with the possibility of voids below, the area around each pile shaft should be tamped and backfilled as necessary. In extreme cases, it may be necessary to remove the upper sloughed-in soil in order to place the backfill properly. These are generally high-capacity foundation units and require full lateral support for stability.

Records.   In addition to the normal records of installed lengths and payment quantities, the inspector should keep an accurate record of the elevation to which each enlarged base was driven. Such elevations must be referenced to a permanent datum.

Installed length.   The installed length of the pile shaft can be determined in accordance with Pile Lengths under Pile Installation. To this length may be added the length or height of the base. However, if the measured length of the pile shaft includes a portion recessed into the base, the amount of the recess should not be included in the length of the precast base to be added.

Pay Quantities.   The pay length of an enlarged-base pile might not include the precast base. Bases might be paid for separately at a unit price.

Load Tests.   The actual capacity of an enlarged-base pile depends upon several factors such as the density, thickness, and physical properties of the bearing stratum, the degree of compaction of the soil resulting from driving the base into it, and the type and bearing capacity of the underlying soil.

  If final penetration resistances vary and if the results of closely spaced reliable test borings are not available, the capacities of the piles may require periodic proof load testing. Piles to be tested should be selected at random by the engineer.

5.11.6 Minipile

Minipiles, also called micropiles, are grouted cast-in-place piles which are 4 to 12 in (10.16 to 30.48 cm) in diameter. The pile is formed in a hole advanced by rotary or rotary percussive drilling methods (with or without temporary casing) or by driving a temporary casing. Typically they are reinforced and a portion or all of the pile is cast directly against soil without permanent casing.

Required Lengths.   Required pile lengths may be specified for each area of similar subsoil conditions. However, additional subsoil data may be required. The inspector should be alert to any changes in the drilling operation which may indicate different subsoil conditions, such as torque required to drill, the rate of advance of the casing or auger, and changes in type or consistency of soil cuttings or washings. Any observed changes should be investigated as to cause in case pile lengths may have to be adjusted.

Equipment.   The 4- to 12-in (10.16- to 30.48-cm) diameter casing is typically installed in 5-ft (1.52-m) lengths. The grout pump should be a positive-displacement piston pump capable of developing pressures at the pile tip during grout pumping in excess of any hydrostatic or lateral soil pressures. A pressure gauge should be mounted in the grout line as close to the drill head as is practical. The grout pump should be equipped at the discharge end with a metering device capable of accurately measuring the volume of grout pumped.

Grout.   The grout should be of proper consistency to be pumped yet not too fluid. If the water-cement ratio is less than 0.45, the grout mix is too dry. If the grout splashes when poured into a container of grout and the mix gets a glossy sheen (sand grains disappear below water and cement surface), the grout is too wet. Strengths will be determined from cube tests, ASTM C91. Admixtures are used to improve the pumpability of the grout and to retard its set.

Drilling.   When drilling to the required depth, the casing should be advancing continuously at a rate which would prevent removing excess soil. If the casing is advancing too slowly, the excess soil may be removed by washing, which could have a detrimental effect on adjacent structures or piles. After reaching the required depth, the grout hose is connected to the drill head.

Pumping Grout and Withdrawing Casing.   The grout is introduced at the bottom of the hole either by a tremie or through the drill casing. After filling the hole with grout, extraction of the casing begins. The casing is removed in stages and the grout inside the casing is normally pressurized at each stage. Pumping pressures are normally reduced during withdrawal of the final sections of casing.

Interruptions and Pressure Drops.   The formation of each pile should be a continuous, uninterrupted operation once grouting has started. If the grouting process is interrupted, the casing should be redrilled to the original tip elevation and the pile re-formed.

Volume of Grout.   The volume of grout will exceed the theoretical volume of the hole created by the drill casing due to the method of drilling and, when the mix has too much water, due to the bleeding of water into the surrounding soil. If a considerable excess of grout is pumped, the cause should be investigated. The excess may be due to excessive pumping pressures in very soft soils or to the grout flowing into fissures, crevices, or solution channels and cavities in limestone formations. Grout may also be lost in underground structures such as sewer lines. The injection of excess grout may be totally wasteful or damaging to underground or adjacent structures.

Forming Pile Butts.   To form and protect the pile butt properly, a steel sleeve should be placed at the top of the pile before grouting that portion of the pile and removing the casing. The steel sleeve should extend from pile cutoff grade or ground surface, whichever is higher, to a point not less than 1 ft (0.3048 m) below pile cutoff grade or the ground surface, whichever is lower, and should be left in place. If a steel sleeve is used, the ground surface adjacent to the pile should be at least 1 ft higher than pile cutoff grade and the hole should be filled with grout to the ground surface; the pile butt is then trimmed to grade. Excess grout should be pumped to displace as much potential laitance as possible.

Placing Reinforcing Steel.   Reinforcing steel commonly consists of single bars conforming to ASTM A615, Grade 60, or high-strength threaded bars with yield strengths of 150 or 160 ksi conforming to ASTM A722. The advantage of threaded bars is ease of splicing. Reinforcing steel is normally placed after the pile is grouted to grade.

Drop in Grout Level.   Completed piles which are still unset should be checked periodically during and after installation of adjacent piles to see if the grout level is maintained. If any drop in grout level occurs in a completed pile during the drilling of adjacent piles, the completed pile should be rejected and replaced. Normally, pile installation is staggered to prevent this occurrence. Another cause of the grout level's dropping is bleeding of water from the grout mix. In this case the grout level should be topped off. This is very important in underpinning applications.

Installed Length.   The installed length of each minipile can be determined by knowing the total length of the casing used and observing the depth or elevation to which the casing tip is drilled before grouting starts. This requires that inspectors keep an accurate count of casing sections and that they know the length of the cutting bit. Depths or elevations should be referenced to a fixed datum. Pile lengths can be calculated from tip and specified butt elevations or distances to the tip and butt from a fixed reference.

Payment Quantities.   Minipiles should be paid for in accordance with Pile Lengths under Pile Installation. If grout is placed above cutoff grade, no extra payment should be made for pile lengths above cutoff or for trimming pile tops back to cutoff grade.

The contract may provide for payment per cubic yard for the volume of grout pumped in excess of the theoretical volume of the piles. In such cases grout metering devices must be accurate, and the inspector must keep records of the actual volume of grout pumped. Waste grout resulting from spillage or excess pumping once the pile hole has been filled should not be included in this pay item.

5.11.7 Monotube Piles

TAPERED SECTIONS

Type	Size Tip Point Diameter x Butt Diameter x Length	Weight Per Foot				Est. Conc. Vol. Cu. Yds.
		9 Ga. .1495"	7 Ga. .1793"	5 Ga. .2092"	3 Ga. .2391"	
F Taper .14 Inch Per Foot	8½" x 12" x 25' 8" x 12" x 30' 8½" x 14" x 40' 8" x 16" x 60' 8" x 18" x 75'	17 16 19 20 -	20 20 22 24 26	24 23 26 28 31	28 27 31 33 35	0.43 0.55 0.95 1.68 2.59
J Taper .25 Inch Per Foot	8" x 12" x 17' 8" x 14" x 25' 8" x 16" x 33' 8" x 18" x 40'	17 18 20 -	20 22 24 26	23 26 28 30	27 30 32 35	0.32 0.58 0.95 1.37
Y Taper .40 Inch Per Foot	8" x 12" x 10' 8" x 14" x 15' 8" x 16" x 20' 8" x 18" x 25'	17 19 20 -	20 22 24 26	24 26 28 31	28 30 33 35	0.18 0.34 0.56 0.86

EXTENSION SECTIONS

Type	Diameter x Length	9 Ga. .1495"	7 Ga. .1793"	5 Ga. .2092"	3 Ga. .2391"	Cu.Yd./Ft.
N 12	12"x 12"x 20'/40'	20	24	28	33	.026
N 14	14"x 14"x 20'/40'	24	29	33	38	.035
N 16	16"x 16"x 20'/40'	27	32	38	44	.045
N 18	18"x 18"x 20'/40'	-	37	43	49	.058

PHYSICAL PROPERTIES

Steel Thickness	Tips		Butts of Pile Sections																
	8 In.	8½ In.	12 In.				14 In.				16 In.				18 In.			
	A In.²	A In.²	A In.²	I In.⁴	S In.³	r In.	A In.²	I In.⁴	S In.³	r In.	A In.²	I In.⁴	S In.³	r In.	A In.²	I In.⁴	S In.³	r In.
9 Gauge (.1495")	3.63	3.93	5.81	102	16.3	4.18	6.75	159	22.0	4.86	7.64	232	28.3	5.50	-	-	-	-
7 Gauge (.1793")	4.40	4.77	6.97	122	19.5	4.18	8.14	194	26.7	4.89	9.18	278	33.9	5.51	10.4	404	43.5	6.23
5 Gauge (.2092")	5.19	5.61	8.18	145	23.0	4.21	9.50	227	31.0	4.88	10.8	329	39.9	5.53	12.2	478	51.2	6.26
3 Gauge (.2391")	5.87	6.58	8.96	148	24.2	4.07	10.6	239	33.6	4.77	12.0	348	43.1	5.40	13.6	504	55.4	6.10
Concrete Area (In.²)	42.3	47.3	101				136				176				224			

1. Choice of tapered section is usually determined by estimated pile length.
2. Longer lengths are obtained by splicing extension sections (N) into tapered sections (F) (J) or (Y).
3. Extension sections are essentially constant in diameter, and are furnished one foot longer than nominal length to provide for telescopic splice.
4. All tapered sections include forged steel conical nose factory-attached.
5. Monotube tabulated areas and thicknesses are minimums and nominals would be 5 percent greater; unlike pipe whose minimum thickness can be 12.5 percent less than nominal.

1. Use a driving head (or cap) compatible with both the hammer and the pile. Concentric alignment and proper dimensioning will transfer energy more efficiently and minimize damage to the butt of the pile. A tapered cóne (6"- 12" long) providing a close internal fit facilitates entry and holds the pile butt centered for axial alignment.

2. Maintain axial alignment of adjoining sections. Monotube tapered and extension sections can be
• shop assembled to custom lengths (subject to shipping limitations)
• rack-assembled on the job site
• spliced in the leads. When splicing in the leads, the weight of the hammer is usually enough to assure a snug fit prior to welding. (Never fire the hammer!) The self-aligning telescopic Monotube splice shown below facilitates axial alignment.

3. Use welding electrodes compatible with chemical and physical requirements of the joint. E6010 and E6011 rods are recommended for adequate penetration of the root pass. Alternately, an E7018 rod may be preferred by a welder experienced in its use. Quality welds are critical to the ultimate performance of the driven pile.

4. Closely monitor resistance when approaching bearing on high capacity piles requiring final blow counts on the order of 10 per inch or more. Prolonged driving at rapidly escalating blows can damage both the pile and the hammer.

VISUAL INSPECTION

With Monotubes, there is no guess-work. After being driven to final penetration, each individual Monotube can be visually inspected its full length to verify the integrity of the pile. No unknown distortion or reduction of area, no unknown corkscrew or out of plumb piles, no tips curled, broomed or bent, or other uncertainties ever unknowingly become part of a Monotube installation. Actual visual inspection is an important and satisfying feature.

Pile contractor splicing up to 126 Monotubes per day to keep two rigs driving Monotubes. A total of 3,940 Monotube piles were driven at plant site in Mississippi.

Typical field splicing of Monotube pile in leads of hammer.

Figure 5.11.7   *(cont.)*

Over the course of 30 years, Monotube piles have been the pile of choice on five major projects at the Port of Wilmington in Delaware. To date, over a quarter million lineal feet of Monotubes have been driven at the port site. This major project, West Dock Expansion Phase 2, has utilized 502 - 18-in. diameter Monotubes driven to average tip elevations of 80-ft. below water line. A primary consideration was an unsupported length approaching 35-ft. above the mud line while carrying a design load of 110 tons per pile.

SPLICING

Extensions can be welded to tapered sections or to other extensions as required. N extension cut-offs can be reused to lengthen other Monotube piles on the job. N sections are furnished longer than nominal lengths to provide for a telescopic splice.

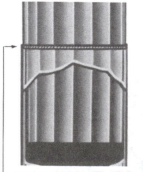

CONTINUOUS FILLET WELD

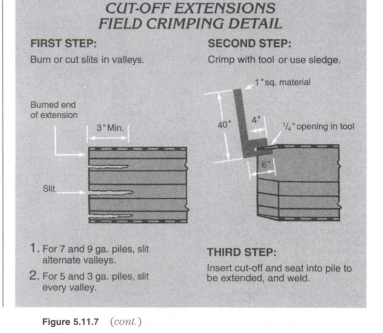

CUT-OFF EXTENSIONS FIELD CRIMPING DETAIL

FIRST STEP:
Burn or cut slits in valleys.

Burned end of extension

3" Min.

Slit

1. For 7 and 9 ga. piles, slit alternate valleys.
2. For 5 and 3 ga. piles, slit every valley.

SECOND STEP:
Crimp with tool or use sledge.

1" sq. material

40"   4"   1/4" opening in tool

6"

THIRD STEP:
Insert cut-off and seat into pile to be extended, and weld.

Figure 5.11.7   (*cont.*)

5.11.8 Pulldown Micropiles

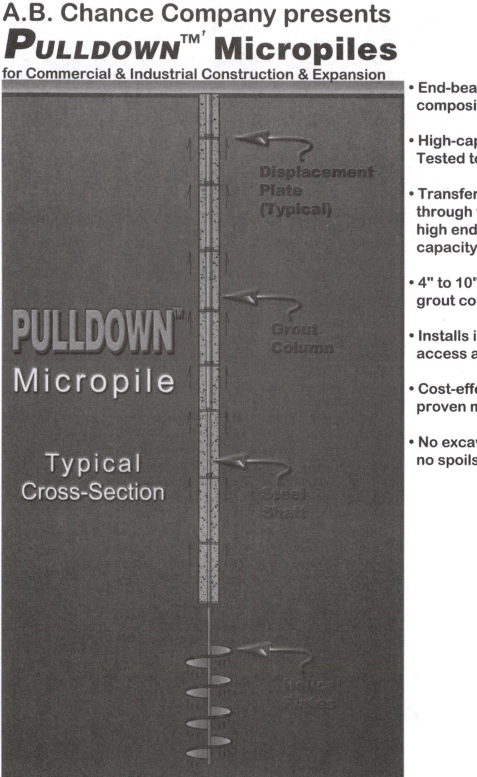

A.B. Chance Company presents
PULLDOWN™† Micropiles
for Commercial & Industrial Construction & Expansion

- End-bearing/friction composite pile

- High-capacity: Tested to 300 kips

- Transfers load through weak soils to high end-bearing-capacity helical plates

- 4" to 10" diameter grout columns

- Installs in limited-access areas

- Cost-effective, proven method

- No excavation, no spoils

PULLDOWN™ Micropile

Typical Cross-Section

Displacement Plate (Typical)

Grout Column

Steel Shaft

Helical Plates

5.11.9 Pin Piles

Nicholson Construction is acknowledged nationwide as an industry leader in geotechnical construction. Nicholson developed its PIN PILES service to provide a cost-effective foundation support system for structures being renovated, and to be a reliable construction technique in difficult ground.

What are PIN PILES?

Pin Piles are high capacity drilled piles usually in the range of 5-12" diameter. Pile depths can extend to 200 ft. and working loads up to 175 tons can be achieved, depending on pile size and ground conditions. A central reinforcing element - either high strength steel bar or tubing - is secured in place with durable cement grout. This grout is usually injected under pressure to improve bonding with the surrounding ground. Despite their advanced design, PIN PILES are constructed with conventional materials and installed using proven drilling and grouting equipment.

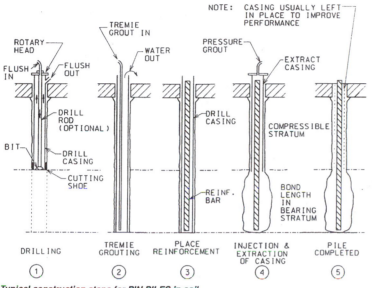

Typical construction steps for PIN PILES in soil

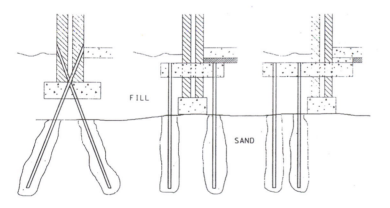

Underpinning options for existing structures.

When Do We Use PIN PILES?

- To replace deteriorated foundation systems.
- To provide extra support for structures during renovation.
- To provide pile foundations where access, geology or environment prevent the use of other methods.
- To support structures affected by adjacent excavation, tunneling or dewatering activities.
- To provide a fast, effective alternative to more traditional underpinning methods.

What Are The Advantages of PIN PILES?

- Can be *installed through virtually any ground condition*, obstruction and foundation and at any inclination.
- Construction methods ensure *minimum vibration or other damage* to foundation and subsoil.
- Can be installed in *as little headroom as 6'*, as close as 8" to existing walls.
- Facility operations can be maintained during construction.
- Can be designed to *resist compressive, tensile or lateral loads*, or combinations of all three.
- Provide *impressively high load capacity* at extremely low total and permanent settlement.
- *Connection to existing and new structures is simple and economical.*
- *Can be preloaded* to working load before connecting to particularly sensitive structures.
- Excellent performance can be cited from dozens of projects nationwide. PIN PILES have been tested to 250 tons in soil and 350 tons in rock.

(cont.)

5.11.10 Pin Piles in Soil and Rock

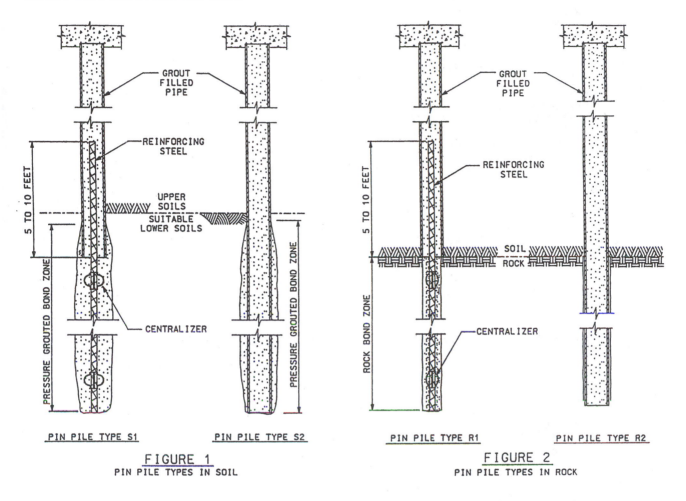

FIGURE 1
PIN PILE TYPES IN SOIL

FIGURE 2
PIN PILE TYPES IN ROCK

5.11.12 Soil Grouting

- *SOIL GROUTING* — three basic types are available:
  - *Compaction grouting:* very low slump cement-based grout is injected to form "bulbs" which compress and densify soft, loose or disturbed soil.
  - *Permeation grouting:* fluid grouts are injected into existing pore spaces in the soil. A wide range of grouts from fine cement to chemicals are used, depending on soil classification and the intended application.
  - *Jet grouting:* in this new technique, thin high-pressure jets of cement grout are discharged sideways into the borehole wall to simultaneously excavate and then mix with the soil (RODIN-JET®). The grout's cutting action can be further enhanced by the simultaneous application of compressed air (RODINJET 2) or air and water (RODINJET 3). The result is a column of modified soil with low permeability and improved strength.
- *ROCK GROUTING* — typically a cement-based, fluid grout is in-

jected at significant pressure into pre-existing fissures. This process seals the fissures to limit water seepage or to improve stability of the rock mass. Our new MPSP^SM system is particularly effective for extremely fractured rock masses.

- *BULK INFILL* — the filling of large cavities, either natural (limestone karsts) or artificial (abandoned mine works, tunnels, sewers). In this procedure, stable, cement-based grouts are normally injected into the cavity at low pressures and high flow rates. In certain conditions, support can be provided by forming special grout columns at preselected locations.
- *STRUCTURAL GROUTING* — high-strength, cement-based grouts injected as part of other geotechnical techniques such as anchors, PIN PILES^SM or INSERT WALL^SM techniques. Such grouts can also be used for slabjacking, and to provide a stable base for machinery. The special RODUR® epoxy resin technique is used for the sealing and rebonding of major concrete structures such as dams.

When Do We Use Rock and Soil Grouting?
- To reduce seepage
- To increase strength
- To reduce deformability
- To reduce uplift
- To fill voids
- To lift or level structures
- To rehabilitate and reinforce major structures

What is Grouting?

Grouting is a procedure whereby a fluid is injected into pores, fissures or voids to improve specific properties of soil, rock or concrete once the grout sets.

Grouts must be carefully selected to provide cost-effective placement and long-term benefits such as reduced permeability, increased strength or decreased deformability. We offer a variety of grouting materials and placement techniques which can be used to satisfy a wide range of applications and geology.

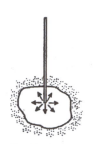

Compaction Grouting to increase density by displacing soil grains.

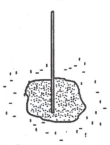

Permeation Grouting to fill existing soil pores.

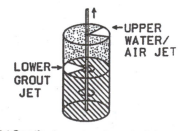

Jet Grouting to create soil-grout mixture in place.

Figure 5.11.12 *(cont.)*

5.12.0 Pile Load Testing

See Figs. 5.12.0.1 and 5.12.0.2. Piles are load-tested to develop design data, to confirm the required load capacity of the pile (proof test), to establish installation criteria, or for a combination of these reasons. Since critical decisions are often made on the basis of load-test results, careful inspection and accurate recording and reporting of the data are essential. Piles can be tested in bearing (compression), in uplift (tension), under lateral loading, or in a combination of such loads. Piles may be tested individually or in groups.

Applicable ASTM standards include D1143, Piles under Static Axial Compressive Load; D3689, Individual Piles under Static Axial Tensile Load; D3966, Piles under Lateral Loads; and D4945, High-Strain Dynamic Testing of Piles.

Test-Pile Data

The pile to be tested should be properly identified (see Identification of Piles under Pile Installation) and adequately described as to material, type, size, length, weight, and date installed and concreted, if applicable. The elevation of the pile butt should be referenced to a fixed and permanent datum.

Driving Record

A complete record of the installation of the test pile, including the driving log, descriptions of the hammer cushion, pile cushion (if used), drive cap, follower (if used), drill or jet (if used), and a record of depths predrilled or jetted should be made. Any interruptions or unusual occurrences during pile installation should be clearly and completely described. See Driving Log: Vibratory Driving under Pile Installation.

Reaction Piles

If reaction piles or other types of anchors are used, they should be described as to type, size, and lengths. Each reaction pile should have a unique designation. The driving log and the location of each reaction pile with reference to the test pile should be reported.

Test-Boring Data

The location of the closest boring with reference to the test pile should be recorded. The boring must be identified, and the boring log should accompany the test report. Ground surface elevations for all test borings must be shown and referenced to the same datum used for the piles.

Instrumentation to Measure Pile-Head Movement

The movement of the pile head under load can be measured by dial gauges, by a wire-and-scale system, or by a remotely stationed surveying level or transit reading a scale attached to the pile or a tar-

Figure 5.12.0.1   Load testing of piles. (*Courtesy of Norwalk Marine Corp., Norwalk, Conn.*)

Figure 5.12.0.2   Load testing to develop design data. (*Courtesy of Norwalk Marine Corp., Norwalk, Conn.*)

get rod held on the pile. Both a primary and a secondary measuring system should be used to assure a check on all readings and continuity of readings in case one of the systems malfunctions or requires resetting. All instrumentation must be properly mounted and installed and be functioning accurately. For tests on pile groups the instrumentation or reference points will normally be on the pile cap instead of on the test piles, as described below. Instrumentation and its supporting system should be protected from wind, extreme temperature variations, and accidental disturbance.

Dial Gauges.   At least two dial gauges, mounted on opposite sides of the pile, should be used. The required sensitivity of the gauges may be specified, but normally readings to 0.01 in (0.254 mm) are sufficient. Gauges should be mounted so as to measure movements at the sides of the pile near the butt relative to an independent reference system. Dial gauge stems should travel freely and bear against a smooth surface.

Wire and Scale.   The scale should preferably be mounted on a mirror, which in turn is fixed to the side of the pile. Consistent readings are thus assured by lining up the wire with its image. The wire should be stretched across the face of the scale, kept taut at all times and supported independently of the test setup.

Level or Transit.   The level or transit should be stationed a sufficient distance from the test pile so as not to be influenced by pile or ground movement during testing. If readings are taken on a scale, it should be fixed to the side of the test pile. If a target rod is used, readings should be taken on a fixed point on the side of the test pile. Readings should be referenced to a fixed benchmark, or the instrument can be mounted on a fixed object such as a pile for consistent readings.

Test Loads

For bearing tests, loads can be applied directly to the test pile or group with objects of known weight or by a hydraulic jack acting against a suitable reaction such as a weighted platform or anchored frame (Fig. 5.12.0.1). For uplift tests and tests on batter piles, the load is generally applied with a hydraulic jack or some type of suitable pulling system. The amount of load applied to the pile during testing should be known within an accuracy of 5 percent.

Direct Load.   The amount of load applied to the pile can be checked by the volume and unit-weight method for the material involved (Fig. 8.10). If necessary, the test-load material can be weighed beforehand to determine the load being applied. Sometimes standard test weights are available. The weight of any test beam, platform, box, or tank should be included in the first load increment. The test load must be balanced on the pile and allowed to act without restraint.

Hydraulic Jack.   A calibration certificate should be submitted for each complete jacking system including the ram, pump, and pressure gauge. If multiple jack rams are used, they should be of equal piston size, connected via a manifold to a single pump and gauge, and if possible, calibrated as a single system. If this is impractical because of the total jacking capacity or the limitations of available testing equipment, each jack ram should be calibrated separately with the pump and gauge. Adjustment factors may have to be applied for the complete system. Calibrations should be furnished for both increasing and decreasing loads.

Load Cell.   A load cell may be used to determine the amount of test load being applied. Load cells must be properly designed and constructed and accurately calibrated for both increasing and decreasing loads. They should be equipped with spherical bearing plates.

Dynamometer.   If a pulling system is used for lateral loading, a calibrated dynamometer should be installed in the system. If the load is applied with multiple-part reaving, the location of the dynamometer within the system should be carefully recorded along with a correlation of the dynamometer readings versus actual loads applied.

Reaction System

The minimum distances between the test pile and the anchor piles or supports for the reaction load will generally be specified. It is recommended that arrangements be made to take readings of the movements of anchor piles or other load reaction systems during the test. This can be done with suitably mounted dial gauges, a surveying instrument reading scales attached to the reaction system, or a target rod held on fixed points.

Test-Pile Instrumentation

If special instrumentation of the test pile, such as strain gauges or telltales, is required, such instrumentation should be installed according to specifications. The installation of electronic strain gauges and subsequent readings of telltales or strain rods is often performed by the contractor, and subsequent readings are made with suitably mounted dial gauges or some other type of measuring system. Telltale readings should generally be referenced to the top of the pile to give direct measurements of elastic shortening under load. Measurements of telltale movements would normally be made as other time-load-movement data are recorded.

Testing Arrangement

Steel test plates having a minimum thickness of 2 in (50.8 mm) should be used on top of the test pile, the jack ram, and the load cell (if used). Eccentric loading must be avoided. All compression and testing loads must be applied directly along the longitudinal axis of the test pile. For lateral tests, the loads must be applied in line with the central axis of the pile. For group tests the loads should act on the center of the group. This requires careful positioning and alignment of all applied loads, loading devices, and special instrumentation such as load cells.

Testing Procedures

Normally the overall testing procedures including rate of load application, holding times, etc., will be specified. There are many different types of testing procedures, and there should be a clear understanding by all parties concerned as to what procedures will be used before testing begins. Testing procedures commonly used include the maintained load, the quick load, and the constant rate of penetration. The inspector should be familiar with the requirements of the testing procedure to be used.

Lateral Load Tests

The lateral load capacity of the pile or group should be determined under service loading conditions with the permanent dead load acting on the pile or group during the test. The test pile or group must not be restrained by the vertical load from free lateral movement. This can be achieved by applying the compression load with the direct load method or by using suitable steel rollers between the test pile or group and the axial load.

Concrete Strength

For tests on cast-in-place concrete piles, the concrete must be of sufficient age (strength) to carry the test loads without failure or excessive creep.

5.13.0 The Basic Parts of a Typical Pile-Driving Rig

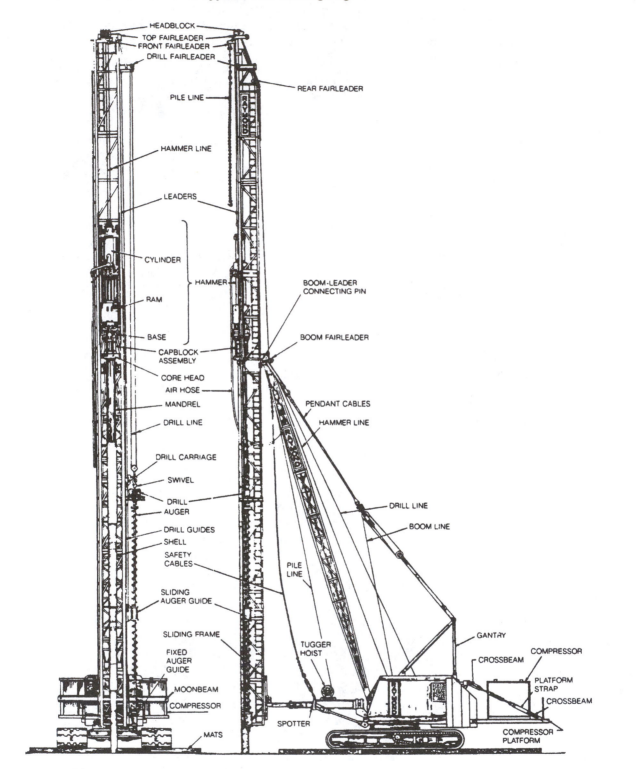

5.13.1 Pile-Driving Rig with Fixed-Lead System

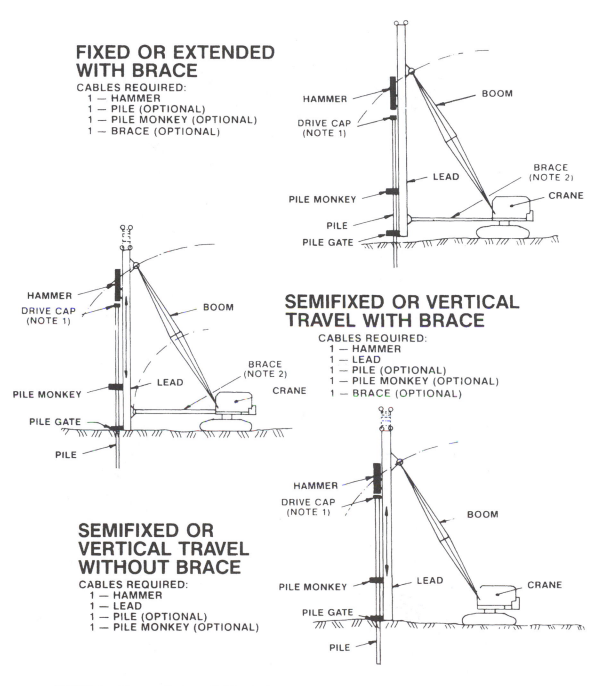

FIXED OR EXTENDED WITH BRACE

CABLES REQUIRED:
1 — HAMMER
1 — PILE (OPTIONAL)
1 — PILE MONKEY (OPTIONAL)
1 — BRACE (OPTIONAL)

SEMIFIXED OR VERTICAL TRAVEL WITH BRACE

CABLES REQUIRED:
1 — HAMMER
1 — LEAD
1 — PILE (OPTIONAL)
1 — PILE MONKEY (OPTIONAL)
1 — BRACE (OPTIONAL)

SEMIFIXED OR VERTICAL TRAVEL WITHOUT BRACE

CABLES REQUIRED:
1 — HAMMER
1 — LEAD
1 — PILE (OPTIONAL)
1 — PILE MONKEY (OPTIONAL)

NOTE 1. Also called anvil block, bonnet, cap, driving head, follow cap, helmet, hood, rider cap.

NOTE 2. Also called A-frame, apron, bottom brace, bottom strut, kicker, parallelogram, platform, spider, spotter, spreader, spreader bars.

5.13.2 Pile-Driving Rig with Swing-Lead System

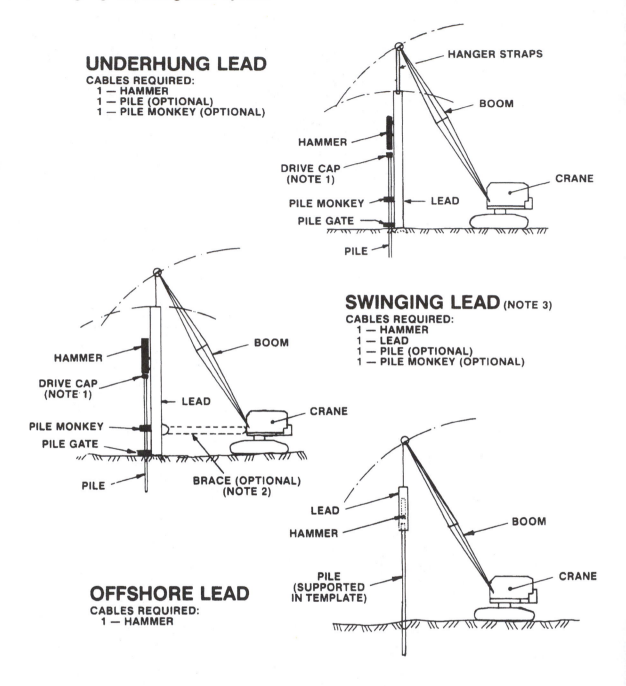

UNDERHUNG LEAD
CABLES REQUIRED:
1 — HAMMER
1 — PILE (OPTIONAL)
1 — PILE MONKEY (OPTIONAL)

SWINGING LEAD (NOTE 3)
CABLES REQUIRED:
1 — HAMMER
1 — LEAD
1 — PILE (OPTIONAL)
1 — PILE MONKEY (OPTIONAL)

OFFSHORE LEAD
CABLES REQUIRED:
1 — HAMMER

NOTE 1. Also called anvil block, bonnet, cap, driving head, follow cap, helmet, hood, rider cap.
NOTE 2. Also called A-frame, apron, bottom brace, bottom strut, kicker, parallelogram, platform, spider, spotter, spreader, spreader bars.
NOTE 3. Also called hanging lead.

5.13.3 Lead Types

H-BEAM

Also Called Spud Lead. Wide Flange. Monkey Stick.
European Lead

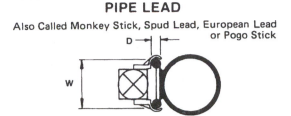

SQUARE TUBE   ROUND TUBE

GUIDE PROFILES & APPROX. WEIGHTS:
10" W — 50 lb./ft.
12" W — 60-80 lb./ft.
14" W — 80-120 lb./ft.
21" W — 145 lb./ft.
3" D x 15¾" W — 80 lb./ft.
2¾" Dia. x 15¾" W — 80-130 lb./ft.

BOX LEAD

Also Called "U" Lead. Steam Lead

WITH OR WITHOUT PLATFORM

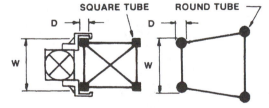

GUIDE PROFILES & APPROX. WEIGHTS:
6", 7" or 8" D x 20" W — 65-90 lb./ft.
8" or 9" D x 26" W — 80-160 lb./ft.
8", 9" or 10" D x 32" W — 130-210 lb./ft.
10" or 11" D x 37" W — 200-280 lb./ft.
11" D x 56" W — 210-300 lb./ft.

TRUSS LEAD

Also Called Monkey Stick. Spud Lead. European Lead

SQUARE TUBE   ROUND TUBE

GUIDE PROFILES & APPROX. WEIGHTS:
3" D x 15¾" W — 60-100 lb./ft.
3" D x 21½" W — 60 lb./ft.
3" D x 28½" W — 100 lb./ft.
5" D x 28½" W — 150-190 lb./ft.
2" Dia. x 11" W — 30 lb./ft.
2¾" Dia. x 15¾" W — 70-125 lb./ft.

PIPE LEAD

Also Called Monkey Stick, Spud Lead, European Lead
or Pogo Stick

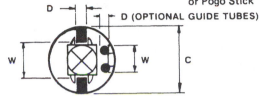

GUIDE PROFILES & APPROX. WEIGHTS:
2" Dia. x 11" W — 30-60 lb./ft.
2¾" Dia. x 15¾" W — 70-125 lb./ft.

OFFSHORE LEAD

Also Called Chuck Lead, Can, Rope Suspended Lead
or Pogo Stick

D (OPTIONAL GUIDE TUBES)

GUIDE PROFILES & APPROX. WEIGHTS:
Diesel Hammers —
6" D x 20" W x 24" C — 4.000 lb.
8" D x 26" W x 36" C — 6.500 lb.
10" D x 32" W x 48" C — 10.000 lb.
2" Dia. x 11" W x 24" C — 4.000 lb.
2¾" Dia. x 15¾" W x 36" C — 6.500 lb.
Air/Steam Hammers —
10" D x 54" W x 48" C — 16.850 lb.
14" D x 80" W x 72" C — 33.700 lb.
19" D x 88" W x 72" C — 68.000 lb.
22" D x 144" W x 120" C — 137.000 lb.
Weights can vary due to hammer size and
thickness of drive plate.

TRIANGULAR LEAD

Also Called Monkey Stick. Spud Lead. European Lead

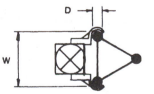

GUIDE PROFILES & APPROX. WEIGHTS:
2" Dia. x 11" W — 20 lb./ft.
2¾" Dia. x 15¾" W — 40 lb./ft.

5.13.4 Pile-Driving Rig in Batter Configurations

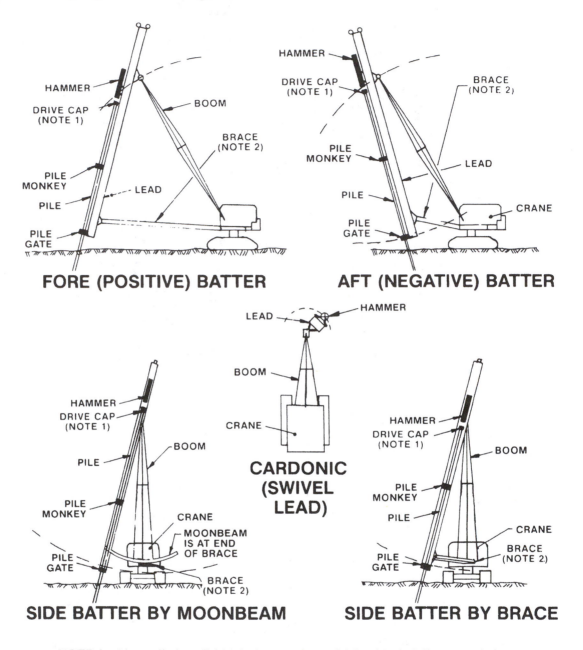

FORE (POSITIVE) BATTER

AFT (NEGATIVE) BATTER

CARDONIC (SWIVEL LEAD)

SIDE BATTER BY MOONBEAM

SIDE BATTER BY BRACE

NOTE 1. Also called anvil block, bonnet, cap, driving head, follow cap, helmet, hood, rider cap.

NOTE 2. Also called A-frame, apron, bottom brace, bottom strut, kicker, parallelogram, platform, spider, spotter, spreader, spreader bars.

5.13.5 Pile-Driving Hammer Types

1. *The drop hammer*   Rarely used, except for installing compacted-concrete piles.

2. *Single-acting hammers*   Powered by steam or air pressure, which is used to raise the hammer ram for each down stroke. Gravity and the weight of the hammer deliver the kinetic energy required to drive the pile.

3. *Double-acting hammers*   Generally powered by compressed air or hydraulics, which provides the power to raise the hammer ram and accelerate its fall.

4. *Vibratory hammers*   Paired, oscillating rotating weights connected to the pile delivers anywhere from 0 to 2000 vibrations per minute at low frequency or from 0 to 8000 vibrations per minute for high-frequency hammers to drive the pile to design depth. This type hammer is effective only in granular or cohesiveless soils.

5.14.0 Sheet Piles

When deep excavations are required in the near vicinity of existing structures, steel sheet piling or soldier beams with wood lagging are generally used to create a stable excavate. Where soil conditions permit, the sheet piling or soldier beams are driven to sufficient depth to allow for ample toe-in to support the sheeting or soldier beams once excavation reaches the required depth. If insufficient depth of soil is available, often in the case of soldier beams, they will be drilled into rock and supported with bracing.

5.14.1 Introduction to Piling Products

Steel sheet piling consists of a series of sections with interlocking connections along each edge of the section. Each pile is engaged with a pile previously driven and then driven itself as nearly as possible to the same depth. Sheet piles can be either temporary or permanent. Temporary sheet piling is usually in the form of retaining structures to control or exclude earth and/or water and allow permanent works to proceed. Steel sheet piles in permanent applications function in similar ways to temporary piles. Examples of usage include river training and flood protection, retaining walls, coastal protection, harbor walls, cofferdams, dock structures, membrane wall systems, cut and cover tunnels and underpasses.

TXI Chaparral Steel first produced PS or flat-web sections on the Large Section Mill in Midlothian, Texas late 1997 and is currently the only producer of the PS Sections in North America. The Richmond, Virginia rolling mill, commissioned mid 1999, produces the entire range of PZ sections. This is the first mill designed specifically to roll sheet pile and has the capability to supply the entire North American sheet pile market. TXI Chaparral Steel produces many of the H Pile sections at our Midlothian, TX facility, but has expanded production to the full range at the Richmond, VA facility.

PS Piling Products

This section has strong interlocks with little beam strength and is typically used for filled cell construction. New designs of open cell construction are providing new opportunities for this product which is discussed in more detail later.

PZ Piling Products

The Z sections, with their optimum distribution of material, are the most efficient sheet piles available for bending strength. With the interlocks located at the outer fibers on the wall rather than at the centerline, the designer is assured of the published section modulus. With their strong bending capacity, these sheet piles are commonly used for cantilevered, tied, or King Pile retaining systems as well as load bearing bridge abutments.

H Piling Products

H Pile sections are special WF beam shapes with the web and flanges the same thickness. These piles are typically used for deep foundations for various types of structures. H piles are also used in conjunction with sheet piling to add lateral stiffness and bending capacity where loads exceed the capacity of the sheet piling alone.

Figure 5.14.1   *(By permission: Chaparral Steel, Midlothian, Texas)*

5.14.2 Steel-Making Process for Sheet Piling

Recycled steel scrap provides the raw material for the Electric Arc Furnaces (EAF). After the scrap is loaded into the EAF, electrodes are lowered through the retractable roof into the EAF near the scrap metal charge. Electricity transfers from one electrode to the scrap metal charge, then back to another electrode. Heat to melt the scrap metal charge is developed by resistance of the metal to the flow of the massive amount of electricity and by the heat of the arc itself. Oxygen is injected into the EAF to speed up the melting process. Fluxes and alloys may be added to EAF at the end of its melt cycle or at the ladle upon tapping the EAF to establish the chemistry of the heat of steel.

Molten steel is tapped from the EAF to a ladle. The ladle is then sent to a ladle metallurgy refining station (LMF). While the ladle is at LMF the chemistry of the heat of steel is verified to determine that proper alloy additions were made during the tapping of the heat. Additional alloys and fluxes may be added at the LMF if required. Homogenization of temperature and chemistry of the heat of steel is accomplished by the bubbling of inert gas through the ladle. Upon completion of LMF treatment the ladle is then sent to the continuous casting operation.

The continuous casting or strand casting operation produces billets, beam blanks, and near-net shape profiles. These semi-finished products are utilized in the rolling mills to produce structural shapes. Molten steel is poured through a slide gate controlled nozzle on the bottom of the ladle into a tundish. The tundish acts as a reservoir and releases the molten steel in a continuous stream through a series of nozzles in its base. The steel flows out at a steady rate into oscillating molds. The hollow interior of the oscillating molds have inside dimensions corresponding to the width and thickness of the billets, beam blanks, and near-net shape profile that is being cast. The molds are contained within water cooled jackets that direct water flow around the outside walls of the mold. As the surface of the metal begins to solidify, a thin skin is formed on the outer edges. During this freezing operation, the metal in the mold continuously moves downward as the mold oscillates up and down to keep the metal from sticking. As the billet leaves the mold, it enters a secondary spray cooling process. The metal solidifies from the outside skin toward the center as it continues to move through the process. When the metal has solidified, the billets, beam blanks and near-net shape profiles are torch cut to length and may be direct charged to a rolling mill or placed in inventory for rolling at a later date.

Figure 5.14.2   *(By permission: Chaparral Steel, Midlothian, Texas)*

5.14.3 Metallurgical Aspects of Sheet Piling

TXI Chaparral Steel produces Carbon and High Strength Low Alloy (HSLA) steels for its structural products.

Carbon Steel - Steel is considered to be carbon steel when no minimum content is specified or required for chromium, columbium (niobium), molybdenum, nickel, titanium, vanadium, or any other element, to obtain a desired alloying effect and when the specified minimum for copper does not exceed 0.40 percent, or when the maximum content specified for any of the following elements does not exceed the following percentages: manganese 1.65, silicon 0.60, copper 0.60.

High Strength Low Alloy (HSLA) Steel – is comprised of a group of steels with chemical composition developed to impart higher mechanical property values and, in certain grades of these steels, greater resistance to atmospheric corrosion than is obtainable from carbon steel.

Common Elements found in steel

Carbon –	is used as a strengthening element. It is maintained at a level consistent with weldability and ductility.
Manganese –	contributes to strength and hardness.
Phosphorus –	results in increased strength and hardness, and improves resistance to atmospheric corrosion.
Sulfur –	higher levels decrease ductility, toughness and weldability, aids machinability.
Silicon –	is one of the principal deoxidizers used in steelmaking and contributes to strength and hardness.
Copper –	contributes to strength and enhances resistance to atmospheric corrosion.

Other elements such as chromium, columbium (niobium), molybdenum, nickel, nitrogen, titanium, and vanadium may be added singly or in combination for their beneficial effects on strength, toughness, atmospheric corrosion resistance, and other desirable properties.

Incidental or Residual Elements – In all steels small quantities of certain elements are unavoidably retained from raw materials. This is especially true of steel produced from recycled steel scrap. The elements of copper, nickel, chromium, molybdenum and tin are always present as residual elements in steel produced from a scrap base. Unless the chemical composition of the steel specifies a minimum and maximum of these elements, they and others that may be present are considered as incidental and are commonly referred to as residual.

Carbon Equivalent – To a good approximation, the weldability of steel can be estimated from its chemical composition. The most significant alloying element effecting weldability is carbon. The effects of other elements can be estimated by equating them to an additional amount of carbon. The total alloy content has the same effect on weldability as an equivalent amount of carbon. Several empirical formulas have been devised to calculate this carbon equivalent. The result of the calculation is expressed as C_{eq} or CE. The most commonly used formula comes from the International Institute of Welding and is listed in the supplementary requirements of ASTM A6/A6M and in Appendix XI of the AWS D1.1 Structural Welding Code.

Figure 5.14.3   (*By permission: Chaparral Steel, Midlothian, Texas*)

Heat Analysis – applies to the chemical analysis representative of a heat of steel as reported to the purchaser. The analysis is determined by analyzing test samples obtained during the pouring of the steel for the elements that have been specified, or for elements that are required to be reported. Samples representing the first, middle, and last portion of the heat are used to survey uniformity. The average of the three samples is reported on the Certified Material Test Report.

Product Analysis – A product analysis is a chemical analysis of the finished steel to determine conformance to the specification requirements and is typically performed by the purchaser. The chemical composition determined shall conform to the listed chemistry specified in the product specification subject to the product analysis tolerances in ASTM A6/A6M. Product analysis tolerances are not to be applied to the chemistry reported on a Certified Material Test Report.

Killed Steel – Carbon and oxygen react in the steelmaking process to form a gas. If the oxygen available for this reaction is not removed (deoxidized) prior to casting, the gaseous products continue to evolve during solidification. De-oxidation of the steel is accomplished by the addition of silicon. The result of this practice is "silicon killed steel". The requirement for killed steel is met when the percentage of silicon in the steel is 0.10 or greater as shown on the Certified Material Test Report for the heat of steel.

Grain Size - Grain size is usually expressed as the average diameter or as a numeric value representing the quantity of grains per unit area or volume. Two types of grain sizes are commonly expressed:
1) "Austenitic" (high temperature) grain size as determined by the McQuaid-Ehn test (ref. ASTM E-112).
2) various forms of room temperature grain sizes which are called as-rolled or ferritic/pearlitic grain size.

TXI Chaparral produces all A992, A572, A588, and Chaparrals A36/A572-50 multi-cert steel to a fine grain practice utilizing Columbium / Niobium or Vanadium. This practice has shown to provide a Fine Austenitic Grain Size when the McQuaid-Ehn tests have been performed; however this test is not performed on all material rolled and inventoried due to the time and cost involved in performing the test.

Charpy V-Notch Impact Testing – A charpy impact test is a dynamic test in which a selected specimen is struck and broken by a single blow in a specially designed testing machine. The energy absorbed in breaking the specimen is measured. The energy values determined are qualitative comparisons on a selected specimen and although frequently specified as an acceptance criterion, they cannot be converted into energy figures that would serve for engineering calculations. Test specimens for structural shapes and test frequency are selected in accordance with ASTM A673 "Sampling Procedures for Impact Testing of Structural Steel." An impact test consists of three specimens taken from a single test coupon. The average of which shall comply with the specified minimum with not more than one value below the specified minimum, but in no case below either two thirds of the specified minimum or 5 ft.-lbf., whichever is greater. The impact properties of steel can vary within the same heat and piece. The purchaser should be aware that testing of one shape does not provide assurance all shapes of the same heat will be identical in toughness with the product tested.

Figure 5.14.3 *(cont.)*

Impact testing is performed only when required by the material specification or by request on the initial purchase order from a customer. TXI Chaparral Steel's standard production is typically capable of meeting a minimum impact value of 25 ft-lbf at test temperatures of 10°F and above. Impact test values can not be certified or guaranteed without the actual tests being performed. Requirements for lower test temperatures require review on an individual basis.

Imperfections – may be present on the surface of structural shapes. Imperfections which do not affect the utility of these products are not considered injurious. ASTM specification A6 makes allowance for these types of imperfections in Section 9. Quality Section 9 states;" The material shall be free of injurious defects and shall have a workmanlike finish." Note 2 states; "Unless otherwise specified, structural quality steels are normally furnished in the as-rolled condition and subjected to visual inspection by the manufacturer. Non-injurious surface, or internal imperfections, or both may be present in the steel delivered and may require conditioning by the purchaser to improve appearance of the steel or in preparation for welding, coating, or further processing."

Limitations of Inspection – There are a number of intrinsic features in the production of structural shapes that affect the properties or conditions of the finished products. Those effects cannot always be precisely known. It is not possible at the present time to identify any reasonable or practical methods of testing or inspection that will ensure the detection and rejection of every piece of steel that varies from the specified requirements with regard to dimensional tolerances, mechanical properties, surface or internal conditions. Therefore, it is technically impossible to give unconditional certification of complete compliance with all prescribed requirements. This fact is manifest to those having a technical knowledge of the subject and is recognized in applying a factor of safety in the design of structural steel.

Figure 5.14.3 *(cont.)*

5.14.4 Typical Sheet Pile Sections

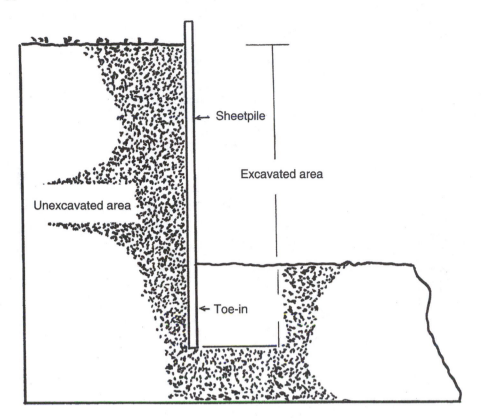

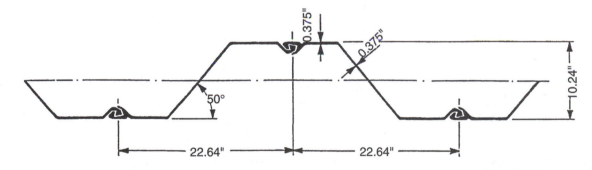

5.14.5 Steel H-Piles (Typical Sections and Specifications)

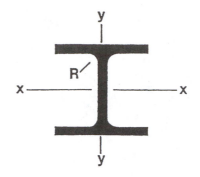

Properties for Design:

Desig-nation and Nominal Size	Weight Per Foot	Area	Depth	Flange		Web Thick-ness	Fillet Rad. R	Sur-face Area	Axis x-x			Axis y-y		
				Width	Thick-ness				I	S	r	I	S	r
inch.	lb.	inch²	inch	inch	inch	inch	inch	ft.²/ft.	inch⁴	inch³	inch	inch⁴	inch³	inch
HP14 14 x 14½	117	34.4	14.21	14.885	.805	.805	.60	1024	1220	172	5.96	443	59.5	3.59
	102	30.0	14.01	14.785	.705	.705	.60	1017	1050	150	5.92	380	51.4	3.56
	89	26.1	13.83	14.695	.615	.615	.60	1010	904	131	5.88	326	44.3	3.53
	73	21.4	13.61	14.585	.505	.505	.60	1002	729	107	5.84	261	35.8	3.49
HP12 12 x 12	84	24.6	12.28	12.295	.685	.685	.60	860	650	106.0	5.14	213	34.6	2.94
	74	21.8	12.13	12.215	.610	.605	.60	850	569	93.8	5.11	186	30.4	2.92
	63	18.4	11.94	12.125	.515	.515	.60	844	472	79.1	5.06	153	25.3	2.88
	53	15.5	11.78	12.045	.435	.435	.60	838	393	66.8	5.03	127	21.1	2.86
HP10 10 x 10	57	16.8	9.99	10.225	.565	.565	.50	707	294	58.8	4.18	101	19.7	2.45
	42	12.4	9.70	10.075	.420	.415	.50	696	210	43.4	4.13	71.7	14.2	2.41
HP8 8 x 8	36	10.6	8.02	8.155	.445	.445	.40	565	119	29.8	3.36	40.3	9.88	1.95

H-Pile Specifications:

ASTM Grades	Yld. Point	Ten. Str.	Advantages
A-36	36,000	70,000	Basic Specification.
A-572 GR50	50,000	65,000	Higher yield.
A-690	50,000	70,000	2-3 times corrosion resistance in splash zone.

5.14.6 HP Shapes—Typical Dimensions and Properties

HP Shapes

Dimensions & Properties

Produced by Chaparral Steel

Designation	Area	Depth	Web Thickness		Flange		Distance		Elastic Properties					
					Width	Thick			X-X Axis			Y-Y Axis		
	A	d	t_w	$t_w/2$	b_f	t_f	k	k_1	I_x	S_x	r_x	I_y	S_y	r_y
	in.^{2}	in.	in.	in.	in.	in.	in.	in.	in.^{4}	in.^{3}	in.	in.^{4}	in.^{3}	in.
HP14x117	34.4	14.21	0.805	0.403	14.885	0.805	1.5000	1.0625	1220	172	5.96	443	59.5	3.59
x102	30.0	14.01	0.705	0.353	14.785	0.705	1.3750	1.0000	1050	150	5.92	380	51.4	3.56
x89	26.1	13.83	0.615	0.308	14.695	0.615	1.3125	0.9375	904	131	5.88	326	44.3	3.53
x73	21.4	13.61	0.505	0.253	14.585	0.505	1.1875	0.8750	729	107	5.84	261	35.8	3.49
HP12x84	24.6	12.28	0.685	0.343	12.295	0.685	1.3750	1.0000	650	106	5.14	213	34.6	2.94
x74	21.8	12.13	0.605	0.303	12.215	0.610	1.3125	0.9375	569	93.8	5.11	186	30.4	2.92
x63	18.4	11.94	0.515	0.258	12.125	0.515	1.2500	0.8750	472	79.1	5.06	153	25.3	2.88
x53	15.5	11.78	0.435	0.218	12.045	0.435	1.1250	0.8750	393	66.8	5.03	127	21.1	2.86
HP10x57	16.8	9.99	0.565	0.283	10.225	0.565	1.1875	0.8125	294	58.8	4.18	101	19.7	2.45
x42	12.4	9.70	0.415	0.208	10.075	0.420	1.0625	0.7500	210	43.4	4.13	71.7	14.2	2.41
HP8x36	10.6	8.02	0.445	0.223	8.155	0.445	0.9375	0.6250	119	29.8	3.36	40.3	9.88	1.95

Figure 5.14.6 *(By permission: Chaparral Steel, Midlothian, Texas)*

HP Shapes

Dimensions & Properties

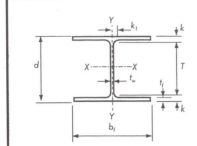

Produced by Chaparral Steel

Designation	Plastic Modulus		Compact Section Criteria												Torsional Properties					
			ASD				LRFD													
	Z_x	Z_y	$b_f/2t_f$	F_y'	d/t_w	F_y'''	h/t_w	F_y'''	r_T	d/A_f	X_1	$X_2 \times 10^6$	J	C_w	W_{no}	S_w	Q_f	Q_w
	in.³	in.³		ksi		ksi		ksi	in.		ksi	(1/ksi)²	in.⁴	in.⁶	in.²	in.⁴	in.³	in.³
HP14x117	194	91.4	9.20	49.4	17.7	-	14.2	-	4.00	1.19	3870	659	8.02	19900	49.9	149	38.5	97.2
x102	169	78.8	10.5	38.4	19.9	-	16.2	-	3.97	1.34	3400	1090	5.40	16800	49.2	128	33.5	84.3
x89	146	67.7	11.9	29.6	22.5	-	18.5	-	3.94	1.53	2960	1840	3.60	14200	48.5	110	29.1	72.9
x73	118	54.6	14.4	20.3	27.0	-	22.6	-	3.90	1.85	2450	3880	2.01	11200	47.8	88.0	23.8	59.2
HP12x84	120	53.2	9.00	52.5	17.9	-	14.2	-	3.29	1.46	3860	670	4.24	7160	35.6	75.0	23.5	59.8
x74	105	46.6	10.0	42.1	20.0	-	16.0	-	3.26	1.63	3440	1050	2.98	6170	35.2	65.5	20.8	52.7
x63	88.3	37.7	11.8	30.5	23.2	-	18.9	-	3.23	1.91	2940	1940	1.83	4990	34.6	54.1	17.5	44.2
x53	74.0	32.2	13.8	22.0	27.1	-	22.3	-	3.20	2.25	2500	3650	1.12	4090	34.2	44.7	14.7	37.0
HP10x57	66.5	30.3	9.00	51.6	17.7	-	13.9	-	2.74	1.73	3920	631	1.97	2240	24.1	34.8	13.1	33.2
x42	48.3	21.8	12.0	29.4	23.4	-	18.9	-	2.69	2.29	2920	1970	0.81	1540	23.4	24.7	9.64	24.2
HP8x36	33.6	15.2	9.20	50.3	18.0	-	14.2	-	2.18	2.21	3840	685	0.77	578	15.4	14.0	6.62	16.8

Figure 5.14.6   (*cont.*)

5.14.7 PS and PSA Piling—Typical Dimensions and Properties

Dimensions & Properties   **US Customary**

PS, PSA Piling

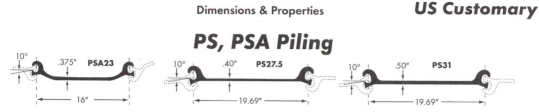

Properties and Weights

Section Designation	Area sq. in.	Nominal Width, in.	Weight in Pounds		Moment of Intertia in.⁴	Section Modulus, in.³		Surface Area, sq. ft. per lin. ft. of bar	
			Per lin. ft. of bar	Per sq. ft. of wall		Single Section	Per lin. ft. of wall	Total Area	Nominal Coating Area
PSA23	8.99	16.00	30.7	23.0	5.5	3.2	2.4	3.76	3.08
PS27.5	13.27	19.69	45.1	27.5	5.3	3.3	2.0	4.48	3.65
PS31	14.96	19.69	50.9	31.0	5.3	3.3	2.0	4.48	3.65

PS sections available now from Texas mill; PSA sections available late 1999

Figure 5.14.7 *(By permission: Chaparral Steel, Midlothian, Texas)*

5.14.8 Z Piling—Typical Dimensions and Properties

Z-Piling

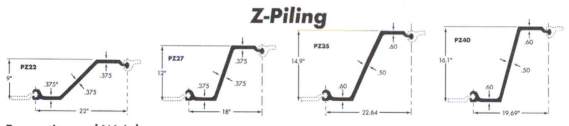

Properties and Weights

Section Designation	Area sq. in.	Nominal Width, in.	Weight in Pounds		Moment of Intertia in.⁴	Section Modulus, in.³		Surface Area, sq. ft. per lin. ft. of bar	
			Per lin. ft. of bar	Per sq. ft. of wall		Single Section	Per lin. ft. of wall	Total Area	Nominal Coating Area
PZ22	11.86	22.00	40.3	22.0	154.7	33.1	18.1	4.94	4.48
PZ27	11.91	18.00	40.5	27.0	276.3	45.3	30.2	4.94	4.48
PZ35	19.41	22.64	66.0	35.0	681.5	91.4	48.5	5.83	5.37
PZ40	19.30	19.69	65.6	40.0	805.4	99.6	60.7	5.83	5.37

PZ sections available mid 1999 from Virginia mill

Grades: ASTM A328
ASTM A572 Gr. 50 & 60
ASTM A690

Notes: All dimensions given are nominal.

PS27.5 and PS31 interlock only with each other.

All Z-sections interlock with one another and with PSA23.

Interlock Strength:
PSA23, when properly interlocked, develops a minimum ultimate interlock strength of 12 kips per inch. However, excessive interlock tension results in web extension for section PSA23. Therefore, the interlock tension for this section should be limited to a maximum working load of 3 kips per inch.
PS27.5 and PS31, when properly interlocked, develop a minimum ultimate interlock strength of 16 kips per inch. Higher interlock strengths are available.

Figure 5.14.8 *(By permission: Chaparral Steel, Midlothian, Texas)*

5.14.9 PS, PSA, and Z Piling Metric Dimensions

Metric   Dimensions & Properties

PS, PSA Piling

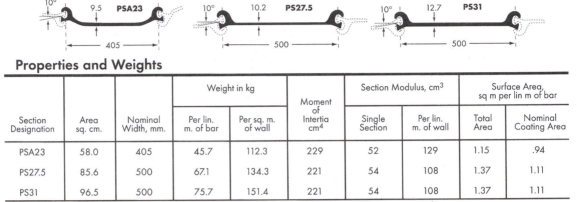

Properties and Weights

Section Designation	Area sq. cm.	Nominal Width, mm.	Weight in kg		Moment of Intertia cm⁴	Section Modulus, cm³		Surface Area, sq m per lin m of bar	
			Per lin. m. of bar	Per sq. m. of wall		Single Section	Per lin. m. of wall	Total Area	Nominal Coating Area
PSA23	58.0	405	45.7	112.3	229	52	129	1.15	.94
PS27.5	85.6	500	67.1	134.3	221	54	108	1.37	1.11
PS31	96.5	500	75.7	151.4	221	54	108	1.37	1.11

PS sections available now from Texas mill; PSA sections available late 1999

Z-Piling

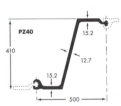

Properties and Weights

Section Designation	Area sq. cm.	Nominal Width, mm.	Weight in kg		Moment of Intertia cm⁴	Section Modulus, cm³		Surface Area, sq m per lin m of bar	
			Per lin. m. of bar	Per sq. m. of wall		Single Section	Per lin. m. of wall	Total Area	Nominal Coating Area
PZ22	76.5	560	60.0	107.4	6440	542	973	1.51	1.37
PZ27	76.8	455	60.3	131.8	11500	742	1620	1.51	1.37
PZ35	125.2	575	98.2	170.9	28370	1500	2610	1.78	1.64
PZ40	124.5	500	97.6	195.3	33520	1630	3260	1.78	1.64

PZ sections available mid 1999 from Virginia mill

Grades: ASTM A328
ASTM A572 Gr. 50 & 60
ASTM A690

Notes: All dimensions given are nominal.

PS27.5 and PS31 interlock only with each other.

All Z-sections interlock with one another and with PSA23.

Interlock Strength:
PSA23, when properly interlocked, develops a minimum ultimate interlock strength of 2100 KN/m. However, excessive interlock tension results in web extension for section PSA23. Therefore, the interlock tension for this section should be limited to a maximum working load of 525 KN/m.
PS27.5 and PS31, when properly interlocked, develop a minimum ultimate interlock strength of 2800KN/m. Higher interlock strengths are available.

Figure 5.14.9 (*By permission: Chaparral Steel, Midlothian, Texas*)

5.14.10 Open Cell, Flat-Web Sheet Piling

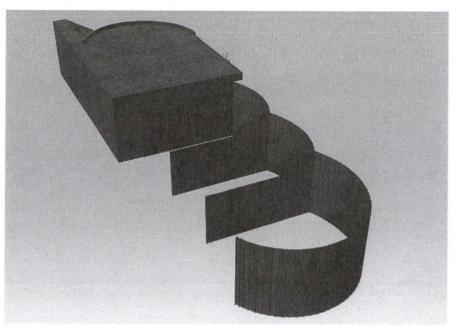

Typical "Open Cell" layout

The need for low-cost, high load capacity docks, bridge abutments, and retaining structures has resulted in the development of a new type of soil/friction structure using modifications to the existing flat-web sheet-piling design technology. Cellular flat-web sheet-pile structures have been successfully used for many years for a wide variety of structures including cofferdam and docks. These types of structures are usually associated with high construction costs due to the difficulty in closing each individual cell. By modifying the closed cell to an open cell concept, a large number of benefits have occurred resulting in a minimum of **25% overall cost savings** in most applications.

Benefits:

- Multiple templates for construction alignment not required.
- No close-tolerance connections required.
- Closing of individual cells not required.
- Toe/Ground support not critical to design.
- Shallower driving depths required for stability.
- Easier access for soil compaction.
- Large axial load capacity.
- All Land-Based Construction; therefore eliminating the need for cumbersome, Barge-Based Construction.
- One quarter less cost than other bulkhead type construction.
- One half of the cost of heavy-duty pile supported docks.

To 1999 about 30,000 flat-web sheet piles have been installed in over 80 open cell structures throughout the country with savings to the owners in the millions of dollars. This concept not only provides a cost-effective solution, but also provides a superior performing wall or dock structure. Existing open cell wall designs have experienced and withstood the forces and ground motion associated with large earthquakes with no major damage.

Figure 5.14.10   *(By permission: Chaparral Steel, Midlothian, Texas)*

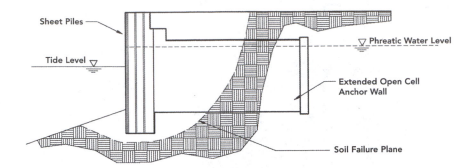

Typical Section "Open Cell" Mass Stability

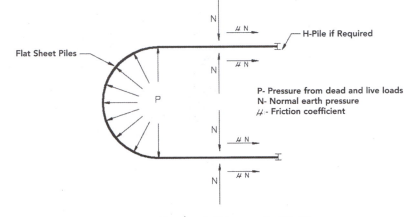

Typical "Open Cell" Plan

Peratrovich, Nottingham, & Drage (PN&D), Inc., a civil engineering consulting firm specializing in bridge and marine structures, was challenged by a general contractor to develop a more cost effective means of construction suitable for dock structures. PN&D had been interested in soil/friction structures which had their formulation in ancient times and which had become the basis for various new adaptations. It seemed probable that flat-web sheet piles could be used as vertical soil friction anchors for membrane wall systems. The system envisioned was the "Open Cell" functioning as a horizontally tied membrane.

Because of the absence of technical information necessary to accurately estimate anchor wall resistance, PN&D began a test program using sand and full size sheet pile sections in a test box, complete with control of vertical pressure. Using a hydraulic jack system, a series of tests were performed under varying soil densities, pile orientation and vertical pressure conditions.

Findings confirmed that pile interlock size had a significant effect in increasing friction resistance. Test pile orientation, both horizontal and normal to vertical pressure, helped verify other variables such as available lateral pressure as related to overburden depth and soil density. Friction resistance often approached two times that suggested in other references.

Figure 5.14.10   *(cont.)*

5.14.11 Plastic Section Modulus—"W" Shapes

Weight Economy Selection Table

Sections in bold are "Weight Economy Sections"

Shape	Z_x in.³	Shape	Z_x in.³	Shape	Z_x in.³	Shape	Z_x in.³
W36X848	3830	W24X408	1250	**W40X183**	781	W27X161	512
W36X798	3570	W27X368	1240	W33X201	772	W24X176	511
W36X650	2840	W30X326	1190	W27X235	769		
W40X593	2760	W40X278	1190	W36X194	767	**W36X135**	509
W40X503	2300	W14X550	1180	W18X311	753	W30X148	500
W36X527	2270	W36X280	1170	W30X211	749	W18X211	490
W40X466	2050	W33X291	1150	W24X250	744	W14X257	487
		W40X264	1130	W14X370	736	W12X279	481
W40X431	1950					W21X182	476
W27X539	1880	**W40X249**	1120	**W36X182**	718	W24X162	468
W36X439	1860			**W40X174**	715		
W14X808	1834	**W44X230**	1100	W27X217	708	**W33X130**	467
W30X477	1790	W36X260	1080			W27X146	461
		W30X292	1060	**W40X167**	692	W18X192	442
W40X392	1710	W14X500	1050	W24X229	676	W30X132	437
		W33X263	1040	W18X283	676	W14X233	436
W40X372	1670	W36X256	1040	W30X191	673	W21X166	432
W36X393	1660	W27X307	1020	W14X342	672	W12X252	428
W14X730	1660	W24X335	1020	W36X170	668	W24X146	418
		W36X245	1010	W33X169	629		
W44X335	1620	W40X235	1010	W27X194	628	**W33X118**	415
W24X492	1550					W30X124	408
W27X448	1530	**W40X215**	963	**W36X160**	624	W18X175	398
W36X359	1510	W36X230	943	W18X258	611	W27X129	395
W14X665	1480	W30X261	941	W24X207	606	W14X211	390
W30X391	1430	W33X241	939	W30X173	605	W12X230	386
		W36X232	936	W14X311	603		
W40X331	1430	W14X455	936	W12X336	603	**W30X116**	378
						W21X147	373
W44X290	1420	**W40X211**	905	**W40X149**	597	W24X131	370
W40X321	1420	W14X426	869	W36X150	581	W18X158	356
W33X354	1420			W27X178	567	W14X193	355
W36X328	1380	**W40X199**	868	W24X192	559	W12X210	348
W40X297	1330	W33X221	855	W33X152	559		
W14X605	1320	W27X258	850	W18X234	549	**W30X108**	346
		W30X235	845	W14X283	542	W27X114	343
W44X262	1270	W24X279	835	W12X305	537	W21X132	333
W33X318	1270	W36X210	833	W21X201	530	W24X117	327
W36X300	1260	W14X398	801			W18X143	322
W40X277	1250			**W33X141**	514	W14X176	320

Figure 5.14.11   *(By permission: Chaparral Steel, Midlothian, Texas)*

5.14.12 Elastic Section Modulus—"W" Shapes

Weight Economy
Selection Table

Sections in bold are "Weight Economy Sections"

Shape	S_x	Shape	S_x	Shape	S_x	Shape	S_x
	in.³		in.³		in.³		in.³
W30X99	**269**	W18X76	146	W12X58	78.0	**W12X26**	**33.4**
W27X102	267	W12X106	145	W14X53	77.8	W10X30	32.4
W12X190	263	W14X90	143	W10X68	75.7	W8X35	31.2
W24X104	258	W21X68	140	W16X45	72.7		
W18X130	256	W16X77	134	W12X53	70.6	**W14X22**	**29.0**
W14X159	254			W14X48	70.3	W10X26	27.9
W21X111	249	**W24X62**	**131**			W8X31	27.5
		W12X96	131	**W18X40**	**68.4**		
W30X90	**245**			W10X60	66.7	**W12X22**	**25.4**
W24X103	245	**W21X62**	**127**			W8X28	24.3
W27X94	243	W18X71	127	**W16X40**	**64.7**	W10X22	23.2
W12X170	235	W10X112	126	W12X50	64.7		
W14X145	232	W14X82	123	W14X43	62.7	**W12X19**	**21.3**
W18X119	231	W12X87	118	W8X67	60.4	W8X24	20.9
W21X101	227	W18X65	117	W10X54	60.0	W10X19	18.8
W24X94	222	W16X67	117	W12X45	58.1	W8X21	18.2
W27X84	**213**	**W24X55**	**114**	**W18X35**	**57.6**	**W12X16**	**17.1**
W14X132	209	W14X74	112	W16X36	56.5	W6X25	16.7
W12X152	209	W10X100	112	W14X38	54.6	W10X17	16.2
W18X106	204	W21X57	111	W10X49	54.6	W8X18	15.2
		W18X60	108	W8X58	52.0		
W24X84	**196**	W12X79	107			**W12X14**	**14.9**
W21X93	192	W14X68	103	**W12X40**	**51.9**	W10X15	13.8
W14X120	190	W10X88	98.5	W10X45	49.1	W6X20	13.4
W18X97	188					W8X15	11.8
W12X136	186	**W18X55**	**98.3**	**W14X34**	**48.6**		
		W12X72	97.4	**W16X31**	**47.2**	**W10X12**	**10.9**
W24X76	**176**			W12X35	45.6	W6X16	10.2
W16X100	175	**W21X50**	**94.5**	W8X48	43.3	W5X19	10.2
W14X109	173	W16X57	92.2	W10X39	42.1	W8X13	9.91
W21X83	171	W14X61	92.2			W6X15	9.72
W18X86	166			**W14X30**	**42.0**	W5X16	8.51
W12X120	163	**W18X50**	**88.9**	W12X30	38.6		
W14X99	157	W12X65	87.9			**W8X10**	**7.81**
W16X89	155	W10X77	85.9	**16X26**	**38.4**	W6X12	7.31
				W8X40	35.5		
W24X68	**154**	**W21X44**	**81.6**			**W6X9**	**5.56**
W21X73	151	W16X50	81.0	**W14X26**	**35.3**	W4X13	5.46
		W18X46	78.8	W10X33	35.0	**W6X8.5**	**5.14**

Figure 5.14.12   (*By permission: Chaparral Steel, Midlothian, Texas*)

5.15.0 Typical Soldier Pile Detail (Rock Bearing)

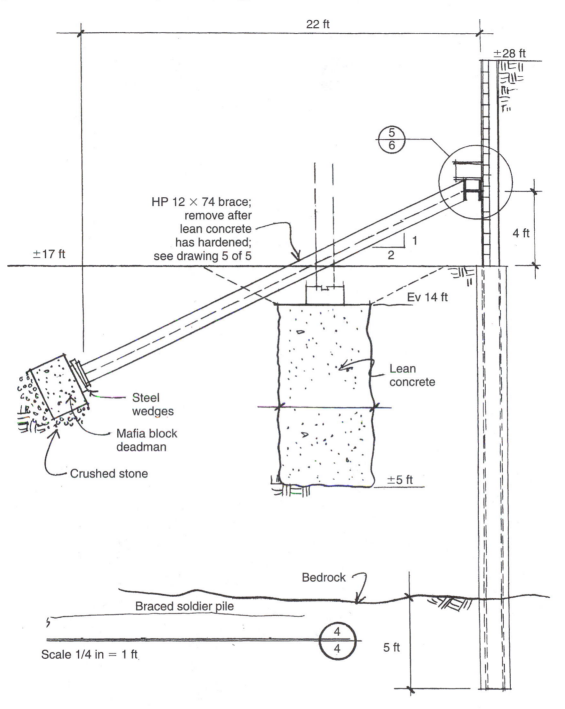

5.15.1 Typical Braced Soldier Pile with Deadman

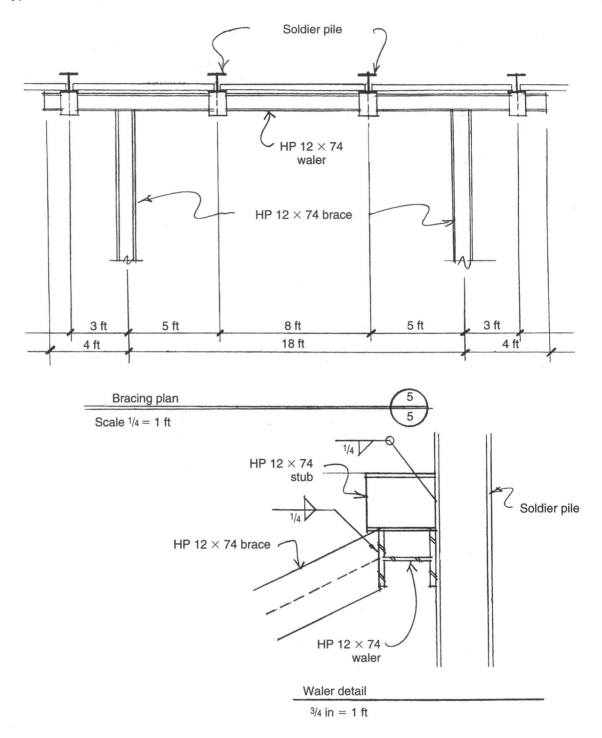

Soldier pile

HP 12 × 74
waler

HP 12 × 74 brace

| 3 ft | 5 ft | 8 ft | 5 ft | 3 ft |

| 4 ft | | 18 ft | | 4 ft |

Bracing plan

5/5

Scale ¼ = 1 ft

HP 12 × 74
stub

¼

¼

HP 12 × 74 brace

Soldier pile

HP 12 × 74
waler

Waler detail

¾ in = 1 ft

5.15.2 Typical Soldier Pile Drilled into Rock (Concrete Filled)

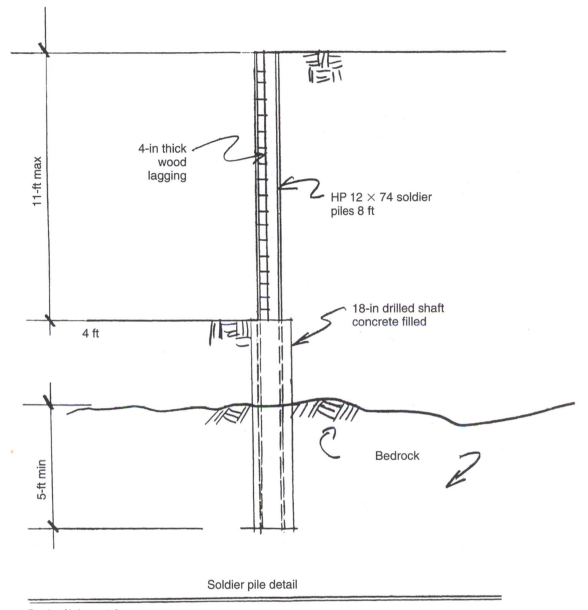

4-in thick wood lagging

HP 12 × 74 soldier piles 8 ft

18-in drilled shaft concrete filled

11-ft max

4 ft

5-ft min

Bedrock

Soldier pile detail

Scale: 1/4 in = 1 ft

5.15.3 Typical Soldier Pile with Wood Lagging, Steel Walers, and Bracing

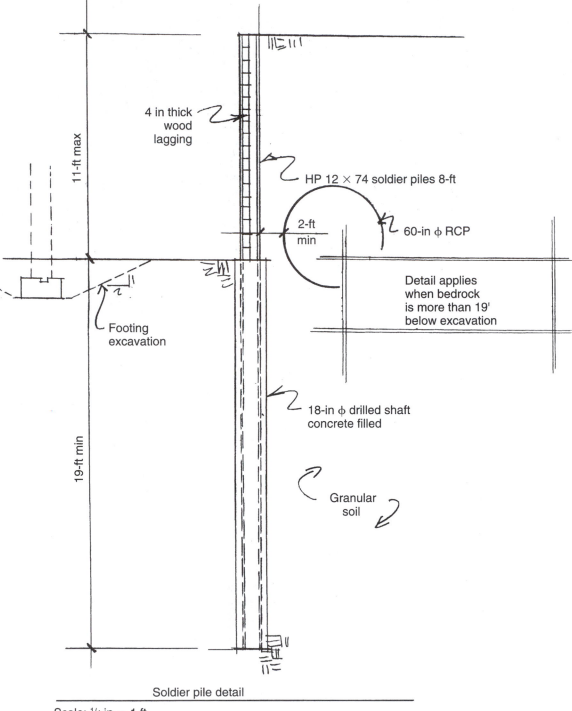

4 in thick
wood
lagging

HP 12 × 74 soldier piles 8-ft

2-ft
min

60-in φ RCP

Detail applies
when bedrock
is more than 19'
below excavation

Footing
excavation

18-in φ drilled shaft
concrete filled

Granular
soil

11-ft max

19-ft min

Soldier pile detail

Scale: 1/4 in = 1 ft

5.16.0 Soil Screws

Description

For retaining projects, Chance screw anchors can be matched to soil and heavy tension loads in the same way HELICAL PIER® Foundation Systems anchors are for compression applications.

Soil Nailing – SOIL SCREW™ Retention Wall System uses screw anchors as bearing devices as compared to grouted anchors which rely on friction. To construct a gravity wall to reinforce the soil, SOIL SCREW anchors have bearing plates spaced along their entire lengths. Anchor size and grid spacing are designed to local soil conditions and load requirements. A shotcrete-reinforced veneer often is applied to the wall face.

Tieback – Chance screw anchors for tiebacks in soldier-pile/waler walls come with shaft sizes and single- or multi-helix plate diameters selected for job-specific requirements. Applications include building sitework, roadways, retaining walls, levees, dams and revetments.

Advantages

- **Competitive installing costs**
- **Immediate loading**
- **Installs in any weather**
- **Speeds site preparation**
- **Installs with available equipment**

- **True helix installs with ease, minimal disturbance**
- **Predictable results**
- **Removable**
- **Less equipment (no concrete trucks or grout pumps)**
- **Labor-saving - keeps crew size small**
- **No spoils to remove**

Installing Tools

For attachment to a torque motor source directly or via a torque-indicating device, SS anchor drive tools are the same basic design.

Catalog No.	Description	Weight
639000I	SS5/SS150 Tool	7 lb.
C303-0195	SS175 Tool	18 lb.
C303-0201	SS200 Tool	30 lb.
C303-0202	SS225 Tool	30 lb.

Components, Parts — *for SOIL SCREW™ uses only on this page*

Rated for 5,500 ft.-lb. maximum installation torque and 70,000 lb. minimum ultimate tension strength, SS5 anchors have 1½"-square steel shafts and 8"-diameter helices and are hot-dip galvanized after assembly.

SOIL SCREW™ Termination Adapters

Both are hot-dip galvanized steel and fit 1½" square shaft SS5 anchors.

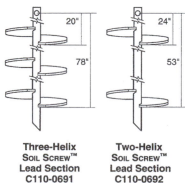
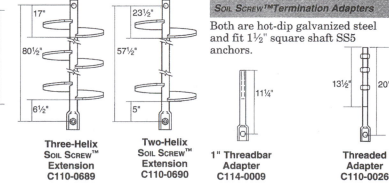

Figure 5.16.0 *(By permission: Atlas Systems, Inc., Independence, Missouri)*

Components, Parts — *for Tieback only on this page*

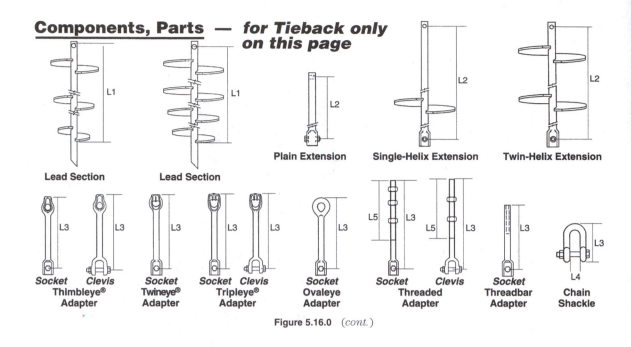

Lead Section **Lead Section** **Plain Extension** **Single-Helix Extension** **Twin-Helix Extension**

Socket *Clevis* **Thimbleye® Adapter** *Socket* **Twineye® Adapter** *Socket* *Clevis* **Tripleye® Adapter** *Socket* **Ovaleye Adapter** *Socket* *Clevis* **Threaded Adapter** *Socket* **Threadbar Adapter** **Chain Shackle**

Figure 5.16.0 *(cont.)*

5.17.0 Theory of Foundation Anchor Design

Soil mechanics

Throughout this discussion we will concern ourselves with the theories of soil mechanics as associated with foundation anchor design. The mechanical strength of the foundations will not be considered in this section as we expect foundations with proper strengths to be selected by the design professional at the time of design. For this discussion, we assume the mechanical properties of the foundations are adequate to fully develop the strength of the soil in which they are installed. Although this discussion deals with the foundation anchor, the design principles are basically the same for either a tension or compression load. The designer simply uses soil strength parameters above or below a helix, depending on the load direction.

Shallow and deep foundation anchors

Two modes of soil failure may occur depending on helix depth: One is a shallow failure mode and the other is a deep failure mode. Foundations expected or proven to exhibit a specific mode are often referred to as "shallow" or "deep" foundations. The terminology "shallow" or "deep" refers to the location of the bearing plate with respect to the earth's surface. By definition, "shallow" foundations exhibit a brittle failure mode with general eruption of the soil all the way to the surface and a sudden drop in load resistance to almost zero. With "deep" foundations, the soil fails progressively, maintaining significant post-ultimate load resistance, and exhibits little or no surface deformation. The dividing line between shallow and deep foundations has been reported by various investigators to be three to eight times the

foundation diameter. Chance Company uses five diameters as the break between shallow and a deep foundation anchors. The five-diameter depth is the vertical distance from the surface to the top helix. The five-diameter rule is often simplified to 5 feet (1.5 m), minimum.

Any time a foundation anchor is considered, it should be applied as a deep foundation. A deep foundation has two advantages over a shallow foundation:

1. Provides an increased ultimate capacity.

2. Failure will be progressive with no sudden decrease in load resistance after the ultimate capacity has been achieved.

Bearing capacity theory

This theory suggests that the capacity of a foundation anchor is equal to the sum of the capacities of individual helices. The helix capacity is determined by calculating the unit bearing capacity of the soil and applying it to the individual helix areas. Friction along the central shaft is not used in determining ultimate capacity. Friction or adhesion on extension shafts (but not on lead shafts) may be included if the shaft is round and at least 3½" (8.9 cm) in diameter.

Graphic representation of individual compression bearing pressures on multi-helix foundation anchor.

A necessary condition for this method to work is that the helices be spaced far enough apart to avoid overlapping of their stress zones. Chance Company manufactures foundations with three-helix-diameter spacing, which has historically been sufficient to prevent one helix from significantly influencing the performance of another.

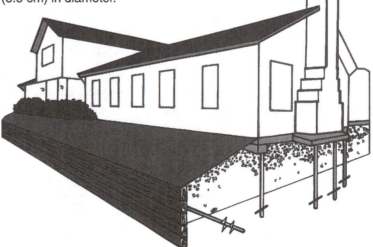

Figure 5.17.0 *(By permission: Atlas Systems, Inc., Independence, Missouri)*

5.17.1 Foundation Anchor Design—Product Specification

Lead and extension section lengths

HELICAL PIER® Foundation Systems standard lead-section lengths are 5, 7, and 10 ft. (1.5, 2 and 3 m). The standard extension section lengths are 3½, 5, 7, and 10 ft. (1, 1.5, 2 and 3 m). These combinations of leads and extensions provide for a variety of installed foundation anchor lengths.

Helix areas

Standard diameters for helices manufactured by the Chance Company are:

    6 in. = 26.7 sq. in.
    (15 cm = 0.0172 m²)
    8 in. = 48.4 sq. in.
    (20 cm = 0.0312 m²)
    10 in. = 76.4 sq. in.
    (25 cm = 0.0493 m²)
    12 in. = 111 sq. in.
    (30 cm = 0.0716 m²)
    14 in. = 151 sq. in.
    (35 cm = 0.0974 m²)

Helix configuration

Standard helices are ⅜ inch (0.95 cm) thick steel plates with outer diameters of 6, 8, 10, 12 and 14 inches (15, 20, 25, 30 and 35 cm). The lead section, or first section installed into the soil always contains helix plate(s). Extensions may be plain or helixed. Multihelix foundations have more than one helix arranged in increasing diameters from the foundation tip to the uppermost helix. The nominal spacing between helix plates is three times the diameter of the next lower helix. For example, a HELICAL PIER® Foundation Systems anchor with an 8-, 10-, and 12-inch (20, 25 and 30 cm) helix combination has a 24-inch (61 cm) space between the 8- and 10-inch (20 and 25 cm) helix and a 30-inch (76 cm) space between the 10- and 12-inch (25 and 30 cm) helix. Extensions with helix plates can be added to the foundation if more bearing area is required. They should be installed immediately after the lead section.

Capacities listed in the Ratings Table above, are mechanical ratings. One must be aware that the actual installed load capacities are dependent on actual soil conditions at each specific project site. Therefore, the design professional should use the bearing capacity method in designing anchor foundations. The number of helices, their size, and depth below grade is determined by obtaining soil

Minimum anchor type required based on mechanical ratings	
Design Load, kips (kN)	Minimum HELICAL PIER® Foundation Systems Anchor Required
0 to 25 (0 to 110)	SS5
25 to 35 (110 to 150)	SS150
35 to 50 (150 to 220)	SS175 or HS

Note: This chart uses a factor of safety vs. ultimate capacity = 2.

shear strength factors, cohesion (c) and angle of internal friction (Ø), and applying them as outlined in Theory of Foundation Anchor Design. The anchor family specified is based on the rated load carrying capacities for the specific foundation shaft size and installation torque required to install the foundation. The shaft sizes are 1½- or 1¾-inch (3.8 or 4.5 cm) square solid steel or 3½-inch (8.9 cm) OD heavy-wall steel pipe.

Chance Company is available to aid the design professional in determining the best helix combination/foundation anchor family for a given application. Additional design considerations are as follows:

Corrosion

Corrosion of foundation anchors is a major consideration in permanent structures. That is why foundation components are hot-dip galvanized per ASTM A153. The zinc coating will add between 5% and 20% to the life of HELICAL PIER® Foundation Systems anchors. The Federal Highway Administration (FHWA-SA-96-072) has established, from an extensive series of field tests on metal pipes and sheet steel buried by the National Bureau of Standards, maximum corrosion rates for steel buried in soils exhibiting the electrochemical index properties shown in the table:

The corrosion rates shown below are suitable for

Property	Criteria	Test Method
Resistivity	>3000 ohm-cm	AASHTO T-288-91
pH	>4.5<9	AASHTO T-289-91
Chlorides	<100 PPM	AASHTO T-291-91
Sulfates	<200 PPM	AASHTO T-290-91
Organic Content	1% max.	AASHTO T-267-86

designs for screw anchor foundations. These rates of corrosion assume a mildly corrosive in-situ soil environment having the electrochemical property limits that are listed in table below. The design corrosion rates, per FHWA-SA-96-072, are:

For Zinc
    15 µm/year (first 2 years)
    4 µm/year (thereafter)

For Carbon Steel
    12 µm/year

Figure 5.17.1 *(By permission: Atlas Systems, Inc., Independence, Missouri)*

5.18.0 Assessing Corrosion Potential for Underground Bare Steel Structures

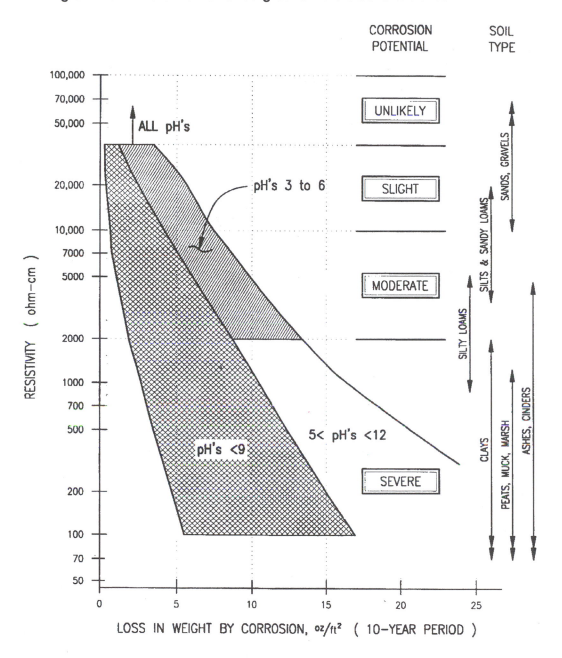

TECHNIQUE FOR ASSESSING CORROSION POTENTIAL
FOR UNDERGROUND BARE STEEL STRUCTURES[1,2]

Figure 5.18.0   *(By permission: Atlas Systems, Inc., Independence, Missouri)*

5.19.0 Rock Anchor Design Considerations

LATERAL FORCE, OVERTURNING AND UPLIFT APPLICATIONS

Most types of engineering structures generate a loading system that requires the foundation system to support bearing (compressive) forces in either soil or rock formations with only slight eccentricities or inclinations of bearing loads. In those cases conventional foundation systems, including the use of **Atlas Piers** and **Atlas-*Helical* Foundation Products**, provide adequate foundation supporting capacity with an adequate Factor of Safety. In a large number of applications, however, foundation systems are required where the forces of the structure require tension (pullout) capacity in rock formations. In those cases **Atlas Rock Anchor Products** can provide a suitable solution.

I.   ROCK MASS

The earth mass into which foundation systems for structures are set consists of either **soil** or **rock**. Soil is generally categorized as uncemented or weakly cemented material overlying the harder rock formations. Soils consist of a wide range of particle sizes and are formed by the decomposition of rock insitu (termed residual soils) or are transported by wind, ice, water or gravity to their present location. Rock formations are generally classified into three categories as follows:

- **Igneous** - Rocks that are formed by the solidification of molten magma ejected to the surface of the earth by volcanic eruption. Some of the more common igneous rocks include granite, basalt, gabbro, prophyrite, dolomite, rhyolite, pumice and diorite.

- **Sedimentary** - Rocks that are formed by compaction and cementation of soil deposits formed by weathering, or by various chemical processes. Among the sedimentary type rocks are shale, sandstone, mudstone, limestone, gypsum and dolomite.

- **Metamorphic** - Rocks that are formed by undergoing a change in form by heat or pressure. Some of the more common metamorphic rocks include gneiss, marble and slate.

It is important to recognize that rock masses are discontinuous and often jointed mass. Because of these discontinuities which include bedding planes, joints, faults and fracture zones, the analysis and design of rock anchors requires careful analysis and design based on the actual engineering properties of the rock formation. Table 1 provides some representative mechanical properties of various rocks.

Figure 5.19.0   *(By permission: Atlas Systems, Inc., Independence, Missouri)*

5.19.1 Elevation and Cross Section of Standard Load-Bearing Pier

ATLAS PIERS®
Resistance Product Designator

Atlas Pier Systems can be installed in either an interior or exterior location. Every Atlas Pier System provides for a two-stage system of initially driving the manufactured piers to load bearing support then, using hydraulics, restoring the structure to the desired elevation. Atlas Piers not only stop settlement, but also actually raise the structure, closing cracks and correcting other structural flaws caused by the settlement and/or ground movement. Design should involve professional engineering input. Specific information involving the structures, soil characteristics and foundation conditions must be used for the final design.

It is recommended that the design be conducted by a Registered Professional Engineer.

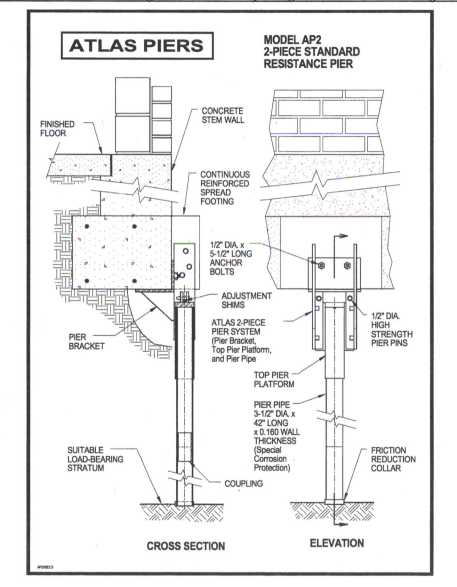

Elevation and cross sectional view of AP2 Standard Pier.

Figure 5.19.1   *(By permission: Atlas Systems, Inc., Independence, Missouri)*

5.20.0 Helical Anchors—Light Duty

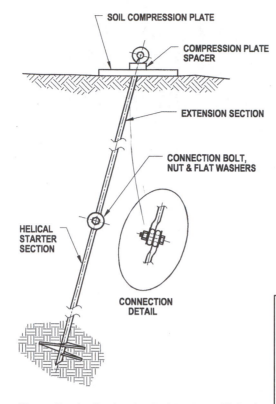

Figure 5. Application detail of the **Atlas-Helical** Light Duty Anchor.

Atlas offers a light duty anchor for mounting and tie down applications up to 5,000 pounds. The anchor system is fabricated from 3/4 inch diameter solid steel bar stock. Attachment and coupling is accomplished with formed and welded "eyes". These anchors may be installed by hand using a torque bar or with a hydraulic torque motor.

Lateral support and tensioning of the anchor is accomplished with a 12 inch square Compression Plate, which is fabricated from one inch thick plate. The plate is quickly attached to the anchor with the slot provided. The assembly is then secured with the Compression Plate Spacer. This spacer also is provided with a slot for easy installation. It is installed with the slot at 90^0 or 180^0 to the Compression Plate.

AHS*-RD0075 /PLATE SIZE/ [5'-6"] STARTER SECTION

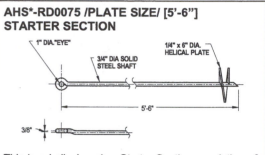

This is a helical anchor Starter Section consisting of a 3/4" solid steel bar with a 2-1/4" pitch on one end and a 1" diameter welded "eye on the other.
AVAILABLE 1/4" THICK HELICAL PLATE SIZES:
      4", 6" or 8" Diameter

AHE*-RD0075 [5'-6"] EXTENSION SECTION

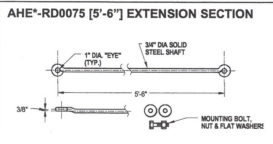

This is a helical anchor Extension Section consisting of a 3/4" solid steel bar with 1" diameter welded "eyes" on both ends. Two 2-1/2" diameter flat washers, bolt and nut are included.

*   Indicate Coating required by entering appropriate letter: **P, G, E, or C** (See Page E13)

AHCP*-RD0075-1"x12"x1'-0" SOIL COMPRESSION PLATE

1" x 12" square steel plate with 13/16" diameter slot for attachment to the top of the anchor.

AHCP*-RD0075-1"x 2"x 3" COMPRESSION PLATE SPACER

1" x 1-1/2" square steel plate with 13/16" diameter slot for attachment the anchor and the Soil Compression Plate.

Figure 5.20.0   *(By permission: Atlas Systems, Inc., Independence, Missouri)*

5.20.1 Distressed Wall with Active Pressure Acting Against Wall

When a helical tieback is installed and anchored in place, two possible options are available as follows:

A.   A portion of the soil is removed, the anchor is used to restore the wall toward its original position and the soil is back filled against the wall, or

B.   The tieback is merely locked in position with no restoration of the wall.

In either case, the soil will continue to exert an "active" pressure against the wall.

The installed helical tieback anchor develops anchoring resistance capacity through development of "passive[2]" earth pressure against the helical plate. Thus it is necessary that the helical tieback anchor be installed properly to insure the ability to develop full "passive" pressure resistance.

It is very important that the basement wall repair should also include remedial drainage work in order to prevent any future condition of soil saturation and resulting water pressure against the wall and/or take into account the full effect of water pressure against the wall in the tieback design. (See *Figure 16.*)

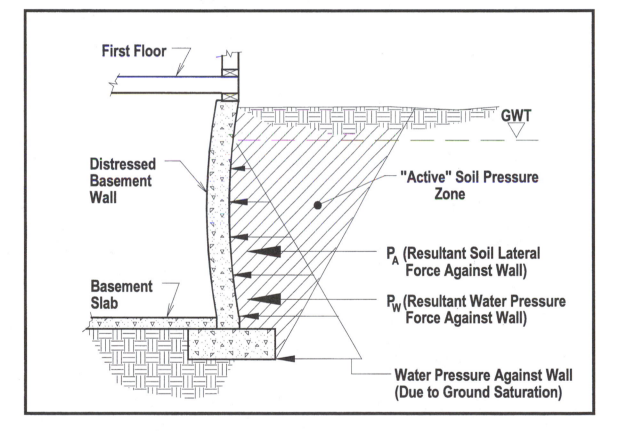

Distressed Basement Wall with "Active" Soil Pressure and Water Pressure Acting Against the Wall.

---

[2] Passive earth pressure is defined as the structure (the anchor) exerting pressure directly on the earth and causing the structure (helical plates) to move in the direction of the soil mass.

Figure 5.20.1   *(By permission: Atlas Systems, Inc., Independence, Missouri)*

5.20.2 Typical Soil Anchor Retaining Wall Tieback

Recommended Factors of Safety for Design

Recognizing the variability of soil conditions that may exist at a site, the varied nature of loading on structures and how these loads are transferred through foundations, Atlas Systems, Inc. recommends that specifiers, contractors, etc. use an appropriate Factor of Safety in design for use of **Atlas-*Helical* Foundation Products**. Generally this Factor of Safety is a minimum of 2:1 on all permanent loading conditions and a minimum of 1.5:1 for any temporary load situation. National and local building code regulations may require more stringent Factors of Safety on certain projects.

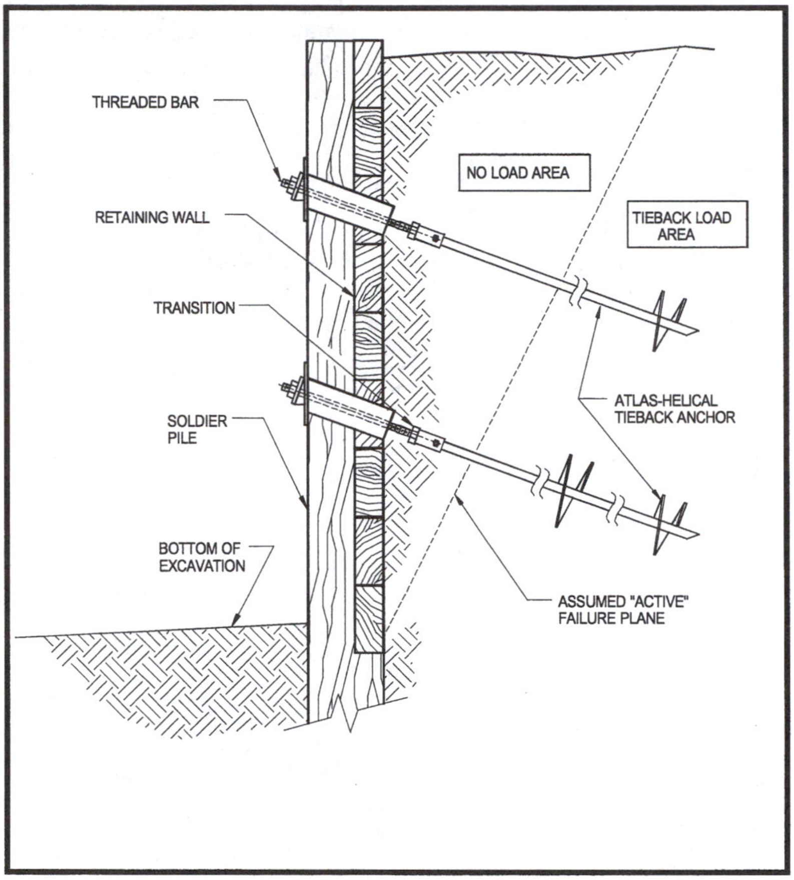

THREADED BAR

RETAINING WALL

TRANSITION

SOLDIER PILE

BOTTOM OF EXCAVATION

NO LOAD AREA

TIEBACK LOAD AREA

ATLAS-HELICAL TIEBACK ANCHOR

ASSUMED "ACTIVE" FAILURE PLANE

Typical retaining wall tieback configuration.

Figure 5.20.2   (*By permission: Atlas Systems, Inc., Independence, Missouri*)

II. Location and Placement of Tiebacks

Every wall tieback situation is unique, but there are some aspects that merit attention. The placement of the anchor is influenced by the height of the soil backfill against the wall. *Figure 5.20.2.1* shows this condition and a guideline for setting the location and minimum length of installation of the tieback. Experience indicates that the tieback should be located close to the point of maximum wall bulge and/or close to the most severe transverse crack. In cases where concrete block walls or severely cracked walls occur, a vertical and/or transverse steel channel (waler) or plate must be used to maintain wall integrity.

For other types of wall distress such as multiple cracking or differential settlement induced cracking, the tieback placement location, must be selected on a case by case basis.

Another factor to consider is the height of soil cover over the helical anchor. *Figure 5.20.2.1* also indicates that the minimum height of cover is 6 times the diameter of the largest helical plate. Finally, the helical anchor must be installed a sufficient distance away from the wall in order that the helical plate(s) can fully develop an anchoring capacity by "passive" pressure as shown in *Figure 5.20.2.1*. This requires the length of installation to be related to the height of soil backfill as shown in *Figure 5.20.2.1*.

III. Estimating Tieback Load Requirements

Estimating the lateral loads acting against basement walls or retaining walls as exerted by the earth requires knowledge of: 1.) The soil type and condition, 2.) The structural dimensions of the retaining structure, 3.) Other geotechnical conditions (e.g. ground water table). *Figures 5.20.2.1–5.20.2.3* were prepared for preliminary design assistance for estimating the tieback load requirements when no ground water (**GWT**) is present at the site. *If hydrostatic water pressure is present, the magnitude of this pressure is determined and added to the tieback load requirement from the earth pressure.* In those cases where the soil and subsurface drainage conditions are not known, it should be assumed in the **design that water pressure will be present.** As a guideline in preparing tieback load requirement estimates, one tieback row was used for walls of 15 feet of height or less and two tieback rows for walls ranging in height from 15 feet to 25 feet. Individual project conditions and design considerations can cause changes in these guidelines.

**IN ALL CIRCUMSTANCES, A REGISTERED PROFESSIONAL ENGINEER
SHOULD CONDUCT AND APPROVE THE FINAL DESIGN.**

Figure 5.20.2   *(cont.)*

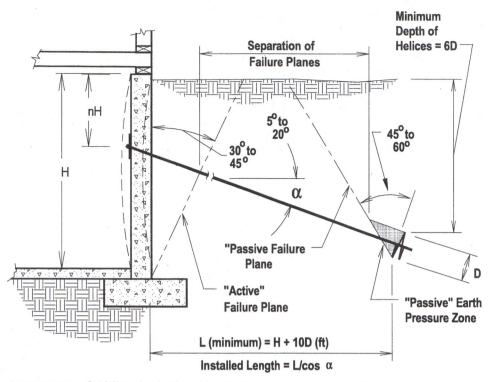

Figure 5.20.2.1   Guidelines for depth and length for a typical installation with helical tiebacks. (*By permission: Atlas Systems, Inc., Independence, Missouri*)

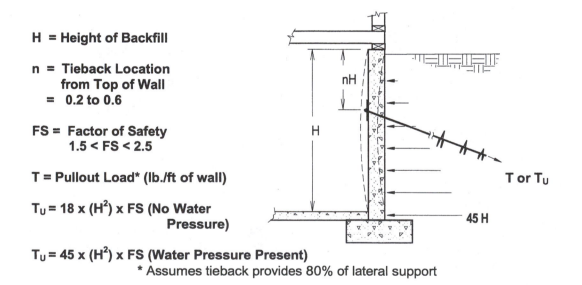

H = Height of Backfill

n = Tieback Location
     from Top of Wall
   = 0.2 to 0.6

FS = Factor of Safety
      1.5 < FS < 2.5

T = Pullout Load* (lb./ft of wall)

T_U = 18 x (H^2) x FS (No Water
                            Pressure)

T_U = 45 x (H^2) x FS (Water Pressure Present)
        * Assumes tieback provides 80% of lateral support

January 2000
2000v1.4

Figure 5.20.2.2   Estimate of tieback force requirement for basement applications. (*By permission: Atlas Systems, Inc., Independence, Missouri*)

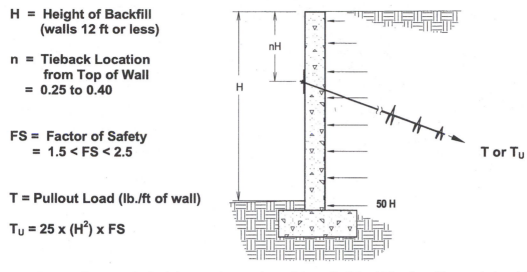

H = **Height of Backfill**
    (walls 12 ft or less)

n = **Tieback Location**
    from Top of Wall
    = **0.25 to 0.40**

FS = **Factor of Safety**
    = **1.5 < FS < 2.5**

T = **Pullout Load (lb./ft of wall)**

$T_U = 25 \times (H^2) \times FS$

Figure 5.20.2.3 Estimate of tieback force requirement for retaining walls 15 feet high or less. (*By permission: Atlas Systems, Inc., Independence, Missouri*)

H = **Height of Backfill**
    (walls 15 to 25 ft.)

n = **Top Tieback Location**
    from Top of Wall
    = **0.20 to 0.30**

m = **Lower Tieback Location**
    from Top of Wall
    = **0.50 to 0.75**

FS = **Factor of Safety**
    = **1.5 < FS < 2.5**

T = **Pullout Load (lb./ft of wall)**

$T_{NU} = 12 \times (H^2) \times FS$

$T_{MU} = 18 \times (H^2) \times FS$

Figure 5.20.2.4 Estimate of tieback force requirement for retaining walls 15 feet to 25 feet. (*By permission: Atlas Systems, Inc., Independence, Missouri*)

5.20.3 Tieback Configuration for Sheet Piles

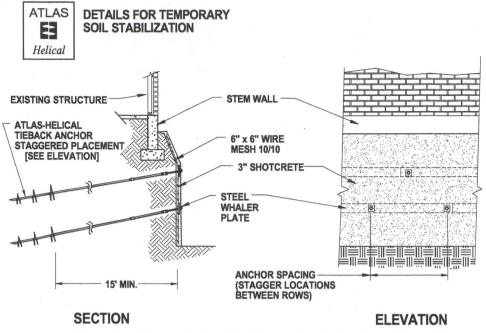

Figure 5.20.3 *(By permission: Atlas Systems, Inc., Independence, Missouri)*

5.20.4 Mechanical Properties of Rock

ROCK	Young's Modulus at Zero Load (10^5 kg/cm^2)	Bulk Density (g/cm^3)	Porosity (percent)	Compressive Strength (kg/cm^2)	Tensile Strength (kg/cm^2)
Granite	2 - 6	2.6-2.7	0.5-1.5	1,000-2,500	70-250
Microgranite	3 - 8				
Syenite	6 - 8				
Diorite	7-10			1,800-3,000	150-300
Dolerite	8-11	3.0-3.05	0.1-0.5	2,000-3,500	150-350
Gabbro	7-11	3.0-3.1	0.1-0.2	1,000-3,000	150-300
Basalt	6-10	2.8-2.9	0.1-1.0	1,500-3,00	100-300
Sandstone	0.5-8	2.0-2.6	5 - 25	200-1,700	40-250
Shale	1-3.5	2.0-2.4	10 - 30	100-1,000	20-100
Mudstone	2 - 5				
Limestone	1 - 8	2.2-2.6	5 - 20	300-3,500	50-250
Dolomite	4-8.4	2.5-2.6	1 - 5	800-2,500	150-250
Coal	1 - 2			50-500	20-50
Quartzite		2.65	0.1-0.5	1,500-3,000	100-300
Gneiss		2.9-3.0	0.5-1.5	500-2,000	50-200
Marble		2.6-2.7	0.5-2	1,000-2,500	70-200
Slate		2.6-2.7	0.1-0.5	1,000-2,000	70-200

Note: 1. For the igneous rocks listed above Poisson's ratio is approximately 0.25.
2. For a certain rock type, the strength normally increases with increase in density and increase in Young's modulus. (After Farmer, 1968)
3. Taken from *"Foundation Engineering Handbook"* by Winterkorn and Fong, Van Nostrand Reinhold, pg 72.

Figure 5.20.4 *(By permission: Atlas Systems, Inc., Independence, Missouri)*

5.20.5 Method of Rock Anchors

Rock Anchors are structural members that transmit tension forces into the rock mass. As shown in *Figure 5.20.5*, Rock Anchors typically consist of a steel tendon, which is inserted into a borehole and then bonded to the rock formation by grout, other chemicals, or by mechanical means. Rock Anchors are somewhat similar in principle to soil anchors, but in most cases provide a much greater load carrying capacity due to the greater strength of the rock mass. As noted in *Figure 5.20.5*, the steel tendon transfers the tension load into the rock mass. Thus, the shear strength of the surrounding material is required to resist this tensile force. Reference to *Figure 5.20.5* indicates that an effort is made to fasten the anchor (tension load transfer) to competent rock at a location well away from the structure itself. This approach is accomplished by creating two zones along the steel tendon anchor as follows:

A.   Fixed Anchor Length

Fixed Anchor Length is that portion of the tendon, which is farthest away from the structure over which the tension load is transferred to the surrounding rock mass. Free Anchor Length

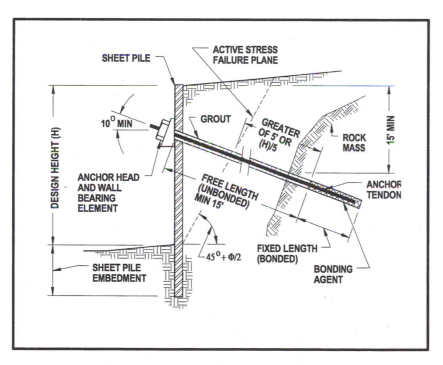

Rock Anchor Setup and Technical Terms (Sheet Pile Application).

Figure 5.20.5   *(By permission: Atlas Systems, Inc., Independence, Missouri)*

is that portion of the tendon between the top of the fixed anchor length and the structure over which none of the tension load is transferred to the surrounding rock mass.

5.20.6 Typical Applications

The following provides some general applications in which Rock Anchors can provide an effective solution for the transfer of structural loads to rock masses.

A.   Uplift Applications

    1.   High Ground Water Table

    The excavations for structures and buildings at waste water treatment plants or at cofferdams for bridge abutments often occur in areas where the ground water table (GWT) is well above the bottom of the excavation. In those cases, water pressure exerted against the base slab of the structure may require anchoring the structural floor slab to an underlying rock mass.

    2.   WIND LOADS

    A second type of application is where wind loads on roof structures lead to a net uplift loading on certain foundation elements. The use of Rock Anchors in an underlying rock mass can be used to transmit these wind load induced uplift forces into the rock mass.

B.   Overturning Applications

There are a number of structures, including communication towers, which are loaded in a manner so as to induce tension loads into a foundation system. In those cases where a rock formation exists in close proximity to the ground surface elevation, the use of rock anchors can provide a cost effective foundation system for these tension loads.

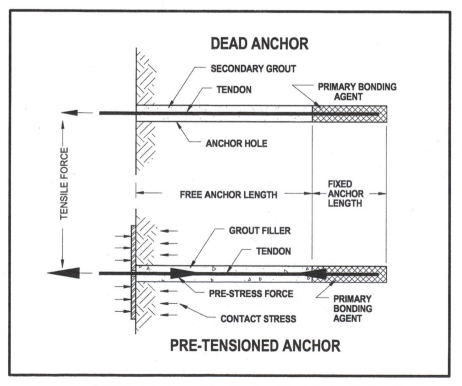

Dead Rock Anchor and Pre-Tensioned Rock Anchor Configuration.

Figure 5.20.6   *(By permission: Atlas Systems, Inc., Independence, Missouri)*

C.   Lateral Force Applications

Earth retaining structures such as sheet piling, retaining walls or reinforced earth may require the use of some form of tieback in order to insure the stability of the structure. These tiebacks are designed to provide some or all of the restraining force against the lateral pressures from the retained earth and water. To provide the required anchor (pullout) support and to limit the amount of lateral movement of the retaining structure, Rock Anchors can be designed and used when competent rock mass is within a reasonable distance of the retaining structure. These anchors are typically installed as the excavations takes place and are installed at a downward angle to the horizontal. In cases of remedial action, rock anchors can be used to provide added stability to an existing structure that has experienced some lateral movement and requires added support and/or restoration.

5.20.7 Dead and Pretensioned Rock Anchors

Since Rock Anchors generally have a long installation length, the steel tendon may undergo considerable elastic movement or deformation once the tension load from the structure is applied when the anchor acts only in a dead capacity.

5.20.8 Average Ultimate Bond Stress at Grout/Rock Interface

Table 3 provides some average ultimate bond stress for a rock-grout interface. Table 4 gives similar values when using epoxy as the chemical bonding agent. The result of laboratory unconfined compression testing of rock cores should be used in the final design of Rock Anchors. In using the equation above to determine the fixed bonded length, it is recommended that a minimum length used be 15 feet and that the beginning of the fixed bonded length be beyond the failure shear plane of the retaining structure for lateral force applications.

It is also recommended that load tests be conducted on a certain number of the installed Rock Anchors in order to verify the load carrying capacity in tension of the Rock Anchor System and to assist in establishing the magnitude of the pre-tension load. In cases where the rock may be fractured, pressure grouting can be used to increase the rock bond values.

In addition to the above, the following are some general guidelines to follow in the design of rock anchors;

- Anchors should have a minimum inclination of 10 degrees below horizontal
- A minimum of 15 feet of cover should exist between any part of the anchor and the ground surface
- The minimum horizontal spacing between Rock Anchors should be three times the diameter of the bonded zone or 4 feet, whichever is larger.

Table 3.	AVERAGE ULTIMATE BOND STRESS AT GROUT/ROCK INTERFACE[1]	
ROCK	AVERAGE ULTIMATE BOND STRESS AT ROCK/GROUT INTERFACE	
	(MPa)[2]	(psi)
Granite & Basalt	1.7 - 3.1	250 - 450
Dolomitic Limestone	1.4 - 2.1	200 - 300
Soft Limestone	1.0 - 1.4	150 - 200
Slates & Hard Shales	0.8 - 1.4	120 - 200
Soft Shales	0.2 - 0.8	30 - 120
Sandstones	0.8 - 1.7	120 - 250
Weathered Sandstones	0.7 - 0.8	100 - 120
Chalk	0.2 - 1.1	30 - 155
Weathered Marl	0.15 - 0.25	25 - 35
Concrete	1.4 - 2.8	200 - 400

1. Taken from Williams Form Engineering Corporation literature.
2. (MPa) – Mega Pascal

Figure 5.20.8 (*By permission: Atlas Systems, Inc., Independence, Missouri*)

5.20.9 Average Ultimate Bond Stress and Epoxy/Rock Interface

Corrosion Protection of Rock Anchors

Rock Anchor steel tendons, anchor heads and all other anchor hardware should be protected against corrosion consistent with the corrosive environment of the ground and water conditions at the site. All Rock Anchors for permanent applications should be double corrosion protected. In most instances this would involve a coating on the steel tendon and coverage with a bonding agent.

TABLE 4.	AVERAGE ULTIMATE BOND STRESS AT EPOXY/ROCK INTERFACE[1]	
ROCK	COMPRESSIVE STRENGTH (psi)	AVERAGE ULTIMATE BOND STRESS ROCK / EPOXY (psi)
Granites	12,000 - 15,000	450 - 550
Limestones	6,000 - 8,000	200 - 300
Shales/Sandstones	1,800 - 2,200	150 - 250
Mudstones/Siltstones	700 - 800	100 - 150

1. Taken from Wiliams Form Engineering Corporation literature.

Figure 5.20.9   *(By permission: Atlas Systems, Inc., Independence, Missouri)*

5.20.10 Corrosion Protection of Rock Anchors

In designing and using sacrificial anode systems, the soil profile conditions as to the type of soil, resistivities, soil pH and location of the groundwater table (if present) must be determined. Among the design considerations for the system:

• Use of wire type or canister type anode

• Selection of the appropriate anode material (magnesium, titanium, etc.)

• Designing the ground bed (location, dimensions, horizontal vs. vertical, depth of placement, type of backfill, etc.)

• Determining the number of piers per anode

• Type, size and connections between pier(s) and the sacrificial anode

• Installation normally 3 feet below the surface and 3 to 7 feet from pier

Impressed Current

In areas of the most severe corrosion potential, where a larger current is required and/or in high resistance electrolytes, an impressed current system is used which requires a power source, rectifier and a ground bed of impressed current anodes. These systems require a continuous external power source.

The majority of applications where Atlas Foundation Support Products may be specified will not require an active corrosion protection system. In those cases where the combination of soil and electrolyte conditions require an "active system", the sacrificial anode protection system will likely be the most economical approach.

Active cathodic protection systems must be individually designed to the specific application. The major variables are soil moisture content, resistivity of soil and pH. Each of these items influences the final selection of the cathodic protection system. Typical design life for the cathodic protection is 10 to 20 years, depending upon the size and length of the anode canister.

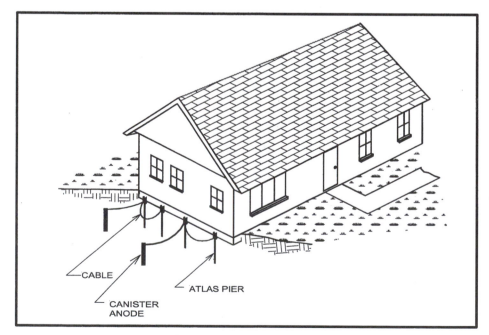

Sacrificial anode cathodic protection system.

Figure 5.20.10   (*By permission: Atlas Systems, Inc., Independence, Missouri*)

5.20.11 Residential Buildings with Concrete Slab Floor

BUILDING CONSTRUCTION	BUILDING DIMENSIONS, (ft.)								
	20' x 20'	20' x 30'	20' x 40'	30' x 30'	30' x 45'	30' x 60'	40' x 40'	40' x 60'	40' x 80'
	ESTIMATED DEAD LOAD at FOUNDATION, DL (lb/ft)								
One Story -- Wood/Metal/Vinyl Walls with Wood Framing on Footing	725	742	753	742	758	768	776	797	810
One Story -- Masonry Walls with Wood Framing on Footing	975	992	1003	992	1008	1018	1026	1047	1060
Two Story -- Wood/Metal/Vinyl Walls with Wood Framing on Footing	965	1004	1012	1004	1040	1063	1082	1129	1160
Two Story -- 1st Floor Masonry, 2nd Wood/Metal/Vinyl with Wood Framing on Footing	1215	1254	1280	1254	1290	1313	1332	1379	1410
Two Story -- Masonry Walls with Wood Framing on Footing	1465	1504	1530	1504	1540	1563	1582	1629	1660

BUILDING CONSTRUCTION	BUILDING DIMENSIONS, (ft.)								
	20' x 20'	20' x 30'	20' x 40'	30' x 30'	30' x 45'	30' x 60'	40' x 40'	40' x 60'	40' x 80'
	ESTIMATED LIVE LOAD at FOUNDATION, LL (lb/ft)								
One Story -- Residential on Slab	N/A								
One Story -- Residential on Basement One Story -- Over Crawl Space	250	300	333	300	346	375	400	461	500
Two Story -- Residential on Slab	250	300	333	300	346	375	400	461	500
Two Story -- Residential on Basement Two Story -- Over Crawl Space	500	600	667	600	692	750	800	923	1000
One Story -- Commercial on Slab	N/A								
One Story -- Commercial on Basement Two Story -- Commercial on Slab	450	540	600	540	623	675	720	831	900
Two Story -- Commercial on Basement	900	1080	1200	1080	1246	1350	1440	1662	1800

Figure 5.20.11   *(By permission: Atlas Systems, Inc., Independence, Missouri)*

5.20.12 Estimating Snow Loads (SL)

The required Snow Load Factor (**S**) can be determined from the locally approved building code. This factor will be given in pounds per square foot. To determine the Snow Load along the perimeter of the structure use the following:

SL = **S** × [Width of Building × Length of Building / 2 × (Width of Building + Length of Building)]

5.20.13 Estimating Foundation Soil Load (W)

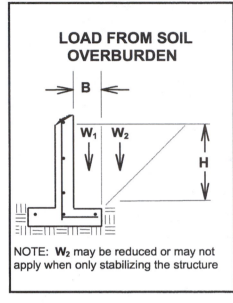

Footing Toe Width,	Height of Soil Overburden,	SOIL TYPE			
		Cohesive		Granular	
B (in.)	H (ft.)	W_1	W_2	W_1	W_2
6 Inches	2 feet	110	220	125	240
	4 feet	220	880	250	960
	6 feet	330	1980	500	2160
	8 feet	440	3520	625	3840
12 Inches	2 feet	220	220	250	240
	4 feet	440	880	500	960
	6 feet	660	1980	1000	2160
	8 feet	880	3520	1250	3840
24 Inches	2 feet	440	220	500	240
	4 feet	880	880	1000	960
	6 feet	1320	1980	2000	2160
	8 feet	1760	3520	2500	3840
36 Inches	2 feet	660	220	750	240
	4 feet	1320	880	1500	960
	6 feet	1980	1980	2250	2160
	8 feet	2640	3520	3000	3840

LOAD FROM SOIL OVERBURDEN

NOTE: W_2 may be reduced or may not apply when only stabilizing the structure

Figure 5.20.13   *(By permission: Atlas Systems, Inc., Independence, Missouri)*

5.20.14 Residential Buildings with Basement

BUILDING CONSTRUCTION	BUILDING DIMENSIONS, (ft.)								
	20' x 20'	20' x 30'	20' x 40'	30' x 30'	30' x 45'	30' x 60'	40' x 40'	40' x 60'	40' x 80'
	ESTIMATED DEAD LOAD at FOUNDATION, DL (lb/ft)								
One Story -- Wood/Metal/Vinyl Walls with Wood Framing on Basement and Footing	1060	1092	1114	1092	1121	1140	1156	1195	1220
One Story -- Masonry Walls with Wood Framing on Basement and Footing	1310	1342	1364	1342	1371	1390	1406	1445	1470
Two Story -- Wood/Metal/Vinyl Walls with Wood Framing on Basement and Footing	1300	1354	1390	1354	1403	1435	1462	1528	1570
Two Story -- 1st Floor Masonry, 2nd Wood/Metal/ Vinyl with Wood Framing, Basement & Footing	1550	1604	1640	1604	1653	1685	1712	1778	1820
Two Story -- Masonry Walls with Wood Framing on Basement and Footing	1800	1854	1890	1854	1903	1935	1962	2028	2070

Figure 5.20.14   *(By permission: Atlas Systems, Inc., Independence, Missouri)*

5.20.15 Commercial Buildings

BUILDING CONSTRUCTION	BUILDING DIMENSIONS, (ft.)								
	20' x 20'	20' x 30'	20' x 40'	30' x 30'	30' x 45'	30' x 60'	40' x 40'	40' x 60'	40' x 80'
	ESTIMATED DEAD LOAD at FOUNDATION, DL (lb/ft)								
One Story -- Precast Concrete Walls on Footing with Slab Floor	2150	2175	2192	2175	2198	2213	2225	2255	2275
One Story -- Precast Concrete Walls and Basement on Footing	3130	3175	3205	3175	3217	3243	3265	3320	3355
Two Story -- Precast Concrete Walls on Footing with Slab Floor	3425	3475	3508	3475	3521	3550	3611	3636	3675
Two Story -- Precast Concrete Walls and Basement on Footing	4490	4560	4607	4560	4624	4665	4700	4786	4840

Figure 5.20.15   *(By permission: Atlas Systems, Inc., Independence, Missouri)*

5.21.0 Introduction to Caissons

Caissons are typically drilled or augered holes, into which a metal casing is installed; caissons are often referred to as *drilled* or *bored* piers. Typically, caissons fall into three groups: straight shaft, belled, or rock socketed. Installing caissons involves excavation, generally by a boring machine, to the depth required to meet either end bearing or friction, lowering a metal casing into the excavated hole, and filling the casing with reinforced concrete. When a belled caisson is constructed, a special belling tool is inserted in the bottom of the excavated shaft to create an enlarged base. In the case of a rock-socketed caisson, the boring machine will have a cutting shoe on the end to bore into the rock-bearing surface. A rock socket is then drilled into the rock to accept an H pile before filling the caisson with concrete.

The slurry method of installing caissons, utilizing bentonite, is also used when proper soil and ground-water conditions prevail.

5.21.1 Caisson Types

The general types of caissons an inspector is likely to encounter include straight-shafted, belled, and rock-socketed. Caissons are typically augered or drilled and may go by a variety of names, including drilled shafts, drilled piers, and bored piles. In some cases a steel casing is driven with impact methods followed by drilling of a rock socket. Typical caisson types are shown in Fig. 5.21.1.1. Caisson shaft diameters typically vary between 36 and 72 in (91.44 and 182.88 cm), although much smaller diameter caissons may be used (e.g., for sign foundations) and larger diameters (over 10 ft; 3.04 m) may be used in special cases.

Because of the expense of the pneumatic (compressed-air) method of caisson construction and the advent of modern caisson construction technology such as slurry methods, pneumatic caissons are rarely encountered and thus are not specifically covered in this text.

5.21.2 Construction Procedures

Construction procedures vary as a function of the contractor's equipment and experience, local soil and groundwater conditions, and type of caisson. Up until World War II caissons were advanced primarily with manual excavation. Today truck- and crane-mounted equipment is available to power drilling buckets, single- and double-flight augers, steel casing, core barrels, and belling buckets. Figures 5.21.2.1 through 5.21.2.10 depict typical rotary drilling equipment and procedures.

The basic technique involves excavating the caisson to the desired depth, verifying that the geologic conditions are as anticipated, and filling the shaft with concrete. For belled caissons a belling tool is inserted to the bottom of the excavated shaft to enlarge the base. Prior to acceptance and concreting the bottom of the excavation may be trimmed and cleaned by hand labor followed by a final "down-the-hole" inspection. In cases where it is deemed safe and necessary to enter the hole for cleaning or inspection, applicable OSHA regulations and appropriate safety precautions should be adhered to. Where human entry is not safe or practical, observations may be made from the ground surface.

Down-Hole Inspection

For "dry" holes floodlights may be used to facilitate topside observations. Video can be used for "dry" or "wet" holes. For slurry caissons, specially fabricated pipe probes can be used to sound the base and feel for collapse or soft sediments. One such pipe probe which has been used for caissons (and some slurry walls) is constructed with aircraft-grade aluminum pipe sections with watertight rubber O-ring joints. The bottom of the pipe is fitted with a weight. The rod is designed so that when inserted in slurry it has a "neutral" buoyancy for maneuverability and a stiff response, thus providing a sensitive "feel" when sounding the bottom.

Entering the caisson for construction or inspection should be done only if necessary and if proper and complete safety procedures are established *in advance*. Inspectors should refer to *Recommended Procedures for the Entry of Drilled Shaft Foundation Excavations* for guidelines on down-hole entry (available through ADSC at the above referenced address). A safety plan for entering the caisson should include at a minimum:

1. Personal protective equipment including safety harnesses, lines, cages, etc.

2. Methods and equipment for verifying down-hole air quality including hazardous vapors, oxygen levels, and percent explosion levels. The potential exposure to contaminated soil and groundwater, organic soils, and heavy exhaust gases should be assessed beforehand.

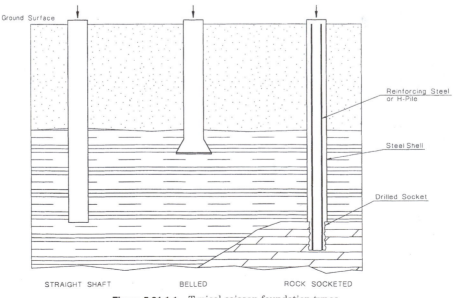

Figure 5.21.1.1 Typical caisson foundation types.

Figure 5.21.2.1   Truck-mounted rotary drilling machine with drilling bucket. (*New England Foundation Company, Andover, Mass.*)

Figure 5.21.2.2   Plumbing Kelly bar prior to excavating. (*New England Foundation Company, Andover, Mass.*)

Figure 5.21.2.3   Spoil removal prior to installation of casing. (*New England Foundation Company, Andover, Mass.*)

Figure 5.21.2.4   Installation of first section of temporary casing. (*New England Foundation Company, Andover, Mass.*)

(*Courtesy: McGraw-Hill: Field Inspection Handbook; Brock, Levy, Sutcliffe*)

Figure 5.21.2.5 Pushing casing with Kelly bar. (*New England Foundation Company, Andover, Mass.*)

Figure 5.21.2.6 Drop chisel to break obstructions. (*New England Foundation Company, Andover, Mass.*)

3. Methods for ventilating the shafts (e.g., blowers).

4. Protective casing.

5. Free and clear access and egress.

6. Experienced top worker and crane operator.

7. Method for two-way communication between bottom and top.

8. Method for safely lowering tools and equipment into hole *with or in advance of* personnel, and removing tools *with or after* personnel.

9. Dewatering as necessary.

For some projects "down-the-hole" field testing may be required. The inspector should have such equipment calibrated and in working order with a safe storage facility. Provision of backup equipment may be prudent to avoid delay in the event of loss or damage.

5.21.3 Shaft Construction

Shafts are commonly constructed by the dry method, casing method, or "wet" method with slurry. In some locales the soil is competent and dry enough where there is little potential for caving or excessive groundwater flow. Where such favorable conditions do not exist, temporary or permanent casing may be used to maintain the excavation and/or control groundwater.

An alternative method of stabilizing the shaft is with a slurry (e.g., bentonite). The density and height of the slurry column must be adequate to maintain a positive head against groundwater pressure. Slurry procedures and characteristics are normally outlined in the specifications. The inspector should be prepared to measure and monitor the following slurry properties in accordance with American Petroleum Institute (API) test specifications: density (pounds per square foot), Marsh fun-

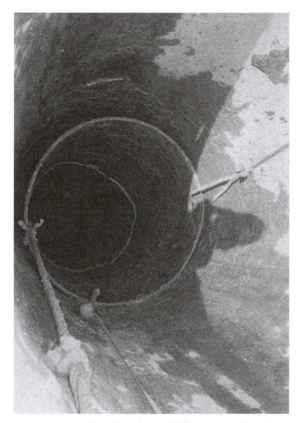

Figure 5.21.2.7   View into shaft with "telescoping" sections of casing. (*New England Foundation Company, Andover, Mass.*)

nel viscosity (seconds), sand content by volume (percent), and pH. Slurry procedures are covered in further detail in the following section on slurry wall construction.

After a caisson location is marked with a survey stake (and offset stakes for as-built centerline measurements) the contractor will set an auger above the stake and level the rig. A carpenter's level is normally used to plumb the Kelly bar which turns the auger. Project specifications set location and alignment tolerances.

For shafts not under water or slurry verticality may be checked with a plumb bob. For shafts under water or slurry approximate plumbness can be checked by measuring the Kelly bar alignment when extended to the bottom of the hole.

Soil cuttings from the auger flights should be observed by the inspector to assess actual soil conditions and elevation of bearing stratum. Inspection personnel should keep a safe distance from operating augers and drill equipment. This includes the aboveground operation of cleaning the spoil from the auger flights.

The inspector should carefully monitor shaft excavation, noting any caving, squeezing, obstructions, or difficulties in advancing the hole. Where temporary steel casing is used, it may be "telescoped" to stabilize the hole to the required depth.

Cleaning of the caisson bottom is accomplished by various methods including clean-out or muck buckets, bailing buckets, airlifts, and where safe and practicable, manual down-hole excavation.

5.21.4 Rock Sockets

Caissons deriving their support from rock sockets are constructed by first driving or drilling a heavy steel casing with a cutting shoe to effect a seal into the top of rock. Rock sockets are then drilled or chiseled, using core barrels, rock augers, churn drills, or chopping bits.

Figure 5.21.2.8 Concrete placement. (*New England Foundation Company, Andover, Mass.*)

Rock sockets usually depend upon a combination of sidewall friction and end bearing to provide design capacity. Desired roughness (if specified) and cleanliness of the sidewalls and socket bottom are conditions the inspection team should assess in accordance with design specification requirements. For slurry caissons a mud cake can develop in permeable rock and reduce the frictional capacity of sidewalls. This condition can be controlled by keeping the slurry agitated and limiting the time of slurry use, or by reaming the sidewalls with a special wirebrush attachment to the auger.

5.21.5 Placement of Concrete

In addition to careful manufacture, transportation, and testing of concrete, special attention must be given to concrete placement within the caissons. Project specifications should address the method of placement, such as free-fall, drop-chute, or tremie methods. A recent research report to the Federal Highway Administration (Baker et al.) on the effects of free-fall concrete in drilled shafts concludes that "The free-fall placement method of concrete into properly constructed clean and dry shafts can be performed to depths of 120 feet or greater without meaningful loss of strength or segregation of concrete aggregate." However, this conclusion may not be immediately reflected in specifications which govern a given project.

The field inspector should carefully observe concrete placement and in all slurry displacement cases should plot a curve which compares concrete volume placed to theoretical volume. Since drilling procedures tend to cut a hole of greater diameter than the minimum specified, some excess concrete "take" (volume) over the theoretical can be anticipated. The plots can be used to detect irregularities and help facilitate assessment of concrete integrity.

In cases where temporary casing is to be withdrawn, special care must be taken to assure that adequate concrete height is present in the shaft to compensate for filling the annular space between the casing and larger-diameter hole. The required concrete level should be computed and verified prior to casing removal. The risk is collapse of unsupported soil into fresh concrete and a defective caisson. In addition an adequate head of concrete must be maintained to counter balance squeezing in of soft soil strata or groundwater flow (and concrete segregation) from pervious water-bearing strata. Information about tremie concrete placement in caissons is contained in the following section covering placement of concrete in slurry walls.

Figure 5.21.2.9   Belling bucket. (*New England Foundation Company, Andover, Mass.*)

(*Courtesy: McGraw-Hill: Field Inspection Handbook; Brock, Levy, Sutcliffe*)

Figure 5.21.2.10   Core barrel. (*New England Foundation Company, Andover, Mass.*)

Tolerances

Tolerances and acceptance criteria are specified in the contract documents. These relate to construction quality control. Shaft location and alignment (plumbness), positioning and minimum cover of reinforcement, concrete strength and slump, bentonite slurry characteristics including density, viscosity, and pH, and concrete placement methods are principal quality control parameters governing acceptance. Location tolerance is usually set to within 3 in (7.62 cm) of plan and alignment to not more than 1 to 2 percent of shaft length.

5.21.6 Caisson Bell Volume Chart

Caisson Bell Volume Chart

Note: Volumes shown in cubic feet for various size shaft
and bell combinations. Bell height includes 6" pad
(STANDARD 60 DEGREES BELLS)

Max. Machine Bell = 3 x Shaft Dia.

Shaft Dia.	Bell Dia.	Bell Vol.	Bell Ht.	Shaft Dia.	Bell Dia.	Bell Vol.	Bell Ht.	Shaft Dia.	Bell Dia.	Bell Vol.	Bell Ht.
12"	1.6	1	0.93	30"	6.0	43	3.53	42"	10.0	197	6.13
	2.0	2	1.37		6.5	56	3.95		10.5	233	6.56
	2.5	4	1.80		7.0	72	4.40		11.0	272	7.00
	3.0	8	2.23		7.5	91	4.83		11.5	316	7.43
	3.5	12	2.67		8.0	112	5.26	48"	4.5	2	0.93
	4.0	18	3.10	36"	3.6	2	0.93		5.0	7	1.37
18"	2.0	1	0.93		4.0	5	1.37		5.5	13	1.80
	2.5	3	1.37		4.5	10	1.80		6.0	21	2.23
	3.0	6	1.80		5.0	17	2.23		6.5	31	2.67
	3.5	10	2.23		5.5	25	2.67		7.0	44	3.10
	4.0	16	2.67		6.0	35	3.10		7.5	59	6.53
	4.5	23	3.10		6.5	48	3.53		8.0	77	3.96
	5.0	31	3.53		7.0	63	3.96		8.5	98	4.40
	5.5	42	3.96		7.5	81	4.40		9.0	122	4.83
24"	2.5	2	0.93		8.0	101	4.83		9.5	150	5.26
	3.0	4	1.37		8.5	125	5.26		10.0	180	5.70
	3.5	7	1.80		9.0	151	5.70		10.5	214	6.13
	4.0	12	2.23		9.5	181	6.13		11.0	253	6.56
	4.5	19	2.67	42"	4.0	2	0.93		11.5	294	7.00
	5.0	27	3.10		4.5	6	1.37		12.0	340	7.43
	5.5	37	3.53		5.0	11	1.80		12.5	391	7.86
	6.0	49	3.96		5.5	19	2.23	54"	5.0	3	0.93
	6.5	63	4.40		6.0	28	2.67		5.5	8	1.37
	7.0	80	4.83		6.5	40	3.10		6.0	14	1.80
30"	3.0	2	0.93		7.0	54	3.53		6.5	23	2.23
	3.5	5	1.37		7.5	70	3.96		7.0	34	2.67
	4.0	9	1.80		8.0	89	4.40		7.5	48	3.10
	4.5	14	2.23		8.5	111	4.83		8.0	65	3.53
	5.0	22	2.67		9.0	137	5.26		8.5	84	3.96
	5.5	31	3.10		9.5	166	5.70		9.0	107	4.40

Shaft Dia.	Bell Dia.	Bell Vol.	Bell Ht.	Shaft Dia.	Bell Dia.	Bell Vol.	Bell Ht.	Shaft Dia.	Bell Dia.	Bell Vol.	Bell Ht.
54"	9.5	132	4.83	60"	15.5	722	9.59	72"	10.0	105	3.96
	10.0	162	5.26	66"	6.0	3	0.93		10.5	133	4.40
	10.5	195	5.70		6.5	9	1.37		11.0	163	4.83
	11.0	231	6.13		7.0	17	1.80		11.5	200	5.26
	11.5	273	6.56		7.5	27	2.23		12.0	238	5.70
	12.0	317	7.00		8.0	41	2.67		12.5	281	6.13
	12.5	365	7.43		8.5	57	3.10		13.0	331	6.56
	13.0	420	7.86		9.0	76	3.53		13.5	383	7.00
	13.5	477	8.29		9.5	98	3.96		14.0	439	7.43
	14.0	540	8.73		10.0	124	4.40		14.5	504	7.86
60"	5.5	3	0.93		10.5	153	4.83		15.0	570	8.29
	6.0	8	1.37		11.0	187	5.26		15.5	642	8.73
	6.5	15	1.80		11.5	224	5.70	78"	7.0	5	.93
	7.0	25	2.23		12.0	264	6.13		7.5	6	1.37
	7.5	37	2.67		12.5	312	6.56		8.0	8	1.80
	8.0	52	3.10		13.0	361	7.00		8.5	9	2.23
	8.5	70	3.53		13.5	414	7.43		9.0	11	2.67
	9.0	91	3.96		14.0	476	7.86		9.5	12	3.10
	9.5	115	4.40		14.5	539	8.29		10.0	14	3.53
	10.0	142	4.83		15.0	609	8.73		10.5	18	3.96
	10.5	175	5.26		15.5	684	9.16		11.0	141	4.40
	11.0	209	5.70		16.0	764	9.59		11.5	172	4.83
	11.5	248	6.13		16.5	850	10.03		12.0	212	5.26
	12.0	292	6.56	72"	6.5	4	0.93		12.5	253	5.70
	12.5	339	7.00		7.0	10	1.37		13.0	299	6.13
	13.0	390	7.43		7.5	18	1.80		13.5	350	6.56
	13.5	448	7.86		8.0	29	2.23		14.0	405	7.00
	14.0	508	8.29		8.5	44	2.67		14.5	465	7.43
	14.5	574	8.73		9.0	61	3.10		15.0	531	7.86
	15.0	646	9.16		9.5	81	3.53		15.5	602	8.29

5.22.0 Slurry Wall Construction Procedures

Structural slurry walls are constructed with varying plan configurations using a variety of construction tools. Fundamental to the construction of slurry walls is the use of a slurry as a means of lateral support of the panel excavation during construction. Slurry walls are constructed in segmental units commonly referred to as panels. These panels can have various configurations in plan (Fig. 5.22.0.1) and usually require special jointing techniques between panels (Fig. 5.22.0.2). Construction tools and procedures will vary depending upon soil and groundwater conditions, the contractor's experience and equipment, and local work rules. Slurry walls were first used in Europe in the late 1950 and were developed primarily for water-cutoff purposes in dams. They are commonly referred to as diaphragm walls in Europe and Asia.

Slurry wall construction requires the excavation of a slot in the ground to a desired depth, length, and width. The slot is kept open by the substitution of a slurry for the removed soil. The use of bentonite slurry as a means of hole support is common to caisson and other drilling operations; however, polymers are sometimes used as a substitution for bentonite. The slot in the ground is usually excavated by a cable-operated clamshell bucket (Fig. 5.22.0.3). Some contractors have developed hydraulic-activated buckets to improve excavation rates in homogeneous and softer materials. Drilling and milling machines are occasionally used to excavate both soil and rock (Fig. 5.22.04). Chisels and rotary drilling machines are used to excavate rock. Figure 5.22.0.5 lists various tools and equipment used in the construction of slurry walls as well as comparisons of their effectiveness in various soil conditions and environments. After completion of the excavation, the slurry in the trench is usually cleaned. All excavation deris is removed from the panel prior to placement of reinforcement. Panel

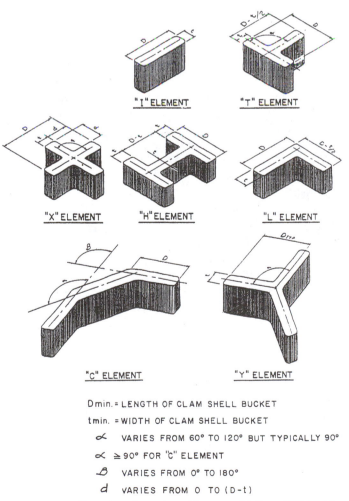

"I" ELEMENT   "T" ELEMENT

"X" ELEMENT   "H" ELEMENT   "L" ELEMENT

"C" ELEMENT   "Y" ELEMENT

Dmin. = LENGTH OF CLAM SHELL BUCKET

tmin. = WIDTH OF CLAM SHELL BUCKET

α VARIES FROM 60° TO 120° BUT TYPICALLY 90°

$\alpha \geq$ 90° FOR "C" ELEMENT

β VARIES FROM 0° TO 180°

d VARIES FROM 0 TO (D-t)

Figure 5.22.0.1 Typical configurations. (*Courtesy: McGraw-Hill: Field Inspection Handbook; Brock, Levy, Sutcliffe*)

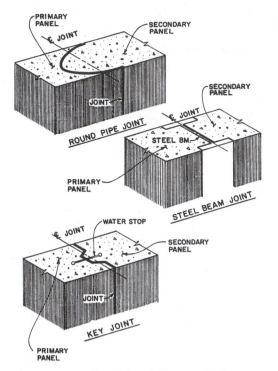

Figure 5.22.0.2   Panel joints. (*Courtesy: McGraw-Hill: Field Inspection Handbook; Brock, Levy, Sutcliffe*)

reinforcing can consist of steel reinforcing bar cages, steel beams, and/or a combination of reinforcing bars and beams.

Concrete is placed into the panel by tremie methods. The concrete expels the slurry as the panel is filled. Special jointing details are necessary at the ends of panels in order to form the ends of the panel and to permit the excavation of the adjacent panels. Because of overbreak during excavation and the difficulty of maintaining close tolerances on vertical and longitudinal alignment of the panel, it is necessary to have proper jointing between panels to permit the removal of concrete which may extend into adjacent panels. All work and measurement of the panel is performed in the blind, inasmuch as the panel is always filled with slurry.

5.22.1 Preconstruction Preparation

Shallow guide walls set along the alignment of the panel excavation are installed prior to the construction of slurry walls. Guide walls range in thickness from 6 in to 2 ft (15.24 to 60.96 cm) and depths from 3 to 6 ft (0.912 to 1.82 m). The guide walls provide a baseline for measurement of line and grade during construction, retain surface soils and prevent localized collapse during excavation, support the reinforcement cage prior to the placement of concrete, and support the steel tremie pipe during concrete placement. Guide walls are usually separated a nominal thickness greater than the width of the excavating tools in order to accommodate the tool during excavation. Guide walls must be blocked and temporarily supported laterally during excavation. Some small lateral inward movement of the guide walls should be expected during construction. The lateral and vertical position of the guide walls and the marks for the location of panel joints should be verified prior to placement of the panel reinforcing. All measurements of the slurry wall excavation will be made from these guide walls. Clear distinct marks should be placed on the guide walls to permit recognition of panel locations and elevation measurement points (Fig. 5.22.1.1).

Figure 5.22.0.3   Cable-operated clamshell bucket. (*Courtesy: McGraw-Hill: Field Inspection Handbook; Brock, Levy, Sutcliffe*)

Figure 5.22.0.4 Milling machine.(*Courtesy: McGraw-Hill: Field Inspection Handbook; Brock, Levy, Sutcliffe*)

		EQUIPMENT TYPE								
		CABLE HUNG CLAMSHELL		KELLY BAR CLAMSHELL		REVERSE CIRCULATION ROTARY DRILL			PERCUSSION	
		MECH	HYDR	MECH	HYDR	ROLL BIT	GANG DRILL	HYDRO MILL	DROP CHISEL	REVER CIRC.
NATURE	SOFT COHESIVE	GOOD	GOOD	GOOD	GOOD	----	POOR	POOR	----	----
	MED GRANULAR	GOOD	GOOD	GOOD	GOOD	----	GOOD	GOOD	----	----
OF	HARD PAN	GOOD	FAIR	FAIR	POOR	GOOD	POOR	GOOD	GOOD	GOOD
SOIL	BOULDER	FAIR	----	POOR	POOR	FAIR	----	FAIR	GOOD	GOOD
	ROCK SOFT	POOR	----	----	----	GOOD	----	FAIR	GOOD	GOOD
	ROCK HARD	----	----	----	----	GOOD	----	----	FAIR	GOOD
DEPTH	0'–50'	GOOD	GOOD	GOOD	GOOD	GOOD	GOOD	----	GOOD	GOOD
OF	50'–100'	GOOD	FAIR	FAIR	FAIR	GOOD	GOOD	GOOD	GOOD	GOOD
EXCAVATION	> 100'	GOOD	POOR	POOR	POOR	GOOD	POOR	GOOD	GOOD	GOOD
SITE	URBANIZED	GOOD	GOOD	GOOD	GOOD	FAIR	FAIR	FAIR	FAIR	GOOD
CONSTRAINT	LOW HEAD ROOM	GOOD	FAIR	----	----	----	FAIR	----	FAIR	GOOD
LABOR	STRICT RULES	GOOD	FAIR	GOOD	FAIR	FAIR	POOR	FAIR	GOOD	GOOD
CONDITIONS	LACK OF SKILL	GOOD	FAIR	FAIR	POOR	FAIR	POOR	POOR	GOOD	FAIR

Figure 5.22.0.5 Performance of excavation tools. (*Courtesy: McGraw-Hill: Field Inspection Handbook; Brock, Levy, Sutcliffe*)

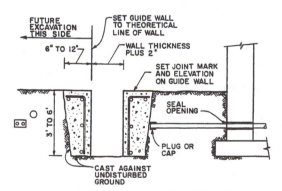

Figure 5.22.1.1 Typical guide wall details.

5.22.2 Slurry Wall Types

General types of slurry walls that inspectors are likely to encounter include:

1. Conventional reinforced-concrete wall with round or other type of slip-formed ends (Fig. 5.22.2.1).

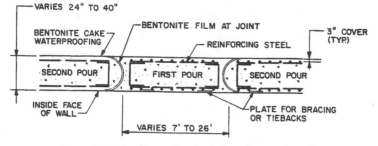

Figure 5.22.2.1   Conventional reinforced-concrete wall.

2. Steel beam reinforced walls with supplemental steel bar reinforcement. The steel beams serve as both reinforcement for the wall and end stops for forming panel ends (Fig. 5.22.2.2).

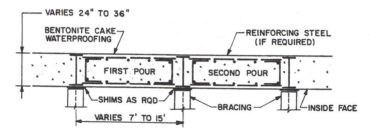

Figure 5.22.2.2   Soldier beam and concrete lagging wall.

Special types of slurry wall not commonly used are:

1. Posttensioned concrete reinforced panel with slip-formed ends (Fig. 5.22.2.3).

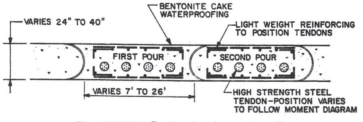

Figure 5.22.2.3   Posttensioned concrete wall.

2. Precast reinforced-concrete panels set into self-hardening cement-bentonite slurry (Fig. 5.22.2.4).

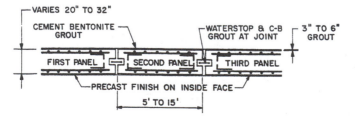

Figure 5.22.2.4   Precast-concrete panel wall. (*Courtesy: McGraw-Hill: Field Inspection Handbook; Brock, Levy, Sutcliffe*)

3. Steel sheet pile set into self-hardening cement-bentonite slurry (Fig. 5.22.2.5).

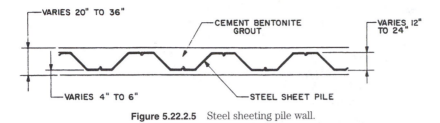

Figure 5.22.2.5 Steel sheeting pile wall.

Very few of these special types of wall have been constructed within the United States. Special skills and a detailed knowledge of the slurry wall technique are necessary for the proper inspection of these unique types of wall.

Conventional reinforced-concrete or steel beam reinforced wall systems are the predominant type of wall used in the United States, and a general methodology for their inspection has been developed.

(Courtesy: McGraw-Hill: Field Inspection Handbook; Brock, Levy, Sutcliffe)

5.22.3 Panel Dimensions and Arrangements

Panel dimensions and arrangements are dependent on the size and type of equipment to be used. Panel widths are established by the width of available buckets, which are conventionally 24, 30, or 36 in (60.96, 76.2, or 91.44 cm) wide. Some wider special buckets range from 40 to 60 in (101.6 cm to 1.52 m) in width. Bucket lengths vary from 7 to 13 ft (2.13 to 3.95 m). Obviously panels cannot be shorter than one bucket length. Multiple bites of the clamshell bucket can be used to extend panel lengths up to 30 ft (9.12 m). Longer panel lengths may require special concreting procedures. Panels can be constructed to various plan configurations using multiple bites in plan. Jointing on curves or to odd plan configurations can be a problem. Careful consideration should be given to jointing details when irregular plan configurations are contemplated.

Panel depths are limited only by the equipment available. Slurry wall elements have been constructed to depths in excess of 300 ft (91.2 m). Panel depths are usually determined by the need to key the bottom of the wall into an impervious material for water cutoff purposes or into a suitable bearing material for lateral or vertical support of the bottom of the wall. Keys cut into an impervious stratum such as stiff clay or rock are common. The depths of keys in rock are usually limited by cost and time considerations. Minimum keys in rock would be in the range of about 2 ft (60.96 cm). Shallower keys can be utilized if careful cleaning of the rock socket is performed prior to placement of concrete.

5.22.4 Bentonite Slurry

Bentonite slurries are obtained by mixing bentonite powder with water. The standard mix consists of one 100-lb (45.35-kg) bag of bentonite powder mixed with 1 yd^3 (0.764 m^3) of water. This mix provides approximately 6 percent of bentonite by weight of water. Bentonite slurries are normally mixed and then stored to permit hydration. Storage up to 24 hours may be necessary to obtain specified properties of specific gravity, viscosity, filtrate loss, and pH, per test API methods. A small bentonite quality control laboratory should be established on site to monitor slurry properties on a daily basis. Once the bentonite has hydrated and the required slurry properties are achieved, slurry is pumped to the trench to replace soil during the excavation process. Flash mixers may be used if careful control is maintained on the slurry prior to use. Bentonite slurries are reused in several panels prior to disposal. The slurry is removed from the panel during concreting and pumped to other panels under excavation or to storage. Bentonite slurries are usually cleaned with a sand separator unit prior to the placement of concrete and/or prior to storage or reuse. Sand content should be measured per API methods and monitored in accordance with specified limits to verify sufficient cleaning to permit concrete placement. Proper cleaning procedures in noncontaminated ground permits multiple reuses of bentonite slurries. Contaminated ground conditions or poor cleaning may require the disposal of bentonite slurries after one use.

Bentonite slurries can be disposed of on site by chemical treatment, separating the clay from the water, by stiffening with cement or lime, or by off-site disposal at a landfill or a sewage treatment plant. Bentonite slurry should not be disposed of in sewers or other drainage systems; upon dilution the suspended solids precipitate out of the slurry and eventually clog the drainage facility.

5.22.5 Polymer Slurries

Polymer slurries are becoming increasingly more common in slurry wall construction because of ease of disposal and the tendency of the slurry to permit suspended soil particles to settle out in the panel, facilitating cleaning of the panel by clamshell bucket. Because of its low unit weight relative to bentonite slurries polymer slurries should be used with caution at locations where there is a high groundwater table, where high surcharge loads abut an open panel, when longer than usual panels are kept open for a long time, and where groundwater chemistry is uncertain.

5.22.6 Excavation

Panel excavation is performed in segments using buckets or drilling tools (Fig. 5.22.6.1). Panel excavation can be underway at various locations on site simultaneously. Excavation is initially performed with clamshell bucket. Initial excavation by clamshell bucket is necessary if special drilling or milling machines are to be used. Occasionally, manufactured or natural obstructions are encountered within the excavation and need to be removed. Special grappling tools, drills, or chisels may be required. On a few occasions explosives are used to split boulders or massive masonry obstructions. Rock is usually removed by drilling and/or chiseling operations (Fig. 5.22.6.2).

Overbreaks in excavation should be expected in areas of obstructions. Chiseling and grappling efforts usually lead to localized collapse of the sidewalls of the trench. These collapses are eventually filled with concrete, forming bulges on the face of the wall. Bulges on the interior face of the wall may have to be removed at a later time. Inspectors should keep careful records of soils encountered during excavation and specific problems which occurred during the excavation operation. These records should be kept on a profile drawing of the wall, maintained by the inspector. These records will prove to be useful during general excavation, when the wall is exposed, to help identify where problems should be anticipated and what corrective action needs to be taken.

The contractor's specialized personnel should periodically measure the verticality of the panel during excavation. The verticality is measured by holding the lift line of the clamshell bucket vertical [using a 4-ft (1.2-m) level] and measuring the position of the cable relative to the guide walls. This measurement should be conducted at 15-ft (4.56-m) intervals during excavation. A more precise

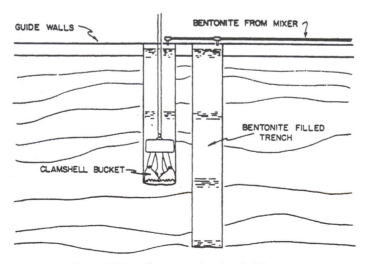

Figure 5.22.6.1 Excavation by clamshell bucket.

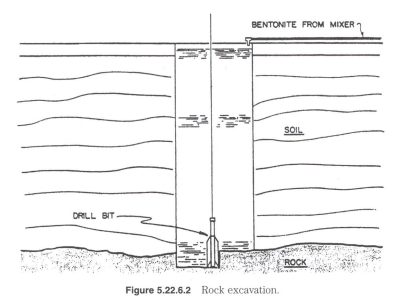

Figure 5.22.6.2   Rock excavation.

measurement can be obtained by attaching steel wires to the ends of the open clamshell jaws and measuring the position of the jaws at 15-ft (4.56-m) depth intervals. The first method of measuring the clamshell cable locates the center of the clamshell bucket but does not identify if the bucket is "corkscrewing" as the panel excavation proceeds downward. The second method, using wires on the tips of the clamshell jaws, offers a more precise means of determining the position of the open clamshell and can determine if "corkscrewing" is occurring.

More precise measurements can be made by sophisticated sonic devices. These devices are used after the panel excavation is completed and therefore report only the position of the panel after the excavation is completed. Contractors have used special expandable spiders which expand within the excavation to measure the position of the two walls of the excavation at any particular location. Information from the spider devices can be misleading, inasmuch as the spider may expand into a void left from an obstruction or boulder, indicating a greater width of trench and a misaligned panel centerline.

5.22.7 Placement of Reinforcement

Reinforcement of the panel is placed as soon as practical after the panel is cleaned (Figs. 5.22.7.1 and 5.22.7.2). Reinforcing cages are constructed of reinforcing bars or steel beams and are usually fabricated on the ground, sometimes on special frames. Reinforcing cages are normally tied with one or two twists of tie wire rather than welded. Welding is normally not accepted, since the steel would be modified by welding and the welded connections are usually very brittle and will fail as the cages are lifted for insertion into the trench (Fig. 5.22.7.3). Long lightweight cages are very flexible and will normally have to be lifted by two lifting lines. The first line is set at the head of the cage where the spreaders are connected to specially reinforced sections of the cage. A second set of lifting cables is attached between midheight and one-third the distance down the cage, to assist in picking up the cage and keeping the cage from bending excessively during lifting. Once the cage is in the vertical position, if the pick point is correctly located, the cage will act as a plumb bob and will straighten itself out. The cage is then placed into the bentonite-slurry-filled trench and supported above the bottom (normally 6 in to 1 ft; 15.24 to 30.48 cm). Special rollers or spacers are normally attached to the face of the cage to keep the cage from coming in contact with the soil. These rollers or skids are at least 3 in (7.62 cm) thick and provide the minimum cover necessary for reinforced-concrete structures cast against the ground (Fig. 5.22.7.4).

In many cases special attachments are placed onto the cage prior to lifting. Special attachments include tieback trumpets or sleeves; bearing plates with embedded anchors for future framing or bracing; blockout keys for future beams, floor slabs, or walls; pipe sleeves for future utility penetrations; and in some cases blockouts for future doorways, tunnels, or shafts. Blockouts or keys are usually

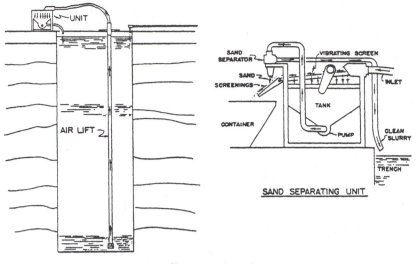

Figure 5.22.7.1   Cleanup with sand separator unit.

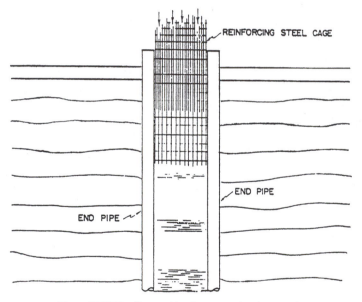

Figure 5.22.7.2   Preparation for concrete placement.

Figure 5.22.7.3   Lifting reinforcing steel cage.

constructed of plywood and Styrofoam sandwiches which are bolted or wired to the reinforcing cage. The blockouts and keys are exposed as the general excavation proceeds downward. The blockout material is then removed and disposed of. The total volume of blockouts and voids should displace no more slurry than the weight of the cage.

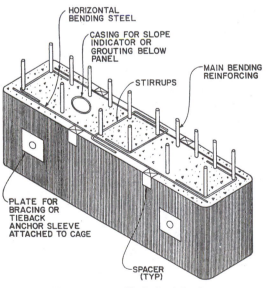

Figure 5.22.7.4   Typical reinforcing.

5.22.8 Placement of Concrete

Concrete is placed within the slurry-filled panel using tremie methods. Tremie pipes are installed in sections and a tremie hopper is set on the top of the pipe (Fig. 5.22.8.1). The tremie pipe joints should be inspected to assure that there is no leakage at the joint. All joints should be watertight and the hopper should be arranged to permit lifting and lowering of the tremie pipe if necessary to free blockage and improve the flow of the tremie concrete. Concrete is placed at slumps in the range of

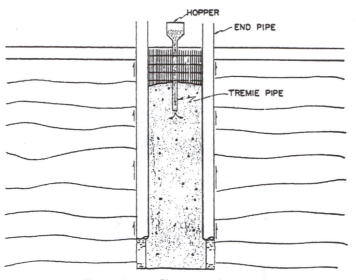

Figure 5.22.8.1 Placement of concrete.

7 to 9 in (17.78 to 22.86 cm), 8 in (20.3 cm) being preferred. The concrete-mix design should be such that there is sufficient cement, sand, and other fluidifying agents to make as workable a mix as possible while still achieving the specified strengths. At the start of placement concrete is usually placed very quickly into the hopper and permitted to free-fall down the slurry-filled chute. This contaminated concrete is slowly brought to the surface as the concrete rises through the panel. "Go-devils" or other types of plugs (such as "nerf-ball rabbits"), although used in tremie caisson work, are rarely used in slurry wall construction.

Sections of tremie pipe are removed as the concrete rises. The bottom of the tremie pipe is kept 5 to 15 ft (1.52 to 4.56 m) below the top of the rising concrete surface. Further immersion within the concrete will reduce the placement speed and lead to blockage of the tremie pipe. Concrete is never vibrated in the panel. Sound concrete is brought above the theoretical top of the panel. The top of the panel is trimmed to the design elevation upon exposure. Measurements of the rise of concrete are made following placement of each truck of concrete. Measurements are plotted against the theoretical rise of concrete to verify that the concrete placement is proceeding as expected (Fig. 5.22.8.2). After placement the ground around the panel is cleaned and the top of the wall is roughly cleaned. The cleaning operation is difficult to perform if the top of the slurry wall is set some distance

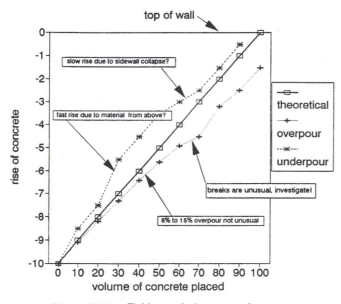

Figure 5.22.8.2 Field record of concrete placement.

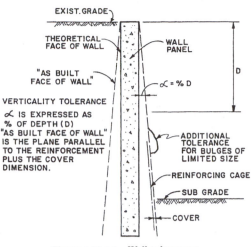

Figure 5.22.9.1 Wall tolerances.

below the guide walls. It is preferred to set the guide walls at the elevation of the top of the slurry wall and to overpour the concrete, letting the contaminated concrete pour off onto the ground to be carted away.

5.22.9 Accuracy

Cages are normally suspended within the panel within 2 in (5.08 cm) of their theoretical location. More rigorous tolerances may be required in some cases. All reinforcing and inserts should be properly positioned prior to the lifting and setting of the cage so that it is only necessary to set the top of the cage accurately on the guide walls in order to assure that the reinforcing and inserts have been properly located. The reinforcing cage should be set vertically within 1 percent of depth. Lumps on the face of the wall should not extend more than 3 in (7.62 cm) beyond the theoretical face of the concrete plus the verticality tolerances (Fig. 5.22.9.1). Other tolerances may be specified by the designer.

5.22.10 Finishes

The finishes of concrete walls cast in slurry-filled panels are rather rough and are direct reflections of the types of material passed through during excavation. For example, relatively smooth surfaces can be obtained in clays, silts, and fine sands while very irregular surfaces are to be expected when passing through fill, boulder, and gravelly materials. Obstructions removed from sidewalls of the trench will leave a void in the excavation. The void will be filled with concrete which will protrude from the face of the wall and may have to be removed when exposed during general excavation. The face of the slurry wall is usually scraped, trimmed, and cleaned as the general excavation proceeds downward. Occasionally high-pressure air, sandblasting, or high-pressure water is used to clean the surface. Water cleaning may lead to softening of the subgrade and a messy excavation if proper drainage is not maintained along the line of the wall.

Joints between panels may have occasional leaks which will have to be sealed by the contractor by grouting from behind the wall, grouting from inside the wall, or removal of the defect and patching with cement mortars. Surficial patches are usually useless and serve only to hide a defect which will leak at a later date. Extreme caution must be taken when patching leaks where the wall is resisting high water pressure and where erodible soils are present. Specifications often permit moisture (like beads of perspiration) on the face of the wall; a flowing water condition is not acceptable.

Sanitary—Storm Water and Site Electrical Structures

Contents

6.0.0 Precast Concrete Manholes

The proper functioning of a sewer system depends to a large degree on the performance of its appurtenances, with manholes being one of the most important. Precast concrete manholes offer significant savings in labor cost over poured-in-place concrete, masonry or brick manholes and are universally accepted for use in sanitary or storm sewers. Precast reinforced concrete manhole sections are available throughout the United States and Canada, and are generally manufactured in accordance with the provisions of American Society for Testing and Materials Specification C478.

The typical precast concrete manhole consists of riser sections, a top section, grade rings and, in many cases, precast base sections or tee sections. The riser sections are usually 48 inches in diameter, but are also available up to 72 inches and larger. They are of uniform circular cross section, and a number of sections may be jointed vertically on top of the base or junction chamber. Most precast manholes employ an eccentric or concentric cone section instead of a slab top. These reinforced cone sections effect the transition from the inside diameter of the riser sections to the specified size of the top opening. Flat slab tops are normally used for very shallow manholes and consist of a reinforced circular slab at least 6 inches thick for up to 48 inches in diameter and 8 inches thick for larger sizes. The slab which rests on top of the riser sections has an access opening cast into it.

Precast grade rings, which are placed on top of either the cone section or flat slab top, are used for close adjustment of top elevation. Cast iron manhole cover assemblies are normally placed on top of the grade rings.

The entire manhole assembly may be furnished with or without steps inserted into the walls of the sections as specified. Reinforcement required by ASTM Specification C478 is primarily designed to resist handling stresses incurred before and during installation. Such stresses are more severe than those encountered in the vertically installed manhole. In such normal installations, the intensity of the earth loads transmitted to the manhole risers is only a fraction of the intensity of the vertical pressure.

To determine the allowable depth to which precast concrete manholes can be placed, it is necessary to consider the forces acting on the manhole ring.

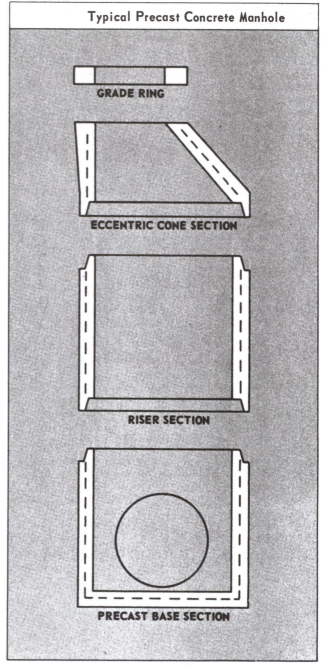

Typical Precast Concrete Manhole

GRADE RING

ECCENTRIC CONE SECTION

RISER SECTION

PRECAST BASE SECTION

6.1.0 Typical Manhole Assembly Combinations

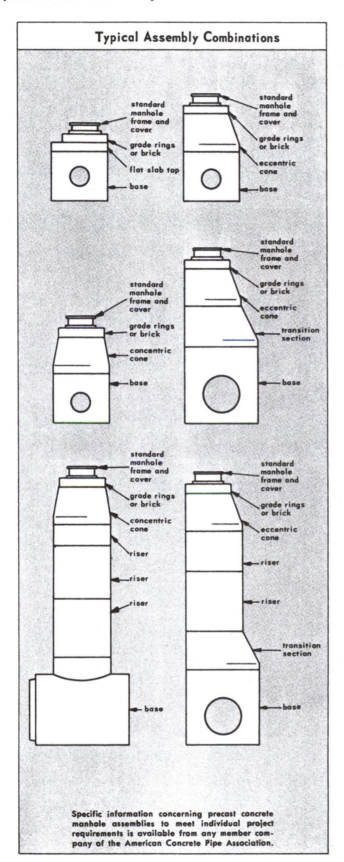

Typical Assembly Combinations

standard manhole frame and cover
grade rings or brick
flat slab top
base

standard manhole frame and cover
grade rings or brick
eccentric cone
base

standard manhole frame and cover
grade rings or brick
concentric cone
base

standard manhole frame and cover
grade rings or brick
eccentric cone
transition section
base

standard manhole frame and cover
grade rings or brick
concentric cone
riser
riser
riser
base

standard manhole frame and cover
grade rings or brick
eccentric cone
riser
riser
transition section
base

Specific information concerning precast concrete manhole assemblies to meet individual project requirements is available from any member company of the American Concrete Pipe Association.

6.2.0 Storm and Sanitary Manhole Schematics with Sections

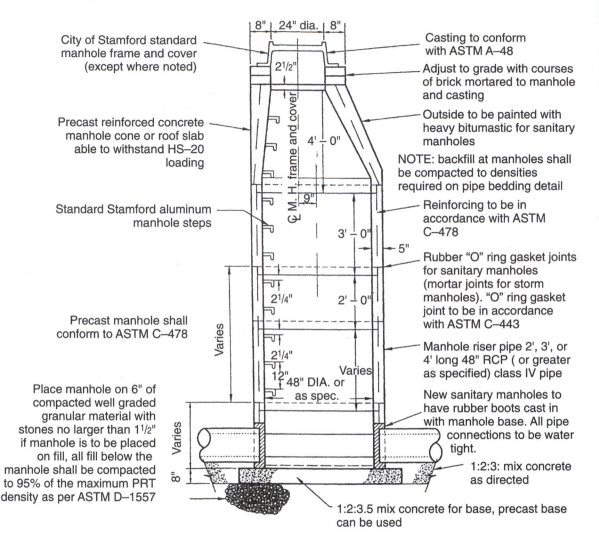

City of Stamford standard manhole frame and cover (except where noted)

Precast reinforced concrete manhole cone or roof slab able to withstand HS–20 loading

Standard Stamford aluminum manhole steps

Precast manhole shall conform to ASTM C–478

Place manhole on 6" of compacted well graded granular material with stones no larger than 1½" if manhole is to be placed on fill, all fill below the manhole shall be compacted to 95% of the maximum PRT density as per ASTM D–1557

8" 24" dia. 8"

2½"

¢ M. H. frame and cover

4' – 0"

9"

3' – 0"

5"

2¼"   2' – 0"

Varies

2¼"

12"   Varies

48" DIA. or as spec.

Varies

8"

Casting to conform with ASTM A–48

Adjust to grade with courses of brick mortared to manhole and casting

Outside to be painted with heavy bitumastic for sanitary manholes

NOTE: backfill at manholes shall be compacted to densities required on pipe bedding detail

Reinforcing to be in accordance with ASTM C–478

Rubber "O" ring gasket joints for sanitary manholes (mortar joints for storm manholes). "O" ring gasket joint to be in accordance with ASTM C–443

Manhole riser pipe 2', 3', or 4' long 48" RCP (or greater as specified) class IV pipe

New sanitary manholes to have rubber boots cast in with manhole base. All pipe connections to be water tight.

1:2:3: mix concrete as directed

1:2:3.5 mix concrete for base, precast base can be used

6.2.1 Storm Sewer Manhole Components

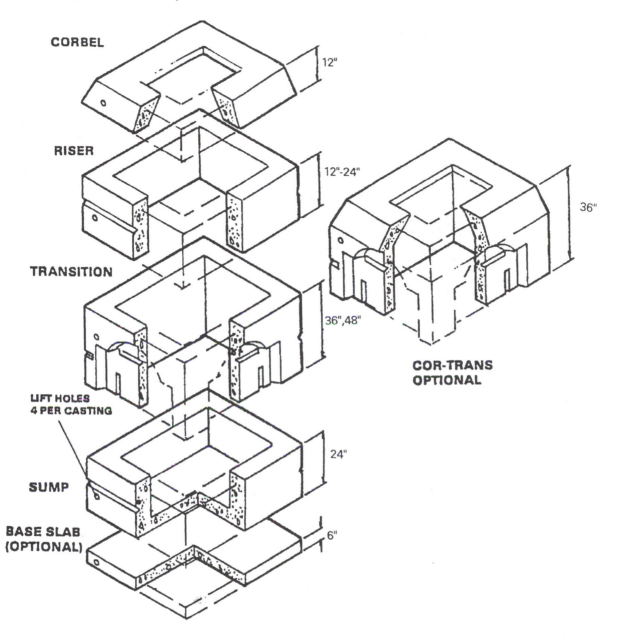

CORBEL

12"

RISER

12"-24"

TRANSITION

36",48"

36"

COR-TRANS
OPTIONAL

LIFT HOLES
4 PER CASTING

24"

SUMP

BASE SLAB
(OPTIONAL)

6"

6.2.2 Forces Acting on a Manhole

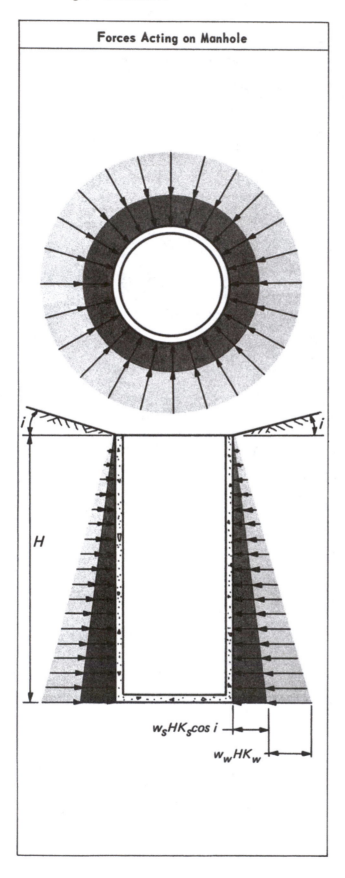

6.3.0 Typical Sewer Manhole and the Effect of Hydrogen Sulfide

A Typical Sewer Manhole

This sewer manhole schematic depicts how hydrogen sulfide (H_2S) comes out of solution, particularly in areas of turbulence. Through the action of bacteria, the sulfur in this hydrogen sulfide becomes sulfuric acid (H_2SO_4), which is what attacks and destroys concrete surfaces especially in the crown areas of sewer lines.

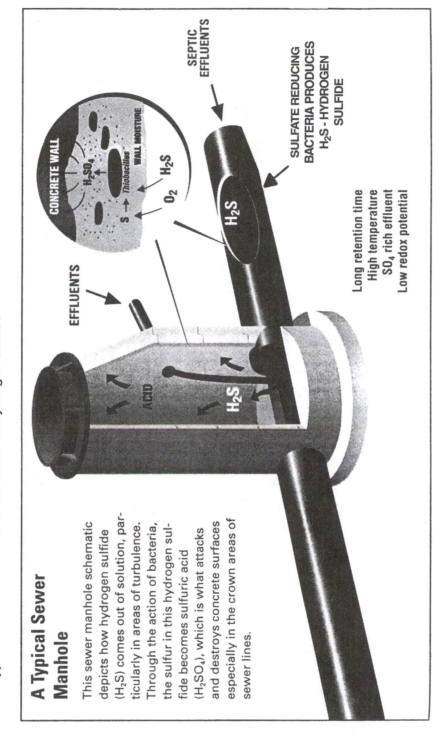

EFFLUENTS

CONCRETE WALL

WALL MOISTURE

H_2SO_4 ← Thiobacillus

S → O_2 → H_2S

H_2S

ACID

H_2S

SEPTIC EFFLUENTS

SULFATE REDUCING BACTERIA PRODUCES H_2S - HYDROGEN SULFIDE

Long retention time
High temperature
SO_4 rich effluent
Low redox potential

6.4.0 Castings for Sanitary and Storm Manholes

Gray iron exhibits excellent corrosion resistance, as well as excellent compressive strength.

Ductile iron is the material of choice when gray iron castings do not have enough load-bearing capacities or where impact resistance is a factor. Ductile iron castings are used where loads greater than H20 are required, such as locations subject to fork truck traffic, airports, and container ports.

Mating ductile iron lids with Class-35B gray iron frames is often a cost-effective method of providing covers in most situations other than H20.

CASTINGS FOR DIFFERING TRAFFIC CONDITIONS

FOUR STRENGTH CATEGORIES

Extra Heavy Duty: For airport and concentrated loads. These applications require special castings to accommodate uniformly distributed loads of 100 to over 225 psi. In selecting castings for heavy concentrated loads, advise loading conditions: total wheel load, tire or wheel contact area, and tire pressure. We will confirm your choice as being suitable for the purpose intended.

Heavy duty: Castings in this category are generally suitable for highway traffic or H20 wheel loads of 16,000 lb. Many of the castings are suitable for much heavier loading.

Medium duty: Use these castings for driveway, parks, ramps, and similar installations where wheel loads will not exceed 2000 lb.

Light Duty: These castings are recommended for sidewalks, terraces, and very light traffic.

Typical Specifications and Mechanical Properties of Gray Iron and Ductile Iron

Gray Iron

Class no.	Tensile strength, psi	Specifications
30	30,000	ASTM A48-83
35	35,000	ASTM A48-83 AASHTO M105-82
40	40,000	ASTM A48-83
45	45,000	ASTM A48-83

Ductile Iron

Grade	Tensile strength, psi	Yield strength, psi	Elongation %	Specifications
60-40-18	60,000	40,000	18 min.	ASTM A536-80 SAE J434C
65-45-12	65,000	45,000	12 to 20	ASTM A536-80 SAE J434C
80-55-06	80,000	55,000	6 to 12	ASTM A536-80 SAE J434C
100-70-03	100,000	70,000	3 to 10	ASTM A536-80 SAE J434C
CLASS A	60,000	45,000	15 min.	MIL-1-24137 CLASS A

Specifications and Mechanical Properties for Nonferrous Metals

Aluminum

ASTM no.	Alloy no.	Physical properties		
		Tensile, psi	Yield, psi	Elongation, %
B26	713.0	32,000	22,000	3
B26	319.0	23,000	13,000	1.5

Bronze

ASTM no.	Alloy no.	Physical properties		
		Tensile, psi	Yield, psi	Elongation, %
B584	C87200-12A	45,000	18,000	20
B584	C86300	110,000	60,000	12

6.4.1 Typical Manhole Adjusting Rings

R-1979 Series
Manhole Adjusting Rings

Heavy Duty

For raising manhole covers when pavement is resurfaced.

Adjusting rings or extension rings, as they are sometimes called, elimi- nate the necessity of raising or replacing existing manhole frame on resurfacing projects. The cover recess is the same depth and diameter as the existing manhole frame. Amount of raise coincides with thick- ness of resurfacing material, thereby bringing the manhole lid to new position flush with refinished roadway surfaces.

Manhole adjusting rings can be placed rapidly and are a permanent installation. Use of adjusting rings may reduce raising costs to as little as 20% of digging up the old manhole frame and resetting it to new grade.

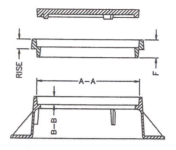

CAUTION-WHEN ORDERING.

Be sure existing lid diameters are smaller than "AA" dimension as shown in table. "AA" dimen- sion in table is opening size in extension ring. Your lid should measure $1/4$" to $3/8$" less in diameter than this dimension.

Adjusting Ring Installed

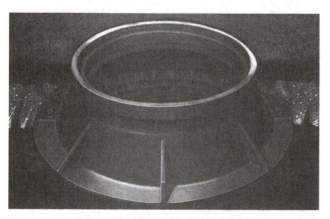

Cutaway View of Adjusting Ring Installation

Dimensions in inches					
A-A	B-B	F	RISE	Reference No.	Previous Catalog No.
9 1/4	1	2 1/2	1 1/2	1979-0247	
9 1/4	1	3	2	1979-0248	
9 5/16	5/8	2 1/8	1 1/2	1979-0245	
9 5/16	5/8	2 5/8	2	1979-0246	
18 3/8	1	3	1 15/16	1979-0166	
18 3/8	1	4	2 15/16	1979-0167	
21 1/2	1	3	2	1979-0008	R-1979-A1
21 1/2	1	3 1/2	2 1/2	1979-0115	R-1979-A2
21 1/2	1	4	3	1979-0022	R-1979-A3
21 7/8	1 1/4	3 1/8	2	1692-7200	
22 1/8	1 1/8	3 1/8	2	1979-0223	
22 1/4	1 1/2	3 1/2	2	1979-0035 *	
22 1/4	1 1/2	4	2 1/2	1979-0036 *	
22 1/4	1 1/2	4 1/2	3	1979-0119 *	
22 5/16	1 1/2	3	1 1/2	1979-0270 *	
22 3/8	1	3	2	1979-0258	
22 3/8	1 1/2	3 1/2	2	1979-0001	R-1979-B4
22 3/8	1 1/2	4	2 1/2	1979-0009	R-1979-B5
22 3/8	1 1/2	4 1/2	3	1979-0021	R-1979-B6
22 1/2	1	2 1/2	1 1/2	1979-0014	
22 1/2	1	5	4	1979-0215	
22 1/2	1 1/2	3 1/2	2	1979-0034	R-1979-C1
22 1/2	1 1/2	4 1/2	3	1979-0116	R-1979-C3
22 3/4	1 1/2	3 1/2	2	1979-0047	R-1979-D1
22 3/4	1 1/2	4	2 1/2	1979-0131	R-1979-D2
22 3/4	1 1/2	4 1/2	3	1979-0019	R-1979-D3
22 3/4	1 3/4	3 3/4	2	1979-0129 *	
22 3/4	1 3/4	4 3/4	3	1979-0170 *	
23	1 1/2	3	1 1/2	1661-7150	
23	1 1/2	3 1/2	2	1661-7200	R-1979-E1
23	1 1/2	4	2 1/2	1979-0004	R-1979-E2
23	1 1/2	4 1/2	3	1979-0005	R-1979-E3
23	1 3/4	2 3/4	1	1713-7100	
23	1 3/4	4	2 3/16	1979-0177	
23	1 3/4	4	2 1/2	1713-7250	

Figure 6.4.1 *(Courtesy: Neenah Foundry Company, Neenah, Wisconsin)*

6.4.2 Typical Manhole Cast Iron Step Inserts

R-1980-1-2 Series
Cast Iron Manhole Steps

Cast iron manhole steps shown in this series are designed for use in conical shaped sanitary sewer manholes equipped with standard cast iron frames and covers.

The **R-1982-J** and the **R-1982-W** steps can be used as a fixed ladder.

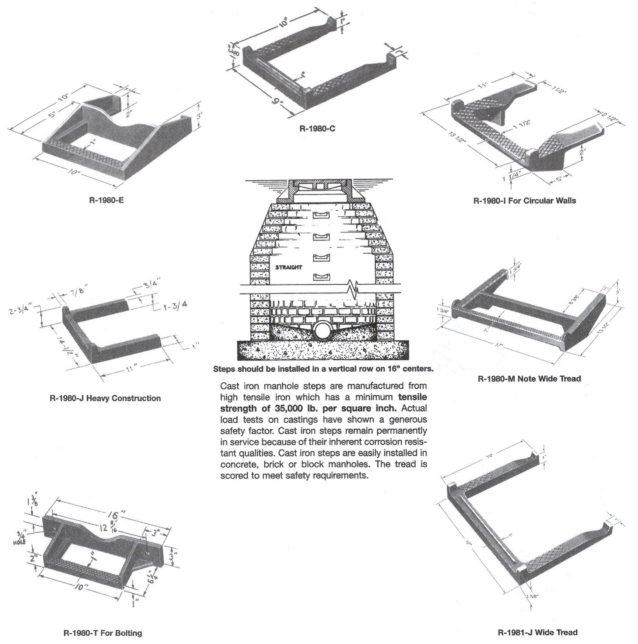

R-1980-C

R-1980-E

R-1980-I For Circular Walls

R-1980-J Heavy Construction

Steps should be installed in a vertical row on 16" centers.

Cast iron manhole steps are manufactured from high tensile iron which has a minimum **tensile strength of 35,000 lb. per square inch.** Actual load tests on castings have shown a generous safety factor. Cast iron steps remain permanently in service because of their inherent corrosion resistant qualities. Cast iron steps are easily installed in concrete, brick or block manholes. The tread is scored to meet safety requirements.

R-1980-M Note Wide Tread

R-1980-T For Bolting

R-1981-J Wide Tread

Figure 6.4.2   *(Courtesy: Neenah Foundry Company, Neenah, Wisconsin)*

6.5.0 Choosing the Proper Inlet Grill

Adopting an inlet grate design requires an analysis of the functions the grate should perform. A designer must review the drainage requirements and then specify which functions will be necessary for satisfactory grate and all other performance requirements of each application. For example, if the only consideration is for the grate to efficiently intercept large quantities of storm water from the gutter - then a grate with the proper geometry and flow through area is the one to choose. Although capacity is an important function, there are other considerations that must be evaluated to satisfy the requirements of a functional, correctly located, safe grate. Some considerations are listed below. Designers must note that there may be other considerations in addition to those listed below that should be evaluated prior to making a choice.

NOTE: ALSO SEE "CAUTIONS" ON PAGE 6 AND "STORM WATER MANAGEMENT WITH NEENAH GRATES" ON PAGE 110.

CONSIDERATIONS FOR HANDICAPPED PERSONS, BICYCLES AND PEDESTRIANS
Design and placement of the grate must be suitable for use in areas where it is possible for handicapped persons, bicycles, and pedestrians to be present

THE ADOPTION OF THE AMERICANS WITH DISABILITIES ACT (ADA), THE PROMINENCE OF NARROW TIRED BICYCLES AND CONCERN FOR PEDESTRIAN SAFETY DICTATES THAT DESIGNERS CAREFULLY CONSIDER THE SAFETY OF THE TYPE OF GRATINGS SELECTED FOR VARIOUS APPLICATIONS. IF THE INSTALLATION IS CHANGED IN CONCEPT, SCOPE OR USAGE PATTERN AT SOME FUTURE DATE, DESIGNERS MUST AT THAT TIME RECONSIDER THE TYPES OF GRATINGS USED AND DETERMINE IF THEY STILL MEET THEIR SAFETY REQUIREMENTS. IF NOT, DESIGNERS MUST REPLACE THESE EXISTING GRATES WITH MORE APPROPRIATE STYLES.

GRATES FOR BICYCLES The designer must determine the appropriate grate design and inlet location for safe use by bicycle traffic, having in mind the specific details of location and foreseeable use. The following are some important considerations:
1.   The style of grate employed should allow bicycle travel from all accessible directions. Grates located in open areas, or adjacent to driveways, or in other locations where they are not adjacent to barrier curbs, require special attention in this regard.
2.   In general, we suggest consideration of our Type L vane style grates. We think certain grates shown in our catalog are generally not suitable for use where bicycle traffic is a consideration, and those grates are so designated in this catalog. The absence of such a designation, however, does not mean that a grate would necessarily be safe in any specific location, and the designer is responsible for making the appropriate grate selection.
3.   If slotted grates or vane-type grates other than our Type L are considered, they should have sufficiently narrow openings, and/or should have appropriately spaced transverse (cross) bars, to ensure that any foreseeable width and diameter of bicycle tire cannot drop down into openings to an unsafe extent.
4.   Gutter slopes should not be substantially swaled into the curb so as to create a disturbance in the roadway which might affect the ability of bicycles or other traffic to traverse them.

GRATES FOR PEDESTRIANS AND HANDICAPPED Various regulations or customs may dictate the use of specific types of grates: such as gratings with longitudinal slot openings no greater than 1/4" or 1/2". Designers must carefully evaluate the location and placement of grates against applicable specifications to be sure they are using grates that will satisfy the specification's requirement and provide a pathway (accessible route) for pedestrians and handicapped persons. The Americans With Disabilities Act (ADA) provides grating specification guidelines that can be used in creating accessible routes for handicapped individuals.

HYDRAULIC EFFICIENCY
The ability of a grate to intercept storm water or its hydraulic efficiency is an important function. The main considerations in design are the geometry and the flow through area of the openings for each individual grate. NEENAH has recognized the designer's need for hydraulic information on individual grates. We studied individual grate performance by establishing our own hydraulic testing program and are now capable of supplying grate capacity information on most of the grates shown in this catalog. Please call us for specific charts and information or to request your copy of Neenah's Inlet Grate Capacities Manual.

Figure 6.5.0 *(Courtesy: Neenah Foundry Company, Neenah, Wisconsin)*

SCREENING OUT HARMFUL DEBRIS
An inlet grate must act as a strainer and prevent harmful debris from entering sewer lines. Objects such as branches, large sheets of semi rigid material, sticks, chunks of wood, etc. which are easily passed by large curb openings (open throat) are conveniently prevented from entering the catch basin by a well chosen grate.

If large debris gets into the catch basin, it can often float or wash into the lines, clogging them at inaccessible locations and making the drainage system ineffective.

Should clogging of the grate occur at the upstream end, grates of longer lengths usually provide the extra flow through capacity necessary to accommodate the gutter flow as well as some of the side flow.

ABILITY TO PASS UNOBJECTIONABLE DEBRIS
Organic material, such as grass clippings, leaves, small stones, scraps of paper and even small twigs should be passed into the catch basin. Because of their size and configuration they are not a hazard for the sewer lines.

Grates designed with closely spaced bars for strength and safety become easily clogged from very small but always present debris. The problem is magnified even more as this debris is packed into the openings by passing traffic creating a hard, semi solid surface. The velocity of the gutter flow is insufficient to dislodge the packed debris and so the grate becomes ineffective.

The answer then is to provide grate openings wide enough, of suitable length or of special design to pass this debris and still meet the other requirements. Neenah has many sizes and styles that can be implemented into your designs.

STRENGTH
Inlet grates placed in roadways must be designed to withstand heavy traffic loads. A common designation for standard highway loads is described as H20-44 loading for single axle trucks and HS20-44 for tandem axle trucks (tractor semi-trailer units). The maximum axle loading in both cases is 32,000 lbs. Or 16,000 lb. for each set of dual wheels. There may be cases where more extreme design loads are necessary such as airports, commercial applications, industrial sites or installations where extra heavy loads and/or extremely hard tires are present. (See page 4 for a list of published specifications available to assist you in your design.)

PERMANENCY A gutter inlet grate should be designed to match or exceed the expected life of the installation. The inherent rust resisting properties of unpainted cast iron and its time-tested performance insures long life. The strength of cast iron grates resists earth and pavement pressures-qualities one should question in other non-cast iron materials.

Figure 6.5.0 *(cont.)*

6.5.1 Typical Heavy Duty Round Drainage Grates

R-4370 Series
Heavy Duty Round Drainage Grates

Most grates listed can be furnished with lightweight cast iron angle frames when specified. Frame is of 1/2" metal as shown on page 266.

Catalog No.	Dimensions in Inches				
	A	B	G	H	Type
R-4370-1	6 3/8	1 1/4	1 x 1	1/2	C
R-4370-2	9 1/2	2 1/2	1 1/2 x 3 1/2	5/8	G
R-4370-3§	15	1 1/4	1 x 12 1/4	1	E
R-4370-4	15	1 1/4	3/4 x 2 1/4	3/4	G
R-4370-5	18	1 1/2	1 x 4 1/2	1 1/4	G
R-4370-6	19 1/2	1 1/2	3/4 x 5 1/4	3/4	G
R-4370-7	20	1 3/8	2 x 2	1	G
R-4370-8	21	1 1/4	3/4 x 19 1/2	3/4	E
R-4370-9	22	1 1/2	1 1/8	1	C
R-4370-10	22	1 1/2	1 1/4	1	D
R-4370-12	22	1 1/2	1 3/8 x 5	1	F
R-4370-13*	22	1 1/2	2 x 6	1 1/4	G
R-4370-15	22 1/2	1 3/4	1 1/8	1	D
R-4370-17	22 3/4	1 3/4	1 1/2	1	D
R-4370-18§	22 3/4	1 3/4	1 1/2	7/8	E
R-4370-21	23	1 1/2	1 1/4	1	D
R-4370-22	23	1 3/4	1	1 1/8	D
R-4370-23	24	1 1/2	2 1/4 x 2 1/4	1 1/2	Special
R-4370-25	29	3	3/4	1	G
R-4370-26	33	2	1 1/2	1	G
R-4370-27A	38	1 1/2	1	1	G

§Not recommended for bicycle traffic.
*Rated as Light Duty - not Heavy Duty.

Type C

Type D

Type E

Type G

Type C

Type D

Type E

Type G

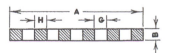

Figure 6.5.1   (*Courtesy: Neenah Foundry Company, Neenah, Wisconsin*)

6.5.2 Typical Rectangular Frames and Solid Lids

R-1800 Series
Rectangular Frames, Solid Lids

Heavy Duty

Catalog No.	Dimensions in Inches				
	A	B	C	E	F
R-1801	11 1/2 x 14 3/4	1 1/4	10 x 13	19 x 22	4
R-1803	14 1/2 x 21 1/2	2	13 x 20	23 x 30	5 1/2
R-1815	18 x 30	2	16 x 28	26 x 38	8 1/2
R-1817	18 x 36	2	16 x 34	28 x 46	7
R-1830	23 1/2 x 35	2	22 x 33	34 x 46	6 1/4
R-1833-A	23 7/8 x 35 3/4	1 3/4	22 x 34	36 x 48	9
R-1848-A2	36 x 48	1 3/4	33 x 45	42 x 54	5 1/2*

*Lid in two pieces. (Also frame)

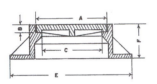

Illustrating Type C Lid

R-1860-A
Large Frame with Double Lid

Heavy Duty

R-1868
Street Type Electric Vault Frame, 2-Piece Top Lid

Heavy Duty

Specify:

1. Lettering as shown or no lettering.

R-1869
Sidewalk Type Electric Vault Frame, 2-Piece Top Lid

Heavy Duty

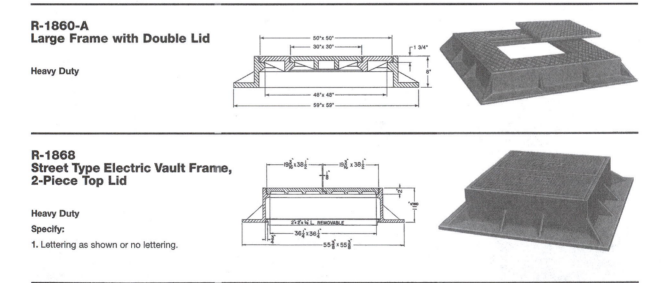

Figure 6.5.2   (*Courtesy: Neenah Foundry Company, Neenah, Wisconsin*)

6.5.3 Typical Rectangular Drainage Grates

SQUARE AND RECTANGULAR DRAINAGE GRATES

Heavy Duty (unless noted otherwise)

The gratings in this series are rated heavy duty when supported on all four sides.

Other special sizes quoted on request in the event none of the standards shown in this series meet your requirements. In ordering replacement grates to be used on existing catch basins, be sure to specify the exact size of opening in which the grate will be used.

Many of the grates in this series can be adapted to trench frames with support on two sides as shown on pages 282 and 283. Most are qualified as heavy duty when the short dimension spans the trench.

Advise loading requirements so we can assist you with your choice.

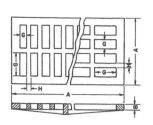

Type A

Type B

Type C

Type P

The above schematic drawing identifies basic dimensions only and does not apply to all grate designs. Bar and rib depths, plate thicknesses, and seating widths, may vary on different sizes and styles.   If your project has design restrictions, ask for approval drawings.

Free open areas for most grates in this section are listed on pages 326 to 330.

Catalog No.	A	B	G	H	Grate Type
Square — Heavy Duty					
R-4400	8x8	$1\,1/4$	1x5	1	B †
R-4401	8x8	1	$3/8$x3	$3/8$	A †
R-4408	$10\,1/4$x$10\,1/4$	1	$7/8$x$8\,1/4$	1	B †
R-4441-1	12x12	1	$3/8$x$4\,1/2$	1	P †
R-4511	14x14	$1\,1/2$	$1\,1/2$x$5\,1/8$	$1\,1/8$	A †
R-4550	15x15	$1\,1/4$	$3\,1/2$x$3\,1/2$	$3/4$	C †
R-4552	16x16	$1\,3/4$	$1\,1/4$x$6\,1/4$	$1\,1/4$	A †
R-4557	16x16	$1\,1/4$	$1\,3/16$x$3\,3/8$	$3/4$	A †
R-4558	$17\,5/8$x$17\,5/8$	$1\,1/4$	1x5	$1/2$	C †
R-4660	18x18	$1\,1/2$	3x$4\,1/2$	$1\,1/4$	C †
R-4662	$19\,3/4$x$19\,3/4$	$2\,1/2$	$1\,3/8$x$8\,1/2$	1	A
R-4720	20x20	$1\,3/4$	$1\,3/8$x$3\,1/2$	$1\,1/4$	C †
R-4721-A	21x21	$1\,3/4$	$1\,1/2$x8	$1\,1/4$	A †
R-4725	21x21	2	3x3	1	C †
R-4760	22x22	$1\,3/4$	3x$5\,3/4$	1	C †
R-4765	23x23	2	1x$6\,1/4$	1	A
R-4808-A	24x24	2	$3/8$x$6\,1/8$	$1\,5/8$	P †
R-4810	24x24	2	1 3/4x6	$1\,1/4$	C †
R-4820	24x24	2	$1\,3/4$x$5\,7/8$	$1\,3/8$	C † •
R-4826	24x24	2	1x6	$1\,1/4$	A †
R-4832	24x24	$1\,5/8$	$1\,1/16$x$4\,3/4$	1	C †
R-4832-B	26x26	$1\,3/4$	$1\,1/16$x4	1	C †
R-4833	26x26	2	2x7	1	A †
R-4850	27x27	2	$2\,1/2$x$7\,1/2$	1	C †
R-4852	27x27	2	$1\,1/2$x5	1	C †
R-4852-A	27x27	1	$1\,1/2$x5	1	C †
R-4859-C	$27\,1/2$x$27\,1/2$	1	2x7	1	A † Ω
R-4871	$27\,1/2$x$27\,1/2$	1	2x24	1	B † §

Catalog No.	A	B	G	H	Grate Type
Rectangular— Heavy Duty					
R-4389-0	6x12	$1\,1/4$	$3/4$x4	1	B
R-4390	6x24	1	$3/4$x$4\,1/2$	1	B
R-4403	8x14	1	$1\,1/2$x$2\,1/8$	$1/2$	A
R-4404-C	8x24	1	$1\,1/16$x$5\,7/8$	1	B
R-4406	8x24	$1\,3/4$	$1\,5/16$x6	1	B
R-4406-1	8x24	$1\,3/4$	1x$2\,5/8$	$3/4$	A
R-4406-2	8x24	$1\,3/4$	1x5	1	C
R-4406-C	$9\,1/4$x$23\,7/8$	1	1x$2\,7/8$	1	A
R-4407-2A	$9\,1/2$x24	$1\,3/4$	1x$7\,1/2$	1	B
R-4409	10x$17\,3/4$	$1\,1/2$	1x$3\,1/2$	1	A
R-4409-A	10x24	$1\,3/4$	$1\,1/4$x$4\,3/4$	1	C
R-4409-C	10x24	$1\,3/4$	1x7	1	B †
R-4409-E	10x24	$1\,3/4$	1x$3\,1/2$	1	A †
R-4409-G	10x24	$1\,1/2$	2x8	1	B
R-4410	10x29	$1\,1/2$	2x8	1	B
R-4421	10x40	$1\,1/8$	$1\,1/4$x$8\,1/2$	1	C †
R-4423-A	$10\,1/4$x48	$1\,5/8$	$1\,1/2$x$3\,1/8$	1	A
R-4424	$10\,1/2$x23 1/8	$1\,3/4$	2x2	1	A
R-4430-A	11x24	$1\,1/2$	$3/4$x$4\,1/8$	$3/4$	A
R-4430-B	$11\,1/2$x24	$1\,1/2$	1x$6\,5/8$	1	C
R-4435-1	$11\,7/8$x14	2	1x$5\,1/2$	1	C
R-4443	12x14	$1\,3/4$	$1\,1/2$x$5\,1/4$	$1\,1/4$	A
R-4449	12x24	$1\,1/2$	$1\,1/2$x$4\,1/4$	1	A †
R-4450	12x24	$1\,3/4$	1x$4\,1/2$	1	A †
R-4450-A	12x24	$1\,3/4$	2x2	1	A †
R-4451	12x24	2	$1\,1/2$x$3\,3/4$	1	C †
R-4454	12x26	$1\,1/2$	$1/2$x$11\,1/2$	$1/2$	C
R-4460	12x30	2	$1\,1/4$x4	1	A †

• Convex
Ω Light Duty
⋆ Grate in two pieces.
† Angle frame available.
§ Not recommended for bicycle traffic.

Figure 6.5.3   (*Courtesy: Neenah Foundry Company, Neenah, Wisconsin*)

SQUARE AND RECTANGULAR DRAINAGE GRATES (CONTINUED)

Catalog No.	A	B	G	H	Grate Type
Dimensions in inches					
Square — Heavy Duty					
R-4880	28x28	2	1 $\frac{1}{2}$x7 $\frac{1}{4}$	1	C †
R-4880-C	30x30	1 $\frac{1}{2}$	1 $\frac{1}{4}$x5 $\frac{7}{8}$	1 $\frac{1}{4}$	C †
R-4882	34x34	2	2x14 $\frac{1}{2}$	1 $\frac{1}{8}$	A † §
R-4882-A	34x34	2	1x9	1 $\frac{1}{8}$	A †
R-4884-A	36x36	1 $\frac{3}{4}$	2x4	1 $\frac{3}{8}$	C † *
R-4540	14 $\frac{1}{2}$x26 $\frac{1}{2}$	2	2 $\frac{1}{4}$x4	1	C
R-4541	14 $\frac{1}{2}$x28	1 $\frac{3}{4}$	1 $\frac{1}{4}$x5	1 $\frac{1}{4}$	A †
R-4544	14 $\frac{1}{2}$x37 $\frac{3}{4}$	1 $\frac{3}{4}$	2x5 $\frac{3}{8}$	1 $\frac{1}{4}$	A
R-4545	14 $\frac{1}{2}$x48	1 $\frac{3}{4}$	1 $\frac{1}{4}$x5	1 $\frac{1}{4}$	A
R-4548	15x20 $\frac{1}{2}$	1 $\frac{1}{2}$	2x2	$\frac{3}{4}$	A
R-4570-1	15x23	1 $\frac{3}{4}$	1 $\frac{1}{2}$x5 $\frac{1}{2}$	1 $\frac{1}{8}$	A
R-4570-2	15x23 $\frac{7}{8}$	1 $\frac{1}{2}$	1x6	1	A
R-4573	15x30	2 $\frac{1}{2}$	1 $\frac{1}{8}$x5 $\frac{1}{2}$	1	A †
R-4575-A	15x36	1 $\frac{3}{4}$	$\frac{13}{16}$x6 $\frac{1}{2}$	$\frac{3}{4}$	A †
R-4579	15 $\frac{1}{8}$x23	1 $\frac{1}{2}$	1 $\frac{3}{8}$x5 $\frac{1}{2}$	1 $\frac{3}{8}$	A
R-4583	15 $\frac{1}{2}$x22 $\frac{1}{4}$	1 $\frac{3}{4}$	1 $\frac{5}{8}$x5 $\frac{1}{2}$	1 $\frac{3}{8}$	A
R-4584	15 $\frac{1}{2}$x36	1 $\frac{3}{4}$	1 $\frac{1}{2}$x5 $\frac{1}{4}$	1 $\frac{3}{8}$	A
R-4585	15 $\frac{1}{2}$x39	1 $\frac{3}{4}$	1 $\frac{7}{8}$x5 $\frac{7}{8}$	1 $\frac{1}{4}$	A
R-4586	15 $\frac{3}{4}$x36	1 $\frac{3}{4}$	1 $\frac{1}{2}$x5 $\frac{1}{4}$	1 $\frac{3}{8}$	A †
R-4600	16x23	2	1 $\frac{1}{4}$x3 $\frac{3}{4}$	1	A
R-4603-A	16x24	1 $\frac{1}{8}$	$\frac{3}{4}$x5 $\frac{1}{2}$	1	A †
R-4604	16x24	1 $\frac{1}{2}$	2x4	$\frac{7}{8}$	C
R-4604-C	16x24	2	2x5	2	A †
R-4604-D	16x24	2 $\frac{1}{2}$	2 $\frac{1}{2}$x6 $\frac{5}{8}$	$\frac{3}{4}$	A
R-4608	16 $\frac{1}{2}$x19 $\frac{3}{4}$	1 $\frac{1}{2}$	1 $\frac{1}{4}$x6	1	A
R-4610	16 $\frac{1}{2}$x22	1 $\frac{3}{4}$	1 $\frac{1}{2}$x4	1	A
R-4620	17x19	2	1 $\frac{1}{4}$x4 $\frac{1}{4}$	1 $\frac{1}{8}$	A
R-4630	17x20	1 $\frac{1}{2}$	1 $\frac{1}{4}$x6 $\frac{1}{2}$	1	C
R-4632	17x21	2	1 $\frac{3}{8}$x8 $\frac{3}{4}$	1	C
R-4640	17x21 $\frac{1}{2}$	1 $\frac{3}{4}$	2 $\frac{1}{2}$x4 $\frac{1}{4}$	1	A
R-4641-A	17x24	2	1x7	$\frac{7}{8}$	A †
R-4641-C	17x24	2	1x6 $\frac{1}{2}$	1	C †
R-4641-F	17x34	1 $\frac{1}{2}$	1 $\frac{1}{2}$x6 $\frac{3}{4}$	1	A †
R-4649	17 $\frac{1}{4}$x24	1 3/4	1x6 $\frac{1}{2}$	1	A
R-4649-1	17 $\frac{1}{4}$x43	1 3/4	1 $\frac{3}{4}$x3 $\frac{7}{8}$	1 $\frac{1}{4}$	C
R-4652	17 $\frac{3}{4}$x29 $\frac{3}{4}$	2	1 $\frac{1}{2}$x4 $\frac{1}{2}$	1 $\frac{1}{8}$	A †
R-4670-A	18x24	1 $\frac{1}{2}$	1 $\frac{1}{8}$x6 $\frac{3}{4}$	1 $\frac{1}{8}$	A †
R-4671	18x24	2	1x7 $\frac{1}{2}$	1 $\frac{1}{16}$	A †
R-4672	18x24	2	1 $\frac{3}{8}$x5 $\frac{3}{4}$	1 $\frac{1}{8}$	C †
R-4689	18x27 $\frac{1}{4}$	2	2x6 $\frac{3}{4}$	1	C †
R-4692-A	18x34	2 $\frac{1}{4}$	1 $\frac{7}{16}$x8	1	C
R-4698	18x36	1 $\frac{3}{4}$	2x4	1 $\frac{3}{8}$	A †
R-4710	18x36	2	1 $\frac{3}{4}$x6 $\frac{1}{4}$	1 $\frac{1}{4}$	C †
R-4711	18x39	2	1 $\frac{1}{2}$x5 $\frac{1}{2}$	1 $\frac{1}{4}$	C †
R-4718	18 $\frac{3}{4}$x36 $\frac{3}{4}$	2	2x8	1	A
R-4730	19 $\frac{1}{2}$x32	1 $\frac{1}{2}$	2x5	1	A Ω

Catalog No.	A	B	G	H	Grate Type
Dimensions in inches					
Rectangular — Heavy Duty					
R-4462	12x35 $\frac{3}{4}$	1 $\frac{1}{2}$	$\frac{3}{4}$x9	1	B †
R-4470	12x48	1 $\frac{3}{4}$	1x4 $\frac{1}{2}$	1	A †
R-4525	14x24	2	1x5 $\frac{1}{2}$	1	A †
R-4530	14 $\frac{1}{2}$x22	2	1 $\frac{1}{2}$x5 $\frac{1}{2}$	1 $\frac{1}{4}$	A
R-4531	14 $\frac{1}{2}$x23 $\frac{7}{8}$	1	1x5	1	A
R-4731	19 $\frac{1}{2}$x39	1 $\frac{1}{2}$	2x4 $\frac{1}{2}$	1	C Ω
R-4732	19 $\frac{1}{2}$x39	1 $\frac{3}{4}$	1x7 $\frac{1}{2}$	1	A
R-4736	20x24	2	1 $\frac{1}{2}$x2 $\frac{1}{2}$	1 $\frac{1}{4}$	C †
R-4738	20x24	2	1x5	1	A †
R-4739	20x24	2	1 $\frac{1}{8}$x4 $\frac{3}{4}$	1	C †
R-4740	20x28 $\frac{1}{2}$	2	1 $\frac{3}{4}$x8	1	C
R-4750	20x30	2	1 $\frac{3}{4}$x6	1	C †
R-4750-1	20x48	2	1 $\frac{1}{2}$x5	1	A
R-4751	20 $\frac{1}{2}$x27	2	1 $\frac{3}{4}$x5 $\frac{1}{2}$	1	C
R-4752	20 $\frac{1}{2}$x36	1 $\frac{7}{8}$	1 $\frac{5}{8}$x5 $\frac{3}{4}$	1 $\frac{1}{8}$	C
R-4755-B	21x42	1 $\frac{1}{2}$	1 $\frac{1}{2}$x5 $\frac{1}{2}$	1	A †
R-4755-C	21x42	1 $\frac{1}{2}$	1 $\frac{1}{2}$x5 $\frac{7}{8}$	1	C †
R-4759	22x23	1 $\frac{3}{4}$	1 $\frac{1}{2}$x6	1	C
R-4762	22x30	1 $\frac{1}{2}$	2x6 $\frac{1}{4}$	1	C †
R-4780	22x37	2	2x8	1	C †
R-4781	22x48	2 $\frac{1}{2}$	1 $\frac{5}{8}$x5 $\frac{3}{8}$	1 $\frac{1}{4}$	C †
R-4795	22 $\frac{1}{2}$x45	2	2x6	1	A †
R-4798	23 $\frac{1}{4}$x36 $\frac{1}{2}$	2 $\frac{3}{4}$	1 $\frac{1}{2}$x7 $\frac{1}{4}$	1	C
R-4821-A	23 $\frac{7}{8}$x26	2	1 $\frac{9}{16}$x4 $\frac{15}{16}$	1 $\frac{1}{4}$	C †
R-4825	24x30	1 $\frac{1}{2}$	1 $\frac{1}{2}$x4 $\frac{1}{4}$	1 $\frac{1}{4}$	C †
R-4825-A	24x30	2 $\frac{1}{4}$	1 $\frac{1}{4}$x4 $\frac{3}{4}$	1	A †
R-4825-B	24x30	2 $\frac{1}{4}$	1 $\frac{1}{2}$x6	1	C †
R-4828	24x33	2 $\frac{1}{4}$	1x4 $\frac{1}{2}$	1	A †
R-4829	24x33	2 $\frac{1}{4}$	1 $\frac{1}{2}$x6 $\frac{1}{2}$	1 $\frac{1}{4}$	C †
R-4837	24x35	2 $\frac{1}{2}$	1 $\frac{1}{2}$x6 $\frac{1}{2}$	1 $\frac{1}{2}$	A
R-4839	24x35 $\frac{7}{8}$	1 $\frac{3}{4}$	$\frac{7}{8}$x9 $\frac{3}{4}$	1	A †
R-4840	24x36	2	2x4 $\frac{3}{4}$	1	C †
R-4843	24x36	1 $\frac{3}{4}$	1 $\frac{3}{4}$x6	1	A †
R-4853	24x39	2 $\frac{1}{2}$	1 $\frac{3}{4}$x5 $\frac{3}{4}$	1 $\frac{5}{8}$	A †
R-4853-A	24x45	3	1 $\frac{1}{2}$x6 $\frac{1}{4}$	1 $\frac{1}{4}$	A
R-4853-B1	24x51	3	1 $\frac{1}{2}$x5	1 $\frac{1}{4}$	C †
R-4855	26x30	2	1 $\frac{1}{2}$x5	1 $\frac{1}{4}$	A †
R-4856	26x32 $\frac{3}{8}$	4	3 $\frac{3}{8}$x5 $\frac{5}{8}$	$\frac{7}{8}$	C
R-4857	26x48	2	1 $\frac{1}{2}$x5 $\frac{1}{2}$	1	A
R-4890	28x37	2	1 $\frac{3}{4}$x7 $\frac{5}{8}$	1 $\frac{1}{4}$	C †
R-4891	30x48	1 $\frac{1}{2}$	1 $\frac{1}{2}$x4 $\frac{1}{4}$	1 $\frac{1}{4}$	A*
R-4893	30x78	2	1 $\frac{1}{2}$x5	1 $\frac{1}{4}$	C**
R-4893-B	32x34	2	1x2	1	C †
R-4894	33x36	2	1 $\frac{1}{2}$x6 $\frac{1}{2}$	1 $\frac{1}{4}$	C †
R-4895-2	36x66	2	1 $\frac{1}{2}$x6 $\frac{1}{2}$	1 $\frac{1}{4}$	A †*

• Convex
Ω Light Duty
* Grate in two pieces.
† Angle frame available.
** Grate in three pieces.

Figure 6.5.3 *(cont.)*

6.5.4 Narrow Slotted Grates

R-1881 Series
Narrow-Slotted Grates

Heavy Duty

Catalog No.	Dimensions in inches.				
	A	B	G	H	TYPE
Square					
R-1881-A*	12 x 12	1	$3/8$ x $4 1/2$	1	Q
R-1881-B	24 x 24	2	$1/2$ x $4 1/2$	1	Q
R-1881-C	$25 3/4$ x $25 3/4$	$1 1/2$	$1/2$ x $4 1/2$	1	Q
Rectangular					
R-1881-D	8 x $23 7/8$	$1 1/2$	$1/4$ x $2 15/16$	1	P
R-1881-E	10 x $23/78$	$1 1/2$	$1/4$ x $3 15/16$	1	P
R-1881-F*	$11 1/2$ x $23 7/8$	$3/4$	$1/4$ x $4 1/16$	1	P
R-1881-G	$11 1/2$ x $23 7/8$	$1 1/2$	$1/4$ x $4 1/16$	1	P
R-1881-H	12 x $23/7/8$	$1 1/2$	$1/4$ x $4 15/16$	1	P
R-1881-J	14 x $23/7/8$	$1 1/2$	$1/4$ x $5 15/16$	1	P
R-1881-K	17 x $23 7/8$	$1 1/2$	$1/4$ x $6 5/8$	$1 13/16$	P
R-1881-L	20 x $23 7/8$	$1 1/2$	$1/4$ x $8 5/16$	1	P
R-1881-M	$23 13/16$ x $26 1/2$	$2 1/2$	$1/2$ x $4 1/2$	1	Q
R-1881-N	$23 7/8$ x 26	$1 1/2$	$1/2$ x $4 1/2$	1	Q

* Light Duty

Type P Grate Openings   **Type Q Grate Openings**

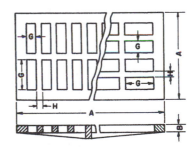

Figure 6.5.4 *(Courtesy: Neenah Foundry Company, Neenah, Wisconsin)*

6.6.0 Catch Basin Inlets—Heavy Duty

Inlet Grate Types

Grate type "D" is not recommended for bicycle traffic.

Grate types "B" and "R" are not considered bicycle-safe when used without a curb box.

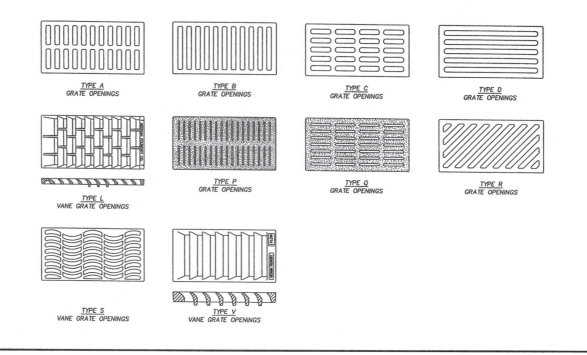

TYPE A
GRATE OPENINGS

TYPE B
GRATE OPENINGS

TYPE C
GRATE OPENINGS

TYPE D
GRATE OPENINGS

TYPE L
VANE GRATE OPENINGS

TYPE P
GRATE OPENINGS

TYPE Q
GRATE OPENINGS

TYPE R
GRATE OPENINGS

TYPE S
VANE GRATE OPENINGS

TYPE V
VANE GRATE OPENINGS

Curb Face Types

The templates shown illustrate a few of the many types of curb sections which can be properly matched with the catch basin inlets shown on the following pages.

SQUARE   VERTICAL 1" 8 2" RAD.   "D" 6" RAD.   TAPER 1" 8 2" RAD.   SEMI ROLL 3" RAD.   ROLL 6" RAD.

Figure 6.6.0 *(Courtesy: Neenah Foundry Company, Neenah, Wisconsin)*

6.7.0 Fastening Devices for Manhole Lids

Fastening Devices

The standard types of fastening devices shown below can be adapted to most of the manhole lids and frames shown in the foregoing pages.

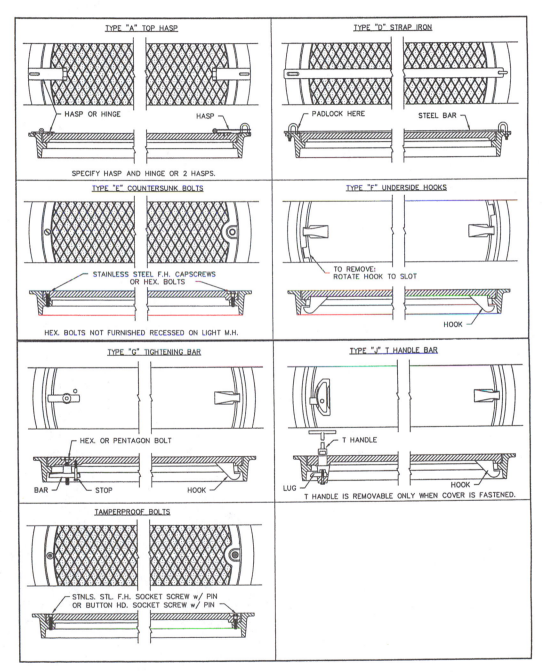

Figure 6.7.0   *(Courtesy: Neenah Foundry Company, Neenah, Wisconsin)*

6.8.0 Water Meter Vault Frames and Lids

R-1910 to R-1912 Series
Water Meter Frame, Solid Lid

Light Duty

Outdoor location of water meters offers advantages to the Water Department and to the consumer. Metering ahead of service lines records all water used, including leakage and wastage. Meter reading is simplified both as to time and accuracy. Meters can be changed without inconvenience, and an unauthorized person cannot gain access to a consumer's home by posing as a meter reader.

Specify:

1. Catalog number.

2. Special lettering if required.

3. Special lift handle if required.

Furnished standard without lettering.

Furnished with as-cast bearing surfaces.

Catalog No.	Dimensions in Inches				
	A	B	C	E	F
R-1910 Series Round					
R-1910-A	18	1	17	25	4 1/2
R-1910-B	15 5/8	1 3/8	15	23	4 3/4
R-1910-C	15 5/8	1 3/8	15	21	4 3/4
R-1910-D	15 5/8	1 3/8	15	22	3
R-1910-E	15 1/4	3/8	14 5/8	20 1/2	4 3/4
R-1911 Series Round					
R-1911-A	26	1 1/2	21	30 1/2	5
R-1911-B	15	1	11 3/4	27 1/4	4
R-1912 Series Square and Rectangular					
R-1912-1*	49 1/2 x 25 1/2	3/4	48 x 24	54 x 30	4
R-1912-A1	38 x 32	5/8	36 x 30	43 x 36	4
R-1912-A2	38 x 26	5/8	36 x 24	42 x 30	4
R-1912-B	32 x 32	3/4	30 x 30	36 x 36	4
R-1912-B1	32 x 26	5/8	30 x 24	36 x 30	4
R-1912-C	30 x 30	3/4	28 x 28	34 x 34	4
R-1912-C1	26 x 14	3/4	24 x 12	30 x 18	4
R-1912-D	25 1/2 x 25 1/2	3/4	24 x 24	30 x 30	4
R-1912-D1	22 x 18	1/2	21 x 17	27 x 23	3
R-1912-E	20 x 20	3/4	18 x 18	24 x 24	4

*Lid in two pieces.

R-1910   R-1911   R-1912

Furnished standard without lettering.

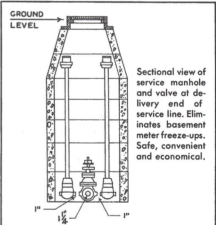

GROUND LEVEL

Sectional view of service manhole and valve at delivery end of service line. Eliminates basement meter freeze-ups. Safe, convenient and economical.

R-1910   R-1911   Type R-1912

Figure 6.8.0   *(Courtesy: Neenah Foundry Company, Neenah, Wisconsin)*

6.9.0 Airport Castings

Airport construction castings should be selected to sustain the loading requirements of a Boeing 727-200 and Lockheed L1011-500 aircraft. Although not the heaviest aircraft, the load is concentrated on their gear configurations making them effectively as heavy a loading to which construction castings will be subjected. Individual wheel loads of various commercial aircraft are shown in the table at right.

The FAA indicates for heavier dual gear aircraft, the assumed spacing of 34" between centerline of the tires is reasonable. Based on this assumption, construction castings with clear openings greater than 34" can be subjected to loadings greater than that of one wheel. However, due to the distribution of this loading on the casting, the effective loading may not necessarily be as critical as a single concentrated wheel load in the center of the casting.

Notes to project designer:

- All items shown in Neenah's Airport Series are capable of withstanding minimum 100,000-pound loads as called for in FAA Advisory Circular AC 150/5320-6D Appendix 3 Item 2.d. (1).

- All items shown in Neenah's Airport Series will be furnished with covers or grates fastened to frames as called for in FAA Advisory Circular AC 150/5370-10A.

- All items shown in Neenah's Airport Series will be furnished unpainted per FAA Advisory Circular AC 150/5370-10A.

- Castings furnished in Austempered ductile iron per ASTM A 897 are furnished per FAA Advisory Circular AC 150/5370-10A.

- Slotted vane drain shown on page 205 is furnished per FAA Advisory Circular AC 150/5370-10A.

Military installations may have heavier loadings than commercial airports, in which case, contact us regarding the tire pressures, contact area and wheel spacing.

The Boeing 727-200 is considered because of the possibility that the spacing of wheels is such that 2 wheels may be concentrated on one casting.

Aircraft	Wheels Per Strut	VMG* Pounds	Max. Load Per-Tire Pounds
727-100	2	76,900	38,450
727-200 STD	2	79,900	39,950
727-200 ADV	2	96,800	48,400
737-300	2	62,200	31,100
737-400	2	70,600	35,300
737-500	2	61,800	30,900
737-700	2	71,500	35,750
737-800	2	82,100	41,050
737-900	2	84,300	42,150
L1011-100	4	204,120	51,030
L1011-500	4	230,000	57,500
747-100B/300	4	174,000	43,500
747-400F	4	204,600	51,150
757-300	4	125,500	31,375
767-200ER	4	180,000	45,000
767-300ER	4	188,200	47,050
777	6	297,500	49,583
DC-10-10	4	212,535	53,134
DC-10-40	4	210,532	52,633
MD-11	4	242,000	60,500
MD-80-83	2	76,280	38,140
MD-90-30	2	75,740	37,870

*Maximum static load per strut at the most aft center of gravity.

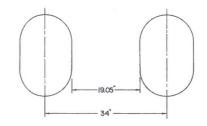

Boeing 727-200 Gear Configuration
Contact Area 290 Sq. In. Tire Pressure 167 psi

Figure 6.9.0 *(Courtesy: Neenah Foundry Company, Neenah, Wisconsin)*

6.10.0 Suggested Forming Procedures for Drainage Systems (Neenah)

Suggested Forming Procedures For Installing Neenah Drainage Structures

Heavy Duty

For those who are not experienced in the installation of Neenah drainage structures (R-4990 and R-4999 series), the following procedures are one way of achieving desirable results.

Materials

Under normal situations it would be customary to use a good grade of weather resistant $3/4$ inch plywood for forming walls. Construction grade 2 x 4's are suitable for studs, plates, bracing and spreaders. The amount and position of the bracing, studs and spreaders to assure a safe working environment is a function of site conditions along with the depth and width of the trench. A typical installation is shown in Figure 1. Details shown and suggestions are based on using the Neenah Foundry Type X frame.

Forming Procedures

Once the floor slab of the trench has been poured and cured according to the plans and/or specifications, begin the forming procedure. The width of the forming (see Figure 1), measured from the outside edges of the forms, corresponds to the "C" dimension in the catalog and on Figure 6. During the entire forming procedure, continually verify that the forms are PLUMB, STRAIGHT, SOLID and LEVEL.

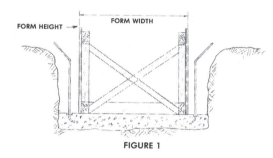

FIGURE 1

The height of the form corresponds to the final grade elevation when installing the non-bolted frame and grates. Extend the spreaders beyond the edge of the forms as shown in Figure 2a to provide a stop for the frame and seat form. Once the elevation has been verified and the forms are level, begin attaching the frames to the formwork.

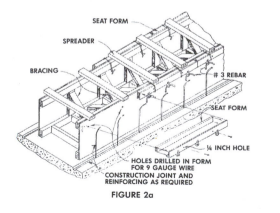

FIGURE 2a

To attach the C.I. frame to the forming, the use of a "seat form" is recommended to assure that the frame is at the proper elevation and level. The seat form has the same dimensions as the frame, with the height corresponding to the frame height (the "B" dimension in the catalog), and the width the same as the seat width of the frame. The seat width should be field measured to assure a proper fit. All Neenah frames have a slight radius at the corner of the seat and vertical face so the seat form should be beveled to accommodate this radius and assure that solid contact is made along the entire length of the C.I. frame.

Several methods may be used to attach the C.I. frames - two are shown here. One involves attaching the seat to the frame using the holes in the face of the frame to nail it in place (Figure 2a) and the other requires nailing the seat form to the framework (Figure 2b).

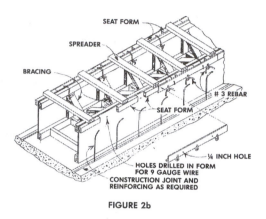

FIGURE 2b

Both then are held in place by using a 9 gauge wire to force the frame into the proper position (Figure 3a & 3b). Frames should butt together snugly, leaving as little gap as possible to eliminate any "creep" if the installations are long. Place a #3 bent rebar through the holes in the anchor lugs to provide additional anchorage in the concrete. Prior to pouring the concrete, verify the annular space allowance (space between the edge of the grate and frame) so grates will fit properly in the space allowed. This should be $3/16$ inch (Figure 6).

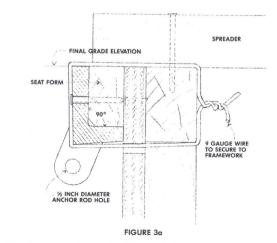

FIGURE 3a

Figure 6.10.0   *(Courtesy: Neenah Foundry Company, Neenah, Wisconsin)*

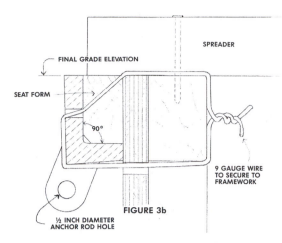

FIGURE 3b

Once the elevation has been verified and the forms are level and braced, begin installing the frame and grate units on the forms. Be sure to keep the sections tight up against one another to eliminate creep due to spacing voids. When the sections are in the proper position, wire them to the bracing as shown in Figure 5. Place a #3 bent rebar through the holes in the anchor lugs to provide additional anchorage in the concrete.

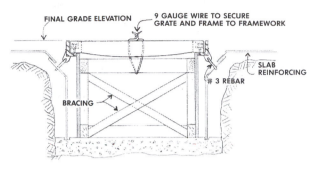

FIGURE 5

BOLTED UNITS

When bolted frames and grates are furnished, they are shipped assembled (see Figure 4a) and therefore require different forming procedures. AT NO TIME SHOULD THE UNITS BE DISASSEMBLED DURING INSTALLATION! DO CHECK THAT THE 3/16" GAP HAS NOT CHANGED IN TRANSPORT.

Once the concrete has been poured and cured, strip the forms and remove any exposed wire which was used to secure the grate or frames to the forms. The completed installation should resemble the illustration shown in Figure 6.

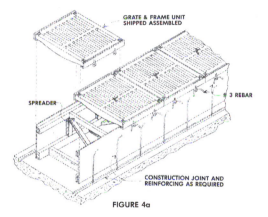

FIGURE 4a

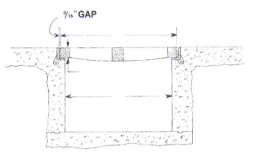

FIGURE 6

The height of the side of the form is such that the top of the form is the final grade minus the seat depth of the C.I. frame (this is actually the same as the "B" dimension referred to in the R-4990 series shown on page 285. Figure 4b illustrates the positioning of the casting on the forms.

General Comments
NOTE: All frame sections are furnished in standard manufactured lengths. It is the responsibility of the installer to cut frame pieces to the proper length and to miter corners where applicable. In cases where trench direction must change, special drawings will be furnished by our Engineering Services Department. These prints will show special grate lengths and cuts, as well as other essential information. Forming procedures, however, are basically the same.

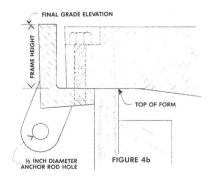

FIGURE 4b

Figure 6.10.0 *(cont.)*

6.11.0 Typical Utility Company Transformer Location Requirements

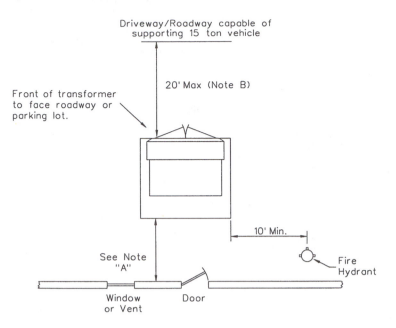

Notes:

A) 2 Ft. Min. To masonry fire resistant walls of buildings with no openings.

   20 Ft. Min. To any flammable building wall. A min. diagonal dist. of 20 feet
   from top of transformer is required if placed beneath window, unless barrier wall
   is constructed according to BGE standards (can be supplied by BGE).

   20 Ft. Min. On any opening in a building wall including: Doors, windows,
   ventilating exhaust, intake ducts or any fire escape.

B) 3 phase Transformer to be no further than 20 feet from a paved access road.
   (Single Phase can be 30' away) to allow vehicular access for future
   transformer/cable maintenance.

C) Traffic protection is required on all sides of transformer that are within
   8' of a roadway or parking lot. (See Traffic Protection Standard)

D) The above are suggested minimum clearances between the transformer foundation
   and windows, doors, fire escapes, entrances and ventilating ducts. It shall be
   the customer's responsibility to see that applicable National Electrical Code,
   municipality and/or insurance regulations and requirements are met.

E) Tranformer location must have 5' horizontal clearance from any undergound facilities

BGE00547

Storm water run-off shall be directed away from transformers and other equipment. Transformers shall also be placed to prevent flooding of the building and/or electrical conduits. Transformers and conduits shall never be placed so that they drain toward a building, unless adequate drainage and run-off protection has been installed.

Transformer shall be placed on a level area with a minimum of eight feet of clear and level operating space in front of the transformer pad. The access road/paving area must be capable of supporting the weight of a 15-ton vehicle.

6.11.1 Typical Transformer Location Requirement Guidelines

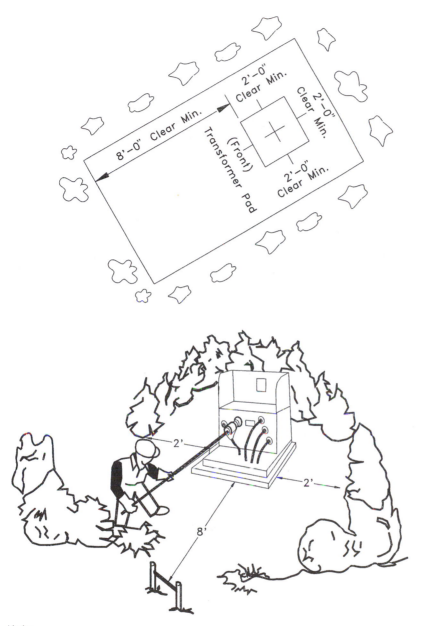

Note:
Minimum clearances are from the edge of mature plants, not
from the stem of planting stock and should be coordinated
with the nursey providing the plants.

BGE 00518

6.11.2 Schematic of Typical Utility Company Precast Electrical Manhole Requirement

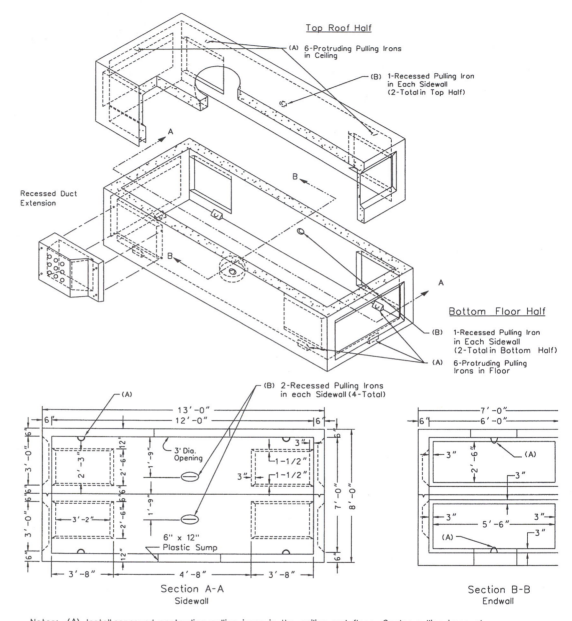

Section A-A
Sidewall

Section B-B
Endwall

Notes: (A) Install approved protruding pulling irons in the ceiling and floor. Center pulling irons at endwalls and at sidewalls opposite duct knockouts (12-total).

(B) Install approved recessed pulling irons in bottom half and top half of manhole. Center in sidewall (4-total).

Installation Depth

Minimum - top of structure 1 ft. below ground level.
Maximum - top of structure 5 ft. below ground level.

Structural design must be approved by the Underground Standards Engineering Unit, Distribution Engineering Department before structure is acceptable for use on the underground distribution system. BGE designation •6.12.7.

6.11.3 Schematic of Typical Utility Company Electrical Manhole Frame and Cover

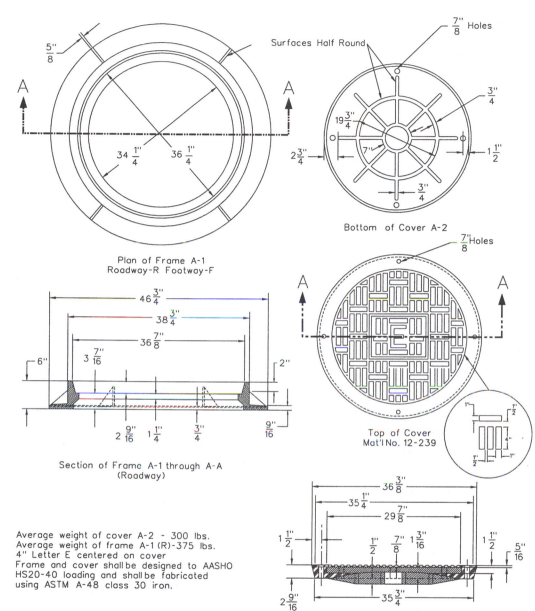

Plan of Frame A-1
Roadway-R Footway-F

Bottom of Cover A-2

Top of Cover
Mat'l No. 12-239

Section of Frame A-1 through A-A
(Roadway)

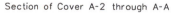

Section of Cover A-2 through A-A

Average weight of cover A-2 - 300 lbs.
Average weight of frame A-1 (R)-375 lbs.
4" Letter E centered on cover
Frame and cover shall be designed to AASHO
HS20-40 loading and shall be fabricated
using ASTM A-48 class 30 iron.

6.11.4 Typical Utility Company Electrical Splice Box Specifications for Roadway Use

Splice Box/Handhole
for roadway use H-20 (4'x 4'x 4')

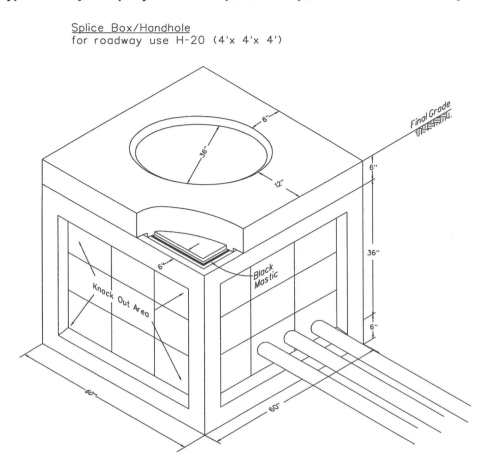

Concrete: 5000 psi @ 28 days
Steel Reinforcement
BGE can supply further specification as necessary.

Limit 6 Sets Secondary 500 kcmil Cable
Designated Enclosed Space (OSHA) - All Safety Procedures Must Be Met

Internal Requirements:
    Sump - 12"dia x 4-1/2" deep
    Pulling Irons in floor
    Ground Rod -$\frac{1}{2}$" Cu-coated

6.11.5 Typical Utility Company Electrical Splice Box Installation Guidelines for Nonroadway Use

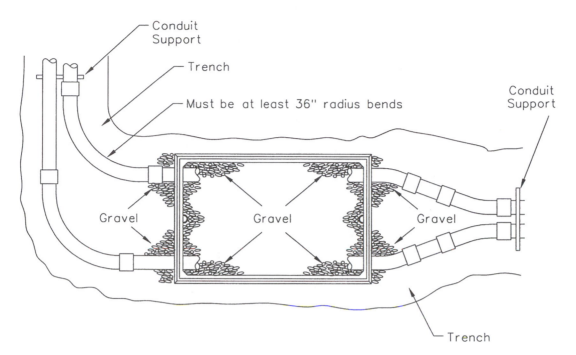

Notes:

Top of splice box should match finished grade. Box does not have to be perfectly level.

All stacked conduits require concrete encasement

BGE00591

6.11.6 Handhold Conduit Capacities

The chart below shows the number of service conduits that the customer must install, depending on the disconnect nameplate capacity. The actual configuration of the duct bank, whether vertical, horizontal or square, must be approved by BGE prior to construction. For large disconnects not addressed by this chart, or if future expansion is planned, please contact your BGE representative. Handholes/manholes may be required if the duct routing contains more than 180 degrees of bend. See specifications for further detail.

Disconnect Nameplate Capacity (Amperes)	Required No. of Service Conduits	Minimum Size Handhole - if required***
Single Phase Services		
0-200	2	4'x4'
201-400	3	4'x4'
401-1200	6**	4'x4'
Three Phase Services		
0-200	2	4'x4'
201-400	3	4'x4'
401-700	4*	4'x4'
701-1000	6**	4'x4'
1001-1600	8**	Manhole Required
1601-2400	10**	Manhole Required
2401-3000	12**	Manhole Required

Gas Service *(no concrete encasement)*	Required Conduit
½" CTS	1 - 4"
1" CTS	1 - 4"
1¼" CTS	1 - 4"
1¼" IPS	1 - 4"
2" IPS	1 - 4"

Notes:

   * **Must be concrete-encased if installed in a 2x2 duct bank.**

   ** **Must be concrete-encased.**

   *** **Method of securing cable to wall must be provided.**

6.11.7 Duct Bank Configurations and Concrete Encasement Schematics

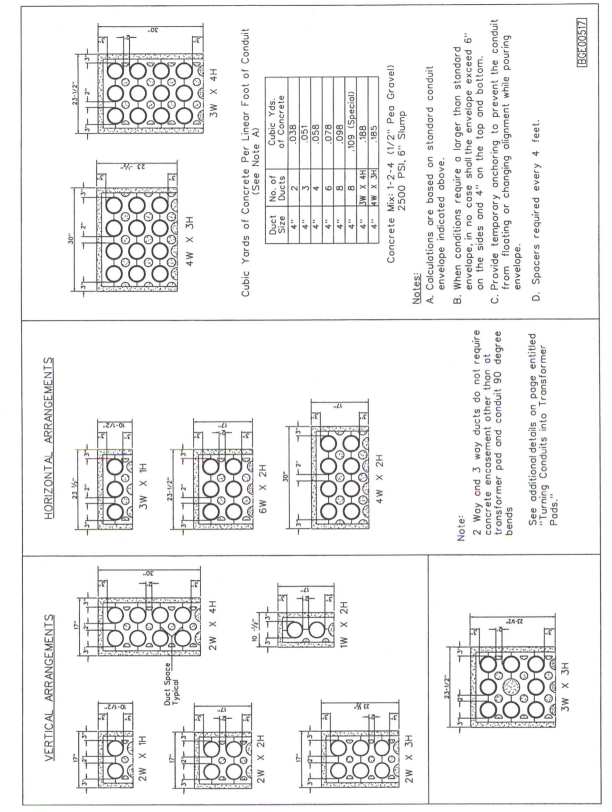

Cubic Yards of Concrete Per Linear Foot of Conduit (See Note A)

Duct Size	No. of Ducts	Cubic Yds. of Concrete
4"	2	.038
4"	3	.051
4"	4	.058
4"	6	.078
4"	8	.098
4"	8	.109 (Special)
4"	3W X 4H	.188
4"	4W X 3H	.185

Concrete Mix: 1-2-4 (1/2" Pea Gravel) 2500 PSI, 6" Slump

Notes:

A. Calculations are based on standard conduit envelope indicated above.

B. When conditions require a larger than standard envelope, in no case shall the envelope exceed 6" on the sides and 4" on the top and bottom.

C. Provide temporary anchoring to prevent the conduit from floating or changing alignment while pouring envelope.

D. Spacers required every 4 feet.

HORIZONTAL ARRANGEMENTS

3W X 1H   6W X 2H   4W X 2H

3W X 4H   4W X 3H

Note:
2 Way and 3 way ducts do not require concrete encasement other than at transformer pad and conduit 90 degree bends

See additional details on page entitled "Turning Conduits into Transformer Pads."

VERTICAL ARRANGEMENTS

2W X 4H   1W X 2H

Duct Space Typical

2W X 1H   2W X 2H   2W X 3H

3W X 3H

Site Details and Landscaping

Contents

7.0.0 Concrete Curb and Gutter Detail

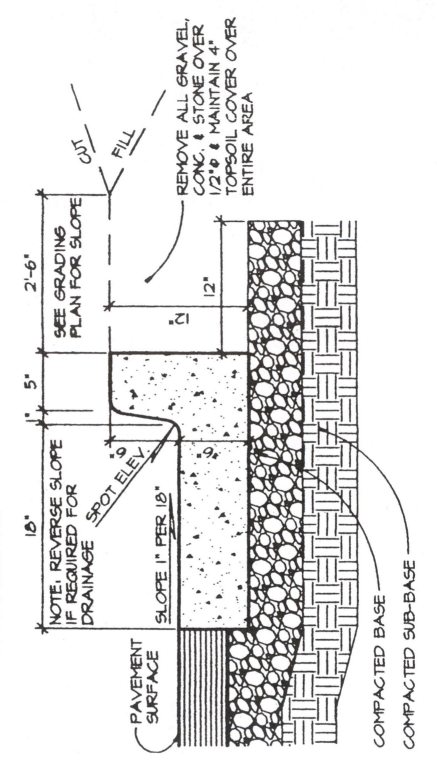

CUT

FILL

REMOVE ALL GRAVEL, CONC. & STONE OVER 1/2"Ø & MAINTAIN 4" TOPSOIL COVER OVER ENTIRE AREA

SEE GRADING PLAN FOR SLOPE

2'-6"

12"

12"

5"

18"

NOTE; REVERSE SLOPE IF REQUIRED FOR DRAINAGE

SPOT ELEV.

6"

6"

SLOPE 1" PER 18"

PAVEMENT SURFACE

COMPACTED BASE

COMPACTED SUB-BASE

7.1.0 Typical Concrete Curb Detail

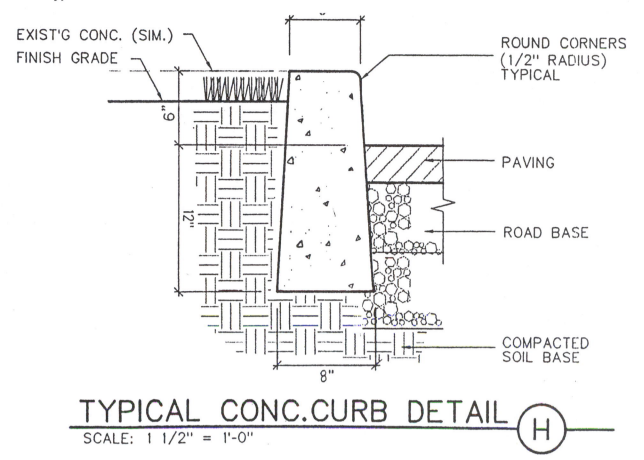

EXIST'G CONC. (SIM.)

FINISH GRADE

ROUND CORNERS
(1/2" RADIUS)
TYPICAL

PAVING

ROAD BASE

COMPACTED
SOIL BASE

6"

12"

8"

TYPICAL CONC.CURB DETAIL Ⓗ

SCALE: 1 1/2" = 1'-0"

7.2.0 On-Site Concrete Sidewalk Detail

On–Site Concrete Sidewalk Detail

Not To Scale

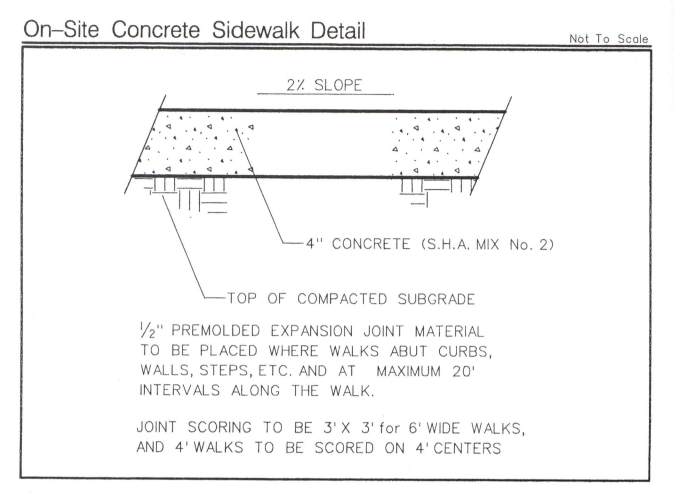

2% SLOPE

4" CONCRETE (S.H.A. MIX No. 2)

TOP OF COMPACTED SUBGRADE

½" PREMOLDED EXPANSION JOINT MATERIAL
TO BE PLACED WHERE WALKS ABUT CURBS,
WALLS, STEPS, ETC. AND AT   MAXIMUM 20'
INTERVALS ALONG THE WALK.

JOINT SCORING TO BE 3' X 3' for 6' WIDE WALKS,
AND 4' WALKS TO BE SCORED ON 4' CENTERS

7.3.0 Concrete Paving Detail

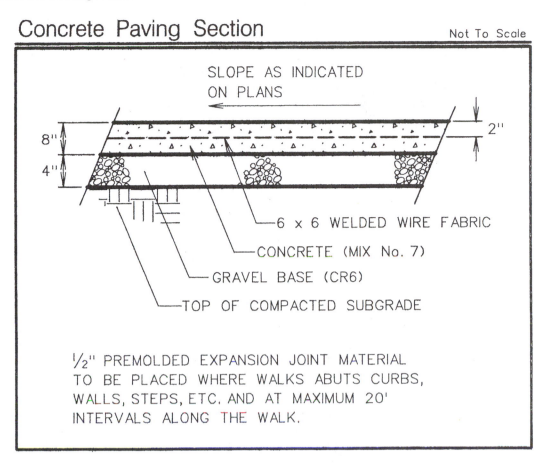

Concrete Paving Section

Not To Scale

SLOPE AS INDICATED
ON PLANS

8"

4"

2"

6 x 6 WELDED WIRE FABRIC

CONCRETE (MIX No. 7)

GRAVEL BASE (CR6)

TOP OF COMPACTED SUBGRADE

1/2" PREMOLDED EXPANSION JOINT MATERIAL
TO BE PLACED WHERE WALKS ABUTS CURBS,
WALLS, STEPS, ETC. AND AT MAXIMUM 20'
INTERVALS ALONG THE WALK.

7.4.0 Brick Plaza Paving Detail

Brick Plaza Paving

Not To Scale

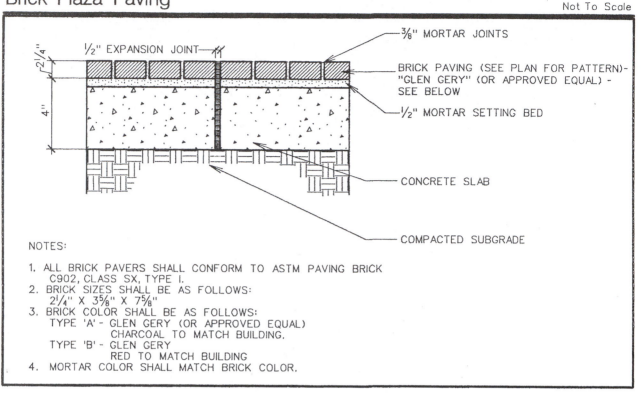

NOTES:
1. ALL BRICK PAVERS SHALL CONFORM TO ASTM PAVING BRICK
   C902, CLASS SX, TYPE I.
2. BRICK SIZES SHALL BE AS FOLLOWS:
   $2\frac{1}{4}$" X $3\frac{5}{8}$" X $7\frac{5}{8}$"
3. BRICK COLOR SHALL BE AS FOLLOWS:
   TYPE 'A' - GLEN GERY (OR APPROVED EQUAL)
               CHARCOAL TO MATCH BUILDING.
   TYPE 'B' - GLEN GERY
               RED TO MATCH BUILDING.
4. MORTAR COLOR SHALL MATCH BRICK COLOR.

7.5.0 Brick Paver Details with Planting Bed

Brick Planter – Section

Scale: 1"=1'

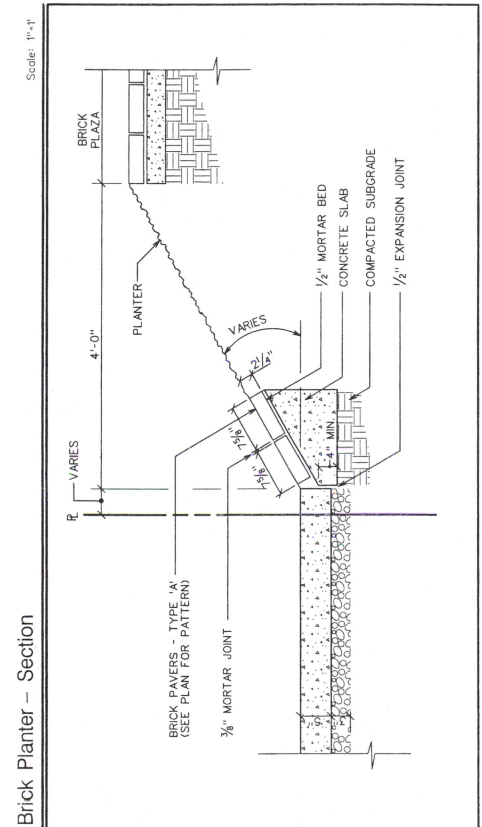

BRICK PLAZA

PLANTER

4'-0"

VARIES

VARIES

2¼"

7⅝"

7⅝"

4" MIN.

½" MORTAR BED

CONCRETE SLAB

COMPACTED SUBGRADE

½" EXPANSION JOINT

BRICK PAVERS - TYPE 'A'
(SEE PLAN FOR PATTERN)

⅜" MORTAR JOINT

7.6.0 Silt Fence Installation Details for Erosion Control

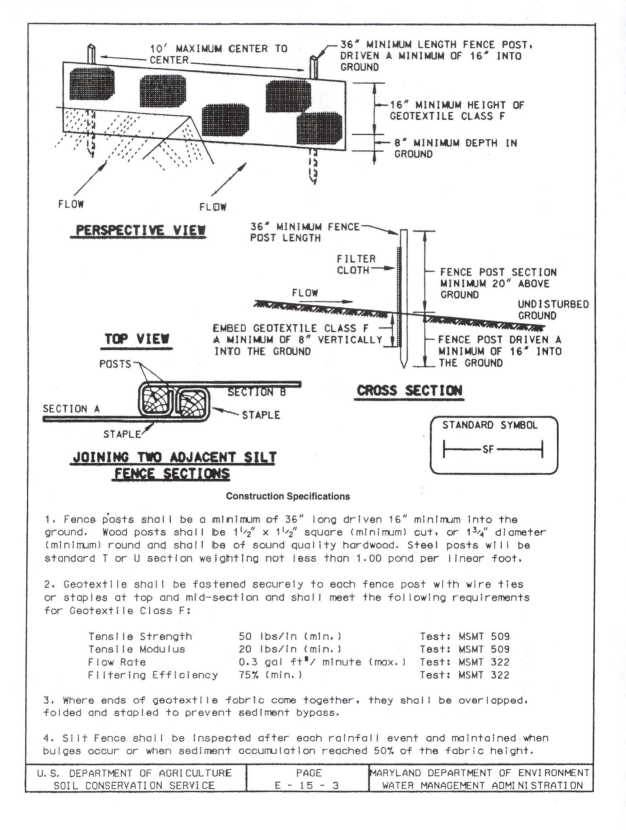

Construction Specifications

1. Fence posts shall be a minimum of 36" long driven 16" minimum into the ground. Wood posts shall be 1½" x 1½" square (minimum) cut, or 1¾" diameter (minimum) round and shall be of sound quality hardwood. Steel posts will be standard T or U section weighting not less than 1.00 pond per linear foot.

2. Geotextile shall be fastened securely to each fence post with wire ties or staples at top and mid-section and shall meet the following requirements for Geotextile Class F:

Tensile Strength	50 lbs/in (min.)	Test: MSMT 509
Tensile Modulus	20 lbs/in (min.)	Test: MSMT 509
Flow Rate	0.3 gal ft^8/ minute (max.)	Test: MSMT 322
Filtering Efficiency	75% (min.)	Test: MSMT 322

3. Where ends of geotextile fabric come together, they shall be overlapped, folded and stapled to prevent sediment bypass.

4. Silt Fence shall be inspected after each rainfall event and maintained when bulges occur or when sediment accumulation reached 50% of the fabric height.

U.S. DEPARTMENT OF AGRICULTURE SOIL CONSERVATION SERVICE	PAGE E - 15 - 3	MARYLAND DEPARTMENT OF ENVIRONMENT WATER MANAGEMENT ADMINISTRATION

7.7.0 Sump Pit for Filtering Dewatering Operations

SUMP PIT
NOT TO SCALE

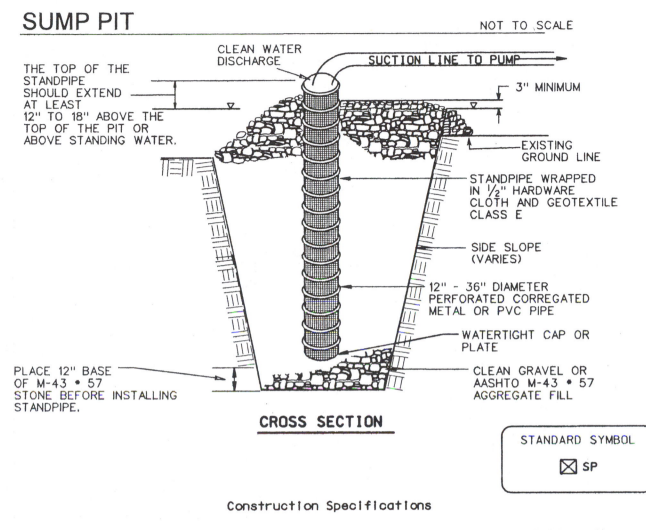

THE TOP OF THE STANDPIPE SHOULD EXTEND AT LEAST 12" TO 18" ABOVE THE TOP OF THE PIT OR ABOVE STANDING WATER.

CLEAN WATER DISCHARGE

SUCTION LINE TO PUMP

3" MINIMUM

EXISTING GROUND LINE

STANDPIPE WRAPPED IN 1/2" HARDWARE CLOTH AND GEOTEXTILE CLASS E

SIDE SLOPE (VARIES)

12" - 36" DIAMETER PERFORATED CORREGATED METAL OR PVC PIPE

WATERTIGHT CAP OR PLATE

CLEAN GRAVEL OR AASHTO M-43 • 57 AGGREGATE FILL

PLACE 12" BASE OF M-43 • 57 STONE BEFORE INSTALLING STANDPIPE.

CROSS SECTION

STANDARD SYMBOL

⊠ SP

Construction Specifications

1. Pit dimensions are variable, with the minimum diameter being 2 times the standpipe diameter.

2. The standpipe should be constructed by perforating a 12" to 24" diameter corrugated or PVC pipe. Then wrapping with 1/2" hardware cloth and Geotextile Class E. The perforations shall be 1/2" x 6" slits or 1" diameter holes.

3. A base of filter material consisting of clean gravel or #57 stone should be placed in the pit to a depth of 12". After installing the standpipe, the pit surrounding the standpipe should then be backfilled with the same filter material.

4. The standpipe should extend 12" to 18" above the lip of the pit or the riser crest elevation (basin dewatering only) and the filter material should extend 3" minimum above the anticipated standing water elevation.

7.8.0 Heavy Duty Stabilized Construction Entrance

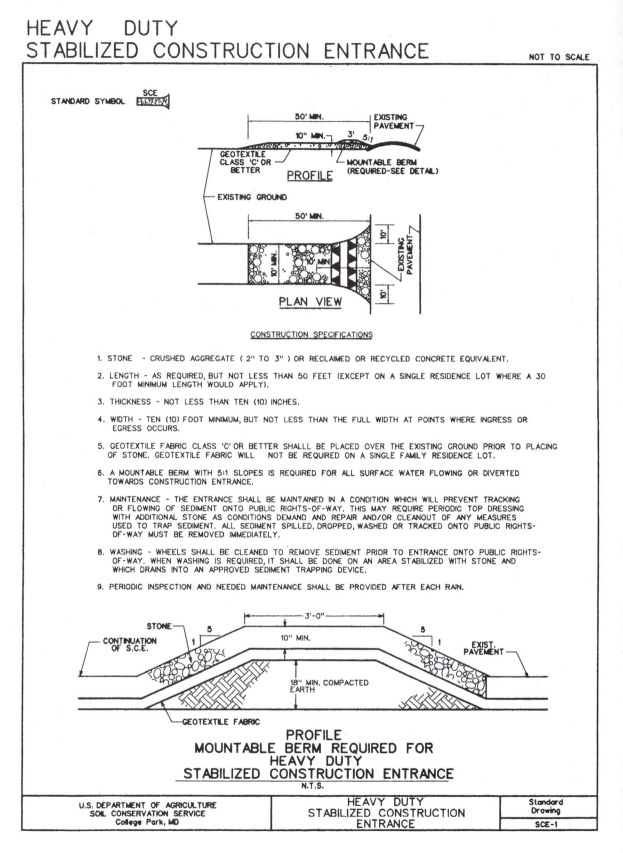

HEAVY DUTY
STABILIZED CONSTRUCTION ENTRANCE

NOT TO SCALE

STANDARD SYMBOL   SCE

50' MIN.

EXISTING PAVEMENT

10" MIN.   3'   5:1

GEOTEXTILE CLASS 'C' OR BETTER

PROFILE

MOUNTABLE BERM (REQUIRED-SEE DETAIL)

EXISTING GROUND

50' MIN.

10' MIN.   10' MIN.

EXISTING PAVEMENT

10'

10'

PLAN VIEW

CONSTRUCTION SPECIFICATIONS

1. STONE - CRUSHED AGGREGATE (2" TO 3") OR RECLAIMED OR RECYCLED CONCRETE EQUIVALENT.

2. LENGTH - AS REQUIRED, BUT NOT LESS THAN 50 FEET (EXCEPT ON A SINGLE RESIDENCE LOT WHERE A 30 FOOT MINIMUM LENGTH WOULD APPLY).

3. THICKNESS - NOT LESS THAN TEN (10) INCHES.

4. WIDTH - TEN (10) FOOT MINIMUM, BUT NOT LESS THAN THE FULL WIDTH AT POINTS WHERE INGRESS OR EGRESS OCCURS.

5. GEOTEXTILE FABRIC CLASS 'C' OR BETTER SHALL BE PLACED OVER THE EXISTING GROUND PRIOR TO PLACING OF STONE. GEOTEXTILE FABRIC WILL   NOT BE REQUIRED ON A SINGLE FAMILY RESIDENCE LOT.

6. A MOUNTABLE BERM WITH 5:1 SLOPES IS REQUIRED FOR ALL SURFACE WATER FLOWING OR DIVERTED TOWARDS CONSTRUCTION ENTRANCE.

7. MAINTENANCE - THE ENTRANCE SHALL BE MAINTAINED IN A CONDITION WHICH WILL PREVENT TRACKING OR FLOWING OF SEDIMENT ONTO PUBLIC RIGHTS-OF-WAY. THIS MAY REQUIRE PERIODIC TOP DRESSING WITH ADDITIONAL STONE AS CONDITIONS DEMAND AND REPAIR AND/OR CLEANOUT OF ANY MEASURES USED TO TRAP SEDIMENT. ALL SEDIMENT SPILLED, DROPPED, WASHED OR TRACKED ONTO PUBLIC RIGHTS-OF-WAY MUST BE REMOVED IMMEDIATELY.

8. WASHING - WHEELS SHALL BE CLEANED TO REMOVE SEDIMENT PRIOR TO ENTRANCE ONTO PUBLIC RIGHTS-OF-WAY. WHEN WASHING IS REQUIRED, IT SHALL BE DONE ON AN AREA STABILIZED WITH STONE AND WHICH DRAINS INTO AN APPROVED SEDIMENT TRAPPING DEVICE.

9. PERIODIC INSPECTION AND NEEDED MAINTENANCE SHALL BE PROVIDED AFTER EACH RAIN.

3'-0"

STONE

CONTINUATION OF S.C.E.

5
1

10" MIN.

5
1

EXIST. PAVEMENT

18" MIN. COMPACTED EARTH

GEOTEXTILE FABRIC

PROFILE
MOUNTABLE BERM REQUIRED FOR
HEAVY DUTY
STABILIZED CONSTRUCTION ENTRANCE

N.T.S.

| U.S. DEPARTMENT OF AGRICULTURE SOIL CONSERVATION SERVICE College Park, MD | HEAVY DUTY STABILIZED CONSTRUCTION ENTRANCE | Standard Drawing SCE-1 |

7.8.1 Mountable Berm for Construction Entrance

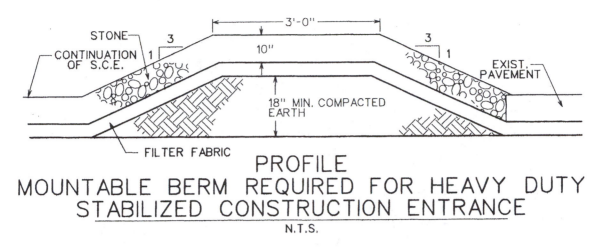

PROFILE
MOUNTABLE BERM REQUIRED FOR HEAVY DUTY
STABILIZED CONSTRUCTION ENTRANCE
N.T.S.

7.9.0 Tree and Shrub Planting Guidelines

Tree Planting Detail

Not to Scale

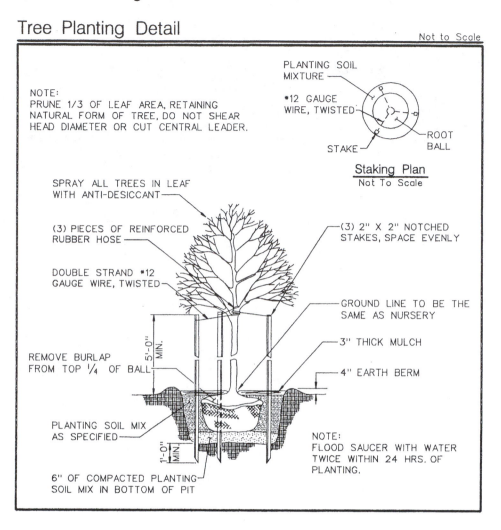

NOTE:
PRUNE 1/3 OF LEAF AREA, RETAINING NATURAL FORM OF TREE, DO NOT SHEAR HEAD DIAMETER OR CUT CENTRAL LEADER.

PLANTING SOIL MIXTURE

#12 GAUGE WIRE, TWISTED

ROOT BALL

STAKE

Staking Plan
Not To Scale

SPRAY ALL TREES IN LEAF WITH ANTI-DESICCANT

(3) PIECES OF REINFORCED RUBBER HOSE

DOUBLE STRAND #12 GAUGE WIRE, TWISTED

5'-0" MIN.

REMOVE BURLAP FROM TOP 1/4 OF BALL

PLANTING SOIL MIX AS SPECIFIED

1'-0" MIN.

6" OF COMPACTED PLANTING SOIL MIX IN BOTTOM OF PIT

(3) 2" X 2" NOTCHED STAKES, SPACE EVENLY

GROUND LINE TO BE THE SAME AS NURSERY

3" THICK MULCH

4" EARTH BERM

NOTE:
FLOOD SAUCER WITH WATER TWICE WITHIN 24 HRS. OF PLANTING.

Shrub Planting Detail

Not to Scale

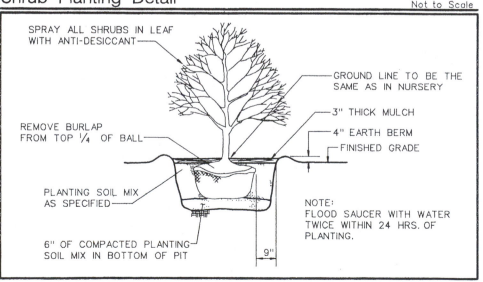

SPRAY ALL SHRUBS IN LEAF WITH ANTI-DESICCANT

REMOVE BURLAP FROM TOP 1/4 OF BALL

PLANTING SOIL MIX AS SPECIFIED

6" OF COMPACTED PLANTING SOIL MIX IN BOTTOM OF PIT

GROUND LINE TO BE THE SAME AS IN NURSERY

3" THICK MULCH

4" EARTH BERM

FINISHED GRADE

NOTE:
FLOOD SAUCER WITH WATER TWICE WITHIN 24 HRS. OF PLANTING.

9"

7.10.0 Grass and Legume Plant Characteristics

Common Name	Redtop	Rye	Ryegrass Italian	Sweet Clover	Sudangrass	Crown Vetch	Lespedeza Korean	Lespedeza Sericea	Ryegrass Perennial	Birdsfoot Trefoil
Botanical Name	Argostis Alba	Secale Cereals	Lolium Multiflorum	Lolium Perenne	Melilotus Alba Officinalis	Sorghum Sudanese	Coronilla Varia	Lespedeza Etipulacea	Lolium Perenne	Lotus Corniculatis
Germination Time (Days)	5 - 10	4 - 7	5 - 14	10	4 - 10	14 - 21	5 - 14	7 - 28	5 - 14	10
Growth Habitat	P, SL, B	A	A	B, 1	A	P, L, R	A	P, L, R	P, S, B	P, L
Seasons — Cool	X	X	X	X		X			X	X
Seasons — Warm				X	X	X	X	X	X	X
Dry, Not Droughty	X	X		X	X	X	X	X		X
Well Drained	X	X	X	X	X	X	X	X	X	X
Drainage Class — Moderately Well Drained	X	X	X	X	X	X	X	X	X	X
Drainage Class — Somewhat Poorly Drained	X		X		X		X	X	X	X
Drainage Class — Poorly Drained	X									
Annual Cover — Winter		X	X	X	X		X		X	X
Annual Cover — Summer				X	X		X			
pH Range	4.0 - 7.5	5.5 - 7.5	5.5 - 7.5	6.5 - 7.5	4.5 - 7.5	5.5 - 7.5	5.5 - 7.5	5.5 - 7.5	5.5 - 7.5	5.0 - 7.5
Flooding Tolerance	X		X				X	X	X	X
Erodable Areas	X	X	X		X		X	X	X	X
Waterways and Channels	X									
Shade Tolerance			X			X			X	
Foot Traffic	X									
Playgrounds, Athletic Fields, Lawns	X									
Beautify				X		X			X	
Levels of Maintenance — High				X						
Levels of Maintenance — Medium	X	X	X	X		X		X	X	X
Levels of Maintenance — Low		X	X	X	X	X	X	X	X	X

7.10.1 Permanent Seeding for Low-Maintenance Areas

MIX	SEED MIX (USE CERTIFIED MATERIAL IF AVAILABLE)	PLANTING LBS/AC.	PLANTING LBS/1000 SQ.FT.	SITE CONDITIONS	USDA HARDINESS ZONES	3/1-5/15	3/15-6/1	5/16-8/14	6/2-7/31	8/1-10/1	8/15-10/1	8/15-10/15	
1 SECOND CHOICE	TALL FESCUE (75%), CANADA BLUEGRASS (10%), KENTUCKY BLUEGRASS (10%), REDTOP (5%)	150	3.4	MOIST TO DRY	5b		X			X			A
					6a		X			X			
					6b	X					X		
					7a	X						X	
					7b	X						X	
3 FIRST CHOICE	TALL FESCUE (85%), PERENNIAL RYEGRASS (10%), KENTUCKY BLUEGRASS (5%)	125 / 15 / 10	2.9 / .34 / .23	MOIST TO DRY	5b		X			X			C
					6a		X			X			
					6b	X					X		
					7a	X						X	
					7b	X						X	
5 THIRD CHOICE	TALL FESCUE (85%)OR PERENNIAL RYEGRASS (50%) PLUS CROWNVETCH OR FLATPEA	110 / 20 / 20	2.5 / .46 / .46	MOIST TO DRY	5b		X			X			D
					6a		X			X			
					6b	X					X		
					7a	X						X	
					7b	X						X	

A - USED BY SHA ON SLOPED AREAS. ADD A LEGUME FOR SLOPES > 3:1.
C - POPULAR MIX - PRODUCES PERMANENT GROUNDCOVER QUICKLY. BLUEGRASS THICKENS STAND.
D - BEST USE ON SHADY SLOPES NOT ON POORLY DRAINED CLAYS.

7.11.0 USDA Plant Hardiness Zone Map

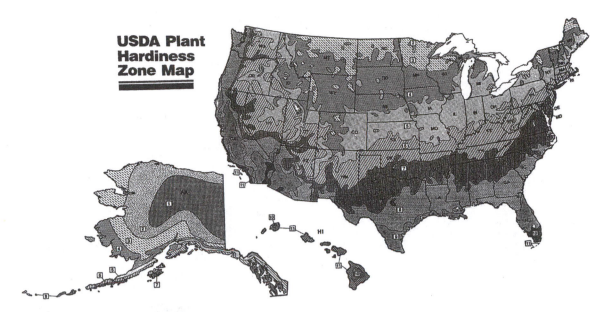

(SOURCE: Agricultural Research Service, USDA)

Figure 7.11.0   *(By permission: McGraw-Hill, The Site Calculations Pocket Reference, Ed Hannon)*

7.12.0 Shade and Ornamental Trees—Botanical and Common Names

Common Name	Botanical Name	Hardiness Zone	Height (ft.)	Spread (ft.)	Shape	Growth Rate	Foliage Texture	Fall Foliage	Fruit	Bark	Unusual Characteristics
Almond, Halls	*Prunus amygdalus* "Halls"	4	15	12	Oval	Fast	Medium	None	Peach-like	Smooth	Nut fruit
Ash, Autum Applause	*Fraxinus americana* "Autumn Applause"	3	50	30	Oval	Fast	Medium	Purple	Winged	Rough dark	Fall color
Ash, Autumn Purple	*Fraxinus americana* "Autumn Purple"	3	50	30	Oval	Fast	Medium	Red-purple	Seedless	Rough dark	Fall color
Ash, Black Hawk Mountain	*Sorbus aucuparia* "Black Hawk"	3	20	20	Oval	Medium	Medium	Gold	Orange	Gray brown	Orange fruit
Ash, Champaign County	*Fraxinus americana* "Champaign County"	3	50	30	Oval	Fast	Medium	Red	None	Dark brown	Glossy foliage
Ash, European Mountain	*Sorbus aucuparia* "European Mountain"	3	30	15	Oval	Medium	Fine	None	Orange	Brown	Orange fruit
Ash, Green	*Fraxinus pennsylvanica*	2	60	40	Oval	Fast	Medium	Yellow	Winged	Rough dark	Shiny green foliage
Ash, Marshall's Seedless	*Fraxinus pennsylvanica* "Marshall's Seedless"	3	60	40	Oval	Fast	Medium	Yellow	None	Rough dark	Seedless
Ash, Patmore	*Fraxinus pennsylvanica* "Patmore"	2	50	40	Oval	Fast	Medium	Yellow	None	Rough dark	Seedless
Ash, Rosehill	*Fraxinus americana* "Rosehill"	5	50	30	Oval	Fast	Medium	Bronze	None	Dark brown	Seedless
Ash, Summit	*Fraxinus pennsylvanica* "Summit"	3	60	40	Oblong	Fast	Medium	Yellow	Very few	Rough dark	Straight growth habit
Bald Cypress	*Taxodium distichum*	5	70	30	Long	Medium	Medium–fine	Russet	None	Rough	Stately, hardy
Birch, Cutleaf Weeping	*Betula pendula* "Gracilis"	2	30	25	Oval	Medium	Fine	Golden	Cone-like	Peeling	White bark
Birch, Clump European White	*Betula pendula* (clump form)	2	30	25	Oval	Medium	Fine	Golden	Cone-like	Peeling	White bark
Birch, European White	*Betula pendula*	2	50	25	Oval	Medium	Fine	Golden	Cone-like	Peeling	White bark
Birch, Clump Paper	*Betula papyrifera* (Clump)	2	50	30	Oval	Medium	Fine	Yellow	Cone-like	Peeling	White bark
Birch, Paper	*Betula papyrifera*	5	50	20	Oval	Medium	Fine	Yellow	Cone-like	Peeling	White bark
Birch, Clump Monarch	*Betula maximowicziana* (Clump)	5	50	20	Oval	Medium	Fine	Yellow	Cone-like	Peeling	Borer resistant
Birch, Monarch	*Betula maximowicziana*	5	50	20	Oval	Medium	Fine	Yellow	Cone-like	Peeling	Borer resistant
Birch, Clump River	*Betula nigra* (Clump)	4	50	40	Oval	Medium	Fine	Yellow	Cone-like	Red-brown peeling	Borer resistant
Birch, River	*Betula nigra*	4	50	40'	Oval	Medium	Fine	Yellow	Cone-like	Red-brown peeling	Borer
Birch, Whirespire	*Betula Platphylla japonica* "Whitespire"	3	50	25	Oval	Medium	Fine	Golden	Cone-like	Peeling	Very borer resistant
Catalpa, Northern	*Catalpa speciosa*	4	60	20	Irregular	Medium	Coarse	None	Pod	Gray brown	White blooms
Catalpa, Umbrella	*Catalpa bungei*	4	10	10	Umbrella	Fast	Coarse	Yellow	Pod	Rough dark	Umbrella form

Figure 7.12.0 (*By permission: McGraw-Hill, The Site Calculations Pocket Reference, Ed Hannon*)

Common Name	Botanical Name	Hardiness Zone	Height (ft.)	Spread (ft.)	Shape	Growth Rate	Foliage Texture	Fall Foliage	Fruit	Bark	Unusual Characteristics
Cherry, Flowering Akebono	*Prunus xyedoensis* "Akebono"	5	20	20	Spreading	Medium	Medium	Red	Orange red	Red-brown, smooth	Double flowers
Cherry, Flowering Canada Red	*Prunus virginia* "Canada Red"	2	25	20	Oval	Medium	Medium	Red	None	Red-brown, smooth	Double flowers
Cherry, Flowering Sargent Columnar	*Prunus sargenti* "Columnaris"	4	25	10	Vase	Medium	Medium	Amber	Red	Brown	Deep blush pink flowers, columnar habit
Cherry, Flowering Kwanzan	*Prunus serrulata* "Kwanzan"	5	25	15	Vase	Medium	Medium	Red-gold	Small-yellow	Brown	White blooms
Cherry, Flowering Mt. Fuji	*Prunus serrulata* "Mt. Fuji"	5	25	20	Drooping	Medium	Medium	Reddish	Orange-red	Dark	Drooping habit
Cherry, Flowering Sargent	*Prunus, sargenti*	4	25	10	Round	Medium	Medium	Amber	Red	Dark	Pale pink flowers, vigorous
Cherry, Flowering Yoshino	*Prunus xyyeddensis* "Yoshino"	5	20	20	Spreading	Medium	Medium	Red-gold	Orange-red	Red, smooth	Fragrant whitish pink blossoms
Cootonwood, Great Plains	*Populus sargentii* "Great Plaims"	3	80	60	Vase	Rapid	Medium	Yellow-green	None	Light-yellow	Rapid growth
Cottonwood, Narrowleaf	*Populus angustifolia* "Narrowleaf"	2	75	50	Round	Rapid	Medium	Golden	None	Light-brown	Drought tolerant
Cottonwood, Nor'Easter	*Populus x canadensis* "Nor'Easter"	2	60	40	Spreading	Rapid	Medium	Golden	None	Light-brown	Canker resistant
Cottonwood, Siouxland	*Populus deltoides* "Siouxland"	2	80	50	Spreading	Rapid	Medium	Golden	None	Light Buff	Rust resistant
Crab, American Beauty	*Malus* "American Beauty"	4	15	12	Round	Medium	Fine	Diverse	Purple	Smooth	Double red blossoms
Crab, Brandywine	*Malus* "Brandywine"	4	25	20	Round	Fast	Fine	Deep-purple	Green	Smooth	Double rose blossoms
Crab, Candied Apple	*Malus* "Candied Apple"	4	15	12	Weeping	Medium	Fine	Diverse	Red	Smooth	Purple pink blossoms
Crab, Centennial	*Malus* "Centennial"	3	18	18	Oval	Medium	Fine	Diverse	Red	Smooth	White blossoms
Crab, Centurian	*Malus* "Centurian"	4	25	20	Columnar	Fast	Fine	Diverse	Red	Smooth	Rosy red blossoms
Crab, Dolgo	*Malus* "Dolgo"	3	25	20	Round	Medium	Fine	Diverse	Crimson	Smooth	Large white blossoms
Crab, Eleyi	*Malus* "Eleyi" "Floribunda"	4	20	15	Upright	Medium	Fine	Diverse	Red	Smooth	Single pink blossoms
Crab, Floribunda	*Malus* "Floribunda"	4	25	20	Round	Medium	Fine	Diverse	Yellow-red	Smooth	Fragrant deep pink blossoms
Crab, Halls Parkmani	*Malus* "Halls Parkmani"	4	15	12	Vase	Medium	Fine	Diverse	Red	Smooth	Double rose blossoms
Crab, Hopa	*Malus* "Hopa"	4	25	20	Upright	Medium	Fine	Diverse	Red	Smooth	Dense rose blossoms
Crab, Indian Magic	*Malus* "Indian Magic"	4	20	15	Round	Medium	Fine	Diverse	Red	Smooth	Pink blossoms

Figure 7.12.0 *(cont.)*

Common Name	Botanical Name	Hardiness Zone	Height (ft.)	Spread (ft.)	Shape	Growth Rate	Foliage Texture	Fall Foliage	Fruit	Bark	Unusual Characteristics
Crab, Liset	*Malus* "Liset"	4	15	15	Spreading	Medium	Fine	Diverse	Red-purple	Smooth	Purple blossoms
Crab, Makamik	*Malus* "Makamik"	4	15	15	Spreading	Medium	Fine	Diverse	Red	Smooth	Purple rose
Crab, Pink Perfection	*Malus* "Pink Perfection"	4	25	20	Spreading	Medium	Medium	Diverse	Red	Smooth	Double pink blossoms
Crab, Pink Spires	*Malus* "Pink Spires"	3	12	10	Pyramidal	Medium	Fine	Mahogany	Maroon-red	Smooth	Light pink blossoms
Crab, Profusions	*Malus* "Profusions"	4	12	10	Shrub	Medium	Fine	Diverse	Oxblood red	Smooth	Reddish purple blossoms
Crab, Red Jade	*Malus* "Red Jade"	4	12	10	Weeping	Medium	Fine	Diverse	Red	Smooth	Blush white blossoms
Crab, Red Silver	*Malus* "Red Silver"	4	20	15	Upright	Medium	Fine	Diverse	Purple	Smooth	Fruit makes excellent jelly, rose blossoms
Crab, Red Splendor	*Malus* "Red Splendor"	3	18	12	Spreading	Medium	Fine	Diverse	Small red	Smooth	Rosy pink blossoms
Crab, Robinson	*Malus* "Robinson"	4	25	15	Upright	Medium	Fine	Diverse	Dark red	Smooth	Deep pink blossoms
Crab, Royal Ruby	*Malus* "Royal Ruby"	4	15	10	Upright	Fast	Fine	Diverse	Red	Smooth	Double reddish pink blossoms
Crab, Royalty	*Malus* "Royalty"	4	15	10	Uprigh	Medium	Fine	Brilliant purple	Dark red	Smooth	Crimson purple blossoms
Crab, Sargents	*Malus* "Sargenti"	4	8	6	Shrub	Medium	Fine	Diverse	Dark red	Smooth	White blossoms
Crab, Snowcloud	*Malus* "Snowcloud"	4	15	8	Upright	Fast	Fine	Diverse	Red	Smooth	Pure white blossoms
Crab, Snowdriff	*Malus* "Snowwdriff"	3	20	15	Upright	Medium	Fine	Diverse	Orange red	Smooth	White blossoms
Crab, Sparkler	*Malus* "Sparkler"	4	10	6	Horizontal	Medium	Fine	Diverse	Small red	Smooth	Bright pink blossoms
Crab, Spring Snow	*Malus* "Spring Snow"	4	20	15	Oval	Medium	Fine	Diverse	None	Smooth	Prolific white blossoms
Crab, Strathmore	*Malus* "Strathmore"	3	15	8	Pyramidal	Medium	Fine	Red	Purplish-red	Smooth	Pink blossoms
Crab, Vanguard	*Malus* "Vanguard"	4	12	8	Vase	Medium	Fine	Diverse	Bright red	Smooth	Rose pink blossoms
Crab, White	*Malus* "White Angel"	4	15	8	Upright	Medium	Fine	Diverse	Glossy red	Smooth	Pure white blossoms
Crab, White Candle	*Malus* "White Candle"	4	14	6	Columnar	Medium	Fine	Diverse	Few	Smooth	White with pink tint blossoms
Crab, Zumi	*Malus sieboldi* var. "Zumi Calocarpa"	5	15	8	Bushy	Medium	Fine	Yellow	Red-orange	Smooth	White blossoms
Dogwood, Cherokee Chief	*Cornus florida* "Cherokee Chief"	5	20	12	Horizontal	Slow	Medium	Scarlet	Red	Tan smooth	Ruby red flowering
Dogwood, Cherokee Princess	*Cornus florida* "Cherokee Princess"	5	20	12	Horizontal	Slow	Medium	Scarlet	Red	Tan smooth	Light pink flowering
Dogwood, Cloud Nine	*Cornus florida* "Cloud Nine"	5	20	12	Horizontal	Slow	Medium	Scarlet	Red	Tan smooth	White flowering
Dogwood, Kousa	*Cornus florida* "Kousa"	5	20	12	Horizontal	Slow	Medium	Scarlet	Pinkish-red	Tan smooth	White flowering
Dogwood, Pink	*Cornus florida* "Pink"	5	20	12	Horizontal	Slow	Medium	Scarlet	Pinkish-red	Tan smooth	Pink flowering

Figure 7.12.0 *(cont.)*

Common Name	Botanical Name	Hardiness Zone	Height (ft.)	Spread (ft.)	Shape	Growth Rate	Foliage Texture	Fall Foliage	Fruit	Bark	Unusual Characteristics
Dogwood,	Cornus florida "Rainbow"	5	20	12	Horizontal	Slow	Medium	Scarlet	Pinkish-red	Tan smooth	Varigated foliage, white flower
Dogwood, Red	Cornus florida "Red"	5	20	12	Horizontal	Slow	Medium	Scarlet	Pinkish-red	Tan smooth	Red flowering
Dogwood, White	Cornus florida	5	20	12	Horizontal	Slow	Medium	Scarlet	Pinkish-red	Tan smooth	White flowering
Golden Raintree	Koelereutaria paniculata	5	40	30	Spreading	Medium	Medium	Yellow	Brown	Light brown ridged	Yellow flowering in mid summer
Hackberry	Celtis occidentalis	2	70	60	Oval	Medium–fast	Medium-coarse	Yellow-green	Orange-red to purple	Corky grayish brown	Hardy, tolerant tree
Hawthorn, Crimson Cloud P.P.#2679	Crataegus laevigata "Crimson Cloud"	4	15	12	Oval	Medium	Medium	None	Scarlet	Brown	Red flowers
Hawthorn, Paul's Scarlet	Crataegus laevigata "Paulii"	5	20	12	Oval	Slow	Fine	None	Scarlet	Flaky	Double scarlet flowers
Hawthorn, Toba	Crataegus x mordensis "Toba"	3	15	12	Oval	Medium	Medium	None	Scarlet	Brown	Fragrant, double white flower
Hawthorn, Washington	Crataegus phaenopyrum "Washington"	4	30	25	Columnar	Medium-slow	Medium-fine	Red-purple	Glossy-red	Brown	White flowers in spring
Honey Locust, Thornless	Gleditsia triacanthos inermix "Thornless"	4	70	60	Spreading	Fast	Medium—Fine	Yellow-green	Reddish-brown	Grayish-brown	Fragrant white flowers
Honey Locust, Golden	Gleditsia triacanthos inermis "Golden"	4	60	40	Oval	Fast	Medium—Fine	Yellow-green	None	Brown	Deligate golden foliage
Honey locust, Green glory	Gleditsia triacanthos inermis "Green Glory"	4	60	40	Pyramidal	Fast	Medium—Fine	Yellow-green	None	Brown	Fragrant white flowers
Honey Locust, Moraine	Gleditsia triacanthos inermis "Moraine"	4	60	40	Spreading	Fast	Medium—Fine	Yellow-green	None	Brown	Old favorite
H.cust, Rubylace	Gleditsia triacanthos inermis "Rubylace"	4	40	30	Spreading	Fast	Medium—Fine	Yellow-green	None	Brown	Purplish-bronze foliage
Honey Locust, Shademaster	Gleditsia triacanthos inermis "Shademaster"	4	60	40	Spreading	Fast	Medium—Fine	Yellow-green	None	Brown	Very dark green foliage
Honey Locust, Skyline	Gleditsia triacanthos inermis "Skyline"	4	60	60	Pyramidal	Fast	Medium—Fine	Yellow-green	None	Brown	Very dark green foliage
Honey Locust, Sunburst	Gleditsia triacanthos inermis "Sunburst"	4	40	30	Pyramidal	Fast	Medium—Fine	Yellow-green	None	Brown	Golden foliage on tips of branch

Figure 7.12.0 *(cont.)*

Common Name	Botanical Name	Hardiness Zone	Height (ft.)	Spread (ft.)	Shape	Growth Rate	Foliage Texture	Fall Foliage	Fruit	Bark	Unusual Characteristics
Linden, American	*Tilia americana*	3	75	50	Pyramidal	Medium	Medium	None	Nutlet	Rough-gray	Pyramidal habit
Linden, Glenleven	*Tilia cordata* "Glenleven"	3	50	35	Pyramidal	Medium	Medium	None	Nutlet	Rough-gray	Pyramidal habit
Linden, Greenspire	*Tilia cordata* "Greenspire"	4	50	35	Pyramidal	Medium	Medium	None	Nutlet	Rough-gray	Spicy fragrant flowers
Linden, June Bride	*Tilia cordata* "June Bride"	3	50	35	Pyramidal	Medium	Medium	None	Nutlet	Rough-gray	Prolific flowering
Linden, Littleleaf	*Tilia cordata*	4	50	35	Oval	Medium	Medium	None	Nutlet	Rough-gray	Fragrant flowers
Linden, Redmond	*Tilia enchora* "Redmond"	4	40	30	Pyramidal	Medium	Medium	None	Nutlet	Rough-gray	Reddish bark in winter
Magnolia, Saucer	*Magnolia soulangiana*	5	30	25	Spreading	Medium	Coarse	None	Pod	Gray	Large white to pinkish flowers
Magnolia, Southern	*Magnolia grandiflora*	6	80	50	Pyramidal	Slow	Medium—coarse	None	Pod	Dark	Evergreen, large white fragrant flowers
Magnolia, Star	*Magnolia stellata*	4	20	15	Oval	Slow	Medium	Yellow-brown	Pod	Brown	Double white fragrant flowers
Maple, Amur	*Acer ginnala*	2	20	20	½ round	Medium	Fine	Scarlet	Winged	Thin, dark	Extremely hardy
Maple, Autumn Flame	*Acer rubrum* "Autumn Flame"	3	50	35	Round	Fast	Medium	Scarlet	Winged	Gray	Fall color
Maple, Blair	*Acer saccharinum* "Blair"	4	80	50	Round	Fast	Medium	Pale yellow	Winged	Gray	Strong branching
Maple, Crimson, King	*Acer platanoides* "Crimson King"	4	40	30	Oval	Medium	Medium	None	Winged	Rough, dark	Purple foliage
Maple, Columnar Norway	*Acer platanoides* "Columnar"	4	50	20	Columnar	Medium	Medium	None	Winged	Rough-dark	Columnar
Maple, Emerald Lustre P.P.#4837	*Acer platanoides* "Pond"	3	50	30	Oval	Fast	Medium	Yellow	Winged	Rough, dark	Glossy foliage
Maple, Emerald Queen	*Acer platanoides* "Emerald Queen"	3	40	25	Ascending	Medium	Medium	Yellow	Winged	Rough, dark	Cutleaf
Maple, Japanese Dwarf	*Acer palmatum*	6	10	10	Round	Slow	Fine	Yellow	Winged	Reddish-brown	Delicate appearance
Maple, Norway	*Acer platanoides*	3	50	30	Oval	Medium	Medium	None	Winged	Rough, dark	Dense foliage
Maple, Northwood Red	*Acer rubrum* "Northwood"	3	50	35	½ round	Fast	Medium	Scarlet	Winged	Gray	Fall foliage
Maple, October Glory	*Acer rubrum* "October Glory"	4	50	35	½ round	Fast	Medium	Scarlet	Winged	Gray	Fall foliage
Maple, Oregon Pride	*Acer platanoides* "Oregon pride"	4	40	25	Oval	Medium	Medium	None	Winged	Rough, dark	Red foliage
Maple, Royal Red	*Acer platanoides* "Royal Red"	4	40	25	Oval	Medium	Medium	None	Winged	Rough, dark	Dark-red foliage
Maple, Red	*Acer rubrum*	4	60	35	½ round	Fast	Medium	Scarlet	Winged	Gray	Fall foliage
Maple, Red Sunset	*Acer rubrum* "Red sunset"	3	50	30	½ round	Fast	Medium	Scarlet	Winged	Gray	Fall color
Maple, Schwedler	*Acer platanoides* "Schwedler"	4	60	40	Round	Medium	Medium	None	Winged	Rough, dark	Purplish-red foliage
Maple, Silver	*Acer saccharinum*	4	100	60	Round	Rapid	Medium	Pale yellow	Winged	Gray	Fast growth
Maple, Silver Queen	*Acer saccharinum* "Silver Queen"	4	60	50	Round	Rapid	Medium	Pale yellow	Seedless	Gray	Seedless

Figure 7.12.0 *(cont.)*

Common Name	Botanical Name	Hardiness Zone	Height (ft.)	Spread (ft.)	Shape	Growth Rate	Foliage Texture	Fall Foliage	Fruit	Bark	Unusual Characteristics
Maple, Sugar	*Acer saccharinum*	3	75	50	Oval	Medium	Medium	Yellow-scarlet	Winged	Dark brown	Fall color
Maple, Bonfire Color	*Acer saccharinum* "Bonfire"	3	60	40	Oval	Medium	Medium	Brilliant red	Winged	Dark brown	Fall color
Maple, Sugar Green Mt.	*Acer saccharinum* "Mt. Green"	3	60	40	Oval	Medium	Medium	Orange-red	Winged	Dark brown	Fall color
Maple, Variegated Norway	*Acer platanoides* "Variegatum"	5	40	25	Oval	Medium	Fine	None	Winged	Rough, dark gray	Green leaves edged in white
Maple, Wiers	*Acer saccharinum* "Weiri"	3	70	50	Weeping	Fast	Medium	Pale yellow	Winged	Gray	Cutleaf
May Day	*Prunus padus commutata*	2	20	30	Round	Medium	Medium	None	Black	Dark brown	Large fragrant white flowers
Mimosa	*Albizia julibrissin*	6	30	20	Vase	Fast	Fine	None	Pod	Greenish	Foliage and brush-like pink flowering
Mt. Ash, Black Hawk	*Sorbus aucuparia* "Black Hawk"	3	20	25	Oval	Medium	Medium	Gold	Orange	Gray-brown	Orange fruit
Mt. Ash, Cardinal	*Sorbus aucuparia* "Cardinal"	3	25	20	Ovate	Medium	Medium	Gold	Apricot color	Gray-brown	Apricot color fruit
Mt. Ash, Cardinal Royal	*Sorbus aucuparia* "Card. royal"	3	25	20	Oval	Medium	Medium	Gold	Orange-red	Gray-brown	Orange-red fruit
Mt. Ash, Cherokee	*Sorbus aucuparia* "Cherokee"	3	25	20	Oval	Medium	Medium	Gold	Red	Gray-brown	Red fruit
Mt. Ash, Decora	*Sorbus decora*	2	15	20	Oval	Medium	Medium	Gold	Red	Gray-brown	Hardy
Mulberry, Common	*Morus alba*	4	30	20	Rounded	Fast	Coarse	None	Flesh-Purpl	Brown	Fruiting
Mulberry, Fruitless	*Morus alba* "Fruitless"	6	50	50	Rounded	Fast	Coarse	None	None	Brown	Fruitless
Mulberry, Weeping	*Morus alaba urbana*	6	50	50	Rounded	Fast	Coarse	None	None	Brown	Weeping
Oak, Pin	*Quercus palustris*	4	70	40	Pyramidal	Slow	Medium	Scarlet	Acorn	Dark	Fast growth, fall color
Oak, Red	*Quercus borealis*	4	70	50	Oval	Slow	Medium	Dark red	Acorn	Rough, Dark	Fast growth, fall color
Oak, Schumard	*Quercus schumardi*	6	80	60	Oval	Medium	Medium	Crimson	Acorn	Rough, dark	Fast growth, fall color
Oak, Water	*Quercus nigra*	6	60	40	Oval	Medium	Medium	None	Acorn	Smooth	Holds leaves into winter
Peach, Cardinal	*Prunus persica* "Cardinal"	5	15	15	Rounded	Medium	Medium	Yellow	Red blush	Red-brown	Double red pink flowers
Pear, Aristocrat	*Pyrus calleryna* "Aristocrat"	4	30	20	Oval	Medium	Medium	Purple-Red	Small yellow	Brown	Fall color, white flowers
Pear, Bradford	*Pyrus calleryna* "Bradford"	4	30	20	Conical	Medium	Medium	Red-gold	None	Brown	Fall colors, white flowers
Pear, Capital	*Pyrus calleryna* "Capital"	5	50	15	Narrow upright	Medium	Medium	Red-purple	Few	Brown	Fall color, white flowers
Pear, Red Spires P.P.#3815	*Pyrus calleryna* "Red Spires"	5	30	20	Conical narrow	Medium	Medium	Crimson purple	None	Brown	Fall color, white flowers
Pear, Whitehouse	*Prunus calleryana* "whitehouse"	5	50	20	Columnar	Medium	Medium	Reddish purple	Few	Brown	Fall color, white flowers
Plum, Blireiana	*Prunus x blireiana*	5	15	10	Round	Medium	Medium	Purple	Dark red	Dark	Double fragrant flowers

Figure 7.12.0   *(cont.)*

Common Name	Botanical Name	Hardiness Zone	Height (ft.)	Spread (ft.)	Shape	Growth Rate	Foliage Texture	Fall Foliage	Fruit	Bark	Unusual Characteristics
Plum, K.V.	*Prunus cerasifera* "Krauter Vesuvius"	6	20	15	Oval	Fast	Medium	Purple	None	Dark red	Purple black foliage
Plum, Newport	*Prunus cerasifera* "Newport"	3	15	15	½ round	Fast	Medium	Purple red	Dark red	Rough, dark	Reddish purple
Plum, Thundercloud	*Prunus cerasifera* "Thundercloud"	4	15	15	½ round	Medium	Medium	Purple	Dark purple	Dark-red	Purple foliage
Poplar, Bolleana	*Populus alba* "Pyramidalis"	3	60	6	Columnar	Rapid	Medium	Golden	None	Smooth	Columnar
Poplar, Lombardy	*Populus nigra* "Thevestina"	3	60	6	Columnar	Rapid	Medium	None	None	Gray	Columnar
Poplar, Robusta	*Populus x canadensis* "Robusta"	2	50	15	Oval	Rapid	Medium	Golden	None	Light buff	Rapid growth
Poplar, Sliver	*Populus alba*	3	70	40	Rounded	Rapid	Medium–coarse	None	None	Gray-green	Silver green foliage
Quaking Aspen	*Populus tremuloides*	2	60	30	Vase	Rapid	Medium	Yellow	None	Greenish white	Fall foliage
Red Bud, Eastern	*Carcis canadensis*	4	20	20	Round	Slow	Coarse	Yellow	Small pods	Dark gray	Rosy pink flowers
Red Bud, Forest Panzy	*Cercis canadensis* "Forest Panzy"	5	20	20	Round	Slow	Coarse	Yellow	Small pods	Dark gray	Purple foliage
Red Bud, Oklahoma	*Cersis canadensis* "Oklahoma"	6	20	20	Round	Slow	Coarse	Yellow	Small pods	Dark gray	Glossy foliage
Red Bud, White	*Cersis canadensis alba*	6	20	20	Round	Slow	Coarse	Yellow	Small pods	Dark gray	White flowe
Russian Olive	*Elaeagnus angustifolia*	2	30	20	Spreading	Fast	Fine	None	Berries	Silver-brown	Silver gray foliage
Sweet Gum	*Liquiddambar styraciflua*	5	80	50	Pyramida	Medium	Medium	Yellow red	Winged	Gray-brown	Fall foliage
Sycamore, American	*Platanus occidentalis*	4	100	100	Oval	Fast	Coarse	None	Balls	Brown-white	Adaptable
Sycamore, Bloodgood	*Platanus x acerifolia* "Bloodgood"	5	80	100	Oval	Fast	Coarse	None	Balls	Brown-white	Disease resistant
Smoke Tree	*Cotinus coggygria*	5	15	15	Spreading	Medium	Fine	Purple	Small pods	Brown-purple	Smoky pink flowers
Smoke Tree, Velvet Cloak	*Cotinus coggygria* "Velvet Cloak"	5	15	15	Spreading	Medium	Fine	Reddish purple	Small Pods	Brown purple	Purple folia
Tulip Tree	*Liriodendron tulipifera*	5	90	50	Pyramidal	Fast	Coarse	Golden yellow	Winged	Brownish	Tulip shape flowers
Willow, Corkscrew	*Salix matsudana* "Tortuosa"	4	30	12	Upright	Rapid	Twisted	None	Cottony	Greenish	Oriental loo
Willow, Globe	*Salix matsudana* "Globosa"	4	15	8	Round	Rapid	Fine	None	Cottony	Greenish	Globe shape
Willow, Niobe	*Salix alba tristis*	3	50	30	Weeping	Rapid	Fine	None	Cottony	Greenish	Golden colo
Willow, Wisconsin Weeping	*Salix x blanda*	4	40	35	Weeping	Rapid	Fine	None	Cotton	Greenish branches	Pendulous
Wisteria Tree	*Wisteria sinensis*	4	30	15	Weeping	Fast	Fine	None	Pods	Grayish	Fragrant violet flowers

Figure 7.12.0 *(cont.)*

7.12.1 Ornamental Shrubs—Botanical and Common Names

Common Name	Botanical Name	Hardiness Zone	Height (ft.)	Spread (ft.)	Shape	Growth Rate	Foliage Texture	Fall Foliage	Fruit	Bark	Unusual Characteristics
Althea, Rose of Sharon	*Hibiscus syriacus*	5	10	5	Upright	Medium	Medium	None	Brown	Gray	Purple, red, white, pink flowers in summer
Almond, Pink Flowering	*Prunus glandulosa "Rosea"*	4	6	5	Upright	Medium	Medium	None	None	Dark	Pink, double flowers in early spring
Barberry, Crimson Pygmy	*Berbiris thunbergii atropurpurea "Crimson Pgymy"*	4	2	3	Mound	Slow	Fine	Red	None	Thorns	Extremely dwarf, deep crimson color
Barberry, Golden	*Berberis thunbergii aurea*	4	2	3	Mound	Slow	Fine	Gold	None	Thorns	Bright golden foliage
Barberry, Red Leaf	*Berberis thunbergii aurea atropurpurea*	4	5	5	Compact	Medium	Fine	Red	Red	Thorns	Mahogany red foliage
Barberry, Rosy Glow	*Berberis thunbergii "Rosy Glow"*	4	4	5	Compact	Medium	Fine	Red	None	Thorns	Red leaves with tinge of white and rose
Barberry, William Penn	*Berberis x gladwynensis*	5	1	3	Spreading	Medium	Medium	None	None	Thorns	Evergreen dwarf, bright yellow spring flowers
Burning Bush, Dwarf	*Euonymusalatus "Compacta"*	3	6	4	Compact	Slow	Medium	Red	Red	Winged	Brilliant red foliage in fall
Cotoneaster, Cranberry	*Cotoneaster apiculatus*	4	2	3	Mound	Slow	Medium	None	Red	None	Masses of bright red berries in fall
Cotoneaster, Spreading	*Cotoneaster divaricatus*	4	2	3	Spreading	Medium	Medium	None	Red	None	Bright red berries hold into winter
Cotoneaster, Rock	*Cotoneaster horizontalis*	4	2	3	Spreading	Medium	Medium	Red	Red	None	Masses of bright red berries in fall
Crape Myrtle	*Lagerstroemia indica*	6	25	15	Upright	Rapid	Medium	Yellow red	None	Smooth	Deciduous shrub or small tree, prolific cluster of flowers in a wide range of colors in summer
Dogwood, Baileyi	*Cornus stolonifera*	3	8	6	Spreading	Medium	Coarse	Red	None	Red	Bright red twigs
Dogwood, Red Twig	*Cornus sericea*	3	8	6	Spreading	Medium	Coarse	Red	None	Red	Red twigs, small white flowers in summer
Dogwood, Variegated	*Cornus alba "Elegantissima"*	3	6	6	Upright	Slow	Coarse	None	White	None	Foliage is green with cream colored fringe

Figure 7.12.1   *(By permission: McGraw-Hill, The Site Calculations Pocket Reference, Ed Hannon)*

Common Name	Botanical Name	Hardiness Zone	Height (ft.)	Spread (ft.)	Shape	Growth Rate	Foliage Texture	Fall Foliage	Fruit	Bark	Unusual Characteristicsi
Dogwood, Yellow Twig	*Cornus stolonifera* "Flaviramea"	3	8	5	Spreading	Medium	Coarse	None	White	Yellow	Bright yellow branches white flowers in late spring
Forsythia, Lynnwood Gold	*Forsythia x intermedia* "Lynnwood Gold"	4	5	6	Spreading	Rapid	Medium	None	Capsule	Dull yellow	Brilliant yellow flowers in early spring
Forsythia, Sunrise	*Forsythia x intermedia* "Sunrise"	4	4	5	Spreading	Medium	Medium	None	None	Dull yellow	Shorter growing
Golden Elder	*Sambucus canadensis* "Aurea"	3	10	6	Open	Rapid	Coarse	Yellow	Red	Tan	Golden yellow leaves, clusters of tiny white flowers
Golden Nine bark	*Physocarpus opulifolius aureus*	4	8	4	Round	Medium	Medium	None	Capsule	Orange brown	Golden colored leaves
Honeysuckle, Arnold's Red	*Lonicera tatarica* "Arnold Red"	3	8	8	Upright	Medium	Medium	None	Red	Gray	Red flowers in spring followed by red berries
Honeysuckle, Clavey's	*Lonica xylosteum* "Claveyi"	4	6	4	Compact	Slow	Medium	None	Red	Gray	Dwarf, fragrant creamy white flowers
Honeysuckle, Goldflame	*Lonicera x heckrottii*	4	2	10	Spreading	Fast	Soft	None	None	None	Brilliantly colored coral red flowers, yellow foliage
Honeysuckle, Hall's	*Lonicera japonica* "Halliana"	4	2	10	Spreading	Fast	Soft	None	None	None	Tubular pure white twig, yellow flowers, fragrant
Honeysuckle, Purpleleaf	*Lonicera japonica* "Purpurea"	5	2	10	Spreading	Fast	Soft	Purple	None	None	Foliage turns deep purple in winter
Honeysuckle, Zabelt	*Lonicera tatarica* "Zabelii"	3	8	8	Upright	Medium	Medium	None	Red	Gray	Bright red blooms in spring
Hydrangea, P.G.	*Hydrangea paniculata* "Grandiflora"		8	5	Stiff	Medium	Coarse	None	Capsule	Buff	Profuse bloomers with large cluster of white flowers
Lilac, Old Fashioned	*Syringa vulgaris*	3	12	10	Upright	Medium	None	None	Capsule	Drab	Clusters of lavender blossoms in spring

Figure 7.12.1 *(cont.)*

Common Name	Botanical Name	Hardiness Zone	Height (ft.)	Spread (ft.)	Shape	Growth Rate	Foliage Texture	Fall Foliage	Fruit	Bark	Unusual Characteristics
Lilac, Purple Persian	*Syringa x chinensist*	3	8	6	Upright	Medium	Medium	None	Capsule	Dark gray	Profusion of lilac purple blooms in spring
Lilac, french Hybrid	*Syringa* french hybrids	3	15	8	Upright	Medium	Coarse	None	Capsule	Dark	Various attractive flowers in late spring
Mockorange, Minnesota Snowflake	*Philadelphus x virginalis* "Minnesota Snowflake"	3	8	6	Upright	Medium	Medium	None	None	Gray	Double white fragrant flowers, very hardy
Mockorange, Sweett	*Philadelphus coronarius*	3	10	6	Mound	Rapid	Medium	None	Capsule	Brown	Very fragrant single white flowers
Potentilla, Abbottswood	*Potentilla fruticosa* "Abbottswood"	3	3	3	Bushy	Medium	Medium	None	Capsule	Flaking	White flowers in summer
Potentilla, Coronation	*Potentilla fruticosa* "Coronation Triumph"	3	4	3	Compact	Medium	Fine	Golden	Seed pods	Brown	Brilliant yellow flowers in summer until frost
Potentilla, Goldfinger	*Potentilla fruticosa* "Goldfingers"	3	3	3	Round	Medium	Fine	None	Seed pods	Brown	Golden yellow flowers from spring until fall, very hardy
Potentilla, Jackmani	*Potentilla fruticosa* "Jackmani"	3	4	3	Upright	Slow	Fine	None	Seed pods	Brown	Dwarf with golden yellow flowers in mid summe
Potentilla, Katherine Dykes	*Potentilla fruticosa* "Katherine Dykes"	3	3	3	Spreading	Medium	Fine	None	Seed pods	Brown	Soft yellow flowers, very hardy
Privet, Golden Vicary	*Ligustrum x vicaryi*	5	6	5	Upright	Slow	Fine	None	Black	Gray	Bright yellow flowers, very hardy
Privet, AR North	*Ligustrum amurence*	4	10	5	Upright	Medium	Medium	None	None	None	Very hardy hedge
Privet, AR South	*Ligustrum sinence*	6	10	5	Upright	Medium	Medium	None	None	None	Semi-evergreen hedge
Pussy Willow, Black	*Salix discolor*	4	15	8	Upright	Medium	Coarse	None	None	Brown	Pearly woolly catkins in spring, very hardy
Pussywillow, French Pink	*Salix caprea*	4	15	12	Upright	Medium	Medium	None	None	Brown	Large pink catkins in spring, very hardy
Quince, Red Flowering	*Chaenomeles japonica* "Rubra"	5	6	4	Spreading	Medium	Medium	None	Green	Brown	Pinkish-red, flowers in early spring
Sand Cherry Purpleleaf	*Prunus x cistena*	4	15	5	Bushy	Medium	Medium	Purple Red	Black	Rough	Deep purple red foliage
Spirea, Anthony Waterer	*Spirea x bumalda* "Anthony Waterer"	4	3	3	Mound	Slow	Medium	Red	None	Red-brown	Cluster of ros blossoms summer/fall

Figure 7.12.1   (*cont.*)

Common Name	Botanical Name	Hardiness Zone	Height (ft.)	Spread (ft.)	Shape	Growth Rate	Foliage Texture	Fall Foliage	Fruit	Bark	Unusual Characteristicsi
Spirea, Billiardi	*Spirea x billiardii*	4	6	7	Upright	Medium	Medium	None	None	None	Bright pink flowers in july/august
Spirea, Blue Mist	*Caryopteris incana*	4	4	3	Mound	Medium	Medium	None	None	None	Powdery blue-fringed flowers in summer
Spirea, Froebeli	*Spirea x bumalda "Froebeli"*	4	4	4	Spreading	Medium	Medium	Red	None	Red-brown	Clusters of bright pink flowers in may/june
Spirea, Gold Flame	*Sprirea x bumalda "Goldflame"*	3	3	2	Round	Slow	Medium	Red	None	Brown	Foliage is golden red, clusters of flowers in summer
Spirea, Little Princess	*Spirea japonica "Little Princess"*	4	3	2	Mound	Slow	Medium	None	None	None	Mint green foliage clusters of rose crimson flowers in summer
Spirea, Snowmound	*Spirea niponica "Snowmound"*	3	4	3	Round	Medium	Fine	Yellow	None	Red-brown	Covered with white clusters in early spring
Spirea, Vanhoutte	*Spirea x vanhouttei*	4	10	8	Recurving	Medium	Fine	Orange red	None	Red-brown	Arching branches covered with pure white blooms in late spring
Viburnum, Burkwood	*Viburnum x burkwoodii*	5	10	5	Upright	Medium	Coarse	None	Red to black	Hairy tan	Clusters of fragrant flowers in early spring
Virburnum, Dilatatum	*Viburnum dilatatum*	5	10	5	Upright	Medium	Medium	None	Red	None	Gray-green leaves, red fruit
Virburnum, Mariesi	*Viburnum plicatum totemtosum "Mariesii"*	5	8	6	Spreading	Medium	Coarse	None	Red to black	Brown	White flowers with flat heads in early spring
Virburnum, Snowball	*Viburnum opulus "Sterilis"*	3	10	15	Rounding	Medium	Medium	Purple	None	None	Masses of pure white snowball-like flowers in summer
Virburnum, Pragense	*Viburnum x pragnese*	4	8	6	Upright	Medium	Glossy	None	Red	None	Creamy white flowers in spring
Virburnum, Willowwood	*Viburnum rhytidophylloides "Willowwood"*	4	8	6	Upright	Medium	Glossy	None	Black-blue	Brown, smooth	Clusters of white flowers, foliage turns burgundy in fall

Figure 7.12.1 *(cont.)*

Common Name	Botanical Name	Hardiness Zone	Height (ft.)	Spread (ft.)	Shape	Growth Rate	Foliage Texture	Fall Foliage	Fruit	Bark	Unusual Characteristicsi
Weigela, Rosea	*Weigela florida*	5	6	6	Spreading	Medium	Medium	None	None	None	Showy rosy-pink trumpet shaped flowers in May/June
Weigela, Vanicek	*Weigela florida "Vanicek"*	5	6	5	Spreading	Medium	Medium	None	Brown	Gray	Abundant red flowers in late May
Weigela, Variegated	*Weigela florida "Variegata"*	5	4	4	Compact	Medium	Medium	None	None	Gray	Pink ball-shaped flowers in spring, green/white leaves
Wisteria, Purple	*Wisteria sinensis*	4	5	35	Spreading	Rapid	Medium	None	None	None	Vine with long grape-like clusters of blossoms in spring

Figure 7.12.1   (*cont.*)

7.12.2 Fruit Trees—Varieties and Characteristics

Variety	Fruit Color	Characterictics	Pollinator Required	Hardiness Zone
Apple, Anna	Greenish-yellow	Low chill, good storage	Yes	7
Apple, Beverly Hills	Pale yellow	Low chill, cooking	Yes	8
Apple, Cortland	Greenish-yellow	Large fruit, aromatic	Yes	5
Apple, Delicoius Red	Red	Eating variety	Yes	5
Apple, Delicoius Yellow	Yellow	Keeps well	Yes	5
Apple, Ein Shemer	Red striped	Low chill, juicy	Yes	8
Apple, Golden Delicious	Yellow	Large, crisp, sweet	No	5
Apple, Granny Smith	Green	Large fruit, keeps well	Yes	5
Apple, Grimes golden	Yellow	Medium fruit, spicy	No	5
Apple, Gravestein	Green, red striped	Medium fruit, cooking	Yes	5
Apple, Haralson	red striped	Keeps well	Yes	5
Apple, Honey gold	Yellow	Hardy and sweet	Yes	4
Apple, Hyslop	Red	Tart, good jelly	Yes	5
Apple, Jonadel	Red	Eating variety, keeps well	Yes	5
Apple, Jonathan	Dark red	Crisp, keeps well	Yes	5
Apple, Keepsake	Red	Cooking variety	Yes	5
Apple, Lodi	Yellow	Large fruit, cooking variety	Yes	4
Apple, Mcintosh	Red	Large fruit, aromatic	Yes	4
Apple, Red Baron	Red	Medium fruit, eating variety	Yes	5
Apple, Regent	Red	Dessert variety	Yes	3

Figure 7.12.2   (*By permission: McGraw-Hill, The Site Calculations Pocket Reference, Ed Hannon*)

Variety	Fruit Color	Characterictics	Pollinator Required	Hardiness Zone
Apple, Rome Beauty	Red	Cooking variety	Yes	4
Apple, State Fair	Red	Crisp, keeps well	Yes	3
Apple, Stayman Winesap	Red	Cooking and eating variety	Yes	5
Apple, Sweet Sixteen	Red striped	Medium fruit, crisp	Yes	3
Apple, Wealthy	Red	Tender Dessert and cooking variety	Yes	5
Apple, Whitney Crab	Red	Juicy, Sweet flavor	Yes	5
Apple, Winesap Crimson	Red	Wine flavored, cooking and keeps well	Yes	6
Apple, Yellow Transparent	Yellow	Medium fruit, cooking variety	Yes	5
Apricot, Chinese	Gold	Fuzzless, freestone	No	5
Apricot, Early Golden	Orange	Fuzzless	No	5
Apricot, Moongold	Gold	Early Bearer	No	4
Apricot, Moonpark	Yellow	Large fruit, freestone	No	5
Apricot, Royal Blenheim	Orange-red	Large fruit, low chill	No	8
Apricot, Scout	Yellow	Hardy	No	3
Apricot, Sungold	Bright gold	Mild and sweet	No	4
Apricot, Tilton	Gold	Canning, drying variety	No	5
Cherry, Early Richmond	Red	Sour variety, highly productive	No	5
Cherry, Montmorency	Red	Sour variety, cooking	No	5
Cherry, Meteor	Red	Dwarf sour variety, cooking	No	4
Cherry, North Star	Red	Dwarf sour variety, large fruit, hardy	No	3
Cherry, Bing	Red	Good fresh or frozen, highly productive	Yes	6
Cherry, Black Tartarian	Purplish/black	Good pollinator	Yes	5
Cherry, Kansas Sweet	Red	Large fruit, highly productive	Yes	4
Cherry, Lambert	Black	Large fruit, fast grower	Yes	5
Cherry, Rainier	Yellow	Keeps well, highly productive	Yes	5
Cherry, Royal Anne	Yellow	Large fruit, sweet and juicy	Yes	5
Cherry, Stella	Black	Good pollinator	No	6
Cherry, Windsor	Black	Highly productive, fast grower	Yes	5
Cherry, Van	Black	Large fruit, sweet and juicy	Yes	5
Cherry, Yellow Glass	Yellow	Hardiest of all sweet cherries	Yes	4
Fig, Brown Turkey	Mahogany brown	Few seeds, best eaten fresh	No	7
Fig, Texas Ever Bearing	Purple	Large fruit, sweet and juicy	No	7
Nectarine, Autumn Delight	Red	Large fruit, highly productive	No	5
Nectarine, Fantasia	Yellow	Large fruit, freestone	No	5
Nectarine, Flavortop	Red	Large fruit, freestone	No	5
Nectarine, Goldmine	Yellow blush	Large fruit, freestone, low chill	No	8
Nectarine, Golden Prolific	Yellow	True genetic dwarf, freestone	No	5
Nectarine, Redgold	Yellow	Juicy, highly productive	No	5
Nectarine, Sunred	Red	Semi-freestone, low chill	No	8
Nectarine, Surecrop	Yellow	Smooth skin, freestone	No	5
Peach, Bell of Georgia	White	Freestone	No	5
Peach, Bonanza	Red blush	True genetic dwarf, freestone	No	5
Peach, Desert Gold	Yellow	Highly productive, low chill	No	8

Figure 7.12.2 *(cont.)*

Variety	Fruit Color	Characterictics	Pollinator Required	Hardiness Zone
Peach, Early Elberta	Yellow	Freestone, good canning	No	5
Peach, Elberta	Yellow	Freestone, good canning	No	5
Peach, Elbertita	Yellow	True genetic dwarf, large fruit	No	5
Peach, Flordabelle	Yellow	Freestone, low chill	No	8
Peach, Flordasun	Yellow	Semi-freestone, low chill	No	8
Peach, Golden Jubilee	Yellow	Freestone, juicy	No	5
Peach, Golden Glory	Golden	True genetic dwarf, freestone	No	5
Peach, Halehaven	Yellow	Freestone	No	4
Peach, J.H. Hale	Yellow	Freestone, keeps well	Yes	5
Peach, Polly	White	Freestone, good canning	No	4
Peach, Redhaven	Yellow	Freestone, highly productive	No	5
Peach, Reliance	Yellow	Freestone, juicy	No	4
Peach, Richhaven	Red blush	Freestone, juicy	No	5
Pear, Bartlett	Golden red	Sweet, eating and canning variety	Yes	5
Pear, Bartlett Max Red	Red blush	Eating and canning variety	Yes	5
Pear, Clapp's Favorite	Red blush	Sweet eating variety	Yes	5
Pear, Dessert	Yellow	Large fruit, tart flavor	Yes	5
Pear, Douglas	Red blush	Blight resistant	Yes	5
Pear, Duchess	Green	Highly productive	No	5
Pear, Kieffer	Green	Blight resistant	No	4
Pear, Luscious	Yellow	Blight resistant	Yes	3
Pear, Moonglow	Yellow	Blight resistant	Yes	5
Pear, Parker	Red blush	Very prolific	Yes	5
Pear, Patton	Yellow	Blight resistant	Yes	5
Pear, Red Sensation	Red	Sweet eating and canning variety	Yes	5
Pear, Seckel	Yellow-brown	Blight resistant, spicy flavor	Yes	5
Plum, Abundance	Red	Light yellow flesh	Yes	5
Plum, Blue Damson	Purple	Excellent preserves	Yes	5
Plum, Bruce	Wine red	Rich juicy flavor	Yes	5
Plum, Burbank	Red	Yellow flesh, canning variety	Yes	5
Plum, Green Gage	Green	Juicy, cooking variety	Yes	5
Plum, Hanska	Red-blue	Yellow fresh, extra sweet	Yes	5
Plum, Italian Prune	Purple	Large fruit, yellow flesh	Yes	5
Plum, La Crescent	Yellow	Extra juicy	Yes	4
Plum, Monitor	Red	The largest red plum	Yes	4
Plum, Mount Royal	Blue	Extra hardy	Yes	4
Plum, Ozark Premier	Red	Yellow flesh	No	4
Plum, Pipestone	Red	Large fruit, yellow flesh	Yes	4
Plum, Santa Rosa	Purple-red	Very prolific, compact growing	Yes	5
Plum, Sapa	Green	Small fruit, good sauce variety	Yes	4
Plum, Shropshire Damson	Purple	Yellow flesh, tart flavor	Yes	#
Plum, Stanley Prune	Purple	Freestone eating or canning variety	Yes	5
Plum, Superior	Red	Yellow flesh, very prolific	Yes	4
Plum, Toka	Red	Good pollinator, yellow flesh	Yes	4
Plum, Underwood	Red	Large fruit, freestone	Yes	4
Plum, Waneta	Red	Yellow flesh, extremely sweet	Yes	4

Figure 7.12.2   (*cont.*)

7.12.3 Nut Trees—Varieties and Characteristics

Variety	Nut Size	Characterictics	Pollinator Required	Hardiness Zone
Pecan, Cherokee	Small to medium	Very vigorous and productive	Yes	6
Pecan, Cheyenne	Small to medium	Highly productive	Yes	6
Pecan, Choctaw	Medium	Unusually thin shelled	Yes	6
Pecan, Colby	Small to medium	Early maturity	Yes	5
Pecan, Comanche	Small	Vigorous grower	Yes	6
Pecan, Desirable	Small to medium	Very thin shelled, highly productive	Yes	6
Pecan, Mahan	Large	Disease resistant (largest of all pecan nuts)	Yes	6
Pecan, Mohawk	Large	Very thin shelled	Yes	6
Pecan, Major	Medium	Vigorous grower	No	5
Pecan, Peruque	Medium	Thin shelled	No	5
Pecan, Shawnee	Medium	Vigorous grower	Yes	6
Pecan, Stuart	Large	Scab resistant, highly productive	No	6
Pecan, Success	Large	Scab resistant, highly productive	No	6
Pecan, Western Chley	Large	Vigorous growth, highly productive	No	6
Pecan, Wichita	Large	Early maturity, highly productive	Yes	6
Walnut, Carpathian	Large	Vigorous growing, disease free	Yes	3
Walnut, Thomas Black	Medium	Vigorous growing, thin shelled	Yes	4

7.12.4 Conifers—Botanical and Common Names

Common Name	Botanical Name	Hardiness Zone	Height	Spread (ft)	Shape	Growth Rate	Foliage Texture	Unusual Characteristics
Arborvitae, Berckman's Gold	*Thuja orientalis* "Aurea Nana"	6	6–8'	4	Egg shape	Moderate	Soft	Dwarf egg shaped evergreen, soft green foliage with bright golden tips
Arborvitae, Excelsa	*Thuja orientalis* "Excelsa"	6	10–12'	4	Pyramid	Rapid	Soft	Bright green foliage arranged in vertical planes, compact in habit
Arborvitae, Pyramidal	*Thuja occidentalis* "Pyramidalis"	3	20–30'	6	Pyramid	Moderate	Soft lacy	Soft lacy green foliage, minimal trimming required
Arborvitae, Emerald Green	*Thuja occidentalis* "Smaragd"	3	15–25'	4	Pyramid	Moderate	Soft lacy	Neat appearance, narrow pyramidal form, dense foliage, emerald green color
Arborvitae, Techni	*Thuja occidentalis* "Techney"	3	15–25'	6	Pyramid	Moderate	Soft lacy	Open pyramidal form, green foliage
Arborvitae, Woodwardi Globe	*Thuja accidentalis* "Woodwardii"	3	4–5'	4	Glove	Slow	Soft lacy	Small dense rounded evergreen, bright green foliage, slow growing
Juniper, Admiral	*Juniperus scopulorum* "Admiral"	4	20–25'	6	Pyramid	Moderate	Plume-like	Broad based and bushy growth habit, gray-green color
Juniper, Ames	*Juniperus chinensis* "Ames"	4	4–6'	4	Pyramid	Slow	Medium	Dwarf dense pyramidal juniper, broad base, green color
Juniper, Andorra	*Juniperus horizontalis* "Plumosa"	2	1–2'	10	Pyramid	Moderate	Plume-like	Low growing, flat top look, green foliage
Juniper, Andorra Youngstown	*Juniperu horizontalis* "Plumosa Youngstown"	2	1–2'	10	Low compact	Moderate	Plume-like	Densely branched prostrate grower, flattened branch structure

Figure 7.12.2 *(By permission: McGraw-Hill, The Site Calculations Pocket Reference, Ed Hannon)*

Common Name	Botanical Name	Hardiness Zone	Height	Spread (ft)	Shape	Growth Rate	Foliage Texture	Unusual Characteristics
Juniper, Armstrong	*Juniperus chinensis* "Armstrong" *Juniperus horizontalis*	4	4'	4	Medium compact	Slow	Lacy	Medium size, light green lacy juniper, dense branching habit
Juniper, Bar Harbor	*Juniperus horizontalis* "Bar Harbor"	3	1'	10'	Ground cover	Moderate	Feathery	Feathery blue foliage, plum color in winter, good ground hugger
Juniper, Blue Chip	*Juniperus horizontalis* "Blue Chip"	4	1'	6'	Ground cover	Moderate	Medium	Silver-blue low mounding juniper, excellent ground cover
Juniper, Blue Danube	*Juniperus sabina* "Blue Danube"	3	2'	4	Low compact	Moderate	Soft	Semi-erect branching, greenish blue foliage, good spreader
Juniper, Blue Haven	*Juniperus scopulorum* "Blue Haven"	4	20'	5	Pyramid	Moderate	Medium	Neat compact narrow blue pyramid, likes exposed locations
Juniper, Blue Pacific	*Juniperus scopulorum* "Blue Pacific"	6	1–2'	6–8	Ground cover	Rapid .	Medium	Blue foliage, superior plant, doesn't brown out during summer heat
Juniper, Blue Pfitzer	*Juniperus chinensis* "Pfitzeriana Glauca"	4	5–6'	10	Upright spreader	Rapid	Medium	Spreading compact habit of growth, silvery blue foliage
Juniper, Blue Point	*Juniperus chinensis* "Blue Point"	4	10–15'	4	Pyramid	Slow	Needle-like	Blue-gray foliage, dense pyramidal form, tolerant to heat and poor soil
Juniper, Blue Rug	*Juniperus horizontalis* "Wiltoni"	4	5"	10	Ground cover	Moderate	Medium	Dense short branchlets on long trailing branch, intense blue color
Juniper, Blue Sargent	*Juniperus chinensis* "Sargenti Glauca"	4	8–10"	6–8	Ground cover	Moderate	Medium	Broad spreading, heavily branched, ground hugging form, rich blue-green color
Juniper, Blue Star	*Juniperus squamata* "Blue Star"	4	3'	5	Irregular mounding	Slow	Needle-like	Dwarf form, irregular growth habit, mounding steel blue needle-like foliage
Juniper, Broadmoor	*Juniperus sabina* "Broadmoor"	3	1–2'	10'	Mounding	Moderate	Soft	Low growing, attractive mounding, green foliage, takes full exposure
Juniper, Buffalo	*Juniperus sabina* "Buffalo"	3	1'	8	Ground cover	Moderate	Feathery	Very hardy, feathery branches create unusual spreading form, bright green color
Juniper, Burki	*Juniperus virginiana* "Burki"	4	10–20'	6–8'	Pyramid	Moderate	Medium	Dense steel blue foliage, broad based, responds well to trimming, very hardy
Juniper, Calgary Carpet	*Juniperus sabina* "Calgary Carpet"	3	1'	6'	Ground cover	Moderate	Soft	Attractive foliage is soft green, low growth habit.
Juniper, Canaerti	*Juniperus virginiana* "Canaerti"	4	30'	8–10'	Pyramid	Slow	Medium	Slow growing, wide based, rich dark green foliage with silver berries in the fall
Juniper, Cologreen	*Juniperus scopulorum* "Cologreen"	3	20'	6–8'	Pyramid	Rapid	Medium	Compact upright grower, broad based forest green foliage.
Juniper, Columnaris	*Juniperus chinensis* "Hetzi Columnaris"	3	12–15'	4'	Pyramid	Rapid	Medium	Rapid growing, narrow based pyramid, good green color
Juniper, Compact Irish	*Juniperus communis* "Hibernica"	6	12–20'	4–6'	Pyramid	Rapid	Needle-like	Dark green foliage, very narrow base, erect grower
Juniper, Compact Pfitzer Nicks	*Juniperus chinensis pfitzeriana* "Compact Nicks"	4	2'	4–6'	Medium compact	Moderate	Medium	Gray-green color, graceful branching, very hardy, compact in growth
Juniper, Dundee	*Juniperus virginiana* "Hillii"	4	10'	4–5'	Pyramid	Slow	Soft	Blue green foliage, compact habit of growth, changes color in fall
Juniper, Goldtip Pfitzer	*Juniperus chinensis pfitzerianna* "Aurea"	3	4'	4–6'	Compact spreader	Rapid	Soft	Medium size plant, greenish yellow foliage with golden yellow tips

Figure 7.12.2 *(cont.)*

Common Name	Botanical Name	Hardiness Zone	Height	Spread (ft)	Shape	Growth Rate	Foliage Texture	Unusual Characteristics
Juniper, Gray Gleam	*Juniperus scopulorum* "Gray Gleam"	4	10–20'	4–6'	Pyramid	Slow	Soft	Very dense columnar with gray blue foliage
Juniper, Hetzi	*Juniperus chinensis* "Hetzi Glauca"	4	15'	15'	Medium	Rapid	Medium	Rapid growing, medium to large plant, bright blue green foliage
Juniper, Hollywood	*Juniper chinensis* "Torulosa"	6	15'	4–5'	Upright	Moderate	Medium	Rich green foliage, upright with twisted branching, fine accent plant
Juniper, Hughes	*Juniperus horizontalis* "Hughes"	4	2'	6–8'	Medium compact	Moderate	Feathery	Blue green foliage, requires little shearing, good for ground cover & slopes
Juniper, Keteleeri	*Juniperus chinensis* "Keteleeri"	5	15–20'	4–5'	Pyramid	Moderate	Scalelike	Vigorous pyramid with compact ascending branches, bright green foliage
Juniper, Medora	*Juniperus scopulorum* "Medora"	3	20'	6–8'	Pyramid	Moderate	Medium	Requires very little trimming, silver blue, upright
Juniper, Montana Green	*Juniperus scopulorum* "Montana Green"	3	10–15'	6'	Pyramid	Moderate	Soft	Gray green foliage, tends to have a wide base with age
Juniper, Mount Batten	*Juniperus scopulorum* "Montana Green"	3	12'	6–8'	Pyramid	Moderate	Needle-like	Gray green in color, broad columnar with needlelike foliage
Juniper, Old Gold	*Juniperus chinensis* "Armstrong Aurea"	4	4'	4'	Compact spreader	Slow	Medium	Rich gold coloring, dense compact growth, excellent mid-size plant
Juniper, Parsoni	*Juniperus squamata* "Parsonii"	4	1'	3–4'	Ground cover	Rapid	Soft	Grows full and thick, hugging fashion, lovely gray green foliage
Juniper, Pathfinder	*Juniperus scopulorum* "Pathfinder"	3	15–20'	3	Pyramid	Rapid	Medium	Showy blue-green foliage, likes full sun and low humidity, desirable for dry soil
Juniper, Pfitzer	*Junifer chinensis* "Pfitzeriana"	3	5'	12	Large spreader	Rapid	Medium	Rapid growing, open spreading shrub of medium height
Juniper, Prince Of Wales	*Juniperus horizontalis* "Prince of wales"	3	8"	8–10	Ground cover	Rapid	Soft	Bright green foliage snakes over ground or cascades over walls and ledges
Juniper, Procumbens Green mound	*Juniperus procumbens* "Greenmound"	4	8"	6	Ground cover	Slow	Needle-like	Ground cover, light green, foliage, short needle-like foliage, likes full sun, compact
Juniper, Procumbens	*Juniperus procumbens*	4	1–2'	6	Ground cover	Moderate	Needle-like	Gray-green in color, feathery foliage, on strong spreading branches
Juniper, Procumbens Dwarf	*Juniperus procumbens* "Nana"	4	1'	4–5	Ground cover	Slow	Needle-like	Low spreader with heavy compact ascending branches
Juniper, Variegated Procumbens	*Juniperus procumbens* "Variegata"	6	1–2'	6	Ground cover	Moderate	Needle-like	Low spreading with gray-green foliage, ascending branches
Juniper, Prostrate	*Juniperus communis* "Prostrata"	5	18"	6	Ground cover	Moderate	Scale-like	Stiff branching with dark green scale-like foliage
Juniper, Pyramidalis Green	*Juniperus scopulorum* "Pyramidalis"	4	15–30'	6	Pyramid	Moderate	Needle-like	Foliage narrow pyramid, broadening with age
Juniper, San Jose	*Juniperus japonica* "San Jose"	4	1–2'	6–8	Ground cover	Moderate	Needle-like	Thicky branched full growing, excellent ground cover and bonsai training
Juniper, Scandia	*Juniperus sabina* "Scandia"	3	18"	8	Low spreader	Rapid	Feathery	Good green color year round, attractive low growing spreader
Juniper, Sea Green	*Juniperus chinensis* "Sea Green"	4	4'	6	Vase	Moderate	Medium	Rich green coloring, arching branches, great eye appeal plant
Juniper, Shore	*Juniperus conferta*	6	1'	6–8	Ground cover	Rapid	Medium	Excellent prostrate juniper, does well in poor, sandy soil, bluish green foliage
Juniper, Sneedi	*Juniperus scopulorum* "Sneedi"	4	15–20'	4–6	Pyramid	Rapid	Medium	Columnar and rather coarse growing foliage, compact gray-blue

Figure 7.12.2 *(cont.)*

Common Name	Botanical Name	Hardiness Zone	Height	Spread (ft)	Shape	Growth Rate	Foliage Texture	Unusual Characteristics
Juniper, Spartan	*Juniperus chinensis densaerecta* "Spartan"	4	15-20'	6	Pyramid	Rapid	Needle-like	Rapid growth with a densely branched tall pyramidal form, rich green foliage
Juniper, Spiney Green	*Juniperus excelsa stricta*	5	10–15'	6–8	Pyramid	Rapid	Spiney	Short cone shaped evergreen, dense gray-green foliage, erect branching
Juniper, springtime	*Juniperus scopulorum* "Springtime"	4	15–20'	4–6	Pyramid	Moderate	Soft	Good dark green color, very dense, pyrqamidal
Juniper, Sutherland	*Juniperus scopulorum* "Sutherland"	4	20'	6	Pyramid	Moderate	Medium	Strong growing broad upright juniper, silver-green in color, requires light trimming
Juniper, Tamarix	*Juniperus sabina* "Tamariscifolia"	4	2'	10	Mounding	Moderate	Medium	Medium spreading ground cover, good green color year round
Juniper, Von Ehron	*Juniperus sabina* "Von Ehron"	3	6–8'	10–15	Vase	Rapid	Soft feathery	Rapid growing evergreen with loose upright branching, greenish color
Juniper, Welchi	*Juniperus scopulorum* "Welchi"	3	15–20'	4–5	Pyramid	Rapid	Soft	Dense growing, silver-green foliage, outstanding narrow upright
Juniper, Whicita Blue	*Juniperus acopulorum* "Whicita Blue"	4	20'	4-5	Pyramid	Moderate	Medium	Bright blue foliage, wide based upright
Cedar, Deodara	*Cedrus deodara*	7	40–60'	20–30	Upright	Moderate	Fine-needle	Dense growth habit, widely spreading branches which droop at the tips, needles 1 ½" long, green to blue in color
Pine, Austrian	*Pinus Nigra*	3	90'	25	Upright	Moderate	Needle	Grows practically anywhere, dark gren needles 4–6" long, withstands adverse conditions, ideal for windbreak or as a specimen tree
Pine, Japanes Black	*Pinus Thunbergi*	5	80–100'	25	Upright	Moderate	Needle	Rugged look, irregular branching, good specimen tree, picturesque
Pine, Loblolly	*Pinus taeda*	6	80–100'	30	Upright	Rapid	Needle	Soft light green 6–9" needles, winter color is pale green to yellow brown
Pine, Mugho	*Pinus mugo mighus*	3	6–8'	10–15	Mounded	Slow	Needle	Bright green dwarf mounded pine, good in garden type landscape
Pine, Ponderosa	*Pinus ponderosa*	3	100–125'	40	Upright	Slow	Needle	Gray-green, 6–16" long needles, open growing, adapts to adverse conditions
Pine, Scotch	*Pinus sysvestris*	3	75'	25	Upright	Moderate	Needle	Irregular shaped, needles 2–3" long and twisted, orange brown bark, resists drought
Pine, Slash	*Pinus elliotti*	7	80–100'	30–40	Upright	Rapid	Needle	Narrowly pyramidal when young, more oval with age, needles 10" long medium green
Pine, White	*Pinus strobus*	3	80–100'	25–30	Upright	Moderate	Soft needle	Soft green graceful tree, when young it is pyramidal in habit, but as it grows older becomes open and picturesque
Spruce, Alberta Dwarf	*Picea glauca* "Conica"	3	6'	2'	Cone	Slow	Small needle	Slow dense conifer that grows in a cone shape without shearing, new growth is bright green, may take 10 years to reach 3' in height.
Spruce, Black Hills	*Picea glauca densata*	3	60'	25'	Upright	Slow	Needle	Slow growing, short needled tree, blue green in color.

Figure 7.12.2   *(cont.)*

Common Name	Botanical Name	Hardiness Zone	Height	Spread (ft)	Shape	Growth Rate	Foliage Texture	Unusual Characteristics
Spruce, Colorado Blue	*Picea pungens glauca*	3	100'	30'	Upright	Slow	Needle	Most popular spruce, often steel blue in color, desirable specimen.
Spruce, Nest	*Pices abies* "Nidiformis"	3	2'	2'	Semi-globe	Slow	Needle	Main branches arch outward, leaving a central depression resembling a birds nest, slow growing, ideal for rock gardens.
Yew, Dark Green Spreader	*Taxus x media* Dark Green Spreader	3	3–4'	4–5'	Medium spreader	Slow	Needle	Hardiest variety grown, ¾" dark green needles, excellent foundation plant.
Yew, Densiformis	*Taxus x media* "Densiformis"	3	3–4'	4–5'	Medium spreader	Slow	Needle	Most versatile and popular yew, 3/4" needles excellent foundation plant.
Yew, Hicksi	*Taxus x media* "Hicksi"	3	6'	4'	Upright	Slow	Needle	Upright growing yew, deep green needles, bright red berries in fall, extremely hardy in north, good for hedging
Yew, Browni	*Taxus x media* "Brownii"	3	3'	4'	Globe	Slow	Needle	Compact dense with medium dark green needles, can be sheared to a formal globe shape, very hardy

Figure 7.12.2   (*cont.*)

7.12.5 Broadleaf Trees—Botanical and Common Names

Common Name	Botanical Name	Hardiness Zone	Height	Shape	Growth Rate	Foliage Texture	Fall Foliage	Fruit	Unusual Characteristics
Abelia, Grandifora	*Abelia grandiflora*	6	4–8'	Compact	Rapid	Medium	None	None	Semi-green, dark purple flowers in spring
Aucuba, Gold Dust Plant	*Aucuba japonica* "Variegata"	6	6'	Compact	Rapid	Smooth	Berries	Red berries	Golden spotted, dark green glossy leaves
Aucuba, Picturata	*Aucuba japonica* "Picturata"	7	6'	Compact	Rapid	Smooth	None	None	Deep green leaves with golden yellow center
Azalea, Exbury	*Azalea exburry* hybrids	Varies	Varies	Upright	Moderate	Medium Fine	None	None	Extremely winter hardy, deciduous
Azalea, Kurume	*Azalea kurume*	6	2–4'	Shrubby	Moderate	Leathery	None	None	Evergreen
Boxwood, Japanese	*Buxus japonica*	7	4'	Shrubby	Slow	Smooth	None	None	Very effective hedge
Boxwood, Korean	*Buxus microphylla* "Koreana"	5	18–24'	Compact	Moderate	Smooth	None	None	Excellent hedge plant
Boxwood, Winter Gem	*Buxus microphylla asiaticum* "Winter Gem"	6	2–3'	Oval	Moderate	Smooth	None	None	Excellent hedge plant
Elaeagnus, Fruitlandi (silverberry)	*Eleagnus pungens* "Fruitlandi"	7	10–12'	Upright	Rapid	Coarse	Fragrant flowers	None	Dainty fragrant flowers in fall
Euonymus, Boxleaf	*Euonymus japonica* "Microphylla"	6	2'	Compact	Moderate	Smooth	None	None	Excellent hedge or border
Euonymus, Coloratus	*Euonymus fortunei* "Colorata"	4	6"	Low mounding	Rapid	Coarse	Plum color	None	Good ground cover, plum color foliage in fall
Euonymus, Emerald Gaiety	*Euonymus fortunei* "Emerald Gaiety"	5	4'	Erect	Moderate	Smooth	Purple	None	White margin on green rounded foliage
Euonymus, Emerald 'N Gold	*Euonymus fortunei* "Emerals 'N Gold"	5	3'	Mounding	Medium	Smooth	None	None	Deep green foliage with small green center

Figure 7.12.2   (*By permission: McGraw-Hill, The Site Calculations Pocket Reference, Ed Hannon*)

Common Name	Botanical Name	Hardiness Zone	Height	Shape	Growth Rate	Foliage Texture	Fall Foliage	Fruit	Unusual Characteristics
Euonymus, Golden	*Euonymus japonica* "Microphylla" "Aureo Marginata"	6	5'	Upright	Moderate to rapid	Smooth	None	None	Large golden foliage with small green center
Euonymus, Gold Spot	*Eunymus japonica* "Aureo Variegata"	6	3–4'	Upright	Moderate to rapid	Smooth	None	None	Large green foliage, gold in center
Euonymus, Green	*Eunymus japonica*	6	5'	Upright	Moderate to rapid	Smooth	None	None	Excellent hedge
Euonymus, Green Crusher	*Euonymus fortunei* "Green Crusher"	4	5'	Upright	Moderate to rapid	Coarse	None	None	Similar to sarcoxie in appearance, lighter colored foliage
Euonymus, Manhattan	*Eunoymus patens* "Manhattan"	6	5–6'	Upright	Rapid	Glossy	None	None	One of the best semi-evergreen shrubs
Euonymus, Pauli	*Eunoymus patens* "Pauli"	5	5–6'	Upright	Rapid	Smooth	None	None	Hardiest of patens varieties
Euonymus, Sarcoxie	*Euonymus fortunei* "Sarcoxie"	5	3'	Upright	Vigorous	Glossy	None	None	Hardy and useful as a foundation shrub
Euonymus, Sheridan Gold	*Euonymus fortunei* "Sarcoxie" "Sheridan Gold"	4	3'	Spreading	Moderate	Smooth	None	None	Deep green foliage with yellow color planted in full sun
Euonymus, Silver King	*Euonymus japonica* "Silver King"	6	8'	Upright	Rapid	Leathery	None	None	Large green foliage edged in creamy white
Euonymus, Vegetus	*Euonymus fortunei* "vegetus"	4	2'	Vine or semi-shrub	Moderate	Coarse	Orange fruits	None	Pink fruit capsules open in fall to reveal orange fruit
Grass, Mondo	*Ophiopogon japonicum*	6	10"	Clumpy	Fast	Slow	Grass-like	None	Used as bed edging and a ground cover
Grass, Pampas Pink	*Cortaderia selloana* "Rosia"	6	10'	Upright	Fast	Rapid	Serrated leaves	None	Produces beautiful pink plumes in fall
Grass, Pampas White	*Cortaderia selloana*	6	10'	Upright	Fast	Rapid	Serrated leaves	None	Produces white plumes in fall
Hawthorn, Indian	*Raphiolepis indica*	7	5'	Upright	Moderate	Leathery	None	None	White to pink flowers in summer
Holly, Blue Prince P.P. #3517	*Iiex x meserveae* "Blue Prince"	5	15'	Pyramidal	Rapid	Crinkled	None	None	Male pollinator for the two female varieties
Holly, Blue Angel P.P. #3662	*Iiex x meserveae* "Blue Angel"	5	6–8'	Shrubby	Rapid	Crinkled	Berries	Red berries	Hardy, vigorous, dense branching
Holly, Blue Princess P.P. #3675	*Iiex x meserveae* "Blue Princess"	5	15'	Pyramidal	Rapid	Crinkled	Berries	Red berries	Excellent for a high, dense hedge
Holly, China Boy P.P. #4803	*Iiex x meserveae* "China Boy"	5	6–8'	Upright	Rapid	Glossy	None	None	Shears and shapes well
Holly, China Girl P.P. #4878	*Iiex x meservea* "China Girl"	5	6–8'	Upright	Rapid	Glossy	None	Red berries	Shears and shapes well
Holly, Burford	*Iiex cornuta* "Burfordi"	6	6'	Upright	Moderate	Glossy	None	Red berries	Deep glossy foliage
Holly, Dazzler	*Iiex cornuta* "Dazzler"	6	8'	Upright	Moderate	Spiney	None	Red berries	Good hedge or specimen
Holly, Dwarf Burford	*Iiex cornuta* "Burfordi Nana"	6	3'	Upright	Slow	Glossy	None	Red berries	Smaller leaves than Burford
Holly, Dwarf Chinese	*Iiex cornuta* "Rotunda"	6	3–4'	Mounding	Medium	Spiney leaves	None	None	Mounding with spiney, glossy green foliage
Holly, Dwarf Yaupon	*Iiex vomitoria* "Nana"	6	2–3'	Compact	Slow	Smooth	None	None	Compact with small gray-green foliage
Holly, Fosteri	*Iiex opaca (attenuata)* "Fosteri"	6	20'	Pyramidal	Medium	Smooth	None	Large red berries	Fruits heavily with male
Holly, Green lustre	*Iiex crenata* "Green Lustre"	6	4'	Compact	Medium	Shiny	None	None	Ideal for specimen or foundation plantings
Holly, Helleri	*Iiex crenata* "Helleri"	6	3'	Compact	Slow	Smooth	None	None	Tiny dark green foliage tp ⅜" long
Holly, Japanese Compact	*Iiex crenata* "Compacta"	6	3–4'	Oval	Moderate	Glossy	None	Blue-black berries	Minutes shiny green foliage
Holly, Japanese Convex	*Iiex crenata* "Convexa"	6	4–6'	Mounding	Moderate	Smooth	None	None	Round leaves that cup downward

Figure 7.12.2   *(cont.)*

Common Name	Botanical Name	Hardiness Zone	Height	Shape	Growth Rate	Foliage Texture	Fall Foliage	Fruit	Unusual Characteristics
Holly, Japanese Hetzi	*Ilex crenata* "Hetzi"	6	5–6'	Upright	Vigorous	Glossy	Berries	Black berries	Dark green convex foliage
Holly, Needlepoint	*Ilex cornuta* "Needlepoint"	6	10'	Upright	Moderate	Lustrous	Berries	Dark red berries	Long narrow dark green foliage
Holly, Nellie R. Stevens	*Ilex hybrid* "Nellie R. Stevens"	6	8–10'	Pyramidal	Rapid	Smooth	Berries	Red berries	Early fall berries
Holly, Rotundifolia	*Ilex crenata* "Rotundifolia"	6	3'	Upright	Moderate	Glossy	None	None	Round foliage to ¼" long
Ivy, English	*Hedera helix*	5		Spreading	Rapid	Leathery	None	None	Favorite ground cover
Jasmine, Carolina	*Gelsemium sempervirens*	7		Vine or mounding	Rapid	Smooth	None	None	Very showy, yellow fragrant flower in early spring
Jasmine, Nudiflorum	*Jasminum nudiflorum*	7		Viney	Moderate	Smooth	None	Yellow flowers	Use as a trailing or cascading planting
Ligustrum, Waxleaf	*Ligustrum texanum*	7	8–10'	Upright	Rapid	Glossy	None	None	White flowers in spring
Liriope, Green	*Liriope muscari*	6	18"	Clumpy	Fast	Moderate	Grass-like	None	Excellent for borders Purple flowers
Liriope, Variegated	*Liriope muscari* "variegata"	6	18"	Clumpy	Fast	Moderate	Grass-like	None	Foliage is yellow striped on outer margin Purple flowers
Nadina, Dwarf Purpurea	*Nandina domestica* "Nana Purpurea"	6	2'	Mounding	Slow	Smooth	Scarlet	None	Excellent for planter boxes
Nandina Domestica	*Nandina domestica*	6	5–6'	Compact	Rapid	Lacy	Red	Red berries	Foliage turns several shades of red in fall
Nandina, Harbour Dwarf	*Nandina domestica* "Harbour Dwarf"	6	18–20"	Mound	Moderate	Lacy	Red	Red berries	Red foliage in fall
Photinia, Freseri	*Photinia fraseri*	6	12–15'	Upright	Moderate	Glossy	Red	None	New foliage
Pittosporum, Tobira	*Pittosporum tobira*	8	6–10'	Upright	Moderate	Leathery	None	White to yellow flowers	Best for screens, massing or individual as free standing small tree
Pittosposporum, Variegated	*Pittosporum tobira* "Variegata"	8	4–5'	Upright	Moderate	Leathery	None	None	Makes an attractive trimmed hedge
Pittosporum, Wheeler's Dwarf	*Pittosporum tobira* "Wheeler's Dwarf"	8	4–5'	Compact	Moderate	Leathery	None	None	Fragrant white flowers
Pyracantha, Kasan	*Pyracantha coccinea* "Kasan"	5	6–10'	Upright	Rapid	Smooth	Berries	Orange-red berries	Retain fruit for a long period, white flower in spring
Pyracantha, Lelandi	*Pyracantha coccinea* "Lelandi"	5	12–15'	Upright	Rapid	Smooth	Berries	Orange-red berries	Strong growing
Pyracantha, Mohave	*Pyracantha, hybrid* "Mohave"	6	8–10'	Upright	Rapid	Smooth	Berries	Orange-red berries	Produces orange-red berries in the fall
Pyracantha, Victory	*Pyracantha Koidzummi* "Victory"	7	6–10'	Upright	Rapid	Smooth	Berries	Orange-red berries	Fire blight resistant, produces orange-red berries in fall
Pyracantha, Wyatti	*Pyracantha coccinea* "Wyattii"	5	6'	Mounding	Moderate	Smooth	Berries	Orange berries	Foundation & hedge plantings, produces orange berries in fall
Rhododendron	*Rhododendron*	Varries	2–4'	Upright	Moderate	Leathery	None	None	Acid soil and good drainage are required
Yucca, Adam's Needle	*Yucca filamentosa*	4	12'	Upright	Moderate	Spiney leaves	None	None	Produces bold, stiff, swordlike rosettes

Figure 7.12.2 *(cont.)*

Contents

8.0.0 Introduction to the 1975 Metric Conversion Act

As the federal government moves to convert the inch-pound units to the metric system, in accordance with the 1975 Metric Conversion Act, various parts of the construction industry will begin the conversion to this more universal method of measurement.

Metric units are often referred to as *SI units*, an abbreviation taken from the French: le Système International d'Unités. Another abbreviation that will be seen with more frequency is ISO—the International Standards Organization charged with supervising the establishment of a universal standards system. For everyday transactions it may be sufficient to gain only the basics of the metric system.

Name of metric unit	Symbol	Approximate size (length/pound)
meter	m	39½ inches
kilometer	km	0.6 mile
centimeter	cm	width of a paper clip
millimeter	mm	thickness of a dime
hectare	ha	2½ acres
square meter	m²	1.2 square yards
gram	g	weight of a paper clip
kilogram	kg	2.2 pounds
metric ton	t	long ton (2240 pounds)
liter	L	one quart and two ounces
milliliter	mL	⅕ teaspoon
kilopascal	kPa	atmospheric pressure is about 100 Pa

The Celsius temperature scale is used. Instead of referring to its measurement as *degree centigrade*, the term *degree Celsius* is the correct designation. Using this term, familiar points are

- Water freezes at 0 degrees
- Water boils at 100 degrees
- Normal body temperature is 37 degrees (98.6 F)
- Comfortable room temperature 20 to 35 (68 to 77 F)

8.1.0 What Will Change and What Will Stay the Same?

What will change:

- The basic building module, from 4 inches to 100 mm.
- The planning grid, from 2' × 2' to 600 × 600 mm.

What will stay the same:

- A module and grid based on rounded, easy-to-use dimensions. The 100 mm module is the global standard.

Drawings

What will change:

- Units, from feet and inches to millimeters for all building dimensions and to meters for site plans and civil engineering drawings. Unit designations are unnecessary: if there is no decimal point, it is millimeters; if there is a decimal point carried to one, two, or three places, it is meters. In accordance with ASTM E621, centimeters are not used in construction because (1) they are not consistent with the preferred use of multiples of 1000, (2) the order of magnitude between a millimeter and centimeter is only 10 and the use of both units would lead to confusion and require the use of unit designations, and 93) the millimeter is small enough to almost entirely eliminate decimal fractions from construction documents.

- Drawing scales, from inch-fractions-to-feet to true rations. Preferred metric scales are:

1:1 (full size)

1:5 (close to 3" = 1'-0")

1:10 (between 1" = 1'-0" and 1½" = 1'-0")

1:20 (between ½" = 1'-0" and ¾" = 1'-0")

1:50 (close to ¼" = 1'-0")

1:100 (close to ⅛" = 1'-0")

1:200 (close to ¹⁄₁₆" = 1'-0")

1:500 (close to 1" = 40'-0")

1:1000 (close to 1" = 80'-0")

As a means of comparison, inch-fraction scales may be converted to true ratios by multiplying a scale's divisor by 12; for example, for ¼" = 1'-0", multiply the 4 by 12 for a true ratio of 1:48.

- Drawing sizes, to ISO "A" series:

A0 (1189 × 841 mm, 46.8 × 33.1 inches)

A1 (841 × 594 mm, 33.1 × 23.4 inches)

A2 (594 × 420 mm, 23.4 × 16.5 inches)

A3 (420 × 297 mm, 16.5 × 11.7 inches)

A4 (297 × 210 mm, 11.7 × 8.3 inches)

Of course, metric drawings can be made on any size paper.

What will stay the same:

- Drawing contents

Never use dual units (both inch-pound and metric) on drawings. It increases dimensioning time, doubles the chance for errors, makes drawings more confusing, and only postpones the learning process. An exception is for construction documents meant to be viewed by the general public.

Specifications

What will change:

- Units of measure, from feet and inches to millimeters for linear dimensions, from square feet to square meters for area, from cubic yards to cubic meters for volume (except use liters for fluid volumes), and from other inch-pound measures to metric measures as appropriate.

What will stay the same:

- Everything else in the specifications

Do not use dual units in specifications except when the use of an inch-pound measure serves to clarify an otherwise unfamiliar metric measure; then place the inch-pound unit in parentheses after the metric. For example, "7.5 kW (10 horsepower) motor." All unit conversions should be checked by a professional to ensure that rounding does not exceed allowable tolerances.

For more information, see the July–August 1994 issue of *Metric in Construction.*

Floor Loads

What will change:

- Floor load designations, from "psf" to kilograms per square meter (kg/m^2) for everyday use and kilonewtons per square meter (kN/m^2) for structural calculations.

What will stay the same:

- Floor load requirements

Kilograms per square meter often are used to designate floor loads because many live and dead loads (furniture, filing cabinets, construction materials, etc.) are measured in kilograms. However, kilonewtons per square meter or their equivalent, kilopascals, are the proper measure and should be used in structural calculations.

Construction Products

What will change:

- Modular products: brick, block, drywall, plywood, suspended ceiling systems, and raised floor systems. They will undergo "hard" conversion; that is, their dimensions will change to fit the 100 mm module.
- Products that are custom-fabricated or formed for each job (for example, cabinets, stairs, handrails, ductwork, commercial doors and windows, structural steel systems, and concrete work). Such products usually can be made in any size, inch-pound or metric, with equal ease; therefore, for metric jobs, they simply will be fabricated or formed in metric.

What will stay the same:

- All other products, since they are cut-to-fit at the jobsite (for example, framing lumber, woodwork, siding, wiring, piping, and roofing) or are not dimensionally sensitive (for example, fasteners, hardware, electrical components, plumbing fixtures, and HVAC equipment). Such products will just be "soft" converted—that is, relabeled in metric units. A 2¾" × 4½" wall switch face plate will be relabeled 70 × 115 mm and a 30 gallon tank, 114 L. Manufacturers eventually may convert the physical dimensions of many of these products to new rational "hard" metric sizes but only when it becomes convenient for them to do so.

"2-By-4" Studs and Other "2-By" Framing (Both Wood and Metal)

What will change:

- Spacing, from 16" to 400 mm, and 24" to 600 mm.

What will stay the same:

- Everything else.

"2-bys" are produced in "soft" fractional inch dimensions so there is no need to convert them to new rounded "hard" metric dimensions. 2-by-4s may keep their traditional name or perhaps they will eventually be renamed 50 by 100 (mm), or, more exactly, 38 × 39.

Drywall, Plywood, and Other Sheet Goods

What will change:

- Widths, from 4'-0" to 1200 mm.
- Heights, from 8'-0" to 2400 mm, 10'-0" to 3000 mm.

What will stay the same:

• Thicknesses, so fire, acoustic, and thermal ratings will not have to be recalculated.

Metric drywall and plywood are readily available but may require longer lead times for ordering and may cost more in small amounts until their use becomes more common.

Batt Insulation

What will change:

• Nominal width labels, from 16" to 16"/400 mm and 24" to 24"/600 mm.

What will stay the same:

• Everything else.

Batts will not change in width, they will just have a tighter "friction fit" when installed between metric-spaced framing members.

Doors

What will change:

• Height, from 6'-8" to 2050 mm or 2100 mm and from 7'-0" to 2100 mm.
• Width, from 2'-6" to 750 mm, from 2'-8" to 800 mm, from 2'-10" to 850 mm, from 3"-0" to 900 mm or 950 mm, and from 3'-4" to 1000 mm.

What will stay the same:

• Door thicknesses.
• Door materials and hardware.

For commercial work, doors and door frames can be ordered in any size since they normally are custom-fabricated.

Ceiling Systems

What will change:

• Grids and lay-in ceiling tile, air diffusers and recessed lighting fixtures, from 2' × 2' to 600 × 600 mm and from 2' × 4' to 600 × 1200 mm.

What will stay the same:

• Grid profiles, tile thicknesses, air diffuser capacities, fluorescent tubes, and means of suspension.

On federal building projects, metric recessed lighting fixtures may be specified if their total installed costs are estimated to be more than for inch-pound fixtures.

Raised Floor Systems

What will change:

• Grids and lay-in floor tile, from 2' × 2' to 600 × 600 mm.

What will stay the same:

• Grid profiles, tile thicknesses, and means of support.

HVAC Controls

What will change:

- Temperature units, from Fahrenheit to Celsius.

What will stay the same:

- All other parts of the controls.

Controls are now digital so temperature conversions can be made with no difficulty.

Brick

What will change:

- Standard brick, to 90 × 57 × 190 mm.
- Mortar joints, from ⅜" and ½" to 10 mm.
- Brick module, from 2' × 2' to 600 × 600 mm.

What will stay the same:

- Brick and mortar composition.

Of the 100 or so brick sizes currently made, 5 to 10 are within a millimeter of a metric brick so the brick industry will have no trouble supplying metric brick.
For more information, see the March–April 1995 issue of *Metric in Construction.*

Concrete Block

What will change:

- Block sizes, to 190 × 190 × 390 mm.
- Mortar joints, from ½" to 10 mm.
- Block module, from 2' × 2' to 600 × 600 mm.

What will stay the same:

- Block and mortar composition.

On federal building projects, metric block may be specified if its total installed cost is estimated to be more than for inch-pound block. The Construction Metrication Council recommends that, wherever possible, block walls be designed and specified in a manner that permits the use of either inch-pound or metric block, allowing the final decision to be made by the contractor.

Sheet Metal

What will change:

- Designation, from "gage" to millimeters.

What will stay the same:

- Thickness, which will be soft-converted to tenths of a millimeter.

In specifications, use millimeters only or millimeters with the gage in parentheses.

Concrete

What will change:

• Strength designations, from "psi" to megapascals, rounded to the nearest 5 megapascals per ACI 318M as follows:

2500 psi to 20 MPa

3000 psi to 25 MPa

3500 psi to 25 MPa

4000 psi to 30 MPa

4500 psi to 35 MPa

5000 psi to 35 MPa

Depending on exact usage, however, the above metric conversions may be more exact than those indicated.

What will stay the same:

• Everything else.

For more information, see the November–December 1994 issue of *Metric in Construction.*

Rebar

What will change:

• Rebar will not change in size but will be renamed per ASTM A615M-96a and ASTM A706M-96a as follows:

No. 3 to No. 10   No. 9 to No. 29

No. 4 to No. 13   No. 10 to No. 32

No. 5 to No. 16   No. 11 to No. 36

No. 6 to No. 19   No. 14 to No. 43

No. 7 to No. 22   No. 18 to No. 57

No. 8 to No. 25

What will stay the same:

• Everything else.

For more information, see the July–August 1996 issue of *Metric on Construction.*

Glass

What will change:

• Nominal pipe and fitting designations, from inches to millimeters.

What will stay the same:

• Pipe and fitting cross sections and threads.

Pipes and fittings are produced in "soft" decimal-inch dimensions but are identified in nominal-inch sizes as a matter of convenience. A 2-inch pipe has neither an inside nor an outside diameter of

2 inches, a 1-inch fitting has no exact 1-inch dimension, and a ½-inch sprinkler head contains no ½-inch dimension anywhere; consequently, there is no need to "hard" convert pipes and fittings to rounded metric dimensions. Instead, they will not change size but simply be relabeled in metric as follows:

⅛" = 6 mm	1½" = 40 mm
³⁄₁₆" = 7 mm	2" = 50 mm
¼" = 8 mm	2½" = 65 mm
⅜" = 10 mm	3" = 75 mm
½" = 15 mm	3½" = 90 mm
⅝" = 18 mm	4" = 100 mm
¾" = 20 mm	4½" = 115 mm
1" = 25 mm	1" = 25 mm for all larger sizes
1¼" = 32 mm	

For more information, see the September–October 1993 issue of *Metric in Construction*.

Electrical Conduit

What will change:

• Nominal conduit designations, from inches to millimeters.

What will stay the same:

• Conduit cross sections.

Electrical conduit is similar to piping: it is produced in "soft" decimal-inch dimensions but is identified in nominal-inch sizes. Neither metallic nor nonmetallic conduit will change size; they will be relabeled in metric units as follows:

½" = 16 (mm)	2½" = 63 (mm)
¾" = 21 (mm)	3" = 78 (mm)
1" = 27 (mm)	3½" = 91 (mm)
1¼" = 35 (mm)	4" = 103 (mm)
1½" = 41 (mm)	5" = 129 (mm)
2" = 53 (mm)	6" = 155 (mm)

These new metric names were assigned by the National Electrical Manufacturers Association.

Electrical Wire

What will change:

• Nothing at this time.

What will stay the same:

• Existing American Wire Gage (AWG) sizes.

Structural Steel

What will change:

• Section designations, from inches to millimeters and from pounds per foot to kilograms per meter, in accordance with ASTM A6M.

• Bolts—to metric diameters and threads per ASTM A325M and A490M.

What will stay the same:

• Cross sections.

Like pipe and conduit, steel sections are produced in "soft" decimal-inch dimensions (with actual depths varying by weight) but are named in rounded-inch dimensions so there is no need to "hard" convert them to metric units. Rather, their names will be changed to metric designations, and rounded to the nearest 10 mm. Thus, a 10-inch section is relabeled as a 250-mm section and a 24-inch section is relabeled as a 610-mm section.

8.2.0 How Metric Units Will Apply in the Construction Industry

	Quantity	Unit	Symbol
Masonry	length	meter, millimeter	m, mm
	area	square meter	m²
	mortar volume	cubic meter	m³
Steel	length	meter, millimeter	m, mm
	mass	megagram (metric ton) kilogram	Mg (t) kg
	mass per unit length	kilogram per meter	kg/m
Carpentry	length	meter, millimeter	m, mm
Plastering	length	meter, millimeter	m, mm
	area	square meter	m²
	water capacity	liter (cubic decimeter)	L (dm³)
Glazing	length	meter, millimeter	m, mm
	area	square meter	m²
Painting	length	meter, millimeter	m, mm
	area	square meter	m²
	capacity	liter (cubic decimeter) milliliter (cubic centimeter)	L (dm³) mL (cm³)
Roofing	length	meter, millimeter	m, mm
	area	square meter	m²
	slope	percent ratio of lengths	% mm/mm, m/m
Plumbing	length	meter, millimeter	m, mm
	mass	kilogram, gram	kg, g
	capacity	liter (cubic decimeter)	L (dm³)
	pressure	kilopascal	kPa
Drainage	length	meter, millimeter	m, mm
	area	hectare (10 000 m2) square meter	ha m²
	volume	cubic meter	m³
	slope	percent ratio of lengths	% mm/mm, m/m
HVAC	length	meter, millimeter	m, mm
	volume (capacity)	cubic meter liter (cubic decimeter)	m³ L (dm³)
	air velocity	meter/second	m/s
	volume flow	cubic meter/second liter/second (cubic decimeter per second)	m³/s L/s (dm³/s)
	temperature	degree Celsius	°C
	force	newton, kilonewton	N, kN
	pressure	pascal, kilopascal	Pa, kPa
	energy	kilojoule, megajoule	kJ, MJ
	rate of heat flow	watt, kilowatt	W, kW
Electrical	length	millimeter, meter, kilometer	mm, m, km
	frequency	hertz	Hz
	power	watt, kilowatt	W, kW
	energy	megajoule kilowatt hour	MJ kWh
	electric current	ampere	A
	electric potential	volt, kilovolt	V, kV
	resistance	milliohm, ohm	mΩ, Ω

8.3.0 Metrification of Pipe Sizes

Pipe diameter sizes can be confusing because their designated size does not correspond to their actual size. For instance, a 2-inch steel pipe has an inside diameter of approximately 2⅛ inches and an outside diameter of about 2⅝ inches.

The *2 inch* designation is very similar to the 2" × 4" designation for wood studs, neither dimensions are "actual," but they are a convenient way to describe these items.

Pipe sizes are identified as *NPS (nominal pipe size)* and their conversion to metric would conform to ISO (International Standards Organization) criteria and are referred to as *DN (diameter nominal)*. These designations would apply to all plumbing, mechanical, drainage, and miscellaneous pipe commonly used in civil works projects.

NPS size	DN size
⅛"	6 mm
³⁄₁₆"	7 mm
¼"	8 mm
⅜"	10 mm
½"	15 mm
⅝"	18 mm
¾"	20 mm
1"	25 mm
1¼"	32 mm
1½"	40 mm
2"	50 mm
2½"	65 mm
3"	80 mm
3½"	90 mm
4"	100 mm
4½"	115 mm
5"	125 mm
6"	150 mm
8"	200 mm
10"	250 mm
12"	300 mm
14"	350 mm
16"	400 mm
18"	450 mm
20"	500 mm
24"	600 mm
28"	700 mm
30"	750 mm
32"	800 mm
36"	900 mm
40"	1000 mm
44"	1100 mm
48"	1200 mm

NPS size	DN size
52"	1300 mm
56"	1400 mm
60"	1500 mm

For all pipe over 60-inches nominal, use 1 inch equals 25 mm.

8.4.0 Metrification of Standard Lumber Sizes

Metric units: ASTM Standard E-380 was used as the authoritative standard in developing the metric dimensions in this standard. Metric dimensions are calculated at 25.4 millimeters (mm) times the actual dimension in inches. The nearest mm is significant for dimensions greater than $1/8$ inch, and the nearest 0.1 mm is significant for dimensions equal to ore less than $1/8$ inch.

The rounding rule for dimensions greater than $1/8$ inch: If the digit in the tenth of mm position (the digit after the decimal point) is less than 5, drop all fractional mm digits; if it is greater than 5 or if it is 5 followed by at least one nonzero digit, round one mm higher; if 5 followed by only zeroes, retain the digit in the unit position (the digit before the decimal point) if it is even, or increase it one mm if it is odd.

The rounding rule for dimensions equal to or less than $1/8$ inch: if the digit in the hundredths of mm position (the second digit after the decimal point) is less than 5, drop all digits to the right of the tenth position; if greater than or it is 5 followed by at least one nonzero digit, round one-tenth mm higher; if 5 followed by only zeros, retain the digit in the tenths position if it is even or increase it one-tenth mm if it is odd.

In case of a dispute on size measurements, the conventional (inch) method of measurement shall take precedence.

8.5.0 Metric Rebar Conversions

A615 M-96a & A706M-96a Metric Bar Sizes	Nominal Diameter	A615-96a & A706-96a Inch-Pound Bar Sizes
#10	9.5 mm/0.375"	#3
#13	12.7 mm/0.500"	#4
#16	15.9 mm/0.625"	#5
#19	19.1 mm/0.750"	#6
#22	22.2 mm/0.875"	#7
#25	25.4 mm/1.000"	#8
#29	28.7 mm/1.128"	#9
#32	32.3 mm/1.270"	#10
#36	35.8 mm/1.410"	#11
#43	43.0 mm/1.693"	#14
#57	57.3 mm/2.257"	#18

8.6.0 Metric Conversion of ASTM Diameter and Wall Thickness Designations

Metric conversion of ASTM diameter designations

in	mm	in	mm	in	mm	in	mm
6	150	30	750	57	1425	96	2400
8	200	33	825	60	1500	102	2550
10	250	36	900	63	1575	108	2700
12	300	39	975	66	1650	114	2850
15	375	42	1050	69	1725	120	3000
18	450	45	1125	72	1800	132	3300
21	525	48	1200	78	1950	144	3600
24	600	51	1275	84	2100	156	3900
27	675	54	1350	90	2250	168	4200

Metric conversion of ASTM wall thickness designations

in	mm	in	mm	in	mm	in	mm
1	25	3-1/8	79	5	125	8	200
1-1/2	38	3-1/4	82	5-1/4	131	8-1/2	213
2	50	3-1/2	88	5-1/2	138	9	225
2-1/4	56	3-3/4	94	5-3/4	144	9-1/2	238
2-3/8	59	3-7/8	98	6	150	10	250
2-1/2	63	4	100	6-1/4	156	10-1/2	263
2-5/8	66	4-1/8	103	6-1/2	163	11	275
2-3/4	69	4-1/4	106	6-3/4	169	11-1/2	288
2-7/8	72	4-1/2	113	7	175	12	300
3	75	4-3/4	119	7-1/2	188	12-1/2	313

8.7.0 Metric Conversion Scales (Temperature and Measurements)

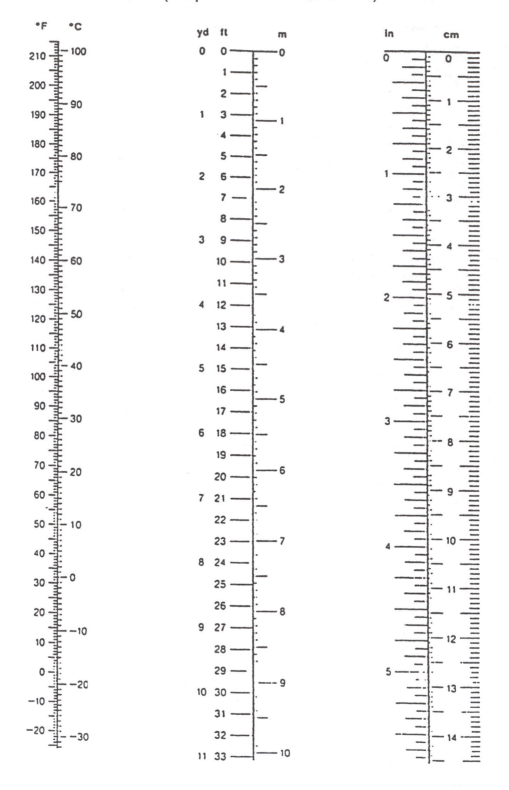

8.8.0 Approximate Metric Conversions

Symbol	When You Know	Multiply by	To Find	Symbol
LENGTH				
mm	millimeters	0.04	inches	in
cm	centimeters	0.4	inches	in
m	meters	3.3	feet	ft
m	meters	1.1	yards	yd
km	kilometers	0.6	miles	mi
AREA				
cm^2	square centimeters	0.16	square inches	in^2
m^2	square meters	1.2	square yards	yd^2
km^2	square kilometers	0.4	square miles	mi^2
ha	hectares $(10,000\ m^2)$	2.5	acres	
MASS (weight)				
g	grams	0.035	ounces	oz
kg	kilograms	2.2	pounds	lb
t	metric ton $(1,000\ kg)$	1.1	short tons	
VOLUME				
mL	milliliters	0.03	fluid ounces	fl oz
mL	milliliters	0.06	cubic inches	in^3
L	liters	2.1	pints	pt
L	liters	1.06	quarts	qt
L	liters	0.26	gallons	gal
m^3	cubic meters	35	cubic feet	ft^3
m^3	cubic meters	1.3	cubic yards	yd^3
TEMPERATURE (exact)				
°C	degrees Celsius	multiply by 9/5, add 32	degrees Fahrenheit	°F

°C -40 -20 0 20 37 60 80 100
°F -40 0 32 80 98.6 160 212

water freezes   body temperature   water boils

(U.S. Department of Commerce Technology Administration. Office of Metric Programs, Washington, DC 20230.)

8.8.0 Approximate Metric Conversions (Continued)

Symbol	When You Know	Multiply by	To Find	Symbol
LENGTH				
in	inches	2.5	centimeters	cm
ft	feet	30	centimeters	cm
yd	yards	0.9	meters	m
mi	miles	1.6	kilometers	km
AREA				
in^2	square inches	6.5	square centimeters	cm^2
ft^2	square feet	0.09	square meters	m^2
yd^2	square yards	0.8	square meters	m^2
mi^2	square miles	2.6	square kilometers	km^2
	acres	0.4	hectares	ha
MASS (weight)				
oz	ounces	28	grams	g
lb	pounds	0.45	kilograms	kg
	short tons (2000 lb)	0.9	metric ton	t
VOLUME				
tsp	teaspoons	5	milliliters	mL
Tbsp	tablespoons	15	milliliters	mL
in^3	cubic inches	16	milliliters	mL
fl oz	fluid ounces	30	milliliters	mL
c	cups	0.24	liters	L
pt	pints	0.47	liters	L
qt	quarts	0.95	liters	L
gal	gallons	3.8	liters	L
ft^3	cubic feet	0.03	cubic meters	m^3
yd^3	cubic yards	0.76	cubic meters	m^3
TEMPERATURE (exact)				
°F	degrees Fahrenheit	subtract 32, multiply by 5/9	degrees Celsius	°C

(United States Department of Commerce Technology Administration. National Institute of Standards and Technology, Metric Program. Gaithersburg, Maryland 20899.)

8.9.0 Quick Imperial (Metric Conversions)

Distance

Imperial		Metric		Metric		Imperial	
1 inch	= 2.540	centimetres		1 centimetre	=	0.3937	inch
1 foot	= 0.3048	metre		1 decimetre	=	0.3281	foot
1 yard	= 0.9144	metre		1 metre	=	3.281	feet
1 rod	= 5.029	metres			=	1.094	yard
1 mile	= 1.609	kilometres		1 decametre	=	10.94	yards
				1 kilometre	=	0.6214	mile

Weight

1 ounce (troy)	=	31.103 grams	1 gram	=	0.032 ounce (troy)
1 ounce (avoir)	=	28.350 grams	1 gram	=	0.035 ounce (avoir)
1 pound (troy)	=	373.242 grams	1 kilogram	=	2.679 pounds (troy)
1 pound (avoir)	=	453.592 grams	1 kilogram	=	2.205 pounds (avoir)
1 ton (short)	=	0.907 tonne*	1 tonne	=	1.102 ton (short)

*1 tonne = 1000 kilograms

Capacity

Imperial			U.S.		
1 pint	=	0.568 litre	1 pint (U.S.)	=	0.473 litre
1 gallon	=	4.546 litres	1 quart (U.S.)	=	0.946 litre
1 bushel	=	36.369 litres	1 gallon (U.S.)	=	3.785 litres
1 litre	=	0.880 pint	1 barrel (U.S.)	=	158.98 litres
1 litre	=	0.220 gallon			
1 hectolitre	=	2.838 bushels			

Area

1 square inch	= 6.452 square centimetres
1 square foot	= 0.093 square metre
1 square yard	= 0.836 square metre
1 acre	= 0.405 hectare*
1 square mile	= 259.0 hectares
1 square mile	= 2.590 square kilometres
1 square centimetre	= 0.155 square inch
1 square metre	= 10.76 square feet
1 square metre	= 1.196 square yard
1 hectare	= 2.471 acres
1 square kilometre	= 0.386 square mile

*1 hectare = 1 square hectometre

Volume

1 cubic inch	= 16.387 cubic centimetres
1 cubic foot	= 0.0283 cubic decimetres
1 cubic yard	= 0.765 cubic metre
1 cubic centimetre	= 0.061 cubic inch
1 cubic decimetre	= 35.314 cubic foot
1 cubic metre	= 1.308 cubic yard

8.10.0 Metric Conversion Factors

The following list provides the conversion relationship between U.S. customary units and SI (International System) units. The proper conversion procedure is to multiply the specified value on the left (primarily U.S. customary values) by the conversion factor exactly as given below and then round to the appropriate number of significant digits desired. For example, to convert 11.4 ft to meters: 11.4 × 0.3048 = 3.47472, which rounds to 3.47 meters. Do not round either value before performing the multiplication, as accuracy would be reduced. A complete guide to the SI system and its use can be found in ASTM E 380, Metric Practice.

To convert from	to	multiply by
Length		
inch (in.)	micron (μ)	25,400 E°
inch (in.)	centimeter (cm)	2.54 E
inch (in.)	meter (m)	0.0254 E
foot (ft)	meter (m)	0.3048 E
yard (yd)	meter (m)	0.9144
Area		
square foot (sq ft)	square meter (sq m)	0.09290304 E
square inch (sq in.)	square centimeter (sq cm)	6.452 E
square inch (sq in.)	square meter (sq m)	0.00064516 E
square yard (sq yd)	square meter (sq m)	0.8361274
Volume		
cubic inch (cu in.)	cubic centimeter (cu cm)	16.387064
cubic inch (cu in.)	cubic meter (cu m)	0.00001639
cubic foot (cu ft)	cubic meter (cu m)	0.02831685
cubic yard (cu yd)	cubic meter (cu m)	0.7645549
gallon (gal) Can. liquid	liter	4.546
gallon (gal) Can. liquid	cubic meter (cu m)	0.004546
gallon (gal) U.S. liquid**	liter	3.7854118
gallon (gal) U.S. liquid	cubic meter (cu m)	0.00378541
fluid ounce (fl oz)	milliliters (ml)	29.57353
fluid ounce (fl oz)	cubic meter (cu m)	0.00002957
Force		
kip (1000 lb)	kilogram (kg)	453.6
kip (1000 lb)	newton (N)	4,448.222
pound (lb) avoirdupois	kilogram (kg)	0.4535924
pound (lb)	newton (N)	4.448222
Pressure or stress		
kip per square inch (ksi)	megapascal (MPa)	6.894757
kip per square inch (ksi)	kilogram per square centimeter (kg/sq cm)	70.31
pound per square foot (psf)	kilogram per square meter (kg/sq m)	4.8824
pound per square foot (psf)	pascal (Pa)†	47.88
pound per square inch (psi)	kilogram per square centimeter (kg/sq cm)	0.07031
pound per square inch (psi)	pascal (Pa)†	6,894.757
pound per square inch (psi)	megapascal (MPa)	0.00689476
Mass (weight)		
pound (lb) avoirdupois	kilogram (kg)	0.4535924
ton, 2000 lb	kilogram (kg)	907.1848
grain	kilogram (kg)	0.0000648

To convert from	to	multiply by
Mass (weight) per length		
kip per linear foot (klf)	kilogram per meter (kg/m)	0.001488
pound per linear foot (plf)	kilogram per meter (kg/m)	1.488
Mass per volume (density)		
pound per cubic foot (pcf)	kilogram per cubic meter (kg/cu m)	16.01846
pound per cubic yard (lb/cu yd)	kilogram per cubic meter (kg/cu m)	0.5933
Temperature		
degree Fahrenheit (°F)	degree Celsius (°C)	$t_C = (t_F - 32)/1.8$
degree Fahrenheit (°F)	degree Kelvin (°K)	$t_K = (t_F + 459.7)/1.8$
degree Kelvin (°K)	degree Celsius (C°)	$t_C = t_K - 273.15$
Energy and heat		
British thermal unit (Btu)	joule (J)	1055.056
calorie (cal)	joule (J)	4.1868 E
Btu/°F · hr · ft²	W/m² · °K	5.678263
kilowatt-hour (kwh)	joule (J)	3,600,000. E
British thermal unit per pound (Btu/lb)	calories per gram (cal/g)	0.55556
British thermal unit per hour (Btu/hr)	watt (W)	0.2930711
Power		
horsepower (hp) (550 ft-lb/sec)	watt (W)	745.6999 E
Velocity		
mile per hour (mph)	kilometer per hour (km/hr)	1.60934
mile per hour (mph)	meter per second (m/s)	0.44704
Permeability		
darcy	centimeter per second (cm/sec)	0.000968
feet per day (ft/day)	centimeter per second (cm/sec)	0.000352

*E indicates that the factor given is exact.
**One U.S. gallon equals 0.8327 Canadian gallon.
†A pascal equals 1.000 newton per square meter.

Note:
One U.S. gallon of water weighs 8.34 pounds (U.S.) at 60°F.
One cubic foot of water weighs 62.4 pounds (U.S.).
One milliliter of water has a mass of 1 gram and has a volume of one cubic centimeter.
One U.S. bag of cement weighs 94 lb.

The prefixes and symbols listed below are commonly used to form names and symbols of the decimal multiples and submultiples of the SI units.

Multiplication Factor	Prefix	Symbol
1,000,000,000 = 10^9	giga	G
1,000,000 = 10^6	mega	M
1,000 = 10^3	kilo	k
1 = 1	—	—
0.01 = 10^{-2}	centi	c
0.001 = 10^{-3}	milli	m
0.000001 = 10^{-6}	micro	μ
0.000000001 = 10^{-9}	nano	n

Useful Tables, Charts and Formulas

Contents

9.0.0 Nails: Penny Designation ("d") and Lengths (U.S. and Metric)

Nail—penny size	Length in inches	Length in millimeters
2d	1	25.40
3d	1 1/4	31.75
4d	1 1/2	38.10
5d	1 3/4	44.45
6d	2	50.80
7d	2 1/4	57.15
8d	2 1/2	63.50
9d	2 3/4	69.85
10d	3	76.20
12d	3 1/4	82.55
16d	3 1/2	88.90
20d	3 3/4	95.25
30d	4 1/2	114.30
40d	5	127.00
50d	5 1/2	139.70
60d	6	152.40

9.1.0 Stainless Steel Sheets (Thickness and Weights)

Gauge	Thickness inches	mm.	Weight lb/ft^2	kg/m^2
8	0.17188	4.3658	7.2187	44.242
10	0.14063	3.5720	5.9062	28.834
11	0.1250	3.1750	5.1500	25.6312
12	0.10938	2.7783	4.5937	22.427
14	0.07813	1.9845	3.2812	16.019
16	0.06250	1.5875	2.6250	12.815
18	0.05000	1.2700	2.1000	10.252
20	0.03750	0.9525	1.5750	7.689
22	0.03125	0.7938	1.3125	6.409
24	0.02500	0.6350	1.0500	5.126
26	0.01875	0.4763	0.7875	3.845
28	0.01563	0.3970	0.6562	3.1816
Plates				
3/16"	0.1875	4.76	7.752	37.85
1/4"	0.25	6.35	10.336	50.46
5/16"	0.3125	7.94	12.920	63.08
3/8"	0.375	9.53	15.503	75.79
1/2"	0.50	12.70	20.671	100.92
5/8"	0.625	15.88	25.839	126.15
3/4"	0.75	19.05	31.007	151.38
1"	1.00	25.4	41.342	201.83

9.2.0 Comparable Thicknesses and Weights of Stainless Steel, Aluminum, and Copper

STAINLESS STEEL				ALUMINUM				COPPER		
Thickness (Inch)	Gauge (U.S. Standard)	Lb/sq ft		Thickness (Inch)	Gauge (B&S)	Lb/sq ft		Thickness (Inch)	Oz sq ft	Lb/sq ft
.010	32	.420		.010	30	.141		.0108	8	.500
.0125	30	.525		.0126	28	.177		.0121	9	.563
								.0135	10	.625
.0156	28	.656		.0156		.220		.0148	11	.688
				.0179	25	.253		.0175	13	.813
.0187	26	.788								
.0219	25	.919		.020	24	.282		.021	16	1.000
.025	24	1.050		.0253	22	.352				
								.027	20	1.250
.031	22	1.313		.0313	—	.441		.032	24	1.500
.0375	20	1.575		.032	20	.451		.0337	28	1.750
				.0403	18	.563		.0431	32	2.000
				.0453	17	.100				
.050	18	2.100		.0506	16	.126				

Note that U.S. Standard Gauge (stainless sheet) is not directly comparable with the B&S Gauge (aluminum). A 20-gauge stainless averages .0375" thick; while a 20-gauge aluminum averages .032" thick; and 20-ounce copper is .027" thick. The higher strength of stainless steel permits use of thinner gauges than required for aluminum or copper, which makes stainless more competitive with aluminum on a weight-to-coverage basis and provides stainless with a substantial weight saving compared to copper. For example, 100 sq ft of .032" aluminum will weigh about 45 pounds, .021" (16-ounce) copper will weigh about 100 pounds, and .015" stainless will weigh about 66 pounds.

9.3.0 Wire and Sheetmetal Gauges and Weights

Name of Gage	*United States Standard Gage		The United States Steel Wire Gage	American or Brown & Sharpe Wire Gage	New Birmingham Standard Sheet & Hoop Gage	British Imperial or English Legal Standard Wire Gage	Birmingham or Stubs Iron Wire Gage	Name of Gage
Principal Use	Uncoated Steel Sheets and Light Plates		Steel Wire except Music Wire	Non-Ferrous Sheets and Wire	Iron and Steel Sheets and Hoops	Wire	Strips, Bands, Hoops and Wire	Principal Use
Gage No.	Weight Oz. per Sq. Ft.	Approx. Thickness Inches	Thickness, Inches					Gage No.
7/0's			.4900		.6666	.500		7/0's
6/0's			.4615	.5800	.625	.464		6/0's
5/0's			.4305	.5165	.5883	.432	.550	5/0's
4/0's			.3938	.4600	.5416	.400	.454	4/0's
3/0's			.3625	.3648	.500	.372	.425	3/0's
2/0's			.3310	.3249	.4452	.348	.380	2/0's
1/0			.3065	.2893	.3964	.324	.340	1/0
1			.2830	.2576	.3532	.300	.300	1
2			.2625	.2294	.3147	.276	.284	2
3	160	.2391	.2437	.2043	.2804	.252	.259	3
4	150	.2242	.2253	.1819	.250	.232	.238	4
5	140	.2092	.2070	.1620	.2225	.212	.220	5
6	130	.1943	.1920	.1443	.1981	.192	.203	6
7	120	.1793	.1770	.1285	.1764	.176	.180	7
8	110	.1644	.1620	.1144	.1570	.160	.165	8
9	100	.1495	.1483	.1019	.1398	.144	.148	9
10	90	.1345	.1350	.0907	.1250	.128	.134	10
11	80	.1196	.1205	.0808	.1113	.116	.120	11
12	70	.1046	.1055	.0720	.0991	.104	.109	12
13	60	.0897	.0915	.0641	.0882	.092	.095	13
14	50	.0747	.0800	.0571	.0785	.080	.083	14
15	45	.0673	.0720	.0508	.0699	.072	.072	15
16	40	.0598	.0625	.0453	.0625	.064	.065	16
17	36	.0538	.0540	.0403	.0556	.056	.058	17
18	32	.0478	.0475	.0359	.0495	.048	.049	18
19	28	.0418	.0410	.0320	.0440	.040	.042	19
20	24	.0359	.0348	.0285	.0392	.036	.035	20
21	22	.0329	.0317	.0253	.0349	.032	.032	21
22	20	.0299	.0286	.0226	.0313	.028	.028	22
23	18	.0269	.0258	.0201	.0278	.024	.025	23
24	16	.0239	.0230	.0179	.0248	.022	.022	24
25	14	.0209	.0204	.0159	.0220	.020	.020	25
26	12	.0179	.0181	.0142	.0196	.018	.018	26
27	11	.0164	.0173	.0126	.0175	.0164	.016	27
28	10	.0149	.0162	.0113	.0156	.0148	.014	28
29	9	.0135	.0150	.0100	.0139	.0136	.013	29
30	8	.0120	.0140	.0089	.0123	.0124	.012	30
31	7	.0105	.0132	.0080	.0110	.0116	.010	31
32	6.5	.0097	.0128	.0071	.0098	.0108	.009	32
33	6	.0090	.0118	.0063	.0087	.0100	.008	33
34	5.5	.0082	.0104	.0056	.0077	.0092	.007	34
35	5	.0075	.0095	.0050	.0069	.0084	.005	35
36	4.5	.0067	.0090	.0045	.0061	.0076	.004	36
37	4.25	.0064	.0085	.0040	.0054	.0068		37
38	4	.0060	.0080	.0035	.0048	.0060		38
39			.0075	.0031	.0043	.0052		39
40			.0070		.0039	.0048		40

* U.S. Standard Gage is officially a weight gage, in oz per sq ft as tabulated. The Approx. Thickness shown is the "Manufacturers' Standard" of the American Iron and Steel Institute, based on steel as weighing 501.81 lb per cu ft (489.6 true weight plus 2.5 percent for average over-run in area and thickness).

9.4.0 Weights and Specific Gravities of Common Materials

Substance	Weight Lb per Cu Ft	Specific Gravity	Substance	Weight Lb per Cu Ft	Specific Gravity
METALS, ALLOYS, ORES			**TIMBER, U. S. SEASONED**		
Aluminum, cast, hammered	165	2.55-2.75	Moisture Content by Weight:		
Brass, cast, rolled	534	8.4-8.7	Seasoned timber 15 to 20%		
Bronze, 7.9 to 14% Sn	509	7.4-8.9	Green timber up to 50%		
Bronze, aluminum	481	7.7	Ash, white, red	40	0.62-0.65
Copper, cast, rolled	556	8.8-9.0	Cedar, white, red	22	0.32-0.38
Copper ore, pyrites	262	4.1-4.3	Chestnut	41	0.66
Gold, cast, hammered	1205	19.25-19.3	Cypress	30	0.48
Iron, cast, pig	450	7.2	Fir, Douglas spruce	32	0.51
Iron, wrought	485	7.6-7.9	Fir, eastern	25	0.40
Iron, spiegel-eisen	468	7.5	Elm, white	45	0.72
Iron, ferro-silicon	437	6.7-7.3	Hemlock	29	0.42-0.52
Iron ore, hematite	325	5.2	Hickory	49	0.74-0.84
Iron ore, hematite in bank	160-180	——	Locust	46	0.73
Iron ore, hematite loose	130-160	——	Maple, hard	43	0.68
Iron ore, limonite	237	3.6-4.0	Maple, white	33	0.53
Iron ore, magnetite	315	4.9-5.2	Oak, chestnut	54	0.86
Iron slag	172	2.5-3.0	Oak, live	59	0.95
Lead	710	11.37	Oak, red, black	41	0.65
Lead ore, galena	465	7.3-7.6	Oak, white	46	0.74
Magnesium, alloys	112	1.74-1.83	Pine, Oregon	32	0.51
Manganese	475	7.2-8.0	Pine, red	30	0.48
Manganese ore, pyrolusite	259	3.7-4.6	Pine, white	26	0.41
Mercury	849	13.6	Pine, yellow, long-leaf	44	0.70
Monel Metal	556	8.8-9.0	Pine, yellow, short-leaf	38	0.61
Nickel	565	8.9-9.2	Poplar	30	0.48
Platinum, cast, hammered	1330	21.1-21.5	Redwood, California	26	0.42
Silver, cast, hammered	656	10.4-10.6	Spruce, white, black	27	0.40-0.46
Steel, rolled	490	7.85	Walnut, black	38	0.61
Tin, cast, hammered	459	7.2-7.5	Walnut, white	26	0.41
Tin ore, cassiterite	418	6.4-7.0			
Zinc, cast, rolled	440	6.9-7.2			
Zinc ore, blende	253	3.9-4.2	**VARIOUS LIQUIDS**		
			Alcohol, 100%	49	0.79
			Acids, muriatic 40%	75	1.20
VARIOUS SOLIDS			Acids, nitric 91%	94	1.50
			Acids, sulphuric 87%	112	1.80
Cereals, oats bulk	32	——	Lye, soda 66%	106	1.70
Cereals, barley bulk	39	——	Oils, vegetable	58	0.91-0.94
Cereals, corn, rye bulk	48	——	Oils, mineral, lubricants	57	0.90-0.93
Cereals, wheat bulk	48	——	Water, 4°C. max. density	62.428	1.0
Hay and Straw bales	20	——	Water, 100°C	59.830	0.9584
Cotton, Flax, Hemp	93	1.47-1.50	Water, ice	56	0.88-0.92
Fats	58	0.90-0.97	Water, snow, fresh fallen	8	.125
Flour, loose	28	0.40-0.50	Water, sea water	64	1.02-1.03
Flour, pressed	47	0.70-0.80			
Glass, common	156	2.40-2.60			
Glass, plate or crown	161	2.45-2.72	**GASES**		
Glass, crystal	184	2.90-3.00			
Leather	59	0.86-1.02	Air, 0°C. 760 mm	.08071	1.0
Paper	58	0.70-1.15	Ammonia	.0478	0.5920
Potatoes, piled	42	——	Carbon dioxide	.1234	1.5291
Rubber, caoutchouc	59	0.92-0.96	Carbon monoxide	.0781	0.9673
Rubber goods	94	1.0-2.0	Gas, illuminating	.028-.036	0.35-0.45
Salt, granulated, piled	48	——	Gas, natural	.038-.039	0.47-0.48
Saltpeter	67	——	Hydrogen	.00559	0.0693
Starch	96	1.53	Nitrogen	.0784	0.9714
Sulphur	125	1.93-2.07	Oxygen	.0892	1.1056
Wool	82	1.32			

The specific gravities of solids and liquids refer to water at 4°C, those of gases to air at 0°C and 760 mm. pressure. The weights per cubic foot are derived from average specific gravities, except where stated that weights are for bulk, heaped or loose material, etc.

9.4.0 Weights and Specific Gravities of Common Materials—Continued

Substance	Weight Lb per Cu Ft	Specific Gravity	Substance	Weight Lb per Cu Ft	Specific Gravity
ASHLAR MASONRY			**MINERALS**		
Granite, syenite, gneiss	165	2.3-3.0	Asbestos	153	2.1-2.8
Limestone, marble	160	2.3-2.8	Barytes	281	4.50
Sandstone, bluestone	140	2.1-2.4	Basalt	184	2.7-3.2
			Bauxite	159	2.55
MORTAR RUBBLE MASONRY			Borax	109	1.7-1.8
			Chalk	137	1.8-2.6
Granite, syenite, gneiss	155	2.2-2.8	Clay, marl	137	1.8-2.6
Limestone, marble	150	2.2-2.6	Dolomite	181	2.9
Sandstone, bluestone	130	2.0-2.2	Feldspar, orthoclase	159	2.5-2.6
			Gneiss, serpentine	159	2.4-2.7
DRY RUBBLE MASONRY			Granite, syenite	175	2.5-3.1
Granite, syenite, gneiss	130	1.9-2.3	Greenstone, trap	187	2.8-3.2
Limestone, marble	125	1.9-2.1	Gypsum, alabaster	159	2.3-2.8
Sandstone, bluestone	110	1.8-1.9	Hornblende	187	3.0
			Limestone, marble	165	2.5-2.8
BRICK MASONRY			Magnesite	187	3.0
Pressed brick	140	2.2-2.3	Phosphate rock, apatite	200	3.2
Common brick	120	1.8-2.0	Porphyry	172	2.6-2.9
Soft brick	100	1.5-1.7	Pumice, natural	40	0.37-0.90
			Quartz, flint	165	2.5-2.8
CONCRETE MASONRY			Sandstone, bluestone	147	2.2-2.5
Cement, stone, sand	144	2.2-2.4	Shale, slate	175	2.7-2.9
Cement, slag, etc.	130	1.9-2.3	Soapstone, talc	169	2.6-2.8
Cement, cinder, etc.	100	1.5-1.7			
VARIOUS BUILDING MATERIALS			**STONE, QUARRIED, PILED**		
Ashes, cinders	40-45	------	Basalt, granite, gneiss	96	------
Cement, portland, loose	90	------	Limestone, marble, quartz	95	------
Cement, portland, set	183	2.7-3.2	Sandstone	82	------
Lime, gypsum, loose	53-64	------	Shale	92	------
Mortar, set	103	1.4-1.9	Greenstone, hornblende	107	------
Slags, bank slag	67-72	------			
Slags, bank screenings	98-117	------			
Slags, machine slag	96	------	**BITUMINOUS SUBSTANCES**		
Slags, slag sand	49-55	------	Asphaltum	81	1.1-1.5
			Coal, anthracite	97	1.4-1.7
EARTH, ETC., EXCAVATED			Coal, bituminous	84	1.2-1.5
Clay, dry	63	------	Coal, lignite	78	1.1-1.4
Clay, damp, plastic	110	------	Coal, peat, turf, dry	47	0.65-0.85
Clay and gravel, dry	100	------	Coal, charcoal, pine	23	0.28-0.44
Earth, dry, loose	76	------	Coal, charcoal, oak	33	0.47-0.57
Earth, dry, packed	95	------	Coal, coke	75	1.0-1.4
Earth, moist, loose	78	------	Graphite	131	1.9-2.3
Earth, moist, packed	96	------	Paraffine	56	0.87-0.91
Earth, mud, flowing	108	------	Petroleum	54	0.87
Earth, mud, packed	115	------	Petroleum, refined	50	0.79-0.82
Riprap, limestone	80-85	------	Petroleum, benzine	46	0.73-0.75
Riprap, sandstone	90	------	Petroleum, gasoline	42	0.66-0.69
Riprap, shale	105	------	Pitch	69	1.07-1.15
Sand, gravel, dry, loose	90-105	------	Tar, bituminous	75	1.20
Sand, gravel, dry, packed	100-120	------			
Sand, gravel, wet	118-120	------			
EXCAVATIONS IN WATER			**COAL AND COKE, PILED**		
Sand or gravel	60	------	Coal, anthracite	47-58	------
Sand or gravel and clay	65	------	Coal, bituminous, lignite	40-54	------
Clay	80	------	Coal, peat, turf	20-26	------
River mud	90	------	Coal, charcoal	10-14	------
Soil	70	------	Coal, coke	23-32	------
Stone riprap	65	------			

The specific gravities of solids and liquids refer to water at 4°C, those of gases to air at 0°C and 760 mm. pressure. The weights per cubic foot are derived from average specific gravities, except where stated that weights are for bulk, heaped or loose material, etc.

9.5.0 Useful Formulas

Circumference of a circle = π × diameter or 3.1416 × diameter

Diameter of a circle = circumference × 0.31831

Area of a square = length × width

Area of a rectangle = length × width

Area of a parallelogram = base × perpendicular height

Area of a triangle = ½ base × perpendicular height

Area of a circle = π radius squared or diameter squared × 0.7854

Area of an ellipse = length × width × 0.7854

Volume of a cube or rectangular prism = length × width × height

Volume of a triangular prism = area of triangle × length

Volume of a sphere = diameter cubed × 0.5236 (diameter × diameter × diameter × 0.5236)

Volume of a cone = π × radius squared × ⅓ height

Volume of a cylinder = π × radius squared × height

Length of one side of a square × 1.128 = diameter of an equal circle

Doubling the diameter of a pipe or cylinder increases its capacity 4 times

Pressure (in lb/sq in.) of a column of water = height of the column (in feet) × 0.434

Capacity of a pipe or tank (in U.S. gallons) = diameter squared (in inches) × length (in inches) × 0.0034

1 gal water = 8⅓ lb = 231 cu in.

1 cu ft water = 62½ lb = 7½ gal.

9.6.0 Decimal Equivalents of Inches in Feet and Yards

Inches	Feet	Yards
1	.0833	.0278
2	.1667	.0556
3	.2500	.0833
4	.333	.1111
5	.4166	.1389
6	.5000	.1667
7	.5833	.1944
8	.6667	.2222
9	.7500	.2500
10	.8333	.2778
11	.9166	.3056
12	1.000	.3333

9.7.0 Conversion of Fractions to Decimals

Fractions	Decimal	Fractions	Decimal
1/64	.015625	33/64	.515625
1/32	.03125	17/32	.53125
3/64	.046875	35/64	.546875
1/16	.0625	9/16	.5625
5/64	.078125	37/64	.578125
3/32	.09375	19/32	.59375
7/64	.109375	38/64	.609375
1/8	.125	5/8	.625
9/64	.140625	41/64	.640625
5/32	.15625	21/32	.65625
11/64	.1719	43/64	.67187
3/16	.1875	11/16	.6875
13/64	.2031	45/64	.70312
7/32	.2188	23/32	.71875
15/64	.234375	47/64	.734375
1/4	.25	3/4	.75
17/64	.265625	49/64	.765625
9/32	.28125	25/32	.78125
19/64	.296875	51/64	.796875
5/16	.3125	13/10	.8125
21/64	.328125	53/64	.828125
11/32	.34375	27/32	.84375
23/64	.359375	55/64	.859375
3/8	.375	7/8	.875
25/64	.398625	57/64	.890625
13/32	.40625	29/32	.90625
27/64	.421875	60/64	.921875
7/16	.4375	15/16	.9375
20/64	.453125	61/64	.953125
15/32	.46875	31/32	.96875
31/64	.484375	63/64	.984375
1/2	.50	1″	1.000000

Figure 9.7.0   (*By permission of Cast Iron Soil Pipe Institute.*)

9.7.1 Decimals of a Foot for Each 1/32"

Inch	0	1	2	3	4	5
0	0	.0833	.1667	.2500	.3333	.4167
1/32	.0026	.0859	.1693	.2526	.3359	.4193
1/16	.0052	.0885	.1719	.2552	.3385	.4219
3/32	.0078	.0911	.1745	.2578	.3411	.4245
1/8	.0104	.0938	.1771	.2604	.3438	.4271
5/32	.0130	.0964	.1797	.2630	.3464	.4297
3/16	.0156	.0990	.1823	.2656	.3490	.4323
7/32	.0182	.1016	.1849	.2682	.3516	.4349
1/4	.0208	.1042	.1875	.2708	.3542	.4375
9/32	.0234	.1068	.1901	.2734	.3568	.4401
5/16	.0260	.1094	.1927	.2760	.3594	.4427
11/32	.0286	.1120	.1953	.2786	.3620	.4453
3/8	.0313	.1146	.1979	.2812	.3646	.4479
13/32	.0339	.1172	.2005	.2839	.3672	.4505
7/16	.0365	.1198	.2031	.2865	.3698	.4531
15/32	.0391	.1224	.2057	.2891	.3724	.4557
1/2	.0417	.1250	.2083	.2917	.3750	.4583
17/32	.0443	.1276	.2109	.2943	.3776	.4609
9/16	.0469	.1302	.2135	.2969	.3802	.4635
19/32	.0495	.1328	.2161	.2995	.3828	.4661
5/8	.0521	.1354	.2188	.3021	.3854	.4688
21/32	.0547	.1380	.2214	.3047	.3880	.4714
11/16	.0573	.1406	.2240	.3073	.3906	.4740
23/32	.0599	.1432	.2266	.3099	.3932	.4766
3/4	.0625	.1458	.2292	.3125	.3958	.4792
25/32	.0651	.1484	.2318	.3151	.3984	.4818
13/16	.0677	.1510	.2344	.3177	.4010	.4844
27/32	.0703	.1536	.2370	.3203	.4036	.4870
7/8	.0729	.1563	.2396	.3229	.4063	.4896
29/32	.0755	.1589	.2422	.3255	.4089	.4922
15/16	.0781	.1615	.2448	.3281	.4115	.4948
31/32	.0807	.1641	.2474	.3307	.4141	.4974

9.7.2 Decimals of an Inch for Each ¹⁄₆₄″, with Millimeter Equivalents

Fraction	¹⁄₆₄ths	Decimal	Millimeters (Approx.)	Fraction	¹⁄₆₄ths	Decimal	Millimeters (Approx.)
...	1	.015625	0.397	...	33	.515625	13.097
¹⁄₃₂	2	.03125	0.794	¹⁷⁄₃₂	34	.53125	13.494
...	3	.046875	1.191	...	35	.546875	13.891
¹⁄₁₆	4	.0625	1.588	⁹⁄₁₆	36	.5625	14.288
...	5	.078125	1.984	...	37	.578125	14.684
³⁄₃₂	6	.09375	2.381	¹⁹⁄₃₂	38	.59375	15.081
...	7	.109375	2.778	...	39	.609375	15.478
¹⁄₈	8	.125	3.175	⁵⁄₈	40	.625	15.875
...	9	.140625	3.572	...	41	.640625	16.272
⁵⁄₃₂	10	.15625	3.969	²¹⁄₃₂	42	.65625	16.669
...	11	.171875	4.366	...	43	.671875	17.066
³⁄₁₆	12	.1875	4.763	¹¹⁄₁₆	44	.6875	17.463
...	13	.203125	5.159	...	45	.703125	17.859
⁷⁄₃₂	14	.21875	5.556	²³⁄₃₂	46	.71875	18.256
...	15	.234375	5.953	...	47	.734375	18.653
¹⁄₄	16	.250	6.350	³⁄₄	48	.750	19.050
...	17	.265625	6.747	...	49	.765625	19.447
⁹⁄₃₂	18	.28125	7.144	²⁵⁄₃₂	50	.78125	19.844
...	19	.296875	7.541	...	51	.796875	20.241
⁵⁄₁₆	20	.3125	7.938	¹³⁄₁₆	52	.8125	20.638
...	21	.328125	8.334	...	53	.828125	21.034
¹¹⁄₃₂	22	.34375	8.731	²⁷⁄₃₂	54	.84375	21.431
...	23	.359375	9.128	...	55	.859375	21.828
³⁄₈	24	.375	9.525	⁷⁄₈	56	.875	22.225
...	25	.390625	9.922	...	57	.890625	22.622
¹³⁄₃₂	26	.40625	10.319	²⁹⁄₃₂	58	.90625	23.019
...	27	.421875	10.716	...	59	.921875	23.416
⁷⁄₁₆	28	.4375	11.113	¹⁵⁄₁₆	60	.9375	23.813
...	29	.453125	11.509	...	61	.953125	24.209
¹⁵⁄₃₂	30	.46875	11.906	³¹⁄₃₂	62	.96875	24.606
...	31	.484375	12.303	...	63	.984375	25.003
¹⁄₂	32	.500	12.700	1	64	1.000	25.400

9.8.0 Solutions of the Right Triangle

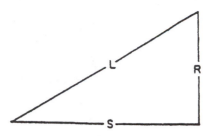

To find side	When you know side	Multiply side	For 45 Ells-By	For 22 1/2 Ells-By	For 67 1/2 Ells-By	For 72 Ells-By	For 60 Ells-By	For 80 Ells-By
L	S	S	1.4142	2.6131	1.08	1.05	1.1547	2.00
S	L	L	.707	.3826	.92	.95	.866	.50
R	S	S	1.000	2.4142	.414	.324	.5773	.1732
S	R	R	1.000	.4142	2.41	3.07	1.732	.5773
L	R	R	1.4142	1.0824	2.61	3.24	2.00	1.1547
R	L	L	.7071	.9239	.38	.31	.50	.866

Figure 9.8.0 (*By permission of Cast Iron Soil Pipe Institute.*)

9.9.0 Area and Other Formulas

Parallelogram	*Area = base × distance between the two parallel sides*
Pyramid	*Area = ½ perimeter of base × slant height + area of base*
	Volume = area of base × ⅓ of the altitude
Rectangle	*Area = length × width*
Rectangular prisms	*Volume = width × height × length*
Sphere	*Area of surface = diameter × diameter × 3.1416*
	Side of inscribed cube = radius × 1.547
	Volume = diameter × diameter × diameter × 0.5236
Square	*Area = length × width*
Triangle	*Area = one half of height times base*
Trapezoid	*Area = one half of the sum of the parallel sides × height*
Cone	*Area of surface = one half of circumference of base × slant height + area of base*
	Volume = diameter × diameter × 0.7854 × one third of the altitude
Cube	*Volume = width × height × length*
Ellipse	*Area = short diameter × long diameter × 0.7854*
Cylinder	*Area of surface = diameter × 3.1416 × length + area of the two bases*
	Area of base = diameter × diameter × 0.7854
	Area of base = volume + length
	Length = volume + area of base
	Volume = length × area of base
	Capacity in gallons = volume in inches + 231
	Capacity of gallons = diameter × diameter × length × 0.0034
	Capacity in gallons = volume in feet × 7.48
Circle	*Circumference = diameter × 3.1416*
	Circumference = radius × 6.2832
	Diameter = radius × 2
	Diameter = square root of = (area + 0.7854)
	Diameter = square root of area × 1.1283

9.10.0 Volume of Vertical Cylindrical Tanks (in U.S. Gallons per Foot of Depth)

Diameter in Feet	Inches	U.S. Gallons	Diameter in Feet	Inches	U.S. Gallons	Diameter in Feet	Inches	U.S. Gallons
1	0	5.875	3	6	71.97	6	0	211.5
1	1	6.895	3	7	75.44	6	3	220.5
1	2	7.997	3	8	78.99	6	6	248.2
1	3	9.180	3	9	82.62	6	9	267.7
1	4	10.44	3	10	86.33	7	0	287.9
1	5	11.79	3	11	90.13	7	3	308.8
1	6	13.22	4	0	94.00	7	6	330.5
1	7	14.73	4	1	97.96	7	9	352.9
1	8	16.32	4	2	102.0	8	0	376.0
1	9	17.99	4	3	106.1	8	3	399.9
1	10	19.75	4	4	110.3	8	6	424.5
1	11	21.58	4	5	114.6	8	9	449.8
2	0	23.50	4	6	119.0	9	0	475.9
2	1	25.50	4	7	123.4	9	3	502.7
2	2	27.58	4	8	127.9	9	6	530.2
2	3	29.74	4	9	132.6	9	9	558.5
2	4	31.99	4	10	137.3	10	0	587.5
2	5	34.31	4	11	142.0	10	3	617.3
2	6	36.72	5	0	146.9	10	6	647.7
2	7	39.21	5	1	151.8	10	9	679.0
2	8	41.78	5	2	156.8	11	0	710.9
2	9	44.43	5	3	161.9	11	3	743.6
2	10	47.16	5	4	167.1	11	6	777.0
2	11	49.98	5	5	172.4	11	9	811.1
3	0	52.88	5	6	177.7	12	0	846.0
3	1	55.86	5	7	183.2	12	3	881.6
3	2	58.92	5	8	188.7	12	6	918.0
3	3	62.06	5	9	194.2	12	9	955.1
3	4	65.28	5	10	199.9			
3	5	68.58	5	11	205.7			

Figure 9.10.0 *(By permission of Cast Iron Soil Pipe Institute.)*

9.11.0 Volume of Rectangular Tank Capacities (in U.S. Gallons per Foot of Depth)

Width Feet	2	2 1/2	3	3 1/2	4	4 1/2	5
2	29.92	37.40	44.88	52.36	59.84	67.32	74.81
2 1/2	—	46.75	56.10	65.45	74.81	84.16	93.51
3	—	—	67.32	78.55	89.77	101.0	112.2
3 1/2	—	—	—	91.64	104.7	117.8	130.9
4	—	—	—	—	119.7	134.6	149.6
4 1/2	—	—	—	—	—	151.5	168.3
5	—	—	—	—	—	—	187.0

Width Feet	5 1/2	6	6 1/2	7	7 1/2	8	8 1/2
2	82.29	89.77	97.25	104.7	112.2	119.7	127.2
2 1/2	102.9	112.2	121.6	130.9	140.3	149.6	159.0
3	123.4	134.6	145.9	157.1	168.3	179.5	190.8
3 1/2	144.0	157.1	170.2	183.3	196.4	209.5	222.5
4	164.6	179.5	194.5	209.5	224.4	239.4	254.3
4 1/2	185.1	202.0	218.8	235.6	252.5	269.3	286.1
5	205.7	224.4	243.1	261.8	280.5	299.2	317.9
5 1/2	226.3	246.9	267.4	288.0	308.6	329.1	349.7
6	—	269.3	291.7	314.2	336.6	359.1	381.5
6 1/2	—	—	316.1	340.4	364.7	389.0	413.3
7	—	—	—	366.5	392.7	418.9	445.1
7 1/2	—	—	—	—	420.8	448.8	476.9
8	—	—	—	—	—	478.8	508.7
8 1/2	—	—	—	—	—	—	540.5

Width Feet	9	9 1/2	10	10 1/2	11	11 1/2	12
2	134.6	142.1	149.6	157.1	164.6	172.1	179.5
2 1/2	168.3	177.7	187.0	196.4	205.7	215.1	224.4
3	202.0	213.2	224.4	235.6	246.9	258.1	269.3
3 1/2	235.6	248.7	261.8	274.9	288.0	301.1	314.2
4	269.3	284.3	299.2	314.2	329.1	344.1	359.1
4 1/2	303.0	319.8	336.6	353.5	370.3	387.1	403.9
5	336.6	355.3	374.0	392.7	411.4	430.1	448.8
5 1/2	370.3	390.9	411.4	432.0	452.6	473.1	493.7
6	403.9	426.4	448.8	471.3	493.7	516.2	538.6
6 1/2	437.6	461.9	486.2	510.5	534.9	559.2	583.5
7	471.3	497.5	523.6	549.8	576.0	602.2	628.4
7 1/2	504.9	533.0	561.0	589.1	617.1	645.2	673.2
8	538.6	568.5	598.4	628.4	658.3	688.2	718.1
8 1/2	572.3	604.1	635.8	667.6	699.4	731.2	763.0
9	605.9	639.6	673.2	706.9	740.6	774.2	807.9
9 1/2	—	675.1	710.6	746.2	781.7	817.2	852.8
10	—	—	748.1	785.5	822.9	860.3	897.7
10 1/2	—	—	—	824.7	864.0	903.3	942.5
11	—	—	—	—	905.1	946.3	987.4
11 1/2	—	—	—	—	—	989.3	1032.0
12	—	—	—	—	—	—	1077.0

Figure 9.11.0 (*By permission of Cast Iron Soil Pipe Institute.*)

9.12.0 Capacity of Horizontal Cylindrical Tanks

% Depth Filled	% of Capacity	% Depth Filled	% of Capacity	% Depth Filled	% of Capacity	% Depth Filled	% of Capacity
1	.20	26	20.73	51	51.27	76	81.50
2	.50	27	21.86	52	52.55	77	82.60
3	.90	28	23.00	53	53.81	78	83.68
4	1.34	29	24.07	54	55.08	79	84.74
5	1.87	30	25.31	55	56.34	80	85.77
6	2.45	31	26.48	56	57.60	81	86.77
7	3.07	32	27.66	57	58.86	82	87.76
8	3.74	33	28.84	58	60.11	83	88.73
9	4.45	34	30.03	59	61.36	84	89.68
10	5.20	35	31.19	60	62.61	85	90.60
11	5.98	36	32.44	61	63.86	86	91.50
12	6.80	37	33.66	62	65.10	87	92.36
13	7.64	38	34.90	63	66.34	88	93.20
14	8.50	39	36.14	64	67.56	89	94.02
15	9.40	40	37.36	65	68.81	90	94.80
16	10.32	41	38.64	66	69.97	91	95.50
17	11.27	42	39.89	67	71.16	92	96.26
18	12.24	43	41.14	68	72.34	93	96.93
19	13.23	44	42.40	69	73.52	94	97.55
20	14.23	45	43.66	70	74.69	95	98.13
21	15.26	46	44.92	71	75.93	96	98.66
22	16.32	47	46.19	72	77.00	97	99.10
23	17.40	48	47.45	73	78.14	98	99.50
24	18.50	49	48.73	74	79.27	99	99.80
25	19.61	50	50.00	75	80.39	100	100.00

Figure 9.12.0 (*By permission of Cast Iron Soil Pipe Institute.*)

9.13.0 Round-Tapered Tank Capacities

$$Volume = \frac{h^3}{3} \frac{[(Area_{Top} + Area_{Base}) + \sqrt{(Area_{Top} + Area_{Base}]}}{231}$$

If inches are used.

$$Volume = \frac{h}{3} [(Area_{Base} + Area_{Top}) + \sqrt{(Area_{Base} + Area_{Top}]} \times 7.48$$

If feet are used.

Sample Problem

Let d be 12" (2 ft)
 D be 36" (3 ft)
 h be 48" (4 ft)
Find volume in gallons.

$$Volume = \frac{48}{3} \frac{[(\pi \times 12^2) + (\pi + 18^2) + \sqrt{\pi\, 12^2 \times 18^2}]}{231}$$

Where dimensions are in inches

$$Volume = \frac{4}{3} [(\pi \times 12^2) + (\pi + 1\frac{1}{2}^2) + \sqrt{(\pi \times 1^2) \times \frac{1}{2}^2)}] \times 7.48$$

Where dimensions are in feet

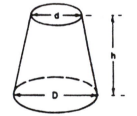

Figure 9.13.0 *(By permission of Cast Iron Soil Pipe Institute.)*

9.14.0   Circumferences and Areas of Circles

	Of One Inch				Of Inches or Feet				
Fract.	Decimal	Circ.	Area	Dia.	Circ.	Area	Dia.	Circ.	Area
1/64	.015625	.04909	.00019	1	3.1416	.7854	64	201.06	3216.99
1/32	.03125	.09818	.00077	2	6.2832	3.1416	65	204.20	3318.31
3/64	.046875	.14726	.00173	3	9.4248	7.0686	66	207.34	3421.19
1/16	.0625	.19635	.00307	4	12.5664	12.5664	67	210.49	3525.65
5/64	.078125	.24545	.00479	5	15.7080	19.635	68	213.63	3631.68
3/32	.09375	.29452	.00690	6	18.850	28.274	69	216.77	3739.28
7/64	.109375	.34363	.00939	7	21.991	38.485	70	219.91	3848.45
1/8	.125	.39270	.01227	8	25.133	50.266	71	223.05	3959.19
9/64	.140625	.44181	.01553	9	28.274	63.617	72	226.19	4071.50
5/32	.15625	.49087	.01917	10	31.416	78.540	73	229.34	4185.50
11/64	.171875	.53999	.02320	11	34.558	95.033	74	232.48	4300.84
3/16	.1875	.58.905	.02761	12	37.699	113.1	75	235.62	4417.86
13/64	.203125	.63817	.03241	13	40.841	132.73	76	238.76	4536.46
7/32	.21875	.68722	.03757	4	43.982	153.94	77	241.90	4656.63
15/64	.234375	.73635	.04314	15	47.124	176.71	78	245.04	4778.36
1/4	.25	.78540	.04909	16	50.265	201.06	79	248.19	4901.67
17/64	.265625	.83453	.05542	17	53.407	226.98	80	251.33	5026.55
9/32	.28125	.88357	.06213	18	56.549	254.47	81	254.47	5153.0
10/64	.296875	.93271	.06922	19	59.690	283.53	82	257.61	5281.02
5/16	.3125	.98175	.07670	20	63.832	314.16	83	260.75	5410.61
21/64	.328125	1.0309	.08456	21	65.973	346.36	84	263.89	5541.77
11/32	.34375	1.0799	.09281	22	69.115	380.13	85	267.04	5674.50
23/64	.35975	1.1291	.10144	23	72.257	415.48	86	270.18	5808.80
3/8	.375	1.1781	.11045	24	75.398	452.39	87	273.32	5944.68
25/64	.390625	1.2273	.11984	25	78.540	490.87	88	276.46	6082.12
13/32	.40625	1.2763	.12962	26	81.681	530.93	89	279.60	6221.14
27/64	.421875	1.3254	.13979	27	84.823	572.56	90	282.74	6361.71
7/16	.4375	1.3744	.15033	28	87.965	615.75	91	258.88	6503.88
29/64	.453125	1.4236	.16126	29	91.106	660.52	92	289.03	6647.61
15/32	.46875	1.4726	.17257	30	94.248	706.86	93	292.17	6792.91
31/64	.484375	1.5218	.18427	31	97.389	754.77	94	295.31	6939.78
1/2	.5	1.5708	.19635	32	100.53	804.25	95	298.45	7088.22

Figure 9.14.0   *(By permission of Cast Iron Soil Pipe Institute.)*

9.14.0 Circumferences and Areas of Circles—Continued

		Of One Inch				Of Inches or Feet			
Fract.	Decimal	Circ.	Area	Dia.	Circ.	Area	Dia.	Circ.	Area
33/64	.515625	1.6199	.20880	33	103.67	855.30	96	301.59	7238.23
17/32	.53125	1.6690	.22166	34	106.81	907.92	97	304.73	7339.81
35/64	.546875	1.7181	.23489	35	109.96	962.11	98	307.88	7542.96
9/16	.5625	1.7671	.24850	36	113.10	1017.88	99	311.02	7697.69
37/64	.578125	1.8163	.26248	37	116.24	1075.21	100	314.16	7853.98
19/32	.59375	1.8653	.27688	38	119.38	1134.11	101	317.30	8011.85
30/64	.609375	1.9145	.29164	39	122.52	1194.59	102	320.44	8171.28
5/8	.625	1.9635	.30680	40	125.66	1256.64	103	323.58	8332.29
41/64	.640625	2.0127	.32232	41	128.81	1320.25	104	326.73	8494.87
21/32	.65625	2.0617	33824	42	131.95	1385.44	105	327.87	8659.01
43/64	.671875	2.1108	.35453	43	135.09	1452.20	106	333.01	8824.73
11/16	.6875	2.1598	.37122	44	138.23	1520.53	107	336.15	1992.02
45/64	.703125	2.2090	.38828	45	141.37	1590.43	108	339.29	9160.88
23/32	.71875	2.2580	.40574	46	144.51	1661.90	109	342.43	9331.32
47/64	.734375	2.3072	.42356	47	147.65	1734.94	110	345.58	9503.32
3/4	.75	2.3562	.44179	48	150.80	1809.56	111	348.72	9676.89
49/64	.765625	2.4050	.45253	49	153.94	1885.74	112	351.86	9853.03
23/32	.78125	2.4544	.47937	50	157.08	1963.50	113	355.0	10028.75
51/64	.796875	2.5036	.49872	51	160.22	2042.82	114	358.14	10207.03
13/16	.8125	2.5525	.51849	52	163.36	2123.72	115	361.28	10386.89
53/64	.828125	2.6017	.53862	53	166.50	2206.18	116	364.42	10568.32
27/32	.84375	2.6507	.55914	54	169.65	2290.22	117	367.57	10751.32
55/64	.859375	2.6999	.58003	55	172.79	2375.83	118	370.71	10935.88
7/8	.875	2.7489	.60123	56	175.93	2463.01	119	373.85	11122.02
57/64	.890625	2.7981	.62298	57	179.07	2551.76	120	376.99	11309 '3
29/32	.90625	2.8471	.64504	58	182.21	2642.08	121	380.13	11499 01
59/64	.921875	2.8963	.66746	59	185.35	2733.97	122	383.27	11689.07
15/16	.9375	2.9452	.69029	60	188.50	2827.43	123	386.42	11882.29
61/64	.953125	2.9945	.71349	61	191.64	2922.47	124	389.56	12076.28
31/32	.96875	3.0434	.73708	62	194.78	3019.07	125	392.70	12271.85
63/64	.984375	3.0928	.76097	63	197.92	3117.25	126	395.84	12468.98

Figure 9.14.0 *(By permission of Cast Iron Soil Pipe Institute.)*

9.15.0   Tap Drill Sizes for Fractional Size Threads

Approximately 65% Depth Thread / AMERICAN NATIONAL THREAD FORM

Tap Size	Threads per Inch	Hole Diameter	Drill	Tap Size	Threads per Inch	Hole Diameter	Drill
1/16	72	.049	3/64	1/2	20	.451	29/64
1/16	64	.047	3/64	1/2	13	.425	27/64
1/16	60	.046	56	1/2	12	.419	27/64
5/64	72	.065	52	9/16	27	.526	17/32
5/64	64	.063	1/16	9/16	18	.508	33/64
5/64	60	.062	1/16	9/16	12	.481	31/64
5/64	56	.061	53	5/8	27	.589	19/32
3/32	60	.077	5/64	5/8	18	.571	37/64
3/32	56	.076	48	5/8	12	.544	35/64
3/32	50	.074	49	5/8	11	.536	17/32
3/32	48	.073	49	11/16	16	.627	5/8
7/64	56	.092	42	11/16	11	.599	19/32
7/64	50	.090	43	3/4	27	.714	23/32
7/64	48	.089	43	3/4	16	.689	11/16
1/8	48	.105	36	3/4	12	.669	43/64
1/8	40	.101	38	3/4	10	.653	21/32
1/8	36	.098	40	13/16	12	.731	47/64
1/8	32	.095	3/32	13/16	10	.715	23/32
9/64	40	.116	32	7/8	27	.839	27/32
9/64	36	.114	33	7/8	18	.821	53/64
9/64	32	.110	35	7/8	14	.805	13/16
5/32	40	.132	30	7/8	12	.794	51/64
5/32	36	.129	30	7/8	9	.767	49/64
5/32	32	.126	1/8	15/16	12	.856	55/64
11/64	36	.145	27	15/16	9	.829	53/64
11/64	32	.141	9/64	1	27	.964	31/32
3/16	36	.161	20	1	14	.930	15/16
3/16	32	.157	22	1	12	.919	59/64
3/16	30	.155	23	1	8	.878	7/8
3/16	24	.147	26	1 1/16	8	.941	15/16
13/64	32	.173	17	1 1/8	12	1.044	1 3/64
13/64	30	.171	11/64	1 1/8	7	.986	63/64
13/64	24	.163	20	1 3/16	7	1.048	1 3/64
7/32	32	.188	12	1 1/4	12	1.169	1 11/64
7/32	28	.184	13	1 1/4	7	1.111	1 7/64
7/32	24	.178	16	1 5/16	7	1.173	1 11/64
15/64	32	.204	6	1 3/8	12	1.294	1 19/64
15/64	28	.200	8	1 3/8	6	1.213	1 7/32
15/64	24	.194	10	1 1/2	12	1.419	1 27/64
1/4	32	.220	7/32	1 1/2	6	1.338	1 11/32
1/4	28	.215	3	1 5/8	5 1/2	1.448	1 29/64
1/4	27	.214	3	1 3/4	5	1.555	1 9/16
1/4	24	.209	4	1 7/8	5	1.680	1 11/16
1/4	20	.201	7	2	4 1/2	1.783	1 25/32
5/16	32	.282	9/32	2 1/8	4 1/2	1.909	1 29/32
5/16	27	.276	J	2 1/4	4 1/2	2.034	2 1/32
5/16	24	.272	I	2 3/8	4	2.131	2 1/8
5/16	20	.264	17/64	2 1/2	4	2.256	2 1/4
5/16	18	.258	F	2 5/8	4	2.381	2 3/8
3/8	27	.339	R	2 3/4	4	2.506	2 1/2
3/8	24	.334	Q	2 7/8	3 1/2	2.597	2 19/32
3/8	20	.326	21/64	3	3 1/2	2.722	2 23/32
3/8	16	.314	5/16	3 1/8	3 1/2	2.847	2 27/32
7/16	27	.401	Y	3 1/4	3 1/2	2.972	2 31/32
7/16	24	.397	X	3 3/8	3 1/4	3.075	3 1/16
7/16	20	.389	25/64	3 1/2	3 1/4	3.200	3 3/16
7/16	14	.368	U	3 5/8	3 1/4	3.325	3 5/16
1/2	27	.464	15/32	3 3/4	3	3.425	3 7/16
1/2	24	.460	29/64	4	3	3.675	3 11/16

9.16.0   Common Material R-Values

R-value is a unit of measure for the rate of heat flow through a given thickness material(s) by conduction. It can include a cavity that incorporates air space reflective insulation. It is measured by the temperature difference between outside surfaces required to cause one **BTU** to flow through one square hour. A **BTU,** (British Thermal Unit), is the amount of heat required to raise temperature of one pound of water 1°F.

MATERIAL	R-value	MATERIAL	R-value	MATERIAL	R-value
1" mineral wool	3.70	3½" fiberglass	13.48	3" honeycomb	2.59
1/2" gypsum	0.45	½" mineral tile	1.19	3" isocyanurate	22.5
1/2" plywood	0.02	1" isocyanurate	7.50	3" polystyrene	12.0
1/8" floor tile	0.05	1" polystyrene	4.00	3" polyurethane	17.6
1/8" hardboard	0.09	1" wood core door	1.96	8" con. block	1.11
3/16" hardboard	0.14	6" fiberglass	19.00	insulated glass	1.65
5/8" gypsum	0.56	1" polyurethane	5.88	single glass pane	0.94

9.17.0   Conversion Factors—Power, Pressure, Energy

Power

Multiply	By	To Get
Boiler hp	33.472	Btu/hr
Boiler hp	34.5	lbs H.O evap. at 212°F
Horsepower	2.540	Btu/hr
Horsepower	550	ft-lb/sec
Horsepower	33.000	ft-lb/min
Horsepower	42.42	Btu/min
Horsepower	0.7457	Kilowatts
Kilowatts	3.415	Btu/hr
Kilowatts	56.92	Btu/min
Watts	44.26	ft-lb/min
Watts	0.7378	ft-lb/sec
Watts	0.05692	Btu/min
Tons refrig.	12.000	Btu/hr
Tons refrig.	200	Btu/min
Btu/hr	0.00002986	Boiler hp
lb H.O evap. at 212°F	0.0290	Boiler hp
Btu/hr	0.000393	Horsepower
ft-lb/sec	0.00182	Horsepower
ft-lb/min	0.0000303	Horsepower
Btu/min	0.0236	Horsepower
Kilowatts	1.341	Horsepower
Btu/hr	0.000293	Kilowatts
Btu/min	0.01757	Kilowatts
ft-lb/min	0.02259	Watts
ft-lb/sec	1.355	Watts
Btu/min	1.757	Watts
Btu/hr	0.0000833	Tons refrig.
Btu/min	0.005	Tons refrig.

Energy

Multiply	By	To Get
Btu	778	ft-lb
Btu	0.000393	hp-hr
Btu	0.000293	kw-hr
Btu	0.0010307	lbs H.O evap. at 212°F
Btu	0.293	Watt-hr
ft-lb	0.3765	Watt-hr
Latent heat of ice	143.33	Btu/lb H$_2$O
lb H.O evap. at 212°F	0.284	kw-hr
lb H.O evap. at 212°F	0.381	hp-hr
ft-lb	0.001287	Btu
hp-hr	2.540	Btu
kw-hr	3.415	Btu
lb H.O evap. at 212°F	970.4	Btu
Watt-hr	3.415	Btu
Watt-hr	2.656	ft-lb
Btu/lb H$_2$O	0.006977	Latent heat of ice
kw-hr	3.52	lb H.O evap. at 212°F
hp-hr	2.63	lb. H.O evap. at 212°F

Pressure

Multiply	By	To Get
atmospheres	29.92	in Mercury (at 62°F)
atmospheres	406.8	in H.O (at 62°F)
atmospheres	33.90	ft. H.O (at 62°F)
atmospheres	14.70	lb/in²
atmospheres	1.058	ton/ft²
in. H.O (at 62°F)	0.0737	in. Mercury (at 62°F)
ft H.O (at 62°F)	0.881	in. Mercury (at 62°F)
ft H.O (at 62°F)	0.4335	lb/in²
ft H.O (at 62°F)	62.37	lb/ft²
in. Mercury (at 62°F)	70.73	lb/ft²
in. Mercury (at 62°F)	0.4912	lb/in²
in. Mercury (at 62°F)	0.03342	atmospheres
in. H.O (at 62°F)	0.002458	atmospheres
ft. H.O (at 62°F)	0.0295	atmospheres
lb/in²	0.0680	atmospheres
ton/ft²	0.945	atmospheres
in. Mercury (at 62°F)	13.57	in. H.O (at 62°F)
in. Mercury (at 62°F)	1.131	ft H.O (at 62°F)
lb/in²	2.309	ft H.O (at 62°F)
lb/ft²	0.01603	ft H.O (at 62°F)
lb/ft²	0.014138	in. Mercury (at 62°F)
lb/in²	2.042	in. Mercury (at 62°F)
lb/in²	0.0689	Bar
lb/in²	0.0703	kg/cm²

Velocity of Flow

Multiply	By	To Get
ft/min	0.01139	miles/hr
ft/min	0.01667	ft/sec
cu ft/min	0.1247	gal/sec
cu ft/sec	448.8	gal/min
miles/hr	88	ft/min
ft/sec	60	ft/min
gal/sec	8.02	cu ft/min
gal/min	0.002228	cu ft/sec

Heat Transmission

Multiply	By	To Get
Btu/in /sq ft /hr/°F	0.0833	Btu/ft /sq ft /hr/°F
Btu/ft /sq ft /hr /°F	12	Btu/in /sq ft /hr/ °F

Weight

Multiply	By	To Get
lb	7.000	grains
lb H.O (60°F)	0.01602	cu ft H.O
lb H.O (60°F)	0.1198	gal H.O
tons (long)	2.240	lb
tons (short)	2.000	lb
grains	0.000143	lb
cu ft H.O	62.37	lb H.O (60°F)
gal H.O	8.3453	lb H.O (60°F)
lb	0.000446	tons (long)
lb	0.000500	tons (short)

Circular Measure

Multiply	By	To Get
Degrees	0.01745	Radians
Minutes	0.00029	Radians
Diameter	3.142	Circumference
Radians	57.3	Degrees
Radians	3.438	Minutes
Circumference	0.3183	Diameter

Volume

Multiply	By	To Get
Barrels (oil)	42	gal (oil)
cu ft	1.728	cu in
cu ft	7.48	gal
cu in	0.00433	gal
gal (oil)	0.0238	barrels (oil)
cu in	0.000579	cu ft
gal	0.1337	cu ft
gal	231	cu in

Temperature

$$F = (°C \times 1.8) + 32$$
$$C = (°F - 32) \div 1.8$$

Fractions and Decimals

Multiply	By	To Get
Sixty-fourths	0.015625	Decimal
Thirty-seconds	0.03125	Decimal
Sixteenths	0.0625	Decimal
Eighths	0.125	Decimal
Fourths	0.250	Decimal
Halves	0.500	Decimal
Decimal	64	Sixty-fourths
Decimal	32	Thirty-seconds
Decimal	16	Sixteenths
Decimal	8	Eighths
Decimal	4	Fourths
Decimal	2	Halves

Gallons shown are U.S. standard.

9.18.0   Useful Engineering Tables—Schedule 40 Pipe Dimensions, Diameter of Circles, and Drill Sizes

Schedule 40 Pipe, Standard Dimensions

Size (in)	Diameters		Nominal Thickness (in)	Circumference		Transverse Areas			Length of Pipe per sq ft		Length of Pipe Containing One Cubic Foot	Nominal Weight per foot		Number Threads per inch of Screw
	External (in)	Approx-imate Internal (in)		External (in)	Internal (in)	External (sq in)	Internal (sq in)	Metal (sq in)	External Surface Feet	Internal Surface Feet	Feet	Plain Ends	Threaded and Coupled	
1/4	0.540	0.364	0.088	1.696	1.114	0.229	0.104	0.125	7.073	10.493	1383.789	0.424	0.425	18
1/4	0.675	0.493	0.091	2.121	1.549	0.358	0.191	0.167	5.658	7.747	754.360	0.567	0.568	18
1/2	0.640	0.622	0.109	2.639	1.954	0.554	0.304	0.250	4.547	6.141	473.906	0.850	0.852	14
3/4	1.050	0.824	0.113	3.299	2.589	0.866	0.533	0.333	3.637	4.635	270.034	1.130	1.134	14
1	1.315	1.049	0.133	4.131	3.296	1.358	0.864	0.494	2.904	3.641	166.618	1.678	1.684	11 1/2
1 1/4	1.660	1.380	0.140	5.215	4.335	2.164	1.495	0.669	2.301	2.767	96.275	2.272	2.281	11 1/2
1 1/2	1.900	1.610	0.145	5.969	5.058	2.835	2.036	0.799	2.010	2.372	70.733	2.717	2.731	11 1/2
2	2.375	2.067	0.154	7.461	6.494	4.430	3.355	1.075	1.608	1.847	42.913	3.652	3.678	11 1/2
2 1/2	2.675	2.469	0.203	9.032	7.757	6.492	4.788	1.704	1.328	1.547	30.077	5.793	5.819	8
3	3.500	3.068	0.216	10.996	9.638	9.621	7.393	2.228	1.091	1.245	19.479	7.575	7.616	8
3 1/2	4.000	3.548	0.226	12.566	11.146	12.566	9.886	2.680	0.954	1.076	14.565	9.109	9.202	8
4	4.500	4.026	0.237	14.137	12.648	15.904	12.730	3.174	0.848	0.948	11.312	10.790	10.899	8
5	5.563	5.047	0.258	17.477	15.856	24.306	20.006	4.300	0.686	0.756	7.198	14.617	14.810	8
6	6.625	6.065	0.280	20.813	19.054	34.472	28.891	5.581	0.576	0.629	4.984	18.974	19.185	8
8	8.625	7.981	0.322	27.096	25.073	58.426	50.027	8.399	0.442	0.478	2.878	28.554	28.809	8
10	10.750	10.020	0.365	33.772	31.479	90.763	78.855	11.908	0.355	0.381	1.826	40.483	41.132	8
12	12.750	11.938	0.406	40.055	37.699	127.640	111.900	15.740	0.299	0.318	1.288	53.600	—	—
14	14.000	13.125	0.437	43.982	41.217	153.940	135.300	18.640	0.272	0.280	1.069	63.000	—	—
16	16.000	15.000	0.500	50.265	47.123	201.050	176.700	24.350	0.238	0.254	0.817	78.000	—	—
18	18.000	16.874	0.563	56.548	52.998	254.850	224.000	30.850	0.212	0.226	0.643	105.000	—	—
20	20.000	18.814	0.593	62.831	59.093	314.150	278.000	36.150	0.191	0.203	0.519	123.000	—	—
24	24.000	22.626	0.687	75.398	71.063	452.400	402.100	50.300	0.159	0.169	0.358	171.000	—	—

Equivalent Length of Pipe to be Added for Fittings—Schedule 40 Pipe

Pipe Size (in)	Length in Feet to be Added Run				
	Standard Elbow	Side Outlet Tee	Gate Valve*	Globe Valve*	Angle Valve*
1/2	1.3	3	0.3	14	7
3/4	1.8	4	0.4	18	10
1	2.2	5	0.5	23	12
1 1/4	3.0	6	0.6	29	15
1 1/2	3.5	7	0.8	34	16
2	4.3	8	1.0	46	22
2 1/2	5.0	11	1.1	54	27
3	6.5	13	1.4	66	34
3 1/2	8.0	15	1.6	80	40
4	9.0	18	1.9	92	45
5	11.0	22	2.2	112	56
6	13.0	27	2.8	136	67
8	17.0	35	3.7	180	92
10	21.0	45	4.6	230	112
12	27.0	53	5.5	270	132

*Valve in full open position

Thermal Expansion of Pipe

*From Piping Handbook, by Walker and Crocker, by special permission. This table gives the expansion from −20°F to temperature in question. To obtain the amount of expansion between any two temperatures take the difference between the figures in the table for those temperatures. For example, if cast iron pipe is installed at a temperature of 80°F and is operated at 240°F, the expansion would be 1.780 − 0.649 = 1.131 in.

Temp (°F)	Elongation in Inches per 100 Ft from −20°F Up			
	Cast Iron Pipe	Steel Pipe	Wrought Iron Iron Pipe	Copper Pipe
−20	0.000	0.000	0.000	0.000
0	0.127	0.145	0.152	0.204
20	0.255	0.293	0.306	0.442
40	0.390	0.430	0.465	0.655
60	0.518	0.593	0.620	0.888
80	0.649	0.725	0.780	1.100
100	0.787	0.898	0.939	1.338
120	0.926	1.055	1.110	1.570
140	1.051	1.209	1.265	1.794
160	1.200	1.368	1.427	2.008
180	1.345	1.526	1.597	2.255
200	1.495	1.691	1.778	2.500
240	1.780	2.020	2.110	2.960
280	2.085	2.350	2.465	3.422
320	2.395	2.690	2.800	3.900
360	2.700	3.029	3.175	4.380
400	3.008	3.375	3.521	4.870
500	3.847	4.296	4.477	6.110
600	4.725	5.247	5.455	7.388

Diameters and Areas of Circles and Drill Sizes

Drill Size	Dia.	Area	Drill Size	Dia.	Area	Drill Size	Dia.	Area	Drill Size	Dia.	Area
3/64	.0469	.00173	27	.1440	.01629	C	.2420	.04600	27/64	.4219	.13920
55	.0520	.00212	26	.1470	.01697	D	.2460	.04753	7/16	.4375	.15033
54	.0550	.00238	25	.1495	.01705	1/4	.2500	.04909	29/64	.4531	.16117
53	.0595	.00278	24	.1520	.01815	E	.2500	.04909	15/32	.4688	.17257
1/16	.0625	.00307	23	.1540	.01863	F	.2570	.05187	31/64	.4844	.18398
52	.0635	.00317	5/32	.1562	.01917	G	.2610	.05350	1/2	.500	.19635
51	.0670	.00353	22	.1570	.01936	17/64	.2656	.05515	33/64	.5156	.20831
50	.0700	.00385	21	.1590	.01986	H	.2660	.05557	17/32	.5313	.22166
49	.0730	.00419	20	.1610	.02036	I	.2720	.05811	9/16	.5625	.24850
48	.0760	.00454	19	.1660	.02164	J	.2770	.06026	19/32	.5937	.27688
5/64	.0781	.00479	18	.1695	.02256	K	.2810	.06202	5/8	.6250	.30680
47	.0785	.00484	11/64	.1719	.02320	9/32	.2812	.06213	21/32	.6562	.33824
46	.0810	.00515	17	.1730	.02351	L	.2900	.06605	11/16	.6875	.37122
45	.0820	.00528	16	.1770	.02461	M	.2950	.06835	23/32	.7187	.40574
44	.0860	.00581	15	.1800	.02545	19/64	.2969	.06881	3/4	.7500	.44179
43	.0890	.00622	14	.1820	.02602	N	.3020	.07163	25/32	.7812	.47937
42	.0935	.00687	13	.1850	.02688	5/16	.3125	.07670	13/16	.8125	.51849
3/32	.0938	.00690	3/16	.1875	.02761	O	.3160	.07843	27/32	.8437	.55914
41	.0960	.00724	12	.1890	.02806	P	.3230	.08194	7/8	.8750	.60132
40	.0980	.00754	11	.1910	.02865	21/64	.3281	.08449	29/32	.9062	.64504
39	.0995	.00778	10	.1935	.02941	Q	.3320	.08657	15/16	.9375	.69029
38	.1015	.00809	9	.1960	.03017	R	.3390	.09026	31/32	.9687	.73708
37	.1040	.00850	8	.1990	.03110	11/32	.3438	.09281	1	1.0000	.78540
36	.1065	.00891	7	.2010	.03173	S	.3480	.09511	1-1/16	1.0625	.88664
7/64	.1094	.00940	13/64	.2031	.03241	T	.3580	.10066	1-1/8	1.1250	.99402
35	.1100	.00950	6	.2040	.03268	23/64	.3594	.10122	1-3/16	1.1875	1.1075
34	.1110	.00968	5	.2055	.03317	U	.3680	.10636	1-1/4	1.2500	1.2272
33	.1130	.01003	4	.2090	.03431	3/8	.3750	.11045	1-5/16	1.3125	1.3530
32	.1160	.01039	3	.2130	.03563	V	.3770	.11163	1-3/8	1.3750	1.4859
31	.1200	.01131	7/32	.2188	.03758	W	.3860	.11702	1-7/16	1.4375	1.6230
1/8	.1250	.01227	2	.2210	.03836	25/64	.3906	.11946	1-1/2	1.5000	1.7671
30	.1285	.01242	1	.2280	.04083	X	.3970	.12379	1-5/8	1.6250	2.0739
29	.1360	.01453	A	.2340	.04301	Y	.4040	.12819	1-3/4	1.7500	2.4053
28	.1405	.01550	15/64	.2344	.04314	13/32	.4062	.12962	1-7/8	1.8750	2.7612
9/64	.1406	.01553	8	.2380	.0449	Z	.4130	.13396	2	2.0000	3.1416

9.19.0   Thermal Expansion of Various Materials

Material	Inches per inch 10^{-6} X per °F	Inches per 100' of pipe per 100°F.	Ratio-assuming cast iron equals 1.00
Cast iron	6.2	0.745	1.00
Concrete	5.5	0.66	.89
Steel (mild)	6.5	0.780	1.05
Steel (stainless)	7.8	0.940	1.26
Copper	9.2	1.11	1.49
PVC (high impact)	55.6	6.68	8.95
ABS (type 1A)	56.2	6.75	9.05
Polyethylene (type 1)	94.5	11.4	15.30
Polyethylene (type 2)	83.3	10.0	13.40

Here is the *actual* increase in length for 50 feet of pipe and 70° temperature rise.

Cast Iron		.261
Concrete		.231
Mild Steel	Building Materials	2.73
Copper	Other Materials	.388
PVC (high Impact)	Plastics	2.338
ABS (type 1A)		2.362
Polyethylene (type 1)		3.990
Polyethylene (type 2)		3.500

9.20.0 Miscellaneous Tables of Weights, Measures, and Other Information

Square Measure

144 inches	1 square foot
9 square feet	1 square yard
30¼ sq. yds. 272¼ sq. ft.	1 square rod
160 square rods	1 acre
640 acres	1 square mile

Cubic Measure

1728 cubic inches	1 cubic foot
1 cubic foot	7.4805 gallons
27 cubic feet	1 cubic yard
128 cubic feet	1 cord

Dry Measure

2 pints	1 quart
8 quarts	1 peck
4 pecks	1 bushel
1 bushel	1.24 cu. feet
1 bushel	2150.42 cu inches

Liquid Measure

4 gills	1 pint
2 pints	1 quart
4 quarts	1 gallon
31½ gallons	1 barrel
2 barrels	1 hogshead

Linear Measure

12 inches	1 foot
3 feet	1 yard
16½ feet	1 rod or pole
5½ yards	1 rod or pole
40 rods or poles	1 furlong
8 furlongs	1 statute mile
320 rods	1 mile
5280 feet	1 mile
4 inches	1 hand
7.92 inches	1 link
18 inches	1 cubit
1.15156 miles	1 knot or 1 nautical mile

Weight – Avoirdupois or Commercial

437.5 grains	1 ounce
16 ounces	1 pound
112 pounds	1 hundredweight
2000 pounds	1 net ton or 1 short ton
20 hundredweight	1 gross or long ton
20 hundredweight	2240 pounds
2204.6 pounds	1 metric ton

Index

Sidney M. Levy is a construction consultant with more than 40 years of experience in the industry. The author of 11 books, including several devoted to international construction and the award winning *Project Management in Construction,* published in English, Spanish and Japanese editions; Mr. Levy is the owner of a construction consulting firm located in Baltimore, Maryland.